Mechanical Engineering Series

Series Editor

Francis A. Kulacki, Department of Mechanical Engineering, University of Minnesota, Minneapolis, MN, USA

The Mechanical Engineering Series presents advanced level treatment of topics on the cutting edge of mechanical engineering. Designed for use by students, researchers and practicing engineers, the series presents modern developments in mechanical engineering and its innovative applications in applied mechanics, bio-engineering, dynamic systems and control, energy, energy conversion and energy systems, fluid mechanics and fluid machinery, heat and mass transfer, manufacturing science and technology, mechanical design, mechanics of materials, micro- and nano-science technology, thermal physics, tribology, and vibration and acoustics. The series features graduate-level texts, professional books, and research monographs in key engineering science concentrations.

More information about this series at https://link.springer.com/bookseries/1161

George Sidebotham

An Inductive Approach to Engineering Thermodynamics

Springer

George Sidebotham
Department of Mechanical Engineering
Cooper Union
New York, NY, USA

ISSN 0941-5122 ISSN 2192-063X (electronic)
Mechanical Engineering Series
ISBN 978-3-030-20432-7 ISBN 978-3-030-20430-3 (eBook)
https://doi.org/10.1007/978-3-030-20430-3

This Springer imprint is published by the registered company Springer Nature Switzerland AG
The registered company address is: Gewerbestrasse 11, 6330 Cham, Switzerland

In this world of information overload, my life partner,
Judy Wirkula,
has an uncanny way of putting just the right book
at just the right time on my nightstand.
She is the right brain to my left.

Note to Students

This book is written for you. My hope is that you will gain confidence and apply the techniques developed here to things that interest you (whether directly related to this subject or not). You can have fun and challenge yourself at the same time. While the content is focused on engineering thermodynamics (energy and devices made to transform it), my firm belief is that the main purpose of an undergraduate education is the broad development of your thinking and feeling brain, achieved through the study of specific content. You're like an athlete that spends time in the gym. The sport may involve putting a ball in a hoop, a puck in a net or swinging a stick to hit a little ball, for example. Yet that is facilitated by lifting weights or running on a treadmill. Or like a musician that practices scales in order to better perform on stage. You spend a lot of time in the gym or "woodshed" before you are put in the game, a form of delayed gratification. Most undergraduate programs aim to find meaningful ways to put you in the game earlier, yet you must have a certain degree of faith that you are being prepared to make important future decisions that can impact the lives of many people. Mistakes now have little consequence, and you can learn from them. Mistakes made in professional practice can have serious negative consequences. Earning money is another example of delayed gratification. It might not feel like it, but the way you apply yourself to day-to-day decisions as an undergraduate influences your ability to land a paying job and more importantly, your ability to advance in your profession for your whole career (whether as a practicing engineer or in some other field). Your choices either build your character (if you work through adversity) or can degrade it (if you are immature about it).

Chapters are designed with "Deeper Dives" and Appendices that you could skip without loss of continuity. They provide depth or more thorough treatments of various topics. You may find you come back to them later on in your careers. So you can follow a reading schedule defined by your instructor, but you can feel free to jump ahead and start in on topics that you think might interest you.

I have attempted as much as possible to present the subject in a way that shows its relevance to the challenges of modern life (such as global warming or the limits to finite resource dependency), without compromising the knowledge base that professional practice requires. In the end, however, the experience gained from applying abstract principles (like energy, entropy, and eXergy) to realistic problem solving is what makes the study of engineering thermodynamics a valuable endeavor for any career.

Note to Instructors

Students in the modern internet-driven world do not learn the way my generation learned engineering content, before the internet, which was strictly textbook based. While there are no easy answers to addressing this challenge, the most important first step is to recognize that the way you or I learned may not be the only way to learn. Therefore, be open to change. Separating wheat from chaff is not easy or clear; some topics and methods are obsolete, others are fundamental. We can evolve to advance our profession, despite the many disagreements that will arise along the way. Yet, ironically, here I am writing a textbook.

My challenge to you is to draw from your experience and apply it using principles developed in this text. If you have taught Thermodynamics many times before, maybe you will see some things in a new light. My advice is to create your own in-class and homework-based group exercises and to allow students to create problem statements and then give them to other students. I occasionally let students write their own exams, administer them to their peers, evaluate (i.e., grade) them, and then I collect the graded exams. I "grade" them primarily based on their problem statements. However, going through the exercise develops to a high level in the Bloom's taxonomy educational model. I find I only really "learn" something when I am put in a position to "teach" it to someone else. So it seems that an obvious approach to education is to create active learning environments in which students "teach" each other.

I have developed a "Workshop-based" formula for how I conduct courses that meet for 3 h weekly. It is a hybrid of a traditional lecture course and "flipped classroom" (where students are expected to watch videos and read a textbook outside class with class time devoted to problem solving). One hour is devoted to team-based "Workshops" during which students work (usually in teams, often on a computer) on a problem that applies topical content. The Workshop is sandwiched between a lecture hour that prepares them for the Workshop, and another lecture that builds on its content. I have received nearly universally favorable feedback on the Workshops and generally mixed reviews about the effectiveness of my lectures.

Traditional engineering textbooks include long lists of "Back of Chapter" problem statements. These represent a treasure of applications, yet in practice, only a small percentage of them can be addressed directly in a course. This textbook has only a few problem statements, as other readily available resources exist for widening the application pool. Also, traditional thermodynamics textbooks contain long Appendices, but all that information is readily available on the internet from a variety of sources. The approach here is to include samples of Tables to explain how to use them and encourage students to obtain more detail from the internet when needed. A good early exercise is to guide students to prepare their own saturation and superheat tables from NIST or other resource.

Note to Lifelong Learners

If you've already taken a course in Thermodynamics and especially if your practice involves it, I hope that this textbook will help you attain a deeper level of understanding and practice, as it reinforces what you are already comfortable with.

Preface for Instructors (and Curious Students)

There are many excellent textbooks on Engineering Thermodynamics. Why another?

This textbook provides an alternative treatment of traditional Engineering Thermodynamics (defined loosely as energy and its transformations in engineering systems). The patient reader is left with virtually all the traditional tools of professional practice that are developed in a first semester in a thermal/fluids program, that is, 1st and 2nd Law for both closed and open systems with various working fluids. An early and repeated emphasis on the abstract concept of eXergy destruction (with the letter X deliberately capitalized) as a measure of energy dissipation sets this textbook apart. eXergy combines 1st Law (energy) and 2nd Law (entropy) concepts and facilitates a distinction between thermal and mechanical energy forms. The approach is to start with energy methods developed in mechanics and transition to thermodynamics by introducing both 1st and 2nd Laws simultaneously and early, and then build content using practical applications as much as possible. That is, an inductive methodology will be favored, in which consideration of a specific example leads to general conclusions.

eXergy, also called Availability in many texts, is a measure of the potential of the thermal energy of a substance to produce useful mechanical work in being brought by a hypothetical process from its current state to the conditions of the local environment. For example, a cold soda placed in a room will naturally warm to the temperature of the room, without any real noticeable change in the room itself. A machine could theoretically (but not practically) be constructed that produces electricity as the soda warms to the room. The eXergy of the soda while it is cold is the maximum possible work that could be extracted, at least in principle. It is only a portion of the inherent energy difference of a hot or cold object.

Just about everyone is familiar with the term "energy conservation" and most recognize that efforts to reduce "energy waste" or "dissipation" in the modern world is a wise thing to do. On the demand side, we ask "Can we use less electricity, gasoline, or other energy input to do what we do without giving up too much of what we like to do?" On the supply side, "Can we increase the use of renewable resources and improve the efficiency of the delivery of useful energy from burning fossil fuels?" However, the 1st Law of Thermodynamics guarantees that energy is ALWAYS conserved and is never wasted. So the term "energy conservation" really means the conservation of particular forms of energy that are considered high value in favor of low value energy forms. eXergy naturally distinguishes the quality of energy, and therefore a more accurate aim of modern society would be to practice "eXergy conservation."

The combined 1st and 2nd Laws show that all real processes destroy eXergy, or dissipate useful energy, to some degree. Therefore, eXergy destruction is an energy measure of "dissipation." The recognition of where, how, and to what extent can reveal opportunities for improvement through engineering design. The 2nd Law shows that entropy is generated by dissipative processes, but it is not an energy measure. However, simply multiplying the entropy generation by the dead state temperature results in an energy measure of dissipation, namely, eXergy destruction.

eXergy destruction is given an unusual nomenclature in this book, namely

eX_D as an energy measure of dissipation (i.e., in SI units of Joules)

$e\dddot{X}_D$ as an energy rate measure of dissipation (i.e., in SI units of Joules/second = Watt)

The triple dots in the rate form are an indication that the effect takes place within the boundary of a defined control volume (not a single dot, which indicates a quantity that crosses a boundary).

There is nothing wrong with the array of traditional textbooks on the subject. They just tend to approach thermodynamics in one particular way, while this textbook approaches the same subject from a different angle. Traditional textbooks develop background (such as property evaluation and defining various energy forms) for several chapters, then spend several chapters on the 1st Law, then several chapters on the 2nd Law, then a few chapters on eXergy (which is often material that is skipped). Practical examples are always included in these chapters (along with long lists of problems statements of which only a few are solved during a course), but their focus is on the general principles first. This material is a full semester's worth of content. When taught that way, eXergy arrives near the end of the semester, the absolute worst time to introduce a new idea. After that, often in a follow-up course, the center of attention is turned to applications.

With eXergy as an afterthought, it becomes something to be feared. This textbook embraces eXergy and develops all that content by introducing an application first, and then applying the 1st and 2nd Laws simultaneously, and combining them. A single semester will be sufficient for a foundation suitable for most Engineering students, and follow-up courses can develop advanced topics such as the fluid dynamics, chemical processing, design and optimization of energy systems, heat transfer, combustion, and compressible flow, to name a few.

One futuristic vision of the future (there are many others) consists of ubiquitous wind and solar (photovoltaic and thermal) farms in regions of high average wind speeds and solar intensities (such as the Great Plains region of the United States).

The electricity generated will be stored by the production of high pressure H_2 gas (from the electrolysis of H_2O), which will be mixed with Argon (an abundant chemical species at 1% of the atmosphere, obtainable from air separation plants) to form an attractive primary transportation fuel. Mixing H_2 with Argon reduces its explosiveness (Argon is an inert gas that absorbs chemical energy released in combustion) and increases the thermal efficiency of heat engines (Argon has a high ratio of specific heats). The O_2 produced in both electrolysis and air separation will be used to improve combustion of low BTU fuels (municipal solid waste, for example) in stationary power plants. It makes much more sense to develop engineering systems that capture the vast renewable resources available (water, wind, solar) to meet human needs rather than deplete the finite fossil fuel resources that raise the CO_2 levels in the atmosphere. That evolution will produce many job opportunities for young engineers that have acquired a fundamentally sound working knowledge of thermodynamics.

Deeper Dives

Most chapters are designed with two levels, an introductory level and a "deeper dive." The deeper dives and Appendices can be skimmed without loss of continuity. Engineering programs that require a single semester of thermodynamics could use the textbook that way. For programs with more than one semester of thermal/fluid coursework, the entire textbook could be used in a first semester focused on the introductory material, and more on the deeper dives in subsequent semesters, which themselves would be launching points for more advanced and focused texts. Alternatively, each chapter could be studied more thoroughly the first time, and the entire text over more than one semester. There is a fair amount of redundancy, by design, in recognition that learning is an iterative process. Each pass on a topic is intended to build depth.

Reading this textbook is a process-oriented endeavor analogous to the way an artist creates a painting. The painter first sketches the entire scene, subject, and background, on the whole canvas. This drawing is critical toward obtaining the correct proportions that will identify the subject and a compelling composition. The artist imagines the final product, and that image will morph as the painting is developed. A higher level of detail, and color, is filled without completing any section of the painting. Different portions are gone over repeatedly and content and detail built with each pass. The entire canvas evolves. The artist does not start in one corner, put in all the detail there, and move on, never to return to that corner. When you look at the painting, you see it all at once. The preliminary sketch of this textbook, Chaps. 1 and 2, introduces the 1st and 2nd Laws, and their combination, eXergy, using basic mechanics examples (Newton's apple falling from a tree and colliding bullets), and then builds content of a more thermal nature throughout. Chapter 1 focuses on mechanical forms of energy (reviewing Mechanics), and

Chap. 2 adds thermal forms. After completion, the reader can look back and see the entire painting; the subject of the painting and the composition that frames it.

Thermodynamics is a subtle and abstract science. Yet thermodynamic phenomena are also a part of normal life. Therefore, part of the challenge of a course in thermodynamics is to change the patterns of thought that a student or other lifelong learner has acquired through common experience. For example, everyone understands that the cold soda that warms too much if you wait too long and won't become cold again without putting it back in the refrigerator (which constantly draws electrical energy to maintain a cold space). However, a shift in your fundamental appreciation of that event is required to identify and quantify the dissipation (or eXergy destruction) associated with the warming and subsequent recooling in the refrigerator. You will learn to identify situations where energy is "wasted" and opportunities to do better. For example, friction brakes in automobiles are very effective at their primary function, namely, stopping a car, but regenerative braking systems convert a fraction of the kinetic energy of the vehicle into stored eXergy that can be used to reaccelerate the vehicle, conserving eXergy (in the form of gasoline). You will also learn that a large portion of the energy of a heating (thermal) source is not "available" to produce useful work. Rejection of energy in the form of heat to the environment is necessary to make the engine do its job.

The primary target audience is a first or second course in Thermodynamics for sophomore or junior Engineering students. The focus is on "simple compressible substances" as working fluids, in which phase changes are important, but chemical reactions and changing composition (e.g., distillation) are not treated formally. Traditional treatments consider combustion as an "equivalent heat addition" which can reinforce a common misconception. An introductory treatment of combustion is given that recognizes it as a conversion of one form of internal energy (chemical) to another (thermal), and not as a heat transfer process. Different disciplines will build on this foundation in follow-up courses.

A word about units is in order. The practicing engineer must be able to adapt to any relevant set of units. The strict SI set of units (kg, m, s, Kelvin) is a convenient set and will be emphasized. Frequent use of SI units with different prefixes will be used, for example, energy in kJ or MJ rather than J. English units will not be used at all, except when necessary to convert published data into SI units. One of the unfortunate aspects of the historical development of science and technology is the inconsistency of units for similar quantities. With regard to energy, an emphasis of this textbook is a distinction between mechanical/electrical and thermal forms of energy. Mechanical energy (force through a distance) is given historically in foot-pounds (ft-lb, English) or Newton-meters (N-m, SI). Thermal energy is given in BTU (British Thermal Units) or in calories. Electrical energy is given in kW-hr. All of these are energy units, yet only the Joule (J) is the strict SI unit. For example, the Energy Efficiency Rating (EER) of air conditioners or refrigerators is given in units of a ratio of the thermal energy delivered to the electrical input, in units of BTU/Watt-hr. This measure is related to a dimensionless quantity, the coefficient of performance (COP), which makes a direct comparison more meaningful. The size of an air conditioner is advertised in units of BTU (an energy unit), but what is really meant is BTU/hr (an energy rate). Also, the price of various energy quantities are listed in terms of price per unit of a convenient measurable quantity, like $/kW-hr for electricity, or $/cubic foot for natural gas, or $/gallon for fuel oil. A comparison of the true cost on a consistent basis requires the analyst to perform unit conversions. Such a comparison is conducted in Chap. 3, to put energy costs of common usage into perspective as students develop their experience with thermodynamic calculations.

In this text, units that are specified on a single line are expressed the way they would be in a line of computer code. For example, the common unit J/kg/K is written, rather than J/kg* K which is incorrect and should really be written J/(kg*K), since K is in the denominator. You will also encounter an unorthodox unit with *CHANGE* in temperature in the denominator. For example J/kg/ΔK = J/kg/Δ^{o}C is used for specific heats, which are based on temperature changes.

Along a similar line, a word about nomenclature is in order. The practicing engineer must be able to navigate nomenclature in which the same symbol is used to mean very different things, sometimes even in the same textbook (like V for volume or velocity). For example, the symbol "U" is used to represent work in courses in Mechanics, internal energy in Thermodynamics, and overall heat transfer coefficient in Heat Transfer. Similarly, the symbol "h" can mean height, enthalpy, convection coefficient, Plank's constant, etc. The only way around this dilemma for the foreseeable future is to insist that the practicing engineer know what is meant from the context, which requires experience and thought. An undergraduate education begins the acquisition process of experience through consistent and guided thinking.

The language of thermodynamics is generally precise, except there are a few misnomers, particularly with regard to the common word "heat." An internet search for the definition of "thermo" returns "related to heat." A better definition of "thermo" would be, "related to thermal energy." Heat is just one of several forms of thermal energy. For example, the term "latent heat" would more precisely be "latent enthalpy" which relates to phase change. An egregious misnomer that is well-entrenched relates to the physical property "specific heat" which is not directly related to heat at all and is not "specific" as defined in thermodynamics as referring to a quantity per unit mass. Furthermore, the properties "specific heat at constant volume" and "specific heat at constant pressure" trip up many students (and faculty!) who interpret that as the appropriate quantity to use for a constant volume or constant pressure process, respectively. It is nothing of the sort. In reality, "specific heat" is a material property that relates two thermodynamic properties (specific internal energy and temperature or specific enthalpy and temperature) for processes that involve a single phase only (no condensing or evaporating). Having failed to think of a better term, and due to the widespread use of these and other misnomers in thermodynamic language, I will point them out, but reluctantly continue to perpetuate their use. I will, however, call c_v the internal energy-specific heat, and c_p the enthalpy-specific heat. The term "heat transfer" instead of just "heat" will be used to help differentiate precise technical language from colloquial usage.

This textbook is intended to cover virtually all of the content expected of a first course in Engineering Thermodynamics at the undergraduate level (in a bit of a different order than in traditional textbooks). It is also intended to be read by practicing professionals that have taken traditional Thermodynamics courses, either recently, or long ago. Learning is a lifelong and iterative process. Having taught Thermodynamics (and Statics, Dynamics, Solid Mechanics, Fluid Mechanics, Heat Transfer, Combustion, Gas Dynamics) to gifted undergraduate students for over 30 years, I find that I learn new things all the time from texts that I've read many times. More importantly to me, I often learn things from students.

Problem-Solving Methodology

The traditional way to develop engineering courses is through problem solving, using paper and pen (or pencil) to work through problems in an organized way, using symbolic notation first, and then using a hand calculator to perform calculations, and transcribing these numbers onto paper. Most traditional textbooks define similar rules for problem solving. These rules are intended to guide the development of important technical communication skills. Many students will know how to get an answer (sometimes correct) but have difficulty putting it down on paper in an orderly, traceable way. I aim to shift focus and introduce guidelines (not rules) to produce electronic deliverables. Pen and paper can be useful for working out ideas (not necessarily to be used as submissions for technical communication, but could be valuable visuals as Appendix material), but the reporting medium is a word processor or presentation program, and a spreadsheet or other computer program to replace the calculator.

I see the hand calculator portion becoming obsolete, replaced by an organized spreadsheet or computer program used for all calculations. I even say (tongue in cheek) in a syllabus "throw away your calculator." The two main advantages gained are the ability to quickly conduct a parametric study (i.e., observe the effect of changing a parameter value) and retaining a high numerical precision, which can be very important in detailed thermodynamic calculations. There is value in working out thoughts with pen and paper analysis. It can be a quick way to represent a physical process schematically, to develop governing equations, and a way to bridge the pictorial or parallel processing nature of what's inside a brain with the linear progression required of technical communication. But it is not a technical communication. It is a form of personal journal entry. Submitted work can be done with a word processing or presentation program, in which the linear nature of technical communication is naturally developed. With some practice, equation editors and sketches with a drawing program do not take much longer than writing it out by hand and can often save time. Think about times when you've written out a careful derivation on paper, and then find an error near the beginning, and then must either start a new one or ross out and rewrite.

So, as you are working out a problem (whether in this course, or another), whenever you find yourself reaching for a calculator, open a spreadsheet or computer program (MATLAB, Python, for example) instead. Create a block of INPUT parameters (straight from the problem statement), a block of DERIVED parameters (which are formulas for other quantities and/or unit conversions that reference the INPUT parameters), and RESULTS for what is being sought. This textbook focuses primarily on spreadsheets, as all applications are readily solved that way. Other computer programming approaches can certainly be followed.

The other advantage of this approach is that a high precision is retained during calculations. For example, an introductory classroom workshop is to drop a golf ball on a hard floor, measure the initial and final heights, and then analyze it as a sequence of processes: falling ball, collision with Earth, rising ball. Initially, the ball has potential energy, but at the end, the potential energy is lower. Where did all that potential energy go? It is "lost," but where? A high precision is needed to calculate the temperature rise associated with the conversion of mechanical to thermal energy.

There are rules, and there is the spirit of the rules. Traditionally, engineering students are taught that the precision of a reported answer should not exceed the uncertainty of input data. For example, if a golf ball has a mass of 0.04 kg (one significant figure, as opposed to 0.0400 to 3 significant figures), then the precision of a reported result should technically not be more than 1 significant figure. Therefore, if a calculated kinetic energy reported by a calculator screen is 0.533664 J after falling a height of 1.36 m, if the rules are followed, it should be reported as 1 J and 1 meter if the mass is specified as 0.04 kg, which is clearly an example where following the rules violates the spirit of the rule. The answer is to use engineering judgment. In this case, it is appropriate to formulate the problem with a mass of 0.04 kg, and then report the kinetic energy as 0.534 J (or even 0.5337 J), with the understanding that with the underlying uncertainty a value like 0.53 J is warranted for making engineering decisions. Therefore, a general GUIDELINE (not "rule") to follow is "while making calculations, carry extra significant figures, and report 3 or 4." Certain types of thermodynamic calculations, especially those that involve 2nd Law analysis, involve reporting an effect that is the difference between two larger numbers. For example, the initial temperature of the golf ball could be 25.0000 °C, and the temperature after colliding with the floor is 25.0032 °C (assuming all thermal energy goes into warming the ball, not into sound or into the floor). The temperature rise is 0.00320 °C to 3 significant figures, but the temperature of the golf ball is the same before and after (25.0 °C to 3 significant figures). However, the "lost" potential energy has made the golf ball hotter, so it is not really the same. It's just that temperature is not a good metric to see it. On the other hand, the "lost" potential energy equals the eXergy destruction for this event. Other detailed examples of this type of numerical calculation are pointed out along the way. Calculations conducted in a spreadsheet template naturally allow for a high precision to be retained, and then cell formatting can be used to report an appropriate number of significant figures for final results. It is too easy to drop precision along the way with hand calculations, and thereby obtain erroneous results for certain types of calculation.

As much as possible, spreadsheet templates are designed with a block of input parameters, which generally come from a problem statement. Input parameters are entered as numbers (with default values). These are followed by a block of derived parameters (and unit conversions) that contain quantities that are fixed but require formulas that reference the input parameters. For example, if the diameter of a sphere is a given input, but its surface area and volume are needed, then the diameter is placed in the input block, and the appropriate formulas for area and volume are given in the derived parameters. If a different value for diameter is input to conduct a parametric study, the area and volume automatically adjust. Additional blocks (typically state tables, process tables, results tables) are developed depending on the specific problem at hand. It actually can be a big time saver to set up conventional problems this way.

As with my first textbook, *Heat Transfer Modeling: An Inductive Approach*, this textbook, targeted to undergraduate engineering students (but accessible to all lifelong learners), is intended to be inexpensive to students' budget and a nurturing challenge to their intellect. It is dedicated to the Cooper Union students (and three semesters of Air Force cadets) that I have had the honor and privilege of serving over the years. I feel blessed to have been given a chance to influence, in some incremental way, hundreds of young women and men on their way to becoming more enlightened and loving human beings (through the intense development of engineering skills).

I am grateful to Michael Luby, senior editor at Springer Science and Business Media, who took a chance with me as a first-time author and has encouraged me to produce this second work, with an eye to further building with textbooks on combustion and gas dynamics (i.e., rocket science). I've been teaching a long time, all the while developing notes and sometimes novel approaches to traditional content. My aim is to organize as much of it as I can into a series of volumes, from which more modern ways of dissemination (i.e. websites) can be drawn going forward. Michael has provided a supportive and professional environment to do that.

Outline of Textbook

The text is broken into two broad parts. In Part I (Chaps. 1, 2, 3, 4, 5 and 6), the main objective is to introduce and begin to practice with the fundamental theoretical tools of thermodynamics. In Part II (Chaps. 7, 8, 9, 10, 11, and 12), these methods are applied to engineering systems in widespread use. Each chapter has its own Appendices (part of the deeper dives), and Appendix 1 for the textbook organizes thermodynamic properties used in the main text into one place. Appendix 2 contains solutions to Workshop problems from the end of each chapter. Appendix 3 lists nomenclature used.

Part I: Fundamental Tools of Thermodynamic Analysis

The main objective of Chap. 1, *Newton's Apple*, is to establish a formal study of Engineering Thermodynamics as a follow-up to Engineering Mechanics by reviewing energy methods. The focus is on mechanical energy forms, namely kinetic and potential as storable forms of energy and work as the transferrable mechanical energy form by investigating the common scenario of a falling object. The 1st and 2nd Laws of Thermodynamics are introduced and combined.

The focus of Chap. 2, *Thermodynamics of Collisions*, is on thermal energy forms, namely internal energy as a storable form of energy and heat as the transferrable form (1st Law measures). Entropy (as a property), entropy transfer and entropy generation are introduced as 2nd Law measures. Combining the 1st and 2nd Laws, eXergy as a pseudo-property emerges as a mechanical form of stored energy and eXergy destruction as an energy form of dissipation. Another mechanics problem, a collision of two high-speed objects, is used to demonstrate how mechanical energy is dissipated into thermal energy, and how the reverse never occurs spontaneously. Thermodynamic properties of solids are naturally introduced.

Chapter 3, *Thermodynamic Cycles and Efficiencies*, introduces thermodynamic efficiencies of engineering systems that involve the conversion of an input form of energy to an output. First Law efficiencies (ratio of output to input) are differentiated from 2nd Law efficiencies (comparison of actual to theoretical maximum performance), and the Carnot efficiency is derived in a general way (without recourse to any specific cycle). A wide array of engineering devices is analyzed in a general way (departing somewhat from an inductive approach). The deeper dives put thermodynamics in a broader societal context.

Chapter 4, *Liquid/Vapor Properties*, develops the important skill of property evaluation for applications that involve phase change of the working fluid, that is, relating one thermodynamic property to another, using published saturation and superheat tables. Compressed liquid tables are shown to be unnecessary, as all properties can be obtained from appropriate use of saturation tables. Several detailed workshops develop a general approach to setting up and analyzing engineering systems.

After an introductory section on the principles of pumps (devices that move compressed liquids), Chapter 5, *Open Systems: Pumps and Turbines*, develops energy and entropy balances for open systems. The emphasis is on applications where the working fluid can be treated as incompressible (no density changes). Pumps and turbines convert one form of mechanical energy into another (shaft work to pressure energy and vice versa). Another goal is to introduce the exchange of fluid in reciprocating engines, using a liquid as the working fluid. There is much overlap of this chapter with courses in Fluid Mechanics.

The main goal of Chap. 6, *Compressors and the Ideal Gas*, is to introduce ideal gas relations and apply them to a situation that has both open and closed system processes, namely, a reciprocating compressor. It is similar to Chap. 5, but the working fluid is a gas, not a liquid.

Part II: Applications of Fundamental Thermodynamic Principles to Engineering Systems

While the main focus of Part I was to develop a methodology for comprehensive thermodynamic analysis (1st and 2nd Laws, with the constraints of mass conservation) with applications and societal context, the goal of Part II is to apply those principles to the most common engineering systems in use.

Chapter 7, *An Idealized Model of a Throttled Four-Stroke Reciprocating Spark Ignition Engine*, is a first step toward detailed modeling of the performance of reciprocating engines, with a gas as the working fluid. The engine is treated as the open-system, internal combustion engine, not as a hypothetical closed system as treated traditionally. Of the seven distinct processes, three act with a confined mass (closed system), while four involve gas exchange, either the intake of fresh fuel/air mixture or exhaust of burned gases.

After an introduction that shows the thermal response of an actual household refrigerator, Chap. 8, *Vapor/Compression Refrigeration*, develops the basic principles of the vapor/compression refrigeration cycle, the workhorse of the refrigeration, air-conditioning, and heat pump industries. An extensive deeper dive analyzes each key component from a fundamental point of view, sticking the toes into the waters of fluid mechanics and heat transfer analysis.

The next three chapters build toward a capstone of a combined cycle, the state-of-the-art in electrical power generation. Chapter 9, *Gas Turbines*, or Brayton-type cycles, develops the other class of internal combustion engines in widespread use for both mobile (i.e., aircraft) and stationary power applications. A gas turbine is an attractive option for the topping cycle of a combined cycle. The deeper dives focus on mobile plants and an introduction to supersonic engines.

In Chap. 10, *Steam Turbine Cycles*, or Rankine-type cycles, are developed as stand-alone power plants, in which the thermal source can come from a variety of means, but mostly necessary from solid fuel sources (i.e., coal) or from nuclear source. However, the use of geothermal or stored solar thermal energy may be important future candidates. The best theoretical thermodynamic performance subject to practical constraints is developed and then compared to realistic expectations.

Chapter 11, *Combined Cycles*, is a capstone project that demonstrates how the gas turbine and steam turbine cycle can be combined to address thermodynamic disadvantages of both, and produce a stationary power plant that approaches the best possible thermodynamic performance. The gas turbine, as an internal combustion engine, has a thermodynamic advantage on the high temperature side but a disadvantage on the low side (high exhaust temperature). The steam turbine cycle, with heating from an external source, has a disadvantage on its high temperature side, but an advantage at the low side (where heat is rejected to the environment). Put two and two together.

The bulk of a first course in Engineering Thermodynamics typically deals with working fluids that have the same chemical composition in all phases (called a pure substance). Chapter 12, *Psychrometrics*, is a first look at the analysis of thermodynamic systems that undergo changes in chemical composition. The change is not due to chemical reaction, but because the gas phase composition consists of air and H_2O vapor of varying relative amounts (humidity changes) while the liquid phase in contact with the gas is pure H_2O. There are important applications (air conditioning and cooling towers, for example) that make it an appropriate study in a first course. More than that, pedagogically, the study of psychrometrics is a good place to start to analyze systems for which chemical changes are important, and this chapter serves as an introduction to those more advanced fields of study. The deeper dives develop a much more comprehensive 2nd Law analysis than is typically considered.

New York, NY, USA George Sidebotham

Contents

About the Author

George Sidebotham is Professor of Mechanical Engineering at the Cooper Union for the Advancement of Science and Art in New York City.

Part I
Fundamental Tools of Thermodynamic Analysis

Chapter 1
Newton's Apple

The subject of mechanics is focused on motion caused by forces. Newton's 2nd Law, expressed by the ubiquitous $F = ma$, forms its theoretical basis. Thermodynamics is focused on energy and its transformations. Energy concepts such as work, kinetic energy, and potential energy arise in the study of mechanics, when forces act through distances. These energy forms can be classified as mechanical. Energy forms that are classified as thermal, such as heat, internal energy, entropy, and an additional mechanical property eXergy, are not treated in mechanics. However, the subject of thermodynamics addresses all forms of energy. The First Law of Thermodynamics will be introduced in this chapter with the basic mechanics problem of Newton's apple falling from a tree. The Second Law of Thermodynamics is introduced briefly and the two laws combined to introduce the concept of eXergy, the work potential of a substance that is in a different state than its environment.

Future chapters expand to applications where thermal energy plays a more central role, the subject of Engineering Thermodynamics. It is a field of supply and demand. On the supply side, energy is available (or stored) in some form that is not directly useful. It is transformed by an engineered device into another form that is capable of meeting a demand. In a fossil-fuel-based automobile, for example, a form of thermal energy is stored in the chemical bonds of the fuel. The engine, in a series of repeating steps (a cycle), converts a portion of the supplied energy into a mechanical form, namely, the movement of a rotating shaft, which in turn converts that energy into the linear motion of the vehicle, meeting the demand to go forward. A 1st Law efficiency is defined loosely as the ratio of the desired useful energy output (demand) to the required source energy input (supply). The Second Law places a restriction on the 1st Law efficiency. A 2nd Law efficiency is defined as the ratio of the actual 1st Law efficiency to the theoretical maximum possible 1st Law efficiency.

There are various sources of energy on the supply side, each of which can be classified as either renewable or nonrenewable (or depletable). Modern society has been built on nonrenewable fossil fuels (coal, oil, and natural gas). To extend the availability of these sources, there is still room for improvement in many important applications. That is, their 2nd Law efficiencies are substantially lower than they

G. Sidebotham, *An Inductive Approach to Engineering Thermodynamics*, Mechanical Engineering Series, https://doi.org/10.1007/978-3-030-20430-3_1

could be with improved engineering design. Meanwhile, the evolutionary shift to renewables (primarily hydroelectric, wind, solar and biomass) has begun. The opportunities for a career in these fields at all stages (research, development, and production) are growing. Since solar and wind are intermittent, there is incentive to develop practical means to store energy in devices, to be used during periods of time when the demand exceeds the available supply. This chapter focuses on mechanical energy storage, and several forms of mechanical energy exchange, or work.

Newton's Apple

As shown schematically in Fig. 1.1, *an apple of mass* $m_{apple} = 0.24$ *kg and volume* $\forall_{apple} = 0.30$ *L falls from a tree directly above a contemplative Sir Isaac Newton at time* $t = 0$ *from an altitude of* $H_1 = 3$ *m above ground on a calm day when the air temperature is* $T_\infty = 25\ ^\circ C$ *and the pressure is 1 atm. At what speed does the apple hit Newton on the head, putting the idea of gravity in it?*

The motion of the apple, neglecting relativistic effects, is governed by Newton's 2nd Law:

$$\sum \vec{F}_{\text{On Apple}} = m_{\text{apple}} \vec{a}_{\text{apple}} = m_{\text{apple}} \frac{d\vec{V}_{\text{apple}}}{dt}$$

A free body diagram (FBD) of the apple prior to its release (when it is static) is shown in Fig. 1.2. Before the apple falls, it experiences three identifiable forces, all acting in the vertical (y) direction: the force of the tree holding the stem, gravity

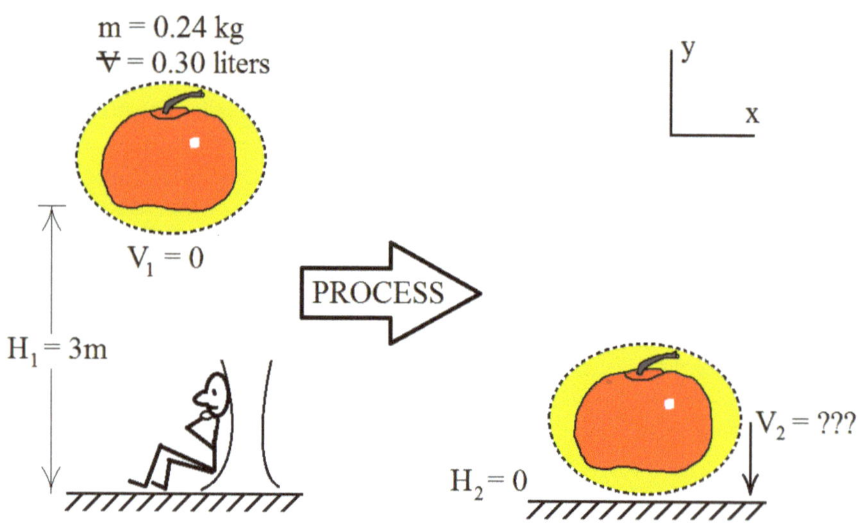

Fig. 1.1 Process schematic of Newton's apple

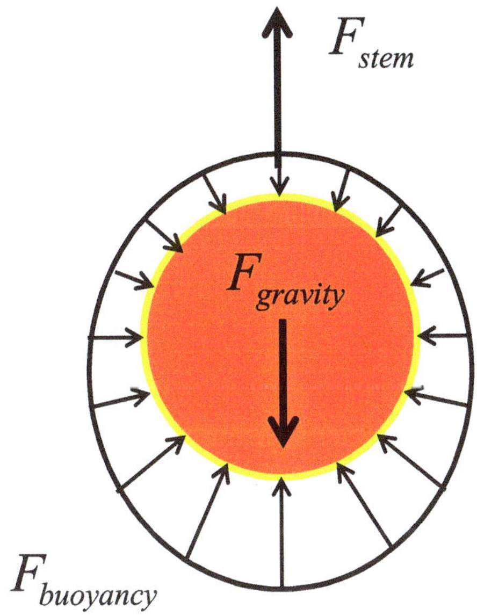

Fig. 1.2 Free body diagram of apple before it falls

acting between the Earth and the apple, and a buoyant force from the air it displaces. The buoyancy force is a net result of the static fluid (air) pressure acting on the surface, a distributed load. If it were a windy day, there would be a fluid force in the x direction. Recognizing that the acceleration is zero prior to release, Newton's 2nd Law is as follows:

$$\vec{F}_{\text{gravity}} + \vec{F}_{\text{buoyancy}} + \vec{F}_{\text{Stem}} = 0 \quad \text{prior to release}$$

The magnitude of the gravitational force between two masses (the apple of mass m) and the Earth (of mass M) is given by

$$F_{\text{Gravity}} = G\frac{Mm}{R^2}$$

where

$G = 0.00000000006672\ \text{m}^3/\text{kg/s}^2$ is the universal gravitational constant (not $g = 9.81\ \text{m/s}^2$)

$M = 5,970,000,000,000,000,000,000,000$ kg is the mass of the Earth

(scientific notation is deliberately not used to emphasize the magnitudes of these values).

$R = R_{\text{Earth}} + H_1 = 6,370,000 + H_1$ m is the distance between the center of masses

Rearranging to observe the effect of altitude (H_1) on gravity, the gravity force is as follows:

$$\vec{F}_{\text{Gravity}} = -G\frac{Mm}{R_{\text{Earth}}^2}\left(\frac{1}{\left(1 + H_1/R_{\text{Earth}}\right)^2}\right)\hat{j} = \frac{-mg_0\hat{j}}{\left(1 + H_1/R_{\text{Earth}}\right)^2},$$

where the familiar "standard gravity value" is introduced:

$$g_0 = \left[\frac{GM_{\text{Earth}}}{R_{\text{Earth}}^2}\right] = \frac{\left[6.672(10)^{-11}\ \text{m}^3/\text{kg/s}^2\right]\left[5.97(10)^{24}\ \text{kg}\right]}{\left[6.37(10)^6\right]^2} = 9.81\ \text{m/s}^2$$

This vector represents the force of gravity acting on the apple, with the unit vector $\hat{j}$ acting in the upward vertical direction. The factor in the denominator takes into account the variation of gravity with elevation and is approximately equal to unity for the vast majority of engineering applications near Earth's surface, in which case $H_1 \ll R_{\text{Earth}}$. This approximation will be invoked in this text unless otherwise stated.

The buoyant force, a hydrostatic effect, is the net vertical pressure force exerted by the stagnant ambient fluid (air) on the solid, and it is equal in value to the weight of the fluid displaced by the apple (Archimedes principle):

$$\vec{F}_{\text{Buoyancy}} = g_0\rho_\infty \forall_{\text{apple}}\hat{j} = m_{\text{air}}g_0\hat{j},$$

where

ρ_∞ is the density of the surrounding air that is displaced by the apple
$\forall_{\text{apple}}$ is the volume of the apple (with the line to distinguish volume from velocity)
m_{air} is the mass of the air displaced by the apple

The density of air is governed by the ideal gas law, which is an excellent approximation for air at nominal room air conditions (discussed at length in a future chapter), is as follows:

$$\rho_\infty = \frac{P_\infty}{R_{\text{air}}T_\infty} = \frac{1.013(10)^5\ \frac{\text{N}}{\text{m}^2}}{\left(\frac{8314}{28.84}\ \frac{\text{N}\cdot\text{m}}{\text{kg}\cdot\text{K}}\right)(273 + 25)\ \text{K}} = 1.18\ \frac{\text{kg}}{\text{m}^3}$$

The mass of air displaced by the apple is

$$m_{\text{air}} = \rho_\infty \forall_{\text{apple}} = \left(1.18\ \text{kg/m}^3\right)\left(0.0003\ \text{m}^3\right) = 0.000354\ \text{kg}$$

Note that in this case, the mass of the air displaced is approximately three orders of magnitude lower than the mass of the apple itself and, therefore, the buoyancy force is much less than the weight of the apple. However, that is not always the case (i.e., consider a helium balloon) and the buoyancy effect is retained here for generality.

Returning to the force balance prior to release, the stem force, or net weight of the apple, becomes the gravity force minus the buoyant force (a static fluid force):

$$\vec{F}_{\text{Stem}} = -\vec{F}_{\text{grav}} - \vec{F}_{\text{buoy}} = -\left(-m_{\text{apple}} g_0 \hat{j}\right) - (m_{\text{air}} g_0)\hat{j} = m_e g_0 \hat{j},$$

where an "effective mass" $m_e = m_{\text{apple}} - m_{\text{air}}$ has been introduced. If the apple were placed on a scale, the force registered by the scale would give an APPARENT weight ($m_e g_0$), which is the actual weight ($m_{\text{apple}} g_0$) minus the buoyant force ($m_{\text{air}} g_0$).

There comes a time in the life cycle of a typical apple that the stem can no longer hold onto the tree, and the apple falls. The stem force suddenly vanishes, the forces on the apple no longer balance, and the apple accelerates in the direction of the net force acting on it. As it picks up speed, it experiences a second type of fluid force, one associated with the relative motion of the apple and the surrounding fluid. These fluid forces are resolved into two components, one in the opposite direction of the velocity of the apple (called drag force) and the other perpendicular to the apple velocity (called a lift force):

$$m_{\text{apple}} \frac{d\vec{V}_{\text{apple}}}{dt} = -\left(m_{\text{apple}} - m_{\text{air}}\right) g_0 \hat{j} + \vec{F}_{\text{Drag}} + \vec{F}_{\text{Lift}}$$

On a windy day, the cross-wind could give rise to a significant lift force (which would be a sideways force in an Earth reference frame). There could also be a lift force if the apple was not symmetrical in shape.

Newton's 2nd Law applies to an inertial reference frame, and Newton is sitting on ground that is rotating with the Earth. The velocity of the apple can be expressed relative to the ground as follows:

$$\vec{V}_{\text{apple}} = \vec{V}_{\text{Ground}} + \vec{V}_{\text{apple/Ground}},$$

where the subscript "apple/ground" means "apple relative to ground." However, the ground moves relative to the center of the Earth and, therefore,

$$\vec{V}_{\text{apple}} = \vec{V}_{\text{Earth}} + \vec{V}_{\text{Ground/Earth}} + \vec{V}_{\text{apple/Ground}}$$

However, the center of the Earth moves relative to the sun, and therefore,

$$\vec{V}_{\text{apple}} = \vec{V}_{\text{Sun}} + \vec{V}_{\text{Earth/Sun}} + \vec{V}_{\text{Ground/Earth}} + \vec{V}_{\text{apple/Ground}} = \ldots$$

However, the Sun moves relative to our galaxy, and therefore...

With this suggestion that true inertial reference frames do not exist, the coordinate system defined by Newton sitting under the tree behaves as an inertial reference frame for engineering purposes, since the time of any event considered is short compared to the time it takes for the Earth to revolve (i.e., 1 day). Therefore, as is reasonable for the majority of engineering applications, the ground is treated as an inertial reference frame. Long-range ballistics studies, however, or modeling the weather, would require a relaxation of that assumption.

Since the day is calm, and assuming a nearly symmetrically shaped apple, there will be no significant side forces, and the apple will move vertically. Treating the apple velocity as a scalar (with "up" considered positive), Newton's 2nd Law, with the approximations discussed and a drag force opposing the motion, becomes

$$m_{\text{apple}} \frac{dV_{\text{apple}}}{dt} = -m_e g_0 \mp F_{\text{Drag}}$$

The drag force varies with speed, so this equation cannot be integrated until the drag force is related to system parameters (a topic for a course in fluid mechanics). Also, the direction of the drag force always opposes the motion (ultimately, a consequence of the 2nd Law of Thermodynamics). If the apple moves upward ($V_{\text{apple}} > 0$, if Newton tosses it up), the drag force will act in the downward direction ($F_{\text{Drag}} < 0$). This dynamic equation will be partially integrated, and investigated from an energy point of view.

Kinetic and Potential Energy: Storable Forms of Mechanical Energy

The height of the apple as it falls, (h_{apple}), will change with time. The rate of change of the height of the apple is given by the kinematic relationship:

$$\frac{dh_{\text{apple}}}{dt} = V_{\text{apple}}$$

Combining this kinematic relation and Newton's 2nd Law by eliminating the differential time (dt) yields a relationship between the change in velocity and change in height:

$$m_{\text{apple}} V_{\text{apple}} \frac{dV_{\text{apple}}}{dh_{\text{apple}}} = -m_e g_0 \mp F_{\text{Drag}} = \sum F_{\text{on apple}}$$

Multiplying through by the differential change in height:

$$m_{\text{apple}} V_{\text{apple}} dV_{\text{apple}} = \left(-m_e g_0 + F_{\text{Drag}}\right) dh_{\text{apple}} = -\left(\sum F_{\text{on apple}}\right) dh_{\text{apple}}$$

The right-hand side is the net force on the apple acting through a distance, and is the definition of the form of mechanical energy called *work*. The SI units of work are as follows:

$$\text{Work } [=] \text{ N} \cdot \text{m } [=] \text{ Joule } [=] \frac{\text{kg} \cdot \text{m}^2}{\text{s}^2}$$

Unit conversions to other systems include

$$1 \text{ J} = 9.478(10)^{-4}\text{BTU} = 2.778(10)^{-7}\text{kW} \cdot \text{hr} = 0.2390 \text{ cal} = 1(10)^7\text{erg}$$

The sign of the work must be carefully defined. In the above expression, the forces are considered to act on the apple by the environment. Therefore, it is the work done by the environment on the apple. This will be given a subscript "in" to be clear. Work "out" would be work done by the apple on the environment.

A source of confusion arises in the transition from the traditional study of mechanics to the study of thermodynamics. Standard practice in both sciences involves the definition of a "system" (the apple) and its "environment" or surroundings. In mechanics, free-body diagrams are used to represent the physical situation, and forces that act ON the system (BY the environment) are indicated. When there is motion in the direction of these forces, the work is a positive quantity. This work is done BY the environment ON the system. In thermodynamic conventions, a control-volume is a representation that defines a boundary between the system and its surroundings and arrows are used to show work crossing this boundary. However, this work is a representation of work done BY the system ON the environment, the opposite of mechanics practice. Therefore, the sign of work must always be carefully interpreted. It is good practice to pause and consider the "physics of the problem" to ensure a proper sign is being applied. Is a particular work term tending to increase or decrease the energy stored? Such a practice also builds intuition.

Three forces act on Newton's apple as it falls: gravity, buoyancy, and the fluid drag force. The fluid drag force is considered to be a "nongravity external force," a term, which will be used periodically. Note that buoyancy is a net pressure force, but is caused by gravity effects that act on the surroundings (air), not on the system (apple). Newton's 2nd Law in the form that relates changes in velocity to changes in height can be integrated (partially) from an initial state when the apple is released from rest (held by the tree), to some point during its fall, when it attains a velocity (V_{apple}) at some elevation (h):

$$\int_0^{V_{\text{apple}}} m_{\text{apple}} V_{\text{apple}} dV_{\text{apple}} = -\int_{H_1}^{h} m_e g_0 dh_{\text{apple}} + \int_{H_1}^{h} F_{\text{ext}} dh_{\text{apple}}$$

The first and second integrals are readily evaluated. The third integral represents the work of the external forces and cannot be integrated without further relating it to elevation (a fluid mechanics application that is not developed here). Evaluating the two integrals with separated variables:

$$\frac{1}{2} m_{\text{apple}} V_{\text{apple}}^2 = -m_e g_0 (h - H_1) + \int_{H_1}^{h} F_{\text{ext}} dh_{\text{apple}}$$

The left-hand side is a form of mechanical energy called *kinetic energy* and is a property of the apple, relative to a specific coordinate frame (Newton sitting under the tree). The first term on the right-hand side is the work done by the apparent weight (actual weight minus buoyant force, what the apple would register on a scale) of the apple. Since the apple falls in the same direction as the force ($h < H_1$), this work is positive, and it tends to increase the kinetic energy. Moving the variable part of the gravity term to the left-hand side:

$$\frac{1}{2} m_{\text{apple}} V_{\text{apple}}^2 + m_e g_0 h = m_e g_0 H_1 + \int_{H_1}^{h} F_{\text{ext}} dh_{\text{apple}}$$

The 2nd term on the left-hand side is a form of mechanical energy called gravitational *potential energy*, and is also a property of the apple (immersed in a fluid in a gravitational field), relative to a fixed point in the coordinate frame (Newton under the tree). It arises from the work associated with movement of the apple through a gravity field, immersed in a fluid. Use of this effective mass is valid for sufficiently small displacements for which the density of the surrounding fluid is constant. Notice that time is not directly involved in this integrated form of the momentum equation, having been eliminated from Newton's 2nd Law using the kinematic relationship between velocity and altitude. Buoyancy effects need to be treated separately for applications where the density of the surrounding fluid changes significantly with the elevation change (e.g., a weather balloon). Also, this form of potential energy is valid when the acceleration due to gravity (g) does not deviate significantly from the standard gravity value, g_0, and can be called "near-planet gravitational potential energy." Appendix 1 is a deeper dive that details the gravitational potential energy when the acceleration due to gravity changes substantially, called "far planet gravitational potential energy."

Kinetic and potential energy are considered to be "storable" forms of energy. In Thermodynamics, they are *properties* of the system (the apple), or equivalently, *state variables*. Work associated with the external forces, on the other hand, is a "transferrable" form of energy that depends on the nature and details of the event. It is a *process variable*. When the apple is first released, it makes sense to say that it has zero kinetic energy and a value $m_e g_0 H_1$ of potential energy associated with it, but it makes no sense to say that the apple has a certain amount of work. On the other hand, work done during a process can change the kinetic and/or potential energy.

Since the mass of the apple is greater than the mass of air displaced by the apple ($m_e > 0$), the apple will fall toward the ground. The speed of the apple when it hits Newton's head can be expressed by setting the altitude (h) to zero and rearranging for the final apple speed:

$$V_2 = \sqrt{2g_0H_1\left(\frac{m_e}{m_{\text{apple}}}\right) + \frac{2}{m_{\text{apple}}}\int_{H_1}^{0} F_{\text{ext}}dh_{\text{apple}}}$$

The integral is the work done by all nongravity forces on the apple. In this case, the only external force is the fluid drag force, which acts in the opposite direction of the motion. That is, if $dh < 0$, $F_{\text{ext}} > 0$ and vice versa making drag a so-called nonconservative force. Mechanical energy is not conserved if the drag force is significant. With the work of the drag force done by the environment being inherently negative (the work done by the system being inherently positive), the final velocity can be expressed as an inequality:

$$V_2 \leq \sqrt{2g_0H_1\left(\frac{m_e}{m_{\text{apple}}}\right)} = \sqrt{2\left(9.81\,\frac{\text{m}}{\text{s}^2}\right)(3\text{ m})\left(\frac{0.24\text{ kg} - 0.000354\text{ kg}}{0.24\text{ kg}}\right)}$$
$$= 7.666\,\frac{\text{m}}{\text{s}}$$

Note that this value becomes 7.672 m/s if buoyancy is neglected, a negligible 0.07% error for a falling apple.

The common experimental observation that two objects released simultaneously from the same height hit the floor at the same time requires both buoyancy and fluid drag to be negligible. That is, the mass of air displaced must be much less than the mass of both objects (or the density of air is much less than the average density of the objects). It is harder to formally justify the neglect of the fluid drag force without considering fluid mechanics. Neglecting these two fluid effects, two objects of different mass will attain the same velocity if they fall the same distance.

A simple demonstration that shows the effect of buoyancy in air is to blow up a balloon and drop it alongside a similar sized basketball. The initial acceleration (when the drag force is zero) is clearly lower for the balloon, so this effect is due to buoyancy. As the balloon picks up speed, the effects of fluid drag will be enhanced also.

What about a helium balloon? In this case, the buoyant force is greater in magnitude than the gravity force. It would appear that, mathematically, the final velocity is an imaginary number (square root of a negative number) and that this theory does not hold. However, retrace the steps. The final velocity was determined by setting the altitude h to zero. The velocity when it reaches a height greater than the initial height (say, hits the ceiling) is needed in that case.

To generalize for a one-dimensional problem, in its traditional form, Newton's 2nd Law is considered to be a momentum equation, in differential form with respect to time. Multiplying through by differential time, the momentum equation becomes an impulse equation:

$$mdV = \sum F \cdot dt,$$

where the left-hand side is the differential change in momentum and the right-hand side is the differential impulse (force multiplied by elapsed time). When time is eliminated with the kinematic relation $V = dh/dt$, Newton's 2nd Law becomes a differential energy equation:

$$mVdV = \sum F \cdot dh$$

The left side is the differential change in kinetic energy, and the right-hand side is a force acting through a distance, or work. The forces consist of gravity directly acting of the object, buoyancy (a fluid force that can be separated from all other external forces and moved to the left-hand side), and all other forces:

$$mVdV + \left(m - m_{\text{displaced}}\right)g_0 dh = F_{\text{ext}} dh$$

With the effective mass defined as $m_{\text{eff}} = m - m_{\text{displaced}}$ this differential equation can be integrated between two states, 1 and 2:

$$\int_{V_1}^{V_2} mVdV + \int_{H_1}^{H_2} m_{\text{eff}} g_0 dh = \int_1^2 F_{\text{ext}} dh$$

In its integrated form, it is considered to be a mechanical energy equation:

$$\left(\frac{1}{2}mV_2^2 - \frac{1}{2}mV_1^2\right) + \left(m_{\text{eff}} g_0 H_2 - m_{\text{eff}} g_0 H_1\right) = -W_{\text{OUT}} + \text{Thermal Effects}$$

Note that evaluation of the second integral on the left requires both $m_{\text{displaced}}$ and g_0 to be treated as constant. The displaced mass responsible for a buoyant force depends on the density of the fluid, which can change substantially for significant elevation changes within Earth's atmosphere. The vague term "*Thermal Effects*" is introduced in anticipation of phenomena not related to the mechanical nature of the process.

A more general form of the energy equation for an event is as follows:

$$\Delta E = \left(E_{\text{FINAL}} - E_{\text{INITIAL}}\right) = -W_{\text{OUT}} + \text{Thermal Effects},$$

where E is the total energy stored in the system.

For a specific event that has defined initial and final states, the mechanical energy equation can be expressed with the changes in the stored energy on the left-hand side and process-dependent quantities (work, using the thermodynamic sign convention) on the right-hand side. A subscript "OUT" is applied to the work, to represent work done by the system on the surroundings. It is possible, but not for the simple falling apple example, that the work out can be negative. That is, nongravity external work can be done on the system and increase its stored mechanical energy. For example, a magnet could be embedded inside an apple and another magnet held above it. Good engineering judgment is often called for to ensure a correct sign for the work term.

In general, there are other forms of storable mechanical energy, important ones being the energy stored in a spring or the kinetic energy stored in a rotating solid. Another important mechanism for work is associated with the motion of electrons that experience forces in an electrical field. That is, the flow of electricity is an important practical form of work. This and other forms of work are discussed shortly.

Consider another event, namely, the apple being released from the tree and then striking Newton on the head, a collision event. In this case, the initial state of the apple is the same as the falling apple, but the final state is defined to be one in which the apple is at rest (momentarily) on Newton's head. There are two stages to this event; the falling of the apple, and the striking of the apple. The elapsed time of the latter is immeasurably short, and there is an immeasurably large force required to decelerate the apple. The mechanical energy equation is as follows:

$$(KE_{\mathrm{FINAL}} - KE_{\mathrm{INITIAL}}) + (PE_{\mathrm{FINAL}} - PE_{\mathrm{INITIAL}}) = -W_{\mathrm{OUT}} + \text{Thermal Effects}$$
$$(0 - 0) + (0 - m_{\mathrm{eff}} g_0 H_1) = -W_{\mathrm{OUT}} + \text{Thermal Effects}$$

While it may be reasonable to neglect the work done by nongravity forces during the fall, it may not be reasonable to neglect it during the collision. However, while the force may be large, it must act through a distance in order to perform work. In fact, the apple (and Newton's head) undergoes a deformation during the collision event. All of the initial potential energy is accounted for either in this work, or in thermal effects. Whatever kinetic energy that was acquired during the fall is suddenly converted into some other form of energy. These effects will be considered by a general study of particle collisions in the next chapter.

Fundamental Tools of Thermodynamics

This section introduces many concepts and terms related to thermodynamics, in a more deductive than inductive manner. The thermodynamic property *temperature* will be used without carefully defining it, as no simple definition exists, yet it has universal awareness.

Fig. 1.3 Schematic defining Newton's apple as the system in a particular state, a snapshot. The dotted line represents the control surface, a boundary that separates the system (contained within a control volume) from the surroundings. The yellow region represents a boundary layer, a thin region of space between the physical system and the boundary

The First Law of Thermodynamics states that there exists a fixed total amount of a property of nature, called energy (E). That is, energy cannot be created or destroyed. However, energy can be converted from one form to another. Application of the First Law generally involves dividing the universe into two: a *system*, and its *surroundings* (or environment), separated by a *boundary*. The system has a definite volume, called the *control volume*, and the boundary, or control surface, has a definite surface area. The boundary may move in time, changing both the surface area and volume. A mathematical expression of the First Law is as follows:

$$E_{\text{Universe}} = E_{\text{System}} + E_{\text{Surroundings}} = \text{constant}$$

The Second Law of Thermodynamics states that there is a property of nature, called *entropy* (S) that always increases (or stays the same) with the passage of time.

$$S_{\text{Universe}} = S_{\text{System}} + S_{\text{Surroundings}} \neq \text{constant}$$

Use of the 2nd Law of Thermodynamics subtly distinguishes thermal energy forms from mechanical energy forms. In due course, consequences of the 2nd Law will be demonstrated such as *heat* flows from hot to cold and mechanical energy tends to be dissipated into thermal energy. Therefore, a mechanical form of energy is considered to be a more valuable form in practice than a thermal one.

Figure 1.3 shows a schematic of an object, called Newton's apple, as a system. The dotted line surrounding the apple defines a *control surface*. The surroundings are contained within the blue region outside the control surface, which extends in all directions to infinity. The apple itself is the key part of the system. However, the control surface is slightly larger than the apple itself, so that the system incorporates a portion of the air surrounding it. This region will be called a *boundary region* and will be shown to facilitate 2nd Law analysis. The outer boundary is chosen so that the *temperature* at the control surface equals that of the immediate surroundings of the apple, called the "*dead state*", which can be different than the temperature of the apple surface (i.e., an apple taken out of a refrigerator). The yellow region represents this extended portion of the system, the boundary region. It is difficult to establish the precise outer location of the boundary region, but it is usually not necessary to do so. It is an important conceptual tool for thorough 1st and 2nd Law thermodynamic analysis.

Systems are often defined so that the total mass of the system remains constant. Such a system is classified as a *closed system*. That is, no mass crosses the boundary of a closed system. For example, the apple of Fig. 1.3 was treated using a closed system in the previous analysis. An *open system* is one for which mass crosses the boundary. An apple is alive, and it undergoes metabolism and respiration. If the process of the apple ripening is of interest, the apple may need to be investigated as an open system with chemical reactions. However, for the purposes of studying the mechanics of an apple falling from a tree, this exchange of mass has a negligible effect. Treating the apple as a closed system for that purpose shows sound engineering judgement. Bad engineering judgement would be to divert attention to investigate the ripening of an apple as it falls from a tree, if the objective of the analysis is to study the motion of the apple.

Figure 1.3 is a representation of the system in a particular *state*. It is like a snapshot. Practical application of the 1st and 2nd Laws considers how the state changes during specific events, or *processes*. Therefore, an equivalent way to express the 1st Law is that the change (Δ) in the total energy of the universe associated with any event is zero:

$$\Delta E_{\text{Universe}} = \Delta E_{\text{System}} + \Delta E_{\text{Surroundings}} = 0$$

or

$$\Delta E_{\text{System}} = -\Delta E_{\text{Surroundings}}$$

If the energy of the system changes, the energy of the surroundings changes by an equal but opposite amount. Energy (specific forms to be discussed) can cross the boundary (in either direction), giving rise to a change in the amount of energy stored within the system and an equal and opposite change in the energy stored in the surroundings. The energy of the system can increase ($\Delta E_{\text{System}} > 0$)or decrease ($\Delta E_{\text{System}} < 0$) depending on the nature of the event.

There are two forms of energy that can cross the boundary of a closed system: heat and work, given symbols Q and W, respectively. Their nature and use will be discussed iteratively going forward. Subscripts are used to define the direction, whether energy enters or leaves. An energy balance for a process applied to a control mass (system) is, therefore,

$$\Delta E_{\text{System}} = Q_{\text{in}} - W_{\text{out}}$$

The 1st Law of Thermodynamics states that this change equals the change in energy of the surroundings. This sign convention ("in" for heat, "out" for work) follows from the early days of thermodynamics, where a primary practical interest was the design of heat engines, in which a thermal input source (heat) was used to produce output in the form of motion against opposing forces (work). Some disciplines use different conventions.

The 2nd Law for an event is expressed as an inequality:

$$\Delta S_{\text{Universe}} = \Delta S_{\text{System}} + \Delta S_{\text{Surroundings}} \geq 0$$

or

$$\Delta S_{\text{System}} \geq -\Delta S_{\text{Surroundings}}$$

Whenever heat crosses the boundary, entropy crosses with it, with an amount given by the magnitude of the heat divided by the absolute temperature, or Q/T_B, where T_B is the temperature of the boundary where heat crosses. Entropy can also be generated internally by processes that are considered to be dissipative, given the symbol Σ. Therefore, an entropy balance is

$$\Delta S_{\text{System}} = \frac{Q_{\text{in}}}{T_{\text{B}}} + \Sigma$$

The 2nd Law of Thermodynamics states that the entropy generation (Σ) must be positive, or in the limit, zero. That is, for any process,

$$\Sigma \geq 0$$

The entropy of a system can decrease ($\Delta S_{\text{System}} < 0$), which does not violate the 2nd Law, as long as the entropy of the surroundings increases by a greater amount due to the event in question. A common mistake is to confuse an entropy change (ΔS) with entropy production (Σ). They are often equal in magnitude (when there is no heat transfer), but they mean different things.

Consider the energy balance (1st Law) and entropy balance (2nd Law) side by side:

$$\Delta E_{\text{System}} = Q_{\text{in}} - W_{\text{out}}$$

$$\Delta S_{\text{System}} = \frac{Q_{\text{in}}}{T_{\text{B}}} + \Sigma$$

The heat transfer (Q_{in}) is the only common term. Eliminating it and rearranging to solve for the output work:

$$W_{out} = -\left(\Delta E_{System} - T_B \Delta S_{system}\right) - T_B \Sigma$$

The term in parentheses represents a change in state of the system and will be shown later to be closely related to (but not quite identical) the change in pseudo-property *eXergy*. This term can be positive or negative, depending on the nature of the process. The 2nd term contains two inherently positive quantities: absolute temperature at the boundary, and the entropy generation (positive from the 2nd Law). This term will be shown to be closely related to the *eXergy destruction* that is a measure of dissipation (or conversion of mechanical to thermal energy) that takes place internally, defined as

$$eX_D \equiv T_\infty \Sigma,$$

where T_∞ is the dead state temperature. The 2nd Law dictates that eXergy destruction is inherently positive (or zero, for reversible processes). Therefore, the output work is always reduced by the eXergy destruction.

The conservation of mass principle is another key component of thermodynamic analysis of engineering systems. It states that there is a property of all material, called mass, that is a constant in the universe:

$$M_{Universe} = M_{system} + M_{surroundings} = \text{constant}$$

For an event, the changes are

$$\Delta M_{Universe} = \Delta M_{System} + \Delta M_{Surroundings} = 0$$

$$\Delta M_{System} = -\Delta M_{Surroundings}$$

If mass crosses the system boundary, the change in mass of the system is equal and opposite to the change in mass of the surroundings.

Einstein postulated that mass is actually a form of energy ($E = Mc^2$) and, therefore, the conservation of mass expressed as a separate Law is not strictly true or necessary. However, for the vast majority of engineering applications, it is a valid and useful construct. Nuclear engineers and astrophysicists need to know when mass conversion to energy must be properly accounted for. This textbook will not formally consider processes for which changes in atomic structure are significant.

It is important in thermodynamic analysis to develop a clear understanding of the difference between a state variable and a process variable. Figure 1.4 shows a schematic of the apple in an initial state on the left, defined by state variables such as color (or temperature, or speed if falling, or ripeness). A process, or event, takes place that causes some change in the state variables, and the system in its final state is shown on the right. Different processes can take place and result in the same final state. For example, if the initial state is an unripe apple, and the final state is a ripe apple, the time it takes to become ripe is a process-dependent variable. An apple in a refrigerator will ripen at a slower rate than an apple left on a countertop. In problem

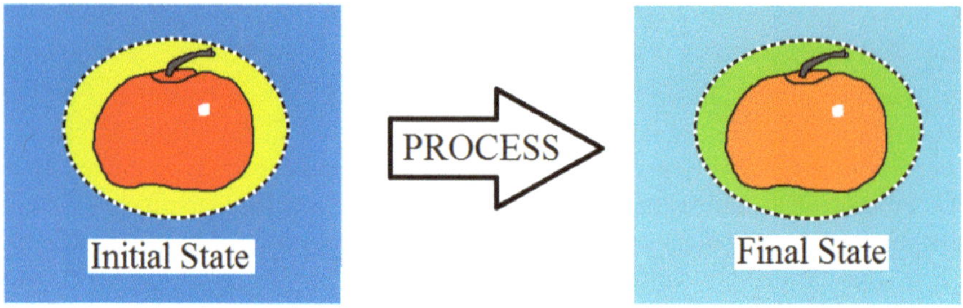

Fig. 1.4 Schematic that distinguishes state variables from process variables

solving, it is often good practice to make a sketch of similar construction to guide the analysis. Information stated in the problem statement is included directly in appropriate places. Word problem statements, a traditional and useful way to develop engineering content, can often be completely redefined by a sketch with no words.

To demonstrate the difference between state and process variables, consider your commute, specifically from the front door of your home to the front door of your place of professional practice (i.e., school, office, work site). Your location is a state variable. At the start of your commute, you are in one state and at the end of your commute, you are in another. However, the time it takes, the exercise you get, and the money you spend are all process-dependent variables. One day you walk, the next you ride your bike, and on another you take the bus. Your change in the state variable (location) is always the same, but the values of the process variables depend on your route.

Many situations naturally lend themselves to analyze specific events, with a before and after. These are considered to be *batch processes*. For example, placing a tray of cookie dough in the oven for, say, 30 min produces a batch of cookies. Other situations behave as *continuous processes*, which do not have defined initial or final states, but rather have inputs and/or outputs of mass and energy. For example, a baker may steadily place cookie dough on a conveyer that sends the dough through an oven, and warm cookies are steadily removed. However, the mathematical expression of a continuous process is derived from a consideration of a continuous series of before and after events. Each cookie spends the same amount of time in the oven, and cookies are produced at a fixed rate. *Transient processes* are those for which one or more variables change with time. Batch processes are inherently transient. The time dependence is often not directly involved in thermodynamics, as only the initial and final states are of interest. Other fields of study, for example, heat transfer, consider the time dependence explicitly. Continuous processes often operate under *steady-state* conditions, defined as processes for which nothing changes with time. Many engineering devices are designed to operate in a steady state. A *quasi-steady-state* situation is one in which the inputs and outputs are not exactly equal and, therefore, some stored quantity must change with time. If the difference between the inputs and outputs is much less than their magnitudes, the change in the stored quantity can be neglected, for engineering purposes.

The mathematics of this textbook requires basic calculus: an understanding of derivatives and integrals. Some vector calculus is used in this and the following introductory chapters only, which is a review of mechanics principles that lead into the study of thermodynamics. Calculus is the mathematics of change, and the following nomenclature is used for different types of quantities associated with change:

Type of quantity	Type of variable	Symbol	Example
Differential change	State	d	dT
Differential amount	Process	δ	δQ
Finite change	State	Δ	ΔE
Rate of flux	Process	$\cdot$	$\dot{Q}$
Internal generation rate	Process	$\cdots$	$\dddot{\Sigma}$

Differential (i.e., small) changes in a state variable are given the symbol "*d*," while small amounts of a process-dependent variable are distinguished by the symbol "δ." A finite (i.e., large) change in a state variable is given the symbol "Δ," and there is no counterpart for a process-dependent variable (like Q). It is common to derive mathematical equations based on the laws of nature using the finite step, and then shrinking the step to zero, resulting in differential equations. For example, the energy balance defined previously, $\Delta E_{\text{System}} = Q_{\text{in}} - W_{\text{out}}$, is in the form of a finite change in the state variable E. If a process that involves a vanishingly small (differential) change were being investigated, the energy balance would be expressed as follows:

$$dE_{\text{System}} - \delta Q_{\text{in}} - \delta W_{\text{out}}$$

The finite change would be obtained by integrating between two states:

$$\int_1^2 dE_{\text{System}} = \int_1^2 \delta Q_{\text{in}} - \int_1^2 \delta W_{\text{out}}$$

or

$$\Delta E_{\text{System}} = Q_{\text{in}} - W_{\text{out}}$$

Process-dependent variables (heat and work) cannot generally be integrated without further specification of the nature of the integrand. An important example where a process-dependent path can be related to differential change in a state variable is the work of an expanding fluid (gas or liquid) along a process-dependent path. Under assumed conditions of quasi-equilibrium, the differential work (δW) can be related to the fluid pressure and a differential change in its volume, a state variable. That is, as will be shown, expansion work is:

$$\delta W = pdV \quad \text{for quasi-equilibrium processes of fluids.}$$

Continuous processes generally involve rates of quantities, which are some quantity per unit time. For example, dividing a differential work by a differential time:

$$\frac{\delta W}{dt} \equiv \dot{W}[=]\frac{\text{Joule}}{\text{sec}} = \text{Watt}$$

The symbol [=] means "has units of.' This quantity is an energy rate, or *power*, expression.

Quantities that cross the boundary on a rate basis (or flux) are indicated by an over-dot '·' above the appropriate symbol. This nomenclature is different from some calculus and other engineering fields, in which the over-dot means a derivative. For example, $\frac{dE}{dt}$ means the rate of change of the quantity E as time proceeds. For steady-state processes, $\frac{dE}{dt} = 0$. The expression $\dot{E}$ would imply something different, namely, a flux of E across the boundary, the amount of E that crosses per unit time. For steady-state continuous processes, it is a nonzero constant. Consider, for example, an open faucet of water. Water is flowing at a steady rate, which could be expressed in gallons per minute. But it is not changing with time. Flux quantities include mass flow rate $(\dot{m})$, heat transfer rate $\left(\dot{Q}\right)$ and work transfer rate, or power $\left(\dot{W}\right)$.

Rate-based quantities that are related to internal processes are indicated by a triple over-dot '...' above the appropriate symbol. The important ones are entropy generation rate and related eXergy destruction rate:

$$e\dddot{X}_{\mathrm{D}} = T_{\infty}\,\dddot{\Sigma}$$

This nomenclature will not be strictly adhered to throughout the text, and these two quantities will usually be indicated with a single over-dot, but it is important conceptually to distinguish quantities that cross the boundary (mass, heat, and work) from quantities that are associated with effects within the control volume (entropy production or eXergy destruction).

Is time a state variable, or process variable, or neither? It could be either, depending on the context. A differential step forward in time is generally given the symbol "*d*" and a large step "Δ," making it a state variable, but the elapsed time of an event is a process-dependent variable.

Other Forms of Stored Mechanical Energy

Mechanical energy can be stored in the form of springs (a potential energy) or rotating masses (kinetic energy). Electrical energy can be stored in batteries and capacitors. At the Engineering device (i.e., macroscopic) level, these devices may

Table 1.1 Forms of Stored Mechanical Energy (include electric energy storage)

Mechanical energy form	Stored energy	Explanation of terms
Linear kinetic energy	$\frac{1}{2}mV^2$	m is the mass of an object (system). V is its linear velocity relative to a coordinate frame (not necessarily Newtonian).
Gravitational potential energy (near planet)	$m_e g_0 (H - H_0)$	m_e is the mass of an object (system) less the mass of displaced fluid. g_0 is the planet surface gravity value, 9.81 m/s^2 for Earth. H is the height relative to a reference height H_0, where the potential energy is defined to have a value of zero.
Gravitation potential energy (far planet)	$\frac{mg_0 h}{1+h/R}$	m is the mass of an object. g_0 is the planet surface gravity value, 9.81 m/s^2 for Earth. h is the altitude from the planet surface. R is the planet radius. The reference height is the planet surface, and the expression reduces to the near-planet case for $h \ll R$.
Elastic solid potential energy (linear spring)	$\frac{1}{2}k(x - x_0)^2$	k is a spring constant (SI unit N/m). $x - x_0$ is the displacement of the spring from its relaxed length (x_0).
Rotational kinetic energy (flywheel)	$\frac{1}{2}I\omega^2$	$I = \iiint r^2 dm$ is the moment of inertia of an object about an axis. ω is the angular speed (in rad/s).
Electrochemical energy (battery)	$(\Psi) * (\mathrm{A} \cdot \mathrm{h}) * \left\lvert \frac{3600\ \mathrm{s}}{\mathrm{h}} \right\rvert$	Ψ is the rated voltage of the battery. $\mathrm{A} \cdot \mathrm{h}$ is the rated capacity of the battery (in Amp-hr units). Note the unit conversion to SI units (Joules).
Electrostatic energy (capacitor)	$\frac{1}{2}C\Psi^2$	C is the capacitance (in Farad). Ψ is the voltage across the capacitor.

play an increasingly important role in a society that relies heavily on renewable, yet intermittent, sources of energy such as solar and wind energy. When the sun is shining and the wind is blowing, the available energy (eXergy) can be converted to a more storable form (like a compressed spring or rotating flywheel), and on a still night, a portion of the stored eXergy can be converted into a desired form, like electricity. These forms of energy storage are discussed in some detail in Appendix 1 of this chapter, a deeper dive, which can be skipped without too much loss in continuity. However, careful reading of the derivations will reinforce the main concept of this chapter, namely, that many mechanics-type problems can be viewed from an energy perspective. The results are summarized in Table 1.1.

At the microscopic level, molecular bonds between atoms in solids can be modeled as mass/spring systems, and the associated vibrations constitute the internal energy of the solid. This stored energy is closely related to temperature: the greater the stored internal energy, the greater the temperature. Also, molecules in gas and liquid phases have molecular bonds that vibrate, and the molecules rotate with stored rotational energy. These energy storage modes, along with the kinetic energy of the center of mass of the molecules, are important components of their internal energy, discussed in future chapters.

Forms of Work

Four forms of work are introduced in this section that are frequently used in the analysis of engineering devices: electrical, shaft, expansion, and flow work. Work associated with friction is touched on in the determination of the energy stored in a spring in Appendix 1, but not considered here. Friction gives rise to a dissipative effect and not usually considered to be a desired input or output of an engineering device.

Electrical Work

Devices that generate or consume electricity exchange energy with the environment in the form of work. The forces that act through distances are at the microscopic level, with each electron that passes through the device moving with, or against, an electromagnetic force. Figure 1.5 shows an electrical circuit with a battery that sustains a voltage difference of $V_1 - V_2$ (units of Volts) that drives a device with an electrical current, I (units of Amperes) passing through both. Figure 1.6 shows a thermodynamic process diagram. The electrons are considered as part of the surroundings. In the battery, work is done on the electrons (by the battery) to move them through the device (electrons flow in the opposite direction of the current). Work is

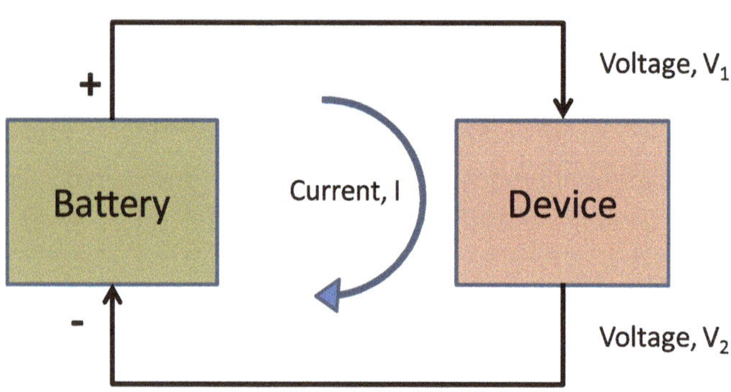

Fig. 1.5 Circuit diagram of a battery driving a device

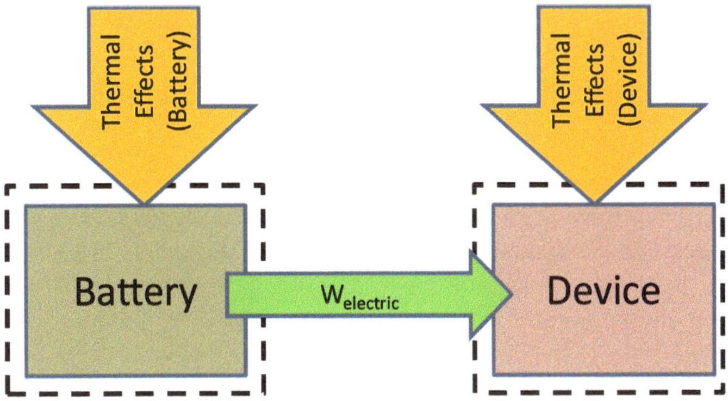

Fig. 1.6 Thermodynamic process diagram of battery and device

done by the electrons on the device as they pass through it. If the voltage and current remain constant over a period of time Δt, the magnitude of the work is as follows:

$$W_{\text{electrical}} = V \cdot (I \Delta t) \quad [=] \quad \text{Volt} \cdot (\text{Amps} \cdot \text{time}) = \text{Joule}$$

The rate at which work is done (power) is obtained by dividing by the time:

$$\frac{W_{\text{electrical}}}{\Delta t} = \dot{W}_{\text{electrical}} = V \cdot I \quad [=] \quad \frac{\text{Joule}}{\text{sec}} = \text{Watt}$$

The sign of the electrical work (or power) depends on whether current flows in (negative work out) or out (positive work out).

Consider an energy and entropy balance on the battery, with energy flows defined by Fig. 1.6, as follows:

$$\frac{dE_{\text{battery}}}{dt} = \dot{Q}_{\text{into battery}} - \dot{W}_{\text{electrical}}$$

$$\frac{dS_{\text{battery}}}{dt} = \frac{\dot{Q}_{\text{into battery}}}{T_{\text{B}}} + \dddot{\Sigma}_{\text{battery}}$$

Any heat exchange takes place between the battery and its local environment, which is at the dead state temperature, T_∞. That is, $T_{\text{B}} = T_\infty$. Combining the energy and entropy balances:

$$\dot{W}_{\text{electrical}} = -\frac{d}{dt}\left(E_{\text{battery}} - T_\infty S_{\text{battery}}\right) - T_\infty \dddot{\Sigma}_{\text{battery}}$$

The quantity that changes with time is defined as the stored eXergy (${}_eX_{\text{battery}} = E_{\text{battery}} - T_\infty S_{\text{battery}}$). The quantity associated with the entropy generation rate (which is inherently positive) is the eXergy destruction rate $e\dddot{X}_{\text{D}} = T_\infty \dddot{\Sigma}$. Therefore, the electrical power produced by the battery is as follows:

$$\dot{W}_{\text{electrical}} = -\frac{d_e X_{\text{battery}}}{dt} - e\dddot{X}_{\text{D,battery}}$$

The eXergy destruction rate is due to dissipative effects within the battery and is inherently positive (or zero for a perfectly designed battery). The stored eXergy, however, is an energy measure of the state of charge of the battery. If the device is being driven by the battery (e.g., a light bulb), the eXergy stored in the battery decreases with time. If the device is a battery charger, then electrical work is being put into the battery, and its stored eXergy increases. The eXergy stored is associated with all forms of stored energy, including thermal and electrochemical energy considered a little more in the deeper dive of Appendix 1.

Consider now the device that receives the electricity for some purpose. This device might be able to store energy or entropy. An energy and entropy balance on a rate basis for any electrical device (i.e., one for which electrical current crosses the control surface into the device), like a light bulb, is as follows:

$$\frac{dE_{\text{device}}}{dt} = \dot{Q}_{\text{into device}} + \dot{W}_{\text{electrical}}$$

$$\frac{dS_{\text{device}}}{dt} = \frac{\dot{Q}_{\text{into device}}}{T_{\text{B}}} + \dddot{\Sigma}_{\text{device}}$$

The boundary across which heat is transferred is taken to be outside the thermal boundary layer of the device, where the temperature equals the dead state temperature. That is, $T_{\text{B}} = T_\infty$. If the device operates at steady state, the stored energy and entropy remain constant ($dE/dt = dS/dt = 0 \quad @\ SS$) and, therefore,

$$\dot{W}_{\text{electrical,in}} = -\dot{Q}_{\text{into device}} \quad \text{from an energy balance at steady state}$$

$$\dot{Q}_{\text{into device}} = -T_\infty \dddot{\Sigma}_{\text{device}} \quad \text{from an entropy balance at steady state}$$

The net mechanical power out must equal the net rate of heat transfer in, a 1st Law consequence. The 2nd Law consequence is that the heat in (and thus work out) must be negative $\left(\dddot{\Sigma} \geq 0 \text{ by the Second Law}\right)$. Combining the energy and entropy balances (by eliminating the heat transfer term) and introducing the eXergy destruction rate:

$$\dot{W}_{\text{electrical,in}} = e\dddot{X}_{\text{D}}$$

That is, thermodynamically, to perform its function, all the electrical input energy is dissipated in the form of heat being transferred to the local environment.

For example, a light bulb (once at steady state, before which its stored thermal energy rises, i.e., it gets hot) provides its desired function, but behaves as an electrical heater. Most electrical devices (like TVs, computers, video projectors, net effect of a refrigerator, blenders, etc.) behave that way thermodynamically. From a 1st Law basis, it is theoretically possible for the same light bulb to receive heat from its environment and generate electricity. However, the 2nd Law analysis shows that in order for the entropy generation rate to be positive, heat must be transferred from the device to the environment.

There has been an important evolution in lighting technology, a shift from candles to incandescent light bulbs to fluorescent to halogen to light-emitting diode (LED). The LED provides similar lighting to an incandescent with a much lower electrical input. Some people argue that incandescents offset heating requirements and are 100% efficient at heating. However, as discussed in the next chapter, since they use an electromechanical input to provide a thermal output, incandescents considered as a heating source dissipate much more eXergy than if the electricity were to be used to drive a heat pump.

Shaft Work

Many engineering applications involve rotational motion: motors, generators, pumps, fans, turbines, compressors, crankshafts. Figure 1.7 shows a flywheel mounted on frictionless bearings. The shaft of the bearing is connected to a simple crank arm of total length L. The ends of the crank are loaded with vertical forces of equal magnitude and opposite sign, creating a couple (no net force). The state/process diagram from an end view (Fig. 1.8) shows the assembly with the crank initially horizontal at time t rotating with an angular velocity ω, and the assembly at time $t + \Delta t$ after it has rotated by an angular displacement $\Delta\theta$ and has accelerated to an angular velocity $\omega + \Delta\omega$. Note that, to be consistent with the process diagram, the angular velocity must be in the direction shown, but the force couple could be of opposite sign, and the change in angular velocity ($\Delta\omega$) could be positive (an acceleration) or negative (deceleration).

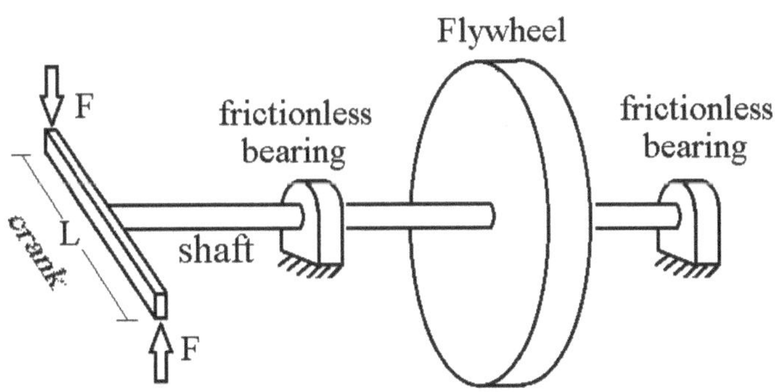

Fig. 1.7 Flywheel mounted on frictionless bearings and turned by a crank

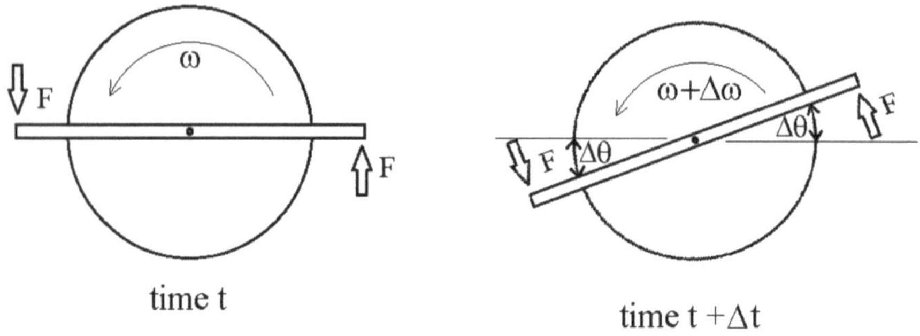

Fig. 1.8 State/process diagram from end view. The forces shown are the external forces acting on the crank. The force exerted by the crank on the surroundings is equal and opposite

Defining the entire assembly (crank, shaft, and flywheel) as the system (plus its boundary layer), the energy balance between the states is as follows:

$$\Delta E = Q_{\text{in}} - W_{\text{out}}$$

Since the center of mass of the assembly does not move, the energy stored in the flywheel (discussed in Appendix 1) consists of internal energy plus rotational energy. That is, $\Delta E = \Delta U + \Delta(I\omega^2/2)$. A formal 1st and 2nd Law analysis would suggest the conditions under which there would be no change in internal energy ($\Delta U = 0$). With these simplifications, the 1st and 2nd Law balances become simply:

$$\Delta\left(\frac{1}{2}I\omega^2\right) = -W_{\text{out}}$$

That is, the change in stored rotational energy equals the negative of the net work done by the system on all external forces. Note that friction on the bearings, or driving another mechanical load (like a piston in a cylinder), would constitute external forces, but those effects are absent in this case. The two forces that make up the couple that act on the crank arm are the only external forces that move through a distance, and there is work associated with that motion.

$$W_{\text{out}} = \begin{pmatrix}\text{Force by}\\ \text{System}\end{pmatrix} \cdot (\text{Displacement}) = 2 * (-F) \cdot (\Delta x) = 2 * (-F) \cdot \left(\frac{L}{2}\Delta\theta\right)$$

$$W_{\text{out}} = -FL\Delta\theta = -\tau\Delta\theta,$$

where the moment of the couple, or its torque ($\tau = FL$), has been introduced. For rotational motion, a torque acting through an angular displacement is analogous to a force acting through a distance in a linear displacement. Note that the torque is the externally applied torque (the opposite of that acting by the system on the surroundings). The energy balance becomes

$$\Delta\left(\frac{1}{2}I\omega^2\right) = \tau\Delta\theta$$

With positive angular velocity, $\Delta\theta$ is positive. Therefore, a positive externally applied torque (as shown) will increase the rotational energy stored, that is, accelerate the flywheel. A retarding, negative, externally applied torque will decelerate the flywheel (with ω still positive). It is very easy to make an error in sign, and once a 1st Law balance is simplified, it is good practice to reflect on whether the signs make sense. The work done by the torque is called "shaft work," and its magnitude is

$$W_{\text{shaft}} = \tau\Delta\theta$$

The rate of work being done (the power) is obtained by dividing the elapsed time, and taking the limit as the time step shrinks to zero:

$$\dot{W}_{\text{shaft}} = \lim_{\Delta t->0} \frac{W_{\text{shaft}}}{\Delta t} = \lim_{\Delta t->0} \tau \frac{\Delta\theta}{\Delta t}$$

That is, $\dot{W}_{\text{shaft}} = \tau\omega$.

*Expansion and Compression Work (***PdV***)*

All fluids exert a pressure, or force per unit area, on the vessels that contain them. If the volume of the container changes, the pressure force acts through a distance, and work is done. If the volume increases, work is done by the fluid on the surroundings, called expansion work. If the volume decreases, work is done on the fluid by the surroundings, called compression work. The general term "expansion work" covers both cases, with the understanding that a negative expansion is a compression. Reciprocating engines rely on a series of compression and expansion processes to convert chemical energy into output work. A free body diagram (FBD) of a common mechanism designed to convert linear motion (a piston moving vertically in a cylinder) to rotational motion (a crankshaft) is shown in Fig. 1.9. The pressure force (PA) acting on the piston is transferred by the connecting rod (of length L) to the crank arm (of length a). The focus of this analysis is on the fluid contained in the cylinder, and a process diagram for a differential change is shown in Fig. 1.10. A more detailed analysis of this mechanism will be conducted in chapter 7 on reciprocating engines.

Treating the contained fluid as the system, the energy balance is

$$dE = \delta Q_{\text{in}} - \delta W_{\text{out}}$$

The work done by the system is

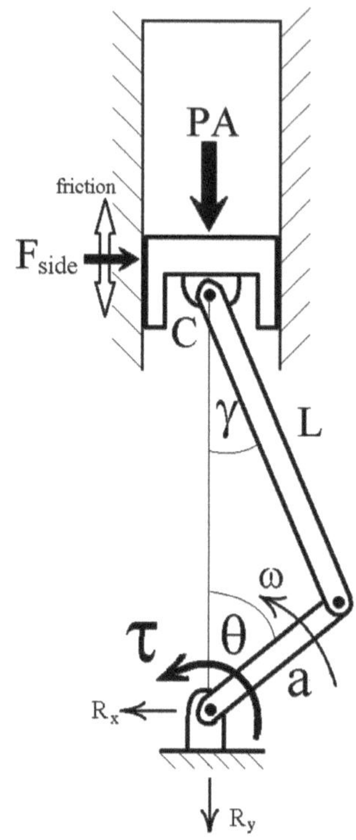

Fig. 1.9 Free body diagram (FBD) of a piston/cylinder and crank mechanism

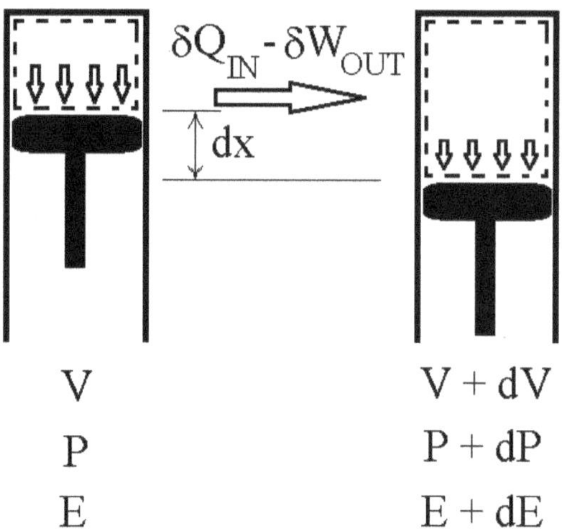

Fig. 1.10 Process diagram of the expansion process of a fluid contained in a piston cylinder

$$\delta W_{\text{out}} = \vec{F}_{\text{On surroundings}} \cdot d\vec{x} \approx (\overline{P}A)dx,$$

where the average pressure $(\overline{P})$ between the initial and final states is used to indicate that the pressure changes during the process. The average pressure during the process is as follows:

$$(\overline{P}) = \frac{1}{2}[P + (P + dP)] = P + \frac{dP}{2} \approx P$$

The latter approximation is valid for differential changes.

The quantity Adx is the change in volume, dV. Therefore, the work done in the differential change is as follows:

$$\delta W_{\text{out}} = PdV$$

Notice that even though both pressure and volume change, the work depends on the value of the pressure, not its change. However, the work depends directly on the change in volume, not on the value of the volume. The work would be the same if the volume were very small, or enormous.

Work is, in general, a process-dependent variable. However, like the work done by gravity, the expansion work is expressed in terms of state variables (pressure and volume), and therefore, the expansion work is an exact differential. The total work for a finite change in volume is as follows:

$$W_{\text{out}} = \int_{\text{state1}}^{\text{state2}} \delta W_{\text{out}} = \int_{\text{state1}}^{\text{state2}} PdV$$

This equation cannot be integrated until an explicit relationship between pressure and volume is determined, as will be done in future chapters. For isobaric (i.e., constant pressure) processes, the pressure can be taken out of the integral and

$$W_{\text{out}} = P \int_{\text{state 1}}^{\text{state 2}} dV = P(V_2 - V_1) \quad \text{for constant pressure processes}$$

The instantaneous rate of expansion work (power) is obtained by dividing by the differential time, dt:

$$\dot{W}_{\text{out}} = \frac{\delta W_{\text{out}}}{dt} = P\frac{dV}{dt}$$

The same expressions for work and power hold for the case where the volume decreases ($dV < 0$). In this case, work is done by the surroundings on the system, resulting in compression work. The 1st Law for the case where expansion work is the only form of work is

$$dE = \delta Q_{\text{in}} - PdV$$

There is a subtle and important assumption underlying this analysis, namely, that the pressure in the cylinder is uniform. Any spatial gradient in pressure within the fluid will tend to equalize itself, the physical mechanism of which is the propagation of sound waves. A uniform pressure will be maintained in practice if the speed of the piston is much less than the speed of sound, which for practical engines is generally a valid approximation. For very-high-speed pistons, the expansion work is still *PdV*, except that *P* represents the pressure acting on the piston face, which may differ from the average pressure in the cylinder. For example, for an expansion, the pressure at the piston face will be less than the average fluid pressure, and for a compression, it will be greater. A more formal treatment of this advanced topic requires a study of compressible flow phenomena.

Flow Work

For open systems, in which a fluid (liquid or gas) crosses the boundary, the fluid that enters or exits experiences a pressure force (from the fluid behind it) that acts through a distance. Therefore, work is associated with mass crossing the boundary, and this work is called flow work. Figure 1.11 shows a process diagram of an open system with a single inlet and single outlet at time *t* (left) and at a differentially later time *t* + *dt*. The dashed line shows a closed system that moves through the device. Fluid enters with an average inlet velocity V_{inlet} and exits with a velocity V_{outlet}. The boundary at the inlet experiences a pressure force $(PA)_{\text{inlet}}$ and moves a distance V_{inlet} dt in the same direction as the force acting on the system. The exit is similar, except that the motion is in the opposite direction as the force.

Treating the system as the mass within the device plus the mass that crosses the boundary, the work done by the system on the surroundings is as follows:

$$\delta W_{\text{flow}} = -\vec{F} \cdot d\vec{x}$$

The force shown in Fig. 1.11 is that of the surroundings on the system and, therefore, the negative sign is needed to express the work done by the system on the

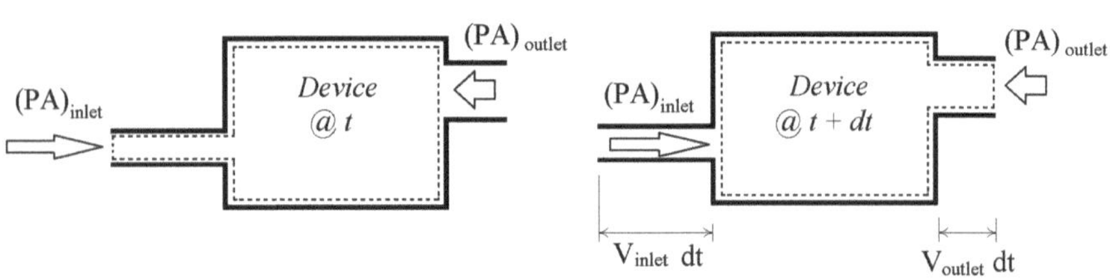

Fig. 1.11 Process diagram of a single inlet, single outlet open system focused on the flow work

surroundings. The displacement is that of the boundary where the mass moves. For mass that enters, the force and displacement are in the same direction, and the work done by the system is negative. For mass that exits, the force and displacement are in opposite directions.

The magnitude of the force is the fluid pressure (force per unit area) multiplied by the cross-sectional area of the inlet or outlet. The displacement is the fluid velocity (V) multiplied by the elapsed time. That is,

$$\delta W_{\text{flow}} = -\left(PA\,\hat{j}\right)_{\text{inlet}} \cdot \left(V_{\text{inlet}}dt\,\hat{j}\right) = -(PAVdt)_{\text{inlet}} \quad \text{for inlet flows}$$

$$\delta W_{\text{flow}} = -\left(PA\,\hat{j}\right)_{\text{outlet}} \cdot \left(V_{\text{outlet}}dt\,\left(-\hat{j}\right)\right) = (PAVdt)_{\text{outlet}} \quad \text{for outlet flows}$$

The quantity $AVdt$ is the volume of the inlet or outlet section, and it will prove convenient to express it as the mass that crosses, by multiplying and dividing the flow work by the fluid density, with the opposite sign and appropriate subscripts for the outlet:

$$\delta W_{\text{flow}} == -\frac{P_{\text{inlet}}}{\rho_{\text{inlet}}}(\rho AVdt)_{\text{inlet}} = -\frac{P_{\text{inlet}}}{\rho_{\text{inlet}}}(m)_{\text{inlet}}$$

Note that the values of all properties (pressure and density) can be different at the inlet and outlet and can also differ from the average property within the device.

It is generally more useful to consider the rate of doing work (power) by dividing by the incremental time step:

$$\dot{W}_{\text{flow}} = \frac{\delta W_{\text{flow}}}{dt} = -\frac{P_{\text{inlet}}}{\rho_{\text{inlet}}}(\rho VA)_{\text{inlet}} = -\frac{P_{\text{inlet}}}{\rho_{\text{inlet}}}\dot{m}_{\text{inlet}},$$

where the mass flow rate has been introduced: $\dot{m} = (\rho VA)$. The quantity VA is the volumetric flow rate, and in fluid mechanics is often given the symbol Q, which will be avoided here, since that symbol is used for heat transfer. The symbol $\dot{V}$ is also common. Note, however, that the symbol V is used for velocity at times, and volume at other times, and, as a general point in reading technical literature, the reader must be careful to consider the context.

A more complete treatment of open systems will be given in chapter 5.

Workshop 1.1: Frequency of a Balloon Pendulum

Working in teams of 3 or 4, weigh a balloon before filling it, then blow it up, and measure its approximate diameter "*d*" and weigh it again. Construct a pendulum by tying it to a string (make it long enough to stand on a stable chair or table). Secure the anchor at a fixed length and gently force an oscillation. Measure and record the time it takes for a number of swings (from which the natural frequency (in Hz) of the pendulum can be determined). Repeat for a wide range of the length of the string.

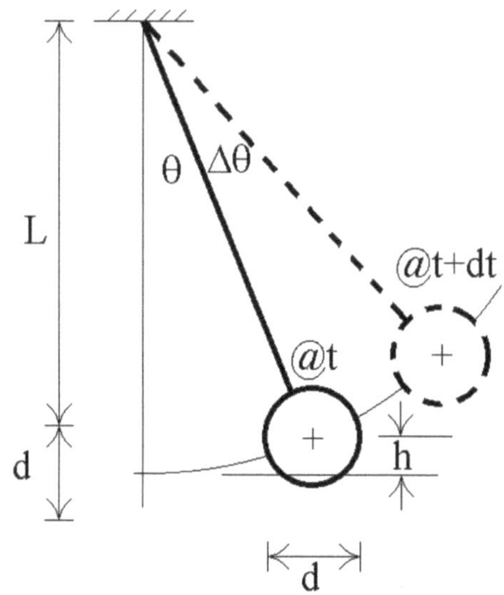

Lab Period Open a spreadsheet and save it as WKSP1_***, where *** is a short team name. Set up a RAW DATA worksheet (similar to the template of Fig. 1.12). Enter numerical values from the experiment directly into the RAW DATA section.

Prepare a well-formatted plot of the frequency (y axis) versus the length (x axis). Show data points as markers only (with no lines or curve fits). Prepare a plot of the frequency (y axis) versus $\sqrt{1/(L+d/2)}$ (x axis). The reason for this will be apparent after analysis. Show data points as markers only, and conduct a linear fit, forced through zero (show the equation in the plot). Strive for enough data points to establish a good experimental trend. Upload workbook with experimental results at the end of the lab period. Move all plots to be separate sheets, not embedded into a worksheet. One file per team.

Homework Follow-up Conduct an analysis on paper using what you learned in Mechanics (a thermodynamic treatment follows). Draw an FBD of the balloon (plus contained gas) and derive a theoretical prediction of the frequency. Neglect the mass of the string in the analysis and any aerodynamic drag force, but include a buoyant force (weight of the displaced fluid). Express the density inside the balloon as ρ_{in} and ambient density as ρ_∞, but in evaluating, it will be assumed that they are equal to that of ambient air at nominal room conditions ($\rho_{\text{in}} \approx \rho_\infty = 1.2\ \text{kg/m}^3$). Note that the inside density of a Helium or hot air balloon would be less than ambient and could result in the buoyancy force exceeding the weight. The schematic and analysis would need to be modified accordingly.

RAW DATA

$M_{unfilled}$		grams, unfilled balloon mass
d		m, filled balloon diameter
M_{filled}		grams, filled balloon mass

trial	L (m)	revs	time (sec)
1			
2			
3			
4			
5			
6			
7			
8			
9			
10			

DATA ANALYSIS

$1/(L+d/2)^{0.5}$	frequency (Hz)

Fig. 1.12 Worksheet template for RAW DATA

Set up a DATA ANALYSIS worksheet (similar to Fig. 1.13) and enter appropriate formulas.

Enter formulas in the THEORETICAL PREDICTION worksheet in the excel file. Prepare a plot that shows the experimental frequency (as markers, no lines) and the theoretical prediction (as lines, no markers) versus string length on the same set of axes. For the theory plot, do not restrict the use of length values to those measured, but make a sufficiently fine spacing of length to yield a smooth curve over the range of the experiment. Upload the updated file (overwrite the original version) before the next class (when we will discuss theoretical aspects of this example). Hand in a single hard-copy of a careful analysis at the start of class (one per team).

Detailed Solution Follows.

Do Not Consult Prior to Attempting Analysis on Your Own

Thermodynamic Analysis

The 1st Law of Thermodynamics (energy balance) applied to the closed system consisting of the balloon (rubber) and the gas inside it for a differential process taking place between the two instants of time shown in the schematic is as follows:

INPUT PARAMETERS

g	9.81	m/s^2	standard gravity value
mrubber	0.0037	kg	mass of empty balloon
rho,in	1.31	kg/m^3	density of gas inside balloon
rho,infinity	1.21	kg/m^3	density of ambient air
d	0.165	m	diameter of filled balloon

DERIVED PARAMETERS

Volume		m^3	volume of balloon
mdisp		kg	mass of displaced air
mtotal		kg	mass of rubber plus gas inside
me		kg	total mass less displaced air

THEORETICAL PLOT

Lmin		m	smallest value of string length in data set
Lmax		m	largest value of string length in data set
n			number of steps to take for plottiing function
step size		m	increment for plotting function

L	ftheory (Hz)
0	

(continue down for "n+1" entries)

Fig. 1.13 Worksheet template for THEORY. Note that the step size for plotting will be a formula (Lmax – Lmin)/n, and there will be $n + 1$ rows copied (with $n = 100$ suggested to give smooth lines). A formula in the L column will increment by the step size. This general way to set up theoretical plots in a spreadsheet to automatically adjust the different ranges can be very useful

$$\Delta E = Q_{\text{in}} - W_{\text{out}}$$

The term ΔE represents the change in the total energy stored within the system, Q_{in} represents heat that crosses into the system through the boundary (to be discussed in future chapters), and W_{out} represents the work done by the system from nongravity forces.

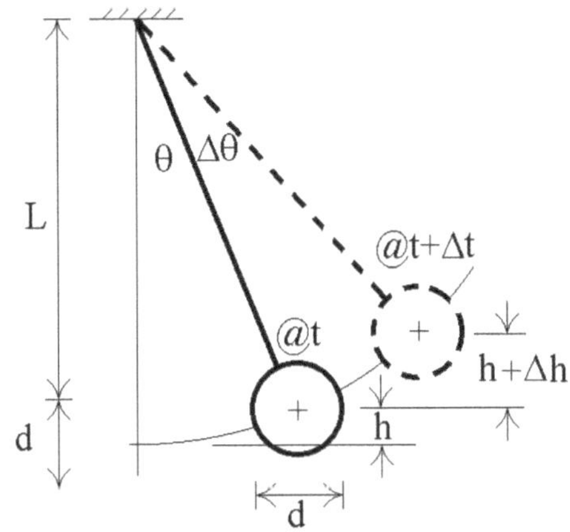

Dividing everything by the elapsed time, Δt:

$$\frac{\Delta E}{\Delta t} = \frac{Q_{\text{in}}}{\Delta t} - \frac{W_{\text{out}}}{\Delta t}$$

The stored energy (E) consists of internal, kinetic, and potential energies:

$$\frac{\Delta E}{\Delta t} = \frac{\Delta}{\Delta t}\left(U + \frac{1}{2}mV^2 + m_e gh\right) = \frac{\Delta U}{\Delta t} + \frac{1}{2}m2V\frac{\Delta V}{\Delta t} + m_e g\frac{\Delta h}{\Delta t},$$

where

U is the total stored internal energy (to be discussed in the next few weeks)
$m = m_{\text{rubber}} + m_{\text{gas}}$ is the total mass (balloon plus contained gas), treated as a point mass
$m_e = m - m_{\text{displaced}}$ is the effective mass needed for potential energy
V is the instantaneous velocity of the center of mass
h is the height of the center of mass above its position at the bottom of the arc
$g = 9.81$ m/s^2 is the standard gravity value

The work out (W_{out}) is work done by the system on nongravity forces. The work of gravity and buoyancy has already been accounted for in the potential energy. The drag force, by definition, is the net force in the direction of relative fluid motion. The work of the drag force is given by

$$\frac{W_{\text{out,drag}}}{\Delta t} = \frac{\vec{F}_{\text{drag}} \cdot d\vec{s}}{\Delta t} = F_{\text{drag}}\frac{\Delta s}{\Delta t} = F_{\text{drag}}V$$

The drag force acting on the system opposes the motion, but the drag force by the system is in the direction of motion. Therefore, the work out is positive (and the friction work decreases the energy stored). When the pendulum moves in the other direction, the sign of the drag force also changes, and therefore, the work done by the system on the environment is always positive. This type of force is a so-called nonconservative force, whereas buoyancy is conservative. Buoyancy work changes sign and the net work of buoyancy for a cycle is zero, so energy is conserved. The 1st Law becomes, upon rearrangement,

$$mV\frac{\Delta V}{\Delta t} + m_e g\frac{\Delta h}{\Delta t} = \left[\frac{Q_{\text{in}}}{\Delta t} - \frac{\Delta U}{\Delta t}\right] - F_{\text{drag}}V$$

Dividing through by mV and letting the time step shrink to a differential time step (i.e., $\Delta t \to dt$, and therefore, all variables expressed as changes become, e.g., $\Delta V \to dV$):

$$\frac{dV}{dt} + \left(\frac{m_e}{m}\right)\frac{g}{V}\frac{dh}{dt} = \frac{1}{mV}\left[\dot{Q}_{in} - \frac{dU}{dt}\right] - \frac{F_{drag}}{m}$$

Note that the heat transfer term has been expressed as

$$\frac{Q}{\Delta t} = \dot{Q}$$

In traditional thermodynamic analysis, a "dot" over a variable signifies a FLUX of the quantity across a control surface, and is not a time derivative as in traditional mechanics or analysis or calculus courses.

Some kinematic relations are used to express velocity and height in terms of angle:

$$V = L_{\text{CM}}\frac{d\theta}{dt} \quad \text{and} \quad \frac{dV}{dt} = L_{\text{CM}}\frac{d^2\theta}{dt^2}$$

$$h = L_{\text{CM}} - L_{\text{CM}}\cos\theta \quad \text{and} \quad \frac{dh}{dt} = L_{\text{CM}}\sin\theta\frac{d\theta}{dt}$$

The 1st Law becomes, upon further rearrangement,

$$\frac{d^2\theta}{dt^2} + \frac{g(m_e/m)}{L_{\text{CM}}}\sin\theta = \left[\frac{\dot{Q}_{in} - dU/dt}{mV}\right] - \left[\frac{F_{\text{drag}}}{mL_{\text{CM}}}\right]$$

The first bracketed term on the right represents "thermal effects," in this case heat transfer (energy driven by a temperature difference) and a rate of change of internal energy. These effects are neglected here, without formal justification, and will be addressed in future chapters.

The second bracketed term represents the effect of fluid drag on the motion. It may seem intuitive that this effect, which changes in magnitude with the instantaneous angle, is important. Its inclusion would require a fluid mechanical analysis, beyond the intended scope of this text. Therefore, it will be neglected, without formal justification, but it is expected (from experience) that drag will have a large effect on amplitude, but a minor effect on the natural frequency of the system.

Therefore, neglecting thermal and fluid drag effects, and invoking the small angle approximation ($\sin\theta \approx \theta$, which yields less than a 1% error for angles up to 15°), the governing differential equation becomes

$$\frac{d^2\theta}{dt^2} + \omega_N^2\theta = 0,$$

where

$\omega_N = \sqrt{\frac{(m_e/m)g}{L_{CM}}}$ is the natural frequency

This second-order linear ordinary differential equation describes a classical harmonic oscillator. The first term is the inertial effect, and the second term is the restoring force.

The general solution is

$$\theta = A\sin(\omega_N t) + B\cos(\omega_N t)$$

If released from rest ($d\theta/dt|_{t=0} = 0$) from an initial angle θ_0, the particular solution is

$$\theta = \theta_0 \sin(2\pi f t)$$

The natural frequency is

$$f = \frac{1}{2\pi}\sqrt{\frac{g(m_e/m)}{(L + d/2)}}$$

A straight line for a plot of the frequency versus $\sqrt{1/(L + d/2)}$ through the origin is predicted (a good test of both the theory and the experiment), and the slope has a value:

$$\text{Slope} = \frac{\sqrt{g(m_e/m)}}{2\pi}$$

Students that Google for the formula of the frequency of a pendulum will likely obtain a result that leaves out the factor m_e/m, which is not negligible in this case, part of the point of the exercise.

Results and Discussion

In the Fall 2015, Fall 2016, Fall 2017 (in Fluids course), and Spring 2018 (Thermo 1 course) semesters, student teams conducted experiments at the Cooper Union. Sample results of one team are shown in Fig. 1.14 (raw data) and Fig. 1.15 (after processing). This team produced a well-spaced set of data with minimal scatter that shows that the frequency decreases with string length. A main point of the exercise is for students to "experience" this effect by measuring it themselves. The shape predicted by theory is confirmed in Fig. 1.15.

Results of all the teams are shown in Fig. 1.16, in which the slopes are plotted versus the balloon diameter. There is good agreement of the shape of this curve between theory and experiment, with the experimental frequencies being approximately 25% lower than the theoretical prediction, and significant scatter. It is likely that the data analysis was done incorrectly for several tests, and reflects that another goal of this exercise is to be an "ice-breaker" exercise in team-based data collection and analysis, with limited time, practice, and instruction. The quality of the data would likely improve if the exercise were repeated.

One effect that was neglected in the model is damping (due to fluid drag) on both the balloon and the string. Inclusion of damping would reduce the predicted natural frequency somewhat, in the wrong direction. Another effect is the treatment of the balloon as a point mass. The kinetic energy of parts of the balloon below its center of

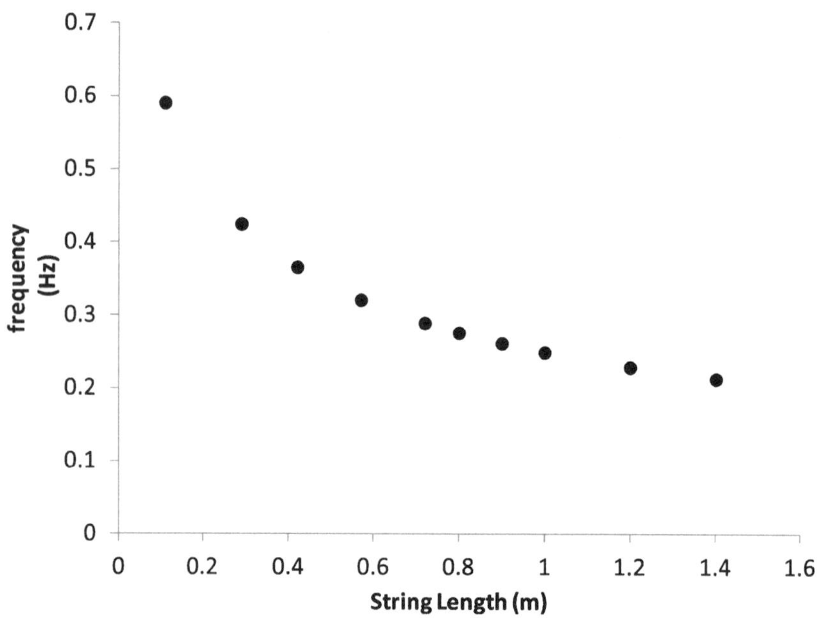

Fig. 1.14 Sample raw experimental data of the frequency of oscillation vs. the string length. The balloon diameter was 0.165 m

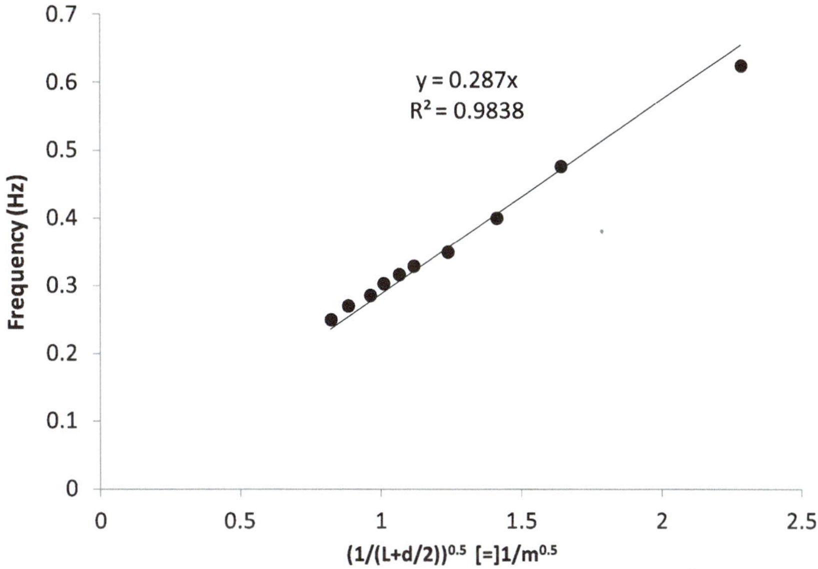

Fig. 1.15 Sample experimental data plotted vs. square root of the effective pendulum length ($L + d/2$). The equation of a linear fit and the correlation coefficient (R^2) values are shown

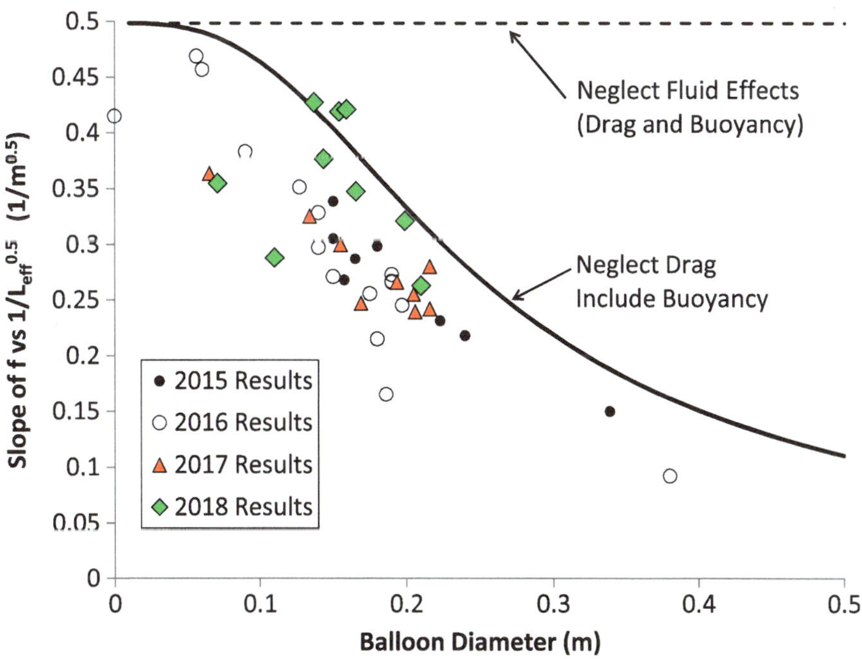

Fig. 1.16 Experimental slope vs. balloon diameter. Theoretical prediction is shown as a solid line. Results from individual teams are shown as symbols

mass is higher than those above, and the effective string length would be larger if this effect were included. The string is also treated as being rigid. No further attempts to include these effects are made.

It is concluded that a simple, inexpensive classroom experiment of a balloon pendulum can be used to tie together concepts from engineering experimentation, mechanics, fluid mechanics, and thermodynamics.

Appendix 1: Other Forms of Stored Mechanical Energy

Mechanical energy can be stored in other modes than linear kinetic energy and near-planet gravitational potential energy. A few of these modes are developed in this Appendix. The results reported in Table 1.1 can be used when needed going forward without going through these derivations, but going through these derivations builds a stronger foundation.

Far Planet Gravitational Potential Energy

For applications in fields such as astrophysics and astronautics, changes in altitude typically occur for which the local acceleration due to gravity (g) does not equal the standard gravity value (g_0). This section considers the potential energy associated with those changes. For a target mass, m, the potential energy equals the work done on the object by a fictitious external force that changes the altitude. The planet has a radius R and mass M. The reference height is taken to be the planet surface. That is,

$$E_{\mathrm{PE}} = \int_{\text{planet surface}}^{\text{altitude, } h} F_{\mathrm{Grav}} dh = \int_{h=0}^{h} G \frac{Mm}{(R+h)^2} dh$$

To integrate, introduce the substitution:

$$u = R + h$$
$$du = dh$$

Inserting, integrating, and manipulating:

$$E_{\mathrm{PE}} = GMm \int_{R}^{R+h} \frac{du}{u^2} = GMm \left[\frac{-1}{u} \right]_{R}^{R+h} = -GMm \left[\frac{1}{R+h} - \frac{1}{R} \right]$$

$$= -GMm \left[\frac{R-(R+h)}{R(R+h)} \right] = \frac{GMm}{R} \frac{h}{R+h}$$

Recall how the standard gravity value relates to the planet mass and radius, namely,

$$g_0 = \left[\frac{GM}{R^2} \right]$$

A few manipulations (multiply and divide by R and rearrange) yield a simple expression for the potential energy:

$$E_{\mathrm{PE}} = \frac{mg_0 h}{1 + h/R}$$

The reference state is the planet surface ($h = 0$), and the potential energy reduces to the near-planet potential energy (mg_0h)when the altitude is much less than the planet radius ($h \ll R$).

Also, there is a distinct maximum potential energy obtained when the altitude is much greater than the planet radius ($h \gg R$), namely, $E_{\mathrm{PE,\ max}} = mg_0R$. The quantity g_0R represents the maximum potential energy per unit mass, which evaluates to 62.5 MJ/kg for Earth. This represents the minimum kinetic energy that a mass launched vertically from Earth's surface would need to escape the effects of Earth's gravity. The corresponding escape velocity is $V_{\mathrm{escape}} = \sqrt{2g_0R}$, which evaluates to 11,180 m/s (25,000 miles per hour) on Earth. It is not easy to launch spacecraft from Earth.

Linear Springs

Most solid materials are elastic, that is, they deform to some degree under an applied load (less than a critical value) and will return to their unloaded state unaltered if the load is removed. If the load exceeds the critical value (called the yield point), the material is no longer perfectly elastic, and permanent deformation may occur or the material will break. Most materials obey Hooke's Law below the yield point, which states that the deformation of a solid is proportional to the applied load, that is, $\Delta L = kF$, where ΔL is the change in dimension from the relaxed state, and F is the applied load (force). The proportionality constant (k) depends on material and geometric properties. A linear spring is an ingenious and common engineering tool in which a solid material in wire form is coiled in such a way that the

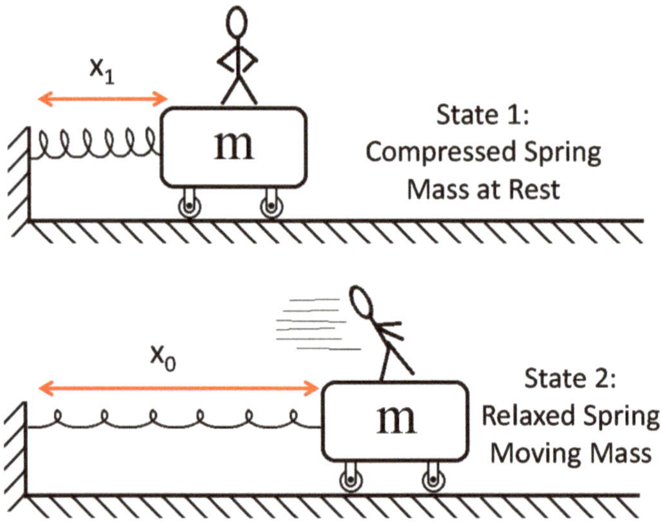

Fig. 1.17 Mass/spring system. The top schematic shows the spring in a compressed condition (with a force acting on the mass to the right), the bottom schematic is in the relaxed position (no force acts), but the mass is moving to the right

deformation is magnified compared to what it would be for a straight material. It also allows for the material to act in both tension and compression (without buckling).

The following example shows a form of potential energy that can be stored in a linear spring:

Example A cart of mass $M = 3.2$ kg with wheels of negligible mass and friction is connected to a linear spring with a spring constant $k = 12$ N/m to a rigid wall as shown in Fig. 1.17 and a relaxed length of $x_0 = 2.7$ m. In state 1, the spring is compressed to a length $x_1 = 0.5$ m by external means (not shown) and released from rest. At state 2, the spring is relaxed. What is the speed of the cart at state 2?

Thermodynamic Analysis A linear spring is one for which the force required to change its length is proportional to the change in length, that is, in the geometry of Fig. 1.17, $F_{\text{spring}} = -k(x - x_0)$. Note that the direction of the spring force is positive if the spring is compressed ($x < x_0$) and negative if the spring is stretched ($x > x_o$). That is, the force is in the direction of the relaxed spring, and is a "restoring force." However, if the mass is in motion, its inertia can prevent it from coming to rest when the spring is relaxed. This mass/spring system can be analyzed with time as the independent variable, but will be treated from a mechanical energy (i.e., thermodynamic) view to establish that a compressed or stretched spring has stored energy that can be classified as another form of potential energy.

The motion is constrained in the x-direction. Newton's 2nd Law in the x-direction applied to the mass is

$$m\frac{dV}{dt} = \sum F_x = -k(x - x_0) \mp F_{\text{friction}},$$

where a friction force acts in the direction opposite to the velocity, which will be minimized in practice by the use of rollers, and neglected here. The kinematic relation between velocity and position is as follows:

$$V = \frac{dx}{dt}$$

Inserting into the momentum equation, and neglecting the friction force,

$$mV\frac{dV}{dx} + k(x - x_0) = 0$$

Multiplying through by the differential displacement, and recognizing that $dx = d(x - x_0)$:

$$mVdV + k(x - x_0)d(x - x_0) = 0$$

The 2nd term is the spring force multiplied by a differential displacement, that is, the work done by the spring. Integrating between two arbitrary states, and assuming a linear spring (that is, constant k),

$$\int_{V_1}^{V_2} mVdV + \int_{x_1}^{x_2} k(x - x_0)d(x - x_0) = 0$$

$$\left(\frac{1}{2}mV_2^2 - \frac{1}{2}mV_1^2\right) + \left(\frac{1}{2}k(x_2 - x_0)^2 - \frac{1}{2}k(x_1 - x_0)^2\right) = 0$$

There are two forms of storable energy on the left side; kinetic energy, and the energy stored in the spring, a form of potential energy:

$$E_{\text{Spring}} = \frac{1}{2}k(x - x_0)^2,$$

where x is the spring length, and $x - x_0$ is distance the spring is displaced from its relaxed length. A concise form of the energy balance is

$$\Delta E = 0,$$

where

$$E = \frac{1}{2}mV^2 + \frac{1}{2}k(x - x_0)^2$$

The sum of the kinetic and potential (spring) energy is conserved in the absence of friction. Friction will tend to degrade this mechanical energy. A formal proof of that will be considered after a more thorough development of 2nd Law concepts in the next chapter. The purpose of this section is to establish the ability to store mechanical energy in linear springs.

For the problem at hand, since the mass starts from rest, the initial energy equals the potential energy stored in the spring at the moment of release. Since the sum of the kinetic and potential energy is constant, the kinetic energy will be maximized when the potential energy is minimized, that is, when the spring is relaxed. Therefore, to find the maximum speed, define state 2 to be when the spring is relaxed. The mechanical energy balance is

$$E_1 = E_2$$

$$\frac{1}{2}k(x_1 - x_0)^2 = \frac{1}{2}mV_2^2$$

Solving for the speed, and recognizing that absolute values should be used when the square root of a squared number is taken:

$$V_2 = |(x_1 - x_0)|\sqrt{\frac{k}{m}}$$

Note that the stiffer the spring and the lighter the mass, the higher the speed. Evaluating for the case at hand:

$$V_2 = |(0.5\ \text{m} - 2.7\ \text{m})|\sqrt{\frac{12\ \text{N/m}}{3.2\ \text{kg}}} = 4.26\ \text{m}\sqrt{\frac{\text{N/m}}{\text{kg}}}$$

Check the units. $\text{m}\sqrt{\frac{\text{N}}{\text{m}\cdot\text{kg}}}[=]\sqrt{\frac{\left(\frac{\text{kg}\cdot\text{m}}{\text{s}^2}\right)\text{m}}{\text{kg}}}[=]\sqrt{\frac{\text{m}^2}{\text{s}^2}}[=]\frac{\text{m}}{\text{s}}$

The final result is

$$V_2 = 4.26\ \frac{\text{m}}{\text{s}}$$

The mass/spring system will oscillate in simple harmonic motion. This behavior can be shown by introducing time by differentiation:

$$\frac{dE}{dt} = 0 = \frac{d}{dt}\left(\frac{1}{2}mV^2 + \frac{1}{2}k(x - x_0)^2\right)$$

$$mV\frac{dV}{dt} + k(x - x_0)\frac{dx}{dt} = 0$$

$$m\frac{dx}{dt}\frac{d^2x}{dt^2} + k(x - x_0)\frac{dx}{dt} = 0$$

$$\frac{d^2(x - x_0)}{dt^2} + \frac{k}{m}(x - x_0) = 0$$

The general solution to this second-order ordinary differential equation is

$$(x - x_0) = A\sin(\omega_n t) + B\cos(\omega_n t),$$

where the natural frequency (in radians per unit time) is

$$\omega_n = \sqrt{\frac{k}{m}}$$

The initial conditions, released from rest when $x = x_1$, are

$$(x - x_0)|_{@t=0} = (x_1 - x_0) \quad \Rightarrow \quad B = (x_1 - x_0)$$

$$\frac{dx}{dt_{@t=0}} = 0 \quad \Rightarrow \quad A = 0$$

The solution to the position versus time is

$$x = x_0 + (x_1 - x_0)\cos(\omega_n t)$$

Or, in simple dimensionless form,

$$\frac{x - x_0}{x_1 - x_0} = \cos(\omega_n t)$$

There are many important applications for macroscopic systems that behave like mass-spring systems that give rise to phenomena such as vibrations. These systems are the focus of more advanced treatments of mechanics. At the microscopic level of molecules in which chemical bonds join atoms, the atoms vibrate as a mass-spring system. The intensity of the vibrations is related to the temperature. The stored energy in both kinetic and potential energy of the atoms (as well as their rotation) contributes to internal energy.

Rotating Flywheel

Consider a small mass, m, attached to a string of length r and being rotated around a central hub with an angular velocity ω.

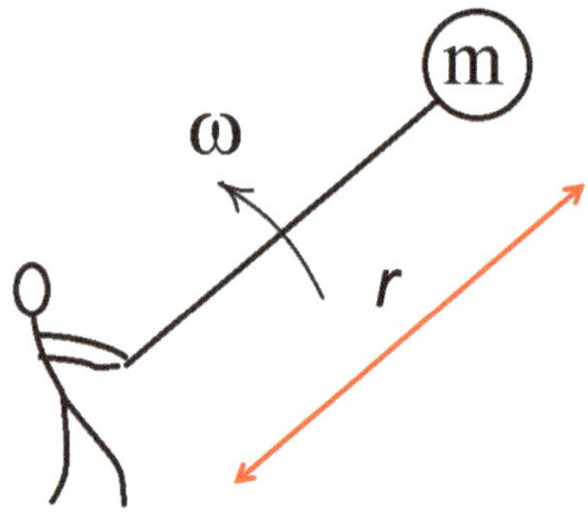

The linear velocity (V) and angular velocity (ω) are related by

$$V_m = \omega r$$

The kinetic energy of the mass is

$$\mathrm{KE} = \frac{1}{2} m V_m^2 = \frac{1}{2} m (\omega r)^2$$
$$\mathrm{KE} = \frac{1}{2} m r^2 \omega^2 = \frac{1}{2} I_m \omega^2,$$

where the quantity mr^2 has been defined as the 2nd mass moment of inertia, I:

$$I_m = mr^2$$

Now consider a symmetrical flywheel that rotates about its central axis with an angular velocity ω. The flywheel can be considered to consist of a large number of individual masses dm, all at different distances r from the hub, all with kinetic energy. The total kinetic energy is

$$KE = \sum \frac{1}{2} dm V_m^2 = \frac{1}{2} \sum dm(\omega r^2) = \frac{1}{2} \left(\sum mr^2 \right) \omega^2$$

The 2nd mass moment of inertia is defined as

$$I = \sum mr^2 = \iiint r^2 dm = \iiint r^2 \rho d\forall$$

For a solid circular disk of uniform composition (density ρ) with a radius R and thickness t, the 2nd mass moment of inertia is

$$I = \int_{r=0}^{R} r^2 \rho(t2\pi r dr) = \frac{\pi \rho t R^4}{2}$$

The total mass is

$$M = \int_{r=0}^{R} \rho d\forall = \rho \int_{r=0}^{R} t2\pi r dr = \rho t \pi R^2$$

Therefore, substituting for the mass into the moment of inertia:

$$I = \frac{MR^2}{2} \quad \text{for a solid circular disk of total mass } M, \text{radius } R$$

The energy stored in the flywheel is

$$E_{\text{Rotation}} = \frac{1}{2} I \omega^2$$

Actual flywheels are important engineering devices in rotating equipment (e.g., automobile engines) where the stored energy helps to maintain steady rotational motion. They also may play an important future role as energy storage devices with renewable energy systems such as solar and wind, where the demand for energy is not synchronized with the available input energy.

At the microscopic level, for gases, the rotation of the molecules contributes to the stored internal energy in addition to the simultaneous vibrational energy.

Batteries

Electrical energy, a high value form of energy like mechanical energy, can be stored in batteries and capacitors. There are certain parallels between gravity and electromagnetism (and much that is fundamentally different). For example, the force between two electrical charges χ_1 and χ_2 separated by a distance r_{12} in a medium with a material property called permittivity ε is

$$F_{12} = \frac{1}{2\pi} \frac{\varepsilon_0}{\varepsilon_r} \frac{\chi_1 \chi_2}{r_{12}^2},$$

where $\varepsilon_0 = 8.85(10)^{-12}$ s^2 · C^2/m^2/kg is a fundamental constant of nature, the permittivity of a perfect vacuum, and $\varepsilon_r = \varepsilon/\varepsilon_0$ is the relative permittivity, also called dielectric constant. Notice that, like gravity, the force between two electrical charges obeys an inverse square law with the distance between them. The symbol χ is used to indicate electrical charge (with SI unit C for Coulomb, the charge of 6.242 $(10)^{18}$ electrons), not the conventional Q, which is reserved exclusively for heat transfer in this text (and is also used in fluid mechanics to indicate volumetric flow rate). In the present analogy, electrical charge is like mass and the gravitational constant G is like $1/(2\pi\varepsilon)$. Unlike gravity, charge can be negative or positive. The force is attractive if the charges are of different sign, and positive if their charge is of like sign. Electrons have negative charge, while protons have positive charge, the sign being an arbitrary convention.

In an electrical conductor, protons are fixed in position, bound by strong molecular bonds in a lattice. They vibrate about this fixed position, and therefore have stored kinetic and potential energy at the molecular level. This stored energy is what constitutes the internal energy of the material and is related to temperature through the specific heat. Electrons are loosely bound outside the nucleus and thus free to move through the lattice, and will do so if subjected to a force. Electrical current (I) in an electrical conductor is the movement of free electrons. Due to the sign convention, if electrons move from left to right, the current moves right to left.

An electric field is established in the space around electrical charges as the force that would be experienced by an infinitesimal test charge. Voltage (given the symbol ψ instead of V, which is reserved for velocity) is an energy measure associated with movement of charge in an electric field, that is, a force acting through a distance. The voltage difference between two locations in space (1 and 2) is defined as the theoretical work that would be required to move a unit test charge from one location to the next in an electric field. The test charge is infinitesimal so as not to alter the electrical field as it moves. That is,

$$\Psi_2 - \Psi_1 = \int_1^2 \frac{\vec{F} \cdot d\vec{x}}{\chi_{\text{test}}} \; [=] \frac{\text{N} \cdot \text{m}}{\text{Coulomb}}$$

The energy stored in batteries is chemical in nature, but as engineering sources of electrical energy, a battery (Fig. 1.18) sends electrical charge at a fixed voltage from its positive terminal (Ψ^+), through an electrical circuit, and returns the same amount of charge to the negative terminal (Ψ^-). The voltage (Ψ) is the difference between positive and negative terminals. The electrical current, I, is the rate at which charge is supplied. The voltage of an ideal battery is not affected by the current. The voltage of an actual battery decreases somewhat with current drawn (due to internal resistance), and there is a clear maximum current that can be drawn.

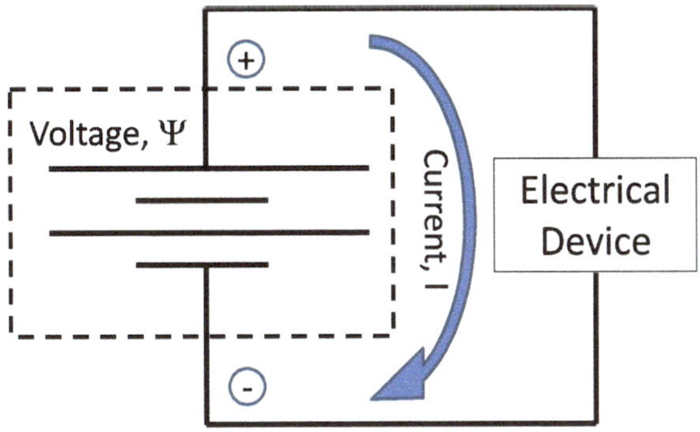

Fig. 1.18 Two cell battery driving an electrical device

The capacity of a battery, listed as "amp-hours" in a specification (spec) sheet, is related to the quantity of charge it can deliver. For example, a typical single AAA battery has a voltage of 1.5 V (Ψ_{cell}) and a capacity of 1.2 amp-hours. The total charge stored in it is therefore

$$\chi_{AAA} = 1.2\left(\frac{\text{Coulomb}}{\text{s}}\right) \cdot 1\,\text{h} \cdot \left|\frac{3600\ \text{s}}{\text{h}}\right| = 4200\ \text{Coulomb}$$

The total stored energy in a fully charged AAA battery is

$$E_{AAA,stored} = \Psi_{cell} \cdot \chi_{AAA} = \left(\frac{1.5\ \text{N}\cdot\text{m}}{\text{Coulomb}}\right)(4200\ \text{Coulomb}) = 6300\ \text{N}\cdot\text{m}$$

Capacitors

Energy storage in capacitors is truly electrical in nature, unlike batteries, which is chemical. Capacitors are used widely in electronic devices and serve a similar role as flywheels in mechanical systems. The basic construction of most capacitors (shown schematically in Fig. 1.19) is that of two conductive plates of surface area A separated by a distance d by a gap of material called a dielectric, which has a permittivity ε. A conducting wire is attached to each plate to allow electrons to flow in or out. The left schematic shows an uncharged state. The middle schematic shows a process in which electrons are forced (green arrows) onto the bottom plate, while the same number of electrons are removed from the top plate. The capacitor on the right is charged. The bottom plate has excess electrons, and a negative charge, while the top plate has been partially depleted of electrons, leaving positively charged protons without their free electrons to balance the net charge of the atom.

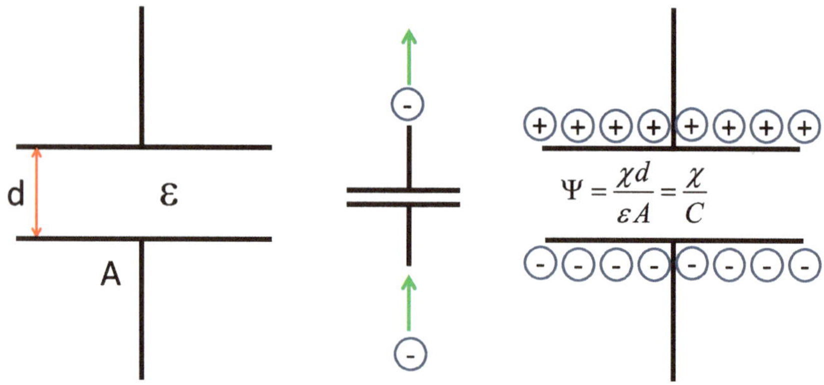

Fig. 1.19 Schematic of a capacitor. Left schematic is of a capacitor in an uncharged state. Middle is during a charging process (with electrons forced into the bottom plate and simultaneously removed from upper plate). Right shows a charged capacitor

The primary metric that defines a capacitor is its capacitance (symbol C), which is a measure of the total charge difference between the plates required to maintain a unit voltage difference between them. That is,

$$C = \frac{\chi}{\Psi} \; [=] \frac{\text{Coulomb}^2}{\text{N} \cdot \text{m}} [=] \text{Farad}$$

Electrostatic theory shows that the capacitance is proportional to the permittivity of the dielectric (ε) and the area of the plates (A) and inversely proportional to the spacing (d). That is,

$$C = \frac{\varepsilon A}{d}$$

In the design of a capacitor, a general goal is to maximize the surface area of the plates, minimize their spacing, and use a dielectric material with as high a permittivity as possible, while having a low electrical conductivity.

An energy balance can be used to relate the electrical energy stored in a capacitor to its capacitance and the voltage difference across its poles. Defining the system to include the physical structure of the capacitor plus the electrons that enter one pole and leave the other, an overall energy balance for a process between two states is

$$\Delta E_{\text{capacitor}} = Q_{\text{in}} - W_{\text{out}}$$

The process is assumed to be adiabatic ($Q_{\text{IN}} = 0$) and isothermal (so that the energy change is due to the electrical energy stored, not thermal energy). In practice, the charging or discharging can be achieved slowly to approximate that behavior.

It takes work to add or remove electrons from the plates. If a differential amount of electrons with charge $d\chi$ leave one plate, the same number of electrons enters the other. If the voltage difference across the plates at the time is ψ, the difference work done by the system is $-\Psi d\chi = -\Psi(Cd\Psi)$ where the negative sign indicates that if net charge is added, work must be done on the system. The total work to change the charge of the capacitor between two states is obtained by integration, and using the relation among voltage, charge, and capacitance:

$$W_{\text{OUT}} = \int_1^2 -\Psi d\chi = -\int_1^2 \Psi C d\Psi = -\frac{1}{2}\left(C\Psi_2^2 - C\Psi_1^2\right)$$

The energy balance becomes

$$\Delta E_{\text{capacitor}} = \frac{1}{2}C\Psi_2^2 - \frac{1}{2}C\Psi_1^2$$

The energy stored in a capacitor is therefore shown to be related to the capacitance and voltage by

$$E_{\text{capacitor}} = \frac{1}{2}C\Psi^2$$

Problem Statements

1. *Flywheel Energy Storage*

This problem is meant to determine whether it would be realistic to use flywheel energy storage for an "off-grid" solar electric installation. Solar Photovoltaic (PV) technology is evolving to the point where the direct conversion of radiant energy available in sunlight to direct current (DC) electricity) is economically attractive. However, the sun does not shine at night. Therefore, use of solar PV by itself ("off-grid," without a utility back-up) requires a means to store energy to meet electricity demand. For example, a system would consist of a PV array directly driving a DC motor, which drives the shaft of the flywheel, which drives a DC generator. When the supply exceeds demand (the sun is shining), the flywheel accelerates. When demand exceeds supply (night), the flywheel decelerates.

Consider a solid cylindrical flywheel with a radius R and thickness t made of a material of density ρ. For a range of thicknesses, calculate the radius, total mass, and material cost needed to store an amount of energy E_{store} when operating between a minimum rotational speed of N_{min} (considered to be discharged) and a maximum of N_{max}. Evaluate for the case where the energy stored equals the equivalent average electrical consumption of a household in the USA, the material is steel, and the minimum and maximum rotational speeds are 1000 and 2000 RPM, respectively. Is it realistic?

2. *Rolling Races*

Two cylinders, each of length L, outer diameter D, and mass M, roll down an inclined plane (of angle θ with the horizontal and initial height H) without slipping. One cylinder is hollow with a wall thickness $t \ll D$. The other cylinder is of uniform construction, but a lower density material so that the masses of the two cylinders are equal. Determine the ratio of their densities (neglect the mass of air in the hollow cylinder), and determine their velocities when they reach the bottom of the hill, using the 1st Law of Thermodynamics. Which one wins the race?

3. *Elastic Collisions*

Two masses, m and M, are on a collision path, one with speed $-v_1$ (moving to the left) and the other with speed V_1 (moving to the right). One of them has a spring with spring constant k attached and aligned to capture the other. The spring compresses, and then recoils with no loss of total mechanical energy. Determine the distance the spring is compressed and their speeds after the spring is relaxed again.

Chapter 2
Thermodynamics of Collisions

Momentum is conserved when two objects collide. However, mechanical energy is not conserved, except for elastic collisions in which case the relative speed between the particles is the same before and after the collision. For inelastic collisions, there is a conversion from high-grade mechanical energy to a form of low-grade thermal energy. This chapter is an investigation of the thermodynamics of collisions, in which any mechanical energy "lost" in collisions appears as thermal energy, first as internal energy and later as heat transfer. All the energy is accounted for (a 1st Law consequence), but the theoretical ability of the system to do work (eXergy) is reduced (a 2nd Law consequence). This chapter introduces thermal energy forms, namely, internal energy (a property that is a storable form of energy, or property) and two forms of transferrable (or process-dependent) thermal energy forms: heat transfer and eXergy destruction. An additional mechanical energy property emerges by combining the energy and entropy balances, namely, eXergy. The relationship between storable forms and temperature for solids and nonphase changing liquids is developed. Future chapters consider phase changes.

Workshop: Colliding Bullets

Two lead bullets of unequal mass ($m = 4$ grams and $M = 8$ grams) travel on a collision course with equal and opposite momentum (state 1). They are at the same temperature as the local environment ($T_\infty = 25\ ^\circ C$) prior to colliding. The smaller mass (m) has a speed $v = 900$ m/s. They collide at a height $H = 10$ m above ground, stick together without breaking apart in any way (state 2′), and fall to the ground (state 2″ just before landing) and come to rest after hitting the ground (state 2). They gradually cool back to local ambient temperature of 25 °C (state 3). Complete the state and process tables in the template of Fig. 2.1, *paying attention to*

G. Sidebotham, *An Inductive Approach to Engineering Thermodynamics*,
Mechanical Engineering Series, https://doi.org/10.1007/978-3-030-20430-3_2

	A	B	C	D	E	F	G	H	I
1	INPUT PARAMETERS								
2	m	0.004	kg, mass of smaller particle						
3	M	0.008	kg, mass of smaller particle						
4	v	900	m/s, initial speed of smaller particle, (relative to the ground)						
5	e	0	coefficient of restitution						
6	H	3	m, height where collision occurs				Color Code		
7	c	128	J/kg/DK, specific heat of both particles				Inputs		
8	Tinf	25	oC, dead state temperature (ambient)				Momentum		
9							1st Law		
10	DERIVED PARAMETERS						2nd Law		
11	V		m/s, initial speed of larger particle (relative to the ground)						
12									
13	STATE TABLE (for combined system, m and M)								
14	STATE	T (oC)	KE (J)	PEg (J)	U (J)	S (kJ/Δ°C)	eX=U-T_∞S (J)		
15	1	25						Prior to collision	
16	2'							immediately after collision	
17	2"							just before hitting ground	
18	2							after coming to rest on the ground	
19	3	25						after cooling to ambient	
20									
21	PROCESS TABLE (all quantities in J)								
22	PROCESS	ΔU	ΔKE	ΔPEg	Qin	Wout	T_∞ΔS	(T_∞/T_B)*Q	eX_D
23	1-2'								
24	2'-2"								
25	2"-2								
26	2-3								
27	net								

Fig. 2.1 Spreadsheet template for the Colliding Bullets Workshop. The appropriate formulas will be added as the analysis evolves

- *The speed of the larger mass before the collision (V)*
- *The average temperature of the two combined masses immediately after the collision (T_2) and after they settle on the ground*
- *The eXergy destruction associated with both the immediate collision and its aftermath*

The first question involves basic mechanics (you probably know the answer already), the second involves the First Law, and the third involves combining the First and Second Laws of Thermodynamics. The complete analysis is organized into a Workshop template (Fig. 2.1). Input parameters are listed with values from the problem statement entered as default values. It is suggested that readers create their own spreadsheet that mimics this structure, and once complete, change input parameters to conduct a parametric study of interest. Blank cells require formulas that reference input parameters. They are color-coded to indicate whether they are based on momentum, 1st Law or 2nd Law calculations. The results of the completed spreadsheet will be presented as appropriate formulas are developed.

A somewhat more general treatment of the colliding bullets problem will be considered, allowing for the two masses to bounce off each other. Consider two masses travelling on a horizontal collision course. They could be two subatomic particles in an atom smasher, two molecules of a gas, two billiard balls, a rubber ball released from rest, a meteor on a collision course with Earth, or, as in this case, two bullets. There will be a gravitational attraction between the two objects, however, that internal force between the two objects does not affect their center of mass. The smaller sphere has a mass of m and the larger a mass M. They will be treated as particles (with no energy stored in rotation) that experience a collision with a central impact. Both particles retain their basic shape (with some permanent deformation like denting possible) after the collision. That is, they do not break apart, nor do they pass through each other.

Momentum Considerations

Appendix 1 (a deeper dive) formally treats collisions from a general mechanics viewpoint using vectors. The final results are summarized in this section. Figure 2.2 shows a collision event in a coordinate frame fixed to the center of mass of the two colliding partners. The larger particle has mass M and speed V, while the smaller has mass m and speed v. Velocities are shown as vectors. Subscripts 1 and 2 refer to before and after the collision, respectively. The masses are in contact for an immeasurably short period of time (Δt) during which they are first brought

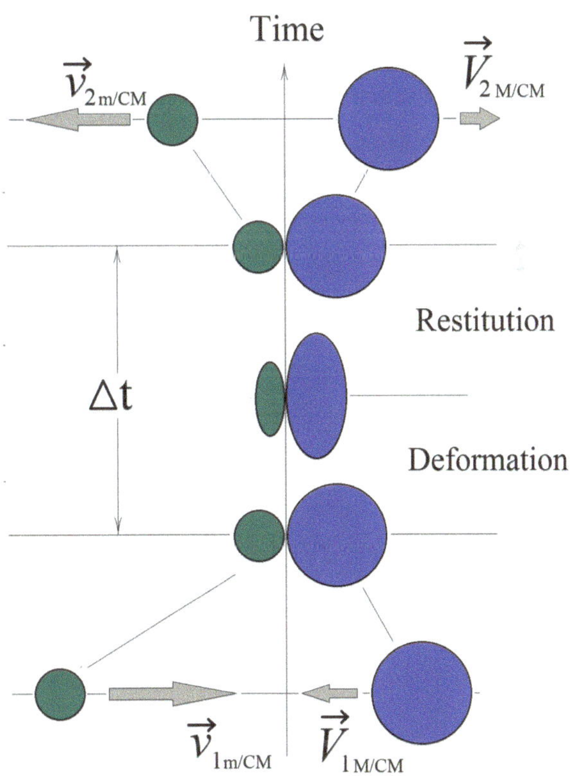

Fig. 2.2 Collision event in coordinates xy fixed to the center of mass. Snapshots are shown before the event, at the start of the event, at the turning point of the event, at the end of the event, and after the event

to rest (period of deformation), then accelerated to their final speeds (period of restitution). The masses experience immeasurably large internal forces during the collision. The two bullets in the problem here do not recoil from each other, a limiting case. Also, the Workshop problem is stated such that the momentum of the two particles is equal in magnitude and opposite in sign in the reference frame of the ground, and therefore, the center of mass of the bullets is motionless in that frame (horizontally).

Conservation of linear momentum in scalar form, with directions accounted for so that all terms are inherently positive, is

$$mv_1 - MV_1 = -mv_2 + MV_2 = 0$$

Given known approach velocities (state 1), this statement of the conservation of momentum provides one equation for two unknowns, the final velocities. For the problem at hand, both final velocities are zero, and the initial speed of M is simply $V_1 = mv_1/M$, a formula, which can be entered into the only momentum calculation of the Workshop. Since the M is twice m, the speed required for it to have the same momentum is half of v.

Another constraint or independent relationship is needed to solve for the two final velocities in the more general case in which the two particles recoil from each other. It is standard to define a coefficient of restitution, ε, as the ratio of impulses of the restitution to the deformation, shown formally in mechanics textbooks and Appendix 1 to equal the ratio of the relative receding speed after the collision ($V_2 + v_2$) to the relative approach speed before the collision ($V_1 + v_1$):

$$\text{Coefficient of Restitution} = \varepsilon = \frac{v_2 + V_2}{v_1 + V_1}$$

It is difficult to predict the value of the coefficient of restitution from first principles. It merely forms a convenient way to express the final velocities in terms of a potentially measurable effectiveness of the collision. Appendix 1 shows that the final speeds are simply related to the initial speeds as follows:

$$v_2 = \varepsilon v_1 \text{ and } V_2 = \varepsilon V_1$$

A perfectly plastic collision ($\varepsilon = 0$) is one in which the particles stick together (as in this case), and their receding velocity is zero (relative to the center of mass). An elastic collision is one where the receding speed equals the approach speed, or $\varepsilon = 1$. A plastic (or inelastic) collision is one for which $0 < \varepsilon < 1$. There is nothing, yet, except for experience, that says ε cannot exceed unity. It is implied that the particles cannot pass through each other, in which case, ε would be negative.

Mechanical Energy Considerations

The collision event will now be analyzed from an energy perspective, as a way to reintroduce the First and Second Laws of Thermodynamics, the real motivation of this chapter. The collision event behaves as if there were a spring between the masses. The spring is at its relaxed length at the point of contact and is compressed during the period of deformation by the momentum of the two masses. From an energy perspective, the kinetic energy of the masses is converted to stored potential energy of the spring. The masses are quickly brought to rest, and then, the stored potential energy in the spring is converted back to kinetic energy during the period of restitution. However, the spring can be damaged (experience permanent deformation) and not fully convert the potential energy to kinetic energy. The mechanical energy "lost" will appear in a thermal form.

Recall, from Newton's apple, the mechanical energy equation applied to a system: in this case, the two mass system consisting of m and M that undergoes a process from an initial state to a final state:

$$\Delta E_{\text{mechanical}} = -W_{\text{out}} + \text{Thermal Effects},$$

where $\Delta E_{\text{mechanical}} = E_{\text{final}} - E_{\text{initial}}$ is the change in stored mechanical energy, W_{out} is the work done by the system on nongravity external forces, and anything not accounted for is considered to be "thermal effects," for the moment. In the case of a collision of particles, the mechanical energy stored is the kinetic energy of the particles plus the potential energy associated with gravity. The kinetic energy associated with rotation is neglected (i.e., the masses are treated as particles). During the collision event, there is no significant change in elevation of the center of mass, and the work done by the nongravity external forces is zero (negligible displacement of the center of mass) and, therefore,

$$\left(\frac{1}{2}mv_2^2 + \frac{1}{2}MV_2^2\right) - \left(\frac{1}{2}mv_1^2 + \frac{1}{2}MV_1^2\right) = \text{Thermal Effects}$$

Introducing the coefficient of restitution, and rearranging:

$$\left(\frac{1}{2}m\varepsilon^2v_1^2 + \frac{1}{2}M\varepsilon^2V_1^2\right) - \left(\frac{1}{2}mv_1^2 + \frac{1}{2}MV_1^2\right) = \text{Thermal Effects}$$
$$\left(\varepsilon^2 - 1\right)\left(\frac{1}{2}mv_1^2 + \frac{1}{2}MV_1^2\right) = \text{Thermal Effects}$$
$$\left(\varepsilon^2 - 1\right)KE_1 = \text{Thermal Effects}$$

There is a change in the kinetic energy of the system, the magnitude of which depends on the coefficient of restitution. There is a decrease in kinetic energy for $\varepsilon < 1$, and an increase for $\varepsilon > 1$ (which will be shown to be a violation of the 2nd Law of Thermodynamics shortly). For perfectly plastic collisions ($\varepsilon = 0$), all the initial kinetic energy is dissipated. For elastic collisions ($\varepsilon = 1$), there is no change in the kinetic energy. That is, the total mechanical energy is conserved in elastic collisions. The 1st and 2nd Laws of Thermodynamics will be introduced and used to demonstrate that the coefficient of restitution must be less than or equal to 1.

First Law of Thermodynamics: Conservation of Energy

Two new forms of energy (one storable, one transferrable) are needed to account for experimental observation: ***internal energy*** (U) and ***heat*** (Q). These will be discussed in some detail, and then inserted into the energy equation that forms the basis for the 1st Law of Thermodynamics.

Internal Energy

Internal energy is a storable form of energy, like kinetic and potential energy, and as such is a property or state variable. It consists of all the kinetic and potential energy that exists at the molecular level. For example, in a solid whose center of mass is at rest, all atoms are in constant vibratory motion. They are bonded together and they vibrate with kinetic energy and a form of stored potential energy associated with the restoring force of stretched molecular bonds (like springs) passing back and forth, even though the average velocity of the center of mass of the solid is zero.

Internal energy is not a measurable quantity, but it can be related to other common measurable properties (like temperature, pressure, volume). The term "property evaluation" in Engineering Thermodynamics involves relating one property to another. Appendix 2, a deeper dive, develops the relationship between internal energy and temperature that applies to solids and liquids that do not undergo a phase change or significant breaking of bonds at the molecular level (chemical reaction).

For a system that has within it material of ***density*** (ρ) that may vary with position, the total mass is given by

$$M = \iiint_{\text{volume}} \rho d\forall$$

The simplified result for a change in internal energy from state 1 to state 2 of a system of mass M, with another property called ***specific heat*** (c, discussed shortly), is as follows:

$\Delta U = U_2 - U_1 = \int_{\text{volume}} \left(\int_{T_1}^{T_2} c dT \right) \rho d\forall$ for solids or incompressible liquids of nonuniform composition and temperature.

Density and specific heat can vary with both temperature and position in the general case, but they can be usually treated as suitably chosen constants in applications that involve a single condensed phase (liquid or solid). If the system has a uniform composition (density does not change with position), and does not change significantly with temperature (reasonable for solids and many liquid applications) the change in total internal energy is expressed as follows:

$$U_2 - U_1 = Mc(T_2 - T_1)$$

or

$\Delta U = Mc\Delta T$ for single phase solids or incompressible liquid

The specific heat is a material property with strict SI units (using the nomenclature [=] to mean "has units of")

$$c[=]\frac{J}{kg \cdot \Delta K}[=]\frac{J}{kg \cdot \Delta^{o}C}$$

It is called the specific heat, which is a misnomer, as it is related to internal energy, not heat (which is a very different thermodynamic property). In the nonintegrated form, the specific heat is an average over the temperature range of the event. It is generally a weak function of temperature, and the useful expression $\Delta U = Mc\Delta T$ treats the specific heat as a constant, which is generally justifiable for engineering purposes, unless clearly stated otherwise. The specific heat can be evaluated at an average temperature if needed for accuracy as explained in Appendix 2. The specific heat simply relates internal energy to temperature (of nonphase changing solids or liquids), and has nothing directly to do with heat, but is a misnomer well entrenched in the technical literature.

The temperature units given here to the specific heat (ΔK or Δ°C) are not traditional, but added to clarify a common source of confusion. The definition of the specific heat is based on temperature differences, not on temperature values and, therefore, the Kelvin scale and Celsius scale are interchangeable in this case. That is, converting temperature between the Kelvin (K) and Celsius (°C) scale, noting that, by convention, how a degree symbol (°) is applied to Celsius but not Kelvin.

$$T(K) = T(^{o}C) + 273$$

Converting between a change in Kelvin and change in Celsius:

$$\Delta T(K) = \Delta T(^{o}C)$$

Some students, who are dutifully paying close attention to units, do the following:

Incorrect Interpretation of Unit Conversion

$$c = 4200\frac{J}{kg \cdot K} = 4200\frac{J}{kg{\cdot}^{o}C}\left|\frac{(T-273)^{o}C}{K}\right| = 4200(T-273)\frac{J}{kg{\cdot}^{o}C}$$

This expression makes no sense.

Correct Interpretation of Unit Conversion

$$c = 4200\frac{J}{kg \cdot \Delta K}\left|\frac{\Delta K}{\Delta^{o}C}\right| = 4200\frac{J}{kg \cdot \Delta^{o}C}$$

The Δ is not normally listed in the units, but they could be and occasionally will be in this text, to avoid this confusion.

The specific heat in Engineering Thermodynamics applications that involve working fluids that are pure, simple compressible substances is most conveniently expressed in mass units, as above. For applications that involve different chemical composition in different phases and/or chemical reactions, it is more convenient to express the specific heat in molar units, which will be discussed further when appropriate.

Returning to the collision process of masses m and M, the total internal energy of the combined system could be expressed as a summation, with the 2 masses (of different materials) with different temperature changes:

$$\Delta U = Mc_M(T_{2M} - T_{1M}) + mc_m(T_{2m} - T_{1m})$$

If the two objects are assumed to be at nominally the same average temperature, $T_{2M} = T_{2m} = T_2$:

$$\Delta U = (Mc_M + mc_m)(T_2 - T_1)$$

If the two objects are also assumed to be of the same material, with $c_M = c_m = c$

$$\Delta U = (M + m)c(T_2 - T_1)$$

It should be recognized that the temperatures at states 1 and 2 represent spatially averaged temperatures, as there will be local hot or cold spots within the system immediately after the collision. The loss of mechanical energy gives rise to a local increase in internal energy (hence temperature) in the vicinity of the point of contact. Subsequently, heat is transferred internally from this hot region to adjacent cooler regions, but not during the collision event itself.

Phase changes (freezing, boiling) and chemical reactions involve energy changes at the molecular level, and therefore, these effects are also forms of internal energy. The term ***latent energy*** (often called latent heat, a misnomer) refers to internal energy associated with phase change, while ***sensible energy*** refers to internal energy associated with changing the temperature (with no phase change). An important class of problems involves the ***humidification*** or dehumidification of air where the composition of the gas mixture changes, as does the latent and sensible energy. Chemical reactions give rise to changes in the composition of the gas mixture, and generally are accompanied by significant conversion of chemical energy into sensible internal energy. ***Combustion*** is the chemical reaction of a fuel with air (or other oxidizer) and is an important source of thermal energy. It is often called a "heat source," but that is also a misnomer. It is a conversion of one form of internal energy (chemical) to another (thermal energy). The effects of phase and chemical changes will be developed in future chapters.

Heat Transfer

Heat transfer (Q) is a process-dependent form of energy, like work, and represents a transfer of energy across the system boundary driven by a spatial temperature difference. The term "heat" is misused rather frequently, even in conventional technical writing or dictionaries as will be pointed out periodically. ***Temperature***, a commonly experienced property, yet one that is awkward to define formally, is related to both internal energy and heat transfer. Temperature is not internal energy, but the higher the temperature of an object, the higher its internal energy. Similarly, temperature is not heat transfer, but if two objects at different temperatures are in contact, heat will flow from one to the other. Temperature is not a form of energy. It is a property of the system that can vary from location to location. Heat flows naturally from hot to cold, a consequence of the 2nd Law of Thermodynamics that will be formally demonstrated at an appropriate point. Heat has a direction, and subscripts (such as "IN" or "OUT") are commonly used to specify the direction of heat transfer.

Heat transfer is fundamentally tied to a temperature difference, and it is also fundamentally a rate-based quantity. That is, "heat transfer takes time." Therefore, for events that take place in a short period of time, such as a collision, heat transfer is negligible, even if there is a substantial temperature difference driving it. Events for which the heat transfer is negligible are termed ***adiabatic*** processes, which are actually quite commonly assumed in practice, and justifiably so. That is,

$Q = 0$ for adiabatic processes

For nonadiabatic processes, the rate of heat transfer is a flux and is indicated by a dot over the Q, and a differential quantity of energy transfer by heat is given by

$$\delta Q = \dot{Q}dt$$

The total heat transfer for an event that occurs over a time span of $\Delta t = t_2 - t_1$ is

$$Q_{IN} = \int_{t_1}^{t_2} \dot{Q}dt \equiv \bar{\dot{Q}}(t_2 - t_1) = \bar{\dot{Q}}\Delta t$$

The bar over the heat transfer rate indicates a time-averaged heat transfer rate over the event. Since the heat transfer is driven by a temperature difference, the heat transfer rate can be expressed in two different ways:

$$\dot{Q}_{\text{IN}} = \frac{T_{\text{surroundings}} - T_{\text{system}}}{\mathfrak{R}} = \Omega A\left(T_{\text{surroundings}} - T_{\text{system}}\right)$$

A thermal resistance has been introduced in the first form:

$$\mathfrak{R}[=]\frac{\Delta K \cdot \sec}{J}[=]\frac{\Delta K}{W}[=]\frac{\Delta^o C}{W}$$

An overall heat transfer coefficient multiplied by the surface area of the defined system boundary has been introduced in the second form. The units of this coefficient are as follows:

$$\Omega[=]\frac{W}{m^2\Delta K}[=]\frac{W}{m^2\Delta^o C}$$

The thermal resistance and overall heat transfer coefficients are topics for a course in heat transfer (where their traditional nomenclature is R and U, which interfere with the gas constant and internal energy nomenclature used in this and other thermodynamics textbooks). They are often modeled as being constant, but there are important applications where they vary with time and/or temperature. The ability for heat to cross the boundary depends on properties such as the thermal conductivity of the material and the nature of any fluid flow in the vicinity of the boundary. The important context emphasized here is that heat transfer is fundamentally tied to a temperature difference.

There are 2 distinct physical mechanisms that drive heat transfer: ***conduction*** and ***radiation***. Conduction involves the transfer of microscopic kinetic energy through direct contact between adjacent molecules. Radiation heat transfer is an electromagnetic phenomenon associated with the energy of photons exchanged between surfaces at different temperatures. ***Convection*** heat transfer is an important mode of heat transfer between solid surfaces and adjacent fluids. Convection is a third mode of heat transfer, but its physical mechanism is conduction that takes place in the thermal boundary layer of a fluid adjacent to a solid surface.

A consequence of the 2nd Law of Thermodynamics, namely, that heat flows from hot to cold, has been implicitly introduced into the sign of the expression that relates heat transfer rate. If the temperature of the surroundings is greater than that of the system, then Q_{IN} will be positive, and vice versa.

Another detail relates to the spatial dependence of heat transfer. It must be recognized that, depending on the system in question, the heat transfer rate at different locations on the boundary can vary considerably. A more formal expression is as follows:

$$Q_{IN} = \int_{state1}^{state2} \dot{Q}dt = \iint_{boundary} \left(\int_{t1}^{t2} q dt \right) dA = \overline{q}_{IN} A \Delta t$$

Lower case q is an intensity of heat transfer: that is, the local heat transfer rate divided by the local area $(q = \dot{Q}/dA)$ and is integrated over the entire surface to obtain the total heat transfer rate per unit area. There is an analogy with the fluid pressure, a force intensity. The inner integral indicates that the total heat transferred is broken up into the sum of small changes over time for the entire event from state 1 to 2. The outer surface integral indicates that heat transfer may not take place uniformly over the entire surface. The average heat transfer per unit area is defined by the last expression.

In most cases of analysis, it is difficult to formally evaluate those integrals. Rather, everything else in the 1st (and/or 2nd) Law is evaluated, and the total heat transfer rate can be inferred, but the details of where and when are not accessible. When that detail is needed, other approaches (i.e., heat transfer theory) must be developed.

In summary, an expression that incorporates the 3 key aspects of heat transfer for the heat transfer (time, temperature difference, resistance) are as follows:

$$Q_{IN} = \int_{t_1}^{t_2} \dot{Q}dt = \left(\frac{1}{\mathfrak{R}}\right)\left(\overline{T}_{surroundings} - \overline{T}_{system}\right)\Delta t = (\Omega A)\left(\overline{T}_{surroundings} - \overline{T}_{system}\right)\Delta t$$

Processes can be modeled reasonably as being adiabatic if one or more of the following conditions apply:

1. There is a large resistance to heat flow due to a well-insulated boundary (low Ω).
2. There is a large resistance to heat flow due to a small contact area (low A).
3. There is minimal temperature difference between the system and the surroundings.
4. The event takes place in a short time period.

Whenever heat crosses a boundary, entropy is transferred with it, quantified by dividing the heat transfer by the absolute temperature at the boundary where it crosses. Formally,

$$\text{Entropy Transfer} = \frac{Q_{IN}}{T_B} = \int_{\text{state1}}^{\text{state2}} \frac{\dot{Q}}{T_B} dt = \iint_{\text{boundary}} \left(\int_{t1}^{t2} \frac{q}{T_B} dt \right) dA = \frac{\overline{q}_{\text{IN}}}{\overline{T}_B} A \Delta t$$

It is often difficult to formally determine the average temperature at the boundary, except in cases where the boundary can be placed outside the thermal boundary layer. Alternatively, a 2nd Law analysis is often facilitated by choosing a system for which heat transfer is internal and the entropy transfer is zero. The internal heat transfer will give rise to internal entropy generation, which can be quantified by an entropy balance.

Extensive Versus Intensive Properties

Various thermodynamic state variables (not process-dependent variables) can be categorized as being ***extensive*** or ***intensive***. The value of an extensive property is proportional to the mass of the object, examples being total volume (V), internal energy (U), and entropy (S). An intensive property does not depend on the mass, examples being temperature (T) and pressure (P). Some extensive properties can be converted to intensive properties by dividing by the mass, and are generally represented by lower case letters (i.e., for volume, $v = V/M$). Storable forms of energy (kinetic, potential, and internal) are extensive properties, while temperature is not. For example, consider Newton's apple of mass M that has a uniform temperature at an altitude H with a velocity V. If it were split in half, both halves would be at the same temperature, but half of the stored energy would be in one half, and half in the other. That is, the total stored energy is expressed as follows:

$$E = U + KE + PE = U + \frac{1}{2} MV^2 + Mg_0 H$$

These energy terms can be converted to intensive properties by dividing by the mass:

$$e = \frac{E}{M} = u + \frac{1}{2} V^2 + g_0 H,$$

where the specific internal energy is defined as follows:

$$u \equiv \frac{U}{M}$$

Note that lower case "*q*" from the heat transfer section case is not an intensive variable (which would be a division by the mass instead of area). Heat transfer is not a property of a system, and not associated with mass. It is a transferable energy form, a process-dependent variable, associated with surface (or boundary) phenomenon.

Defining the System

It is good practice in thermodynamic analysis to clearly define the system in question, aided by a visual sketch of the system. For the collision problem, Fig. 2.3 defines the system to consist of both masses, with a boundary placed far enough away from the particles that the temperature of the boundary (T_B) equals the temperature of the environment surrounding the masses (T_∞). This location is important for the 2nd Law analysis. For particles in a complete vacuum, the boundary would be placed on the surface of the particles. The 2nd Law analysis is a little more difficult in that case. The temperature of the system, a state variable, is in general different than the ambient. This temperature is best considered to be an average temperature, which allows for local hot or cold regions within the system boundary. Energy is exchanged with the surroundings in the form of heat and work, shown as arrows crossing the boundary. By convention, heat added to the system and work done by the system (on nongravity forces) are considered positive quantities, but it is a good habit to use appropriate subscripts. The values of these quantities can be negative. The arrows could be drawn in the other direction (with appropriate subscripts) as long as the application of the 1st and 2nd Laws is consistent. In this problem, the work done by the system on nongravity external forces evaluates to zero, but is shown here for completeness.

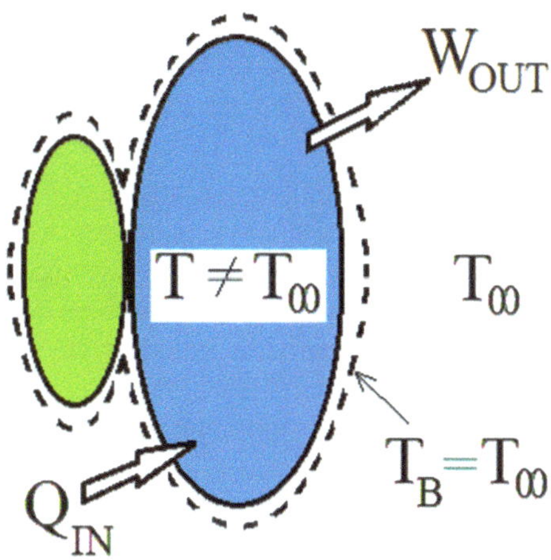

Fig. 2.3 Process diagram of the colliding system during a collision. The system boundary is placed just outside the solid object where the boundary temperature is everywhere equal to the local ambient

Energy Balance

The thermal effects are inserted into the mechanical energy equation in the following way:

$$\text{Thermal Effects} = Q_{\text{IN}} - \Delta U$$

Recall the First Law of Thermodynamics, which states that during any process, energy can be converted from one form to another, but cannot be created or destroyed. That is, considering the Universe to consist of any defined system and its surroundings:

$$\Delta E_{\text{Universe}} = \Delta E_{\text{System}} + \Delta E_{\text{Surroundings}} = 0$$

However, two forms of energy can cross the boundary between the closed system and its surroundings, namely, heat and work. The First Law of Thermodynamics, applied to a system that undergoes a process from an initial state to a final is, therefore,

$$\Delta E_{\text{System}} = Q_{\text{IN}} - W_{\text{OUT}} = -\Delta E_{\text{Surroundings}}$$

That is, the change in the total energy stored within the system is equal to the net energy that enters the system in the form of heat and work, and is equal in magnitude but opposite in sign to the change in the energy stored in the surroundings. The total energy of the system is now considered to consist of 3 forms of storable energy: internal, kinetic, and potential. The 1st Law is

$$\Delta U + \Delta KE + \Delta PE = Q_{\text{IN}} - W_{\text{OUT}}$$

1st Law for Collisions

A process diagram for the collision event is shown in Fig. 2.4. The change in potential energy and work associated with external forces are negligible. Also the collision event is adiabatic for 2 reasons: there is little time for heat transfer to occur and there is no temperature difference driving heat at the defined system boundary (assuming both particles are at the local ambient temperature at the time of the collision). Note that there could be heat transfer taking place WITHIN THE SYSTEM, but that is an internal effect. This effect will be evidenced by the property entropy (S) and the process variable entropy production (Σ), which are shown in Fig. 2.4 and will be discussed in the next section.

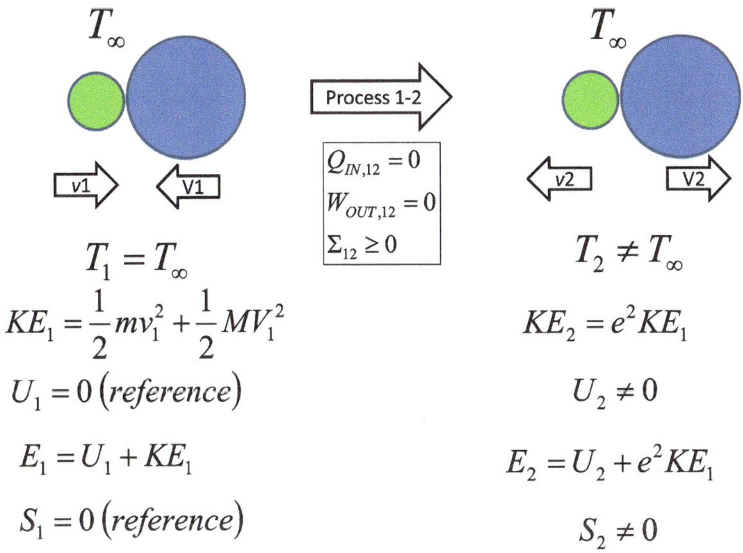

Fig. 2.4 State and Process diagram for the collision event

The First Law reduces to

$\Delta U + \Delta KE = 0$ for collision events

With no energy crossing the boundary of the defined system, the change in total energy is zero. Expressing the internal energy in terms of the temperature (assuming the two particles are of the same material, and acquire the same average final temperature), and expressing the change in kinetic energy in terms of the coefficient of restitution:

$$(M + m)c(T_2 - T_1) + (\varepsilon^2 - 1)KE_1 = 0$$

Rearranging for the final temperature (recall that this is an average temperature):

$$T_2 = T_1 + \frac{(1 - \varepsilon^2)KE_1}{(M + m)c}$$

At this point, the only restriction on the coefficient of restitution is that it be positive ($\varepsilon > 0$). A negative value of ε would mean the particles passed through each other. However, it is possible, from a 1st Law point of view, for the coefficient of restitution to exceed unity, $\varepsilon > 1$, in which case the collision would increase total kinetic energy, and the 1st Law would require a decrease in the internal energy, hence temperature. The particles would become colder, and the receding relative speed of the two particles would be greater than the approach speed, which of course seems counterintuitive. The 2nd Law of Thermodynamics will show formally that the coefficient of restitution must be less than (or equal to) unity.

Workshop 1st Law Results

The results of all 1st Law calculations are shown in Fig. 2.5. The spreadsheet is shown with the rows and columns visible, and the table at the bottom shows the order in which calculations can proceed. For example, the simple momentum calculation cell formula is B2*B4/B3.

Completion of the spreadsheet does not proceed in a linear fashion, but involves going back and forth between the state and process tables, depending on what is known from the problem statement. Many of the formulas can be entered directly from the problem statement PRIOR to formally applying an energy balance. For example, cell E19, the internal energy after the masses have cooled to ambient, can be filled in directly.

Similarly, many process quantities can be entered before conducting an energy balance. For example, there is no work due to nongravity external forces, and the work for all processes can be set to zero. Similarly, the time duration of the collision and the falling of the particles is assumed to be sufficiently short that heat exchange between the particles and the environment is negligible during all but the final cooling process.

	A	B	C	D	E	F	G	H	I
1	INPUT PARAMETERS								
2	m	0.004	kg, mass of smaller particle						
3	M	0.008	kg, mass of smaller particle						
4	v	900	m/s, initial speed of smaller particle, (relative to the ground)						
5	e	0	coefficient of restitution						
6	H	10	m, height where collision occurs				Color Code		
7	c	128	J/kg/Δ°C, specific heat of both particles				Inputs		
8	Tinf	25	oC, dead state temperature (ambient)				Momentum		
9							1st Law		
10	DERIVED PARAMETERS						2nd Law		
11	V	450	m/s, initial speed of larger particle (relative to the ground)						
12									
13	STATE TABLE (for combined system, m and M)								
14	STATE	T (oC)	KE (J)	PEg (J)	U (J)	S (kJ/Δ°C)	eX=U-T_S (J)		
15	1	25	2430	1.18	0			Prior to collision	
16	2'	1607.031	0	1.18	2430.00			immediately after collision	
17	2"	1607.031	1.18	0	2430.00			just before hitting ground	
18	2	1607.798	0	0	2431.18			after coming to rest on the ground	
19	3	25	0	0	0			after cooling to ambient	
20									
21	PROCESS TABLE (all quantities in J)								
22	PROCESS	ΔU	ΔKE	ΔPEg	Qin	Wout	T_ΔS	(T_/T_B)*Q	eX_D
23	1-2'	2430	-2430	0	0	0			
24	2'-2"	0	1.18	-1.18	0	0			
25	2"-2	1.18	-1.18	0	0	0			
26	2-3	-2431.18	0	0	-2431.18	0			
27	net	0	-2430	-1.18	-2431.18	0			

Order of Calculations, 1st Law

C15, D15, E15, C16, D16, D17, C18, D18, C19, D19, E19	States from inputs and process specifications
C23, D23, E23, F23, C24, D24, E24, F24, D25, E25, F25, C26, E26, F26	Process quantities from inputs or process specifications
B23, E16, C24, C17, C25, B25, E17, E18, B26, E26	From energy balances
B16, B17, B18	from property relatioins

Fig. 2.5 Workshop spreadsheet with 1st Law analysis results. The table at the bottom shows the order in which cells were entered, which involves going back and forth within the tables

Next, quantities that involve application of energy balances or follow-up are entered. For example, for the main collision, the changes in kinetic and potential energy, as well as the heat and work exchanges, can be calculated from the input quantities. The change in internal energy is then determined from an energy balance (cell B23), from which the internal energy of state 2′ can be found (cell E16).

Finally, a property relation is used to calculate the average temperatures of the masses. This is the only case in this workshop where locking cells in formulas and copying formulas can be used. That is, the formula entered into cell B16 is

$$\mathrm{B16} = (\mathrm{E16} - \mathrm{E16})/\mathrm{B7}/(\mathrm{B2} + \mathrm{B3})$$

Then this cell can be copied to B17 and B18 to obtain the temperatures from the internal energies in those states. The "$" is used to lock in appropriate rows. A "$" is not needed on the column, since the formula will be copied down, not across.

The results show several things. All the kinetic energy is converted to internal energy and the resulting average temperature of the bullets after the collision is quite high, well above the melting point of lead (327 °C). This case is pretty extreme (and would require incredible marksman-ship). The result suggests that phase change in the property relation that relates internal energy to temperature is not negligible. In practice, perhaps a thin layer of lead at the collision point might melt or soften, and the temperature would be limited to the melting point. However, as the bullets subsequently cool, the energy that went into phase change would be recovered, and the net heat exchange predicted by the single-phase model would be the same.

The results also show that the effects of gravitational potential energy are small compared to the kinetic energy effects. After colliding, the particles fall a substantial height (about 3 stories), and the potential energy is converted to kinetic energy (with the velocity vertical). Then KE is converted to internal energy by the collision with the Earth, but the additional temperature rise is less than 1 °C, much less than that associated with the bullets colliding.

The net effect (row 27) of the collision and its aftermath (falling, then cooling) is that all the original mechanical energy (kinetic and gravitational) is ultimately "lost" as heat to the environment. However, immediately after the collision, before there is time for heat loss to occur, the mechanical energy has been converted into internal energy. Is there a possibility to do something more "useful" with that internal energy rather than letting the hot objects simply cool to the environment (however impractical in this case)? The 2nd Law will tell.

Second Law of Thermodynamics: Entropy Production

In the same way that storable energy is a property central to the First Law, the entropy (S) is a property that is central to the 2nd Law of Thermodynamics, namely, that for any process, the entropy of the Universe must increase (or remain the same).

$$\Delta S_{\text{Universe}} = \Delta S_{\text{System}} + \Delta S_{\text{Surroundings}} \geq 0$$

The increase in the entropy is due to the PRODUCTION of entropy within the system boundary, as opposed to the TRANSFER of entropy between the system and its surroundings. To express the 2nd Law as an equality, the entropy produced is given the Greek symbols for s, the upper case (Σ) for extensive, and the lower case (σ) for the intensive (entropy production per unit mass). Extensive and intensive entropies are related by the mass ($\Sigma = M\sigma$). The total entropy change of the universe is expressed as an equality, then, as:

$$\Delta S_{\text{Universe}} = \Delta S_{\text{System}} + \Delta S_{\text{Surroundings}} = \Sigma_{\text{system}} + \Sigma_{\text{surroundings}}$$

The entropy production takes place internally within both the system and the surroundings. Note that the entropy produced in the surroundings (with an indefinable mass) includes every activity in the universe that takes place during the event under investigation. The entropy produced within the system boundaries is specific to what takes place during the event in question within the finite system boundary. Entropy that crosses the boundary depends on both the heat transfer and the temperature at the location where heat crosses. An entropy balance for any event, in a word statement, is as follows:

$$\begin{bmatrix} \text{Change in Entropy} \\ \text{Stored Within} \\ \text{System Boundary} \end{bmatrix} = \begin{bmatrix} \text{Entropy That} \\ \text{Enters Through} \\ \text{System Boundary} \end{bmatrix} + \begin{bmatrix} \text{Entropy Produced} \\ \text{Within} \\ \text{System Boundary} \end{bmatrix}$$

These effects will be discussed in the following sections. No attempts to "define" entropy will be made, just like it is hard to "define" common quantities like heat, energy, temperature, and love. You know it when you see it. The good engineer knows how to use them in their proper contexts. Virtually all dictionary definitions of these quantities are unsatisfactory, as a quick internet search will reveal.

Entropy Transfer

Entropy that enters through the system boundary is proportional to the heat transferred, and inversely proportional to the absolute temperature of the boundary. It is difficult to simply explain why or how this relationship was "discovered." It is best to accept it, just like $F = ma$ must be simply accepted. Since, in general, the temperature of the boundary varies along the boundary, and can also be different at different times during the event, a general expression of the entropy transfer is as follows:

$$\begin{bmatrix} \text{Entropy That} \\ \text{Enters Through} \\ \text{System Boundary} \end{bmatrix} = \iint\limits_{\text{boundary}} \int_{\text{state1}}^{\text{state2}} \left(\frac{\delta q_{IN}}{T_B} \right) dA = \sum_{\text{boundary}} \frac{Q_{\text{IN}}}{T_B}$$

The first form shows an inner integral between states, and the outer surface integral indicates the spatial variation in both the heat and the boundary temperature (which must be in absolute units).

Heat crosses locally, at a differential surface dA, where the absolute temperature is T_B. Both the rate of heat transfer and the boundary temperature can change during the event in the most general case. Interpretation and use of this entropy transfer will become clearer as applications are accumulated.

The second form (the summation) indicates that the total entropy transferred into the system is the sum of the local entropy transfers around the entire system boundary. Wherever heat crosses the boundary, it carries entropy with it, whose magnitude depends on the local temperature where it crosses. This temperature may be difficult to define. The art of insightful 2nd Law analysis is in choosing appropriate system boundaries so that the temperature where heat crosses is well defined.

The system boundary for the colliding partners in Fig. 2.3 surrounds the two particles and includes just enough air so that the boundary temperature equals ambient. There is entropy generation within this boundary layer of air that cannot be easily distinguished from entropy generation within the colliding particles themselves. For two particles in outer space, where the local ambient is a vacuum, the system boundary could be placed on the surface of the particles, and the integral would not be so easy to evaluate.

In some cases, it is convenient to define a form of heat-transfer-weighted average boundary temperature, even if it is not possible to evaluate it formally:

$\sum\limits_{\text{boundary}} \frac{Q_{\text{IN}}}{T_B} = \frac{Q_{\text{IN}}}{\overline{T}_B}$ when the boundary temperature is not fixed

Any time that $\overline{T}_B$ appears in a subsequent analysis, it means that

- The process in question is not adiabatic.
- The temperature at the boundary where heat crosses is not clearly defined or fixed.
- The analysis is qualitatively correct, but may require a more thorough investigation if more precise calculations are required.

The fundamental difference between heat and work, or between thermal and mechanical energy, is that there is no entropy transfer associated with work.

Entropy Production

The entropy produced within the system is an integral over the volume:

$$\begin{bmatrix} \text{Entropy Produced} \\ \text{Within} \\ \text{System Boundary} \end{bmatrix} = \iiint\limits_{\text{volume}} \left(\int_1^2 \delta\sigma \right) \rho d\forall = M\sigma = \Sigma_{\text{system}}$$

Entropy production is process dependent (hence the 'δ'), it changes with time during the event (the inner integral) and is different at all locations (the outer volume integral). The 2nd Law states that the entropy production must be positive for each incremental step:

$$\delta\sigma \geq 0$$

Note that the differential expression is used to guard against a process that violates the 2nd Law but has a net positive entropy production. During such a fictitious process, the entropy production could decrease initially, and then increase more at a later point, resulting in a net production. Such a process violates the 2nd Law. Entropy production often is localized, such as near the point of contact between the colliding particles, but is contained within the system boundary.

Entropy Balance and the 2nd Law

The entropy balance, expressed mathematically, with a boundary temperature T_B in absolute units (Kelvin or Rankine), applied to a process taking place on a closed system is as follows:

Entropy balance: $\Delta S_{\text{System}} = \sum\limits_{\text{boundary}} \frac{Q_{\text{IN}}}{T_B} + \Sigma$

2nd Law of Thermodynamics: $\Sigma = M\sigma \geq 0$

The equality sign holds for processes that are termed ***reversible***, and all processes for which the inequality holds are termed ***irreversible***. This terminology is explained by the following thought experiment. If a given process is analyzed from an initial to a final state, and all quantities are determined, it is possible to consider whether the process could occur in reverse, by setting the conditions of the initial state to be that of the final, and vice versa. That involves simply changing the signs of all energy and entropy terms, including the entropy production. The only way the reverse process can occur without violating the 2nd Law is if the entropy production is zero.

Expressed in terms of the average boundary temperature, the entropy balance is as follows:

$$\overline{T}_B \Delta S = Q_{\mathrm{IN}} + \overline{T}_B \Sigma$$

This expression shows that the entropy of a system can increase for two reasons. Heat can be added across the surface, or entropy can be produced internally. Since the heat can be a negative quantity, it is possible for the entropy of a system to decrease without violating the 2nd Law. A common mistake is to confuse a change in entropy with entropy production. The entropy of a system can decrease if sufficient heat is transferred out of it. The entropy of the surroundings will increase more in that case. The entropy change and entropy production are only equal in magnitude (but are very different quantities) for adiabatic processes.

2nd Law for Collisions

The collision event, being short in duration, is adiabatic. Therefore,

$$\Delta S = \Sigma \geq 0$$

The relationship between entropy and temperature is developed in the deeper dive of Appendix 3. Assuming both particles are of the same material, and they attain the same average temperature, the entropy change can be expressed in terms of the average temperatures before and after the event:

$$(M + m)c \ln\left(\frac{T_2}{T_1}\right) = \Sigma \geq 0$$

Solving for the temperature after the collision:

$$T_2 = T_1 e^{\frac{\Sigma}{(M+m)c}} \geq T_1$$

In summary, The 1st Law dictates that the change in kinetic energy associated with the collision results in an equal and opposite change in internal energy, but says nothing about whether the kinetic energy can increase or decrease. The 2nd Law dictates that the entropy of the system after the collision must be greater than before the collision. Relating internal energy and entropy to temperature reveals that the temperature and therefore internal energy must increase. Recall, from the energy balance, the final temperature:

$$T_2 = T_1 + \frac{(1-\varepsilon^2)KE_1}{(M+m)c}$$

The entropy change is, therefore,

$$\Delta S = (M+m)c\ln\left(1 + \frac{(1-\varepsilon^2)KE_1}{(M+m)cT_1}\right) = \Sigma$$

Note that $\ln(1 + y) < 0$ if $y < 0$. Therefore, since the mass and specific heat are inherently positive quantities, as are all factors within the natural logarithm term, the only way the final temperature can be greater than the initial temperature is if the coefficient of restitution is less than or equal to unity:

$$\varepsilon \le 1$$

The 2nd Law formally dictates that the coefficient of restitution must be less than or equal to 1. That is, the total kinetic energy after the collision must be less than before it. Then the 1st Law requires a rise in internal energy. There is a conversion from a mechanical energy form (kinetic energy) to a thermal form (internal energy) in a plastic collision, a kind of dissipation.

An elastic collision, $\varepsilon = 1$, results in no entropy generation. The situation in Fig. 2.4 could be reversed; that is, if the velocities of both particles at state 2 were reversed, and the collision resulted in the velocities of state 1 (reversed), this reverse process would not generate entropy. An elastic collision is therefore reversible. On the other hand, collisions for which $0 \le \varepsilon \le 1$ are possible, but if the process were reversed, the entropy production would be negative in violation of the 2nd Law, so plastic collisions are irreversible. Finally, collisions with $\varepsilon \ge 1$ violate the 2nd Law (but not the 1st Law).

Exergy: Combining the First and Second Laws

The 1st Law, which is a statement of the conservation of energy for all events, and the 2nd Law, which is a statement that entropy must be produced, can be combined by simply subtracting a factor $T_\infty \Delta S$ from both sides of the 1st Law energy balance (without fully explaining why):

$$[\Delta U - T_\infty \Delta S] + \Delta KE + \Delta PE = Q_{\text{IN}} - T_\infty \Delta S - W_{\text{OUT}},$$

where T_∞ is the dead state temperature (the temperature of some local ambient) in absolute units.

The first term, in brackets, combines two thermal properties but is not a pure thermodynamic property, because it depends on what the local dead state is. For example, a device operating inside a building experiences a local ambient temperature that is different from the temperature out of doors. Either of these environments could be used as the dead state, depending on the objectives of the particular analysis. This quantity is called ***eXergy*** (the capital "X" being done deliberately to draw attention to it):

$$X = U - T_{\infty}S$$

eXergy can be interpreted as a measure of the usefulness of thermal energy, relative to a local environment, and is therefore a mechanical form of energy. Specifically, eXergy will be shown to be equal to the maximum possible work that could be extracted in a hypothetic process in which a system in a particular state is brought reversibly into thermal and mechanical equilibrium with the dead state. eXergy is an inherently positive quantity (or zero if the system is at the dead state) as shown in Appendix 5 for solids and liquids (but true generally).

Most textbooks incorporate the kinetic and potential energy into the definition of eXergy, and/or call the quantity ***Availability***. In this textbook, the availability will be defined to be the sum of all mechanical forms of stored energy:

$$\text{Availability} = X + \frac{1}{2}MV^2 + MgH$$

The eXergy (X) is the useful portion associated with thermal energy forms. Practitioners must know which definition is being used.

In general, the eXergy of the system changes during a process:

$$\begin{bmatrix}\text{Change in}\\ \text{eXergy}\end{bmatrix} = \Delta X = [\Delta U - T_{\infty}\Delta S]$$

To summarize, for a process on a defined system, the energy balance, entropy balance and 2nd Law are as follows:

Energy balance (modified): $\Delta X + \Delta KE + \Delta PE = Q_{\text{IN}} - T_{\infty}\Delta S - W_{\text{OUT}}$

Entropy balance: $\Delta S = \frac{Q_{\text{IN}}}{T_B} + \Sigma$

2nd Law of Thermodynamics: $\Sigma = M\sigma \geq 0$

Since both heat and work can cross the boundary in either direction, it might be tempting for a clever inventor or engineer to look for a means to increase the stored energy of a system by simply adding heat to it. The energy balance suggests that may be possible. One could certainly apply an external force to raise a weight. Could the same effect be accomplished by merely blowing warm air on it? Of course not. But to prove it, the 2nd Law must somehow be considered.

The common factor between the 1st and 2nd Laws is the heat exchange between the system and surroundings. Combining the energy and entropy balance equations by eliminating the heat transfer, and rearranging:

$$\Delta X + \Delta KE + \Delta PE = -\overline{T}_B\Sigma - \left(T_\infty - \overline{T}_B\right)\Delta S - W_{\mathrm{OUT}}$$

Entropy production (Σ) and the absolute temperature at the boundary are inherently positive (or zero for a reversible process and a boundary at absolute zero temperature). Therefore, anything that PRODUCES entropy internally (i.e., any irreversibility in the process) will decrease the stored energy. The second term on the right side is more subtle, as entropy can increase or decrease and the boundary temperature could be greater than or lower than the dead state, so the possibility might be argued that if heat exchange occurs at a temperature other than at the dead state, it could increase the useful stored energy. Proof that it cannot will be deferred until it is proved that heat must flow from a higher to a lower temperature.

In many cases, it is possible and advantageous to define the system so that the boundary is everywhere at the dead state temperature. In those situations, the combined 1st and 2nd Laws becomes

$\Delta X + \Delta KE + \Delta PE = -T_\infty\Sigma - W_{\mathrm{OUT}}$ for $\overline{T}_B = T_\infty$ and/or adiabatic processes

This expression holds for adiabatic processes, because $\Sigma = \Delta S$.

The first term on the right side, the dead state temperature times the entropy generation, is defined as the ***eXergy destruction***:

$$eX_D = T_\infty\Sigma$$

It has units of energy and is inherently positive. It is considered to be the measure of ***dissipation***, that is, the conversion of stored mechanical energy (capable of doing work) into low-grade thermal energy which has little use, besides possibly keeping something warm. A final form of the combined 1st and 2nd Law, or eXergy balance, is as follows:

$$\Delta X + \Delta KE + \Delta PE = -\left(\frac{\overline{T}_B}{T_\infty}\right)eX_D - T_\infty\left(1 - \frac{\overline{T}_B}{T_\infty}\right)\Delta S - W_{\mathrm{OUT}}$$

The 3 terms that make up the left-hand side are considered to be "available" energy; that is, these forms of energy are capable of producing useful work.

It is important to differentiate what is meant by eXergy (a property), eXergy change (a property change that takes place during a process, which can be positive or negative), and eXergy destruction (a process-dependent variable that must be positive).

Collision Aftermath

A nonelastic collision event results in two particles that are hotter than their immediate surroundings receding from each other at a lower relative speed than their approach speed. Therefore, they are not in a thermal equilibrium with their surroundings, and, given time, there will be heat transfer. Experience dictates that the direction of heat transfer will be from the hot objects to the cold surroundings. The 2nd Law will confirm it. A process sketch of the aftermath is shown in Fig. 2.6, with state 3 being defined when sufficient time has passed that both objects are at the temperature of the surroundings. A process sketch that covers both the collision and the aftermath is shown in Fig. 2.7. This problem statement neglects any fluid forces acting afterward, as the focus is on the heat transfer process. Therefore, the work out (done by the system on nongravity forces) is zero, and the kinetic energy is unchanged (a consequence of $F = ma$).

Overall Process 1–3

An analysis of the overall process is conducted in this section. The collision process was analyzed earlier. State and process tables that summarize the results are presented in Tables 2.1 and 2.2. The final results of the Workshop of the Colliding Bullets are shown in Fig. 2.8.

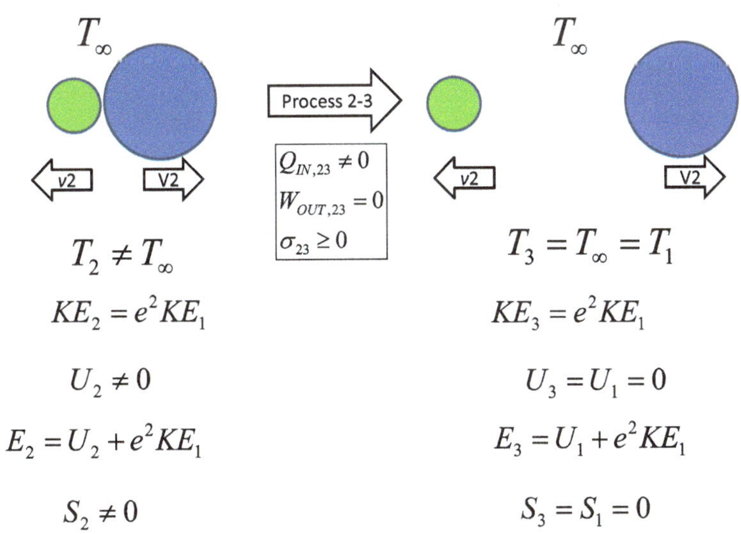

Fig. 2.6 State and process diagram for the aftermath of the collision event

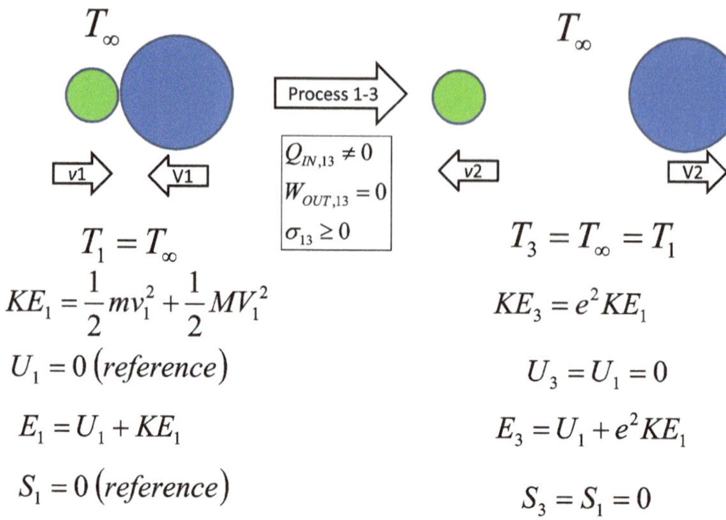

Fig. 2.7 State and process diagram for the collision and its aftermath

Once the 1st Law analysis is complete, the formulas for the 2nd Law states can be entered into the spreadsheet. For example, the entropy formula entered into cell F15 of Fig. 2.8 is as follows:

$$(\text{B2} + \text{B3}) * \text{B7}^* \ln((\text{B25} + 273)/(\text{B8} + 273)).$$

In its cell referencing, this formula uses $ to lock in the rows anticipating the formula being copied down to complete this column in the state table. Note that a $ on the column is not necessary in this case. Locking the column with a $ can be useful in other cases in which formulas are copied across, examples of which will occur later. Note that the temperatures are converted to absolute (Kelvin) units within the formula. A common mistake is to enter the temperatures in Celsius. Since 2nd Law calculations can be very sensitive to small changes, some authors would argue that a higher precision be used for the unit conversion from Celsius to Kelvin (i.e., use 273.15 instead of 273). That can be important when precise property table data is used, but in most cases will not cause significant engineering error.

The eXergy state column can also be entered. Cell G15 is as follows:

$$\text{E15} - (\text{B8} + 273) * \text{F15}$$

This formula is copied down to complete the state table.

The 1st Law applied to both the collision and the aftermath is as follows:

Table 2.1 State Table for collision and its aftermath (neglecting the falling to the ground for the Colliding Bullets Workshop)

state	T	U	KE	S	eX
1	T_∞	0	$\frac{1}{2}mv_1^2 + \frac{1}{2}MV_1^2$	0	0
2	$T_2 = T_\infty + \frac{(1-\varepsilon^2)KE_1}{(M+m)c}$	$(1-\varepsilon^2)KE_1$	$\varepsilon^2 KE_1$	$(M+m)c\ln\left(1 + \frac{(1-\varepsilon^2)KE_1}{(M+m)cT_1}\right)$	$U_2 - T_\infty S_2$
3	T_∞	0	$\varepsilon^2 KE_1$	0	0

Table 2.2 Process Table for collision and its aftermath (neglecting the falling to the ground for the Colliding Bullets Workshop)

Process	Q_{IN}	$T_\infty \Sigma = eX_D$
1–2	0	$(M+m)cT_\infty \ln\left(1+\frac{(1-\varepsilon^2)KE_1}{(M+m)cT_\infty}\right)$
2–3	$-(1-\varepsilon^2)KE_1$	$(1-\varepsilon^2)KE_1 - (M+m)cT_\infty \ln\left(1+\frac{(1-\varepsilon^2)KE_1}{(M+m)cT_\infty}\right)$
1–3 (overall)	$-(1-\varepsilon^2)KE_1$	$(1-\varepsilon^2)KE_1$

INPUT PARAMETERS

m	**0.004**	kg, mass of smaller particle
M	**0.008**	kg, mass of smaller particle
v	**900**	m/s, initial speed of smaller particle, (relative to the ground)
e	**0**	coefficient of restitution
H	**10**	m, height where collision occurs
c	**128**	J/kg/Δ°C, specific heat of both particles
Tinf	**25**	oC, dead state temperature (ambient)

Color Code
Inputs
Momentum
1st Law
2nd Law

DERIVED PARAMETERS

V	450	m/s, initial speed of larger particle (relative to the ground)

STATE TABLE (for combined system, m and M)

STATE	T (oC)	KE (J)	PEg (J)	U (J)	S (kJ/Δ°C)	eX=U-T∞S (J)	
1	**25**	2430	1.18	0	0	0	Prior to collision
2'	1607.031	0	1.18	2430.00	2.829	1586.89	immediately after collision
2"	1607.031	1.18	0	2430.00	2.829	1586.89	just before hitting ground
2	1607.798	0	0	2431.18	2.830	1587.88	after coming to rest on the ground
3	**25**	0	0	0	0	0	after cooling to ambient

PROCESS TABLE (all quantities in J)

PROCESS	ΔU	ΔKE	ΔPEg	Qin	Wout	T∞ΔS	(T∞/T_B)*Q	eX_D
1-2'	2430	-2430	0	0	0	843.11	0	843.11
2'-2"	0	1.18	-1.18	0	0	0	0	0.00
2"-2	1.18	-1.18	0	0	0	0.19	0	0.19
2-3	-2431.18	0	0	-2431.18	0	-843.30	-2431.18	1587.88
net	0	-2430	-1.18	-2431.18	0	0	-2431.18	2431.18

Fig. 2.8 Complete results of Colliding Bullets Workshop

$$\begin{aligned} E_3 - E_1 &= Q_{IN,13} - W_{OUT,13} \\ \left(U_1 + \varepsilon^2 KE_1\right) - \left(U_1 + KE_1\right) &= Q_{IN,13} - 0 \\ Q_{IN,13} &= \left(\varepsilon^2 - 1\right)KE_1 \end{aligned}$$

This overall result demonstrates that all the kinetic energy that is "lost" in the collision is converted into heat transfer between the system and surroundings. The 1st Law does not restrict the coefficient of restitution, however.

The entropy balance applied to both the collision and its aftermath is as follows:

$$S_3 - S_1 = \sum_{\text{boundary}} \frac{Q_{\text{IN},13}}{T_B} + \Sigma_{13}$$

Since the boundary is placed outside the two masses where the local temperature is ambient, the boundary temperature can be taken out of the integral. The entropy returns to its original value of 0. Therefore,

$$0 - 0 = \frac{Q_{IN,13}}{T_\infty} + \Sigma_{13}$$

Rearranging for the entropy production (Σ) multiplied by the dead state temperature yields the eXergy destruction:

$$eX_{D,13} = T_\infty \Sigma_{13} = -Q_{\text{IN},13}$$

Combining the 1st and 2nd Laws by eliminating the heat transfer, and introducing the inequality imposed by the second law:

$$eX_{D,13} = T_\infty \Sigma_{13} = \left(1 - \varepsilon^2\right) KE_1 \geq 0$$

Since kinetic energy is an inherently positive quantity, the consequence of the inequality is that the coefficient of restitution must be less than unity:

$$\varepsilon \leq 1$$

Also, the heat transfer into the system must be negative; that is, heat must leave the system, not enter it:

$$Q_{\text{IN},13} \leq 0$$

There are a few reasons for breaking the collision event into two stages (collision and its aftermath). First, it is good practice going forward to analyze a given situation from different points of view. Second, it is interesting to note that immediately after the collision, all of the kinetic energy "lost" in the collision is converted to internal energy. The eXergy of state 2 is nonzero, but less than the internal energy. Theoretically, the warm particles are capable of doing work in being brought to equilibrium with its environment. In practice, however, the eXergy associated with the warm particles is not of practical use, and that eXergy is destroyed as the particles cool to ambient.

It is re-emphasized that there is a distinction between eXergy and eXergy destruction, just like it is important to distinguish between entropy and entropy production. eXergy (and entropy) are properties. eXergy destruction (and entropy production) are process-dependent quantities related to the degradation of mechanical to thermal energy. The use of eXergy in thermodynamic analysis has the advantage over entropy of being an energy quantity that can be directly compared to other energy quantities.

The numerical results show that immediately after the collision, about 2/3 of the initial kinetic and potential energy is converted into eXergy, and a third is dissipated. This eXergy is subsequently lost as the colliding particles cool during the aftermath.

It is common that 1st Law analysis (use of an energy balance) will yield the most important practical results. The 2nd Law analysis (use of an entropy balance) involves mostly "postprocessing" type calculations, and the results help interpret the nature of the event in question in a deeper, more insightful way. 1st Law analyses are therefore the work horse, and 2nd Law analyses make things more sophisticated.

Exergy as Work Potential of Thermal Energy

The purpose of this section is to show in a general way that eXergy is a measure of the theoretical potential of a system to produce useful work in being brought into equilibrium with the dead state. Consider a system of mass M at a state that differs from a dead state (i.e., the two bullets while they are hot). The system is initially at a uniform temperature T_1, with a velocity V_1, at a height H_1 in a gravitational field. The boundary is placed at the dead state, outside the boundary layer of the object.

$$\text{Energy balance}\quad dE = d\left(U + \frac{1}{2}MV^2 + MgH\right) = \delta Q - \delta W$$

$$\text{Entropy balance}\quad dS = \frac{\delta Q}{T_\infty} + \delta\Sigma$$

Combining (by eliminating the common heat transfer), and rearranging with exact differentials on the left, and inexact differentials on the right:

$$d\left(U - T_\infty dS + \frac{1}{2}MV^2 + Mgh\right) = -\delta W - T_\infty\delta\Sigma$$

Introducing the eXergy, defined as $X = U\text{-}T_\infty S$:

$$d\left(X + \frac{1}{2}MV^2 + Mgh\right) = -\delta W - T_\infty\delta\Sigma$$

The left-hand side consists of 3 exact differential terms. These are path-independent quantities, and they can be directly integrated between 2 states:

$$\int_{\text{state1}}^{\text{state2}} d\left(X + \frac{1}{2}MV^2 + Mgh\right) = (X_2 - X_1) + \left(\frac{1}{2}MV_2^2 - \frac{1}{2}MV_1^2\right) + (MgH_2 - MgH_1)$$

Defining state 2 to be the dead state, where the eXergy, kinetic energy, and potential energy are zero:

$$\int_{\text{state1}}^{\text{Dead State}} d\left(X + \frac{1}{2}MV^2 + Mgh\right) = -\left(X_1 + \frac{1}{2}MV_1^2 + MgH_1\right)$$

On the other hand, there are 2 inexact differentials on the right-hand side that cannot be integrated without specifying the nature of the process. The first is a mechanical work, and the second is dissipation. The work can be positive or negative, while the dissipation is inherently positive (by the 2nd Law). However, if one or the other effect is zero, then the total integrated effect must be equal to the change in the state properties. Therefore, consider 2 limiting cases:

1. No work is extracted, $\delta W = 0$. In this case, the total energy dissipated becomes a path-independent quantity, since it is the only path-dependent property, and all other terms are integrated state properties. The result is as follows:

$$T_\infty \Sigma = \left(X_1 + \frac{1}{2}MV_1^2 + MgH_1\right) = \text{eXergy Destroyed}$$

All of the available energy (eXergy, kinetic, and gravitational) is destroyed if the object is brought to equilibrium with the environment without extracting any useful work.

2. Reversible process, $\delta\Sigma = 0$. Such a process would require a device of some sort that is capable of extracting work from the system. In that case, the total work is as follows:

$$W_{\text{rev}} = \left(X_1 + \frac{1}{2}MV_1^2 + MgH_1\right)$$

That is, all of the eXergy, kinetic energy, and potential energy are extracted as work. A similar analysis will be conducted in the next chapter as well, to reinforce the concept from another point of view.

Appendix 1: Formal Momentum Collision Analysis

A position diagram at an initial time in a fixed reference frame XY is shown in Fig. 2.9. Three position vectors relative to XY are shown, one from the origin to the smaller mass (m), one to the larger (M), and one to the center of mass (CM), whose relationship will be derived. Two relative position vectors are shown: the smaller mass relative to the center of mass $\left(\vec{r}_{m/\mathrm{CM}}\right)$, and the larger mass relative to the center of mass $\left(\vec{R}_{M/\mathrm{CM}}\right)$.

The 2 masses together are defined as the system. The location of its center of mass is defined as the total mass times the position vector to the center of mass, and that equals the sum of the individual masses times their position vectors. That is,

$$(m+M)\ \vec{r}_{\mathrm{CM}} = m\ \vec{r}_m + M\ \vec{R}_M$$

The position vectors to the individual masses can be expressed in terms of the position vector to the center of mass and to the relative position vectors between the center of mass and the individual masses:

$$(m+M)\ \vec{r}_{\mathrm{CM}} = m\ \left(\vec{r}_{\mathrm{CM}} + \vec{r}_{m/\mathrm{CM}}\right) + M\ \left(\vec{r}_{\mathrm{CM}} + \vec{R}_{M/\mathrm{CM}}\right)$$

The terms that involve $\vec{r}_{\mathrm{CM}}$ cancel (by the definition of the center of mass), and therefore:

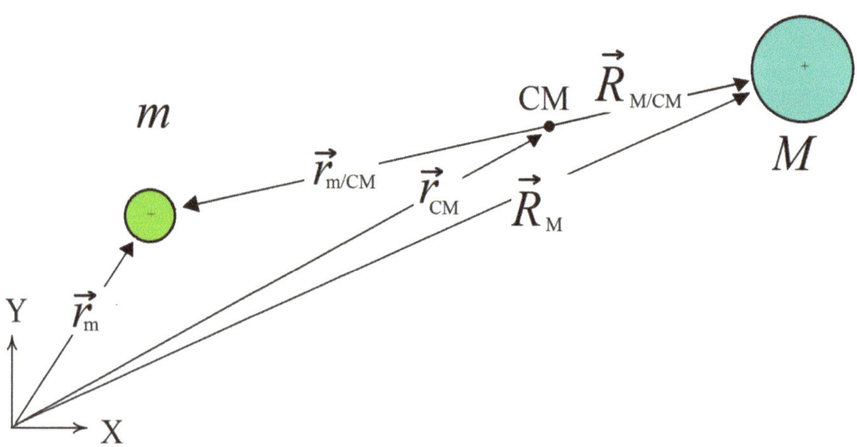

Fig. 2.9 Position vectors of two spheres in a fixed coordinate frame XY

$$0 = m\left(\vec{r}_{m/\mathrm{CM}}\right) + M\left(\vec{R}_{M/\mathrm{CM}}\right)$$

Taking the derivative of position yields the velocity, and recognizing that the coordinate frame is not rotating:

$$0 = m\vec{v}_{m/\mathrm{CM}} + M\vec{V}_{M/\mathrm{CM}}$$

A collision event is shown in Fig. 2.10 in the XY coordinate system, and in Fig. 2.11 in a coordinate system that moves with the center of mass (*xy*), with the x

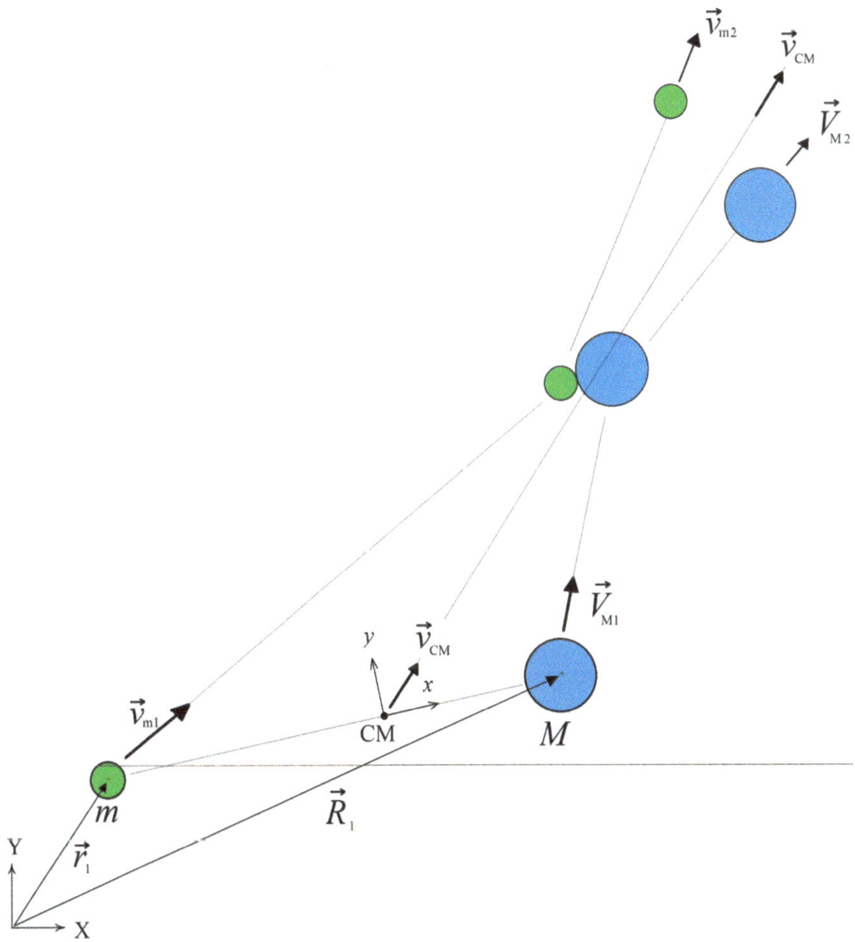

Fig. 2.10 Collision event in fixed coordinates XY. Three snapshots are shown: before, during, and after the collision. The coordinate frame *xy* moves with the center of mass, and its *x*-axis is oriented in the direction between the two masses. The collision takes place along this line, making it one of central impact

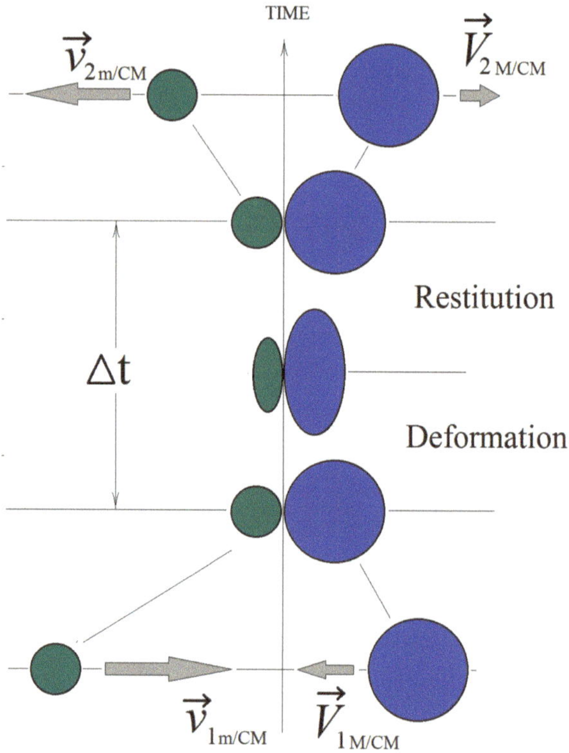

Fig. 2.11 Collision event in coordinates xy fixed to the center of mass. Snapshots are shown before the event, at the start of the event, at the turning point of the event, at the end of the event, and after the event

axis aligned with the direction of the central impact. The particle of mass m moves with a velocity $\vec{v}_{m1}$ before the collision and the larger mass M that moves at $\vec{V}_{M1}$ (in the XY frame). After the collision, the particles move at velocities $\vec{v}_{m2}$ and $\vec{V}_{M2}$ The center of mass is located on a line between the two masses, closer to the larger. In the absence of external forces, its trajectory is not affected by the collision, as will be shown.

The collision takes place in an infinitesimally short time (Δt), during which a period of deformation brings both masses to rest (in the center of mass coordinate frame) while deforming them, followed by a period of restitution during which the masses recoil from each other at new velocities. During the event, there is a large internal force between the two masses (of equal magnitude and opposite sign) that brings the masses to rest, and then accelerates them in the opposite direction (unless they stick together, called a plastic collision). The masses recoil from each other, and may experience some permanent change in shape, but are assumed to remain intact in this scenario.

In the study of mechanics, the principle of conservation of momentum (Newton's 2nd Law) yields a relationship between the initial and final velocities.

$$\frac{\text{(Final Momentum)-(Initial Momentum)}}{\Delta t} = \vec{F}_{\text{ext}}$$

Multiplying through by the time yields the 2nd Law as an impulse form:

$$\text{(Final Momentum)-(Initial Momentum)} = \vec{F}_{\text{ext}}\Delta t$$

The external force multiplied by the time duration is the impulse applied by the external forces. In terms of the velocities:

$$\left(m\vec{v}_2 + M\vec{V}_2\right) - \left(m\vec{v}_1 + M\vec{V}_1\right) = \vec{F}_{\text{ext}}\Delta t$$

Each term on the left-hand side represents the momentum of the center of mass:

$$(m+M)\vec{v}_{2\text{CM}} - (m+M)\vec{v}_{1\text{CM}} = \vec{F}_{\text{ext}}\Delta t$$

Rearranging:

$$\vec{v}_{2\text{CM}} - \vec{v}_{1\text{CM}} = \frac{\vec{F}_{\text{ext}}\Delta t}{(m+M)}$$

External forces (like gravity or friction) can have a significant influence on the center of mass depending on how much time before or after the event is under consideration. However, the collision event is one of very short time duration. The internal forces are very large between the two particles, and they cause each mass to rapidly change speed, but they have no effect on the center of mass. Therefore, the velocity of the center of mass is constant across the collision event (to a good approximation, even if external forces exist). Therefore:

$$m\vec{v}_{m/\text{CM}} + M\vec{V}_{M/\text{CM}} \approx (m+M)\vec{v}_{\text{CM}}$$

In the center of mass reference frame, the velocity of mass m will be expressed as lower case (v), and the velocity of M as upper case (V). Then the momentum equation (in center of mass frame) is

$$m\vec{v} = -M\vec{V}$$

This relationship can be used at any time, for any state before during or after the collision. That is, for example,

$$m\vec{v}_1 + M\vec{V}_1 = m\vec{v}_2 + M\vec{V}_2$$

Since the motion is all along a single direction (x) in the center of mass reference frame, the momentum equation can be written in terms of scalars, being careful of the signs. All scalars in this derivation are inherently positive, with the directions of the velocities defined by Fig. 2.11. The state numbers are retained as subscripts, with the understanding that the reference frame is the center of mass.

$$mv_1 - MV_1 = -mv_2 + MV_2 = 0$$

Given known approach velocities (state 1), this statement of the conservation of momentum provides one equation for two unknowns, the final velocities. Another constraint or independent relationship is needed to solve for the two final velocities. It is standard in mechanics to define a coefficient of restitution, ε, that is the ratio of the internal impulse applied during the period of restitution (where the particles are accelerated in the opposite direction to their approach) to that during the period of deformation (where the particles are decelerated to the center of mass velocity). This ratio is shown formally in mechanics textbooks to be equal to the ratio of the relative speed after the collision, the receding speed ($V_2 + v_2$), to the relative speed before the collision, the approach speed ($V_1 + v_1$):

$$\text{Coefficient of Restitution} = \varepsilon = \frac{v_2 + V_2}{v_1 + V_1}$$

It is difficult to predict the value of the coefficient of restitution from first principles. It merely forms a convenient way to express the final velocities in terms of a potentially empirical measure of the effectiveness of the collision. Eliminating V_2 from the momentum impulse equation:

$$-mv_2 + M(-v_2 + \varepsilon(v_1 + V_1)) = 0$$

Rearranging:

$$v_2(M + m) = M\varepsilon(v_1 + V_1)$$

Eliminating V_1:

$$v_2(M + m) = M\varepsilon\left(v_1 + \left(\frac{mv_1}{M}\right)\right) = \varepsilon v_1(M + m)$$

Or simply:

$$v_2 = \varepsilon v_1$$

The speed of the smaller mass after the collision is the coefficient of restitution times its original speed. Similarly, by eliminating v_2:

$$V_2 = \varepsilon V_1$$

Appendix 2: Relating Internal Energy to Temperature for Solids

The relationship between internal energy and temperature is developed in this Appendix. For solids (and liquids that are not freezing or boiling), the internal energy depends only on the temperature of the solid. That is, $U = U(T)$.

Consider the following hypothetical process. A solid object with a mass M is at an initial temperature of $T\text{-}\Delta T$, where ΔT is a small temperature change. The object is placed ("slowly" so that there is no significant kinetic energy change) into an environment at the temperature T (which is warmer than the object by ΔT). Heat naturally flows into the object until it attains thermal equilibrium. The First Law for this process is

$$\Delta E = \Delta U + \Delta KE + \Delta PE = Q_{\text{IN}} - W_{\text{OUT}}$$

The change in kinetic and potential energy and the work done by the system are all zero, and therefore,

$$\Delta U = Q_{\text{IN}}$$

Multiplying and dividing by the change in temperature:

$$\left(\frac{\Delta U}{\Delta T}\right)\Delta T = Q_{\text{IN}}$$

Dividing both sides by the mass and the temperature change:

$$\left(\frac{\Delta\left(\frac{U}{M}\right)}{\Delta T}\right) = \frac{Q_{\text{IN}}}{M \cdot \Delta T} [=] \frac{J}{kg \cdot \Delta K}$$

The quantity U/M, the internal energy per unit mass, has units of J/kg and is called the "specific" internal energy, and given a lower case u.

$$u = \frac{U}{M}$$

In the limit, as the temperature change shrinks to zero, the change in internal energy per unit mass divided by the change in internal energy is defined as a new thermodynamic property, called the "specific heat" and given the symbol "c."

$$c \equiv \lim_{\Delta T->0}\left(\frac{\Delta\left(\frac{U}{M}\right)}{\Delta T}\right) = \frac{du}{dT} = \frac{Q_{\mathrm{IN}}}{M\cdot\Delta T}[=]\frac{J}{kg\cdot\Delta K}$$

The specific heat can be interpreted as the heat addition required to change the temperature of 1 unit of mass by 1 degree temperature change FOR THIS TYPE OF PROCESS. The reason specific "heat" is a misnomer is because it refers to one way to change the temperature. It could equally be called the "specific work" by considering a process that adds energy to the system by work instead of heat. The specific heat can always be thought of as "change in specific internal energy per change in temperature," which is a lot longer than "specific heat," so the latter will be reluctantly accepted, since it is so engrained in the technical literature.

The specific heat of several solids at nominal room temperature is shown in Fig. 2.12. The internal energy for solids depends only on the temperature, and therefore, the specific heat depends on temperature as well.

In the same way that a reference elevation is defined for potential energy, it is generally convenient to define a reference temperature, $T_{\mathrm{ref},}$ where the specific internal energy is assigned a value of u_{ref} (typically set to zero). Then the specific internal energy at any other temperature (T) can be determined by integrating from the reference temperature to T:

$$u - u_{\mathrm{ref}} = \int_{T_{\mathrm{ref}}}^{T} c dT$$

The change in specific internal energy is related to the change in temperature by integrating between the 2 temperature states.

$$\Delta u = u_2 - u_1 = \int_{T_1}^{T_2} c dT$$

If the specific heat is treated as a constant, which is frequently acceptable, the change in internal energy is proportional to the change in temperature, with the specific heat as the proportionality constant. That is,

$$\Delta u = c\Delta T$$

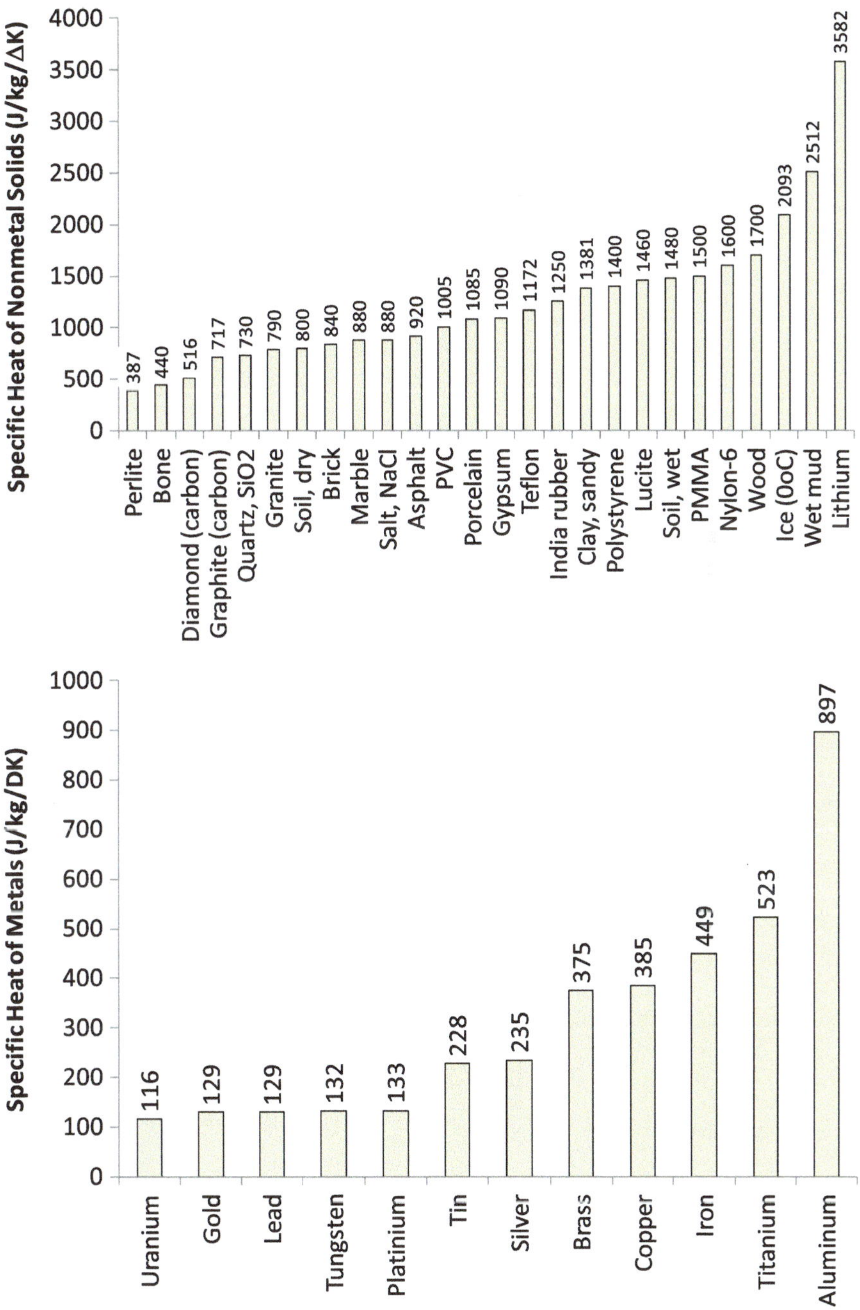

Fig. 2.12 Specific heat of nonmetals (top) and metals (bottom) at nominal room temperature. (Data adapted from http://www.engineeringtoolbox.com/specific-heat-capacity-d_391.html, accessed on 6/26/2015)

Even if the specific heat changes significantly over the temperature range in question, this temperature variation can be handled by either using the average specific heat at the two temperatures or by using the specific heat evaluated at the average temperature:

$$\int_{T_1}^{T_2} cdT \approx \frac{c_{(T_2)} + c_{(T_1)}}{2}(T_2 - T_1) \approx c_{(\overline{T})}(T_2 - T_1)$$

The first form uses the average of the specific heat at the initial and final temperatures. The second form uses the specific heat evaluated at the average temperature $\left(\overline{T} = (T_1 + T_2)/2\right)$. These are both approximate, and usually reasonable for engineering analysis.

Appendix 3: Relating Entropy to Temperature for Solids

It is tempting to define a coefficient like dS/dT as was done for the internal energy, but it is better to combine the 1st and 2nd Laws. Consider the 1st and 2nd Laws applied to an event that gives rise to a differential change in all quantities:

$$dE = \delta Q - \delta W$$
$$dS = \frac{\delta Q}{T_B} + \Sigma$$

Now consider the process to occur with the following restrictions:

- There is no change in kinetic or potential energy ($dE = dU$).
- The system is a nonphase changing solid ($dU = McdT$) and the specific heat is constant.
- The process is reversible ($\Sigma = 0$), and there is no external work ($\delta W = 0$).
- The boundary temperature equals the system temperature (not the dead state), which does not change significantly during the event.

The heat transfer becomes, from the 2nd Law,

$$\delta Q = TdS$$

The heat transfer, under this reversible process, is expressible as an exact differential. The combined 1st and 2nd Law is

$$McdT = TdS$$

Since the specific heat is a function of temperature, the variables can be separated:

$$dS = Mc\frac{dT}{T}$$

This expression is for a differential change in state along a reversible path. A finite change from one state to another can be obtained by then integrating between 2 states:

$$\int_{S_1}^{S_2} dS = \int_{T_1}^{T_2} Mc\frac{dT}{T}$$

$S_2 - S_1 = Mc \ln\left(\frac{T_2}{T_1}\right)$ for nonphase changing solids, and a constant specific heat

There are several restrictions to the use of this relationship between the entropy change of a solid and the temperature change. The most important ones are that it applies to solids with no phase change, and that the specific heat is a constant (chosen to be a suitable average if a large temperature change exists).

It would appear that it only applies to reversible processes. However, that is not the case. The relationship was derived by considering a reversible process, but the final outcome is that two state variables are related at two different states. Temperature and entropy are both path-independent quantities. That argument may seem circular, but it is well founded.

Appendix 4: Relating eXergy to Temperature for Solids

The eXergy (X) of system of mass M at an average temperature T, relative to a dead state at a temperature T_∞, is defined as follows:

$$X = U - T_\infty S$$

The eXergy of the system relative to a reference state (subscript REF), chosen to be at the dead state, is as follows:

$$(X - X_{\text{REF}}) = (U - U_{\text{REF}}) - T_\infty(S - S_{\text{REF}})$$

For solids, with a constant specific heats

$$X - X_{\text{REF}} = Mc(T - T_{\text{REF}}) - McT_\infty \ln\left(\frac{T}{T_{\text{REF}}}\right)$$

Choosing the reference temperature as the dead state, and the eXergy to be zero at the dead state:

$$X = Mc(T - T_\infty) - McT_\infty \ln\left(\frac{T}{T_\infty}\right)$$

Both the internal energy and the entropy will also be zero at the dead state with this choice of reference. Normalizing by the quantity McT_∞

$$\frac{X}{McT_\infty} = \frac{U}{McT_\infty} - \frac{T_\infty S}{McT_\infty} = \left[\frac{T}{T_\infty} - 1\right] - \ln\left(\frac{T}{T_\infty}\right)$$

These normalized quantities are plotted in Fig. 2.13 as a function of the ratio of the absolute temperature to the dead state temperature. Both the internal energy and entropy terms are positive for hot systems (temperatures greater than the dead state) and negative for cold systems. The eXergy, however, is an inherently positive quantity. It has a value of zero only when the system is in thermal equilibrium with the dead state.

It is instructive to consider the Taylor expansion of the entropy term (the natural logarithm). The natural logarithmic function, expanded about $y = 1$ (where $\ln(y) = 0$), is as follows:

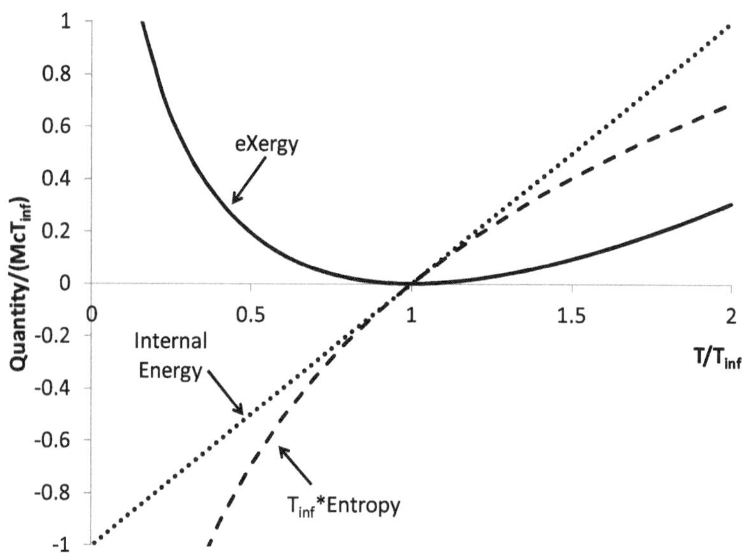

Fig. 2.13 eXergy and its components as a function of absolute temperature for solids with constant specific heats

$$\ln(y) = (y-1) - \frac{(y-1)^2}{2} + \frac{(y-1)^3}{3} - \frac{(y-1)^4}{4} + \ldots + \frac{(-1)^{n+1}(y-1)^n}{n} + \ldots$$

Applying this expansion to the eXergy:

$$\frac{X}{McT_\infty} = \left[\frac{T}{T_\infty} - 1\right] - \left[\left(\frac{T}{T_\infty} - 1\right) - \frac{1}{2}\left(\frac{T}{T_\infty} - 1\right)^2 + \frac{1}{3}\left(\frac{T}{T_\infty} - 1\right)^3 - \ldots\right]$$

The first term of the expansion of the entropy exactly cancels the internal energy, and the dimensionless eXergy of a solid is

$$\frac{X}{McT_\infty} = \left[\frac{1}{2}\left(\frac{T}{T_\infty} \quad 1\right)^2 \quad \frac{1}{3}\left(\frac{T}{T_\infty} \quad 1\right)^3 + \ldots\right]$$

Therefore, only for systems whose temperature deviates substantially from the dead state can be considered to have significant useful eXergy associated with them. Dissipation of mechanical energy tends to only raise the temperature by a small amount. This behavior is shown graphically, in that the internal energy and entropy terms have the same slope when the temperature equals the dead state.

Appendix 5: Friction Work

It is well known that friction opposes motion and has the effect of dissipating energy (destroying eXergy) when there is relative motion (as opposed to a straight static friction case). This Appendix demonstrates both of these points from a fundamental thermodynamic point of view, using $F = Ma$, the 1st Law and the 2nd Law, applied to the application of a block sliding on an inclined plane.

Consider a block of mass M on an inclined plane with an angle θ. The static and kinetic friction coefficients are μ_s and μ_k, respectively, but they are not used directly in this analysis. From statics analysis, the block will slide if the inclined angle exceeds the friction angle $\phi = \tan^{-1}(\mu_S)$. Then, the mechanics of the problem ($F = ma$ applied to the free body diagram of the block) in vector form is

$$M\frac{d\vec{V}}{dt} = M\vec{g} + \vec{F}_{\text{buoyancy}} + \vec{F}_{\text{drag}} + \vec{F}_{\text{friction}}$$

Four forces that act on the block are indicated (gravity, buoyancy, fluid drag, and friction between the block and the sliding plane). Note that these are the forces acting on the block (not the forces of the block on the environment). The focus in this case is on the effect of friction, and the two fluid forces (buoyancy and drag) are neglected going forward, as is commonly done in analyzing this application. This equation of motion can be viewed as a mechanical energy balance by taking the dot product with the velocity, which is equivalent to viewing each term (force or inertia) acting through a distance (work) per unit time (power).

$$M\frac{d\vec{V}}{dt}\cdot\vec{V}=\frac{d}{dt}\left(\frac{1}{2}MV^2\right)=M\vec{g}\cdot\vec{V}+\vec{F}_{\text{friction}}\cdot\vec{V}$$

The left-hand side is recognized as the rate of change of kinetic energy stored in the block. The first term on the right-hand side is the work per unit time (power) associated with the gravity, and the second term is the power associated with the friction between the block and the plane. Recognizing that the block slides along the plane (a kinematic constraint), and the gravity is vertical, the gravity power term is:

$$\vec{V}\cdot\vec{g}=\left(V\hat{i}\right)\cdot g\left(\sin\theta\ \hat{i}+\cos\theta\ \hat{j}\right)=Vg\sin\theta$$

The kinematics of the problem is relatively simple, in that motion is constrained along the plane. Therefore, the rate of change of elevation and velocity of the block (V) are related by

$$\frac{dH}{dt}=-V\sin\theta$$

The mechanical energy equation becomes

$$\frac{d}{dt}\left(\frac{1}{2}MV^2+MgH\right)=\vec{F}_{\text{friction}}\cdot\vec{V}$$

That is, the rates of change of the kinetic and gravitational potential energy equal the power of friction. At this point in traditional mechanics, it is reasonable to simply state that friction acts in the opposite direction of the motion, and therefore, the friction power is inherently negative. This intuitive result is analogous to stating that heat flows from hot to cold. However, it will now be demonstrated that this result is a consequence of the 2nd Law, namely, that the entropy of the universe must increase.

Taking the block as the system and applying the 1st Law of Thermodynamics:

$$\frac{dE}{dt}=\dot{Q}_{\text{in}}-\dot{W}_{\text{out}}$$
$$\frac{d}{dt}\left(U_{\text{block}}+\frac{1}{2}MV^2+MgH\right)=\dot{Q}_{\text{in}}-\dot{W}_{\text{friction}}-\dot{W}_{\text{drag}}-\dot{W}_{\text{bouyancy}}$$

The negative sign on friction here is consistent with the representation of the friction force (from mechanics convention) representing the force acting ON the block. Neglecting drag and buoyancy and rearranging (to isolate thermal from mechanical effects), the 1st Law becomes

$$\frac{d}{dt}\left(\frac{1}{2}MV^2 + MgH\right) = \left[\dot{Q}_{in} - \frac{dU_{\text{block}}}{dt}\right] - \dot{W}_{\text{friction}}$$

The work of friction is the friction force BY the system times the velocity, that is,

$$\dot{W}_{\text{friction}} = \left(-\vec{F}_{\text{friction}}\right) \cdot \vec{V}$$

$$\frac{d}{dt}\left(\frac{1}{2}MV^2 + MgH\right) = \left[\dot{Q}_{\text{in}} - \frac{dU_{\text{block}}}{dt}\right] + \vec{F}_{\text{friction}} \cdot \vec{V}$$

It is recognized that all the mechanical terms cancel, from the mechanics analysis, and, in effect, the thermal and mechanical effects are decoupled. What is left is

$$0 = \left[\dot{Q}_{\text{in}} - \frac{dU_{\text{block}}}{dt}\right]$$

In this case, the heat transfer rate represents heat flowing from the contact surface into the interior of the block. That heat transfer changes the internal energy stored within the block. However, the analysis of the block itself does not demonstrate that the work associated with the friction must be dissipative.

On the other hand, a thermodynamic analysis of the combined block and plane will make the friction effect internal, and demonstrate the dissipative nature of friction. Taking the system to include both the block and the plane and the air immediately around the block:

$$\frac{d}{dt}\left(U_{\text{block}} + U_{\text{plane}} + U_{\text{air}} + \frac{1}{2}MV^2 + MgH\right) = \dot{Q}_{\text{in}} - \dot{W}_{\text{out}}$$

In this case, the boundary of the control volume is taken sufficiently far from the block and plane such that there is no exchange of heat or work with the environment. Therefore, the energy balance of the combined block and plane reduces to

$$\frac{d}{dt}\left(U_{\text{block}} + U_{\text{plane}} + U_{\text{air}} + \frac{1}{2}MV^2 + MgH\right) = 0$$

Introducing the mechanical energy balance ($F = $ ma times velocity), this becomes

$$\frac{d}{dt}\left(U_{\text{block}} + U_{\text{plane}} + U_{\text{air}}\right) = -\vec{F}_{\text{friction}} \cdot \vec{V}$$

This energy balance shows that the work associated with the friction changes the internal energy stored. Any heat transfer would take place in the "aftermath" of the event.

An entropy balance on the combined system is

$$\frac{d}{dt}\left(S_{\text{block}} + S_{\text{plane}} + S_{\text{air}}\right) = \frac{\dot{Q}_{\text{in}}}{T_b} + \dddot{\Sigma}$$

Since the process is adiabatic, and recognizing the 2nd Law,

$$\frac{d}{dt}\left(S_{\text{block}} + S_{\text{plane}} + S_{\text{air}}\right) = \dddot{\Sigma} \geq 0$$

That is, the entropy stored within this adiabatic system must increase with time. To combine the 1st and 2nd Laws, the energy balance is modified by multiplying and dividing by the differential change in entropy:

$$\left[\frac{d\left(U_{\text{block}} + U_{\text{plane}} + U_{\text{air}}\right)}{d\left(S_{\text{block}} + S_{\text{plane}} + S_{\text{air}}\right)}\right]\frac{d}{dt}\left(S_{\text{block}} + S_{\text{plane}} + S_{\text{air}}\right) = -\vec{F}_{\text{friction}} \cdot \vec{V}$$

The derivative term in brackets is inherently a spatial derivative (there are hot and cold spots with the control volume), and is written in shorthand as follows:

$$\left[\frac{d\left(U_{\text{block}} + U_{\text{plane}} + U_{\text{air}}\right)}{d\left(S_{\text{block}} + S_{\text{plane}} + S_{\text{air}}\right)}\right] = \frac{dU}{dS}$$

For a nonphase changing solids, but internal energy and entropy are monotonically increasing functions of temperature, and therefore, this "derivative" is inherently positive. The combined 1st and 2nd laws become

$$\left[\frac{dU}{dS}\right]\dddot{\Sigma} = -\vec{F}_{\text{friction}} \cdot \vec{V} \geq 0$$

And there you have it. The dot product of the friction force acting on the block and the velocity of the block must be negative to satisfy the 2nd Law of thermodynamics. That is, the friction force must oppose the motion.

As is commonly experienced, friction raises the temperature of objects. This "heating" (it is "working" actually) is localized near the surfaces on which friction acts. This thermodynamic analysis, with the control surface located far from the point of action, does not pretend to determine how hot the surfaces get. That is a topic for a heat transfer analysis, not formally developed in this text. It does show formally that friction is an inherently dissipative process, in that a mechanical form of energy is converted to a thermal form.

Workshop 2.1: Bouncing Ball

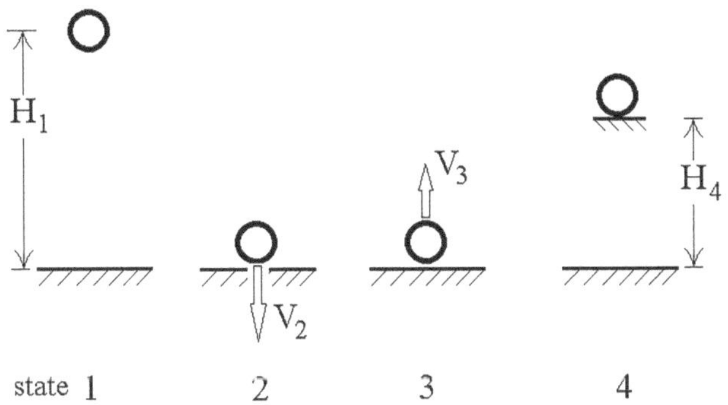

A golf ball (effective mass $M_e = 0.044$ kg, displaced mass $M_d = 5(10)^{-5}$ kg, specific heat

$c = 1500$ J/kg/K) is released from rest from an initial height $H_1 = 1.92$ m (state 1). It collides with the ground with a velocity V_2 (state 2) and acquires an upward velocity V_3 (state 3). It rises to a height $H_4 = 1.56$ m (and zero velocity) where it is caught (state 4). It is allowed to cool to ambient temperature (state 4′, not shown) and then slowly raised back to its initial height to complete a cycle.

Open a spreadsheet and save it with the file name "WKSP1_***.xlsx" where *** is a short name for your team. Create a block at the top with a heading "INPUT PARAMETERS" and enter the data from the problem statement with the values given as the default values. Then prepare a STATE and a PROCESS table using the format below (color-coded as a guide). Enter appropriate formulas into the blank boxes. Use the 1st and/or 2nd Laws of Thermodynamics to analyze each process, that is, $\Delta E = \Delta(U + KE + PE) = Q_{\text{in}} - W_{\text{out}}$ and $\Delta S = Q_{\text{in}}/T_B + \Sigma$ with the ball as the system, noting that internal energy changes for nonphase changing solids are related to temperature changes by $\Delta U = Mc\Delta T$ and $\Delta S = Mc \ln\left(1 + \frac{\Delta T}{T+273}\right)$ with temperatures in Celsius. Pay attention to formatting of the worksheet (i.e., technical communication). Sketch the cycle on the H-V plot provided for the default case. Note: for the temperature change, report enough significant figures to observe a change. Part of the underlying message of this workshop is that "dissipation" of mechanical energy "shows up" as a temperature rise, but the magnitude of the temperature rise is often too small to even be measurable. There are a lot of subtleties in this workshop, and the state and process tables are intertwined.

Work as a team of 3 or 4, and prepare a neat single hard copy submission with the results for the default case and upload your file, both due at the end of the class period. (Keep your own completed individual sheets, bring to next class).

INPUT PARAMETERS

g	9.81	m/s^2, gravity value
me	0.044	kg, effective mass
mdisp	0.00005	kg, displaced mass of air
c	1500	J/kg/K, specific heat
H1	1.92	m, height of release
H4	1.56	m, height of capture
T_∞	25	oC, dead state and initial temperature

COLOR CODE

- INPUTS
- 1st Law
- 2nd Law
- Property Relation
- Process Relation

STATE TABLE

state	Height (m)	Speed (m/s)	Potential Energy (J)	Kinetic Energy (J)	Internal Energy (J)	Temperature (oC)	Entropy (J/K)	eXergy (J)
1								
2								
3								
4								
4'								

note: dead state is reference state for internal energy and entropy

PROCESS TABLE (all in J)

			1st LAW TERMS			2nd LAW TERMS		
Process	Description	Assumptions	E	Q_{in}	W_{out}	$T_\infty \Delta S$	$Q_{in}(T_\infty/T_b)$	$eX_D=T_\infty\sigma$
1-2	Free Fall	Adiabatic, reversible, neglect fluid drag						
2-3	Collision	Adiabatic, unchanged surroundings						
3-4	Free Rise	Adiabatic, reversible, neglect fluid drag						
4=4'	Wait	thermal equilibrium achieved with ambient						
4'-1	Lift	Adiabatic, neglect KE during process						
		NET						

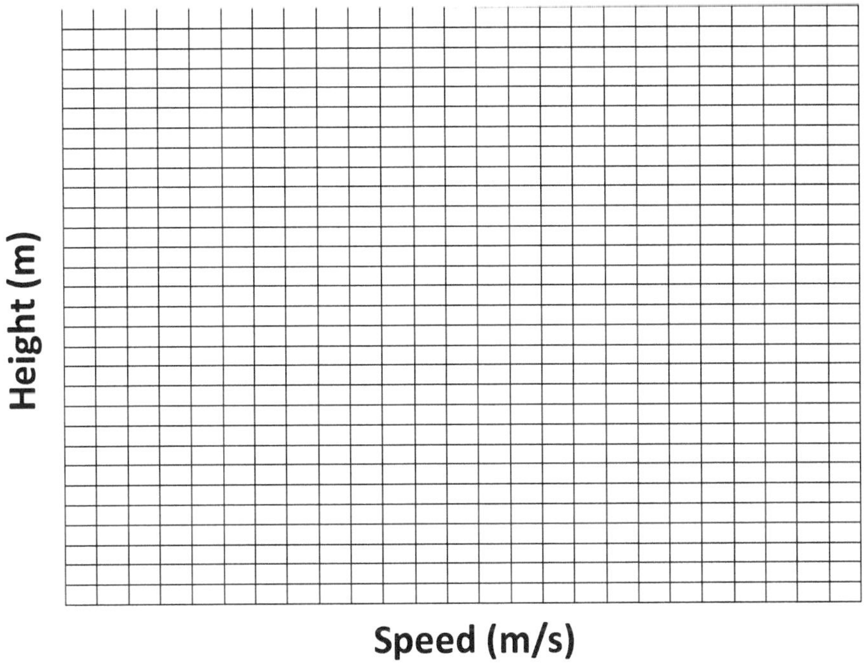

For this plot, base the scale on the largest values of the properties (height and speed), using absolute value for speed (with the origin at the bottom left corner). Show the 4 states clearly (i.e., a dot and a circle around it), and draw curved dotted lines between them as your best guess as to the actual path that is taken by each process. Be aware that this "before and after" thermodynamic analysis does not specifically yield information about those shapes, so use your intuition and experience instead about the shape of the paths.

Workshop 2.2: Measuring the Specific Heat of a Solid

The specific heat of a solid material is to be measured in a calorimeter experiment (a "calorimeter" being a device designed to relate the internal energy of a substance to its temperature change). A sample of material is formed into a sphere of diameter $D = 0.1$ m, embedded with a coiled wire with electrical resistance $R = 100$ Ohm, and a thermistor to measure the temperature near the center. The sample, weighed on a scale, has a mass of $M = 4.0$ kg and is initially at a temperature $T_1 = 25$ °C (equal to ambient). A layer of insulation of negligible mass is placed around the sample. The test consists of the application of a voltage of 12 V (DC) to the wire, and monitoring the temperature and elapsed time. In this case, the temperature reaches $T_2 = 35$ °C in time $\Delta t = 93$ minutes. The insulation is removed, and the solid cools back to ambient (state 3). Complete the derived parameters, state and process tables, with the system defined to include the solid material plus the wires.

The determination of the specific heat of the material involves a simple 1st Law calculation from the measured data. However, the underlying message of this workshop is to pay attention to the subleties of a 2nd Law analysis. In particular, notice that at state 2, the object is warm and has a certain amount of internal energy relative to the dead state. It also has eXergy (the theoretical ability to do work in being brought to the dead state). However, the eXergy is much less than internal energy. Most of the input electrical work results in eXergy destruction. The rest of it is destroyed during process 2-1, when the object cools back to ambient. The net effect is that all of the input electrical work is dissipated. It is a good way to measure specific heat, however.

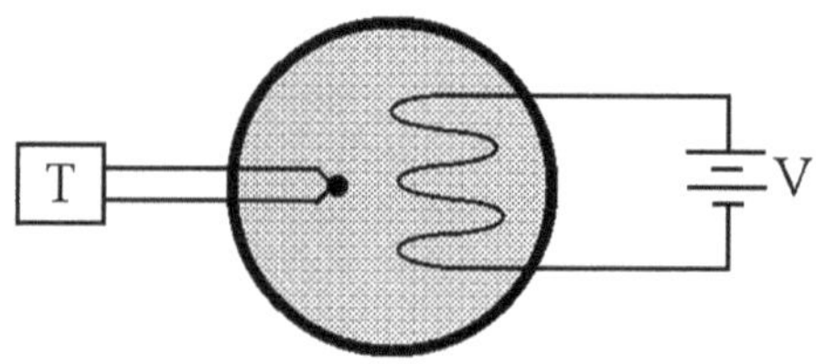

INPUT PARAMETERS

D	0.1	m, diameter of sample
M	4	kg, mass of sample
R	100	Ohms, electrical resistance
V	12	Volts
T1	25	oC, initial and ambient temperature
T2	35	oC, final temperature
t	93	min, elapsed time

DERIVED PARAMETERS

Volume		m^3, volume of material
Wdot,e		Watts, input electrical power
ρ		kg/m^3, density of material
c		J/kg/ΔK, specific heat of material

STATE TABLE

State	T (oC)	u (J/kg)	s (J/kg/K)	eX (J/kg)
1				
2				

note: set reference state for u and s as the dead state

PROCESS TABLE (all in J)

Process	ΔU	Q_{in}	W_{out}	$T_\infty \Delta S$	$Q_{in}(T_\infty/T_B)$	$eX_D = T_\infty \Sigma$
1-2						
2-1						
NET						

Further Comments

- A definition of specific heat (a material property) is as follows:
  "Specific Heat: the change in internal energy of a nonphase changing solid or liquid per unit mass per unit temperature change."
  The term "specific heat" is a misnomer, because, as this example shows, the internal energy can be changed by work. The term "specific internal energy" cannot be used, as that is the total internal energy divided by the mass (i.e., $u = U/M$).
- Some students tend to "hold on" to things learned in High School, or in Freshman Chemistry classes. For example, many will blindly apply the "formula" $Q = MC\Delta T$. This formula applies to a subset of situations. It does NOT apply if there is any phase change, and many such applications going forward.
- With the system defined to include the solid plus the wires, the energy that crosses the boundary is in the form of electrical work. There is no heat transfer. An alternative analysis would treat the solid and the electrical wire as being separate systems. It is a good exercise to analyze those systems separately. A First Law analysis is fine, and you will see that there is in fact heat transfer to the solid, but you will find a complete 2nd Law analysis to be problematic as the temperature of the boundary where heat crosses is ill-defined. Combining the solid and wire eliminates that problem.

First Law of Thermodynamics: $\Delta E = Q_{in} - W_{out}$
Second Law of Thermodynamics: $\Delta S = \frac{Q_{in}}{T_B} + \Sigma$ with the restriction $\Sigma \geq 0$,
where, for a given process,

$\Delta E = \Delta U + \Delta KE + \Delta PE$ is the change in total energy stored within the system (i.e., internal, kinetic, and potential)

For solids that undergo no change in phase: $\Delta U = Mc(T_2 - T_1)$

Q_{in} is the heat transfer into the system during the process

W_{out} is the work done by the system on its surroundings during the process

$\Delta S = Mc \ln\left(\frac{T_2+273}{T_1+273}\right)$ is the change in entropy stored within the boundary (a material property)

T_B is the absolute temperature (K or R) at the system boundary where heat crosses

Σ is the entropy generation within the system boundary. It must be positive (or zero).

$eX_D = T_\infty \Sigma$ is the eXergy destruction (where T_∞ is the dead state temperature, usually ambient). It must be positive (or zero) and is a measure of the "loss" of useful energy.

The eXergy is defined as $eX = U - T_\infty S$ and is work potential of a substance in an environment with a dead state temperature T_∞.

Problem Statements

Saving the Earth

An asteroid of mass $M = 100$ kg is travelling on a collision course with Earth at a relative speed of $V_1 = 25{,}000$ m/s. The nations of the world have developed a plan to intercept the asteroid with a projectile of mass $m = 1$ kg with a velocity of $v_1 = 1000$ m/s, hitting the asteroid from the side to adjust its trajectory to one that will miss the Earth. The hope is that the projectile will bounce off the asteroid to maximize the effect. What is the possible range of the change in perpendicular velocity that the projectile could impart to the asteroid?

http://www.spacedaily.com/reports/Protecting_Earth_Against_Asteroids_999.html, accessed 6/26/2015

Ballistics

A bullet of mass $m = 50$ g fired from a gun is aimed at a block of wood of mass $M = 3$ kg that hangs from a string of length $L = 2$ m. The bullet becomes embedded in the block of wood, and the block swings to an angle $\theta = 45°$. Determine the initial speed of the bullet, and the average temperature of the wood and bullet (assume they are equal) shortly after the event. Assume the bullet is made of lead ($c_{\text{lead}} = 129$ J/kg/ΔK) and $c_{\text{wood}} = 1700$ J/kg/ΔK.

Spring to Action

A novel rocket propulsion application is proposed in which a planned collision between two spacecraft results in a coefficient of restitution greater than 1 by the use of a form of stored energy released by the collision. Both spacecrafts initially orbit Earth at the same altitude, but at different speeds. The lead craft, of mass M, has a stiff spring (spring constant k) that is held in a compressed state x_0. This compression can be done slowly (by astronauts doing exercises, perhaps?) and is in a circular orbit at an altitude h. The orbital speed, V, corresponds to that for which the centripetal acceleration equals gravity. The trailing craft, of mass m, has a greater speed ($v > V$), and when it collides with the lead craft, a mechanism is activated that releases the spring, which pushes the lead craft forward, and the trailing craft backward. Determine the speeds of both crafts after the collision and the apparent coefficient of restitution (which will be greater than 1), using a 1st Law analysis.

Rowing Machine

The main function of an exercise machine is to provide some specific type of workout. The human does thermodynamic work on the machine, and the easy thing to do is design the machine to dissipate the energy. With a little ingenuity, an exercise machine could be redesigned to put that work to use, to charge a battery or supply electrical power where needed. A rowing machine is a perfect example. The action is a pull a bar to which the rower applies a force ($F_{stroke} = 20$ lb, say) that acts through a linear distance ($L_{stroke,}$ about 3 feet). A mechanism is designed to transmit that motion into the rotation of a paddle wheel immersed in a container of water (V_{water}, about 50 liters). All the work goes put in by the rower is dissipated in the water. Estimate the temperature of the water (specific heat $c = 4.2$ kJ/kg/K) after a typical workout ($t_{workout} = 30$ minutes), with a stroke of frequency (f) about one stroke every 2 seconds. Calculate the average power (in Watts) that could be theoretically developed if the machine were re-designed to drive a reversible electric generator (rather than make some warm water).

Chapter 3
Thermodynamic Cycles and Efficiencies

Classically, the 1st Law of Thermodynamics is stated as the conservation of energy principle, while the 2nd Law is stated in the form of a process restriction. The property entropy is a consequence of those restrictions. For example, a classical statement of the 2nd Law is as follows: *"It is impossible for a device to exchange heat with a single source and generate a net work output."* The approach taken in this textbook is to introduce the 2nd Law the same way the 1st Law is introduced, namely, as a conservation principle on a property, in this case, entropy. Unlike energy, however, entropy is not conserved. It is constantly being generated, and a balance equation allows for its quantitative expression. Combining the energy and entropy balance equations results in an eXergy balance, and the entropy production is expressed in energy units as a form of mechanical energy dissipation into thermal energy, or eXergy destruction. The restrictions that pertain to heat and work exchanges are consequences of eXergy destruction. This chapter considers thermodynamic cycles, that is, closed systems that undergo a series of processes that return it to an initial state to demonstrate important thermodynamic principles. First and Second Law efficiencies are defined as performance measures. The distinctions between thermal and mechanical energy forms become clearer, and the limitations of energy conversions are thereby defined. Once fundamental principles are established, a review of energy consumption trends and prices in the United States of America is performed, followed by a case study that compares the energy and cost associated with various home heating alternatives. A novel (or at least not widely available) approach that uses an on-site heat engine to drive a heat pump that pulls heat from the ground is shown to be the most attractive, thermodynamically.

Energy forms are classified in Table 3.1 as being either mechanical (including electrical) or thermal, and as being transferrable or storable. Mechanical energy is generally a more useful form of energy. Entropy production has the effect of degrading a mechanical form of energy into a thermal form, often called dissipation. For example, energy is conserved in a plastic collision, but the thermal energy produced at the expense of the initial kinetic energy is not particularly useful.

G. Sidebotham, *An Inductive Approach to Engineering Thermodynamics*, Mechanical Engineering Series, https://doi.org/10.1007/978-3-030-20430-3_3

Table 3.1 Classification of Energy Forms

	Mechanical	Thermal
Transferrable (associated with energy crossing a system boundary)	Work (force through a distance) or Electrical	Heat
Storable (energy associated with the mass contained within the system boundary)	Kinetic Energy Potential Energy Rotational Energy eXergy	Sensible Internal Energy Latent Internal Energy Chemical Internal Energy

A thermodynamic machine is defined as a device that converts energy from one form to another. Machines that violate the 1st Law are called "perpetual motion machines of the 1st kind." The output of these machines exceeds all the inputs and the energy of the universe changes. Machines that violate the 2nd Law are called "perpetual motion machines of the 2nd kind." The entropy of the universe decreases with these machines. It is often much harder to prove that a machine violates the 2nd Law than it does the 1st Law. Nobody has ever built a perpetual motion machine. Devices can be made to operate continuously from "renewable" sources like solar, wind, wave, or tide and are not perpetual motion machines.

A common misconception of environmentally conscious people is to regard electricity as a completely "green" form of energy, a misconception fueled in part, perhaps, by marketing interests. In fact, electricity is a secondary form of energy produced by a thermodynamic machine that requires a primary form of energy to drive it. The primary sources are dominated by fossil fuels (coal and natural gas) at the time of this writing. Until the electric utility industry can completely switch to renewable sources (a long time down the road), greenhouse gas emissions are an inherent consequence of the use of electricity. These emissions must always be factored in when comparing the environmental impact of a technology that uses electricity (such as automobiles). A broad aim of this chapter is to make a distinction between a primary source of energy and a secondary source.

The analyses in this chapter are conducted as a type of "black box" approach, with no details of what goes on inside the dashed lines that separate system from surroundings, deviating somewhat from the inductive approach. In future chapters, actual devices and their working fluids will be described and analyzed. These machines will always adhere to the general restrictions outlined in this chapter.

First and Second Law Efficiencies

The term "efficiency" is frequently used and often abused. A marketing pitch may tout that a given product is "efficient" as a statement without a clear definition, or what the efficiency is compared to. Many times, a better word would be "effectiveness". The device is good at performing some task. There is often a trade-off between efficiency and effectiveness. For example, a car can be designed for a high acceleration (be effective) at the expense of a high gas mileage (efficiency). Generally, efficiency is a ratio that quantifies the success of converting one quantity into another, or a measure of a desired output given some input. A thermodynamic efficiency pertains to energy conversions, that is, a dimensionless measure of a device that delivers an energy output from an energy input. Gas mileage (miles driven for every gallon of gasoline consumed) is not a thermodynamic efficiency according to that strict definition, as miles driven is not easily expressed as an energy form. It is an output over an input, but it is not dimensionless.

Even within the field of thermodynamics, there can be inconsistency and/or confusion in what is meant by "efficiency." An efficiency can be classified based on whether it compares the output of some device to an input, or whether it is based on the performance compared to a theoretical ideal. An efficiency based on the comparison of outputs to inputs is classified as a First Law efficiency, and can be loosely defined in the following word equation:

$$\eta_I = \begin{pmatrix} \text{First Law} \\ \text{Efficiency} \end{pmatrix} = \frac{\text{"What You Get"}}{\text{"What You Pay For"}} = \frac{\text{Desired Output}}{\text{Required Input}}$$

For example, a 1st Law efficiency applied to consumer products (e.g., in refrigerators, washers, dryers, or air conditioners) is "Energy Efficiency Rating," or EER, which is a ratio of a desired thermal load to a required input electrical load. Conventional units of "thermal" quantities are typically in Btu (for energy) or Btu/hour (for energy rate), and conventional units for electricity are kW-hr (for energy) or Watts (for energy rate). Therefore, the units of EER are Btu/(W-hr).

$$\text{EER} = \frac{\text{Desired Thermal Output}}{\text{Electrical Input}} [=] \frac{\text{Btu}}{W \cdot \text{hr}}$$

It is unfortunate that a complete adoption of modern society to a consistent set of units (like SI) has not been made. Engineering concepts are hard enough to grasp, without the use of inconsistent units that compare apples to oranges. The EER is a ratio of energy output to energy input, but it is not dimensionless. In thermodynamic analysis, it is more convenient to work in consistent units, and the EER is related to the "Coefficient of Performance" or COP:

$$\text{COP} = \frac{\text{Desired Thermal Output}}{\text{Electrical Input}} = \text{EER}\left(\frac{\text{Btu}}{W \cdot \text{hr}}\right) \cdot \left|\frac{W \cdot \text{hr}}{3.798\text{Btu}}\right|$$

That is,

$$\text{COP} = \text{EER}/3.798$$

Typical values of EER are 8–12 for air conditioning units and 4–8 for household refrigerators. Refrigerators have lower EER than air conditioning units in practice because the temperature "hill" they must climb is higher (-10°C to 25°C to cool a freezer compartment sitting in a room compared to 25°C to 35°C to cool a home in summer).

As another example, in the power industry, a measure of the performance of power plants commonly used is the "Heat Rate," which is a ratio of the required added thermal energy to the desired electrical work output. It is an inverse 1st Law efficiency, and it has units of MMBtu/kW-hr, where the unit MMBtu is actually a million Btu. Why not MBtu, for mega Btu? Ask a science historian.

The 1st Law efficiency can be greater than unity, depending on what is defined as the desired output and what is defined as the required input. Some devices are intended to convert a mechanical input (work) into a thermal output (heat), and others are intended to convert a thermal input into a mechanical output. A classification of important engineering devices is presented in Table 3.2. One objective of this chapter is to demonstrate the fundamental restrictions imposed by the 2nd Law in converting a thermal input to an electro/mechanical output, indicated by the listing of the maximum possible 1st Law efficiency. When a work input is converted to a thermal output, it is possible for the desired output to exceed the required input. That does not mean that the conservation of energy is violated, of course, and that will be clearer by the end of this chapter. On the other hand, a thermal input cannot be converted entirely into a mechanical output for any real device.

Therefore, another way to measure the performance of a device is to compare the actual output to what could theoretically be achieved if it were reversible (i.e. no entropy production or equivalently eXergy destruction). A word statement for a 2nd Law efficiency is as follows:

$$\eta_{II} = \begin{pmatrix} \text{Second Law} \\ \text{Efficiency} \end{pmatrix} = \frac{\text{"What You Get"}}{\text{"What You COULD Get"}}$$

A properly defined Second Law efficiency cannot exceed unity. The 2nd Law efficiency is often much harder to determine than the 1st Law efficiency.

Table 3.2 Classification of energy conversion devices and their 1st Law efficiency limits

	Thermal Input	Mechanical Input
Thermal Output	$\eta_{I,\,\mathrm{MAX}} = 100\%$ Heat exchangers (Water heaters, boilers, radiators, etc.)	$\eta_{I,\,\mathrm{MAX}} > 100\%$ Electric heaters, refrigerators, heat pumps, air conditioners
Mechanical Output	$\eta_{I,\,\mathrm{MAX}} < 100\%$ Heat engines (automobiles, power plants, jet engines)	$\eta_{I,\,\mathrm{MAX}} = 100\%$ Electric motors, electric generators, gear boxes

Thermal Reservoirs

The concept of a thermal reservoir is central to thermodynamic analysis. A thermal reservoir is defined (loosely) as any part of the surroundings of a defined system that is able to maintain a stable temperature during the course of an event. This behavior can be achieved by a number of means, in practice, as follows:

The thermal reservoir can be an actual reservoir, in which energy exchange between a system and its surroundings has a much larger effect on the system than the surroundings. For example, if a hot metal part that is created in a manufacturing process needs to be cooled rapidly, it can be placed into a container of water or oil with a much larger thermal mass (M^*c) than the part. The heat exchange between the part and the water will change both temperatures, but the temperature change of the water will be negligible. The dead state is generally a thermal reservoir of this type, and depending on the objective of a particular analysis, it can be a local dead state, like the large container of water in the above example or the temperature of the air inside a building, or a more global dead state, like the outdoor temperature.

Alternatively, a thermal reservoir can be attained if the temperature of the surroundings can be maintained by other simultaneous means. For example, the surroundings for a class of 100 students can be defined as the air in the classroom. The metabolic heating of the students, professor, computer, video projector, laptops, cellphones, etc., tends to raise the temperature of the air, but the cooling system of the building is designed to counter that effect.

Phase change materials can create a thermal reservoir. For example, the hot metal part could be placed into an ice water bath, and the melting of the ice maintains the temperature of the water near its freezing point.

A final way that a thermal reservoir can be considered is for a differential event. Heat exchange between the system and its surroundings may result in a substantial change in the temperature of the surroundings for a finite event, but during a differential change, the temperature of the surroundings is considered to behave like a thermal reservoir. This is a subtle case, and will be discussed as needed, going forward.

An example for which it is difficult to produce a true thermal reservoir is the conversion of chemical to thermal energy (i.e., combustion). Fuel and air can be continuously fed to a device, and their chemical reaction can result in a well-defined energy conversion. However, it is difficult to establish a true reservoir temperature and, therefore, the entropy transfer cannot be clearly established.

Cyclic Processes with a Single Thermal Reservoir

Consider a closed system that consists of an object and its boundary layer separating it from the surroundings at the dead state, shown schematically in Fig. 3.1, with dashed lines indicating the inner and outer edges of the boundary layer. A series of

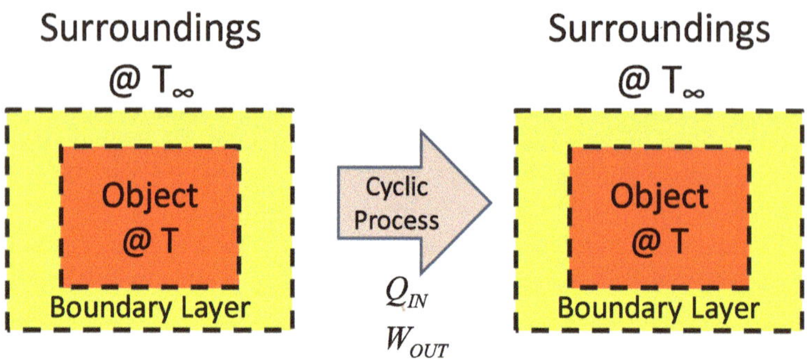

Fig. 3.1 Event acting on an object and its boundary layer. The event is cyclic so that the object is returned to its original state

processes takes place during which energy is exchanged with the environment, but the system is returned to its original state. For example, consider an electrical heater, in which case the object is a wire, subject to a 2-step process. First, electrical current could flow, driven by a voltage difference, constituting an input of work, raising the internal energy (and therefore temperature). Subsequently, heat could flow to the surroundings until the wire returns to its original temperature. Any process (usually broken up into a sequence of distinct processes) that returns a system to its original state is called a ***cycle***. In this section, a cycle will be analyzed in a general way by considering the object and its boundary layer as separate subsystems. Many (if not most) practical devices that operate in a cycle consist of one key process that obtains the desired thermodynamic effect. However, a series of following process is needed to return the working fluid to its starting point in order to repeat the desired process (for batch processes) or to operate continuously (for steady-flow devices).

It is possible for a system to undergo a mechanical cycle, but not a thermal cycle. For example, a piston can move through two strokes and return the volume of the contained fluid to its starting point, but the temperature and pressure will be different. There was a mechanical cycle, but not a thermal cycle.

Analysis of the Object

An energy diagram of just the object (considered to be at a uniform temperature T) is shown in Fig. 3.2, and a corresponding entropy diagram is shown in Fig. 3.3. Work done by the object on the environment and heat added to the object are shown crossing the inner surface of the boundary layer. The directions of the arrows (heat in, work out) are shown by the convention established historically during the development of heat engines (i.e., steam engines) where "heat" in (chemical internal energy actually) was considered the required input, and mechanical work out was the desired output.

An energy balance applied to the process is as follows:

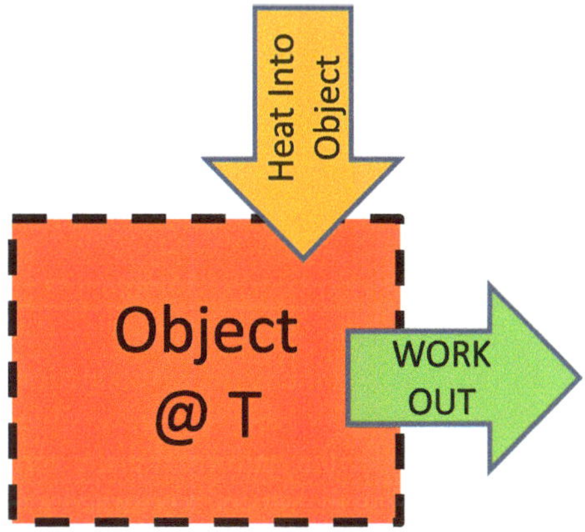

Fig. 3.2 Energy flows associated with object

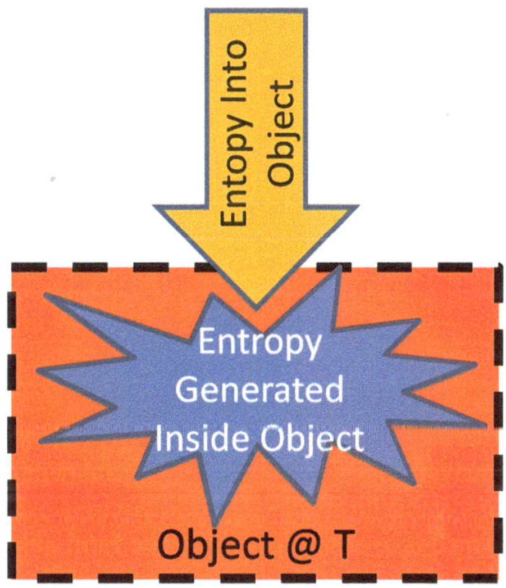

Fig. 3.3 Entropy flows for just the object

$$\Delta U_{\text{OBJECT}} = Q_{\text{IN}} - W_{\text{OUT}}$$

Since the process is cyclic, that is, designed so that average temperature of the object is returned to its original state, the change in internal energy is zero, and

$$Q_{\text{IN}} = W_{\text{OUT}} \text{ for the cycle}$$

The 1st Law dictates that the net heat added to the object must equal the net work output done by the object. Note that heat or work could flow in or out during steps within the overall process. The energy flows indicated here are net flows. Also, the signs of the energy flows can be opposite of those shown. In the example of an electrical heater, the electrical work would be done on the wire, making the work out negative and, therefore, heat must be out of the wire.

It might seem tempting to simply say "add heat to the engine, and it could GENERATE (not consume) electricity and not violate the 1st Law". The 2nd Law will make that impossible. The entropy balance applied to the wire is:

$$\Delta S_{\text{OBJECT}} = \frac{Q_{\text{IN}}}{T} + \Sigma_{\text{OBJECT}}$$

Recall that Σ is the entropy production and the 2nd Law states that it must be a positive quantity (or zero for a reversible process). Also, the absolute temperature must be used for the entropy transfer term. The change in entropy is zero for the cycle process, as the object is returned to its original state. Therefore,

$$T\Sigma_{\text{OBJECT}} = -Q_{\text{IN}} \geq 0$$

Any irreversibility generated internally results in heat transferred out of the object in order to return the object to its original state. That is, the 2nd Law dictates that the direction of heat transfer must be from the object to the surroundings, opposite to the direction shown in Figs. 3.2 and 3.3. According to the 1st Law ($Q_{\text{IN}} = W_{\text{OUT}}$), the work done by the wire must be negative:

$$T\Sigma_{\text{OBJECT}} = -W_{\text{OUT}} \geq 0$$

So given that the system operates in a cycle and exchanges heat with a single thermal reservoir, the 2nd Law requires that work must be supplied to the system from the surroundings, and heat must flow from the system to the surroundings. This analysis confirms one classical statement of the 2nd Law, the ***Kelvin-Planck Statement***, that "it is impossible to devise a cyclically operating device, the sole effect of which is to absorb energy in the form of heat from a single thermal reservoir and to deliver an equivalent amount of work."

Yet another subtle point is that the 1st Law does not prevent the object from being at a lower temperature than its surroundings. Imagine a cold wire. Work added to it would require heat to flow out of wire, from a low temperature to a high one. The analysis of the boundary layer surrounding the wire will make that impossible. The other classical statement of the 2nd Law is the ***Clausius Statement***: "Heat can never pass from a colder to a warmer body without some other change, connected therewith, occurring at the same time"

Analysis of the Boundary Layer

The energy diagram of the boundary layer separating the object from its surroundings is shown in Fig. 3.4. The entropy diagram is shown in Fig. 3.5. Heat enters the

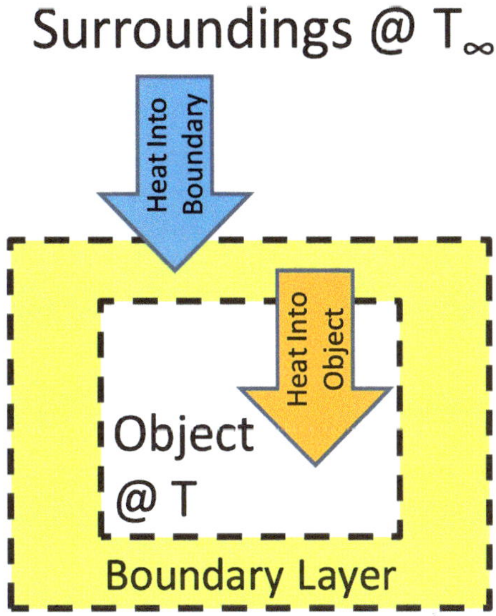

Fig. 3.4 Energy flows associated with the boundary layer

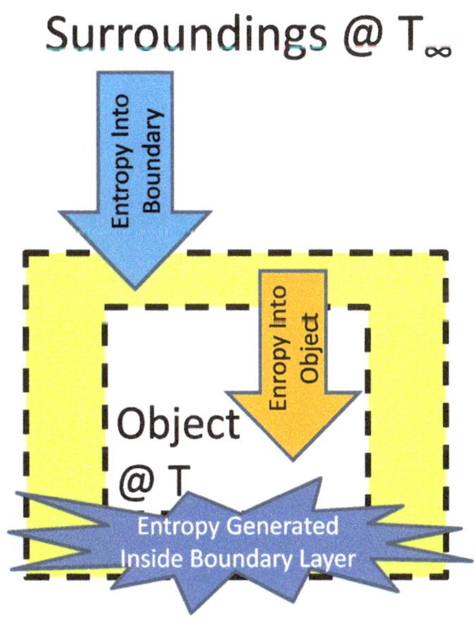

Fig. 3.5 Entropy flows associated with the boundary layer

outer surface directly from the surroundings, and exits the boundary layer across its inner surface into the object. The sign of the heat crossing the inner surface must be consistent with what was assumed for the analysis of the object (even though it is now known that the sign of this heat flow is impossible). There is no work interaction associated with the boundary layer.

The energy balance of the boundary layer (treated as a closed system even though fluid may flow into and out of it, in actuality), consistent with the directions defined in Fig. 3.4 is as follows:

$$\Delta U_{\mathrm{BL}} = Q_{\mathrm{IN(OUTER)}} - Q_{\mathrm{IN(INNER)}}$$

For the cyclic process, the change in internal energy is zero, and therefore,

$$Q_{\mathrm{IN(OUTER)}} = Q_{\mathrm{IN(INNER)}} = Q_{\mathrm{IN}}$$

That is, the heat that enters the boundary layer from the surroundings must equal the heat that enters the wire.

The entropy balance for the boundary layer is, in a word statement, as follows:

$$\begin{bmatrix} \text{Change in} \\ \text{Entropy Stored} \\ \text{in Boundary Layer} \end{bmatrix} = \begin{bmatrix} \text{Entropy that} \\ \text{Enters the} \\ \text{Boundary Layer} \end{bmatrix} - \begin{bmatrix} \text{Entropy that} \\ \text{Exits the} \\ \text{Boundary Layer} \end{bmatrix} + \begin{bmatrix} \text{Entropy} \\ \text{Generated Within} \\ \text{Boundary Layer} \end{bmatrix}$$

$$\Delta S_{\mathrm{BL}} = \frac{Q_{\mathrm{IN}}}{T_\infty} - \frac{Q_{\mathrm{IN}}}{T_{\mathrm{OBJECT}}} + \Sigma_{\mathrm{BL}}$$

While the heat that enters the boundary layer is equal to the heat that exits, the temperatures of these surfaces are different, and therefore, the entropy that crosses is different. For the cyclic process, the change in entropy stored is zero, and therefore, the entropy generated is as follows:

$$\Sigma_{\mathrm{BL}} = (-Q_{\mathrm{IN}})\left(\frac{1}{T_\infty} - \frac{1}{T_{\mathrm{OBJECT}}}\right) \geq 0$$

The analysis of the object requires the heat in to be negative, or $(-Q_{\mathrm{IN}}) \geq 0$. Therefore,

$$\left(\frac{1}{T_\infty} - \frac{1}{T_{\mathrm{OBJECT}}}\right) = \frac{T_{\mathrm{OBJECT}} - T_\infty}{T_{\mathrm{OBJECT}} \cdot T_\infty} \geq 0$$

or

$$T_{\mathrm{OBJECT}} \geq T_\infty$$

In summary, the analysis of the object has demonstrated that heat must flow from the object to the surroundings, and the analysis of the boundary layer requires that the object must be at a higher temperature than the surroundings. That is, heat must flow from high to low temperature, conforming to experience, and confirming the Clausius statement of the 2nd Law.

Combined System

The total entropy generated by the wire and the boundary layer surrounding it is the sum of the entropy generated in both subsystems:

$$\Sigma = \Sigma_{\text{WIRE}} + \Sigma_{\text{BL}} = \frac{(-Q_{\text{IN}})}{T} + (-Q_{\text{IN}})\left(\frac{1}{T_\infty} - \frac{1}{T}\right) = \frac{(-Q_{\text{IN}})}{T_\infty} \geq 0$$

The eXergy destruction is as follows:

$$eX_D = T_\infty \Sigma = (-Q_{\text{IN}}) = (-W_{\text{OUT}}) \geq 0$$

That is, all of the electrical work put into the object is dissipated. Going forward, the system boundary will usually be placed at the outer surface of the boundary layer, and this analysis serves to demonstrate that there can be substantial eXergy destruction within the boundary layer. From an engineering perspective, heat transfer that takes place between machines and their environment that is not part of the thermodynamic objective can generally be minimized but not eliminated and is called ***stray heat***. Stray heat generally detracts from the performance from both a 1st and 2nd Law perspective, and is unavoidable in practice, but can often be designed to have a minor effect.

An important consequence of this analysis is that it is possible (quite easy, actually) to convert input work into heat loss from the device to a single thermal source (the dead state). A simple electrical heater operates this way. It is impossible, however, to run this device in reverse, that is, to add heat to the device from a single source, and generate a net work output. That impossibility, again, is the Kelvin-Planck Statement of the 2nd Law.

Consider an electrical heater, or some other device whose objective is to deliver heat as a desired output from a work input. It is easy to attain a 100% 1st Law efficiency from this device in practice, and it has the appearance of being a good way to heat a living space. However, there is no way to eliminate eXergy destruction if the device exchanges heat with a single reservoir (the heated space). It will be shown, however, that it is theoretically possible to eliminate eXergy destruction, and get more out of the input work if the device operates between two reservoirs. People still will use electric heaters, however, because they have advantages related more to their effectiveness than to their thermodynamic efficiency, such as simplicity, portability, and low capital cost compared to more thermodynamically efficient options.

One practical example that might appear to generate electricity by exchanging heat with a single source (and thereby violate the 2nd Law of Thermodynamics) is the generation of electricity using photovoltaic solar cells. In this case, operating at a quasi-steady-state, the solar energy that enters the device is a form of heat (by the mechanism of radiation, not conduction), and electrical work is extracted. However, the solar panel actually exchanges heat with 2 reservoirs: its immediate surroundings (Earth temperature) and the sun (at a bit higher temperature than Earth). The conversion of solar input cannot be 100%, and the portion of the input not converted into work is converted into thermal energy. From a human perspective, it is an inexhaustible and renewable form of energy we can tap into.

Noncyclic Processes

The analysis of a noncyclic process on a closed system (with its boundary layer) that operates with a single thermal reservoir is worthy of discussion at this point. The object does not return to its original state in this case. The energy balance is as follows:

$$\Delta U = Q_{\mathrm{IN}} - W_{\mathrm{OUT}}$$

This result shows that any work output (a mechanical energy exchange) is accompanied by a combination of two thermal effects: a heat input and/or a decrease in internal energy. The entropy balance is as follows:

$$\Delta S = \frac{Q_{\mathrm{IN}}}{T_\infty} + \Sigma$$

Combining, by eliminating the heat exchange, and rearranging for the work output,

$$W_{\mathrm{OUT}} = -(\Delta U - T_\infty \Delta S) - T_\infty \Sigma$$

Recognizing the eXergy change,

$$W_{\mathrm{OUT}} = -\Delta X - T_\infty \Sigma$$

Net work output is possible if the eXergy decrease associated with the process is greater than the eXergy destruction.

Electric Motors

A motor is a device that converts electrical input work into output shaft work (Figs. 3.6 and 3.7). It can be considered to be a device that interacts with the surroundings through 2 work exchanges (electrical and mechanical) and with a single thermal reservoir (the dead state, defined to be the environment immediately around the motor). The motor will be assumed to undergo a noncyclic process, meaning that the average temperature of the motor is in general different at the end of the process. The case of a cyclic process can be easily obtained if a motor is operated for a period of time, then turned off, and allowed to cool back to its original state. This cycle is analogous to a collision event (which raises the internal energy) followed by its aftermath, the cooling back to ambient. Note that the direction of the energy exchanges deviates from the default signs (heat in, work out) and that subscripts are used to keep the analysis consistent with the assumed directions. Formal 2nd Law analysis will confirm that the heat and two work exchanges for a motor are in the directions indicated in Fig. 3.7.

The system consists of the motor and the boundary layer surrounding it. The 1st Law for this noncyclic process is as follows:

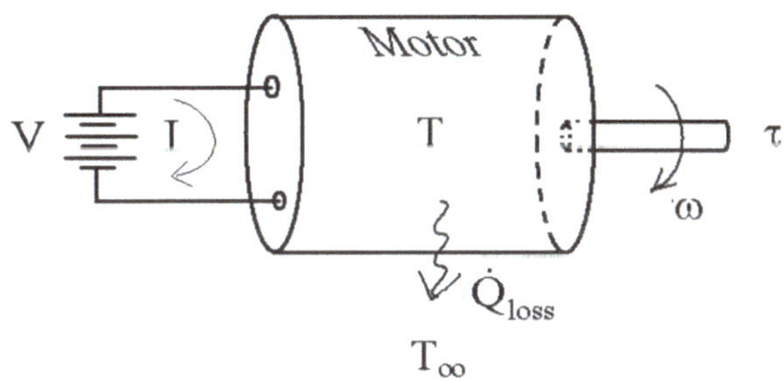

Fig. 3.6 Schematic of a DC electric motor

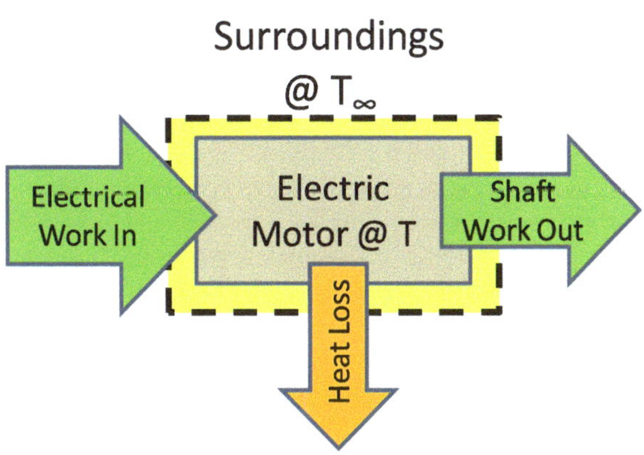

Fig. 3.7 Thermodynamic Process Diagram of an electric motor

$$\Delta U_{\text{Motor}} + \Delta KE_{\text{motor}} = -Q_{\text{OUT}} + W_{\text{ELEC,IN}} - W_{\text{SHAFT,OUT}}$$

Since the motor may contain moving parts, it has kinetic energy, which is neglected going forward (but could easily be retained if, e.g., a form of flywheel were part of the motor design). The entropy balance is as follows:

$$\Delta S_{\text{Motor}} = \frac{-Q_{\text{OUT}}}{T_\infty} + \Sigma$$

Notice that the temperature for heat transfer is taken outside the thermal boundary layer, while the temperature of the motor itself is not well characterized. Rearranging and expressing in terms of the eXergy destruction:

$$eX_D = T_\infty \Sigma = T_\infty \Delta S_{\text{Motor}} + Q_{\text{OUT}}$$

Any irreversibility within the motor (eXergy destruction) results in a combination of an increase in entropy of the motor and heat loss to the environment. Combining the 1st and 2nd Laws and solving for the output shaft work, recognizing that the eXergy change is $\Delta X = (\Delta U - T_\infty \Delta S)$, and eXergy destruction is $eX_D = T_\infty \Sigma$:

$$W_{\text{SHAFT,OUT}} = W_{\text{ELEC,IN}} - \Delta X - eX_D$$

Electrical input work is converted to output shaft work and to the combined thermal effects of an eXergy change and eXergy destruction. This conversion requires, of course, that some physical mechanism or device is constructed that can execute the conversion, a topic of electro-magnetism. The reduction in work is due to the eXergy destruction that takes place within the motor, and this eXergy is dissipated to the environment as low-grade heat loss.

Consider again the 1st Law, rearranged with mechanical terms on the left and thermal on the right:

$$W_{\text{ELEC,IN}} - W_{\text{SHAFT,OUT}} = \Delta U_{\text{Motor}} + Q_{\text{OUT}}$$

Any work imbalance appears as either a change in internal energy (temperature rise) or in heat loss to the environment. This analysis cannot, and is not intended to, distinguish between the two. In practice, when a motor that drives a fixed mechanical load is started (and is initially at the dead state), the response time of the electro-magnetic effects is generally short compared to the thermal response time. That is, for a given input of electricity, the mechanical system will accelerate to its steady-state speed. Therefore, an imbalance of work input and output will go into thermal effects. Initially, the motor is at room temperature, and there is no temperature to drive heat. The work imbalance will increase the internal energy of the motor and, by the property relationship, its temperature. As time proceeds, the temperature of the motor rises, and it loses heat at a greater rate. If left running continuously, the motor will either reach a stable equilibrium temperature, at which point the imbalance in work will be dissipated as heat, or the motor will burn out (the temperature is too high). If the motor is turned off and allowed to cool, it will return to its original state, and the overall process is cyclic. Many applications of motors involve intermittent use, with time elapse between active periods allowing for the motor to cool.

A motor that operates in a cycle (or continuously at a steady state, studied next) is one for which the change in eXergy is zero. A reversible motor is one for which the eXergy destruction is zero, in which case 100% conversion of electrical work to input work is obtained. This analysis shows that it is possible to convert one form of work completely into another form of work, as a theoretical limit. In practice, internal irreversibilities cause a portion of the input work to be converted to thermal energy. The 1st Law efficiency of the motor (operating cyclically) is as follows:

$$\eta_{\text{motor}} = \frac{W_{\text{SHAFT,OUT}}}{W_{\text{ELEC,IN}}} = 1 - \frac{eX_D}{W_{\text{ELEC,IN}}} \leq 1$$

This efficiency is restricted to a value of unity, as both the output and input are both work quantities, and their transfers do not involve entropy transfer.

The 2nd Law efficiency is the same as the 1st Law efficiency, because both desired output (outlet shaft work) and required input (electrical work) are mechanical forms of energy exchange.

Electric Generators

A motor can be operated in reverse, in principle, with shaft work as the input being converted to electrical work as the output (Fig. 3.8). This type of device is an electric generator. The analysis is similar to the motor, but with the signs of the work terms reversed.

The 1st Law for a generator is as follows:

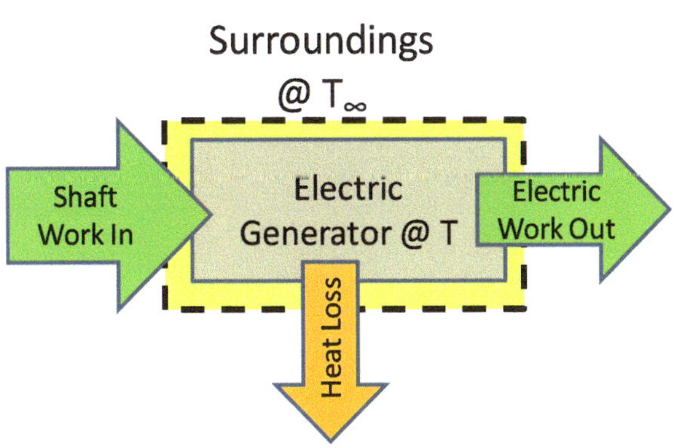

Fig. 3.8 Process diagram of an electric generator

$$\Delta U_{\text{Generator}} = -Q_{\text{OUT}} + W_{\text{SHAFT,IN}} - W_{\text{ELECTRIC,OUT}}$$

The 2nd Law is as follows:

$$\Delta S_{\text{Generator}} = \frac{-Q_{\text{OUT}}}{T_{\infty}} + \Sigma$$

Combining the 1st and 2nd Laws and solving for the output electrical work:

$$W_{\text{ELEC,OUT}} = W_{\text{SHAFT,IN}} - \Delta X - T_{\infty}\Sigma$$

Input shaft work is converted to output electrical work and the remainder is eXergy destruction within the device, which equals heat loss to the environment. This conversion requires, of course, that some physical mechanism or device is constructed that can execute the conversion. The 1st Law efficiency of the generator is defined as follows:

$$\eta_{\text{Generator}} = \frac{W_{\text{ELEC,OUT}}}{W_{\text{SHAFT,IN}}} = 1 - \frac{eX_D}{W_{\text{SHAFT,IN}}} \leq 1$$

A reversible generator is one for which the eXergy destruction is zero, in which case 100% conversion of electrical work to input work is obtained.

As with motors, the 2nd Law efficiency of an electric generator is the same as the 1st Law efficiency, because both desired output (outlet electrical work) and required input (shaft work) are mechanical forms of energy exchange.

Steady-State Continuous Processes

The analysis in the previous section considered the thermodynamics of motors and generators for finite processes. That is, the process was considered to take place over a significant period of time, and the overall effects observed. Another view is to consider a rate basis, or continuous basis, which amounts to analyzing a process that takes place over a short period of time (Fig. 3.9). The heat and work energy transfers

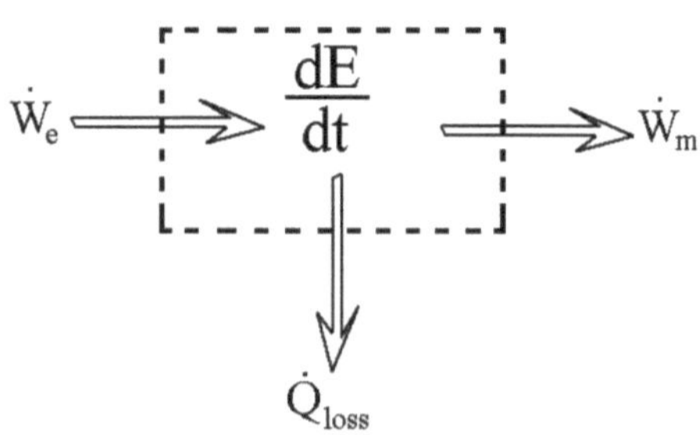

Fig. 3.9 Thermodynamic process diagram of an electric motor in continusous mode

are expressed on a rate basis, and the energy balance for a differential process for the motor (Fig. 3.7) is as follows:

$$\frac{dE_{\text{motor}}}{dt} = -\dot{Q}_{\text{OUT}} + \dot{W}_{\text{ELECTRIC,IN}} - \dot{W}_{\text{SHAFT,OUT}} \quad (= 0@SS)$$

If the device is operated continuously and allowed to reach a steady state, then the energy stored within the system boundary remains constant. There is a net balance between energy inputs and outputs. The desired output, at steady-state, is as follows:

$$\dot{W}_{\text{SHAFT,OUT}} = \left(-\dot{Q}_{\text{OUT}}\right) + \dot{W}_{\text{ELECTRIC,IN}} \quad (@SS)$$

No restriction is placed on the direction of heat transfer by the 1st Law. That is, it might be tempting to attempt a design of a motor that draws heat in from its environment and therefore produces more output power than input power. What does the 2nd Law have to say about that? The entropy balance is as follows:

$$\frac{dS_{\text{motor}}}{dt} = \frac{-\dot{Q}_{\text{OUT}}}{T_{\infty}} + \dddot{\Sigma} \quad (= 0@SS)$$

Rearranging and defining eXergy destruction rate:

$$T_{\infty}\dddot{\Sigma} = e\dddot{X}_D = \dot{Q}_{\text{OUT}} \quad (@SS)$$

This heat loss is a form of "stray heat," that is, heat loss to the environment that is not a desired energy exchange, but is an inevitable part of the operation of a real device.

The 2nd Law requires the eXergy destruction rate to be positive and, therefore, the direction of the heat transfer is constrained to be in the direction assumed in Fig. 3.7. The 1st Law efficiency is as follows:

$$\eta_{\text{motor}} = \frac{\dot{W}_{\text{SHAFT,OUT}}}{\dot{W}_{\text{ELEC,IN}}} = 1 - \frac{e\dddot{X}_D}{\dot{W}_{\text{ELEC,IN}}} \leq 1$$

A similar analysis of an electric generator would show the impossibility of adding heat to extract extra work.

In both motor and generator operating at steady state, there is a conversion of mechanical energy to thermal energy, which is required by the 1st Law to be in the form of heat transfer and to be from the device to ambient by the 2nd Law. Recall the basic heat transfer rate expression:

$$\dot{Q}_{\text{OUT}} = \Omega A\left(T_{\text{system}} - T_{\text{surroundings}}\right),$$

where Ω is an overall heat transfer coefficient (the topic of a course in heat transfer) and A is the surface area of the boundary. That is, the device runs hot. How hot? That is an important practical design consideration. Motors or other devices will fail if their temperature is too high, and manufacturer data sheets usually state the operating conditions. Thermal management of these devices can become problematic, and can be an important consideration in the design of systems that use them. The main design criterion is to ensure the motor operating temperature does not exceed a specific value.

The steady-state operating temperature can be expressed as follows:

$$T_{\text{motor},@SS} = T_\infty + \left(\frac{1-\eta_{\text{motor}}}{\eta_{\text{motor}}}\right)\left(\frac{\dot{W}_{\text{SHAFT,OUT}}}{\Omega A}\right)$$

This result shows a number of practical principles that would factor into their design. First, the temperature of the device runs at a higher temperature than its local ambient. That may seem obvious, but in design practice, it is important to understand the range of ambient conditions a device will be expected to operate under. Second, a higher 1st Law efficiency will result in a lower operating temperature, as the energy is passing from one mechanical form to another, and does not need to be "dissipated" as waste heat. Third, there is the scaling, namely, the higher the output power (offset somewhat by the physical size, A), the higher the operating temperature. Finally, the thermal management will depend on the effectiveness of the heat transfer, all lumped into the overall heat transfer coefficient (Ω). It is common for special systems to be implemented to increase that coefficient. For example, heat transfer "fins" are a common passive means. Coupling fins with the addition of fans to blow ambient air over them makes it a more active strategy. Water cooling may also become necessary.

Appendix 1 is a deeper dive into the performance characteristics of typical DC motors from a thermodynamic viewpoint.

Cyclic Processes with Two Thermal Reservoirs

Many devices can be reasonably modeled to exchange heat with two thermal reservoirs. Engineering devices whose main purpose is to convert energy between work and heat are generally classified as either heat engines or heat pumps. The objective of a heat engine is to produce work from a thermal source. A fundamental understanding of the operation of electrical power plants, automobile engines, and jet engines is facilitated by this analysis, even though the high side temperature source is from a chemical energy conversion (combustion) and is therefore not a strict thermal reservoir, and involves an inherently open system. The objective of a heat pump is to transfer heat from a low temperature to a high temperature, which requires an input of work. Household refrigerators, air conditioners, and heat pumps operate between a cold reservoir and a hot reservoir. Individual components can be analyzed as open systems, but a boundary can be naturally defined as a closed system for the entire device. The basic principles of these devices are the same, but the objectives (inputs and outputs) are different.

Heat Engines

A heat engine is defined as a device with the following characteristics:

1. It operates between two thermal reservoirs: a higher temperature reservoir (at T_H) and a lower one (at T_C). Either reservoir could be the dead state, but a heat engine can operate between any two temperature reservoirs.
2. It undergoes a cycle, so that it returns to its original state.
3. It is a closed system (no mass crosses its boundary).
4. Heat enters from the high-temperature reservoir and exits to the low-temperature one, and net work is produced.

A classic representation of a heat engine is shown in Fig. 3.10. Heat crosses the system boundary (dashed line) in the directions shown: from the high-temperature reservoir to the heat engine and from the heat engine to the low temperature. Work exits the device.

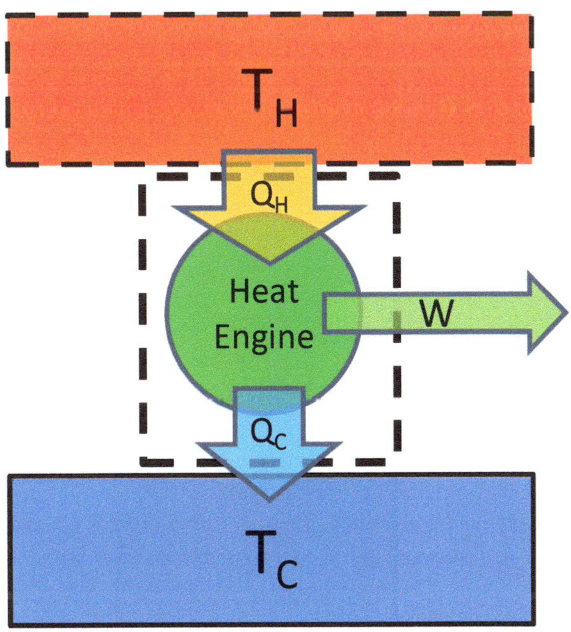

Fig. 3.10 Schematic of a heat engine operating between two temperature reservoirs

The 1st Law, evaluated over an entire cycle, is as follows:

$$\Delta U_{\text{ENGINE}} = 0 = Q_H - Q_C - W_{\text{OUT}}$$

Rearranging,

$$W_{\text{OUT}} = Q_H - Q_C$$

The net work out equals the net heat in. A perpetual motion machine of the 1st kind would violate this principle. Notice that the heat added to the high side must exceed the heat to the low side in order to produce a work output. In practice, the best state-of-the-art utility power plants convert approximately half of the thermal energy in the heating source into work, and simply reject the remainder to the local surroundings (air or river water). This heat rejection requires large, expensive equipment to build and operate. It is common for educated people to say at this point "why not just convert more of the heat into work?" The answer lies in the 2nd Law of Thermodynamics, which demonstrates that the energy associated with the heat source is not fully "available" as follows.

The 2nd Law for the heat engine is

$$\Delta S_{\text{ENGINE}} = 0 = \frac{Q_H}{T_H} - \frac{Q_C}{T_C} + \Sigma$$

Combining the 1st and 2nd Laws by eliminating Q_C,

$$\frac{Q_H}{T_H} - \frac{W_{\text{OUT}} - Q_H}{T_C} + \Sigma = 0$$

Rearranging for the work output:

$$W_{\text{OUT}} = Q_H\left(1 - \frac{T_C}{T_H}\right) - T_C\Sigma$$

The 1st Law efficiency is defined as the net work out divided by the heat input (not the net heat, which equals the output work):

$$\eta_I = \frac{W_{\text{OUT}}}{Q_H} = \left(1 - \frac{T_C}{T_H}\right) - \frac{T_C\Sigma}{Q_H}$$

The low-temperature reservoir is commonly taken to be the dead state, in which case the numerator of the last term is the eXergy destruction that takes place within the engine. A reversible heat engine operating between two temperature reservoirs will yield the maximum output work possible, by setting the entropy generation to zero.

$$W_{\text{OUT,MAX}} = Q_H\left(1 - \frac{T_C}{T_H}\right)$$

The maximum possible 1st Law efficiency is, therefore, as follows:

$$\eta_{1,\text{MAX}} = \frac{W_{\text{OUT,MAX}}}{Q_H} = \left(1 - \frac{T_C}{T_H}\right)$$

This efficiency, called the Carnot efficiency (plotted in Fig. 3.11 for three low side reservoir temperatures as a function of high side reservoir temperature), represents the maximum possible efficiency that can be produced by a heat engine operating between two temperature reservoirs. For a high side reservoir temperature of 100°C and a low side reservoir temperature of 25°C (nominal room temperature), for example, the maximum possible thermal efficiency is 20%, indicating that heat engines that operate between those reservoirs would, at best, convert 20% of a heat input source to useful work, and 80% would be "discarded" to the ambient. This analysis shows that it is not easy to convert heat into work. The Carnot efficiency increases with high side temperature, but with diminishing returns (i.e., the curve is concave down). At a temperature of 600°C, the maximum possible

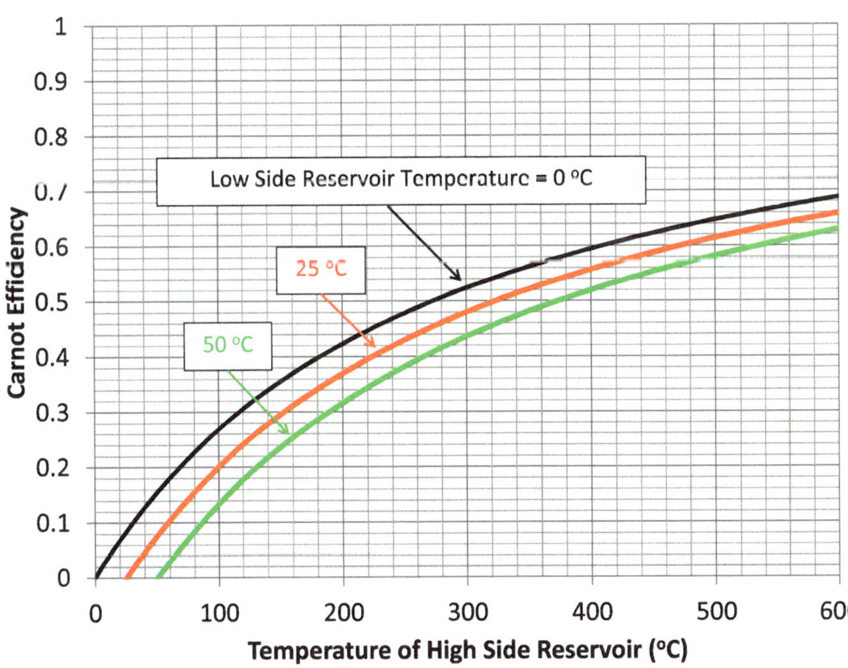

Fig. 3.11 Carnot efficiency (maximum theoretically possible 1st Law efficiency) of a heat engine as a function of high side reservoir temperature, for low side reservoir temperatures of 0, 25 and 50°C

efficiency is 66%. Carnot efficiencies improve with decreasing low side reservoir temperatures.

The derivation conducted here does not restrict the cycle to any specific cycle (such as the Carnot cycle). However, in order for the heat transfer to be reversible, the temperature difference driving it must be zero (or vanishingly small) both at the high side and low side, and these conditions are difficult to achieve in practice. Therefore, any real device will have significant irreversibility associated with it, and the Carnot efficiency represents a theoretical upper limit. In practical heat engines for which the thermal source is chemical energy conversion, it is difficult, if not impossible, to establish a meaningful value for the effective temperature of the high side reservoir. This difficulty is particularly true for internal combustion engines. The principles of the Carnot efficiency are valid in these cases, the main one being that a large fraction of input thermal energy must be rejected to the environment as "waste heat." It is just difficult to quantify from a 2nd Law perspective.

When it is possible to clearly define reservoir temperatures, a 2nd Law efficiency can be defined as the actual work to the maximum possible work:

$$\eta_I = \frac{W_{\mathrm{OUT}}}{W_{\mathrm{OUT,MAX}}} = 1 - \frac{T_H}{(T_H - T_C)} \frac{T_C \Sigma}{Q_H}$$

The 2nd Law efficiency is 100% if there are no irreversibilities within the engine, including the heat transfer processes between each reservoir and the engine (perhaps the hardest irreversibility to eliminate in most cases).

A heat engine that produces no net work represents a situation in which heat flows from a high temperature to a low temperature, as it does in the boundary layer of the electric heater example. In this case, setting the work output to zero:

$$0 = Q_H \left(1 - \frac{T_C}{T_H}\right) - T_C \Sigma$$

Solving for the entropy generation, and multiplying by the dead state temperature (to give the eXergy destruction):

$$eX_D = T_\infty \Sigma = T_\infty Q_H \left(\frac{1}{T_C} - \frac{1}{T_H}\right) = T_\infty Q_H \left(\frac{T_H - T_C}{T_H \cdot T_C}\right)$$

This result reinforces that heat transfer across a finite temperature difference generates entropy (and destroys eXergy).

Appendix 2 revisits the interpretation of eXergy as the maximum possible work that could be extracted using a hypothetical heat engine. Appendix 3 expands on the nature of eXergy destruction associated with heat transfer.

Cogeneration

Heat engines are designed to produce useful work as their primary output and, therefore, the heat rejection to the cold reservoir is considered as "waste heat." However, there are opportunities for a heat engine to be used in a manner in which both the work output and the heat rejected to the cold reservoir are considered to be useful outputs. Such an application is called "cogeneration," the simultaneous production of a useful work and useful heat from a single engine. Opportunities for the use of heat from a heat engine include space heating of residential and commercial buildings, and industrial processes that require a heat addition. Except for some geographical regions that are warm year round, virtually every building in the United States has a heating system designed to maintain the temperature of the interior at nominal room temperature (25°C) when it is colder outside. In principle, rather than reject heat to the environment, a heat engine could use the interior space itself as the cold reservoir, in which case the heat rejected to the cold reservoir would not be "waste heat," it would actually be a desired output. The 1st Law efficiency of such a device could approach 100%. However, if the heat engine was a combustion engine, the exhaust of the engine would need to be safely vented to the environment somewhat warm, an effect that would reduce the 1st Law efficiency.

So why is cogeneration so rarely implemented? Perhaps, because electricity generation historically has been predominantly relegated to large centralized power plants, and distributed to its point of use through a power grid. It would be prohibitively expensive (and nearly impossible), to develop infrastructure that would transfer (e.g., through a network of buried pipes) to the point of use in buildings. It has been done in some cities, notably New York City.

Many industrial processes require a heating source that is at a higher temperature than needed for space heat, but lower than the temperature associated with the high side reservoir. Therefore, many industrial plants use cogeneration. The systems are sized to provide the required thermal load for the process. The electricity produced generally exceeds the needs for electricity on-site, and, by law (google PURPA), arrangements can be made with a local utility for the excess to be supplied to the grid.

More widespread use of cogeneration would require decentralization of power generation. For example, many households use back-up generators to produce electricity in the event of a power outage. These systems reject heat to the environment. With additional components (heat exchangers) and their expense, this heat could be used for space heating of the home, in principle. These systems could be designed in a manner that they operate whenever there is a demand for space heat, and the electricity used either on-site, or returned to the grid. Maybe that will be common in the future, but it is rarely, if ever, done in the early twenty-first century.

Heat Pumps

A heat pump is a device that is used to "pull" heat from a cold reservoir and "pump" it to a hot reservoir, in what appears to violate the Clausius statement of the 2nd Law. However, it is possible to "pump heat uphill," provided a suitable work input is supplied. A heat pump is simply a heat engine operating in reverse, as shown schematically in Fig. 3.12. In this case, the desired output is the heat delivered to the higher temperature (Q_H) and the required input is the work input (W).

The 1st Law is as follows:

$$\Delta U_{\text{HEAT PUMP}} = 0 = Q_C - Q_H + W_{IN}$$

Rearranging for the heat added to the hot reservoir:

$$Q_H = W_{\text{IN}} + Q_C$$

That is, the heat added to the high side is equal to the heat pulled from the cold side plus any work input.

The entropy balance for the heat pump is as follows:

$$\Delta S_{\text{HEAT PUMP}} = 0 = \frac{Q_C}{T_C} - \frac{Q_H}{T_H} + \Sigma$$

Combining the 1st and 2nd Laws by eliminating Q_C:

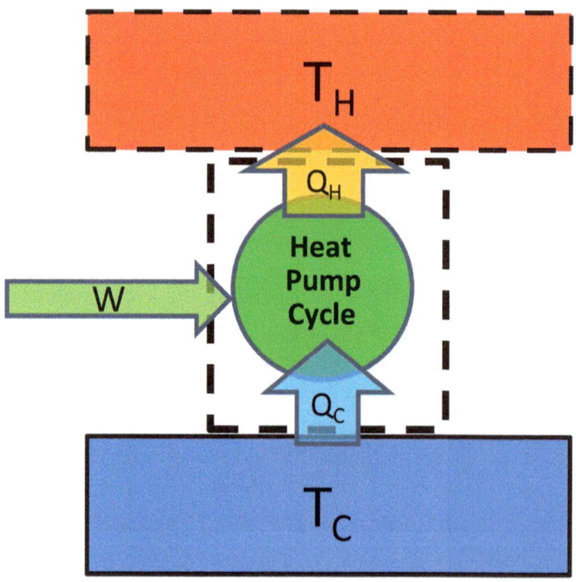

Fig. 3.12 Schematic representation of a heat pump cycle

$$\frac{Q_H - W_{\text{IN}}}{T_C} - \frac{Q_H}{T_H} + \Sigma = 0$$

Rearranging for the high side heat output:

$$Q_H = W_{\text{IN}}\left(\frac{T_H}{T_H - T_C}\right) - \left(\frac{T_H \cdot T_C \cdot \Sigma}{T_H - T_C}\right)$$

The 1st Law efficiency of a heat pump, called by convention the "coefficient of performance," (COP) is defined as the heat delivered to the high side divided by the work input:

$$\eta_I = \frac{Q_H}{W_{IN}} = \left(\frac{T_H}{T_H - T_C}\right) - \left(\frac{T_H \cdot T_C \cdot \Sigma}{(T_H - T_C)W_{\text{IN}}}\right)$$
$$\equiv \text{Coefficient of Performance } (\text{COP}_{\text{HP}})$$

The subscript "HP" indicates the performance of a heat pump, as opposed to a refrigerator, which will be discussed in the next subsection. A reversible heat pump operating between two temperature reservoirs will yield the maximum possible heat delivered, by setting the entropy generation to zero.

$$Q_{H,\text{MAX}} = W_{\text{IN}}\left(\frac{T_H}{T_H - T_C}\right)$$

The maximum possible 1st Law efficiency (COP Carnot coefficient of performance) is therefore

$$\text{COP}_{\text{HP,Carnot}} = \frac{Q_{H,\text{MAX}}}{W_{\text{IN}}} = \left(\frac{T_H}{T_H - T_C}\right)$$

The 2nd Law efficiency of a heat pump is as follows:

$$\eta_{II} = \frac{\text{COP}_{\text{HP}}}{\text{COP}_{\text{HP,Carnot}}} = 1 - \frac{T_C \Sigma}{W_{\text{IN}}}$$

One of the most common applications of a heat pump is in space heating, as shown in Fig. 3.13. Two dashed lines indicate two different systems. The top system considers a living space maintained at a temperature T_H, and the outdoor ambient temperature is considered to be at the dead state temperature. In a more general treatment, there would be additional energy inputs to the living space (electrical work input that is not used to operate the heat pump, heating from metabolism of biological systems such as people, dogs, and plants), and the system would be an open system, with outside air entering and warm air exiting at a different humidity level. In the simplified case considered here, there are only 2 energy transfers: the

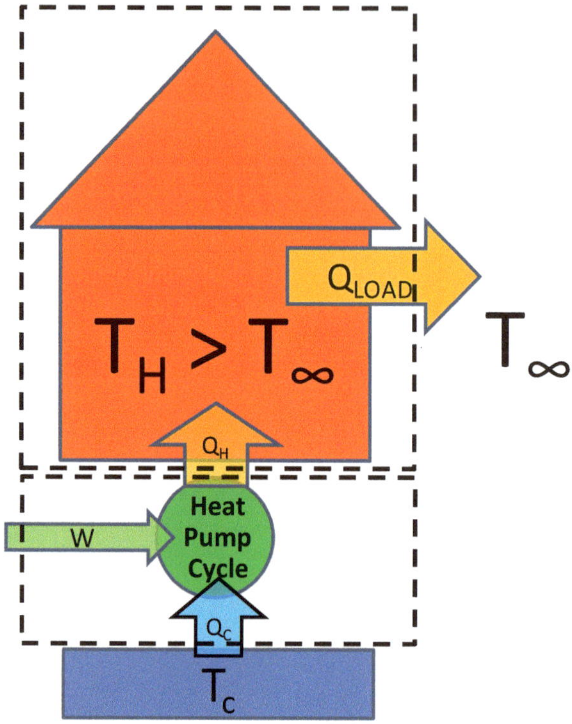

Fig. 3.13 Heat pump used as a space heating source. The dead state is considered to be the outside air temperature. The low side temperature reservoir could be the same as the dead state (an "air source heat pump") or be taken from the Earth's crust (a "ground source heat pump")

heat load (the transfer of heat from the warm interior to the cold ambient), and the heat input from the heat pump. These heating loads are equal in this case (the interior is maintained at T_H).

The heat pump, however, takes in electrical work and pulls heat from a cold reservoir. That cold reservoir could be the dead state (ambient air), and systems that operate this way are called "air source heat pumps." They pull heat from the cold air outside, and pump all that heat, plus the electrical work, into the living space. Alternatively, the temperature of the ground (a few meters below the surface) is at a more stable temperature (nominally 15°C, depending on the location) and systems that extract heat from this reservoir are called "ground source heat pumps."

Refrigerators and Air Conditioners

A refrigerator is the same thermodynamically as a heat pump, in that a work input is used to pump heat from a low-temperature reservoir to a high-temperature reservoir. The difference is only in the definition of what is the desired heat transfer. The basic heat pump cycle is the same as in Fig. 3.12, but the desired heating is from the cold side, as shown in Fig. 3.14, and it is called a refrigerator instead of a heat pump. A household refrigerator is a device that maintains a refrigerated space at a lower

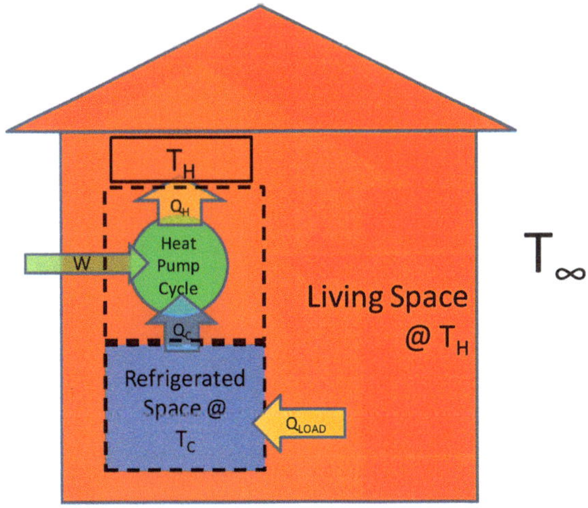

Fig. 3.14 Heat pump cycle used as a refrigerator. The living space could be either greater than the outside temperature (in summer) or less (winter)

temperature than the living space it is in. The dashed line that defines the refrigerated space has a thermal load, which is dominated by the constant flow of heat from the living space to the cold space through the walls and door of the refrigerator. It is a common misconception that the thermal load of a refrigerator is dominated by the placement of warm objects in it (termed an active thermal load) or the opening and closing of the door (a parasitic thermal load). In normal operation, the heating load associated with that effect is small compared to the constant heat pouring into a space maintained at a temperature below its immediate surroundings (a passive thermal load). The refrigeration system removes that heat and pumps it into the room.

Another application where the desired heat load is from the cold space is in space cooling, using air conditioners, shown in Fig. 3.15. In practice, air conditioners also remove water vapor from the interior air (decrease the humidity), and that open-system behavior will be addressed in chapter 12.

It is a good exercise to show that the 1st and 2nd Law combine to yield a relationship between the desired cooling load and the required work input and, therefore, a 1st Law efficiency (coefficient of performance) of:

$$\eta_I = \frac{Q_C}{W_{IN}} = \left(\frac{T_C}{T_H - T_C}\right) - \frac{T_C \cdot T_H \cdot \Sigma}{(T_H - T_C)W_{\mathrm{IN}}}$$
$$\equiv \text{Coefficient of Performance } (\mathrm{COP}_R),$$

where the subscript "R" refers to a refrigerator.

The 2nd Law efficiency of a refrigerator is as follows:

$$\eta_{II} = \frac{\mathrm{COP}_R}{\mathrm{COP}_{R,\mathrm{Carnot}}} = 1 - \frac{T_H \Sigma}{W_{\mathrm{IN}}}$$

The COP of the heat pump and refrigerator are closely related. Recall the 1st Law:

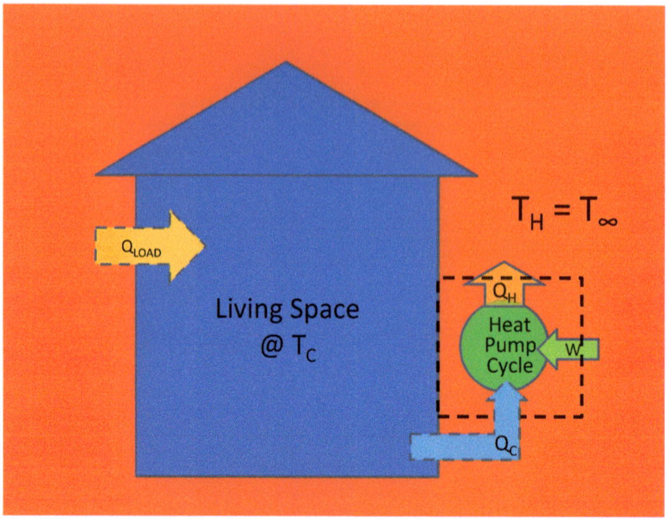

Fig. 3.15 Heat pump cycle used as an air conditioner. Actual air conditioners also modify the humidity of the air, a subject of future chapters

$$Q_H = Q_C + W_{\text{IN}}$$

Dividing each term by the work input reveals the COP definitions:

$$\frac{Q_H}{W_{\text{IN}}} = \frac{Q_C}{W_{\text{IN}}} + 1$$

$$\text{COP}_{\text{HP}} = \text{COP}_R + 1$$

It is noteworthy that the heat pump COP is always greater than (or equal to) unity. The input is a mechanical transfer (electrical work), and the desired output is a thermal transfer (heat). If the heat pump pulls no heat from the low side, it acts as a simple electrical heating element, and a 100% 1st Law efficiency is possible. Also, the COP of a heat pump is larger than the COP of a refrigerator, because the heat pumped to the high side is the sum of the electrical work plus whatever heat is pulled from the low side.

The maximum (Carnot) coefficient of performance for several applications associated with space heating and refrigeration is shown in Fig. 3.16. The vertical dotted line shows the living space temperature, assumed to be nominal room temperature (25°C). The 2 dashed lines are refrigerator COPs, and the solid line is a heat pump COP. For space heating applications, when the living space is maintained at a higher temperature than the outside temperature (i.e., winter), the living space acts as the high side reservoir (T_H), and the outside conditions act as the low side (T_C). As the temperature difference between reservoirs decreases, the maximum COP increases dramatically. The temperature difference between reservoirs is like a "hill to climb."

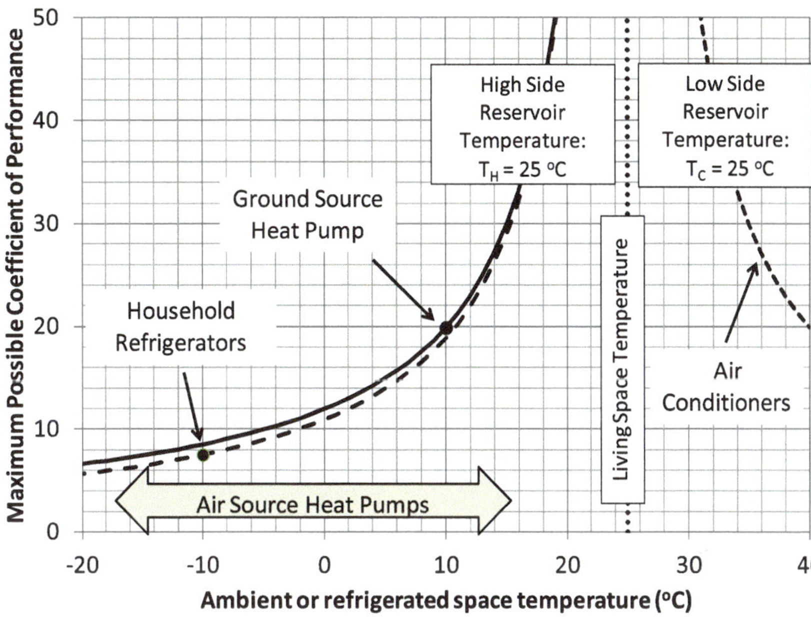

Fig. 3.16 Maximum possible coefficients of Performance of heat pumps, refrigerators, and air conditioners. The two dashed lines are refrigerator COPs, and the solid line is a heat pump COP

In practice, irreversibilities limit the actual COPs to significantly lower values, particularly for applications with low-temperature hills, as will be formally shown shortly. The ground source heat pump application is shown as a case study at a low side temperature of 10°C. On the other hand, air source temperatures vary widely in practice during the course of a heating season. It is well established that air source heat pumps work well in temperate climates, and lose efficiency (and effectiveness) for very cold conditions. In summer, when the living space is maintained at a temperature below the ambient, the living space acts as the low side reservoir, and outside conditions as the high side. The maximum COP of air conditioners decrease strongly with outdoor temperature contributing to the stress placed on electric utilities on hot summer days.

Refrigerators operate year-round with nominally stable reservoir temperatures. The cold space typically is maintained at a nominal -10°C (assuming the unit contains a freezer compartment and the refrigerator compartment is maintained indirectly by suitable thermal design), and the living space at 25°C. The maximum possible COP for a refrigerator is approximately 8, and in practice is more like 2–4 (a 2nd Law efficiency of 25–50%). Air conditioners generally operate more efficiently than refrigerators simply, because the temperature "hill" is much lower. Not apparent in this treatment, however, is that air conditioning also involves dehumidification of the air, and that is generally accomplished by producing much lower temperatures (well below the dew point of the interior space) than the living space temperature to condense liquid water.

Reversible Heat Pump Cycle with Irreversible Heat Transfer

The following analysis demonstrates how difficult in practice it is to produce very high 2nd Law efficiency refrigerators or heat pumps. The schematic of Fig. 3.17 shows a heat pump cycle operating between a high and low reservoir (at T_H and T_C). In this scenario, it is recognized that there must be a finite temperature difference between the system and the surroundings at both reservoirs in order for heat to actually flow. This temperature difference is called ΔT_{HEX}, with "HEX" for heat exchanger. On the high side, the temperature of the location of the system where heat is transferred (@ $T_H + \Delta T_{\text{HEX},\ H}$) must be greater than the high side reservoir temperature (@ T_H), and on the cold side, the temperature of the system (@ $T_C - \Delta T_{\text{HEX},\ C}$) must be lower than the cold reservoir (@ T_C). Therefore, the effective temperature "hill" that must be climbed is the difference in temperature between the reservoirs plus both heat exchanger hills. Considering the heat pump cycle to otherwise be reversible, the Carnot efficiency (for a refrigerator/air conditioner) is as follows:

$$\text{COP}_R = \frac{Q_C}{W} = \frac{T_C - \Delta T_{\text{HEX},C}}{(T_H + \Delta T_{\text{HEX},H}) - (T_C - \Delta T_{\text{HEX},C})}$$

The 2nd Law efficiency can be defined as the ratio of the actual COP (with the heat transfer irreversibilities) to the Carnot COP:

$$\eta_{II} = \frac{\text{COP}}{\text{COP}_{\text{Carnot}}}$$

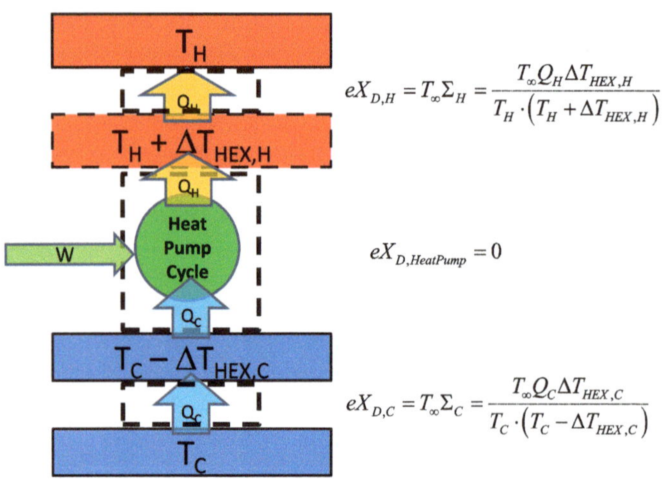

Fig. 3.17 Reversible Heat Pump with Irreversible heat transfer

The eXergy destruction is the sum of the eXergy destruction in the high and low side heat transfers:

$$eX_D = T_\infty(\Sigma_H + \Sigma_C) = T_\infty\left(\frac{Q_H \Delta T_{\mathrm{HEX},H}}{T_H(T_H + \Delta T_{\mathrm{HEX},H})} + \frac{Q_C \Delta T_{\mathrm{HEX},C}}{T_C(T_C - \Delta T_{\mathrm{HEX},C})}\right)$$

Since $Q_H = Q_C + W = Q_C(1 + 1/\mathrm{COP})$, and with $\Delta T_{\mathrm{HEX},\,H} = \Delta T_{\mathrm{HEX},\,C} = \Delta T_{\mathrm{HEX}}$, the eXergy destruction is as follows:

$$\frac{eX_D}{Q_C} = T_\infty\left(\frac{1 + 1/\mathrm{COP}}{T_H(T_H + \Delta T_{\mathrm{HEX}})} + \frac{1}{T_C(T_C - \Delta T_{\mathrm{HEX}})}\right)\Delta T_{\mathrm{HEX}}$$

The effect of irreversibilities on the performance of a heat pump cycle that pumps air from a cold space at 25°C to the outdoor ambient at $T_H = T_\infty$ (like air conditioners) is shown in Figs. 3.18, 3.19, and 3.20, with $\Delta T_{\mathrm{HEX},\,H} = \Delta T_{\mathrm{HEX},\,C} = \Delta T_{\mathrm{HEX}}$, for 3 values of ΔT_{HEX}, as a function of ambient temperature. Realistically, systems require typically 5–10°C temperature differences in order to be sufficiently effective. When the ambient temperature approaches the cold reservoir, the Carnot efficiency is very high. However, the negative effect of the irreversibilities associated with heat transfer becomes more dramatic. For a fixed temperature

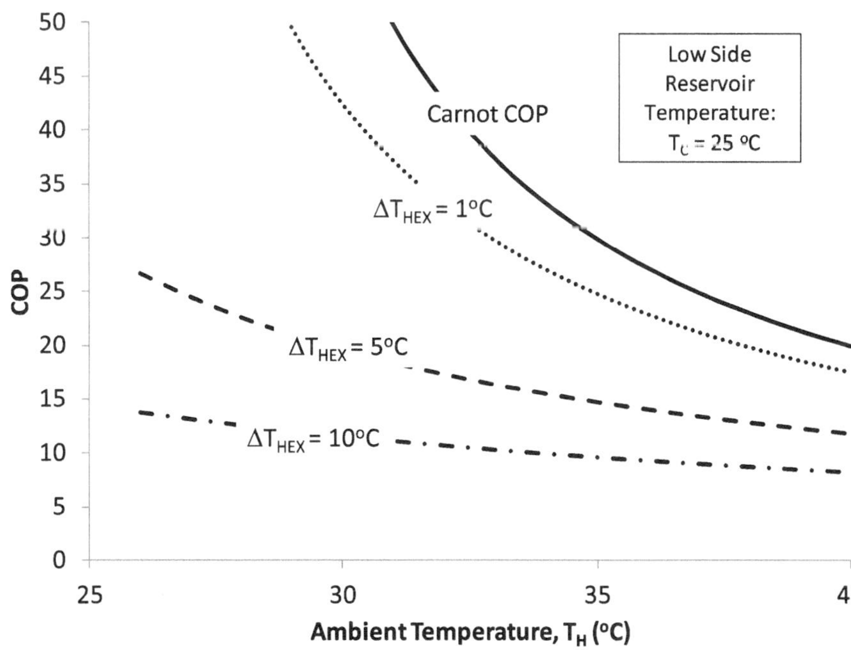

Fig. 3.18 Effect of heat transfer irreversibility on the coefficient of performance of an air conditioner. Each curve represents a scenario with fixed temperature difference across both high side and low side heat exchangers

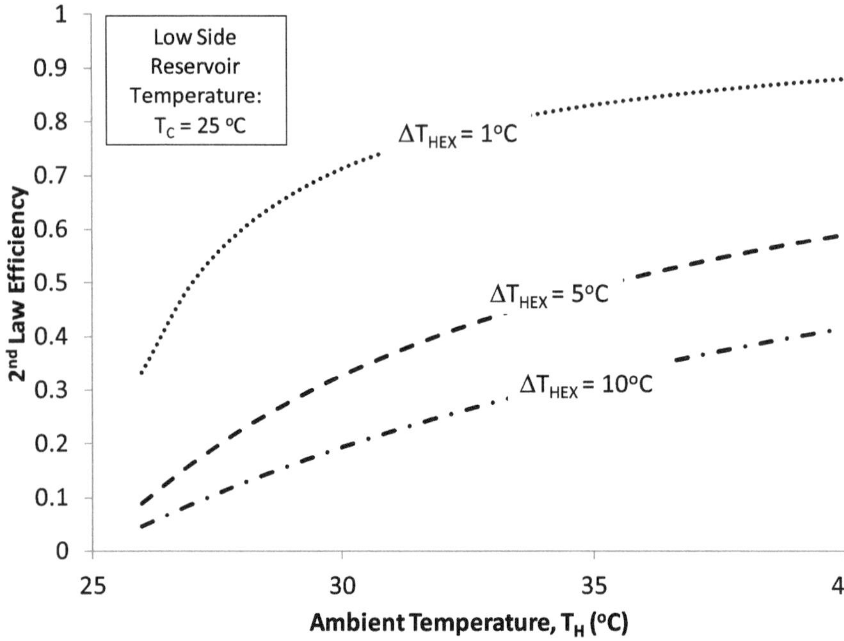

Fig. 3.19 2nd Law efficiency associated with air conditioner with heat transfer irreversibilities

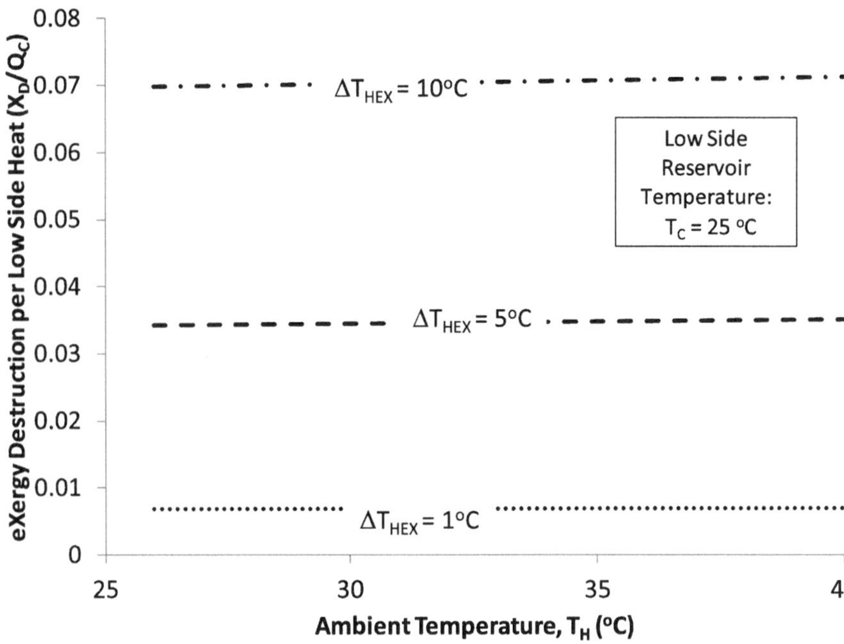

Fig. 3.20 eXergy destruction normalized by the heat extracted from the low side as a function of ambient temperature

difference driving the heat, the 2nd Law efficiencies become quite low as the ambient temperature approaches the cold reservoir temperature. Second Law efficiencies greater than, say, 0.5 are difficult to achieve in practice due to the practical requirements of both heat exchangers. Finally, the eXergy destruction (per unit of cooling heat) is essentially constant for a fixed temperature difference driving the heat (ΔT_{HEX}) and is proportional to that temperature difference.

Deeper Dive: Energy Demand in the USA

The website (www.eia.gov) of the Energy Information Agency (EIA), a division of the Department of Energy (DOE), is an excellent resource for energy-related topics and data. The EIA defines primary energy sources as being from the sources listed in Table 3.3, distinguishing between nonrenewable and renewable sources (energy sources that are considered to be effectively inexhaustible). Electricity is generally considered to be a secondary source of energy, produced from a heat engine in which a primary source of energy acts as a thermal input. The exception is nuclear electricity, classified as a primary energy source by the EIA. A dramatic shift to renewables over the next few decades is expected in an effort to reduce greenhouse gas emissions and thereby address the problem of global warming.

It should be noted that some renewable sources in this classification are not inexhaustible, in particular, wood and other biomass products. In order to be considered renewable, an energy source must be replenished at approximately the same rate that it is consumed.

It is curious that the electricity produced in nuclear power plants is considered to be a primary energy source, which does not make much sense, because electricity is a secondary source, generated from a finite resource. Nevertheless, it is a well-thought-out classification for that special case, in the context of contemporary energy markets.

The demand for primary energy and the population in the USA from 1949 to 2014 is shown in Fig. 3.21. Before 1973, the time of the oil embargo by the Organization of Petroleum Exporting Countries (OPEC), energy demand grew at a much faster rate than the population. The oil shortages of 1979 also had a significant impact.

Table 3.3 Primary energy source classifications of the US Department of Energy

Nonrenewable sources	Renewable sources
Coal	Conventional hydroelectric
Petroleum products	Geothermal electric
Natural gas	Solar thermal
Nuclear electricity	Solar photovoltaic
	Wind
	Wood and wood-derived fuels
	Biomass waste
	Fuel ethanol
	biodiesel

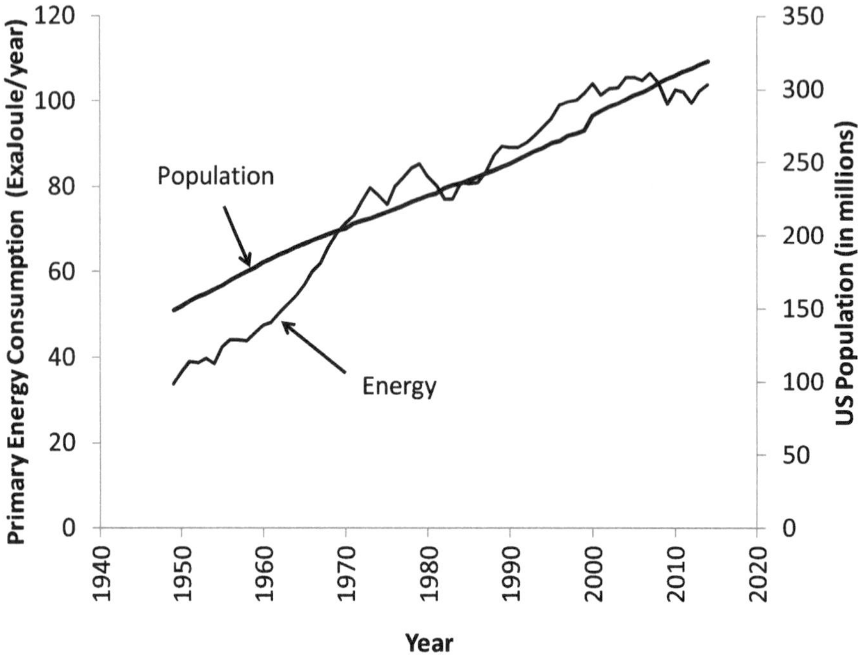

Fig. 3.21 Historical energy consumption and population (1949–2014). Data adapted from www.eia.gov and US census

Between 1973 and 2000, energy consumption and population grew at comparable rates. Since approximately 2000, population has grown at a faster rate than energy consumption.

Fig. 3.22 shows the per capita primary energy consumption by sector (energy consumption divided by the population). In 2014, electricity production accounted for the largest percentage of primary energy consumption, followed by transportation and industrial. Residential and commercial sectors represent the nonelectric energy consumption. The growth in energy consumption prior to 1973 was very high in the electricity section, and to a lesser extent in the transportation section. Per capita energy consumption in the industrial sector has decreased, a reflection in part of manufacturing shifts away from the USA.

Utility Generated Electricity

A common misconception is that electricity is a primary source of energy and is more or less equivalent to a primary fuel source. However, fuel sources constitute a form of thermal energy, and electricity is a form of electro/mechanical energy, which is often produced in a heat engine with a fuel source as the input. They are not

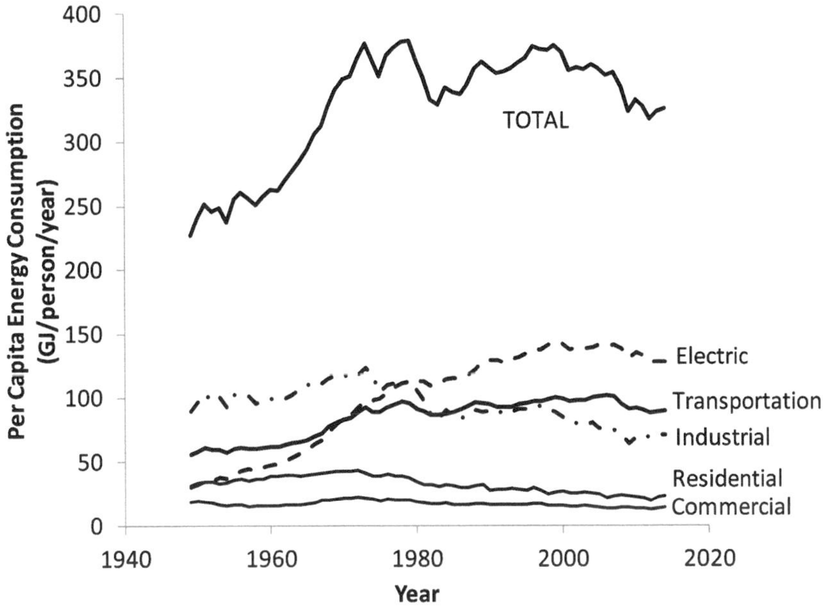

Fig. 3.22 Historical per capita primary energy consumption in the USA. Population data adapted from US Census, and energy data from www.eia.gov

equivalent. Energy conscious people must be mindful that any device that uses electricity is using a secondary form of energy. Energy dissipation and pollutants (local and greenhouse gas) are necessary by-products of electricity production.

A schematic of a utility power plant and the distribution of electricity to the grid are shown in Fig. 3.23. Utility companies use an array of electricity-generating sources and distribute electricity through a vast network of transmission cables. The majority of the generating sources (at the time of this writing) use a form of thermal input to generate electricity as an output, and are therefore modeled as heat engines. The average 1st Law efficiency of the aggregate is obtained from EIA data to be approximately 36%. Some base-load power plants (that operate continuously) are nearly 50% efficient, while peaking power plants (that operate intermittently to meet a rapidly changing demand) are typically 25% efficient. The peaking plants must operate over a range of conditions and have a rapid response, and cannot be built to operate as efficiently as base-load plants. It is difficult to precisely define 2nd Law efficiencies as the heating source is from a chemical energy conversion (combustion); the temperature of the high-side source is not well defined. A great deal of effort has gone into minimizing eXergy destruction in these devices, although significant improvements may yet be forthcoming. On the other hand, the use of renewable resources (hydro, solar, wind, biomass) is expected to increase over time, yet their levelized cost remains high compared to the use of nonrenewables (coal,

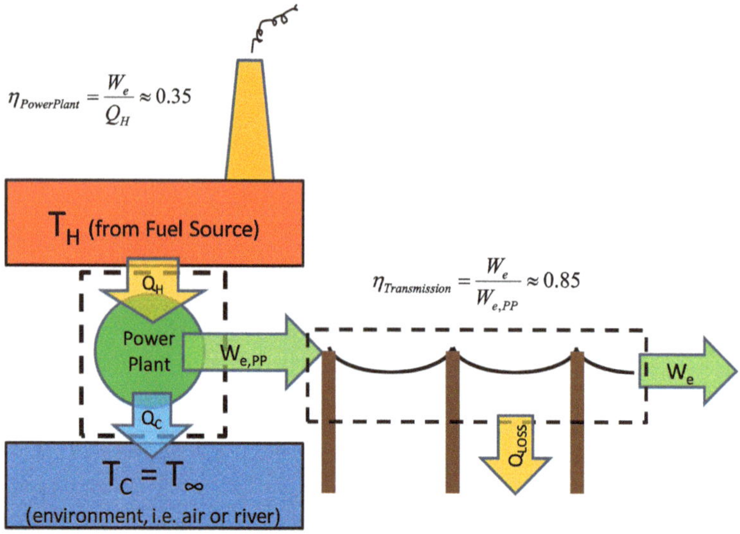

Fig. 3.23 Schematic of a remote utility electricity-generating power plant and its distribution to the grid. Numerical data derived from www.eia.gov

natural gas, nuclear). Social, political and economic breakthroughs are needed to make that shift.

Once electricity is generated at a power plant, it must be distributed over a network of cables before it is ultimately delivered to its point of use. The average 1st Law efficiency of the distribution system is very roughly estimated at 85%, with 15% of the electricity generated being "lost" due to eXergy destruction, which appears as heat loss. The 2nd Law efficiency for transmission is the same as the 1st Law efficiency, as electricity distribution is a conversion of a work input to a work output. The effective efficiency of the generating plant plus transmission is as follows:

$$\eta_{overall} = \frac{W_{e,delivered}}{Q_{IN,fuel}} = \left(\frac{W_{e,generated}}{Q_{IN,fuel}}\right)\left(\frac{W_{e,\text{delivered}}}{W_{e,\text{generated}}}\right) = \eta_{\text{PP}} \cdot \eta_{\text{tranwsmission}} \approx (0.35) \cdot (0.85) = 0.298$$

Seventy percent of the input thermal energy content is simply dissipated to the environment at the power plant or in transmission. The heat rejection at the power plant requires large, expensive equipment to accomplish the task, a significant factor in the inherent cost of generating electricity.

Data for electricity generation by primary source is shown in Fig. 3.24. The total electricity consumption increased by only 0.31% per year between 2004 and 2014, despite the explosive growth of electronic-based technology. Fossil fuels (coal and natural gas) accounted for approximately two-thirds of the source energy for electricity in 2014. Nuclear energy was approximately 20%. These are all thermal sources. Very little petroleum liquid (oil) is used to generate electricity, but that is the primary fuel source for the transportation sector. Renewables (solar and wind, principally) account for a small (7%), but rapidly growing sector (23.8% growth rate per year). Hydroelectricity (another renewable source) can be an economically attractive renewable source, but it is considered to be a somewhat saturated supply source. The number of suitable sites is limited, and those have been tapped, as reflected by the slight decrease between 2004 and 2014.

Coal, due to its low price, is the largest primary source for electricity generation as of 2014. However, there has been a shift from coal to natural gas, which is a cleaner-burning fuel (both from local acid rain and global greenhouse gas pollution), and the price of natural gas has plummeted (probably due to "fracking," a controversial way to increase production of oil and natural gas). While there is a strong shift from coal to natural gas in electricity production, coal will likely remain significant for some time. Natural gas produces approximately half the greenhouse gas emissions as coal, but is still a nonrenewable fossil fuel.

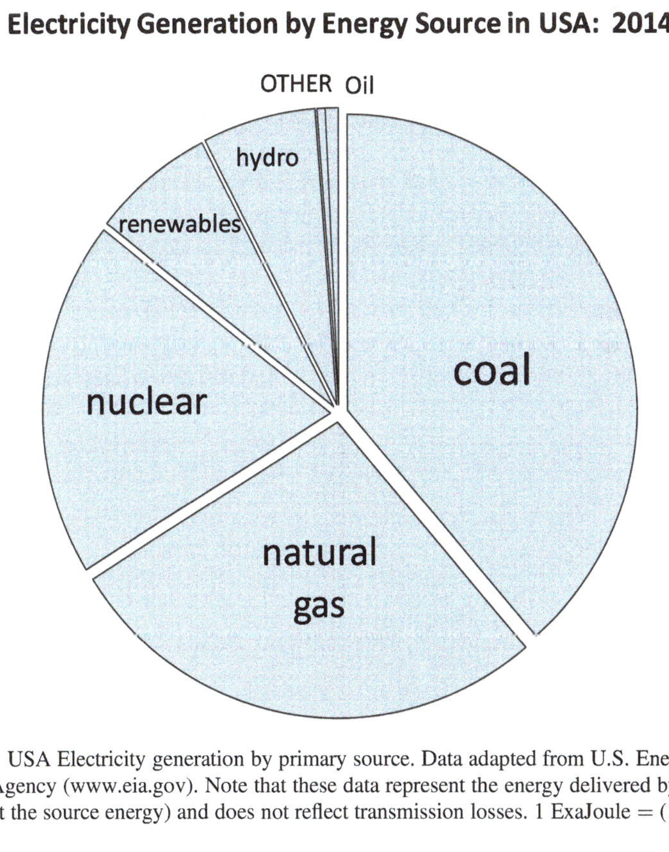

Fig. 3.24 USA Electricity generation by primary source. Data adapted from U.S. Energy Administration Agency (www.eia.gov). Note that these data represent the energy delivered by the power plants (not the source energy) and does not reflect transmission losses. 1 ExaJoule = $(10)^{18}$ Joules

Net Electricity Generation in USA by source (in Exa-Joules)

	2004	%	2014	%	Change/year
TOTAL	14294	100%	14735	100%	0.3%
coal	7122	50%	5709	39%	-2.0%
natural gas	2556	18%	4039	27%	5.8%
nuclear	2839	20%	2869	19%	0.1%
renewables	299	2.1%	1012	6.9%	23.8%
hydroelectric	966	6.8%	931	6.3%	-0.4%
Petroleum Liquids	361	2.5%	67	0.5%	-8.1%
OTHER	150	1.1%	107	0.7%	-2.9%

http://www.eia.gov/electricity
accessed July 21, 2015

Energy Prices

It is instructive, and of general interest, to consider the economic cost of various energy commodities. Consumers routinely buy "energy products" such as gasoline to power their cars, natural gas, or fuel oil to heat their homes, or electricity to run a myriad of devices at home, work and play. However, all these commodities are sold in units that pertain to their physical nature, not to their energy content. Gasoline and fuel oil are liquid fuels, and are sold per gallon, natural gas is sold in cubic feet (or therms, which is an energy unit), and electricity is sold in kW-hr, an energy unit. Therefore, it is difficult to directly compare the true cost of various energy sources, without performing some unit conversions. Figure 3.25 shows typical retail prices, converted to consistent energy units, of several energy sources available to residential customers in the USA for the year 2009. Coal and wood are the cheapest fuels, and electricity (a secondary energy source) is very high. Natural gas has a significantly lower price than oil-based liquid fuels (fuel oil, diesel, gasoline, kerosene). Propane has the highest price of all hydrocarbon fuels in this list, but has a strong niche market as it is stored as a liquid in on-site tanks. These prices are considered to be representative and can vary significantly with geographic location, and shift over time.

Comparison of Space Heating Costs

The heating of living spaces (residential and commercial) constitutes a large economic expense and consumes a great deal of energy resources (Table 3.4). There has been little change in the total energy consumed in residences between 1993 and 2009, but there is a shift from space heating to appliances. The decrease in space heating could be due in part to efforts by homeowners to reduce heat loss from their homes (insulation, weather stripping, new windows). The large increase in

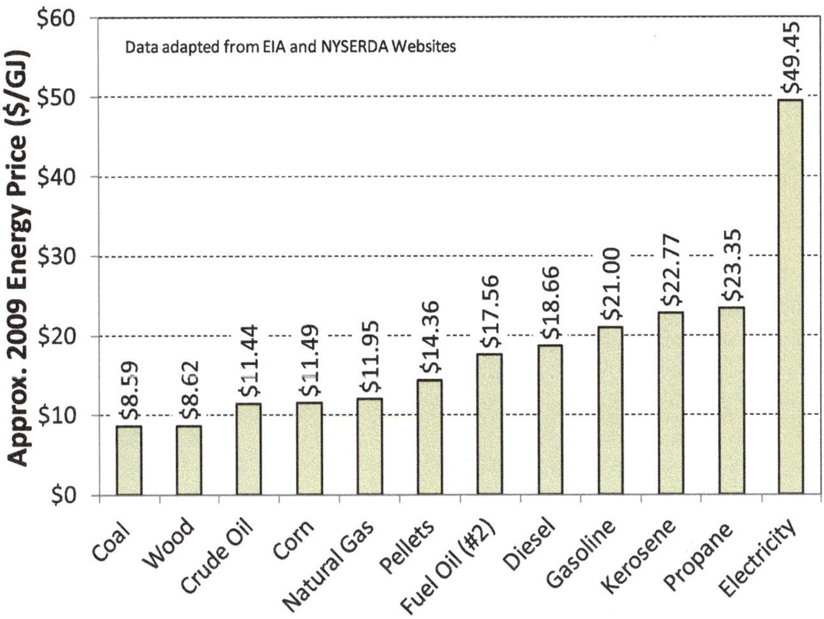

Fig. 3.25 Approximate prices for residential consumers in the USA converted to consistent units

Table 3.4 USA Residential Energy Consumption by End Use

	1993	2009
TOTAL (ExaJoules)	10.56	10.74
Space Heating	53.1%	41.5%
Appliances, Electronics, Lighting	24.0%	34.6%
Water Heating	18.3%	17.7%
Air Conditioning	4.6%	6.2%

Data from www.eia.gov

appliances is likely due to the explosion of electronic-based technology. It is emphasized that space heating and water heating come primarily from direct conversion of a thermal source to a thermal demand (with electric heating a niche market), while appliances and air conditioning require electricity (a secondary source).

The following comparison summarizes many of the associated 1st and 2nd Law principles. Space heating can be accomplished in many ways, 5 of which are compared here. There are other methods (i.e., fireplaces, wood-pellet stoves, propane heaters) and the curious reader could conduct those estimates. Each method is briefly discussed, in the order of increasing overall thermodynamic efficiency, defined by the ratio of the delivery of a unit of space heat to the consumption of a primary fuel source. A quantitative estimate of the required inputs and their cost to a residential or commercial consumer is made as a case study. This analysis is not a complete economic analysis, as it does not consider the capital cost associated with the equipment, or any nonenergy operating costs such as maintenance. The focus is on the energy cost.

Electric Heater

The use of electric heaters is one of the simplest ways to heat a space, assuming of course that a reliable electricity source is available (Fig. 3.26). Portable space heaters are generally inexpensive and easy to operate (and can be somewhat of a fire hazard). More permanent electrical heating systems produce hot water and distribute the thermal energy by circulating it from a central unit to various heat exchange devices located where needed to maintain the temperature of the living space. The conversion of electrical to thermal energy is accomplished by forcing electricity through an element that has an electrical resistance (R), across which Ohm's Law applies $V = I \cdot R$ where V is the voltage difference across the resistor. The electrical work done by the surroundings on the element in a time Δt is as follows:

$$W_e = V^*(I\Delta t) = V^*\left(\frac{V\Delta t}{R}\right).$$

The electrical power developed for continuous operation is as follows:

$$\dot{W}_e = \frac{W_e}{\Delta t} = VI = \frac{V^2}{R}$$

The thermodynamic analysis is simple, as a direct conversion from an electro/mechanical energy form (work) into a thermal energy form is accomplished with a high 1st Law efficiency. However, there is considerable eXergy destruction inherent in the process, and that will ultimately be reflected in a high thermal (fuel) input at the utility power plant, which dissipates approximately 70% of the fuel input to the environment.

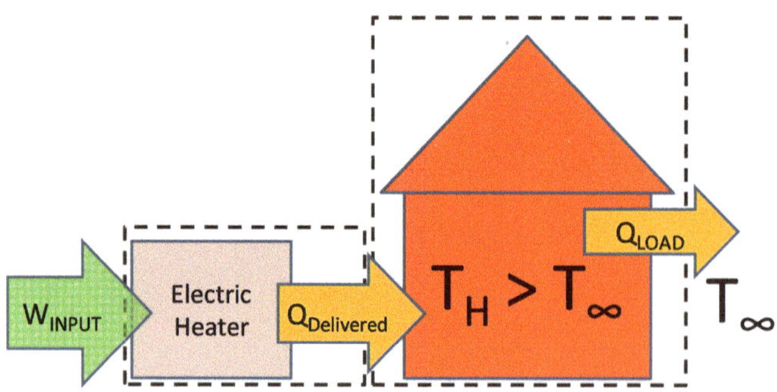

Fig. 3.26 Schematic of space heating using a direct electric heating system

Electrical heating can make economic sense in temperate climates, where space heating is limited to only a few days in the year. Space heaters can also make sense in colder climates if the space heating is localized in only a few rooms of a large dwelling, as opposed to heating many rooms unnecessarily.

Natural-Gas- or Fuel-Oil-Fired Heater

A very common way to heat homes and commercial buildings is through systems that burn a fuel purchased from a utility (natural gas or fuel oil) in a central unit that produces hot air, hot water, or steam to be distributed to individual rooms through a network of ducts or pipes. These devices (Fig. 3.27) take a thermal input (in the form of chemical energy) and deliver a portion of that energy as space heat. However, it is necessary to vent warm combustion products outside the building, and the energy content of this process is termed "stack loss." The 1st Law efficiency, defined as thermal energy delivered to the thermal energy input, is typically 80%.

Air Source Heat Pump

An air source heat pump extracts heat from the cold ambient air and delivers it to the heated space, as discussed previously (Fig. 3.13). This method of heating can be accomplished in principle by simply installing a wall-mounted air conditioner in reverse. In practice, that would not fly, only because currently marketed air conditioners are designed with the control panel on the cold side. Perhaps, someday wall mounted or window heat pumps will be widely marketed.

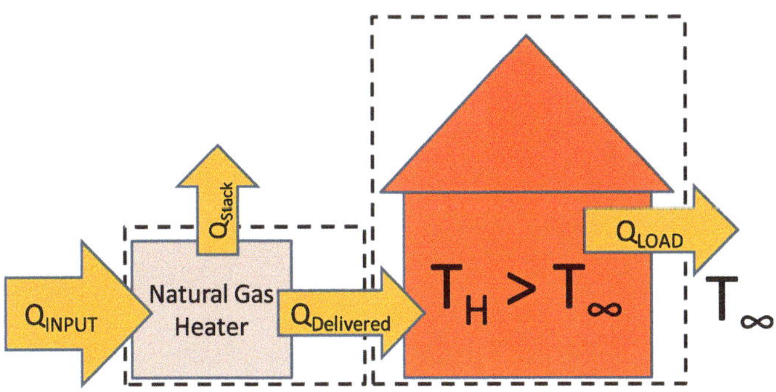

Fig. 3.27 Schematic of space heating using a natural-gas-fired heater (boiler or hot water heater). The stack heat represents the required venting of warm combustion products

Ground Source Heat Pump

A schematic of a typical system that uses a ground source heat pump is shown in Fig. 3.28. The geothermal bed consists of pipes that are buried in the ground. A heat transfer fluid that is colder than the ground temperature is circulated through the pipes and creates the cold side reservoir conditions. Heat flows from the ground into the fluid. In order to work properly, the ground must behave suitably as a thermal reservoir. The heat that is pulled from the ground must be regenerated by conduction heat transfer in the surrounding Earth (a slow process), and that rate of heat transfer depends on the size of the bed (usually quantified as the length of the buried pipe). If the bed is too small, the decrease in temperature of the ground around the pipe due to the heat extraction is too large, and the performance of the heat pump cycle will degrade, and ultimately fail. Therefore, a large capital expense, and available land, is needed to implement a ground-source heat pump, and is one of the major barriers to its widespread implementation. A back-up direct electrical system is shown for times when the geothermal system is not capable of meeting the demand. A well-designed system will not need the back-up system very often.

It should be noted that this form of geothermal heating is fundamentally different from geothermal heating using much deeper wells that tap into the very high temperatures of the Earth's crust well below the surface.

Space heating using heat pumps is a vast improvement, thermodynamically, over direct electrical heating because heat is taken from the environment and added to the

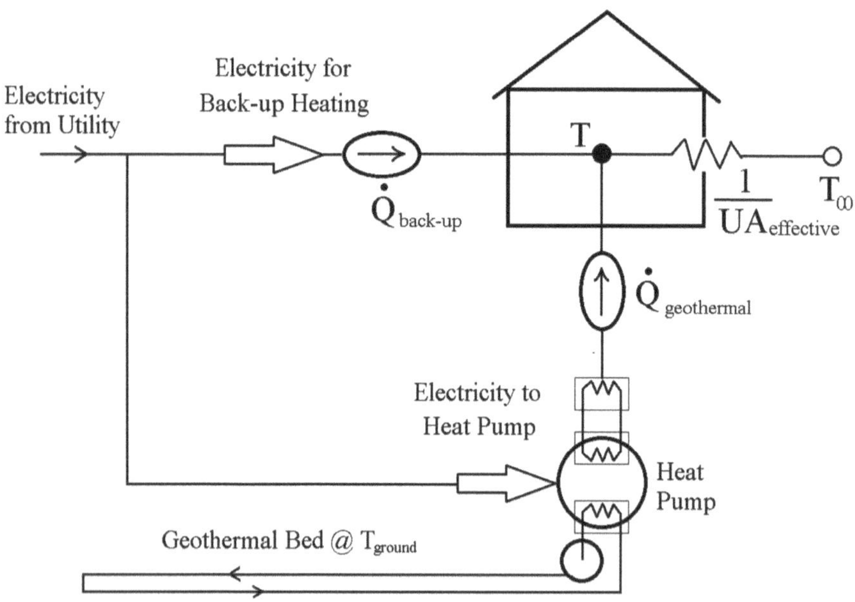

Fig. 3.28 Schematic of a ground source heat pump application. Direct electrical heating is included as an option if the geothermal system is not capable of meeting the demand

space along with the electrical energy. However, the heat pump is driven by electricity and, therefore, there is a large amount of heat dissipated to the environment at the power plant (Fig. 3.23).

Combined Heating Cycle

There is a way around the need to reject most of the thermal energy content of a fuel source to the environment, and that is to produce mechanical shaft power on-site to drive a ground source heat pump, and use the living space as the hot side reservoir. Such a combined heat engine/heat pump system is shown schematically in Fig. 3.29. The heat engine has a 1st Law efficiency of $\eta_{HE} = W_{shaft}/Q_{H,\ HE}$, and an overall efficiency of $\eta_{overall} = (W_{shaft} + Q_{C,\ HE})/Q_{H,\ HE}$. Some heat must be rejected to the stack, in practice. To meet the required heat load, an energy balance on the living space is as follows:

$$Q_{LOAD} = Q_{C,HE} + Q_{H,HP}$$

Expressing the heat transfers in terms of the work:

$$Q_{LOAD} = \left(\eta_{overall} Q_{H,HE} - W_{shaft}\right) + COP_{HP} W_{shaft}$$

Expressing the work in terms of the fuel addition:

$$Q_{LOAD} = \left(\left(\eta_{overall} - \eta_{HE}\right) + COP_{HP} \cdot \eta_{HE}\right) Q_{H,HE}$$

The required fuel input to the heat engine is as follows:

$$Q_{H,HE} = \frac{Q_{LOAD}}{\eta_{overall} + (COP_{HP} - 1) \cdot \eta_{HE}}$$

While the heat engine in the scheme as outlined in Fig. 3.29 uses both the output work and heat rejection, it is not considered to be a cogeneration application, as the work is used only as part of the space heating scheme. Cogeneration would involve using some or all of the shaft power to generate electricity to be sent "to the grid" or used directly on-site to displace utility electricity.

This combined scheme should clearly provide the best possible space heating performance thermodynamically of all the scenarios. A numerical case study that compares the required fuel input (on-site plus utility) and cost for delivering heat is conducted next.

$$Q_{H,HE} = \frac{Q_{LOAD}}{\eta_{overall} + (COP_{HP} - 1) \cdot \eta_{HE}}$$

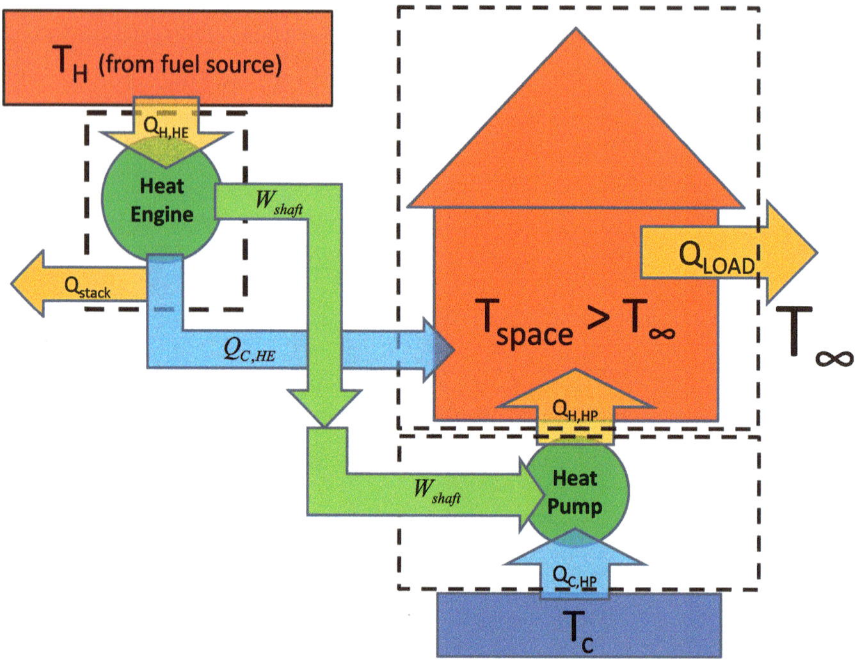

Fig. 3.29 Schematic of a combined heating system that uses a local heat engine to deliver electricity to a ground source heat pump and uses the living space as the cold reservoir

Representative Results

A spreadsheet calculation that compares the energy inputs required to meet a 1 GJ heating demand (Q_{LOAD}) is shown in Fig. 3.30. The required fuel input (local plus utility) and cost to the consumer are shown in Figs. 3.31 and 3.32, respectively. This 1st Law analysis incorporates 2nd Law concepts by the choice of the various component efficiencies, with values taken to be representative of commercially available systems. The energy prices are considered to be typical at the time of writing. The price of electricity (a mechanical form of energy) is more than 4 times that of natural gas (a thermal form) when compared with consistent units (Fig. 3.25). That difference reinforces the notion that electricity is a secondary source of energy, produced mostly from a primary thermal source.

Electric heating is clearly expensive, and should only be used when needed for very localized heating or short heating seasons. Heat pumps show significant advantages from a source and energy cost perspective over fuel-based heating systems. Their capital costs are significantly higher, however, and make the economics either unfavorable or marginally better, as consumers tend to give a higher priority to capital costs than life-cycle costs.

Qload	1	GJ
η_{pp}	0.2975	power plant efficiency
$\eta_{electric\ heater}$	1	electric heater efficiency
$\eta_{gas\ heater}$	0.8	gas heater efficiency
COP, air source	4	air source heat pump COP
COP, ground source	6	ground source heat pump COP
$\eta_{HE,electric}$	0.3	local heat engine efficiency
$\eta_{HE,overall}$	0.8	overall efficiency of heat engine
Electricity Price	$0.15	$/kWhr = $41.67 $/GJ
Fuel Price	$1.00	$/therm = $9.48 $/GJ

	Electric Work Input	Utility Thermal Input	Local Thermal Input	TOTAL FUEL INPUT	ELECTRIC COST	FUEL COST	TOTAL COST
Electric Heating	1	3.361	0	3.36	41.7	0	$41.67
Natural Gas Heating	0	0	1.25	1.25	0	$11.86	$11.86
Air Source Heat Pump	0.25	0.840	0	0.84	10.4	0	$10.42
Ground Source Heat Pump	0.167	0.560	0	0.56	6.9	0	$6.94
Combined Heating System	0	0	0.43	0.43	0	$4.12	$4.12

Fig. 3.30 Spreadsheet calculation of energy inputs and costs for various space heating methods

While the combined heat engine/heat pump system shows a clear advantage from an energy and cost perspective, it is not a readily available technology. Perhaps some ecologically and entrepreneurially oriented readers will help develop it. The thermodynamic advantage is clear due to 2 main effects: the "waste heat" from a heat engine is directly used for space heating, and thermal energy is pulled from the Earth.

Appendix 1: Direct Current (DC) Motor Characteristics

Electric motors can be driven by either direct (DC) or alternating (AC) electrical current. There are many practical considerations that go into the design decision in selecting a DC versus an AC motor for a particular application, and a few different types of each. This section focusses on basic permanent magnet, brushed DC motor behavior because the underlying electromagnetic physics lends itself to a clear and simple thermodynamic analysis. A wire is coiled around an iron core to produce an electromagnet, which can rotate in the magnetic field of fixed magnets. DC voltage is

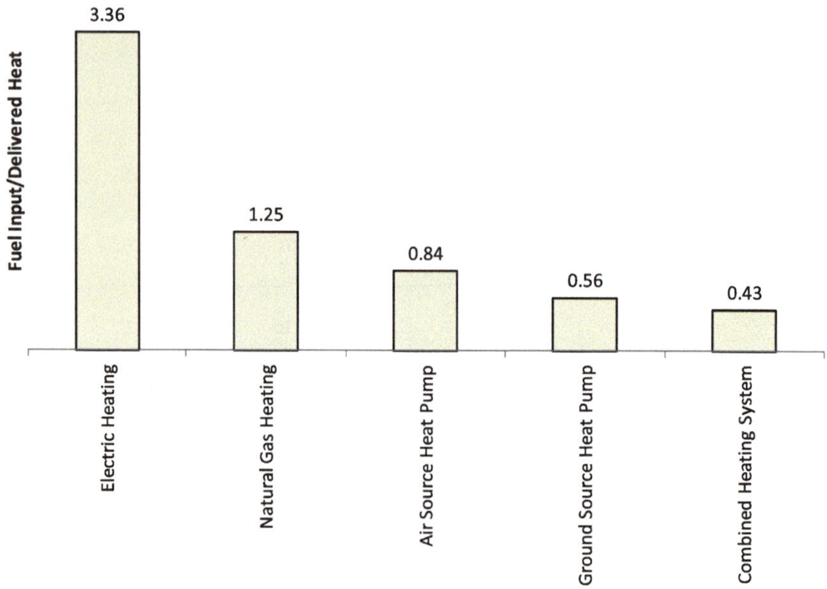

Fig. 3.31 Fuel input required to deliver 1 GJ of space heating for various scenarios

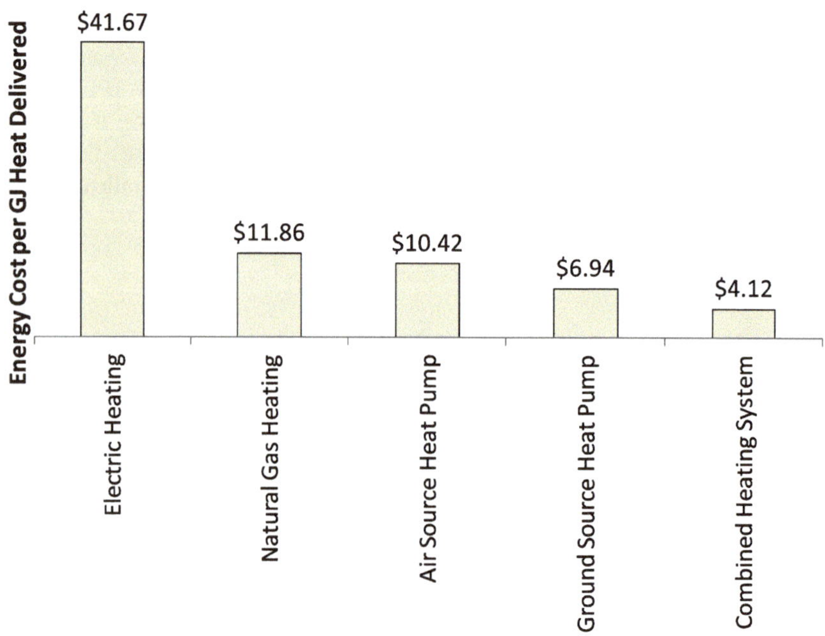

Fig. 3.32 Energy cost to consumer to deliver 1 GJ of space heating for various scenarios

applied across the coiled wire (with a means for reversing the polarity) and the rotating coil drives a mechanical load through an output shaft.

Thermodynamically, a DC motor converts input electrical power $\left(\dot{W}_e = VI\right)$ into output mechanical shaft power $\left(\dot{W}_m = \tau\omega\right)$. The electromechanical principles behind the operation are the Lorentz force (experienced by an electric charge that moves through a magnetic field) and Faraday induction (the electromotive force that opposes a change in the magnetic flux through the coil, called "back emf").

Fig. 3.33 shows the basic response of a typical small-scale brushed DC motor to an applied torque (mechanical load), powered by a nominal 12 V DC source. If the motor is powered without a mechanical load, the motor will rotate freely at a definite maximum speed (no load speed of 5730 rev/min for this motor) and draw a minimum current from the power supply (no load current of 0.25 amps). Faraday induction is the dominant player at low applied torque. The motor operates at a speed for which the applied voltage is almost completely opposed by the "back emf," with a minimum electrical current due to dissipative effects.

If an applied torque is gradually added, the speed will decrease and the current drawn will increase, both linearly with applied torque. The effect of the Lorentz force gradually increases as Faraday induction decreases. Eventually, the speed will drop to zero, and the motor will be stalled at a definite torque value (stall torque of 0.11 N-m) and maximum current (stall current of 7.2 amps). With no motion (no back emf), the motor electrically acts as a simple resistor, for which Ohm's Law applies.

Linear mathematical relationships for speed and current (in dimensionless form) are as follows:

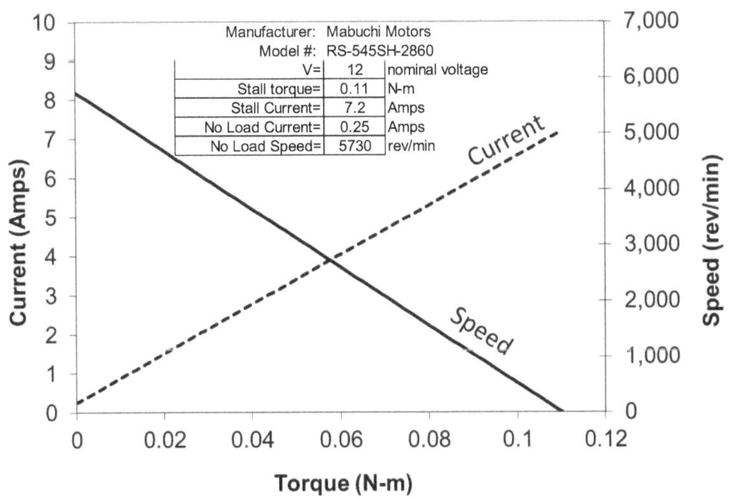

Fig. 3.33 Basic response curve of a typical DC motor to an applied torque. Linear plots are derived from the manufacturer listed parameters shown

$$\frac{N}{N_{\text{no load}}} = 1 - \frac{\tau}{\tau_{\text{stall}}} \quad \text{or, with dimensionless variables :} \quad \widehat{N} = 1 - \widehat{\tau}$$

$$\frac{I}{I_{\text{stall}} - I_{\text{no load}}} = \left(\frac{I_{\text{no load}}}{I_{\text{stall}} - I_{\text{no load}}}\right) + \frac{\tau}{\tau_{\text{stall}}} \; or, \; \widehat{I} = I^* + \widehat{\tau}$$

Other performance parameters are derived directly from these two linear relationships (Fig. 3.34 for energy rates and Fig. 3.35 for efficiency and steady-state motor temperature). The input power is simply voltage times current:

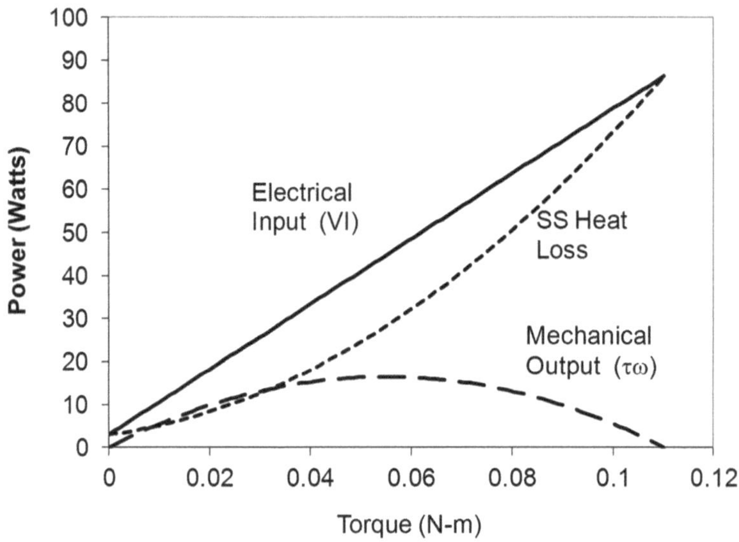

Fig. 3.34 Power terms vs. applied torque for a typical DC motor, derived from basic speed and current functions

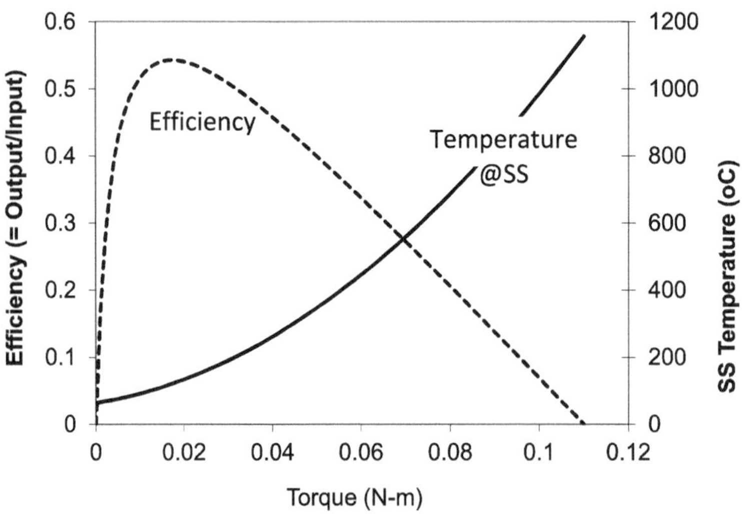

Fig. 3.35 DC motor efficiency and approximate steady-state operating temperature vs. applied torque for a typical DC motor

$$\dot{W}_e = VI = V(I_{\text{stall}} - I_{\text{min}})(I^* + \hat{\tau})$$

The output shaft power is as follows:

$$\dot{W}_m = \tau\omega = 2\pi\tau N = (2\pi\tau_{\text{stall}} N_{\text{no load}})\hat{\tau}(1 - \hat{\tau})$$

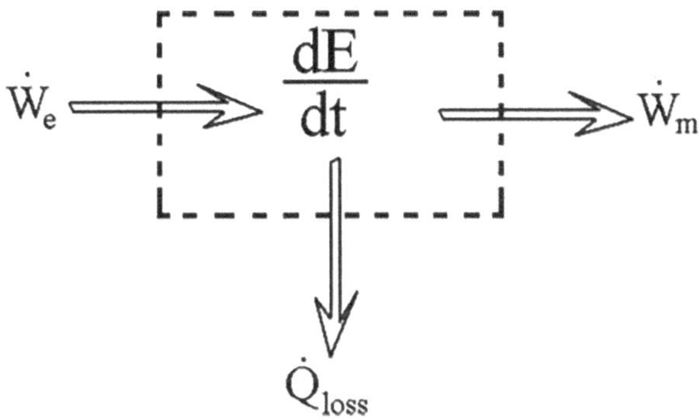

An energy balance applied to the motor yields the effective heat loss rate (which is effectively the eXergy destruction rate of a DC motor):

$$\left(\frac{dE}{dt} + \dot{Q}_{\text{OUT}}\right) = \dot{W}_e - \dot{W}_m$$

The imbalance in the input and output mechanical power is converted into either an increase in stored energy (internal and kinetic) or to heat loss to the environment. An entropy balance is as follows.

$$\frac{dS}{dt} = \frac{-\dot{Q}_{\text{OUT}}}{T_\infty} + \dddot{\Sigma}$$

Combining the energy and entropy balances (by eliminating the heat loss), and expressing stored kinetic energy as rotational energy ($I\omega^2/2$) associated with the inertia of both the motor itself and that of the mechanical load (a flywheel effect):

$$\left(\frac{d(U - T_\infty S)}{dt} + T_\infty \dddot{\Sigma}\right) + \frac{d}{dt}\left(\frac{1}{2}I\omega^2\right) = \dot{W}_e - \dot{W}_m$$

Recognizing the eXergy and the eXergy destruction:

$$\left(\frac{dX}{dt} + e\dddot{X}_D\right) + I\omega\frac{d\omega}{dt} = \dot{W}_e - \dot{W}_m$$

Steady-state operation

If the input is held constant in time and there are no changes in the mechanical system being driven, the system will attain a steady-state operating point:

$$e\dddot{X}_{D,@SS} = \dot{Q}_{OUT@SS} = \dot{W}_e - \dot{W}_m \text{ at steady state}$$

The imbalance between input and output power will arise as heat loss to the environment, equal to the eXergy destruction rate.

The input electrical and output mechanical power terms are generally plotted in a manufacturer spec sheet, but the heat loss rate is not. It is clear that the heat loss rate is important, especially at high loading. The 1st Law efficiency (Fig. 3.35) is the output shaft power divided by input electrical power. A little (not much) differential calculus shows that the maximum output shaft power is attained when the applied torque is exactly half of the listed stall torque, and the corresponding maximum power is a quarter of the stall torque times the no load angular speed. That is,

$$\dot{W}_{m,\text{MAXIMUM}} = \frac{1}{4}\left(2\pi N_{\text{no load}}\tau_{\text{stall}}\right) \quad @\frac{\tau_{\text{stall}}}{2}$$

On the other hand, the efficiency peak requires a little more algebra to attain and is very sensitive to the minimum current drawn for a free spinning motor. A rule of thumb is that the peak efficiency of a small DC motor is approximately 60%, corresponding to an applied torque of approximately 10–20% of the stall torque. Higher efficiencies are obtainable for larger and more expensive motors. Since an electrical input is used to produce a mechanical output, the 1st and 2nd Law efficiencies are identical. Properly designed AC motors can also approach 100% efficiency, but do not lend themselves to such simple analysis (due to very different motor characteristics).

The thermal management of electric devices is an important consideration in the design of engineering systems. An estimate of the steady-state operating temperature of the motor can be made by modeling the heat loss as $\dot{Q}_{OUT@SS} = \Omega A_S(T_{SS} - T_\infty)$. The temperature is as follows:

$$T_{SS} = T_\infty + \frac{\dot{Q}_{OUT@SS}}{\Omega A_S}$$

The case study for Fig. 3.35 is one in which the motor is assumed to be exposed to an ambient room (at 25°C) in still air. The overall heat transfer coefficient (Ω) is estimated to be 10 W/m^2/°C, from heat transfer theory for natural convection in air. The exposed surface area ($A_S = 76\text{cm}^2$) is the external surface area of the housing. Figure 3.35 shows that the predicted steady-state motor temperature is very high for high applied torques, and the motor will surely fail before reaching these predicted temperatures. In practice, a DC motor could be operated for short periods of time under high torque loadings (during a transient thermal period), but would fail if operated continuously. A thermal response time estimate of 13 minutes was made for this motor. Motors are frequently contained in some kind of housing, which has the effect of reducing the overall heat transfer coefficient, resulting in higher operating temperatures.

Transient Operation

If a motor whose shaft is connected to a mechanical system is started from rest, there will be a complicated transient start-up, where the imbalance between electrical input and mechanical output will simultaneously raise the stored internal energy, be lost as heat, and accelerate the motor (i.e., change the kinetic energy). The energy balance, with thermal and mechanical effects separated, is as follows:

$$\left(\frac{dU}{dt} + \dot{Q}_{\mathrm{OUT}}\right) + \frac{d}{dt}(\mathrm{KE}) = \dot{W}_e - \dot{W}_m$$

Relating the electrical input and mechanical output power terms to the speed, and using the property relationship between internal energy and temperature:

$$\left(\mathrm{Mc}\frac{dT}{dt} + \Omega A_S(T - T_\infty)\right) + \frac{d}{dt}(\mathrm{KE}) = VI - \tau\omega$$

To decouple the mechanical from thermal response, it is assumed that the acceleration of the motor and system takes place quickly and the initial condition for the thermal response is a motor operating at a fixed speed, and the imbalance between input electrical power and output shaft power equals the eXergy destruction rate, which is then constant in time. That is,

$$\left(\mathrm{Mc}\frac{dT}{dt} + \Omega A_S(T - T_\infty)\right) = e\dddot{X}_D \quad = \text{constant}$$

Introducing a dimensionless variable for temperature (θ, that starts at a value of 1 when the motor temperature equals ambient and is 0 at steady state) and time:

$\theta = \frac{T - T_{SS}}{T_\infty - T_{SS}}$ where T_{SS} is the temperature at steady state, to be determined
$\widehat{t} = \frac{t}{t^*}$ where t^* is a characteristic time constant, to be determined

The steady-state temperature is determined by setting the derivative of temperature with respect to time to zero. That is,

$$T_{SS} = T_{\infty} + \frac{e\dddot{X}_D}{\Omega A_S}$$

Introducing the dimensionless variables into the governing ODE:

$$\frac{\mathrm{Mc}(T_{\infty} - T_{SS})}{t^*}\frac{d\theta}{d\widehat{t}} + \Omega A_S(\theta(T_{\infty} - T_{SS}) + T_{SS} - T_{\infty}) = e\dddot{X}_D$$

Introducing the steady-state temperature and rearranging:

$$\frac{d\theta}{d\widehat{t}} + \left(\frac{t^*\Omega A_S}{\mathrm{Mc}}\right)\theta = 0$$

The characteristic time (t^*) can now be defined as follows:

$$t^* = \frac{\mathrm{Mc}}{\Omega A_S}$$

$$\frac{d\theta}{d\widehat{t}} + \theta = 0$$

With an initial condition $\theta(0) = 1$, the solution to this first-order ODE is

$$\theta = e^{-\widehat{t}}$$

The temperature approaches the steady-state temperature with a first-order response. The eXergy destruction rate (difference between input electrical and output shaft power) remains constant, but its effect gradually changes over from raising the temperature to heat loss over time. With an overall heat loss coefficient of 10 W/m^2/K (typical of natural convection in air), Table 3.5 shows a spreadsheet template for estimating the response time, which is 13.11 minutes for the case study investigated here. The steady-state temperature depends on the mechanical operating point, as was shown in Fig. 3.35.

Table 3.5 Thermal properties associated with DC motor case study

mass=	0.156	kg
specific heat =	385	J/kg/C, copper
Casing Diameter=	0.0358	m
Casing Length=	0.05	m
Ω=	10	W/m2/C
As=	0.0076	m^2
t^*, response time=	13.11	minutes
Tambient=	25	oC

Appendix 2: eXergy as Maximum Work potential, revisited

As experience and the laws of thermodynamics dictate, an object at some initial temperature placed in an environment at a different temperature will eventually attain thermal equilibrium with its environment by means of irreversible heat transfer to the dead state. It is possible, theoretically, to operate a heat engine between the object and the dead state, and thereby develop useful work (Fig. 3.36). The goal of this analysis is to determine the maximum possible work that could be extracted from a solid object at an initial temperature T that is brought into thermal equilibrium with its surroundings (at a dead state temperature (T_∞).

Consider a system to consist of the object and the heat engine. The object is initially at a temperature above the local dead state (left case of Fig. 3.36). There is internal heat exchange between the object and the engine (Q_H), which does not cross the boundary of the combined system. Work and heat to the environment (the low-temperature side of the engine) exit the system. For a differential change, the 1st and 2nd Law balance equations are as follows:

$$dU_{\text{OBJECT}} + dU_{\text{ENGINE}} = dU_{\text{OBJECT}} = -\delta Q_C - \delta W_{\text{OUT}}$$

$$dS_{\text{OBJECT}} + dS_{\text{ENGINE}} = dS_{\text{OBJECT}} = \frac{-\delta Q_C}{T_\infty} + \delta\Sigma$$

Note that the change in stored internal energy and entropy of the engine is zero, as the heat engine undergoes a cyclic process.

Combining (by eliminating δQ_C) and rearranging for the work output:

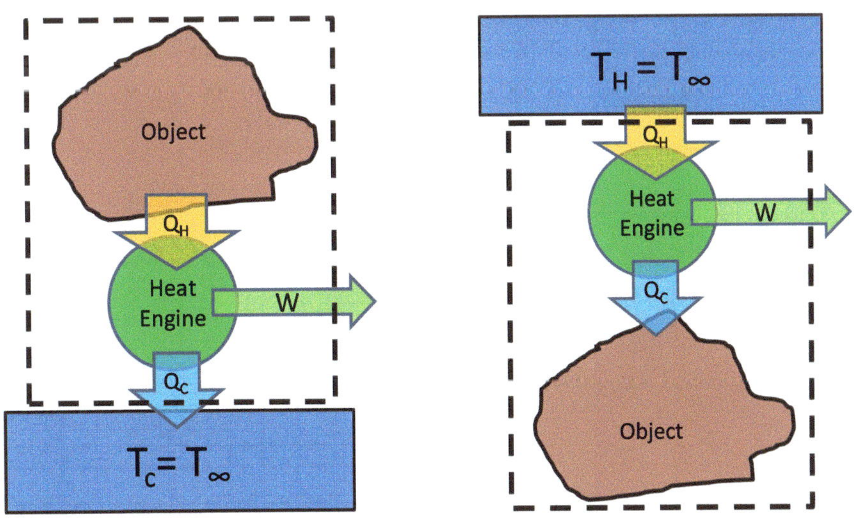

Fig. 3.36 Process diagrams of an object at a different temperature than the dead state being brought to thermal equilibrium with the dead state by a heat engine. The left diagram is for an object that is hotter than the dead state, and the right diagram is for a cold object

$$\delta W_{\text{OUT}} = T_{\infty} dS_{\text{OBJECT}} - dU_{\text{OBJECT}} - T_{\infty} \delta \Sigma$$

Introducing the eXergy ($X = U - T_{\infty} S$)

$$\delta W_{\text{OUT}} = -dX_{\text{OBJECT}} - T_{\infty} \delta \Sigma$$

This equation, with 2 inexact differentials, can only be integrated for the two limiting cases (where one of the differentials is zero). For the case where no work is extracted (i.e., allowing the object to merely cool), entropy production decreases the eXergy of the object. That is,

$$dX_{\text{OBJECT}} = -T_{\infty} \delta \Sigma \qquad \text{SIMPLE COOLING/HEATING, NO WORK EXTRACTED}$$

For the case of an reversible process, there is no entropy production, and

$$\delta W_{\text{OUT,REVERSIBLE}} = -dX_{\text{OBJECT}} \qquad \text{REVERSIBLE PROCESS}$$

The right-hand side can be integrated between states 1 and 2:

$$\int_{\text{state1}}^{\text{state2}} \delta W_{\text{OUT,REVERSIBLE}} = \int_{\text{state1}}^{\text{state2}} -dX_{\text{OBJECT}}$$

$$W_{12,\text{OUT,REVERSIBLE}} = -(X_2 - X_1)$$

For the case where the object is brought into thermal equilibrium with the dead state, $X_2 = 0$, and the object is at some initial state, X:

$$W_{\text{OUT,REVERSIBLE}} = X$$

That is, the eXergy of an object is a measure of the maximum possible work that could be extracted from it in its environment.

The same result ensues if the object is initially colder than the local ambient (right side of Fig. 3.36). A good exercise to practice your analysis and thinking skills…

Appendix 3: eXergy Destruction due to Heat Transfer, Revisited

The eXergy destruction associated with pure heat transfer from a surface at temperature T_H to another surface at temperature T_C has been shown for the case of a heat engine with no work output, and for the heat transfer across a boundary layer of an object to be given by as follows:

$$eX_D = T_\infty \Sigma = T_\infty Q\left(\frac{T_H - T_C}{T_H \cdot T_C}\right)$$

The eXergy destruction is proportional to both the quantity of heat that is transferred and the temperature difference driving the heat. For an event for which both temperatures are maintained that lasts a time Δt, the total heat transferred is also expressed in terms of the temperature difference and a thermal resistance ($\mathfrak{R}$):

$$Q = \dot{Q}\Delta t = \frac{T_H - T_C}{\mathfrak{R}} \Delta t$$

Therefore, the eXergy destruction can be expressed as follows:

$$eX_D = T_\infty \Sigma = \left(\frac{\Delta t \cdot T_\infty}{\mathfrak{R} \cdot T_H \cdot T_C}\right)(T_H - T_C)^2$$

That is, eXergy destruction depends on the temperature difference squared. The derivative of the eXergy with respect to the temperature difference is as follows:

$$\frac{d(eX_D)}{d(T_H - T_C)} = 2\left(\frac{\Delta t \cdot T_\infty}{\mathfrak{R} \cdot T_H \cdot T_C}\right)(T_H - T_C)$$

For "small" temperature differences, not only is the eXergy destruction small, but the rate of change with respect to the temperature difference is small. That subtle mathematical observation means that, in practice, reversible heat transfer can be approached by designing the process in which a relatively small temperature difference is used to drive the heat. However, that requires a combination of either a long time and/or a large surface area. Devices that have a heat transfer as either an input or an output will always have eXergy destruction associated with the heat transfer process. However, that eXergy destruction can be minimized by designing the process with the smallest possible temperature difference driving the heat. There is a trade-off between effectiveness and efficiency. The effectiveness of heat transfer increases with temperature difference, length of contact time, and size of equipment. However, so does the eXergy destruction and the 2nd Law efficiency decreases with temperature difference driving heat transfer.

Chapter 4
Liquid/Vapor Properties

This chapter introduces thermodynamic processes that involve liquids, vapors, and phase change between liquid and vapor (evaporation and condensation). Important applications include steam-based power plants and refrigeration systems, among others. Analyses using the first and second Laws of Thermodynamics require that abstract thermodynamic property values (internal energy, entropy) be determined from property values that can be directly observed and controlled (pressure, temperature, and volume). An example of an electric water heater in a piston/cylinder arrangement is used to develop both a fundamental understanding of the character of phase change, as well as the use of property tables to relate one set of thermodynamic properties to another. This heating process (a working process, actually, if the system includes the electrical heating element plus the water) raises a weight. A series of additional steps are then added to return the working fluid to its initial state, thereby producing a closed cycle. While not practical for use, study of this cycle in a closed system emphasizes the distinctions between thermal (internal energy and entropy) and mechanical (pressure and volume) properties and their behaviors.

Traditional thermodynamics textbooks include tables of property data, usually for water, one or more refrigerants, and possibly others (such as propane or carbon dioxide). There are numerous online sources for data as well. Since different tables may use different reference states, and there may be slight differences depending on the data sources, it is recommended to work a specific problem with a consistent set of tables. This chapter contains a coarse sample of tables with the objective of demonstrating their use.

For example, saturation tables are readily available for many chemicals at the Website of the National Institute of Standards and Testing (NIST):

http://webbook.nist.gov/chemistry/fluid/

In common language, the terms ice, water, and steam generally connote solid, liquid, and vapor, but the word "water" really could refer to any or all of the three phases, depending on the context. In this text, the chemical formula H_2O will usually be used as needed when phase changes are involved.

G. Sidebotham, *An Inductive Approach to Engineering Thermodynamics*,
Mechanical Engineering Series, https://doi.org/10.1007/978-3-030-20430-3_4

Raising a Weight with an Electric Water Heater

Consider an electric water (H_2O) heating apparatus (Fig. 4.1) that consists of a massless, insulated cylinder ($D = 0.1$ m inner diameter) that contains M = 1 kg of H_2O (the working fluid, approximately 1 liter of liquid water at nominal room conditions) with a frictionless, leak-free piston in place. The piston is fitted with a platform onto which weight can be added to control the pressure of the working fluid, which is maintained at a constant value ($P = 10{,}000$ kPa = 100 bar = 98.7 atm) for this example. The weight itself is assumed from the beginning to be thermally insulated from the working fluid, but in an actual device, there could be significant energy exchange between the piston and the working fluid. An electrical resistance element (of negligible mass) immersed in the H_2O is activated with an electrical input power of $\dot{W}_e = 1\ kW$, with the working fluid initially at 99.61 °C (the normal boiling point). The dead state (ambient) temperature is 20 °C. The process continues until the temperature is $T_e = 500$ °C. The system is shown at a series of intermediate stages and will be analyzed from a 1st and 2nd Law point of view. The thermodynamic objective of the device is considered to be the raising of the weight, W, at the expense of input electrical energy. Therefore, a 1st Law efficiency is defined as the potential energy change of the weight divided by the input electrical work. A 2nd Law efficiency can be defined as the ratio of the potential energy raised divided by the potential energy raised plus the total eXergy destruction during the overall process.

The relationship between the pressure of the working fluid and the mass of the piston can be obtained from Newton's 2nd Law ($F = ma$) applied to the piston, namely,

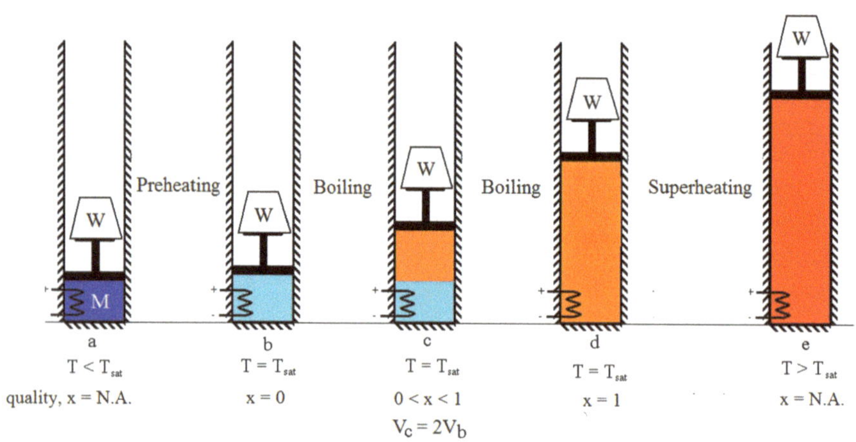

Fig. 4.1 Process sketches for Electric Water Heater

$$M_{\text{piston}} a_y = PA - P_\infty A - F_{\text{friction}} - M_{\text{piston}} g,$$

where a_y is the acceleration in the vertical direction, A is the cross-sectional area of the cylinder, and F_{friction} is the friction acting on the side walls. The forces acting on the piston are the pressure of the working fluid, acting upward, and the downward forces of the atmosphere, friction, and the weight of the piston (which includes the platform and weight W).

Rearranging for the pressure of the working fluid:

$$P = P_\infty + \frac{M_{\text{piston}}(g + a_y) + F_{\text{friction}}}{A}$$

In executing this example, inertial effects will be neglected. That is, the acceleration of the piston will be considered negligible in comparison to the gravitational acceleration. This approximation depends on the electrical input rate and a criterion to justify this assumption could be determined with a more detailed mechanics analysis. Also, friction on the side wall will be neglected, which in practice can be difficult to accomplish while simultaneously specifying a leak-free piston/cylinder arrangement. With these approximations, the pressure remains constant throughout the electrical heating process. The piston mass required to fix a constant pressure can be easily calculated from $M_{\text{piston}} = (P - P_\infty)A/g$.

First and Second Law Analyses

For the purposes of analysis, there are 4 different subsystems that can be considered: the heating element, the water, the walls of the cylinder, and the piston with the weight on it. They will be introduced as separate systems, but will subsequently be evaluated as combinations. Each term in the 1st and 2nd Law balance equations will be introduced and then categorized as being a primary or secondary effect. An explanation and interpretation of each term is given in Table 4.1, which lists terms that are considered primary, and in Table 4.2, which lists terms that are considered secondary. Secondary effects will eventually be neglected in order to evaluate the stated problem numerically.

Energy balances for these four systems are the following:

$$\text{WIRE}: \quad \Delta U_{\text{wire}} = -Q_{\text{wire}} + W_{e,\text{IN}}$$

$$\text{H2O}: \quad (\Delta U + \Delta KE + \Delta PE)_{\text{piston}} = Q_{\text{STRAY,piston}} + W_b - W_\infty - W_{\text{friction}}$$

$$\text{WALLS}: \quad \Delta U_{\text{walls}} = Q_{\text{STRAY,walls}} - Q_\infty$$

Table 4.1 Definition and explanation of terms in the energy balance that are considered to be primary effects

Term	Definition	Comments
Q_{wire}	Heat transfer between wire and water	The key heating effect of an electrical heater. Input electrical energy is converted to thermal energy, making the heating element hotter than the water, and heat is thereby transferred to the water.
W_e	Input electrical energy to heating element	The electrons that pass through the heating element are driven by a voltage difference, and their movement through an electromagnetic force represents a form of work.
ΔU_{H2O}	Internal energy change of the water	A major objective of this chapter is to develop the skill of using thermodynamic tables to relate internal energy, enthalpy, and entropy to temperature, pressure, and volume.
$W_b = P\Delta V_{\text{H2O}}$	Work of expansion	The force associated with the pressurized water that moves through a distance as the water expands. Represents the boundary work done by the water on the piston.
$\Delta PE_{\text{piston}} = W\Delta H$	Potential energy change of weight on the piston	This rise in the heavy piston is the desired thermodynamic effect of this device.
$W_\infty = P_\infty \Delta V_{\text{H2O}}$	Work done by piston on atmosphere	The piston moves against the force exerted by the atmosphere. For a piston undergoing a cycle, the NET work done by this type of work is identically zero, so it is often not included in analyses.

$$\text{WEIGHTED PISTON}: \quad (\Delta U + \Delta KE + \Delta PE)_{\text{piston}} = Q_{\text{STRAY,piston}} + W_b - W_\infty - W_{\text{friction}}$$

The kinetic and potential energy changes of the wire and walls are identically zero, and therefore not included.

Entropy balances for each subsystem are the following:

$$\text{WIRE}: \quad \Delta S_{\text{wire}} = -\iint \frac{\delta Q_{\text{wire}}}{T_{\text{wire/H2O}}} + \Sigma_{\text{wire}}$$

$$\text{H2O}: \quad \Delta S_{H2O} = \iint \frac{\delta Q_{\text{wire}}}{T_{\text{wire/H2O}}} - \iint \frac{\delta Q_{\text{STRAY,piston}}}{T_{\text{H2O/piston}}} - \iint \frac{\delta Q_{\text{STRAY,walls}}}{T_{\text{H2O/walls}}} + \Sigma_{\text{H2O}}$$

$$\text{WALLS}: \Delta S_{\text{walls}} = \iint \frac{\delta Q_{\text{STRAY,walls}}}{T_{\text{H2O/walls}}} - \iint \frac{\delta Q_\infty}{T_\infty} + \Sigma_{\text{walls}}$$

$$\text{WEIGHTED PISTON}: \quad \Delta S_{\text{piston}} = \iint \frac{\delta Q_{\text{STRAY,piston}}}{T_{\text{H2O/piston}}} + \Sigma_{\text{piston}}$$

Table 4.2 Definition and explanation of terms in the energy balance that are considered to be secondary effects. These terms will be neglected in subsequent numerical calculations

Term	Definition	Comments
ΔU_{wire}	Internal energy change of heating element	The heating element (wire) that converts electrical energy into thermal energy has a finite mass and will be at a different temperature (and internal energy level) after each process. When neglected, the heat transfer to the water will equal the electrical input work.
$\begin{pmatrix} \Delta KE \\ +\Delta PE \end{pmatrix}_{H2O}$	Kinetic and potential energy changes of the water	The water will move through natural convection processes, and its center of mass will rise. Therefore, these terms are not identically zero, but will be assumed negligible. It would be easy to justify that for the potential energy, and more difficult to prove formally for the kinetic energy.
$Q_{STRAY,\ piston}$	Stray heat transfer from the water to the piston	As the temperature of the water rises, heat will be transferred to the piston, and perhaps to the heavy weight on it. It is difficult to formally justify the neglect of this term, but its neglect facilitates the main theoretical points to be made.
$Q_{STRAY,\ walls}$	Stray heat transfer from the water to the walls of the cylinder	Heat will be transferred from the water to the walls of the cylinder. Similar arguments made about the stray heat to the piston walls applies.
ΔU_{walls}	Internal energy change of the walls	Similar to the wire, the walls have mass and will rise in temperature, resulting in an internal energy rise.
Q_∞	Heat transfer between the cylinder and the environment	This term represents heat transfer between the outer surface of the cylinder and the environment.
ΔKE_{piston}	Change in kinetic energy of weight on piston	The piston moves and has kinetic energy. It would be relatively easy to justify its neglect, which would depend on the heating rate.
$W_{friction}$	Work of friction between piston and cylinder wall	In practice, there could be considerable friction between the piston and the walls (or there would be substantial leaking), and work would be done by the piston as it moves through a distance. Friction would result in higher (and perhaps variable) pressures during heating.

Each of the entropy transfer terms with heat (the double integrals) is evaluated at a boundary between adjacent systems. Note that the boundary can be located at either extreme between the adjacent systems, and that serves to shift the location of eXergy generation. These terms can be difficult to evaluate, but when two systems are combined, they become internal and make for easier overall eXergy calculations. The double integral is a reminder that the boundary temperature where the differential heat crosses (δQ) may vary with position.

The assumption of an "adiabatic "process for piston/cylinders is frequently made in traditional thermodynamics analysis without much discussion or justification. Here, it is emphasized that an adiabatic piston/cylinder process requires two different assumptions: no heat exchange between the cylinder and the environment $(\dot{Q}_{\mathrm{STRAY,walls}} = 0)$ and thermally massless walls ($\Delta U_{\mathrm{walls}} = 0$). That is, internal heat exchange between the H2O and the cylinder itself is necessary as the walls are heated along with the H2O. The thermal mass of the wall itself is rarely negligible in practice, particularly for processes that involve gases only as the working fluid, as is true for most piston/cylinder devices in engineering practice. While it is useful to develop theoretical models that yield a fundamental understanding of thermodynamic principles, these real effects make piston/cylinder devices difficult to produce in a laboratory, relegating much thermodynamic analysis to the realm of "thought experiment." Recall that an adiabatic process can be approached if the available time for heat transfer during a process is sufficiently short. In reciprocating engines, a formal heat transfer analysis would reveal that the typical engine speeds are sufficiently high that adiabatic processes are justified.

The 1st and 2nd Law balances of the combined wire and H2O are, neglecting potential and kinetic energy changes, and combining the two stray heat terms:

$$\Delta U_{\mathrm{H2O}} + \Delta U_{\mathrm{wire}} = -Q_{\mathrm{STRAY}} + W_{e,\mathrm{IN}} - W_{b,\mathrm{OUT}}$$

$$\Delta S_{\mathrm{H2O}} + \Delta S_{\mathrm{wire}} = -\frac{Q_{\mathrm{STRAY}}}{T_B} + \Sigma_{\mathrm{H2O}} + \Sigma_{\mathrm{wire}}$$

The eXergy destruction is given by

$$eX_D = T_\infty(\Sigma_{\mathrm{H2O}} + \Sigma_{\mathrm{wire}})$$

The source of the eXergy destruction is the internal heat transfer between the heating element and the H_2O. It is difficult to directly evaluate this eXergy transfer, because the temperature of the heating element is not well characterized without a more involved heat transfer analysis, and these temperatures change during the event. However, by combining the heating element and water as a combined system, the total eXergy destruction calculation is simplified.

Rearranging for the input electrical work and eXergy destruction and neglecting all secondary effects (internal energy and entropy change of the wire compared to the H2O and stray heat):

$$W_{e,\mathrm{IN}} = \Delta U_{\mathrm{H2O}} + W_{b,\mathrm{OUT}},$$

$$eX_D = T_\infty \Delta S_{\mathrm{H2O}}$$

The electrical energy added in each process increases the internal energy of the working fluid and overcomes the expansion work of the working fluid. The internal energy change is a thermal effect, while the boundary work is a mechanical effect.

Neglecting all secondary effects, the simplified energy equation for the piston becomes

$$\Delta PE_{\text{piston}} = (P - P_{\infty})\Delta V_{\text{H2O}}$$

This balance shows that the raising of the weight comes from the expansion work of the water, less the work done on the atmosphere. The boundary work done by the working fluid changes the potential energy of the piston and overcomes the work required to move the piston against the atmosphere.

Note that $(P - P_{\infty})$ is the pressure of the working fluid relative to ambient, and is called the gage pressure, whereas P is the absolute pressure (which is inherently positive). It is more common in fluid mechanics applications to work with gage pressure, and more common in thermodynamics to work in absolute pressure. It is important to know the difference and the context.

Finally, consider a combined system that consists of all four systems (heating element, H_2O, walls, and piston). The energy and entropy balances for this system are the following:

$$\Delta E = \Delta U_{\text{H2O}} + \Delta PE_{\text{piston}} = W_e - P_{\infty}\Delta V_{\text{H2O}}$$

$$eX_D = T_{\infty}\Delta S_{\text{H2O}}$$

Rearranging the energy balance

$$W_e = \Delta U_{\text{H2O}} + \Delta PE_{\text{piston}} + P_{\infty}\Delta V_{\text{H2O}}$$

This analysis shows that the net effect of the electrical input work is to raise the weight, increase the internal energy of the H_2O, and overcome some work in expanding against the atmosphere. Any associated entropy rise of the H_2O will result in the destruction of eXergy.

Further thermodynamic analysis requires a means to relate thermodynamic properties (T, u, s) to mechanical properties (P, v). In practice, use of property tables is an important skill, to which attention is now turned, and is the main objective of this chapter.

Property Evaluation

The ***state postulate***, a fundamental thermodynamic principle, is as follows: the specification of any two independent intensive properties is sufficient to uniquely define the state of a simple compressible substance (for which electromagnetic or surface tension effects are absent). An important observation is that pressure and temperature are not independent when a system is undergoing a phase change.

Pressure, volume, and temperature are properties that can be measured, whereas internal energy, enthalpy, and entropy require thermodynamic analysis. Thermodynamic property data is readily available in traditional textbooks and online sources for most substances of engineering relevance. Much scientific work has gone into the construction of these tables, while engineering work is focused on using them. Each set of tables is based on a specific reference point, and if two different sets of tables are used, the reference point must be the same (or adjusted). There can also be some differences between sets of tables (there is uncertainty in each value, higher than the high number of significant figures reported in the tables). Therefore, a consistent set of tables should be used whenever possible.

A ***phase diagram*** is a map on pressure/temperature space that separates regions of different phases. A generic phase diagram for a typical substance is shown in Fig. 4.2. Two points, the ***triple point*** and ***critical point***, define the substance. The 3 phases (solid, liquid, and vapor) can coexist in equilibrium at the triple point. Below the triple point, a saturation line separates regions of solid and vapor. Sublimation occurs in a process that moves from a solid directly to vapor and condensation in the reverse. Other thermodynamic properties such as specific volume, internal energy, and entropy are discontinuous across the saturation line. Above the triple point, there are two saturation lines, one separating solid from liquid, and the other between liquid and vapor. The solid/liquid line is generally nearly vertical, but slopes backward for water and forward for all other substances. Water is unique in that it expands upon freezing (a change from liquid to solid). Within the solid region, there can be many different saturation curves that distinguish various forms of solid with different crystalline structures. The phase energy associated with these changes can be significant, but are not considered in this text, whose primary focus is on working fluids (liquids and gases). Use of the density and specific heat (which relates internal energy to temperature) is considered sufficient for many practical problems that involve thermal processes with solids. Supercritical fluid behavior (shown as a question mark for pressures and temperatures above the

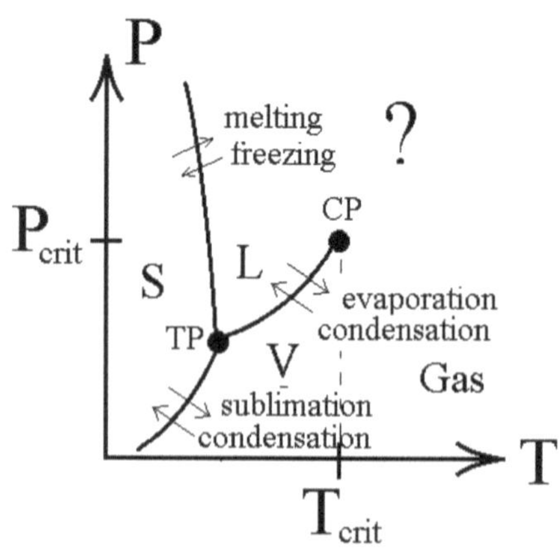

Fig. 4.2 Generic Phase (PT) diagram showing saturation curves associated with the three phases: solid (S), liquid (L), and vapor (V). Supercritical fluids are shown as a question mark, since they are difficult to classify

critical point) is unusual. There are important applications that involve supercritical fluids, but they are not considered further in this text.

There are 2 important types of property tables: saturation tables and superheat tables. The saturation tables pertain to states *b*, *c*, and *d* in the electric heater example. The superheat tables pertain to state *e*. Compressed liquid tables have been constructed for some substances, but they are usually unnecessary since the compressed liquid approximations are generally acceptable for engineering work, as discussed in the next section. That is, all the compressed liquid properties (e.g., for state *a*) can be obtained from the properties of saturated liquids.

Much of the attention of engineering thermodynamics is on processes that involve the liquid/vapor saturation line. The ***saturation curve*** represents the relationship between pressure and temperature at which phase change occurs. Figure 4.3 shows a phase diagram for H_2O that focuses just on the phase boundary between liquid and vapor. Liquids and vapors are the primary working fluids of a large class of engineering devices, so the interpretation of the state of a fluid relative to the liquid/vapor curve is a key skill. Other chemicals have different saturation curves, but share the same qualitative behavior.

The saturation curve in Fig. 4.3 (shown as a semilogy plot because the pressure changes by over 4 orders of magnitude) separates regions of liquid (upper left, i.e. low temperature, high pressure) from regions of vapor (high temperature, low pressure). The term ***compressed liquid*** is used to indicate that the substance is not on

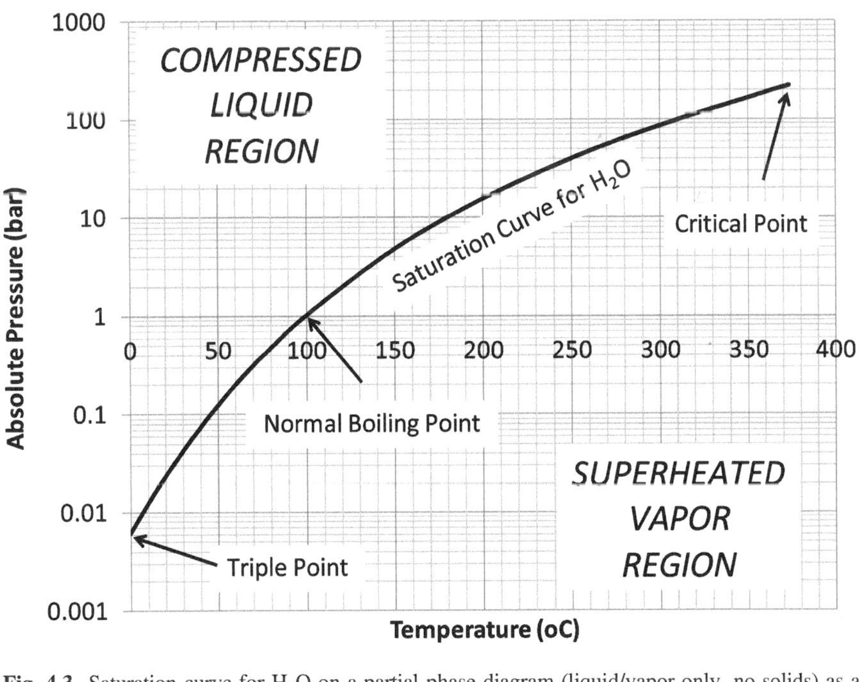

Fig. 4.3 Saturation curve for H_2O on a partial phase diagram (liquid/vapor only, no solids) as a semilogy plot of absolute pressure (in bar) vs. temperature (in °C)

the saturation curve (but to the left), and the term ***superheated vapor*** is used in the region to the right. There is a substantial change in other thermodynamic properties (specific volume, internal energy, entropy) as the saturation curve is crossed from liquid to vapor, and those changes will be observed in other types of thermodynamic plots. Standard nomenclature defines the properties of saturated liquid with a subscript "f" (for fusion) and saturated vapor with a subscript "g" (for gas). For example, the internal energy of a saturated liquid is u_f.

The term "quality," given the symbol x (not to be confused with eXergy, but a standard symbol that will be used here, with another prompt for the reader to always consider the context of a given variable), refers to a liquid/vapor mixture and is defined as the mass fraction of vapor of a liquid/vapor mixture. Specification of a quality demands that the substance be on the saturation curve of a phase diagram, and the relative amounts of liquid and vapor are specified by the value of the quality. That is,

$x = 0.0$ for saturated liquids
$x = 1.0$ for saturated vapors
$x = NA$ (not applicable) for compressed liquids or superheated vapors

It is common for students to report a quality of 0.0 in a case where the substance is a compressed liquid (or a quality of 1.0 for a superheated vapor), and that is a conceptual error that could lead to poor designs. Pressure and temperature are independent properties in those cases.

Three points are identified on the saturation curve. The lower limit to the liquid/vapor curve is the triple point, at which point, all 3 phases can exist in equilibrium. It is a unique thermodynamic point of reference. The upper limit is the critical point, above which there is no clear distinction between liquid and vapor. The ***normal boiling point*** is the temperature at which the phase changes at a standard atmosphere pressure, defined as 1.013 bar. Note that 1 bar = $(10)^5$ N/m^2 and therefore, 1 atm = $1.013(10)^5$ N/m^2 = 1.013 bar. The unit Pascal is commonly used for pressure 1 Pa = 1 N/m^2. The normal boiling point for H_2O is 100 °C. At 100 bar, as in Fig. 4.1, H_2O boils at 311 °C.

Given the saturation curve and a specification of the pressure and temperature, there are two equivalent ways to determine the phase of a substance. For a given pressure, if the temperature is lower than the corresponding saturation temperature (T_{SAT}), then the substance is a compressed liquid. Otherwise, it is a vapor. If the temperature equals the saturation temperature, then the phase is indeterminate. It could be all liquid, all vapor, or any fraction in between, and liquid and vapor are in equilibrium. That is,

$T < T_{SAT}(P)$ for compressed liquids
$T > T_{SAT}(P)$ for superheated vapors
$T = T_{SAT}(P)$ for liquid/vapor mixtures

Alternatively, and equivalently, for a given temperature, if the pressure is higher than the corresponding saturation pressure, then the substance is a compressed liquid. That is,

$P > P_{SAT}(T)$ for compressed liquids
$P < P_{SAT}(T)$ for superheated vapors
$P = P_{SAT}(T)$ for liquid/vapor mixtures

The use of property tables to relate various thermodynamic properties is a skill whose importance cannot be overemphasized. Saturation tables are widely available that list the temperature and corresponding saturation pressure (or vice versa), and then the corresponding properties of saturated liquids and vapors. Tables of compressed liquids (and of superheated vapors) are also published that show subtables of properties as a function of temperature for fixed values of pressure.

Compressed Liquid Property Evaluation

A liquid is a substance that takes the form of an open container into which it is placed. For example, liquid water will fill a cooking pot placed under an open faucet. A direct way to measure the volumetric flow rate (AV) from a faucet is to collect a volume over a measured time. Measurement of the weight of water collected allows for a determination of the specific volume, the volume per unit mass. The specific volume, v, is the inverse of the density, ρ. Of course, if the pot of water is placed on a stovetop, its temperature will rise and eventually begin to boil. Prior to boiling, the liquid is in a state classified as a compressed liquid. Addition of energy (heat or work) to a compressed liquid will not cause it to change phase, but the increase in internal energy will be evidenced by a temperature rise. Once it reaches the boiling point, it is classified as a saturated liquid. Further energy addition will cause phase change without changing the temperature (if the pressure is held constant).

The following examples (one focused on pressure effects, the other temperature) demonstrate, using the 1st law of thermodynamics and the incompressible approximation, that the specific volume, internal energy, and entropy of a pure substance (i.e., H_2O) depend only on temperature, not pressure. Therefore, the saturation tables can be used to determine all thermodynamic properties (specific volume, internal energy, enthalpy, and entropy). Workshop 4.1 is a good early exercise in a thermodynamics course designed to familiarize the student with the structure of compressed liquid tables (which has the same structure as superheat tables). The hand construction of the plots facilitates the interpretation of the results, the other objective of the workshop. Ironically, the real goal of the exercise is to convince the student that compressed liquid tables are unnecessary for the vast majority of engineering applications.

Consider the process shown in Fig. 4.4 in which a unit mass of a saturated liquid in a piston/cylinder arrangement is compressed adiabatically and reversibly by placing a heavy weight on the piston. The initial temperature is specified and, therefore, all other properties of state 1 are obtained from the saturation tables at that temperature. All the unknown thermodynamic properties at the 2nd state (for various pressures) will be first shown from compressed liquid tables for H_2O

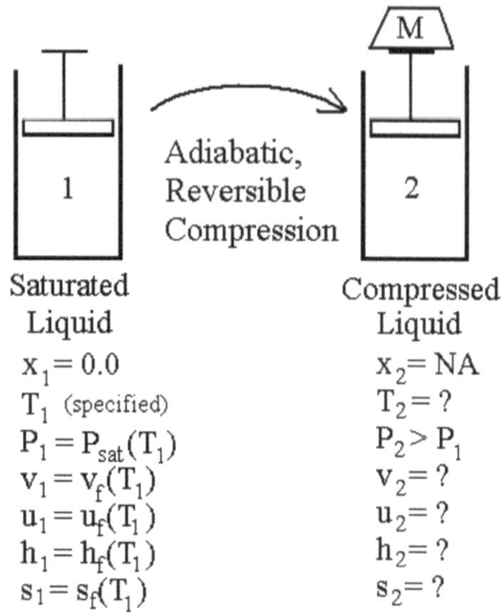

Fig. 4.4 Thermodynamic process diagram of compressing a liquid, initially a saturated liquid

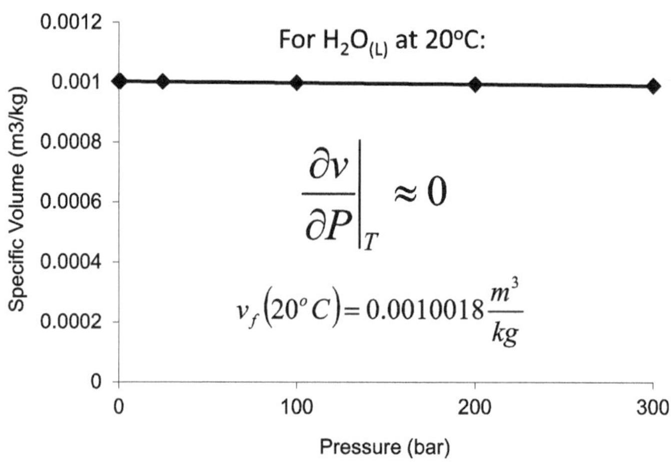

Fig. 4.5 Specific volume of compressed liquid water at a fixed temperature of 20 °C. Data from compressed liquid tables

(obtainable from an internet search). The compressed liquid approximations will be demonstrated from there. Liquids experience a similar change in state when they are pumped in engineering systems (but those are open systems, which will be studied formally in future chapters).

Consider the effect of pressure on specific volume first, shown in Fig. 4.5. A large pressure applied to a liquid will not appreciably reduce its volume. The incompressible approximation states that the volume does not change at all. This behavior should appeal to intuition: merely applying pressure to a liquid is not expected to change its volume much (nor raise its temperature, as will be shown).

In a single-phase region, the specific volume can be considered to be a function of temperature and pressure. That is, $v = v(T, P)$. For a differential change of state, where temperature changes by dT and pressure changes by dP:

$$dv = \left.\frac{\partial v}{\partial T}\right|_P dT + \left.\frac{\partial v}{\partial P}\right|_T dP$$

The results of Fig. 4.5 show that, to an excellent approximation, $\left.\frac{\partial v}{\partial P}\right|_T \approx 0$. Stated another way, the specific volume is a function of temperature only for compressed liquids and, therefore, saturated liquid tables can be used to determine the specific volume of a compressed liquid.

Energy and entropy balances applied to the process are the following:

$$\Delta E = Q_{\text{in}} - W_{\text{out}}$$

$$\Delta S = \frac{Q_{in}}{T_B} + \Sigma$$

The heat transfer is specified to be zero in the problem statement (adiabatic). The compression work (integral of PdV) is approximately zero since the volume does not change appreciably. Finally, there are no kinetic or potential energy changes. The process is considered to be reversible. Therefore, there is no change in the internal energy or entropy.

$$\Delta U = 0$$

$$\Delta S = 0$$

These consequences of the incompressible approximation are confirmed by the compressed liquid table data. The effect of pressure on the specific internal energy (u) is shown in Fig. 4.6 and on specific entropy in Fig. 4.7. Like the specific volume, both the internal energy and entropy are independent of pressure for compressed liquids, and therefore, saturated liquid tables can be used to determine the specific internal energy of a compressed liquid.

If the temperature of a compressed liquid is known, then its specific volume equals that of a saturated liquid at the corresponding temperature. Similar statements are made for internal energy and entropy. Students will sometimes make the mistake of looking up the value for a corresponding pressure. That is, $u(T, P) \neq u_f(P)$. A good way to check this (e.g., in a quiz) is to specify a pressure and an internal energy or entropy (in the compressed liquid region), and determine all other properties.

As shown in Fig. 4.6, there is an appreciable effect of pressure on an important thermodynamic property called ***enthalpy***. Enthalpy, given the symbol h, is a combination property that is convenient to use with open systems and defined as follows:

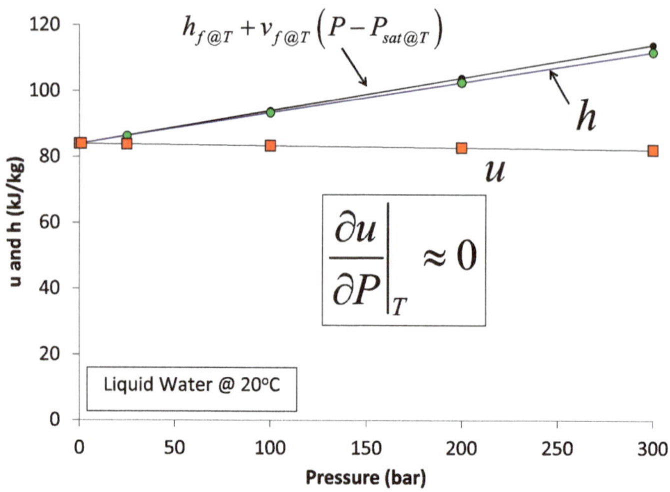

Fig. 4.6 Specific internal energy and enthalpy of compressed liquid water at a fixed temperature of 20 °C. Data from compressed liquid tables, and approximation for enthalpy from saturated liquid data shown

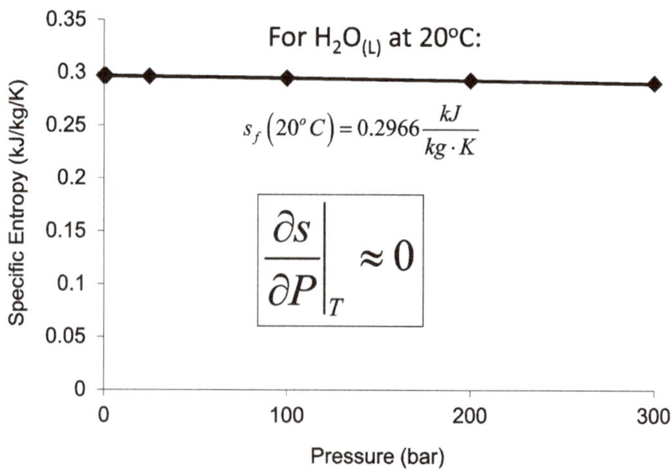

Fig. 4.7 Specific entropy of compressed liquid water at a fixed temperature of 20oC. Data from compressed liquid tables

$$h = u + Pv$$

However, an excellent approximation can be made to account for the pressure effect on enthalpy, as follows. Starting with its definition, for a process from state 1 to state 2:

$$h_2 - h_1 = (u_2 + P_2v_2) - (u_1 + P_1v_1)$$

Since, in the problem at hand, $h_1 = h_f(T)$, $u_2 = u_1 = u_f(T)$, $v_2 = v_1 = v_f(T)$ and$P_1 = P_{SAT}(T)$, the 3rd compressed liquid approximation is as follows:

$$h(T,P) = h_f(T) + v_f(T)(P - P_{\mathrm{SAT}}(T))$$

The enthalpy is therefore a function of both temperature and pressure, but the pressure effect can be calculated using saturated liquid data. In practice, this pressure effect can often, but definitely not always, be neglected. The pressure effect is responsible for the work required to pump a liquid from low to high pressure in an open system.

Therefore, four compressed liquid approximations follow, summarized in Fig. 4.12:

1. The specific volume of a compressed liquid (state 2) equals that of a saturated liquid at the same temperature, or, $v(T,P) = v_{f\ @\ T}$.
2. The internal energy of a compressed liquid equals that of a saturated liquid at the same temperature, or $u(T,P) = u_{f\ @\ T}$.
3. The entropy of a compressed liquid equals that of a saturated liquid at the same temperature, or $s(T,P) = s_{f\ @\ T}$.
4. The enthalpy of a compressed liquid equals that of a saturated liquid at the same temperature plus a pressure correction, or $h(T,P) = h_{f\ @\ T} + v_{f\ @\ T}(P - P_{\mathrm{SAT}\ @\ T})$.

Table 4.3 shows the comparison over the full range of pressure data available from compressed liquid tables at a different temperature, namely, 100 °C. The first row corresponds to the saturated pressure (from which the compressed liquid approximations are obtained). The properties (v, u, h, and s) are entered from the tables. The percentage error of each approximation is listed. The error of all compressed liquid approximations is less than 2% up to a rather high pressure of 300 bar. The specific volume, internal energy, and entropy, all decrease slightly with pressure, while the enthalpy increases due to the $v\Delta P$ effect.

The effect of temperature on the thermodynamic properties of compressed liquids is demonstrated by the process in which electrical work is added to an electrical resistor immersed in a liquid in a well-insulated piston/cylinder with a fixed pressure, the first process of the scenario in Fig. 4.1, and shown in Fig. 4.8. The liquid is initially a compressed liquid ($T < T_{\mathrm{SAT}\ @\ P}$), and work is added until it reaches the saturated liquid state. An alternative means of adding work would be from a paddle wheel that mechanically stirs the liquid.

Figure 4.9 shows the specific volume as a function of temperature that represents the process path that would take place in this experiment, starting with nominal room temperature H2O at an extremely high pressure of 200 bar, just below the critical pressure (220.8 bar). Overall, the volume doubles, but for temperatures well below the saturation temperature, the volume changes very little, and it is often reasonable

Table 4.3 Effect of pressure on compressed liquid H2O. Compressed liquid data compared to compressed liquid approximations. Temperature is fixed at 100 °C

P (bar)	v (m^3/kg)	%ERROR	u (kJ/kg)	%ERROR	h (kJ/kg)	hf(T)+vf(P-Psat)	%ERROR	s (kJ/kg/K)	%ERROR
1.014	1.044E-03	0.00	418.94	0.00	419.04	419.04	0.00	1.3069	0.00
25	1.042E-03	0.12	418.24	0.17	420.85	421.54	0.16	1.3050	0.15
100	1.039E-03	0.48	416.12	0.68	426.50	429.37	0.67	1.2992	0.59
200	1.034E-03	0.95	413.39	1.34	434.06	439.80	1.32	1.2917	1.18
300	1.029E-03	1.41	410.78	1.99	441.66	450.24	1.94	1.2844	1.75

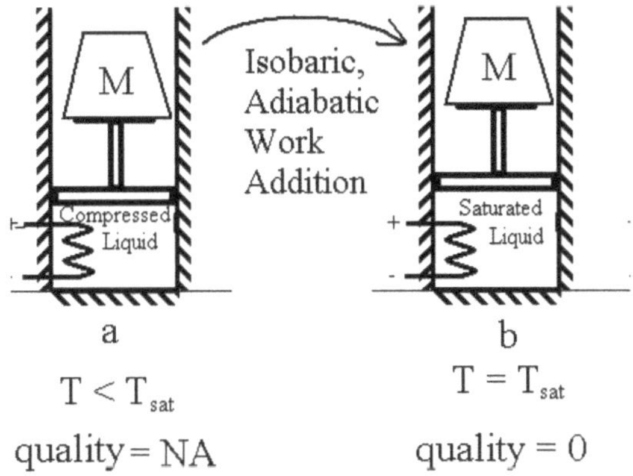

Fig. 4.8 Thermodynamic process diagram for an isobaric electrical heating of a compressed liquid

to assume the specific volume is independent of temperature as well as independent of volume.

Figure 4.10 shows the internal energy for the same process. The internal energy increases linearly with temperature (i.e., with constant slope) well below the saturation point. As the saturation temperature is approached, at which point the slope increases somewhat. The slope of this curve is the specific heat of the liquid.

Figure 4.11 shows a plot of temperature versus entropy, an important type of thermodynamic plot. This plot shows the saturated liquid values (solid line) and the entropy from compressed liquid tables for 3 pressures as data points. The compressed liquids are at very different pressures, but there is no distinction on a Ts plot. That is, all compressed liquid points collapse onto the saturated liquid line, and there is no compressed liquid "region" on a Ts diagram, as is often indicated in the analysis of thermodynamic cycles.

In summary, for pressures and temperatures that place a substance in the compressed liquid region, the specific volume, internal energy, and entropy are all shown to be functions of temperature (not pressure). This situation is similar to that of solids, and use of the specific heat to relate internal energy to temperature is the same as derived in Chap. 2. The enthalpy has both a temperature and pressure dependence, but can still be determined using saturated liquid properties. It is often justified to treat the specific volume as a constant, even if the temperature changes significantly. When that approximation is made, the liquid is considered to be incompressible. However, the internal energy and entropy are strong functions of temperature.

The change of specific volume with temperature is expressed through a fluid property called the "coefficient of thermal expansion, β" and defined by:

$$\beta = \frac{1}{v}\left(\frac{\partial v}{\partial T}\right)_P = \frac{-1}{\rho}\left(\frac{\partial \rho}{\partial T}\right)_P$$

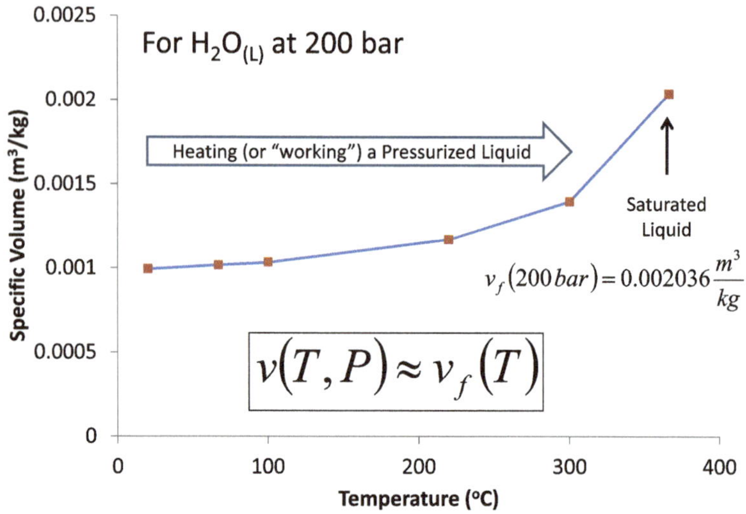

Fig. 4.9 Specific volume of compressed H2O liquid as a function of temperature (fixed pressure of 200 bar)

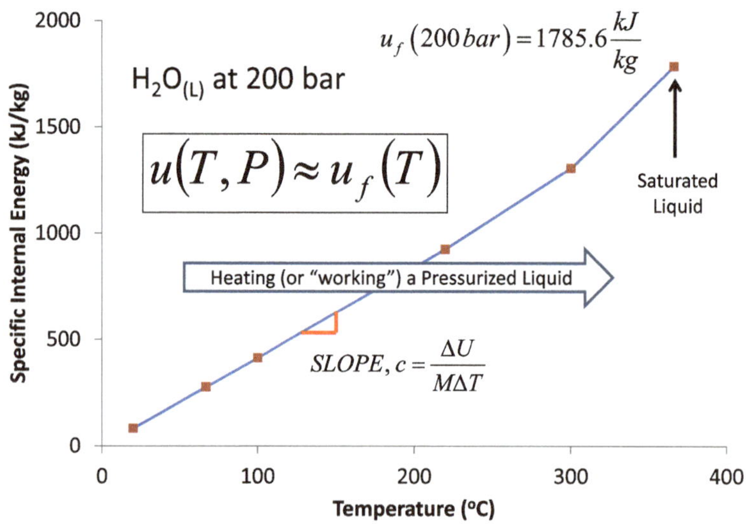

Fig. 4.10 Specific internal energy of compressed H2O liquid as a function of temperature (fixed pressure of 200 bar)

For many applications with liquids only, it is reasonable to neglect the change in specific volume (even if there is substantial temperature change in some cases). That is, the specific volume is treated as a constant. The fluid is modeled as being incompressible. That is,

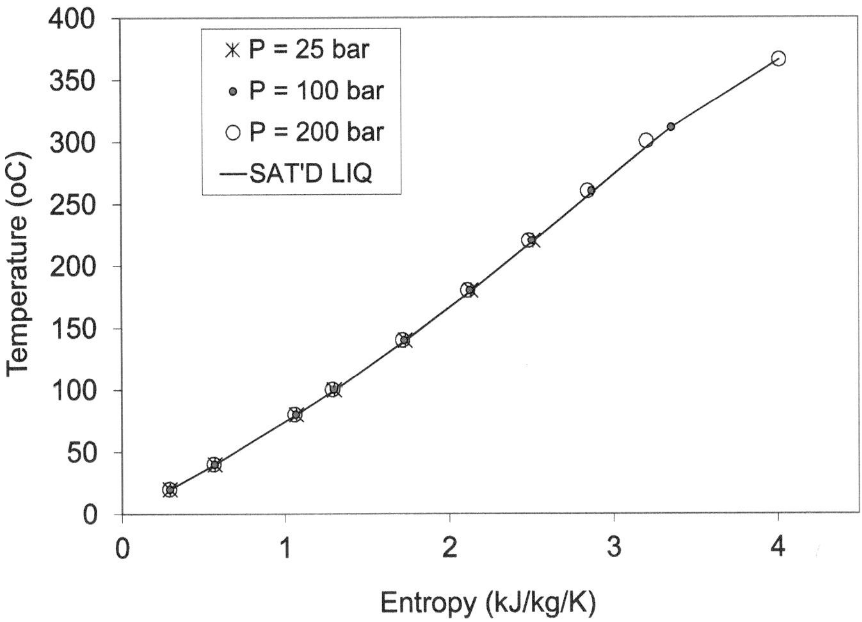

Fig. 4.11 Ts plot for compressed liquid H_2O. Saturated liquid data is shown as a line, and the entropy is shown at 3 pressures from compressed liquid tables which collapse onto the saturation line

$\beta = 0$ for incompressible substances

One case where the incompressible approximation is not good modeling practice is for what is called natural convection in heat transfer modeling. A pot of water on a stove clearly shows that the liquid moves, and this motion is caused by density variations associated with temperature, driven by gravity.

The specific volume, internal energy, and entropy relations are the same as for solids for a change from state 1 to state 2 (both states as compressed liquids), namely,

$$v_2 - v_1 = v_1\beta(T_2 - T_1)$$

$$u_2 - u_1 = c(T_2 - T_1)$$

$$s_2 - s_1 = c\ln(T_2/T_1)$$

A short table of liquid properties (density, specific heat, and coefficient of thermal expansion) and their normal boiling points is provided in Table 4.4. Of particular note is that liquid water has a specific heat nearly double that of most other liquids, one of its unique features. Also, the behavior of liquid water near its freezing point is unique as the density is a maximum at 4 °C, and the density of ice is less than liquid water (so that ice floats, a very unusual characteristic).

Fig. 4.12 Compressed liquid approximations: a summary of how to use saturated liquid properties to determine the state of a compressed liquid, or the change from one compressed liquid state to another. The coefficient of thermal expansion (β) could be used to estimate changes in volume for substantial temperature differences

Compressed Liquid Approximations

$$\text{if} \quad T < T_{SAT@P} \text{ or } P > P_{SAT@T} \quad \text{then}$$

$$v(T,P) = v_{f@T}$$

$$u(T,P) = u_{f@T}$$

$$s(T,P) = s_{f@T}$$

$$h(T,P) = h_{f@T} + v_{f@T} \cdot \left(P - P_{SAT@T}\right)$$

For a process from state 1 to state 2 in which both states are compressed liquids:

$$v_2 - v_1 = \beta v_1 (T_2 - T_1) \approx 0, \text{ often}$$

$$u_2 - u_1 = c(T_2 - T_1)$$

$$s_2 - s_1 = c \ln\left(\frac{T_2}{T_1}\right)$$

$$h_2 - h_1 = c(T_2 - T_1) + v_1 \cdot (P_2 - P_1)$$

Table 4.4 Liquid properties at nominal room temperature and their normal boiling points (sorted by increasing specific heat). Kerosene and engine oil are not well-defined mixtures of components, and any property value is approximate

Liquid	Normal Boiling Point (°C)	Density (kg/m^3)	Specific Heat (J/kg/K)	Coef. of Thermal Expansion (1/K)
Engine oil	>150	~900	~2000	~0.0007
Kerosene	>150	~800	2010	0.00099
Acetone	50.5	784.6	2150	0.00143
n-octane	125.6	699	2150	0.00114
Dodecane	216	754.6	2210	
Ethanol	79	789	2300	0.00109
Ethylene glycol	197	1097	2360	0.00057
Glycerin	290	1259	2430	0.0005
Propylene glycol	187	965	2500	
Water	100	997	4180	0.000257

Temperature Table

T (° C)	P(kPa)	v_f (m3/kg)	v_g (m³/kg)	u_f (kJ/kg)	u_g (kJ/kg)	s_f (kJ/kg/K)	s_g (kJ/kg/K)
0.01	0.6117	0.001000	206.00	0	2374.9	0	9.1562
20	2.3392	0.001002	57.762	83.913	2402.3	0.2965	8.6661
40	7.3851	0.001008	19.515	167.53	2429.4	0.5724	8.2556
60	19.947	0.001017	7.667	251.16	2455.9	0.8313	7.9082
80	47.416	0.001029	3.4053	334.97	2481.6	1.0756	7.6111
100	101.42	0.001043	1.6720	419.06	2506.0	1.3072	7.3542
150	476.16	0.001091	0.392480	631.66	2559.1	1.8418	6.8371
200	1554.9	0.001157	0.127210	850.46	2594.2	2.3305	6.4303
250	3976.2	0.001252	0.050085	1080.7	2601.8	2.7933	6.0721
300	8587.9	0.001404	0.021659	1332.7	2563.6	3.2068	5.7432
350	16529	0.001741	0.008806	1642.4	2418.3	3.7788	5.2114
373.95	22064	0.003016	0.003016	2015.7	2015.7	4.407	4.407

Pressure Table

P(kPa)	T (° C)	v_f (m3/kg)	v_g (m³/kg)	u_f (kJ/kg)	u_g (kJ/kg)	s_f (kJ/kg/K)	s_g (kJ/kg/K)
1	6.97	0.001000	129.19	29.302	2384.5	0.1059	8.9749
10	45.81	0.001010	14.67	191.79	2437.2	0.6492	8.1488
100	99.61	0.001043	1.6941	417.4	2505.6	1.3028	7.3589
1000	179.88	0.001127	0.19436	761.39	2582.8	2.1381	6.5850
10000	311.00	0.001452	0.018028	1393.3	2545.2	3.3603	5.6159

Fig. 4.13 Saturation Property Data for H2O. (Adapted from the Appendices of *Thermodynamics, an Engineering Approach*, 7th ed, by Yunus A. Cengel and Michael A. Boles, McGraw-Hill). These tables are deliberately coarse. More detailed tables are available from a variety of textbook and online sources. Notice that the reference state for these tables is the saturated liquid at the triple point of H2O ($T_{ref} = 0.01$ °C, $P_{ref} = 0.6117$ kPa)

Saturation Tables

Saturation tables list the thermodynamic properties of saturated liquids and saturated vapors at corresponding pressures and temperatures. A coarse version of the saturation tables for H_2O (often called Steam Tables) is shown in Fig. 4.13. There are two tables, which contain similar data. The temperature table lists properties at convenient values of temperature and the pressure table lists convenient pressure values. The two columns temperature and pressure define the saturation line of the phase diagram of Fig. 4.3. Appendix 1 for this chapter dives deeper to develop formulas that can be used to relate saturation pressure to temperature. The next columns list the specific volume, internal energy, and entropy of saturated liquids, which are given the subscript "*f*" (for fusion), and for saturated vapors, subscript "*g*" (for gas). For convenience, most published or online tables also list the enthalpy, and the difference between the vapor and liquid values, given a subscript *fg*. For example, $v_{fg} = v_g - v_f$. They also list values with a finer subdivision of temperature and/or pressure. The purpose here is to gain experience in navigating these tables.

Linear interpolations between points are frequently warranted. For example, if a quantity "*A*" that falls between entries is sought (A_{low} and A_{high}), and the quantity "*B*" is specified, then a linear interpolation formula (a 1st order Taylor expansion) is as follows:

$$A = A_{\mathrm{low}} + \frac{\Delta A}{\Delta B}(B - B_{\mathrm{Low}}),$$

where the sensitivity gradient comes from the table:

$$\frac{\Delta A}{\Delta B} = \frac{A_{\mathrm{high}} - A_{\mathrm{low}}}{B_{\mathrm{high}} - B_{\mathrm{low}}}$$

Note that the term "high" and "low" refer to the location of the entry in the table, not to their value. It is common that several property values are required, and, for hand calculations, it is convenient to calculate and store the interpolation ratio:

$$r = (B - B_{\mathrm{Low}})/\left(B_{\mathrm{high}} - B_{\mathrm{low}}\right)$$

The value for A is therefore as follows:

$$A = A_{\mathrm{low}} + r\Delta A$$

For example, determine the saturation temperature, pressure, and specific volume (quantities "A") of saturated liquid if the internal energy of a saturated liquid (quantity "B") is known to have a value of 215 kJ/kg. From the temperature table, visual inspection reveals that the temperature is between 40 °C and 60 °C, the pressure is between 7.3851 kPa and 19.947 kPa, and the vapor specific volume is between 19.515 m^3/kg and 7.667 m^3/kg). The interpolation ratio is as follows:

$$r = \frac{B - B_{\mathrm{low}}}{B_{\mathrm{high}} - B_{\mathrm{low}}} = \frac{215 - 167.53}{251.16 - 167.53} = 0.5676$$

Therefore,

$$T = 40 + 0.5676\,(60 - 40) = 51.35^{o}C$$

$$P = 7.3851 + 0.5676\,(19.947 - 7.3851) = 14.515\ \ kPa$$

$$v_g = 19.515 + 0.5676\,(7.667 - 19.515) = 12.790\ \ m^3/kg$$

Quality

The real power of the saturation tables is the ability to determine the average properties of liquid/vapor mixtures that lie "under the dome". Consider, once again, state c in the electric heater example, shown in Fig. 4.14. This state falls on the phase boundary of a P-T plot. However, two distinct phases are indicated, liquid

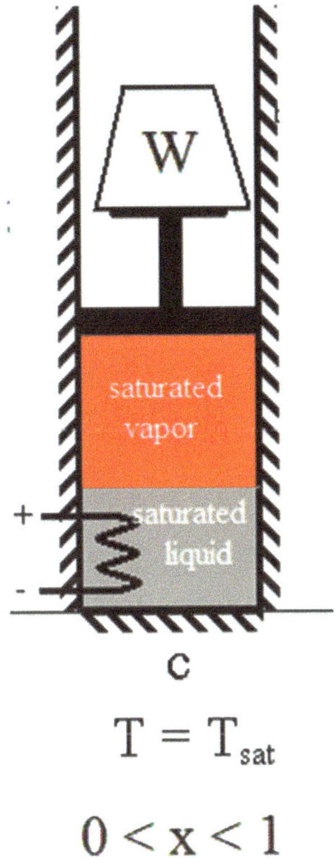

Fig. 4.14 H2O in a mixed liquid/vapor state

and vapor. The two phases are generally assumed to be in a quasi-equilibrium, which means that the liquid is a saturated liquid and the vapor is a saturated vapor, and therefore, their properties are listed in the saturation tables. The total volume, or total internal energy, or total entropy of the working fluid is the sum in both phases. The "quality" of the mixture is simply the fraction of mass that is in the vapor phase. This terminology originated in applications that involved steam, where a high percentage of vapor was desired. Frankly, it should be renamed, but practitioners know what is meant by it. The conventional use of "x" as the symbol for quality will be used, despite its being the same symbol as used for eXergy. As is often the case with inevitable nomenclature conflicts, the practicing engineer must know which quantity is being referred to from the specific context.

To understand quality, consider the following analogy of a piggy bank. For simplicity, assume that the piggy bank owner only places pennies or quarters in the piggy bank. The pennies are like saturated liquid and the quarters like saturated vapor. Suppose, at one point in time, there are $N = 100$ coins in the piggy bank. What is the monetary value of the bank? It depends, of course, on how many coins are pennies, and how many are quarters. If all the coins are pennies, the total value is $1. If they are all quarters, the value is $25. Suppose, however, that 50 coins were pennies and 50 coins were quarters. The total value would be 50*$0.01 + 50*

\$0.25 = \$13. The average value of a coin would be \$13/100 = \$0.13/coin, or 13 cents per coin. Of course, there are no coins in the piggy bank that have a value of 13 cents. Any sample coin taken would have a value of either 1 cent or 25 cent. The quality of the piggy bank can be defined as the fraction of coins that are quarters, given the symbol x (not eXergy). Defining the value of the coins to be $v_{\text{penny}} = \$0.01/$ coin and $v_{\text{quarter}} = \$0.25/\text{coin}$, a formal analysis of the value of the piggy bank is as follows:

$$N = N_{\text{pennies}} + N_{\text{quarters}}$$

That is, the total number of coins is the sum of the pennies and quarters, a form of mass balance. The total value of the bank (\$) is as follows:

$$\$ = N_{\text{pennies}} v_{\text{penny}} + N_{\text{quarters}} v_{\text{quarter}}$$

Using the mass balance and eliminating the number of pennies,

$$\$ = (N - N_{\text{quarters}}) v_{\text{penny}} + N_{\text{quarters}} v_{\text{quarter}}$$

Dividing through by the total number of coins,

$$\frac{\$}{N} = \left(1 - \frac{N_{\text{quarters}}}{N}\right) v_{\text{penny}} + \frac{N_{\text{quarters}}}{N} v_{\text{quarter}}$$

Inserting the definition of the quality of the piggy bank, x:

$$\frac{\$}{N} = (1 - x)\, v_{\text{penny}} + x\, v_{\text{quarter}}$$

The specific value of the piggy bank (the monetary value per total coins) is therefore related to the value of the individual coins and the quality of the piggy bank.

Returning from piggy banks to thermodynamic banks, suppose the total energy of a liquid/vapor mixture is to be expressed in terms of the quality, x.

$$U_{\text{TOTAL}} = U_{\text{liquid}} + U_{\text{vapor}}$$

Expressed in terms of the masses and the specific internal energies:

$$M_{\text{TOTAL}} u_{\text{mixture}} = M_{\text{liquid}} u_f + M_{\text{vapor}} u_g$$

Eliminating the mass of the liquid:

$$M_{\text{TOTAL}} u_{\text{mixture}} = (1 - M_{\text{vapor}}) u_f + M_{\text{vapor}} u_g$$

Dividing by the total mass:

$$u_{\text{mixture}} = \left(1 - \frac{M_{\text{vapor}}}{M_{\text{TOTAL}}}\right)u_f + \frac{M_{\text{vapor}}}{M_{\text{TOTAL}}}u_g$$

Introducing the definition of quality, $x = M_{\text{vapor}}/M_{\text{TOTAL}}$:

$$u_{\text{mixture}} = (1 - x)u_f + xu_g$$

It is often convenient to express the internal energy as:

$$u_{\text{mixture}} = u_f + x(u_g - u_f) = u_f + xu_{fg}$$

Rearranging for the quality:

$$x = \frac{u - u_f}{u_g - u_f} = \frac{u - u_f}{u_{fg}}$$

These calculations can be done similarly for other extensive thermodynamic properties that are made intensive by dividing by the mass (i.e., specific volume, internal energy, enthalpy, entropy). The intensive properties temperature and pressure are not related to quality this way. For liquid/vapor mixtures, they are related by the saturation curve.

In the electric heater example, the problem is formulated such that the specific volume of state c can be determined (set arbitrarily to be twice the volume of state b). Once the quality of state c is known, then the internal energy and entropy can be determined. That is,

$$v_c = 2v_b = 2v_{f@10{,}000kPa} = 2(0.001452) = 0.002904 \ \frac{m^3}{kg}$$

This value lies between the saturated liquid and saturated vapor values for the corresponding pressure and, therefore, state c must be within the "steam dome." That means that the temperature must equal the corresponding saturation temperature (311.00 °C) and that there must be a quality, whose value can be calculated as follows:

$$x_c = \frac{v - v_f}{v_g - v_f} = \frac{0.002904 - 0.001452}{0.018028 - 0.001452} = 0.0876$$

The internal energy and entropy can then be calculated:

Pressure = 100 kPa (T_{sat} = 99.61°C)

T (° C)	v (m³/kg)	u (kJ/kg)	s (kJ/kg/K)
100	1.6959	2506.2	7.3611
200	2.1724	2658.2	7.8356
300	2.6389	2810.7	8.2172
400	3.1027	2968.3	8.5452
500	3.5655	3132.2	8.8362
600	4.0279	3302.8	9.0999
800	4.9519	3665.0	9.5682
1000	5.8755	4055.0	9.9800
1200	6.7988	4470.7	10.3504

Pressure = 10,000 kPa (T_{sat} = 311.00°C)

T (° C)	v (m³/kg)	u (kJ/kg)	s (kJ/kg/K)
400	0.026436	2833.1	6.0433
500	0.032811	3047.0	6.5995
600	0.038378	3242.0	6.9045
700	0.043597	3434.0	7.1693
800	0.048629	3628.2	7.4085
900	0.053547	3826.5	7.6290
1000	0.058391	4029.9	7.8349
1100	0.063183	4238.5	8.0289
1200	0.067938	4452.4	8.2126

Fig. 4.15 Superheated vapor properties of H_2O at 2 pressures. (Adapted from the Appendices of *Thermodynamics, an Engineering Approach*, 7th ed, by Yunus A. Cengel and Michael A. Boles, McGraw-Hill)

$$u_c = (1 - x)u_f + xu_g = (1 - 0.0876)1393.3 + 0.0876\,(2545.2) = 1494.2\frac{kJ}{kg}$$

$$s_c = (1 - x)s_f + xs_g = (1 - .0876)3.3603 + 0.0876\,(5.6159) = 3.5579\frac{kJ}{kg \cdot K}$$

Superheat Tables

Continuing through the stages of Fig. 4.1, once phase change is complete (beyond state *d*), the pressure and temperature become independent (the saturation tables no longer apply), and the working fluid is in the superheated region. A coarse sample of superheated property tables is shown in Fig. 4.15. Superheat tables generally consist of tables at a fixed pressure, with temperature being the main dependent variable, and the corresponding values of specific volume, internal energy, enthalpy, and entropy. Enthalpy is not shown here, as it can be calculated from its definition ($h = u + Pv$). For cases where a known pressure does not correspond to one of those listed, an intermediate table can be formed using linear interpolations, and then to interpolate within that table. That is, two-point linear interpolations are sometimes necessary.

Ideal gas behavior, discussed more fully in future chapters when the working fluid is sufficiently far from saturation, is characterized by two main characteristics, related to mass and energy. The mass relation is defined by the ideal gas law ($Pv = RT$), an equation of state that relates pressure, volume, and temperature. The energy characteristic is that internal energy is a function of temperature only ($u_{(T)} = u_g$ @ T), similar to single-phase processes with solids or liquids. While there are different definitions in the literature, a ***thermally perfect gas*** is one that adheres to both of these characteristics. A ***calorically perfect gas*** obeys the ideal gas law, but is more restrictive on the energy side in that internal energy is a linear function of temperature (i.e., constant specific heats). This approximation simplifies analysis and can often be fully justified. Any deviation from these behaviors of a working fluid in the gas phase is called a ***real gas*** behavior. The behavior of superheated H_2O will be assessed in this regard in the following. While superheated vapors will be shown to adhere to ideal gas behavior over wide areas, there can be considerable quantitative deviation in some regions. Therefore, the accepted practice is to use tables of properties that pertain to the superheated vapor region when they are readily available.

It is instructive to consider several thermodynamic plots of superheated vapor data obtained from superheat tables. These plots are constructed to help identify conditions for which deviations from ideal gas behavior are significant. All the following plots span the entire range of conditions available in the particular set of tables used, which includes conditions that span both the critical temperature and critical pressure.

Deviation from the ideal gas law is indicated by the compressibility factor $Z = Pv/(RT)$, and the dependency of internal energy on pressure. Regarding the ideal gas law ($Z = 1$), Fig. 4.16 shows the specific volume as a function of pressure, for several fixed values of reduced temperature (defined as the temperature normalized by the critical temperature, in absolute units). The ideal gas law predictions are shown as dotted lines, and the markers correspond to values from the superheat tables. The specific volume of an ideal gas behaves inversely with pressure for a constant temperature, which plots as a straight line (with slope − 1) on log-log axes.

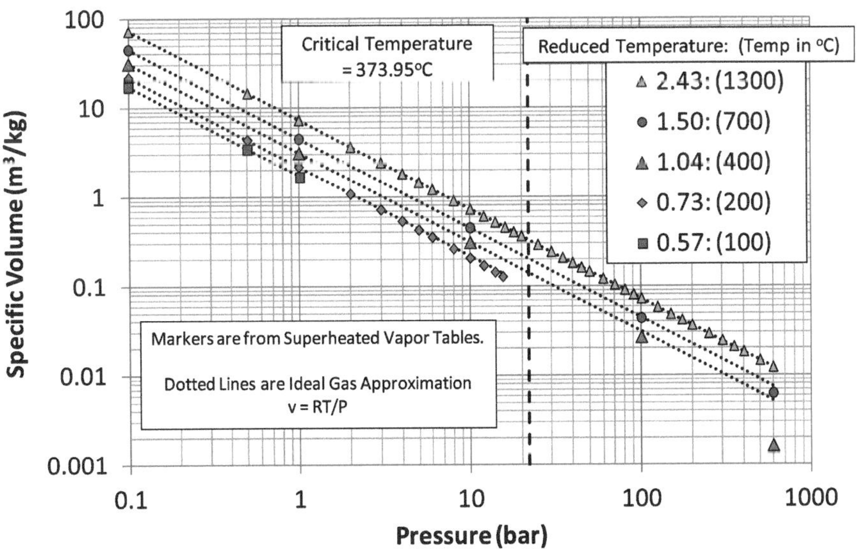

Fig. 4.16 PvT behavior of superheated H2O vapor (log-log axes). Various isotherms are shown for fixed reduced temperature (temperature normalized by the critical temperature, in absolute units), some below and some above the critical temperature. The critical pressure is shown as a vertical dashed line

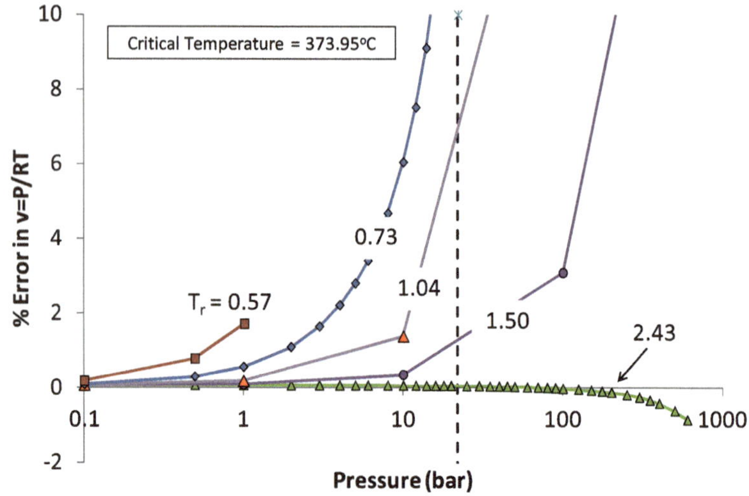

Fig. 4.17 Error in ideal gas approximation for superheated vapors. Percent error is plotted against pressure for several values of reduced temperature. The critical pressure (22.064 bar) is shown as a vertical dashed line. Significant deviation from the ideal gas law occurs as saturated vapor is approached

Ideal gas behavior is closely followed in most regions, with deviations demonstrated in Fig. 4.17, which shows the corresponding percent error of the ideal gas law. Substantial deviations from ideal gas behavior occur as the saturated vapor curve is approached. For reduced temperatures below approximately 1.5, the ideal gas law overestimates the actual specific volume; that is, the compressibility factor Z is less than unity. That behavior is a reflection of intermolecular forces tending to pull molecules together. When the reduced temperature is approximately 2 or greater, pressures much greater than the critical pressure result in the ideal gas law underestimating the volume ($Z > 1$). This effect is associated with molecular crowding, where the volume occupied by the molecules themselves occupies a significant portion of the total volume.

The effect of pressure on the internal energy is shown in Fig. 4.18 for several temperature values. The internal energy is effectively independent of pressure for sufficiently low reduced pressures (i.e., ideal gas behavior). Deviations are significant only for cases where both the reduced pressure is much greater than unity, and the reduced temperature is in the vicinity of unity. It is noteworthy that even with temperatures below the critical temperature (100 °C and 200 °C), the internal energy is independent of pressure, right up to the point of saturation (the far right point for cases above the critical temperature).

A different viewpoint of the effect of pressure on the internal energy effects is shown in Fig. 4.19, in which the internal energy is plotted against temperature for several fixed pressures. For reduced pressures below unity, the internal energy is independent of pressure and increases monotonically with temperature. The internal energy is significantly affected by pressure for reduced pressures above 4.5.

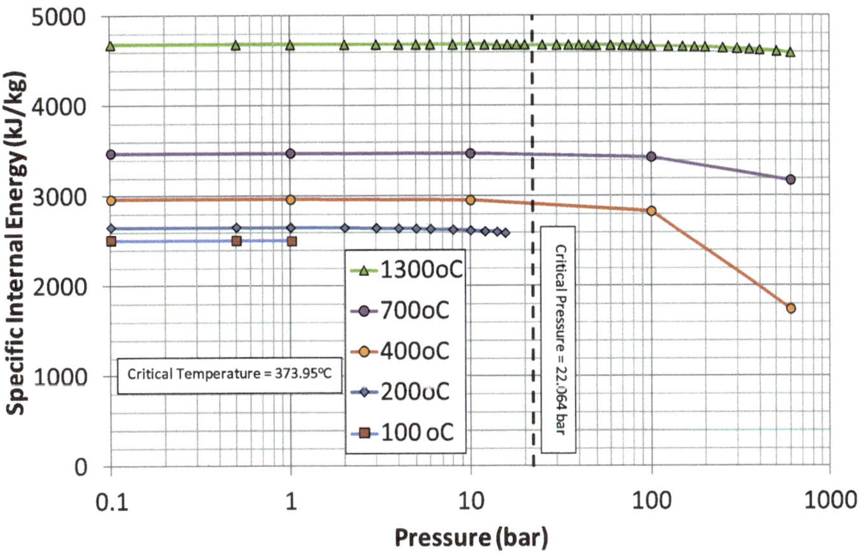

Fig. 4.18 Effect of pressure on internal energy of H_2O superheated vapor. Various isotherms are shown. The critical pressure is shown as a vertical dashed line

To test whether the superheated vapor is calorically perfect or not (i.e., whether the specific heat is a constant), it is tempting to observe the slight curvature of each curve in Fig. 4.19. The specific heat at constant volume is defined as $c_v = \partial u/\partial T|_v$. However, these are isobaric lines (constant pressure), not isochoric lines (constant specific volume). Therefore, Fig. 4.20 shows the first derivative of the enthalpy $c_P = \partial h/\partial T|_P$ for these curves. The derivative is determined as a simple central difference approximation. For pressures for which superheated vapor behaves as an ideal gas to a good approximation (reduced pressures below 0.5), the specific heat is a mildly increasing function of temperature. These vapors are not strictly calorically perfect, but can be treated as such with a suitably chosen average value. It is of interest to note that the value of the specific heat of H_2O superheated vapor is approximately half that of H_2O compressed liquids.

In summary, a general rule of thumb as to whether to use superheated tables or ideal gas approximations is that if, for a given application, the use of the saturation tables is needed at any point, then it is probably best to use superheat tables as well. If a problem involves vapor at some point, but never approaches the steam dome, or if the reduced pressure is always less than 0.5 (11 bar for H_2O), then either tables or ideal gas formulas could be used. Another rule of thumb is that if the conditions under investigation fall within the range of available superheat tables, then use the tables (not the ideal gas formulas). The same reference point must be used if a combination of both tables and formulas is used.

If the pressure of a desired state does not correspond with a tabulated pressure, a 2 point interpolation is required. There are two methods for doing so; a brute force 2

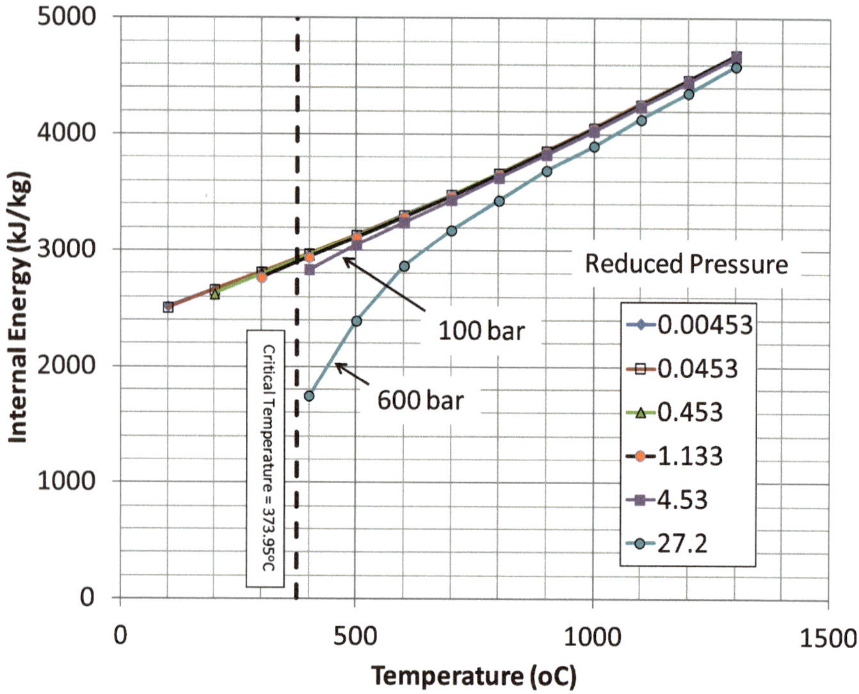

Fig. 4.19 Effect of temperature on internal energy. Various isobars are shown as reduced pressure. (Critical pressure = 221 bar)

point interpolation method, and a more elegant method that uses ideal gas models and a nearby tabulated set of data. These methods are discussed in detail when they arise in applications in future chapters.

Electric Heater Results

These principles are all put into practice in completing the electric heater example. Results of a spreadsheet implementation are shown in Fig. 4.21. The top section contains input parameters from the problem statement. A short block of derived parameters is used. Note that the mass of the piston needed to produce a pressure of 10,000 kPa (= 100 bar) in a 10 cm piston is enormous, almost 8000 kg, or some 20 tons. This example helps build an appreciation of the magnitude of forces associated with pressurized fluids.

The series of events for the electric heater problem are shown on a phase diagram in Fig. 4.22. The phase boundary separates liquid and vapor phases, and is like a cliff for the thermodynamic properties specific volume, internal energy, and entropy,

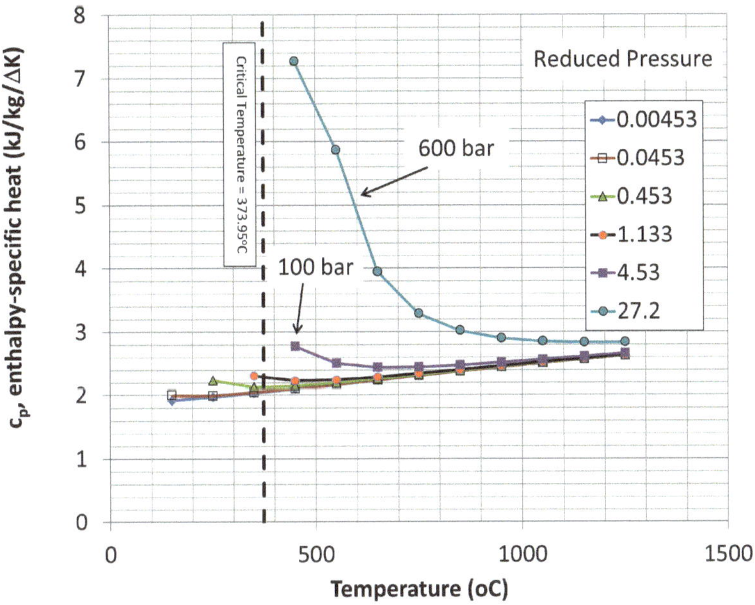

Fig. 4.20 Effect of temperature on enthalpy specific heat (c_p) at various pressures. The specific heat values were obtained from a central difference approximations from the superheat tables

which all change discontinuously as the boundary is crossed. Each intermediate state is defined by a specific event, and the arrows show the process direction.

It is instructive to view property relations and process paths on various plots of thermodynamic properties. Three common ones are temperature versus volume (*Tv*, shown in Fig. 4.23), pressure versus volume (*Pv* in Fig. 4.24), and temperature versus entropy (*Ts* in Fig. 4.25). These plots are shown for H_2O, with the two saturation lines (liquid and vapor) that separate regions of a liquid/vapor mixture from superheated vapor. Compressed liquids lie on the saturation line for *Tv* and *Ts*, and there is a compact compressed liquid region in *Pv* space. The saturation lines define a so-called steam dome in all 3 plots. The term steam dome is commonly used for other substances in practice. Any state under the dome contains a mixture of saturated liquid and saturated vapor. In practice, the phases tend to separate due to their very different densities. However, in some circumstances, particularly as a vapor condenses, a mist or fog may form, with droplets of liquid distributed around a vapor. It is also possible for the mixture to not be in equilibrium, with either a liquid at a temperature above, or a vapor below, the saturation temperature. These nonequilibrium effects, sometimes called supersaturation, are more likely for rapid processes and require a more advanced theory than is attempted here.

The initial state (*a*) is well within the compressed liquid region in the PT phase diagram. The liquid is initially at the normal boiling point, but the system is pressurized, so $P_a > P_{SAT}(T_a)$, the criterion for compressed liquids. Therefore, in

INPUT PARAMETERS

M	1	kg, mass of water
D	0.1	m, diameter of cylinder
P	10000	kPa, pressure maintained on water
Tdead	20	oC, dead state and initial temperature
Pinf	101.3	kPa, ambient pressure
We	1	kW, electrical input power

DERIVED PARAMETERS

Acylinder	0.007854	m^2, cross-sectional area of cylinder
Mpiston	7925	kg, mass of piston weight

STATE TABLE

state	*T(oC)*	*v (m3/kg)*	*u (kJ/kg)*	*s (kJ/kg/K)*	*x,quality*	*H (m)*	*Time (min)*
a	**99.61**	0.001043	417.4	1.3028	NA	0.133	0.0
b	311	0.001452	1393.3	3.3603	**0**	0.185	16.3
c	311	0.002904	1494.2	3.5579	0.0876	0.370	18.3
d	311	0.018028	2545.2	5.6159	**1**	2.295	38.3
e	**500**	0.032811	3047.0	6.5995	NA	4.178	49.1

PROCESS TABLE

Process	*ΔU (kJ)*	*Q_{in} (kJ)*	*$W_{e,in}$ (kJ)*	*$W_{b,out}$ (kJ)*	*eX_D (kJ)*	*ΔPE(kJ)*
a-b	975.9	0	980.0	4.09	602.8	4.0
b-d	1151.9	0	1317.7	165.76	660.9	164.1
d-e	501.8	0	649.6	147.83	288.2	146.3
total	2629.6	0.0	2947.3	317.7	1551.9	314.5

RESULTS

η	0.107	1st Law Efficiency, potential energy of piston over input electrical energy
We,min	1395.3	kJ, minimum theoretical electrical input to heat pump
Qinf	1551.9	kJ, heat pulled from environment by Heat Pump
We,min/We	0.473	a 2nd Law Efficiency

Fig. 4.21 Spreadsheet solution of Electric Heater problem. The system is defined to be the H_2O plus the heating element (whose thermal mass is neglected). Therefore, work is added to the system, not heat

the first process (a-b), called preheating, the rise in internal energy of the working fluid (due to the added energy in the form of electrical work) will be evidenced by a rise in temperature. The internal energy depends on the temperature and is independent of pressure for liquids. The volume will change slightly during this process (and the boundary work will be negligible in comparison with the electrical work). The working fluid appears to move along the saturated liquid line in the *Tv* and *Ts* diagrams, but the fluid is not saturated. The saturated liquid line is like a cliff for pressure with these coordinates. Eventually, the working fluid will reach the phase boundary, and become a saturated liquid at state *b*. Further increase in internal energy will be evidenced by a gradual change of phase, that is, boiling, and the temperature will remain at the saturation temperature that corresponds to the fixed pressure.

The quality (the mass fraction of vapor in a liquid/vapor mixture) is relevant between states *b* and *d*. The quality at state *b* is 0. The term quality pertains only to states that lie on the phase boundary, and therefore is listed as not applicable (N.A.)

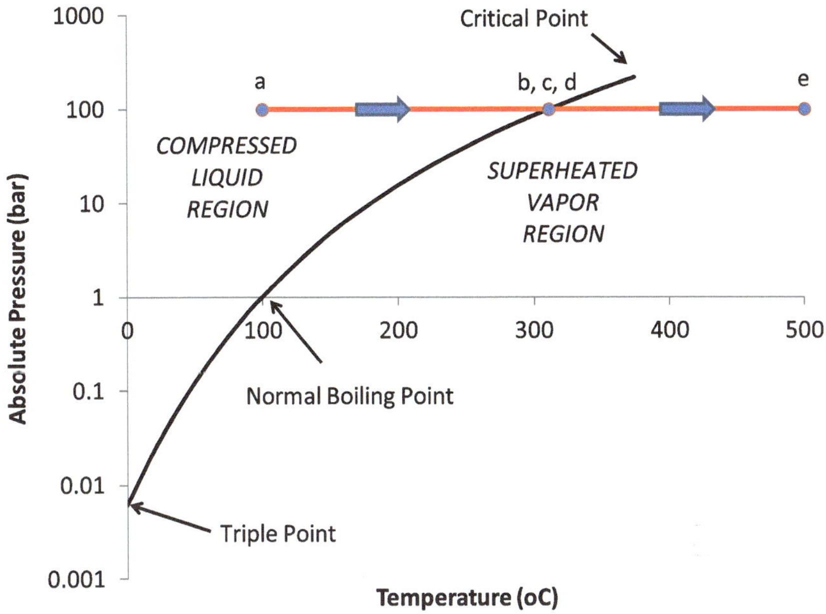

Fig. 4.22 Phase diagram for electric heating problem

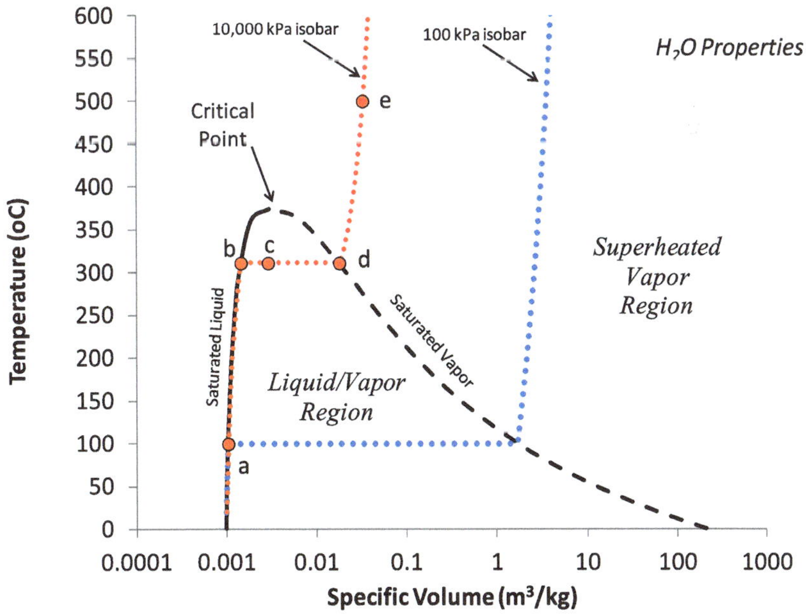

Fig. 4.23 Temperature/Volume curves for H2O on semilogx axes. Two isobars (100 kPa and 10,000 kPa) are shown as dotted lines, as are the points corresponding to the electric heater example

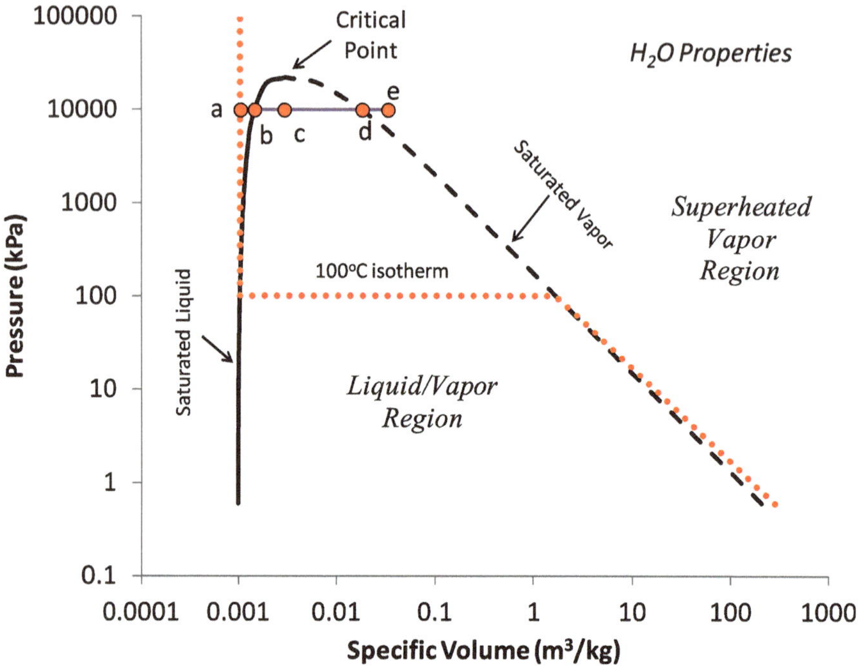

Fig. 4.24 Pressure/Volume curves for H_2O on log-log axes. An isotherm (100 °C) is shown as a red dotted line, as are the points corresponding to the electric heater example

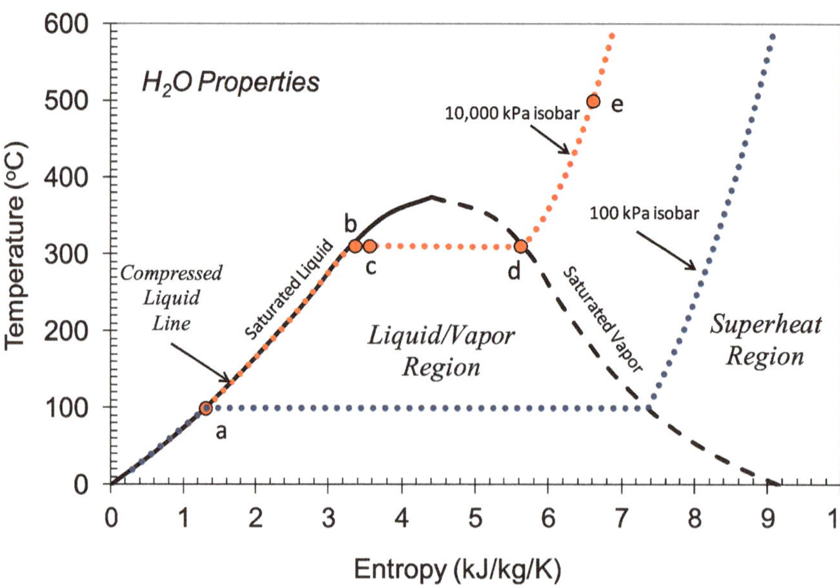

Fig. 4.25 Temperature/entropy curves for H_2O. Two isobars (100 kPa and 10,000 kPa) are shown as dotted lines, as are the points corresponding to the electric heater example

at state *a*, and throughout the preheating process. Once state *b* is reached, further energy addition will cause a change in the state as measured by the quality until phase change is complete, at state *d*. The pressure is fixed by the specification of the process, and the temperature is fixed by its relationship to pressure for liquid/vapor mixtures.

State *c* is defined (somewhat arbitrarily) as a state for which the volume of the working fluid is twice the volume at state *b*. Recall that the quality was evaluated to be 0.0876 in the previous section, determined from the specification of the volume change. Once the quality is known, internal energy and entropy can be determined. As energy is continually added, a gradual conversion from liquid to vapor will occur, along with an increase in quality, until eventually the working fluid will be a saturated vapor with a quality of 1 (state *d*). States *b*, *c*, and d all lie on the same point on a phase diagram, but the working fluid has very different average properties (specific volume, internal energy, and entropy). State *b* is like a piggy bank with all pennies, and state *d* has all quarters. The intermediate state is a piggy bank with both pennies and quarters.

Further electrical energy addition (past state d) results in an increase in internal energy and more boundary work. The working fluid is completely vapor and, therefore, there is a rise in temperature, and the quality is not applicable (reported as NA in Fig. 4.21). This vapor does not strictly behave as an ideal gas, meaning that the compressibility factor is generally not equal to unity and the internal energy is a mild function of pressure. Therefore, the superheat tables are used to determine properties in this region (with linear interpolations often required).

The state table of Fig. 4.21 contains property values obtained from the saturation tables (states *a, b, c*, and *d*) and the superheat tables (state *e*). The quality is listed as "NA" when the state is not a mixed liquid/vapor state. A common conceptual error is to list the quality of state "*a*" as 0, and the quality of state "*e*" as 1. If, for example, in a state table, a quality of 0 is entered, that means that the state is a saturated liquid. State "a" has a much higher pressure than the saturation pressure associated with the input temperature of 20 °C, and therefore has much different properties than a saturated liquid at that temperature. Adding energy (heat or work) to a saturated liquid will commence boiling without changing the temperature, while adding energy to a compressed liquid will raise its temperature.

The height of the piston is obtained simply from the total volume divided by the cross-sectional area of the piston: $H = M_{H2O}v/A$. The elapsed time is obtained from the total electrical energy input (from the process table) divided by the input power: $t = W_e/\dot{W}_e$

The process table, which shows the boiling step as a single process (b-d), shows all the energy terms in the first law balance, namely,

$$\Delta U_{H2O} = Q_{STRAY,in} + W_{e,IN} - W_{b,OUT},$$

as well as the eXergy destruction associated with each process:

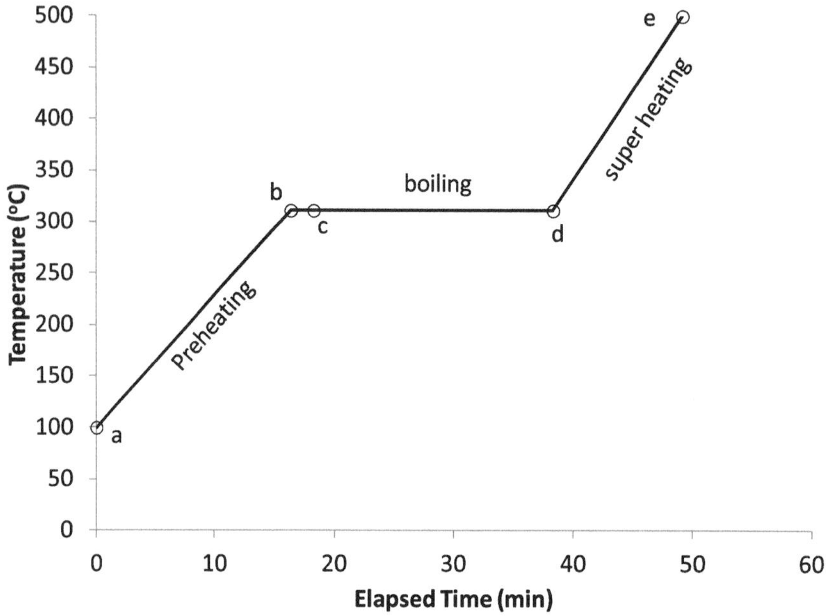

Fig. 4.26 Temperature vs. time for electrical heating

$$eX_D = T_\infty \Sigma = T_\infty \Delta S_{\mathrm{H2O}} + (T_\infty \Delta S_{\mathrm{wire}} + Q_{\mathrm{STRAY}}(T_\infty / T_b))$$

The terms in parentheses (wire and stray heat effects) are neglected, but shown as a reminder of some of the inherent simplifications made, typical of any real engineering analysis.

A change in potential energy of the piston mass is included in the process table:

$$\Delta PE = M_{\mathrm{piston}} gH = WH$$

The change in piston potential energy differs slightly from the boundary work done by the working fluid because work is done by the piston on the atmosphere (which in this case is small in magnitude compared to the work done by the high-pressure fluid).

These results are plotted in Fig. 4.26 (temperature vs. time), Fig. 4.27 (height of piston vs. time), and Fig. 4.28 (energy terms vs. time). The time axis is directly proportional to the electrical energy added. The energy required to bring the compressed liquid to its boiling point is almost equal to the energy required to boil it (a coincidence). For lower pressures, the preheating energy would be less, and the boiling energy greater.

In contrast, the change in height is very small for the preheating (a slight expansion as the specific volume is a weak function of temperature), and then

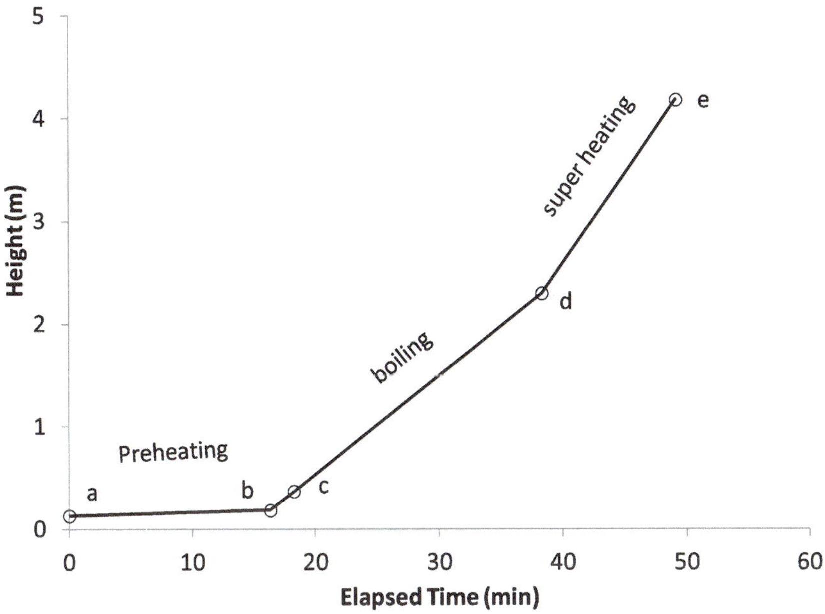

Fig. 4.27 Height of bottom piston face vs. time for electric water heater

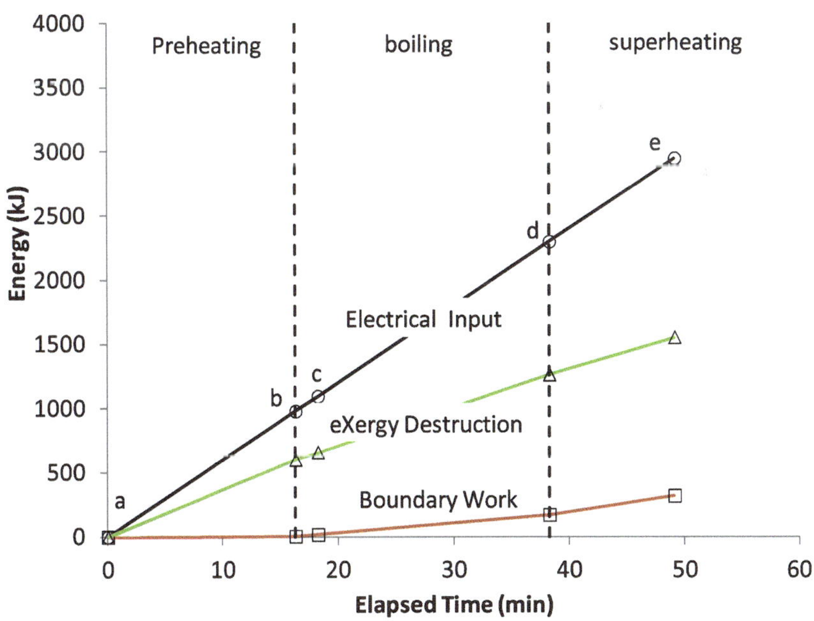

Fig. 4.28 Energy terms in electric heating problem

increases significantly during boiling and superheating, with a total rise of the heavy weight of approximately 4 m.

Finally, input electrical energy is much larger than boundary work done by the working fluid. Most of the input work is used to raise the internal energy of the working fluid. A significant eXergy destruction is associated with the internal electrical heating process.

The 1st Law efficiency, defined as the ratio of the potential energy gained divided by the input electrical work, is a paltry 10.6%. That means that almost 90% of the input power goes to raising the internal energy of the steam, while 10% goes to raising the weight. At the same time, the water is left at a higher entropy state and, therefore, there is a significant eXergy destruction associated with this adiabatic process.

The choice of system to include both the H_2O and the heating element was deliberately made to facilitate both a 1st and 2nd Law analysis. The device as such should really be called an "Electric Worker" instead of an "Electric Heater." If just the H_2O were chosen as the system, then the only work would be boundary work as it expands, and there would indeed be heat exchange between this system and its environment, that is, between the heating element and the H_2O. The 1st Law analysis would be simply done. However, the entropy transfer could not be evaluated easily, because the boundary temperature is not well characterized and it changes with time. Therefore, the full 2nd Law analysis would not be easy. A good exercise is to complete the state and process tables of Fig. 4.21 as much as possible for the case in which the system is defined as just the H_2O.

Since there is so much eXergy destruction, how, in theory, could input electricity be put to better use, thermodynamically?

Deeper Dives

The main objective of this chapter has been met, namely, to explain the use of property tables. The remainder of the chapter reinforces important principles developed to this point.

Raising the Weight with a Heat Pump

The source of all the eXergy destruction in this example is the internal heat transfer between the heating element and the water. By now, it should not be surprising that using an electrical heating element is easy to do in practice, but it comes with a thermodynamic loss of available energy (eXergy). In principle, the electricity could be used to drive a heat pump (Fig. 4.29) rather than directly converting the electrical energy into thermal energy in the heating element. The heat pump would extract heat from the ambient environment and pump it into the water. There would still be an

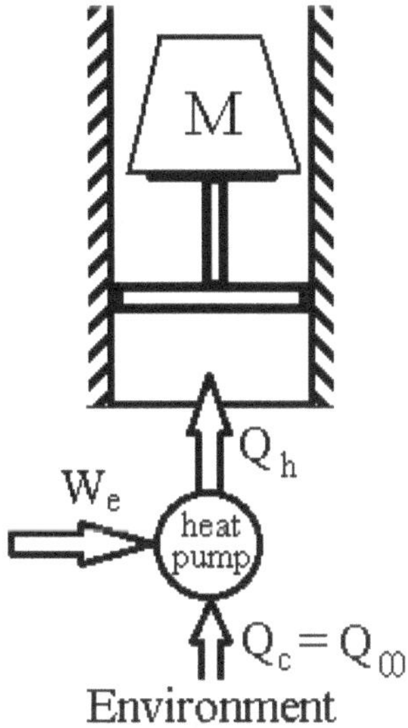

Fig. 4.29 Schematic of piston/cylinder with a heat pump delivering heat to working fluid

increase in entropy of the water, but there would also be entropy transferred into the combined system with the heat absorbed from the environment. The electrical energy input needed to raise the water temperature would be lower than a straight electric heater.

Taking the heat pump, water, walls, and piston as a combined system, the energy and entropy balances are as follows:

$$\Delta U_{\mathrm{H2O}} + \Delta PE_{\mathrm{piston}} = Q_{\infty} + W_e - W_{\infty}$$

$$\Delta S_{\mathrm{H2O}} = \frac{Q_{\infty}}{T_{\infty}} + \Sigma$$

In these balance equations, Q_{∞} represents the heat added to the heat pump from the ambient environment, and W_{∞} is the work against the environment, as before. Combining the energy and entropy balances by eliminating Q_{∞}:

$$\Delta U_{\mathrm{H2O}} + \Delta PE_{\mathrm{piston}} = T_{\infty}\Delta S_{\mathrm{H2O}} - T_{\infty}\Sigma + W_e - P_{\infty}\Delta V_{\mathrm{H2O}}$$

Rearranging for the input electrical work required:

$$W_e = \Delta PE_{\mathrm{piston}} + [\Delta U + P_{\infty}\Delta V - T_{\infty}\Delta S]_{\mathrm{H2O}} + T_{\infty}\Sigma$$

The term in brackets is a form of eXergy change of the water for an isobaric process, which includes the work of expansion. The entropy generation (eXergy destruction) for this combined system equals that in the water plus that which occurs in the heat pump. The minimum theoretical input work required to raise the weight (and bring the working fluid to state d) is obtained by setting the entropy generation (or eXergy destruction) to zero:

$$W_{e,\mathrm{MIN}} = \Delta PE_{\mathrm{piston}} + [\Delta U + P_{\infty}\Delta V - T_{\infty}\Delta S]_{\mathrm{H2O}}$$

Evaluating for this case, taking relevant numbers from the straight electric heater example:

$$W_{e,\mathrm{MIN}} = (314.5)_{\mathrm{piston}} + [2629.6 + 3.2 - 1551.9]_{\mathrm{H2O}}$$

$$W_{e,\mathrm{MIN}} = 1395.3kJ$$

That is a rather dramatic reduction compared to the 2947 kJ required of the straight electric heater. Note that the last term (1551.9 kJ, the entropy change) is equal to the heat that is pulled from the environment by the heat pump, since the boundary temperature where heat enters the heat pump equals the dead state temperature. The boundary work term ($P_{\infty}\Delta V$) is there, but small.

A 2nd Law efficiency for the straight electric heater, defined as the minimum work required divided by the actual work with the electric heater, is as follows:

$$\varepsilon = \frac{W_{e,\mathrm{MIN}}}{W_e} = \frac{1395.3kJ}{2947.3kJ} = 0.473$$

It is impractical to think that a conventional vapor/compression refrigeration cycle (the basis of household refrigeration) as a heat pump would be used to heat a liter of water in a kitchen appliance. However, it is conceivable that thermoelectric modules (based on the Peltier effect) could be designed for that. The heating process would be slower, in all likelihood, than a direct heating element. For industrial scale processes, however, it is likely that there are opportunities for the replacement of straight electric heating processes with some form of heat pump.

Heating in an Oven

Another scenario is to obtain the heating of the working fluid by simply placing the piston/cylinder into an oven set at the final desired temperature (without regard to how the oven conditions are maintained). The energy and entropy balance equations (for H_2O as the system, and neglecting secondary effects) are as follows:

$$\Delta U_{\mathrm{H2O}} = Q_{\mathrm{IN}} - W_b = Q_{\mathrm{IN}} - P\Delta V_{\mathrm{H2O}}$$

$$\Delta S_{\mathrm{H2O}} = \frac{Q_{\mathrm{IN}}}{T_{\mathrm{oven}}} + \Sigma$$

Eliminating Q_{IN} (the heat transferred from the oven air to the working fluid) and rearranging for the eXergy destruction:

$$T_{\infty}\Sigma = T_{\infty}\Delta S_{H2O} - \frac{T_{\infty}}{T_{oven}}(\Delta U_{H2O} + P\Delta V_{H2O})$$

$$T_{\infty}\Sigma = 1551.9 - 1117.1 = 434.8kJ$$

The eXergy destruction associated with heating from the oven is approximately a third of the eXergy destruction associated with the electrical heating. However, the energy required to maintain the oven is not considered in this case. The main point is to just point out that there can be different ways in which a thermodynamic objective can be obtained and that these methods have different 2nd Law consequences.

Creating a Cycle

In the electric heater problem, a heavy weight was raised to a higher elevation. The working fluid, the H_2O, began as a liquid at 100 °C (the normal boiling point temperature) at a high pressure (making it a compressed liquid), and ended as a superheated vapor at the same pressure and high temperature. The weight was raised some 4 meters. If the objective of the device is to deliver the weight to the higher elevation and then repeat the process, a cycle is needed that returns the working fluid to its original state. In the example given here, a series of steps are taken to return to the initial state. For this problem, the system is designed to be the working fluid with the electrical wire (of negligible thermal mass) embedded in it.

The cycle consists of 4 sequential processes, as shown in Fig. 4.30 and discussed in Table 4.5. An analysis of each state follows. A spreadsheet solution for a case study is shown in Fig. 4.31. The input and derived parameter sections are the same as for the electric heater/lifter problem. The state and process tables are color-coded and the order of steps taken to complete it is discussed in detail in Table 4.6. The complete cycle is shown in *Pv* and *Tv* diagrams: Figs. 4.32, 4.33, and 4.34. The path taken for each process is shown as a red line.

The constant volume path taken by the first cooling step (2–3) on the *Ts* diagram is not well defined (without a much more detailed analysis of the process path), and is shown as a curved, dashed line. This method is suggested whenever the end points of a process are well-defined, but the precise path taken is not clear. There is a large decrease in entropy during this process. That is not a violation of the 2nd Law, because there is entropy transfer out of the system.

Note that states 4 (at ambient pressure) and state 1 are coincident on *Tv* and *Ts* diagrams, because the compressed liquid region is collapsed onto the saturated liquid line. Most textbooks show a distinct compressed liquid region, but conceptually, that is only reasonable for *Pv* space. States 3 and 4 have only a slight separation in *Ts* space for this cycle.

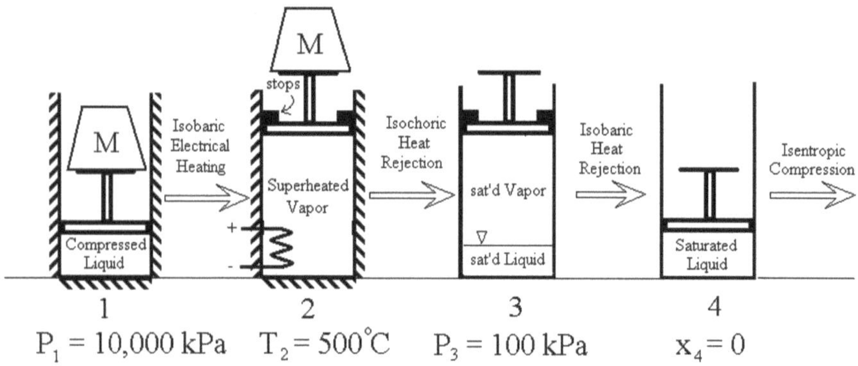

Fig. 4.30 Process diagram of a cycle for an electric heater/lifter. One thermodynamic state variable is specified for each state and one thermodynamic property is held constant during each process, eventually returning the working fluid to its initial state

Table 4.5 Process Summary for Cycle

Process	Type of process	Comments
1–2	Isobaric, adiabatic electrical heating	Electrical work is added to the pressurized water, resulting in the raising of the weight, the desired thermodynamic objective. This process can take considerable time. Stray heat between the cylinder and environment is neglected. The goal of the remaining processes is to return to the initial state so that this process can be repeated.
2–3	Isochoric (constant volume) cooling	The weight is removed from the piston (delivering it to its desired higher elevation). A set of stops prevents further expansion of the working fluid. The insulation is removed, and heat transfer to the ambient (dead state) occurs gradually.
3–4	Isobaric cooling	Once the pressure reaches ambient, continued cooling will result in a decreasing volume. Heat flows from the working fluid to the ambient (at the dead state), and is considered complete when all vapor is completely condensed.
4–1	Isentropic (adiabatic and reversible) compression	A weight is placed on the piston, returning the system to its initial state, thereby completing the cycle. This process takes place quickly, as the weight is simply placed on the platform, and the working fluid is thereby pressurized, with no work being done.

Process 1–2: Isobaric Electrical Heating

This process was studied in detail. However, for an overall thermodynamic analysis, the intermediate steps shown in the first example are not necessary to calculate overall changes. They are necessary to show the process paths on thermodynamic

INPUT PARAMETERS

MH2O	1	kg, mass of water
D	0.1	m, diameter of cylinder
P	10000	kPa, pressure maintained on water
Tdead	20	oC, dead state and initial temperature
Pinf	101.3	kPa, ambient pressure
We	1	kW, electrical input power

DERIVED PARAMETERS

Acylinder	0.007854	m^2, cross-sectional area of cylinder
Mpiston	7925	kg, mass of piston weight

STATE TABLE

state	P (kPa)	T(oC)	v (m3/kg)	u (kJ/kg)	s (kJ/kg/K)	x,quality
1	10000	99.61	0.001043	417.4	1.3028	NA
2	10000	500	0.032811	3047.0	6.5995	NA
3	100	99.61	0.032811	456.6	1.4164	0.0188
4	100	99.61	0.001043	417.4	1.3028	0

COLOR CODE
INPUT PARAMETERS
Process Specification
Direct entry from tables
quality calculation

PROCESS TABLE (all quantities in kJ)

	1st Law Quantities				2nd Law Quantities		
Process	ΔU	Q_{in}	$W_{e,in}$	$W_{b,out}$	$T_\infty \Delta S$	$T_\infty Q_{in}/T_b$	X_D
1-2	2629.6	0	2947.3	317.7	1551.9	0.0	1551.9
2-3	-2590.4	-2590.4	0	0	-1518.6	-2590.4	1071.8
3-4	-39.2	-42.4	0	-3.2	-33.3	-42.4	9.1
4-1	0.0	0	0	0	0.0	0.0	0.0
CYCLE	0.0	-2632.8	2947.3	314.5	0.0	-2632.8	2632.8

COLOR CODE
Process Specification
From State Table
From Energy Balance
From Entropy Balance

Fig. 4.31 Spreadsheet solution for cycle case study. The state and process tables are color-coded to indicate what is needed to fill each cell

Table 4.6 Steps taken to complete state table

step	objective	Explanation of state table calculation
1	Process Spec	P_2 and P_4 entered from the two isobaric process specifications. The other process specifications cannot be implemented before other steps are taken.
2	State 2	The temperature at state 2 (500 °C) is higher than the saturation temperature (311 °C) for the corresponding pressure (10,000 kPa). Therefore, state 2 is a superheated vapor (quality = N.A.), and properties for states 2 can be directly entered from superheat tables. No interpolations are required but are often required for other cases.
3	State 4	State 4 is specified to be a saturated liquid and the pressure has been determined to be 100 kPa from the process specification. Therefore, all properties for state 4 can be directly entered from the saturation pressure table.
4	State 1	s_1 is entered from the isentropic process specification for process 4-1. The entropy (1.3028 kJ/kg/K) is lower than the saturated liquid value (3.36 kJ/kg/K) at the corresponding pressure (10,000 kPa) and, therefore, state 1 is a compressed liquid (specific volume, internal energy, and entropy are all functions of temperature only). Therefore, the only property that changes in process 4-1 is the pressure, and all other values can be entered for state 1 from state 4.
5	State 3	v_3 is entered from the process specification (isochoric heat rejection). The specific volume (0.032811 m^3/kg) is between the saturated liquid (0.001043 m^3/kg) and vapor values (1.6941 m^3/kg) at the corresponding pressure (100 kPa). Therefore, the temperature is the corresponding saturation temperature. The quality is calculated, followed by the internal energy and entropy. These calculations can be done directly into the cells of the worksheet, referring to the saturation tables as needed.

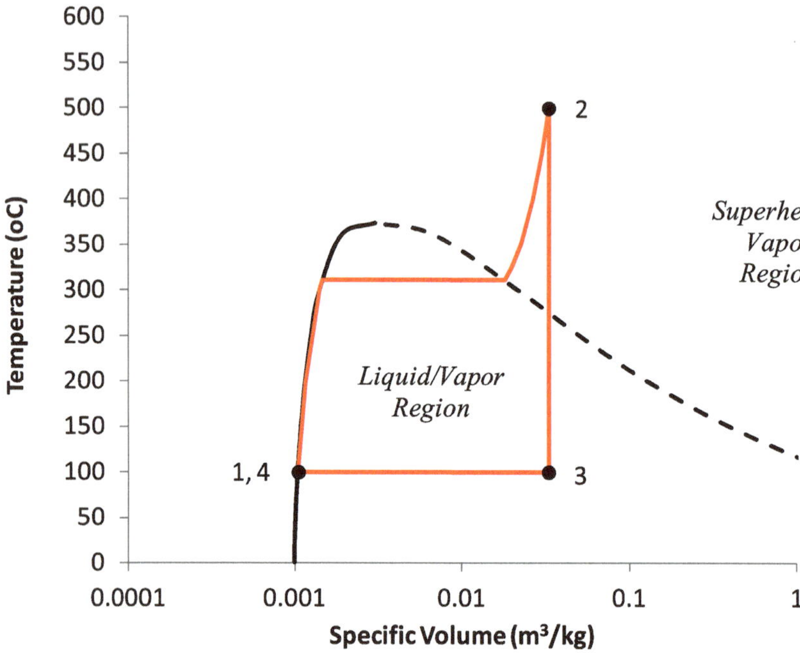

Fig. 4.32 T-V diagram of cycle

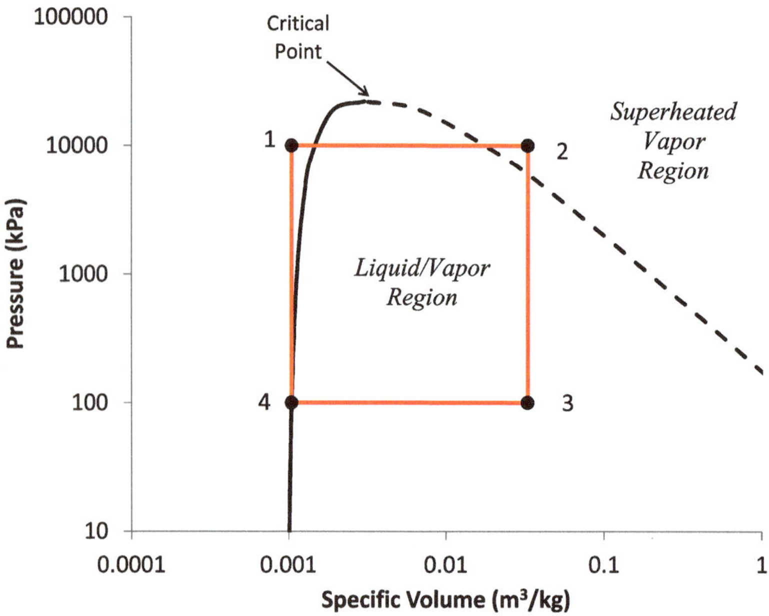

Fig. 4.33 P-V diagram of cycle

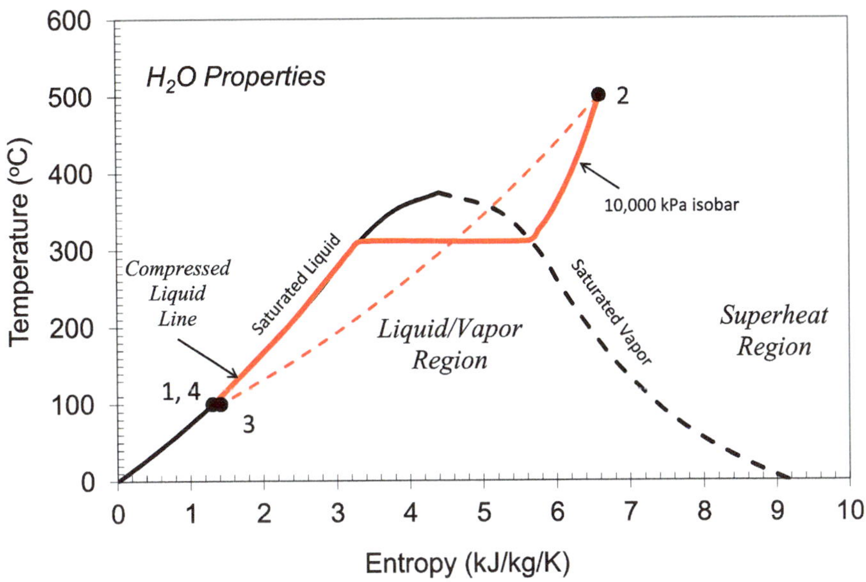

Fig. 4.34 Ts diagram of cycle. The precise path of process 2–3 is not well defined and is shown as a curved, dashed line

coordinates (Figs. 4.32, 4.33, and 4.34). The energy balance, neglecting potential and kinetic energy changes, and with secondary effects shown in brackets, is as follows:

$$\Delta U_{12} + [\Delta U_{\text{wire}}] = \left[-Q_{\text{STRAY},12}\right] + W_{e,12} - P_1 \Delta V_{12}$$

Rearranging for the input electrical work, neglecting secondary effects:

$$W_{e,12} = M_{\text{H2O}}((u_2 - u_1) + P_1(v_2 - v_1))$$

The changes in state properties (internal energy and volume) can be calculated by entering appropriate formulas into the spreadsheet that refer directly to the state table.

$$\Delta S_{12} + [\Delta S_{\text{wire}}] = \left[\frac{-Q_{\text{STRAY},12}}{T_b}\right] + \Sigma_{12}$$

The eXergy destruction is, neglecting secondary effects, as follows:

$$eX_{D,12} = T_\infty \Sigma_{12} = T_\infty M_{\text{H2O}}(s_2 - s_1) = 1551 kJ$$

This large irreversibility is associated with the electrical heating process. As discussed, the cycle could be improved by using a heat pump (Fig. 4.29) instead of direct electrical heating, a good exercise to practice.

Process 2–3: Isochoric Heat Rejection

Prior to this process, the weight is removed (and delivered to its desired elevated state). The force required to maintain the pressure is shifted (with no work) from the weight on the platform to a set of stops on the cylinder wall that prevents further expansion of the high-pressure working fluid. Also, the insulation is removed, and the system then exchanges heat with the environment. The pressure will fall as heat is gradually exchanged. State 3 is under the steam dome. Therefore, the initially superheated vapor will become saturated at some point during the process, and further cooling will result in condensation (which will initiate on the inner walls that must be a bit colder than the working fluid in order for heat to flow). Saturated liquid water will collect on the bottom of the cylinder, and is assumed to be in quasi-equilibrium with the saturated vapor above it. The process ends when the pressure equals ambient (plus the weight per unit area of the unweighted piston, which is neglected). Since there is no work, the energy balance is simply:

$$\Delta U_{23} = M_{\mathrm{H2O}}(u_3 - u_2) = Q_{\mathrm{IN},23}$$

Numerically, the heat addition evaluates to −2590.4 kJ.

The formula for the internal energy change can be entered into the spreadsheet (referencing the state table), and then the 1st Law dictates the heat transferred. Notice that the internal energy decreases and, therefore, the heat in is negative. The entropy balance is as follows:

$$\Delta S_{23} = M_{\mathrm{H2O}}(s_3 - s_2) = \frac{Q_{\mathrm{IN},23}}{T_{B,23}} + \Sigma_{23}$$

The eXergy destruction is as follows:

$$eX_{D,23} = T_\infty \Sigma_{23} = T_\infty M_{\mathrm{H2O}}(s_3 - s_2) - \frac{T_\infty Q_{\mathrm{IN},23}}{T_{B,23}}$$

In this case, the boundary where heat is exchanged is at the dead state ($T_{B,23} = T_\infty$), and therefore, the temperatures will cancel in the heat transfer term when evaluated. It is left in this expression to make the problem more general. It is possible in other cases for the temperature at the heat exchange boundary to differ from the dead state.

Evaluating numerically for the case study:

$$eX_{D,23} = (298K)\,(1kg)\,(0.8182 - 6.5995)\,kJ/kg/K - \left(\frac{293K}{293K}(-2590.4)kJ\right)$$

$$eX_{D,23} = -1518.6 + 2590.4 = 1071.8kJ$$

In this case, the entropy decreases, but the eXergy transfer associated with the heat rejection is greater in magnitude, so that there is a net eXergy destruction, as there must be a need to satisfy the 2nd Law. This process is identified as having a significant eXergy destruction, and that leads to the question as to why, and what could be done to alter the cycle to reduce it, discussed shortly.

Process 3–4: Isobaric Heat Rejection

The force on the stops decreases as the pressure drops during the isochoric heat rejection (process 2–3). When this force drops to zero, the pressure of the working fluid equals ambient (plus the neglected weight per unit area of the piston). Heat will continue to flow, however, as the temperature is still above ambient. The pressure will remain constant (isobaric), and the volume will decrease accordingly as heat transfer continues. Water will continue to condense, until eventually it is completely condensed, and the cylinder is filled with a saturated liquid at ambient pressure.

The energy balance is as follows:

$$\Delta U_{34} = Q_{\mathrm{IN},34} - W_{B,34}$$

The boundary work is the ambient pressure times the change in volume. Rearranging for the heat input (which will evaluate as a negative value):

$$Q_{\mathrm{IN},34} = M_{\mathrm{H2O}}((u_4 - u_3) - P_\infty(v_4 - v_3))$$

For the case study, this heat transfer evaluates to −42.4 kJ, much less than the heat transfer in the isochoric process.

The entropy balance is as follows:

$$\Delta S_{34} = M_{\mathrm{H2O}}(s_3 - s_2) = \frac{Q_{\mathrm{IN},34}}{T_{B,34}} + \Sigma_{34}$$

The eXergy destruction is as follows:

$$eX_{D,34} = T_\infty \Sigma_{34} = T_\infty M_{\mathrm{H2O}}(s_4 - s_3) - \frac{T_\infty Q_{\mathrm{IN},34}}{T_{B,34}}$$

$$eX_{D,34} = -33.3 + 42.4 = 9.1kJ$$

It is common that the entropy generation (or eXergy destruction) calculations involve terms of opposite sign, but comparable absolute values. Therefore, the resulting value is a number that is small compared to the terms that make it up. It is easy and common to make an error in the sign of one of the terms, resulting in a large indicated eXergy destruction (which could be negative, a sign of either an error, or the problem statement is a perpetual motion machine of the 2nd kind). A negative eXergy destruction is a clear sign of an error, and should initiate a check of the analysis. More subtly, an eXergy destruction that seems large compared to an energy change (for example) is also a clue to trace back and search for an execution or sign error. In addition, this is a case in which what you may have been taught about significant figures needs to be modified. A general rule for thermodynamic calculations is to "CARRY EXTRA SIGNIFICANT FIGURES DURING CALCULATIONS TO MINIMIZE ROUND-OFF ERRORS, AND REPORT 3 OR 4 SIGNIFICANT FIGURES FOR FINAL RESULTS, UNLESS MORE IS CALLED FOR." In this case, the result (9.1 kJ) is small compared to the two terms that make it up (−33.3 and 42.4). Other cases can be much more extreme. For example, a calculation might indicate something like $eX_D = -42.398 + 42.407 = 0.009kJ$. If only 3 significant figures were carried, the result would be $eX_D = -42.4 + 42.4 = 0.00kJ$, which could yield a misleading conclusion and possibly bad engineering design decisions.

In fact, for careful 2nd Law analyses that involve eXergy calculations, it should be recognized that the reference state for the steam tables is generally at 0.01 °C = 273.16 K, not 273 K. A good exercise is to analyze a hypothetical reversible isobaric evaporation process starting at the reference state.

Process 4-1: Adiabatic Compression

The final process that completes the cycle is to load the next weight to be raised onto the platform. Starting with a saturated liquid, adding weight simply pressurizes the working fluid, and it becomes a compressed liquid. The volume change is negligible, so there is negligible work, and the process is assumed to be adiabatic (a good approximation, since it takes place quickly). Therefore, the internal energy change is zero. Since the internal energy is a function of temperature only for compressed liquids, the temperature does not change. Since the entropy is also a function of temperature only, the process is isentropic. Since, then, the process is adiabatic and isentropic, the entropy production (and eXergy destruction) is zero. The only change, but an important one, is the pressure. The *Pv* and *Tv* diagrams look different.

Cycle Modifications

The overall First Law efficiency of the cycle, defined as the net boundary work done by the working fluid (H_2O) on the environment (raising the weight) divided by the input electrical work, evaluates to a paltry 0.107. And that is from an electrical input, which itself is a secondary form of energy. This device is not a high performing thermodynamic device (nor is it intended to be). It does raise a weight, however.

Clues to modifying the cycle to improve the thermodynamic performance come from the eXergy destruction associated with each process. In this case, the largest eXergy destruction occurs during the electrical heating step (1–2). The second largest is the constant volume heat rejection step (2–3). A small eXergy destruction occurs during the final cooling (3–4), and the compression process (4–1) involves no eXergy destruction.

To address the large eXergy destruction during the electrical heating, without fundamentally changing the thermodynamic cycle of the working fluid, the electrical heating element can be replaced by a heat pump (recall Fig. 4.29), a theoretical alternative to improving the thermodynamic performance of this cycle by pulling heat from the environment, and pumping it to the water. It is a much harder device to build, but provides a theoretical means to improve the performance. A good exercise is to modify the spreadsheet used to generate Fig. 4.31 for this case study. The overall cycle is analyzed here.

The energy balance for the combined heat pump and cycle is as follows:

$$\Delta E_{\text{cycle}} = Q_{\infty} - Q_{23} - Q_{34} + W_{e,\text{HP}} - W_{b,\text{NET}} = 0$$

There are three heat transfer terms, one is the heat transfer from the environment to the low side of the heat pump, and the other two are the heat rejections (2–3 and 3–4). The heat transfer from the heat pump to the water is an internal process. There are two forms of work, the electrical input work to the heat pump, and the net boundary work of the working fluid.

The entropy balance for the combined heat pump and cycle is as follows:

$$\Delta S_{\text{cycle}} = \frac{Q_{\infty}}{T_{\infty}} - \frac{Q_{23}}{T_{\infty}} - \frac{Q_{34}}{T_{\infty}} + \Sigma_{\text{HP}} + \Sigma_{23} + \Sigma_{34} - 0$$

Note that all three heat transfers occur with the dead state. Combining, and rearranging, for the net work required by the heat pump:

$$W_{e,HP} = W_{b,NET} + T_{\infty}(\Sigma_{\text{HP}} + \Sigma_{23} + \Sigma_{34}) = W_{b,\text{NET}} + (eX_{D,\text{HP}} + eX_{D,23} + eX_{D,34})$$

A 1st Law efficiency for this process could be defined as the ratio of the net boundary work divided by the input electrical power:

$$\eta_I = \frac{W_{b,\text{net}}}{W_{e,\text{HP}}} = \frac{1}{1 + (eX_{D,HP} + eX_{D,23} + eX_{D,34})/W_{b,\text{net}}}$$

Theoretically, with a perfect heat pump drawing heat from the environment, the eXergy destruction within the heat pump is zero, and

$$\eta_{I,\max} = \frac{W_{b,\mathrm{net}}}{W_{e,\mathrm{HP},\min}} = \frac{1}{1 + (eX_{D,23} + eX_{D,34})/W_{b,\mathrm{net}}}$$

This cycle is has a significantly higher First Law efficiency (22.5%) than the cycle with direct electrical heating (10.7%). However, do not forget that electricity is a secondary form of energy, a theme of this textbook. This example is intended to inform basic thermodynamic principles, and not to suggest that such a device would make a good lifting mechanism. If an electrical input were to be used in practice to raise a weight, it would be best to drive a mechanical device, not a thermal one.

The process with the second highest eXergy destruction rate is the constant volume cooling process (2–3). The entropy transfer in this process is large because heat transfer is allowed to occur across a large temperature difference, particularly during the early stages of the process when the temperature of the system is high. Two possible modifications could be introduced. Theoretically, a heat engine could be designed to extract a portion of the internal energy into useful work. Alternatively, the stops could be removed, and the mass removed gradually (if possible), which would raise more mass to a higher elevation. Of course, that would mean breaking the mass up, and allowing a much larger expansion of the cylinder. In this latter case, the cycle would be a closed mass representation of the Rankine cycle, an important cycle that closely models many power generation cycles (but used with open-system devices, rather than the piston/cylinder used here for pedagogical reasons). A good exercise would be to complete the analysis for these two situations, the second of which is laid out in a detailed Workshop problem statement.

Appendix 1: Saturation Pressure and Temperature Formulas

There are times when a closed form expression for the saturation curve (relationship between saturation temperature and pressure) is useful for an investigation, as opposed to the use of tables. This Appendix develops that functional dependency from thermodynamic principles.

Clausius-Clapeyron Heat Engine

The Clausius-Clapeyron relation is a differential thermodynamic relationship between the saturation pressure and saturation temperature of a substance. It is derived in this section by analyzing a reversible heat engine that spans the steam dome, and operates between two temperature reservoirs that differ by a differential temperature. An integrated form of the resulting differential equation between saturation temperature and pressure yields a practical equation called the integrated Clausius/Clapeyron equation that can be used to relate saturation pressures and

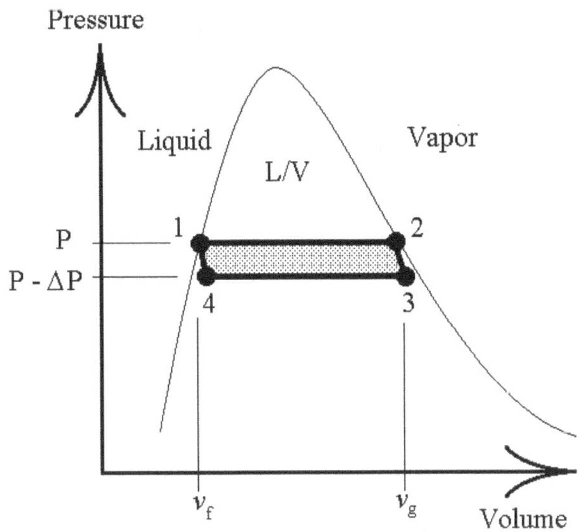

Fig. 4.35 Schematic of PV diagram for the heat engine cycle used to derive the Clausius-Clapeyron relation. Not to scale

temperatures for substances that have limited property data available. For example, the vapor pressure exerted by a liquid fuel at any temperature (below the boiling point) can be estimated given the normal boiling point temperature (i.e., at 1 atm pressure) and the heat of vaporization (at 1 atm) using the integrated Clausius-Clapeyron equation. In practice, the normal boiling point and heat of vaporization at 1 atm may be readily available, but a detailed set of saturation tables may not be.

Consider a piston/cylinder assembly that contains a pure substance that undergoes a cycle between 2 pressure levels, P and P-ΔP where ΔP is a differential change (Fig. 4.35). The pressure is set by placing a weight on the piston, for example. The temperature is set by the saturation conditions. The cycle is shown schematically on a P-v diagram. State 1 is a saturated liquid. Heat is added in the amount Q_H (at constant pressure) from a source at temperature T until the substance is a saturated vapor (state 2). The pressure is reduced by removing mass from the piston. This process is done adiabatically and without friction and is therefore isentropic. The volume increases somewhat, and the fluid enters into the "steam dome." State 3 is at a pressure P-ΔP, and has a high quality (mass fraction of vapor). The temperature of the working fluid changes to a value T-ΔT, which is the saturation temperature that corresponds to the new pressure. Heat is removed in the amount Q_C (in a constant pressure process) to a source at temperature T-ΔT. The working fluid is condensed to a low quality at just the right point (state 4) so that when the weight is then placed back on the cylinder, the substance is compressed isentropically to complete the cycle. The total work of the cycle is $W_{\text{out,cycle}}$ and equals the area within the cycle on a Pv diagram.

The energy balance for the cycle is as follows:

$$\Delta E_{\text{cycle}} = 0 = Q_H - Q_C - W_{\text{out,cycle}}$$

The entropy balance for the cycle is as follows:

$$\Delta S_{\text{cycle}} = 0 = \frac{Q_H}{T} - \frac{Q_C}{T - \Delta T} + \Sigma,$$

where Σ is the entropy generation (irreversibilities) which are zero for this reversible cycle.

Combining the 1st and 2nd Laws by eliminating $Q_{C,}$

$$0 = Q_H - Q_H\left(\frac{T - \Delta T}{T}\right) - W_{\text{out,cycle}}$$

Rearranging,

$$Q_H \frac{\Delta T}{T} = W_{\text{out,cycle}}$$

Expressions for Q_H and $W_{\text{out,cycle}}$ are needed. Analyzing process 1–2:

$$E_2 - E_1 = Q_H - \int_{V_1}^{V_2} PdV$$

Neglecting kinetic and potential energy changes (E→U), and recognizing that P remains constant:

$$Q_H = m[u_2 - u_1] + P(V_2 - V_1) = m[(u_2 + P_2 v_2) - (u_1 + P_1 v_1)] = m(h_2 - h_1)$$

Since state 1 is a saturated liquid and state 2 a saturated vapor, $h_2 - h_l = h_{fg}$, the enthalpy of vaporization (at pressure P).

That is, $Q_H = mh_{fg}$

The work of the cycle is the area under the curve.

$$W_{\text{out,cycle}} = \oint_{\text{cycle}} PdV \approx \Delta P(V_2 - V_1) = \Delta Pm(v_2 - v_1) = \Delta P\left(mv_{fg}\right)$$

This expression is not exact, as the small, nearly triangular areas between 1 and 4 and 2 and 3 do not exactly cancel. This error decreases as ΔP shrinks to zero. Inserting Q_H and $W_{\text{out,cycle}}$ into the combined 1st and 2nd Laws, and cancelling the masses:

$$h_{fg}\frac{\Delta T}{T} \approx v_{fg}\Delta P$$

Rearranging yields the Clausius-Clapeyron equation, which relates the change in saturation pressure and temperature to the heat of vaporization and volume change associated with changing phase. The analysis was done for liquid/vapor phase change, but it holds for any phase change (solid to liquid, solid to vapor, solid crystal form to another solid crystal form):

$$\frac{\Delta P}{\Delta T} = \frac{h_{fg}}{v_{fg}T}$$

This differential Clausius-Clapeyron equation can be integrated once h_{fg} and v_{fg} are expressed in terms of temperature and pressure. A simple approximate integrated form is obtained by recognizing that the vapor volume is much greater than the liquid (valid if the temperature is "much less" than the critical point), and treating the vapor as an ideal gas ($Pv = RT$), where the gas constant is specific to the substance ($R = R_u/$ MW = (8314 J/kmol/K)/MW:

$$v_{fg} = v_g - v_f \approx v_g \approx \frac{P}{RT}$$

Inserting into the Clausius-Clapeyron relation and separating variables:

$$\frac{dP}{P} = \frac{h_{fg}dT}{RT^2}$$

The heat of vaporization is a weak function of temperature. Treating it as a constant and integrating between 2 states (A and B):

$$\ln\left(\frac{P_A}{P_B}\right) = \frac{-h_{fg}}{R}\left(\frac{1}{T_B} - \frac{1}{T_A}\right)$$

The value of h_{fg} could be taken to be an average between the two states if that data were readily available. This approximate method becomes less accurate closer to the critical point, and/or for larger differences between states.

Detailed property tables are not readily available for all substances. However, normal boiling points and enthalpies of vaporation at atmospheric pressure (usually called heats of vaporization) are more widely available. If state A is taken to be atmospheric pressure, then T_A is the normal boiling point of the liquid and the heat of vaporization at 1 atm is $h_{fg@1atm}$. Then, the saturation pressure at any pressure can be approximated as follows:

$$P = e^{\frac{h_{fg@1atm}}{R}\left(\frac{1}{T_{bp}} - \frac{1}{T}\right)}$$

Results of this model for water are shown in Fig. 4.36 for the entire liquid/vapor line (between the triple point and critical point). The solid curve is from the integrated Clausius-Clapeyron equation and the data points are from saturation tables. The percent error is shown in Fig. 4.37. The model gives less than 10% error between 50 °C and 250 °C, which is really extraordinary.

Two Parameter Curve Fit for Saturation Pressure

There are times when a simple formula for the relationship between saturation temperature and pressure is helpful, even if saturation tables are available. For that purpose, the integrated Clausius-Clapeyron equation provides a functional dependency between saturation temperature and pressure, namely,

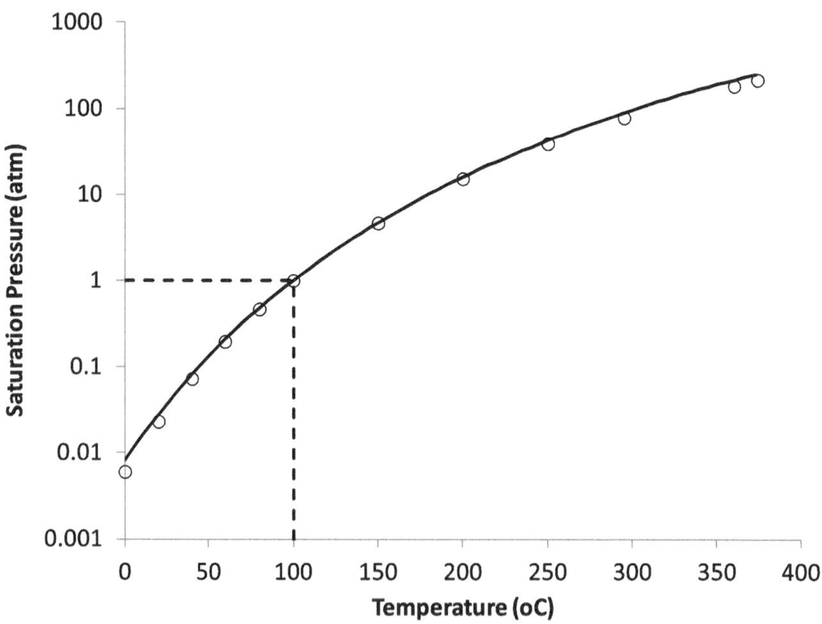

Fig. 4.36 Saturation pressure (vapor pressure) of water. The solid line shows the integrated Clausius-Clapeyron relationship based on the normal boiling point. The open circles are from saturation tables. For water, the normal boiling point (i.e., at 1 atm) is 100 °C, the heat of vaporization (at 1 atm) is 2257 kJ/kg, and the molecular weight is 18

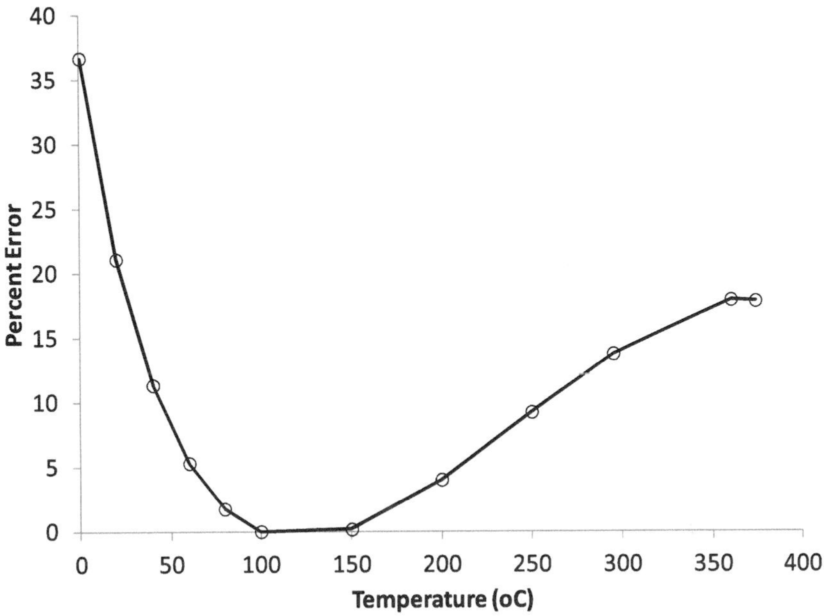

Fig. 4.37 Percent error of the integrated Clausius-Clapeyron equation for water

$$P_{\mathrm{SAT}} = Ae^{\frac{B}{T}},$$

where A and B are parameters that can be determined from 2 entries from saturation tables. Both pressure and temperature are in absolute units.

To determine the parameters, consider two points, 1 and 2:

$$P_{\mathrm{SAT},1} = Ae^{\frac{B}{T_1}}$$

$$P_{\mathrm{SAT},2} = Ae^{\frac{B}{T_2}}$$

Taking the natural logarithm of both sides, subtracting and rearranging yields the parameter B:

$$B = \frac{\ln\left(\frac{P_{\mathrm{SAT},1}}{P_{\mathrm{SAT},2}}\right)}{\frac{1}{T_1} - \frac{1}{T_2}}.$$

Back-substituting yields parameter A (using either point):

CASE 1

PROPERTY DATA

	P (kPa)	Tsat (oC)
1	0.6117	0.01
2	7.3851	40

Clausius/Clapyron Parameters

A	1.7942E+08	kPa
B	-5.3228E+03	K

TEST

T	20	oC
Psat	2.3392	kPa, from table
Psat	2.3133	kPa, from curve
%error	-1.11%	

CASE 2

PROPERTY DATA

	P (kPa)	Tsat (oC)
1	175	116.04
2	18000	356.99

Clausius/Clapyron Parameters

A	3.1933E+07	kPa
B	-4.7130E+03	K

TEST

T	225	oC
Psat	2549.7	kPa, from table
Psat	2478.4	kPa, from curve fit
%error	-2.80%	

Fig. 4.38 Spreadsheet template to determine 2 parameter curve fit, based on Clausius-Clapeyron function. Unit conversions from Celcius to Kelvin are performed in the embedded formulas, as needed

$$A = \frac{P_{\mathrm{SAT},1}}{e^{\frac{B}{T_1}}}$$

A simple spreadsheet template can be created in which the user enters two data points from saturation tables, and the parameters are calculated. An example that considers two cases over different ranges is shown in Fig. 4.38. A test of each case is shown at an intermediate point in the range, which shows the percentage error to be below 3% for both cases.

Antoine Equation: A Three Parameter Curve Fit for Saturation Pressure

It is common practice in the Chemical Engineering field to use the Antoine equation, defined by a 3 parameter curve fit for saturation pressure. It is given by the following fit:

$$\mathrm{Log}_{10} P = A - \frac{B}{T + C}$$

The 3 parameters (A, B, C) are listed in a variety of sources over a variety of temperature ranges (e.g., http://webbook.nist.gov/cgi/cbook.cgi?ID=C7732185&Units=SI&Mask=4). The values depend on the set of units used, of course. Note that if the value of the parameter $C = 273$ K, (and T is in °C), the Antoine Equation is the same as the integrated Clausius-Clapeyron equation.

Workshop 4.1: Compressed Liquid Properties

For each of the grids provided, fill in the tables and prepare a plot from data obtained from compressed liquid tables (internet). Each one demonstrates a compressed liquid approximation.

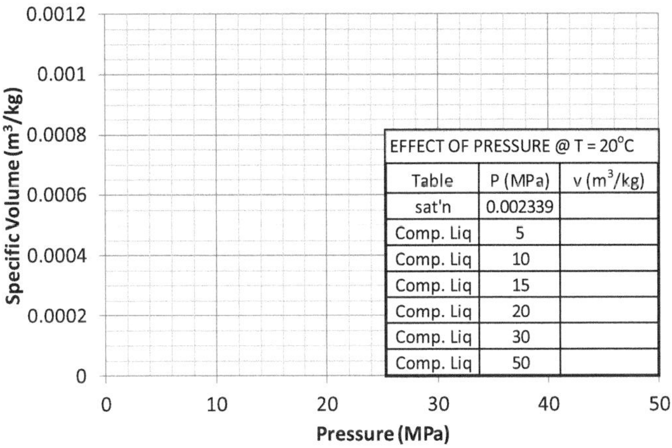

EFFECT OF PRESSURE @ T = 20°C		
Table	P (MPa)	v (m^3/kg)
sat'n	0.002339	
Comp. Liq	5	
Comp. Liq	10	
Comp. Liq	15	
Comp. Liq	20	
Comp. Liq	30	
Comp. Liq	50	

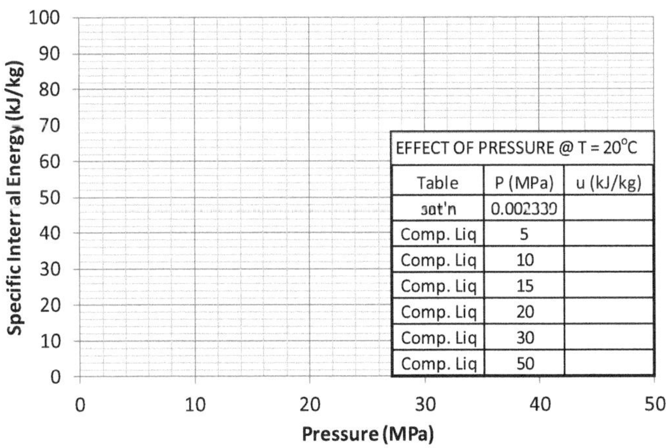

EFFECT OF PRESSURE @ T = 20°C		
Table	P (MPa)	u (kJ/kg)
sat'n	0.002339	
Comp. Liq	5	
Comp. Liq	10	
Comp. Liq	15	
Comp. Liq	20	
Comp. Liq	30	
Comp. Liq	50	

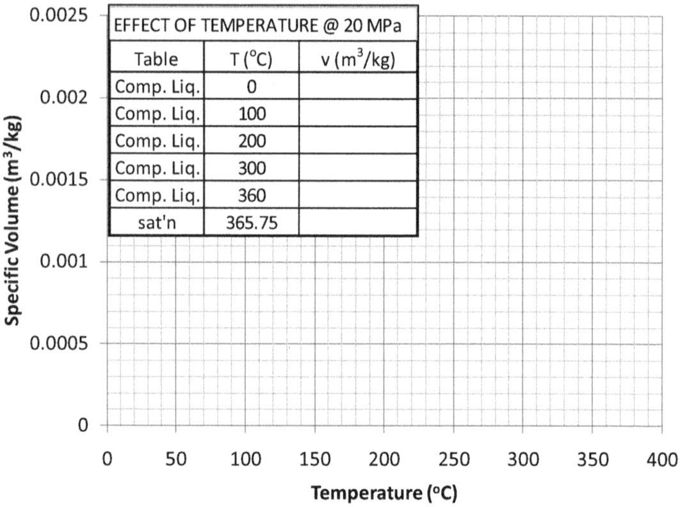

EFFECT OF TEMPERATURE @ 20 MPa		
Table	T (°C)	v (m³/kg)
Comp. Liq.	0	
Comp. Liq.	100	
Comp. Liq.	200	
Comp. Liq.	300	
Comp. Liq.	360	
sat'n	365.75	

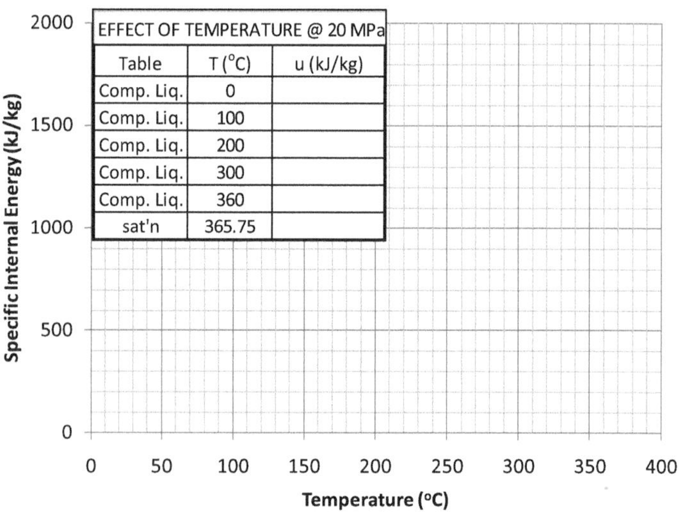

EFFECT OF TEMPERATURE @ 20 MPa		
Table	T (°C)	u (kJ/kg)
Comp. Liq.	0	
Comp. Liq.	100	
Comp. Liq.	200	
Comp. Liq.	300	
Comp. Liq.	360	
sat'n	365.75	

EFFECT OF TEMPERATURE @ Various Pressure				
	5 Mpa	20 Mpa	50 Mpa	SAT LIQ
T (°C)	s (kJ/kg/K)	s (kJ/kg/K)	s (kJ/kg/K)	s (kJ/kg/K)
0				
100				
200				
263.94				
300				
365.75				

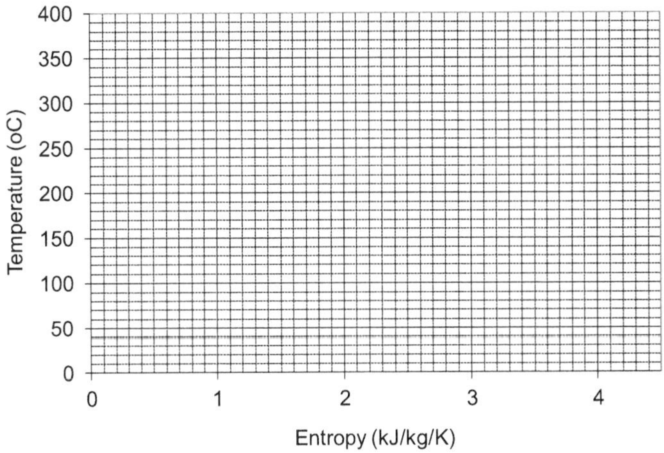

EFFECT OF PRESSURE @ T = 20°C			
P (MPa)	u (kJ/kg)	h (kJ/kg)	$h_{sat}+v_{sat}(P-P_{sat})$
0.002339			
5			
10			
15			
20			
30			
50			

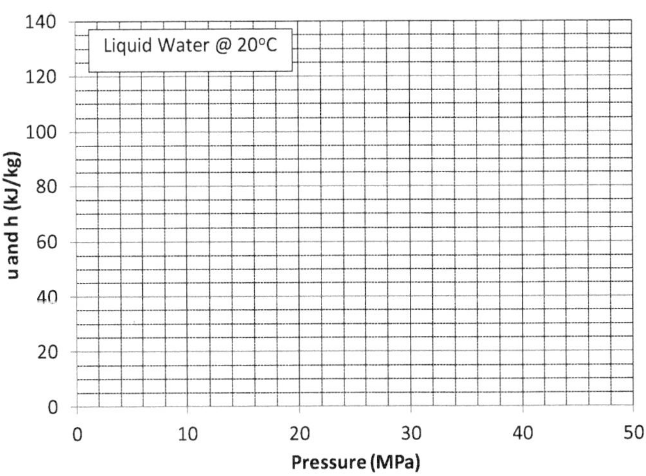

Workshop 4.2: Isothermal Expansion of H2O

The main goals of this workshop are to gain experience with the application of the 1st Law and 2nd Law for closed systems that involve the use of the saturation tables to determine thermodynamic properties of liquid/vapor mixtures. The specific process analyzed is an isothermal expansion of H_2O in a piston/cylinder arrangement, that is, boiling water at room temperature!

Problem Statement

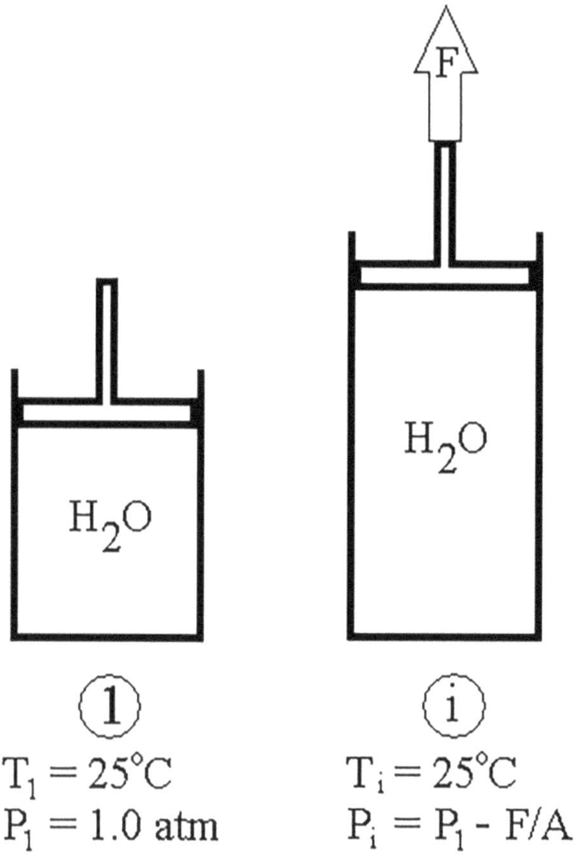

A fixed mass (3.2 kg set as a default) of H_2O is contained in an ideal piston/cylinder (frictionless and leak free), initially at nominal room conditions (the dead state, $T_1 = T_\infty = 25$ °C, $P_1 = P_\infty = 1$ atm). A partial vacuum is created by pulling up on the piston, slowly enough so that the H_2O remains in thermal equilibrium with the environment (i.e., the temperature remains constant), which requires heat to flow from the environment to the working fluid (otherwise, the system would attain a lower temperature). This cooling effect is similar to what takes place in a vapor/

compression refrigeration cycle. Treating the H_2O as the system, a series of processes through intermediate states is indicated in the state and process tables in the spreadsheet template for this workshop. Note that a very specific reference temperature for the tables is indicated (needed to convert temperature into absolute temperature units). Also, the final state will require some thought, and understanding of when and how to use ideal gas approximations.

1st Law

For each process (i→j), an energy balance is $\Delta E = Q_{\text{in}} - W_{\text{out}}$

Changes in kinetic and potential energy are considered to be negligible and, therefore, the change in stored energy becomes $\Delta E = \Delta U = m\Delta u$ where "u" is the specific internal energy (i.e., per unit mass). The expansion work done by the system is its force exerted on the piston (absolute pressure times piston area) times the displacement (dx). Integrated between 2 states:

$$W_{\text{out}} = \int_i^j (PA)dx = \int_i^j PdV$$

2nd Law

For each process (i→j), an entropy balance multiplied by the dead state temperature is as follows: $T_\infty \Delta S = Q_{\text{in}}(T_\infty / T_B) + T_\infty \Sigma$

The eXergy destruction associated with each process is as follows: $eX_D = T_\infty \Sigma$

State Postulate

Specification of 2 independent thermodynamic variables uniquely defines all others for simple compressible substances. In the template, 2 independent thermodynamic variables are specified for each state. A main objective of this workshop is to learn and practice the skill of property evaluation for mixed liquid/vapor states. Note that temperature and pressure are not independent if the system is a liquid/vapor mixture (but are independent for compressed liquids or superheated vapors).

INPUT PARAMETERS

MASS	3.2	kg, mass of H2O
Tref	273.1434706	K, reference temperature for H2O tables

STATE TABLE

state	T(oC)	P(kPa)	v (m^3/kg)	u (kJ/kg)	s (kJ/kg/K)	x(quality)	PHASE (CL, L/V, or SH)
1	**25**	**101.3**					
2	**25**					**0**	
3	**25**		**1.00**				
4	**25**				**4**		
5	**25**					**0.9**	
6	**25**					**1**	
7	**25**		**100**				

PROCESS TABLE (all in kJ)

process	ΔU	Q_{in}	W_{out}	$T_\infty \Delta S$	$Q_{in}(T_\infty/T_B)$	$eX_D = T_\infty \Sigma$
1-2						
2-3						
3-4						
4-5						
5-6						
6-7						
NET						

NOTE: THIS PROBLEM IS SET UP SO THAT ALL PROCESSES ARE REVERSIBLE (NO EXERGY DESTRUCTION). FOR THE 2nd LAW ANALYSIS, THE ABSOLUTE TEMPERATURE OF THE DEAD STATE MUST BE CALCULATED TO A HIGH PRECISION FOR THE EXERGY DESTRUCTION TO EVALUATE TO ZERO. TRY USING 273 INSTEAD OF 273.1434706. IT IS AN ISSUE OF ROUND-OFF ERROR AND RELATED TO THE CONSTRUCTION OF THE STEAM TABLES.

Workshop 4.3: Closed Rankine Cycle

The Rankine cycle is the thermodynamic cycle most closely associated with the operation of steam-turbine-based power plants. In practice, it is used in an open cycle, but it is instructive (and good practice) to analyze the Rankine cycle in a closed cycle, with a unit mass of H_2O ($M = 1$ kg) as the working fluid in the traditional frictionless piston/cylinder arrangement (with a piston diameter $D = 0.05$ m). The objective is to raise a weight by adding heat from a high temperature reservoir (or "source"). Alternatively, an electric heater could be used,

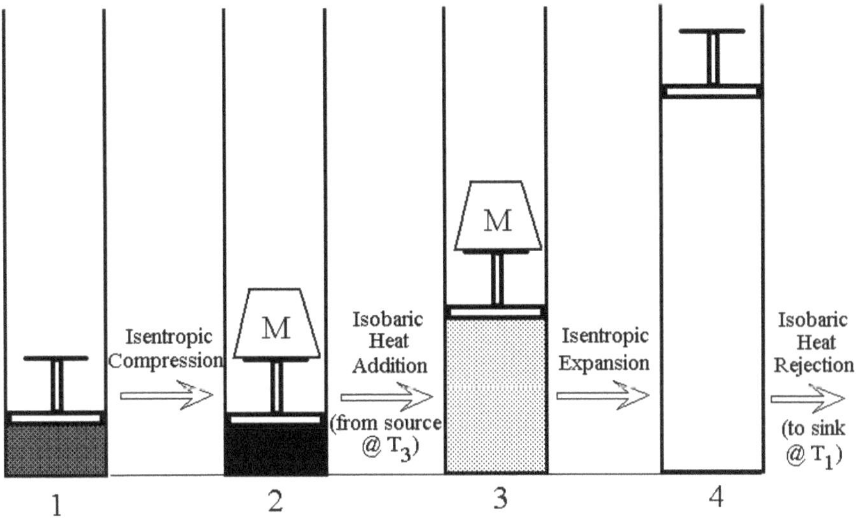

Fig. 4.39 State/Process diagram of closed Rankine cycle

as in the case developed in this chapter. Operation in a cycle requires heat rejection to the local ambient (the dead state, at $T_\infty = 25$ °C).

The basic idealized cycle consists of 4 sequential processes (Fig. 4.39):

1–2Isentropic compression (i.e., "adiabatic, with no entropy generation (frictionless)"). A pump accomplishes this process in an open system (which requires input work). The initial state is specified to be a saturated liquid at a partial vacuum. The partial vacuum could be accomplished by tying a helium balloon to the piston platform (not shown). The process consists of a mass being suddenly placed on the piston platform. The workshop on compressed liquids should help understand what happens in this step.

2–3 Isobaric heating. This step can be accomplished in practice by transferring heat from hot combustion products, or from nuclear reactor core, or focused solar energy, or an electrical heating element, or many other means. For the closed cycle, assume that the piston/cylinder is placed in contact with a thermal reservoir at temperature $T_H = T_3$ (i.e., an oven that is maintained at the highest temperature experienced in the cycle, T_3). The process consists of simply allowing heat to flow until the working fluid attains thermal equilibrium with the oven. Work is done (by the working fluid on the mass), and the mass is raised. In the open system, there will be flow work associated with this step, and that does not become part of the net input or output shaft work.

3–4 Isentropic expansion. A steam turbine (a steady-state open system device) extracts useful mechanical power in an open system. In the closed system, the mass must be gradually removed to accomplish an isentropic expansion (Fig. 4.40). That is, starting at state 3 (where all the mass has already been raised

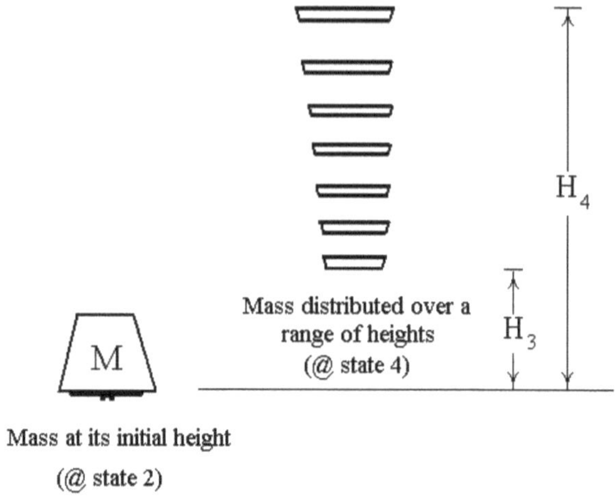

Fig. 4.40 Schematic that shows the state of the mass to be raised after the isentropic expansion (process 3–4)

somewhat), a bit of mass is removed. The pressure on the working fluid is decreased, and the gases expand a little, raising the remaining mass a bit further. Then another bit of mass is removed (at a higher level than the first), and so on until all the mass is removed, and the total mass is distributed across a range of heights. The cylinder is well insulated during this stage (adiabatic process).

4–1 Isobaric cooling. After raising all the weight it can (the system is at the initial low pressure), the working fluid is at a much higher energy state than the initial state. To return to the initial state, heat can be transferred from the working fluid to the ambient (the low temperature reservoir, $T_C = T_\infty$). This step is a "necessary evil" needed to close the cycle. A large percentage of the input thermal energy must be rejected to the local ambient by a condenser, usually the largest physical component of an actual power plant.

To analyze the cycle, perform a 1st and 2nd Law analysis for a unit mass of working fluid as the system for each process (i–j), that is (neglecting potential and kinetic energy changes for the working fluid),

$$\Delta U_{ij} = Q_{ij,\text{IN}} - W_{i,j,\text{OUT}}$$

$$\Delta S_{ij} = \frac{Q_{ij,\text{IN}}}{T_B} + \Sigma_{i,j}$$

The eXergy destruction is $eX_D = T_\infty \Sigma$. Complete the State, Process, and Results Tables. Sketch the cycle on the P-v, T-v, and T-s diagrams (follow the path taken where possible, or draw a dotted curved line where the shape of the path is not clearly defined).

INPUT PARAMETERS

M	1	kg, mass of H2O
D	0.05	m, cylinder diameter
T_∞	25	oC, dead state temperature (ambient)
T_H	400	oC, high temperature source temperature

STATE TABLE

state	P(kPa)	T (° C)	v (m^3/kg)	u (kJ/kg)	s (kJ/kg/K)	quality or phase
1	50					0
2						
3	8000	400				
4						

PROCESS TABLE (all in kJ)

Process	ΔU	Q_{in} (kJ)	W_{out}	$T_\infty \Delta S$	$Q_{in}(T_\infty/T_b)$	$T_\infty \Sigma = eX_D$	Process type
1-2							adiabatic, reversible compression
2-3							Isobaric heating (from "oven")
3-4							Isentropic expansion
4-1							Isobaric cooling (to ambient)
net							

RESULTS TABLE

H_3		m	height of piston at state 3
H_4		m	height of piston at state 4
W_{net}		kJ	Net work done by H2O
Q_{23}		kJ	Heat added in "oven"
eX_D		kJ	Total eXergy destruction
η_1			1st Law Efficiency (defined as net work over heat added)
η_2			a 2nd Law Efficiency (net work over net work + eXergy destruction)
η_{CARNOT}			Carnot efficiency between "reservoirs"

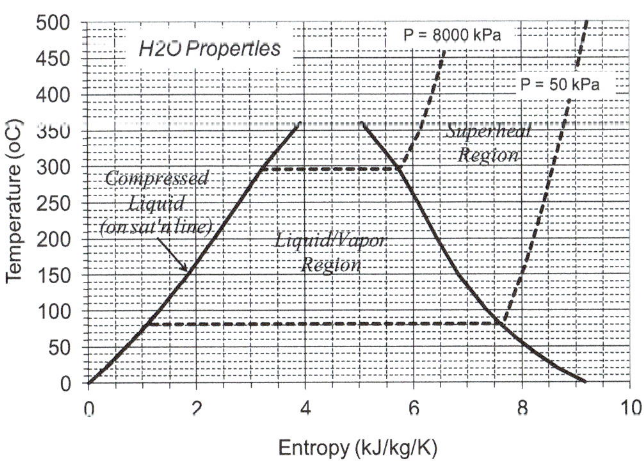

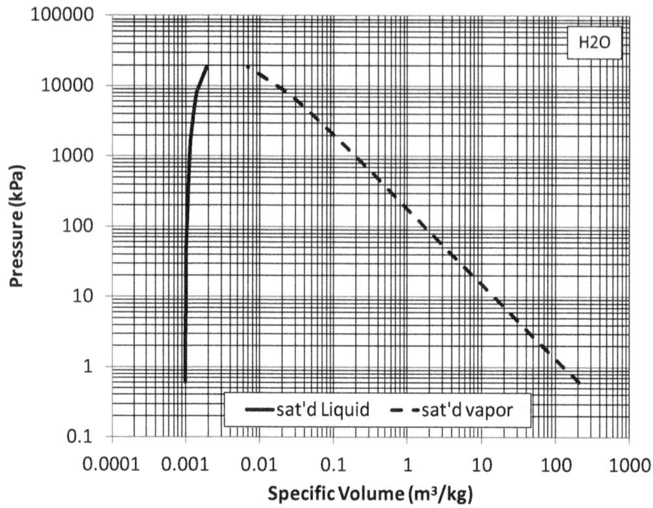

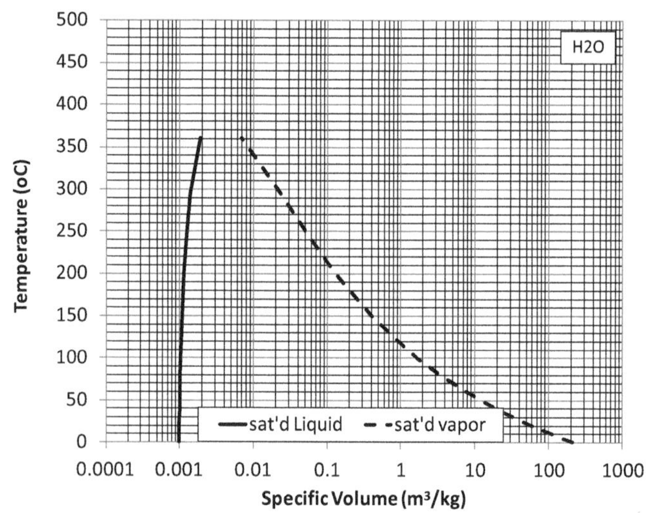

Workshop 4.4: Refrigeration 101

The majority of commercial refrigeration systems are based on vapor/compression refrigeration cycles. This workshop demonstrates the two key sequential processes used to attain a refrigeration effect (remove heat from a space that is at a lower temperature than its surroundings) in a closed system application. In practice, this process would be accomplished in an open system device (a throttling valve), and two components would be needed to create a closed cycle (a compressor to attain a high-pressure vapor, and a condenser to return to the high-pressure liquid starting point). Workshop 4.5 builds on this one by considering one possible cycle for returning the working fluid to the initial state of this workshop.

One kilogram of Refrigerant R134a is contained in a rigid vessel in which a wall partition initially separates the refrigerant (at $P_1 = 1600$ kPa, $u_1 = 90$ kJ/ kg) from a complete vacuum. Figure 4.41 shows the vessel located in a cold space (at 0 °C), a thermal reservoir. Fig. 4.42 shows 3 states of the system (defined as the vessel, assumed to have negligible thermal mass, plus the refrigerant). At state 1, the partition is suddenly removed and the refrigerant fills the evacuated space in an adiabatic (no heat) free expansion (no work) process (1–2). In an open system, a

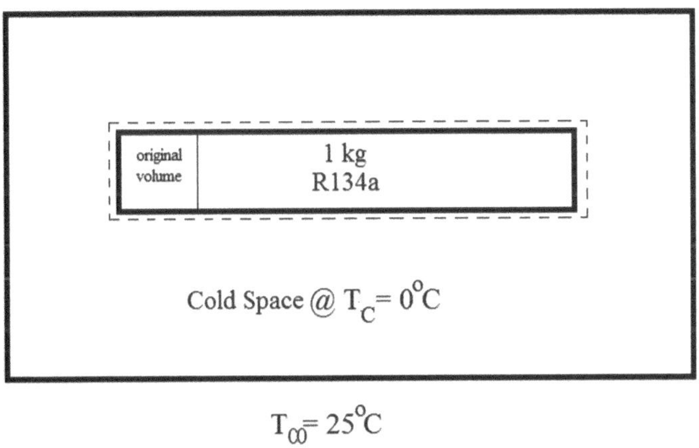

Fig. 4.41 Schematic of a rigid vessel of negligible thermal mass containing 1 kg of refrigerant (R134a). The system to be analyzed consists of the refrigerant and the vessel

Fig. 4.42 Three states of the system

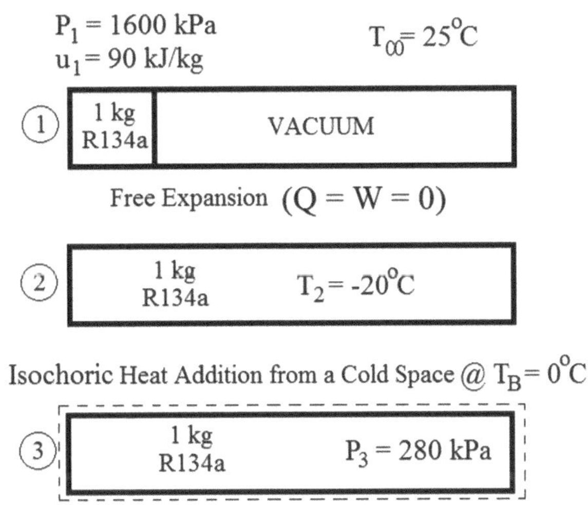

throttling valve is used to achieve this effect, namely, to take advantage of liquid to vapor phase change to create a working fluid that is at a lower temperature without an energy exchange (heat or work).

After equilibrium is established (assumed to be a relatively short time), the temperature at state 2 is −20 °C. Heat then naturally flows (process 2–3) from the cold space until the pressure in the vessel is 280 kPa at state 3. In an open system, a heat exchanger called an evaporator would be used to accomplish this effect, namely, to transfer heat from a cold space to the working fluid.

Complete the state and process tables and sketch the processes on the *Tv*, *Pv*, and *Ts* diagrams. The dead state temperature is 25 °C.

INPUT PARAMETERS

M	1	kg, mass of refrigerant
Tspace	0	oC, temperature of cold space
Tinf	25	oC, dead state temperature

STATE	P (kPa)	T(oC)	V_{R134a} (m³)	u (kJ/kg)	h (kJ/kg)	s(kJ//kg/K)	quality
1	1600			90			
2		-20					
3	280						

all quantities in kJ

PROCESS	ΔU	Q_{in}	W_{out}	ΔS^*T_{inf}	$(Q/T_B)^*T_{inf}$	eX_d
1-2						
2-3						

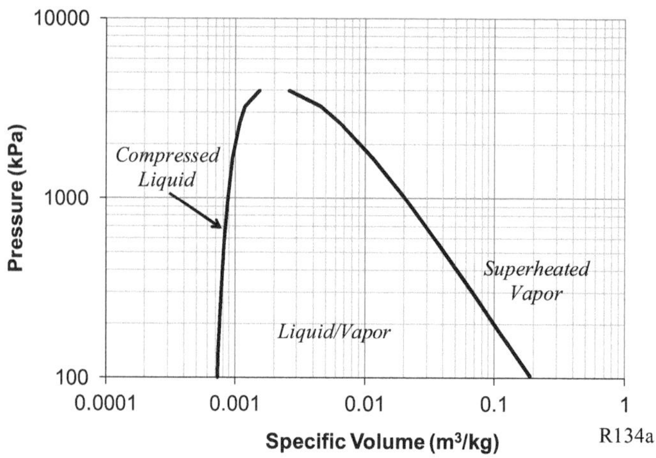

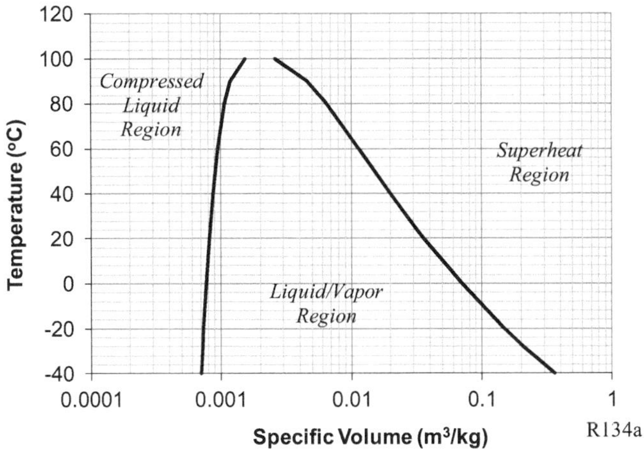

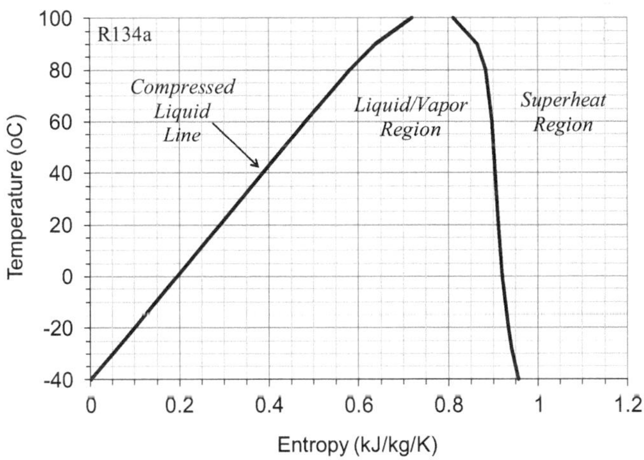

Workshop 4.5: Closed Refrigeration Cycle

THIS PROBLEM BUILDS ON WORKSHOP 4.4, REFRIGERATION 101, BY CREATING A CLOSED CYCLE (WITH A DIFFERENT REFRIGERANT MASS).

A total of 2.5 kg of Refrigerant R134a is contained in a piston/cylinder in which a wall partition initially separates the refrigerant (at $P_1 = 1600$ kPa, $u_1 = 90$ kJ/ kg) from a complete vacuum. The left sketch of Fig. 4.43 shows the piston/cylinder located in a cold space (at 0 °C, a thermal reservoir). Figure 4.44 shows 6 states of the system (defined as the piston/cylinder, assumed to have negligible thermal mass,

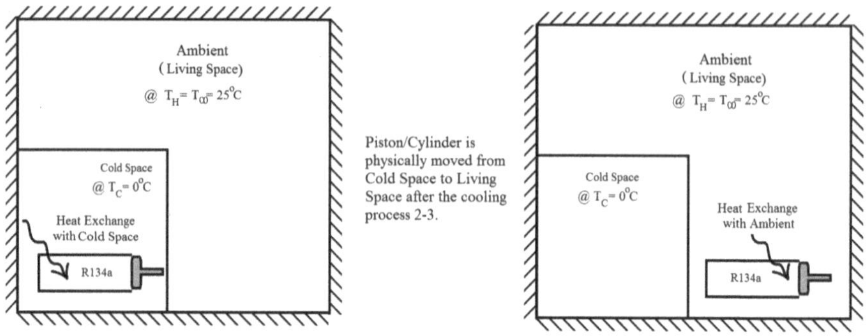

Fig. 4.43 Schematic of a piston/cylinder arrangement (of negligible thermal mass) containing 2.5 kg of refrigerant (R134a). The system to be analyzed consists of the refrigerant and the piston/ cylinder. The left schematic shows the system located in a cold space. The right schematic shows it in the ambient (living space). Heat transfer is shown defined as positive into the working fluid and will evaluate as negative if heat flows out of the working fluid (Fig. 4.44)

plus the refrigerant). At state 1, the partition is suddenly removed and the refrigerant fills the void in an adiabatic (no heat) free expansion (no work) process (1–2). After equilibrium is established, the temperature at state 2 is −20 °C. Heat then naturally flows (process 2–3) from the cold space until the pressure in the vessel is 280 kPa at state 3. These steps were analyzed in the previous Workshop, and the results are included in the spreadsheet template. In an open system, these steps would be accomplished by a throttling valve and a heat exchanger (called an evaporator).

In order to create a cycle that returns the system to its original state, the piston/ cylinder is physically moved out of the cold space into the ambient (living space), shown as the sketch on the right of Fig. 4.43. In process 3–4, heat is allowed to flow between the system and living space (ambient) with the volume remaining constant until the refrigerant is a saturated vapor (state 4). At that point, the piston is used to compress the refrigerant in an isentropic process (adiabatic and no entropy generation, process 4–5). In an open system, these two steps are accomplished by a compressor, although the first heat transfer step does not really occur. The main objective is to achieve the high pressure associated with the initial liquid state.

After compression, the fluid is in a superheated vapor state. Therefore, compression is followed by a constant pressure process (5–6, with heat exchange from the refrigerant to the living space) until the refrigerant is returned to its original state (state 6, same as state 1 for refrigerant).). In the open system, a heat exchanger called a condenser is used for this purpose, and is responsible for the warm air at the back of a refrigerator (since heat is transferred from the hot refrigerant to the ambient air).

At that point (state 6), a removable partition is placed at the inner face of the piston. This contrived mechanism is introduced at this point in order to simplify the theoretical analysis, not because it offers a practical advantage. The piston is then withdrawn, creating a perfect vacuum between the piston and the partition (state 1). Note that the refrigerant properties do not change for process 6–1, but work on the piston against ambient pressure is required.

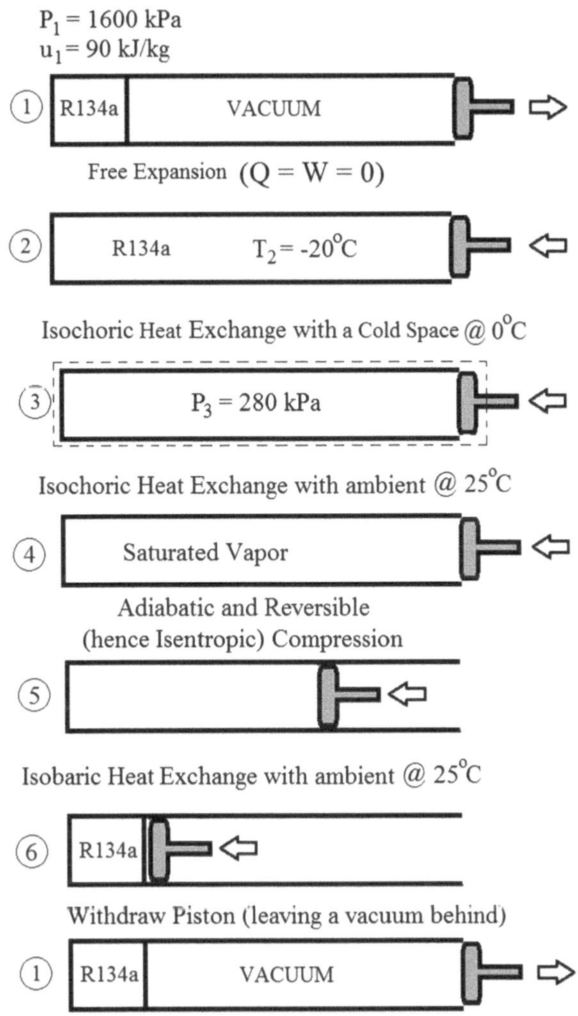

Fig. 4.44 States of the piston/cylinder arrangement. The arrows show the direction of the force that must be applied to the piston to maintain equilibrium (but this force need not be calculated directly)

This type of refrigerator has never been built for practical reasons (sealing problems, low heat transfer rates, complex mechanical design, you name it...). From a theoretical standpoint, however, this closed system cycle demonstrates the fundamental changes the refrigerant undergoes (particularly its volume) and is used here to develop experience with saturated and superheated tables. The open system counterpart used in virtually all household refrigerators in a future chapter will be better understood by going through this exercise.

Complete the state, process, and results tables in the excel template, and sketch the entire process on the *Pv*, *Tv*, and *Ts* diagrams provided on hard copy (show accurate process paths where possible, dashed lines where the process path is not precisely known). Enter values from R134a saturation and superheat tables where appropriate, and enter formulas in cells as needed. Do all calculations in this spreadsheet. Note that there are 2 columns for work. One column represents the work done by the refrigerant on the left side of the piston, called $Wout_{,R134a}$ (the refrigerant side), while the other column represents the work done by the piston on the ambient pressure, called $W_{out,amb}$ (the right side of the piston).

You may interpolate "by eye" where needed. If you have time, perform a linear interpolation for extra credit.

INPUT PARAMETERS

M	**2.5**	kg, mass of refrigerant
Tspace	**0**	oC, temperature of cold space
Pinf	**1**	atm
Tinf	**25**	oC, ambient (and dead state) temperature

R-134a STATE TABLE — results from Workshop 4.4

STATE	P (kPa)	T(oC)	v(m^3/kg)	u (kJ/kg)	s(kJ//kg/K)	quality, or phase
1	**1600**	28	0.000831	**90**	0.33846	NA
2	132.82	**-20**	0.04968	90	0.38552	0.33399
3	**280**	-1.25	0.04968	172.63	0.69972	0.68332
4						**1**
5	**1600**					
6	**1600**	28	0.000831	**90**	0.33846	NA

all quantities in kJ

PROCESS	U	Q_{in}	$W_{out,R134a}$	$W_{ambient}$	ΔS^*T_{inf}	$(Q/T_B)^*\ T_{inf}$	eX_d
1-2	0	0	0	0	35.08	0	35.075
2-3	206.57	206.57	0	0	234.20	225.47	8.726
3-4							
4-5							
5-6							
6-1							
net							

RESULTS

COP		Heat Removed from cold space/Net work In
Carnot Efficiency		= TC/(TH-TC), maximum theoretical COP of a refrigerator between cold space and ambient
2nd Law effy		= COP/CarnotCOP

Chapter 5
Open Systems: Pumps and Turbines

The focus of this chapter is on engineering devices that use a compressed liquid as a working fluid, with pumps and turbines as a case study. The main purpose of a pump, from a thermodynamic point of view, is to convert shaft work into mechanical energy of a flowing liquid. Compressors and fans, considered in future chapters, operate with gases. There are two broad classes of pumps: positive displacement and dynamic. Positive displacement pumps move a definite volume of liquid. Dynamic pumps involve the addition of energy to a fluid by the action of moving blades. The volume of liquid that moves through a dynamic pump is not well characterized. The basic principles behind two examples, reciprocating pumps and centrifugal pumps, are presented first. Turbines are devices that work in reverse, namely, mechanical energy from a fluid is converted into output shaft work (and no classification distinction between liquids and gases is made). Much of this material overlaps with courses on fluid mechanics, but the emphasis is on developing thermodynamic analyses of open systems. Applications that involve the simplest type of working fluid, compressed liquids, are used. The fundamental thermodynamic balance equations are the same for gases and liquid/vapor mixtures, but their property relationships are more complicated. Pump characteristic curves (flow work vs. flow rate) are developed in a general way and show 1st Law efficiency is maximized if a pump is properly matched with a flow system, a form of impedance matching for a flow system. The main goal of this chapter is to develop a fundamental approach to the analysis of open systems and the use of property relations for applications that use liquids as the working fluids.

Two forms of storable mechanical energy (kinetic and potential) and one form of transferable mechanical energy (flow work) relate to flowing fluids and were introduced in Chap. 1. Flow work, the work associated with the pressure force moving a fluid into or out of a device, will be discussed further in this chapter as it is a key to proper thermodynamic analysis of open systems (for which mass crosses the system boundary). The objective of a pump is primarily to pressurize and/or move liquid. The objective of a turbine is to extract useful mechanical power from a fluid stream.

G. Sidebotham, *An Inductive Approach to Engineering Thermodynamics*,
Mechanical Engineering Series, https://doi.org/10.1007/978-3-030-20430-3_5

Hydroelectricity has proven to be an economically viable renewable resource in the locations where it is feasible. Siting locations can be controversial because it generally involves damming or diverting a river to create a large drop in elevation over a short distance. The famed Hoover Dam has been a huge success, and a substantial project is being considered to tie it with other intermittent renewable sources (solar and wind) to act as a means of energy storage. During periods of time when electricity from solar and/or wind exceeds the demand on the grid, the excess electricity will drive pumps that move water from downstream to upstream of the Hoover Dam a form of pumped storage hydropower.

There are other opportunities for renewable energy sources that extract power from liquid water. For example, turbines can be placed in fast-flowing rivers (such as the East River between Manhattan and Brooklyn, Queens, or the Bronx), with minimal disruption of the river. The inherent energy in the rising and falling tides can be extracted. The energy in surface waves can be extracted. Development of these technologies poses significant technical, economic, and political challenges, which can be just the thing to launch a wide range of productive careers for forward-thinking young professionals.

Wind power is a growing field as well, and a deeper dive demonstrates the available power that can be extracted. A fundamental analysis of a wind stream reveals the result of Betz's law, namely, that 59.3% of the inherent power in a wind stream is available for conversion to output shaft power.

A common theme of the engineering devices is that adding or extracting power from a fluid stream requires both a throughput of mass and a conversion of energy. Generally, a low throughput results in a large conversion, and vice versa. Therefore, a compromise results in optimal operating condition in a general form of impedance matching.

Hydrostatics: Pressure Head and Buoyancy

It is instructive to establish the behavior of stationary fluids in a gravitational field (hydrostatics) before diving into fluids in motion. The term "head" is used to indicate the pressure of a liquid, normalized by the specific weight of the fluid (ρg), which gives it units of length. The origin of the term "head" is not obvious, and several Internet searches did not resolve it. The head can be thought of as the equivalent height of a column of liquid. Consider an open cylinder with a cross-sectional area A that contains a stationary liquid with a density ρ filled to a depth H (Fig. 5.1). A force balance in the vertical direction on the liquid is

$$\sum F_y = \mathrm{ma}_y$$

Since there is no motion, the acceleration is zero, and the forces balances. There are three forces: a fluid pressure acting upward on the bottom of the bucket,

Fig. 5.1 Schematic and free-body diagram of a stationary column of liquid of height H in a container of cross-sectional area A

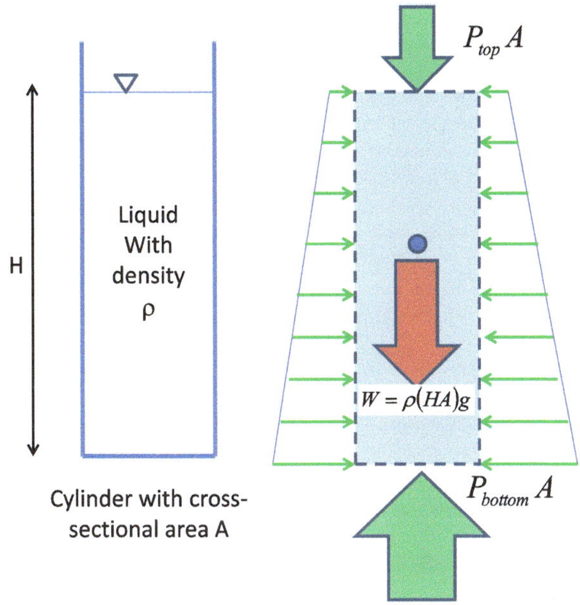

atmospheric pressure acting on the free surface downward, and the weight of the liquid acting downward. That is,

$$P_1 A - \rho A H g - P_\infty A = 0$$

Solving for the pressure at the bottom:

$$P_1 - P_\infty + \rho g H$$

This formula is the basis for pressure measurement using a manometer or atmospheric pressure using a barometer. The pressure force at the bottom opposes atmospheric pressure plus the weight of the fluid in the container. Rearranging for the height, H, and introducing the gage pressure as the pressure relative to local ambient:

$$H = \frac{P_1 - P_\infty}{\rho g} = \frac{P_{1,\text{GAGE}}}{\rho g}$$

It is important in engineering analysis to know whether gage or absolute pressures are being used. In thermodynamics, the absolute pressure is usually used, while gage pressure is more common in fluid mechanics. It is good practice to indicate the difference in the way units are specified. For example, psig versus psia for the English unit pounds per square inch can be used. In SI units, Pa,a versus Pa,g could be used, but seems to be rarely done.

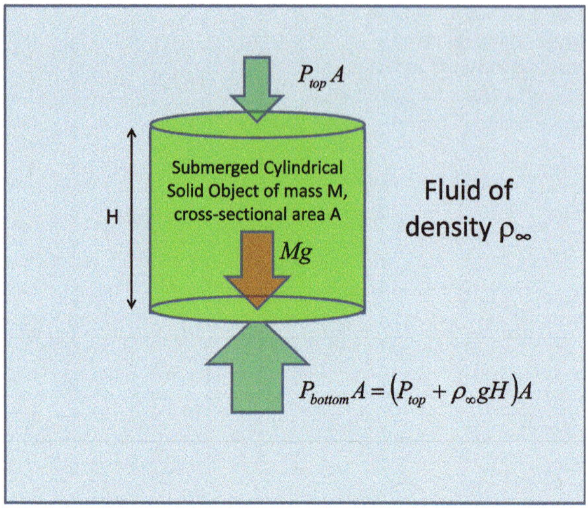

Fig. 5.2 Schematic free-body diagram that demonstrates Archimedes principle. A solid object (cylindrical shape) of cross-sectional area A and height H is submerged in a liquid of density ρ_∞

The pressure at the bottom of the container is expressed as the equivalent height of a column of liquid. In engineering practice, it is common for statements to be made such as "the pump develops 100 feet of head." This means that the pressure rise across the pump is the equivalent of 100 feet of water (whose density is taken to be 1000 kg/m^3, the value at 4°C, at which the density is a maximum). So, 100 feet of water is different than 100 feet of oil due to the different densities.

A similar analysis reveals the physics of buoyancy forces that act on a fluid. Newton's Law of Motion (in the direction of gravity) applied to the free-body diagram of a submerged solid cylinder of mass M (Fig. 5.2) is

$$\sum F_y = \mathrm{m}a_y = P_{\mathrm{bottom}}A - P_{\mathrm{top}}A - Mg$$

The net vertical pressure force represents the buoyant force acting on the submerged object, that is,

$$F_{\mathrm{buoyancy}} = \left(P_{\mathrm{bottom}} - P_{\mathrm{top}}\right)A = \rho_\infty gHA$$

Note that the quantity HA is the volume of the object, and $\rho_\infty HA$ is the mass of the fluid displaced by the object. Finally, $(\rho_\infty HA)g$ is the weight of the displaced fluid. Long before Newton discovered $F = \mathrm{ma}$, Archimedes had it right in 250 BC when he said, "Any object, wholly or partially immersed in a fluid, is buoyed up by a force equal to the weight of the fluid displaced by the object." Note that "equal to" does not mean "the same as." The buoyant force is the net pressure force that acts on a submerged object (and any object in air is a submerged object). Newton's Law of Motion ($F = \mathrm{ma}$) demonstrates that it is equal in magnitude to, but not the same as, the weight of the displaced fluid. The practical result of Archimedes principle is that the magnitude of the buoyant force (the net pressure force on a submerged object) is equal to the weight of the displaced fluid (which is easier to evaluate).

An object submerged in an open container of liquid will accelerate upward if the buoyant force exceeds the weight of the object until it reaches the free surface and then find just the right depth of floating for the buoyant force to balance the weight (i.e., an iceberg). On the other hand, the weight of a stone will exceed the buoyant force and will sink until it reaches the floor. The force exerted by the floor is the difference between the weight of the stone and the buoyant force. A neutrally buoyant object will neither tend to rise nor sink.

Buoyancy also exists in the air we live in. For sufficiently small elevation changes, the density of the air can be treated as a constant. However, the density of the displaced fluid decreases gradually with altitude. A weather balloon released from Earth's surface will rise and experience a gradually decreasing buoyant force as the air pressure at elevation decreases. It will settle at an elevation where the buoyant force equals the weight of the package.

Reciprocating Pumps

Many pumps achieve the objective of moving liquids from a low pressure to a higher pressure by a kind of "bucket brigade." A large amount of liquid can be transported by moving small amounts repetitively. In the early days of firefighting, a line of people would work cooperatively to move water from a source (like a river) to a burning building by passing buckets. On one end of the line, a person (the "pumper") would fill a bucket with water and pass it to the next person who would simultaneously hand the pumper an empty bucket. Each person in the middle would pass a full bucket forward and an empty one back. At the other end, a bucket of water would be thrown onto the fire, and the empty bucket sent back to the source through the line. A steady, but pulsed, stream of water would be delivered to the fire, with a volumetric flow rate (given the nomenclature AV), in a word statement, of

$$\mathrm{AV} = \begin{pmatrix} \text{Volume of} \\ \text{Water Delivered} \\ \text{per unit time} \end{pmatrix} = \frac{(\text{Bucket Volume})}{(\text{time to fill and pass a bucket})} = \frac{V_{\text{bucket}}}{t_{\text{pass}}}$$

This technique was found empirically to be better than a non-cooperative method, that is, with each person walking back and forth from the water source to the fire. The individual effort required is much less and, therefore, the delivery of water could be sustained for a longer period of time.

For example, consider what can realistically be done with the proper number of people and five gallon buckets. Imagine the time needed for the pumper to receive an empty bucket, fill it, and pass it back:

One one thousand (get empty bucket)
Two one thousand (turn)

Three one thousand (dip and fill)
Four one thousand (turn)
Five one thousand (pass full bucket)

That is, 5 seconds. The flow rate for each line would be

$$AV = \frac{V_{\text{bucket}}}{t_{\text{pass}}} = \frac{5\text{gallons}}{5\text{sec}} \times \left| \frac{60\text{sec}}{\text{min}} \right| = 60\frac{\text{gallons}}{\text{min}} \quad \text{using a bucket brigade.}$$

At a steady state, the rate at which the pumper fills buckets will be the rate at which water is delivered to the fire.

Notice that the distance from the water source to the fire (the length of the "human hose") is not involved in the flow rate, provided there are enough people to fill the natural spacing between each person in the line required. At each exchange, each person must step in one direction, pass a full bucket, then step in the other direction and place an empty bucket down. The natural spacing between people would be approximately two paces plus two arm's length, say 4 meters. If the fire were 100 meters from the water source, the line would need to be 25 people long. If fewer people were available, the spacing would grow, and the time needed to pass the buckets (and wear and tear on the pumpers) might be reduced.

Alternatively, using the non-cooperative method, the same 25 people (N_{people}) could work independently, with each person filling a bucket (of volume V_{bucket}) from the source, walking the bucket to the fire (at speed u_{walk} to the fire a distance L away), dumping the water on the fire, and walking (or running) back to the source with an empty bucket.

$$AV = (\#\text{of people}) \times \frac{(\text{Bucket Volume})}{(\text{time to walk cycle})} = N_{\text{people}} \times \frac{V_{\text{bucket}}}{2L/u_{\text{walk}}}$$
$$= \frac{N_{\text{people}} V_{\text{bucket}} u_{\text{walk}}}{2L}$$

For the same example, using the average speed a person walks of 3 miles per hour:

$$AV = 25\text{people} \times \frac{5\text{gallons}}{\text{dump}} \times \frac{3\text{miles}}{\text{hour}} \times \frac{1}{2 \times 100\text{m}} \times \left| \frac{1\text{hour}}{60\,\text{min}} \right| \times \left| \frac{1609\text{m}}{\text{mile}} \right|$$
$$= 50\frac{\text{gallons}}{\text{min}} \quad \text{using a non-cooperative method.}$$

This preliminary analysis suggests, then, that the flow rate of water from both methods is comparable. Therefore, the advantage of the cooperative bucket brigade would be in the lower individual effort required, and therefore, the flow of water to the fire could be sustained longer.

An in-class workshop can serve as a fun, active exercise that stimulates thinking about the flow of a fluid from one point to another, the basis for open thermodynamic system analysis. Take two buckets, one nearly full of water and the other empty, placed at opposite ends of the room. Distribute a cup to each student. The goal is to move all the water from the full bucket to the empty one. First, have every student fill their cup and walk it to the empty one. Continue until all the water is moved. Time the event. Next, have the students stand in a line between the two buckets, and, at the start, pass an empty cup toward the full bucket and a full cup toward the empty one. Time this event. Alternatively, form two lines and make it a race, giving extra points for not spilling.

This principle of delivering a liquid is the basis for many pumps classified as positive displacement pumps. One basic design concept is a reciprocating pump, shown schematically in Fig. 5.3. The pump is built around a piston/cylinder arrangement. The piston moves up and down in a cylinder that forms a chamber whose internal volume changes with time as the crank arm rotates. Check valves, which allow flow in only one direction, automatically control the flow into and out of the chamber. The displacement volume (analogous to the bucket volume) equals twice the crank arm (distance "a"' times the cross-sectional area of the piston, that is,

$$\begin{pmatrix} \text{Displacement} \\ \text{Volume} \end{pmatrix} = V_D = (2a)A_{\text{piston}}$$

The pump operates on a cycle with two strokes (full sweeps of the piston); an intake stroke and an exhaust stroke. The schematic of Fig. 5.3 is shown during the

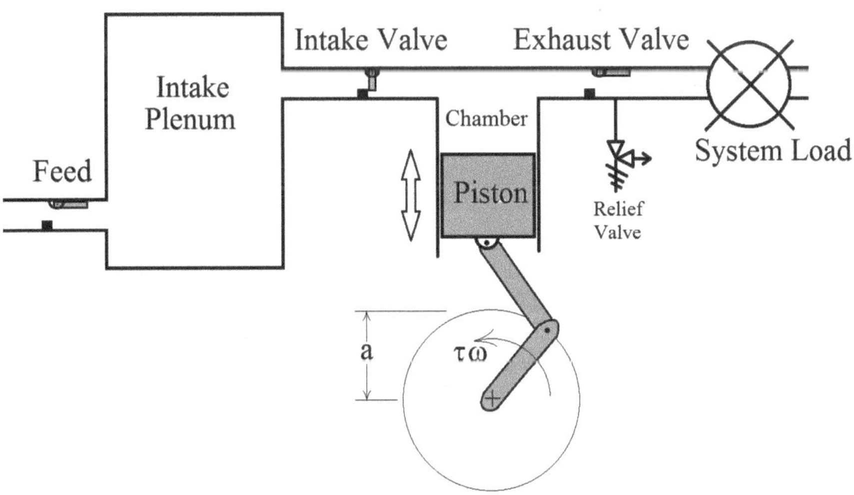

Fig. 5.3 Schematic of a reciprocating pump. The crank turns counterclockwise, while the piston moves up and down, depending on the crank angle. In the orientation shown, the piston is moving upward

exhaust stroke, with the piston moving upward (decreasing the chamber volume). The intake valve is closed and the exhaust valve is open. Liquid is forced through the exhaust valve and fed to a "system load," a term that refers to the flow path taken by the liquid after leaving the pump, shown as a fluid restriction (or impedance). A fire hose could be the system load, for example. During the intake stroke, during which the intake valve is open and the exhaust valve is closed, the piston moves downward (the crank is on the left side) and flow is taken from a plenum and forced through the intake valve to fill the expanding chamber. The plenum (which may not be present as a formal device) acts as a moderator that allows the pump to draw from a stable source and is fed in a manner to maintain an overall balance of flow. A full displacement volume of water is delivered for each stroke. Therefore, the flow rate of the pump (with SI units) is

$$\mathrm{AV} = \begin{pmatrix} \text{Volume of} \\ \text{Liquid Delivered} \\ \text{per unit time} \end{pmatrix} = \begin{pmatrix} \text{Volume} \\ \text{Displaced} \end{pmatrix} \times \frac{(\text{Revolutions})}{(\text{Time})}$$

$$= V_D(m^3) \times \omega\left(\frac{\text{radians}}{\text{sec}}\right) \times \left|\frac{\text{Revolution}}{2\pi \text{ radians}}\right|$$

$$\mathrm{AV} = V_D N = \frac{V_D \omega}{2\pi}$$

where N is the rotational speed (in revolutions per unit time) and ω the angular speed (in radians per unit time).

Pumps act on liquids, whose density does not change appreciably with an applied pressure and, therefore, the flow rate through a displacement pump is very well characterized. It is generally easy to create a high pressure in a liquid without performing work (which requires movement) by the application of a large containment force. Therefore, if the system restriction is too high, the pressures attained could be problematic, and a pressure relief valve can be used for safety and to protect equipment. On the other hand, mechanical work is required to move the liquid from an intake source to the high pressure required to move the liquid through the system. A thermodynamic analysis of the key components will be conducted, and then all the pieces put together in the subsequent sections of this chapter.

In practice, the external torque applied to the crank (τ) to turn it at an angular speed (ω) can come from a variety of sources. It can be turned by human power (a simple hand crank or a bicycle-type mechanism), an electric motor (DC or AC), a windmill, or any other mechanical/electrical source. A pump is generally considered to consist of two parts: a motor (or equivalent) that applies torque to a rotating shaft, and the conversion of the shaft work into fluid energy. The focus of this chapter is on the latter.

Arguably the most important positive displacement-type pump is not an engineering device, but a biological one; the human heart. The anatomy of the heart is shown in Fig. 5.4 during an "intake stroke" when intake valves are open and exhaust valves are closed. A listing that compares the components of the reciprocating pump with its counterpart in the heart is given in Table 5.1. The heart is actually two pumps in a single compact design. The right side (from the point of view of the person, not the viewer) receives "dirty" deoxygenated blood (colored blue) from the body into the right atrium, which acts as a plenum chamber. When the heart dilates, blood

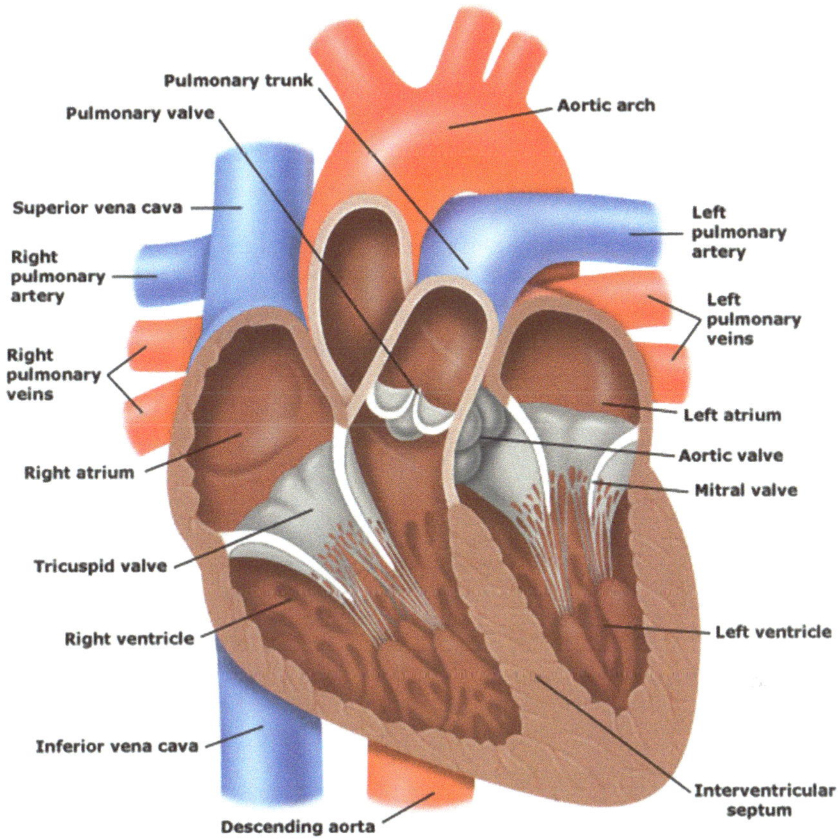

Fig. 5.4 Diagram of the human heart. (Image from http://1.bp.blogspot.com/_-IYNHdaSjEY/TRAVftg324I/AAAAAAAAACA/e2gJbTnOfxA/s1600/Cardiac_anatomy_PI.jpg))

Table 5.1 Reciprocating pump components and corresponding anatomy of the human heart

Reciprocating pump component	Human heart component(s)
Chamber	Left and right ventricle
Inlet plenum	Left and right atrium
Feed	Vena cava, pulmonary veins
Intake valve	Tricuspid valve, mitral valve
Exhaust valve	Pulmonary valve, aortic valve
System load	Left pulmonary artery (to lungs), aortic arch (to body)

flows through the tricuspid valve (a check valve) into the right ventricle. From there, blood is pumped through the pulmonary valve (the tricuspid valve shuts naturally) as the heart contracts and flows to the lungs, the system load of the right side of the heart. The blood that leaves the lungs is oxygenated (red) and serves as the feed to the left atrium (the plenum), then through the mitral valve to the right ventricle, and pumped through the aortic valve to the body. With each heartbeat, a volume of blood is pumped simultaneously to the lungs and to the body. The same volume is sent, but there is greater flow restriction on the body side, so the left ventricle works harder than the right.

Centrifugal Pumps

Unlike positive displacement pumps, the flow rate through dynamic pumps is not well characterized. A good demonstration of the principles of centrifugal pumps (a class of dynamic pumps) is to swing a filled bucket of water around in a vertical plane. It does not spill! The centrifugal force associated with spinning the bucket keeps it in place. Consider Fig. 5.5, a free-body diagram of the water (mass m) in the bucket at the instant it is vertical while being rotated at an angular speed ω and an arm + bucket length of r.

Newton's 2nd Law applied to the water is

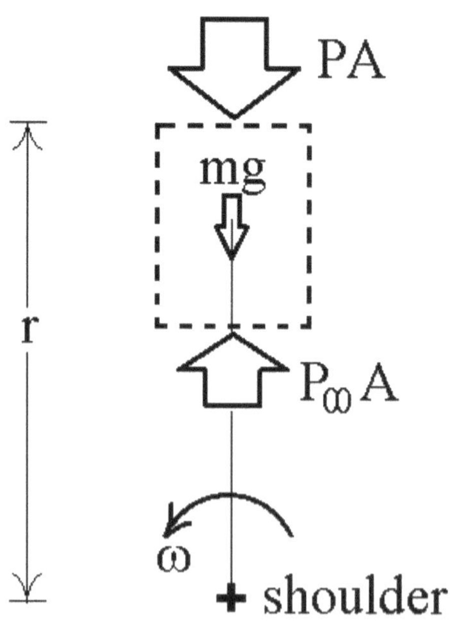

Fig. 5.5 Free-body diagram of water in a bucket being rotated in a vertical plane at its highest point

$$\sum F_y = \mathrm{ma}_y$$

The forces consist of a pressure force on the free surface ($P_\infty A$) acting upward, a pressure acting on the bottom of the bucket (PA) acting downward, the weight of the water (mg) acting down, and the acceleration is the centripetal acceleration. The force balance becomes

$$P_\infty A - PA - mg = -m\frac{V^2}{r} = -m\frac{(\omega r)^2}{r} = -m\omega^2 r$$

The pressure on the bottom of the bucket (when directly overhead) is

$$P = P_\infty + m(\omega^2 r - g)/A$$

A high pressure is created at the bottom of the upside-down bucket. The centrifugal effect is constant around the entire swing. The centripetal acceleration must exceed the gravitational acceleration and, therefore, the minimum angular velocity (below which you would get wet) is $\omega_{\min} = \sqrt{g/r}$. When the bucket is at the bottom of its swing, the sign of the gravity term is positive and the pressure at the bottom of the bucket exceeds the hydrostatic pressure.

Now imagine (do not try, unless you are in a bathing suit) the same demonstration with a hole in the bottom of the bucket. Water will flow out. It is not a good way to move water in a directed line (unless you are watering your lawn), but it shows the basic principle of a centrifugal pump. Water is mechanically forced to flow in a circle, creating a high pressure at the outer extreme, and this high-pressure liquid is bled off. A practical pump provides a means to replace the water (fed at the low pressure of the central hub) and a means to direct the flow.

Centrifugal pumps (and compressors, for gases) generally consist of an impeller contained within a housing (Figs. 5.6 and 5.7). The impeller has a shaft on one side driven by a motor (or equivalent) and vanes on the other that apply a radial force to the fluid, causing it to rotate. Fluid is contained within the housing, and the fluid rotation creates a high pressure on the outside needed to accelerate the fluid toward the axis (centrifugal force). Fluid is fed continuously to the center of the housing and forced to rotate as it flows radially outward. Fluid exits a pipe placed on the outer regions directed in a manner that allows directed flow outward to its intended target.

A big difference between positive displacement and centrifugal pumps is the maximum pressure that can be attained. In a positive displacement pump, if the shaft is rotating, fluid must be moving through the pump; otherwise, the internal pressures will be enormous (and limited by the torque the motor can supply). A properly designed and functioning relief valve will allow for fluid to flow in that case. On the other hand, the impeller of a centrifugal pump can rotate at its rated speed with a completely closed outlet. The input shaft work will be converted to thermal energy of the fluid (internal energy and/or heat transfer) in a dissipative process. That is, eXergy will be destroyed. The fluid will warm and lose heat to the environment.

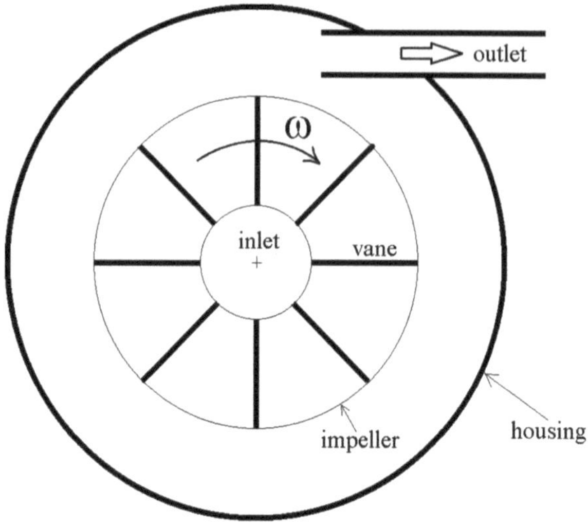

Fig. 5.6 Schematic of the internals of a centrifugal pump

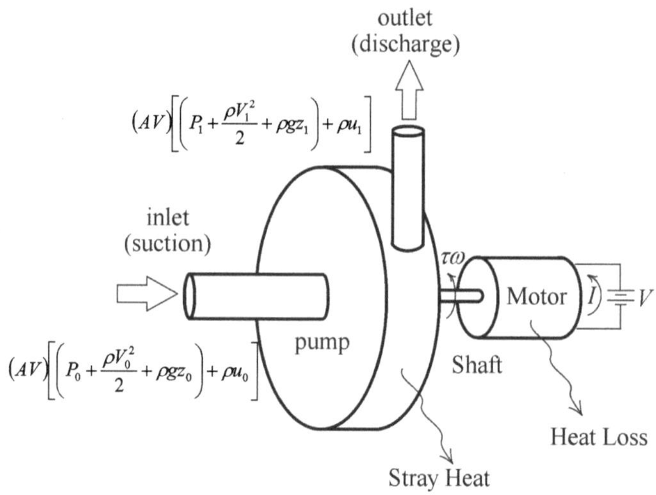

Fig. 5.7 Schematic view of centrifugal pump being driven by a DC motor, with energy transfers indicated

Operating a pump with closed valves can lead to overheating since the flow through the pump often acts as a form of cooling. In short, there is a clearly defined maximum pressure that a centrifugal pump can develop, and an order of magnitude estimate is obtained from the swinging bucket analogy, namely,

$$P - P_0 \leq \rho\omega^2 R^2/2$$

This result (not derived in detail here) is the pressure at the outer radius (R) of a fluid (of density ρ) rotating as a rigid body with an angular speed (ω), with a pressure P_O at the centerline (the feed pressure).

It is harder to establish a definite upper limit to the flow rate passing through, but it is closely related to this pressure and will be reconsidered after some more background and theoretical tools are developed for the analysis of these open systems.

Conservation of Mass for Open Systems

Reciprocating pumps are open systems as fluid enters and exits a device. Their behavior is inherently transient, but cyclical. Therefore, it can be viewed as operating in a steady state over a time span that is long compared to the time of a complete cycle (a heartbeat). The 1st and 2nd laws apply to control masses, but the following derivation of the *Reynolds Transport theorem* is used to obtain a general equation that applies to open systems.

The process diagram of Fig. 5.8 shows a control volume (CV, dashed lines) at two instants of time. The CV encloses a device of some kind whose size may change with time. A process occurs during which time mass will enter (m_i), shown at time t, and mass will exit (m_e), shown at a later time $t + \Delta t$. Heat and work are also exchanged with the surroundings. The combination of the mass inside the control volume at time t plus the mass that *will* enter is defined as a control mass. It is the same mass that is inside the control volume at time $t + \Delta t$ plus that which exited during the process. The properties of the inlet and outlets are different, in general. Mass, energy, and entropy are contained within the control volume, and their amounts can be different before and after the event.

The conservation of mass applied to the system, called a mass balance, is

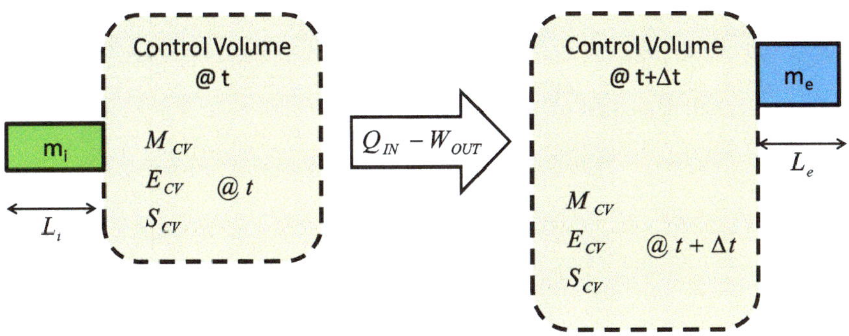

Fig. 5.8 Process diagram for a control volume with a single inlet and single outlet

$$\Delta M = (M_{\text{CV}@t+\Delta t} + m_e) - (M_{\text{CV}@t} + m_i) = 0$$

Rearranging and dividing by the elapsed time:

$$\frac{M_{\text{CV}@t+\Delta t} - M_{\text{CV}@t}}{\Delta t} = \frac{m_i}{\Delta t} - \frac{m_e}{\Delta t}$$

The left-hand side of the mass balance is a first derivative of the mass in the control volume with respect to time:

$$\lim_{\Delta t->0}\left(\frac{M_{\text{CV}@t+\Delta t} - M_{\text{CV}@t}}{\Delta t}\right) = \frac{dM_{\text{CV}}}{dt}$$

The mass of the inlet can be expressed in terms of the fluid properties as the density of the intake fluid (ρ_i) times its volume (A_iL_i), where A_i is the cross-sectional area of the inlet and L_i is the length of the inlet section, which is in turn related to the flow velocity normal to the area ($V_{\perp i}$) and elapsed time. That is,

$$m_i = \rho_i A_i V_{\perp i} \Delta t$$

Similarly, the exit mass is

$$m_e = \rho_e A_e V_{\perp e} \Delta t$$

Note that the fluid may enter at an oblique angle to the inlet area, and it is only the normal component of the velocity that defines the volume of the inlet mass. Also, it is the velocity relative to the control volume as some control surfaces may be moving.

Taking the limit as the time step shrinks to zero, the mass per unit time becomes the mass flow rate or rate of flux of mass:

$$\dot{m}_i = \lim_{\Delta t->0} \frac{m_i}{\Delta t} = \rho_i V_{\perp i} A_i$$

The mass balance equation becomes

$$\frac{dM_{\text{CV}}}{dt} = \dot{m}_i - \dot{m}_e$$

If there are multiple inlet and/or outlet streams, the mass balance equation is

$$\frac{dM_{\text{CV}}}{dt} = \sum_{\text{inlets}} \dot{m}_i - \sum_{\text{outlets}} \dot{m}_e$$

The velocity used is a mass-weighted average velocity across the inlet plane. A more general expression of the mass conservation equation, in integral form, is

$$\frac{\partial}{\partial t}\iiint_{CV} \rho d\forall = -\iint_{CS} \rho \vec{V} \cdot d\vec{A}$$

A line is drawn through the differential volume ($\forall$) to distinguish it from velocity. The left-hand side is a volume integral over the control volume (the rate of change of mass stored) and the right-hand side is a surface integral over the entire control surface and represents the net rate of efflux of mass. The area vector is in the direction of the outward normal, with a magnitude of a differential area (of magnitude dA). The dot product between the fluid velocity vector (relative to the CV) and the area vector gives the normal component of the velocity that crosses the control surface. The sign of inlet and outlet is automatically accounted for by the dot product. A positive dot product means the velocity points outward, and fluid exits, resulting in a decrease in the mass stored within the CV. A negative dot product means mass enters.

The volumetric flow rate is the product of an area and the normal component of fluid velocity that crosses it. It will generally be identified here as (AV), and not the symbol Q as is often used in fluid mechanics texts. The mass flow rate is the density times the volumetric flow rate, that is, $\dot{m} = \rho(\mathrm{AV})$.

First Law for Open Systems

The 1st Law of Thermodynamics applied to the open system of Fig. 5.8, called an energy balance, is

$$\Delta E_{\mathrm{CV}} = (E_{\mathrm{CV}@t+\Delta t} + m_e e_e) - (E_{\mathrm{CV}@t} + m_i e_i) = Q_{\mathrm{IN}} - W_{\mathrm{OUT}}$$

In this expression, the energy of the inlet (e_i) and exit (e_e) represents the total specific energy (i.e., energy per unit mass) of the fluid that enters and exits, respectively, and includes all forms of stored energy (internal, kinetic, and potential):

$$e = \frac{E}{M} = u + \alpha\frac{V^2}{2} + gz$$

The factor α (called the kinetic energy coefficient) before the kinetic energy term reflects that the mass-weighted velocity from mass conservation is different that the kinetic energy-weighted velocity. High-velocity fluid carries a greater kinetic energy than the same mass flux of a low-velocity fluid. A formal derivation for uniform density (density does not vary across the inlet or exit plane) flows shows that

$$\alpha = \frac{A^2 \int u^3 dA}{\left(\int u dA\right)^3}$$

For uniform velocity profiles, the factor $\alpha = 1$. For parabolic velocity profiles (found for laminar flow in pipes or between flat plates), $\alpha = 2.0$. Turbulent flows are more nearly uniform, and it is common to use $\alpha = 1$ as an acceptable engineering approximation.

Rearranging and dividing by the elapsed time:

$$\frac{E_{\text{CV@}t+\Delta t} - E_{\text{CV@}t}}{\Delta t} = \frac{Q_{\text{IN}}}{\Delta t} - \frac{W_{\text{OUT}}}{\Delta t} - \frac{m_e}{\Delta t} e_e + \frac{m_i}{\Delta t} e_i$$

The left-hand side is the first derivative of the total energy stored within the control volume per unit time:

$$\lim_{\Delta t -> 0} \left(\frac{E_{\text{CV@}t+\Delta t} - E_{\text{CV@}t}}{\Delta t} \right) = \frac{dE_{\text{CV}}}{dt}$$

The heat transfer represents heat that crosses the entire control surface. This heat transfer is identically zero for adiabatic devices. For devices that are not intended to exchange heat, but are not at the same temperature as the surroundings, there could be "stray heat." Note that the total heat transfer to the control volume includes the areas of the inlet and exit. A temperature gradient (conduction) or photon flux (radiation) across the inlet or exit plan will act in parallel with the energy that enters with the mass (called "advection" in some quarters, "convection" in others). This effect is commonly not recognized, and can be important, especially when the control volume is considered to be differentially small in size. There is a lot behind this sweeping statement, and the effect is important for computational fluid dynamics and in combustion phenomena. Dividing by the elapsed time and taking the limit yields the total heat transfer rate (a surface integral):

$$\dot{Q}_{\text{in}} = \lim_{\Delta t -> 0} \frac{Q_{\text{IN}}}{\Delta t}$$

The work represents all forms of work, including flow work (derived in Chap. 1). It is common and convenient to separate the work into flow work and all others, for reasons that will be apparent. To wit:

$$W_{\text{OUT}} = W_{\text{flow}} + W_{\text{non-flow}} = W_{\text{flow}} + W_{\text{electrical}} + W_{\text{shaft}} + W_{\text{expansion}} + \ldots$$

For brevity, it is common to call all nonflow work "shaft work" even if there is no shaft involved.

Recall from Chap. 1 that flow work is the work required to push fluid in and push fluid out, and is related to the inlet and exit masses, and the pressures and densities (typical of fluid mechanics analysis):

$$W_{\text{flow}} = m_e \frac{P_e}{\rho_e} \Delta t - m_i \frac{P_i}{\rho_i} \Delta t$$

Expressed in terms of specific volumes (typical of thermodynamic analysis):

$$W_{\text{flow}} = m_e P_e v_e \Delta t - m_i P_i v_i \Delta t$$

The rate of doing work out (power) is obtained by dividing by the elapsed time and taking the limit:

$$\dot{W}_{\text{OUT}} = \lim_{\Delta t -> 0} \frac{W_{\text{OUT}}}{\Delta t} = \dot{m}_e \frac{P_e}{\rho_e} - \dot{m}_i \frac{P_i}{\rho_i} + \dot{W}_{\text{SHAFT}}$$

It is understood that shaft power includes all nonflow work exchanges, and is the work done by the control volume on the surroundings (and can be a negative value, such as for a pump). The final form of the open system energy equation is

$$\frac{dE_{\text{CV}}}{dt} = \dot{Q}_{\text{IN}} - \dot{W}_{\text{SHAFT}} + \dot{m}_i \left(e_i + \frac{P_i}{\rho_i} \right) - \dot{m}_e \left(e_e + \frac{P_e}{\rho_e} \right)$$

It is common and sometimes convenient to introduce a combination property, the enthalpy, which combines the internal energy with the flow work. That is,

$$\text{Enthalpy} = h = u + \frac{P}{\rho} = u + Pv$$

The 1st Law is, writing out the energy in terms of its components,

$$\frac{d(U_{\text{CV}} + KE_{\text{CV}} + PE_{\text{CV}})}{dt} = \dot{Q}_{\text{IN}} - \dot{W}_{\text{SHAFT}} + \dot{m}_i \left(h_i + \alpha_i \frac{V_i^2}{2} + gz_i \right) - \dot{m}_e \left(h_e + \alpha_e \frac{V_e^2}{2} + gz_e \right)$$

If there are more than one inlet and/or outlet streams, the energy balance (1st Law) equation is

$$\frac{dE_{\text{CV}}}{dt} = \dot{Q}_{\text{IN}} - \dot{W}_{\text{SHAFT}} + \sum_{\text{inlets}} \dot{m}_i \left(e_i + \frac{P_i}{\rho_i} \right) - \sum_{\text{outlets}} \dot{m}_e \left(e_e + \frac{P_e}{\rho_e} \right)$$

A more general integral expression of the 1st Law is

$$\frac{\partial}{\partial t}\iiint\limits_{CV} \rho e d\forall = \dot{Q}_{\text{IN}} - \dot{W}_{\text{SHAFT}} - \iint\limits_{CS} \rho\left(e + \frac{P}{\rho}\right) \vec{V} \cdot d\vec{A}$$

The left-hand side is a volume integral over the control volume (the rate of change of energy stored). The right-hand side contains a surface integral over the entire control surface that represents the net rate of efflux of energy, plus the flow work required to move the mass into and out of the control volume.

Many devices involve a single inlet and a single outlet and operate at a steady state. The energy equation for these devices, rearranged to be in the form of energy in = energy out, reduces to

$$\left(h_i + \alpha_i \frac{V_i^2}{2} + gz_i\right) + q_{\text{in}} = w_{S,\text{OUT}} + \left(h_e + \alpha_e \frac{V_e^2}{2} + gz_e\right)$$

for single-inlet, single-exit devices@SS. The heat and work per unit mass are introduced and defined by

$\frac{\dot{Q}_{\text{IN}}}{\dot{m}} = q_{\text{in}}$ heat addition per unit of mass flow

$w_{s,\text{OUT}} = \frac{\dot{W}_{\text{SHAFT,OUT}}}{\dot{m}}$ shaft work output per unit of mass flow

While it is convenient in many cases to use the combined thermodynamic property enthalpy, it combines a mechanical property (Pv) with a thermal property (u). It is instructive to view the energy equation with them separated:

$$\left(\frac{P_i}{\rho_i} + \alpha_i \frac{V_i^2}{2} + gz_i\right) = \left(\frac{P_e}{\rho_e} + \alpha_e \frac{V_e^2}{2} + gz_e\right) + w_{s,\text{out}} - [(u_e - u_i) + q_{\text{in}}]$$

for single-inlet and exit devices @SS.

This equation is a form of the celebrated Bernoulli equation. There are three properties of a flow that are mechanical in nature: pressure divided by density (from flow work), kinetic energy, and potential energy. The sum of these properties can be changed by either a mechanical device (pump or turbine) or by thermal effects (internal energy and heat transfer). Again, for engineering calculations, the velocity profile shape factor α is almost always set to a value of unity.

Second Law for Open Systems

The entropy balance applied to the open system of Fig. 5.8 is

$$\Delta S_{\mathrm{CV}} = (S_{\mathrm{CV}@t+\Delta t} + m_e s_e) - (S_{\mathrm{CV}@t} + m_i s_i) = \iint\limits_{\mathrm{CS}} \frac{\delta Q_{\mathrm{IN}}}{T_B} + \iiint\limits_{\mathrm{CV}} \delta\Sigma$$

In this expression, the specific entropy of the inlet (s_i) and exit (s_e) represents the entropy per unit mass of the mass that enters and exits, respectively. The entropy transfer (first term on right side) is shown as a surface integral of the heat transfer (divided by the corresponding boundary temperature). Entropy transfer is associated with heat that crosses the boundary (control surface). The entropy generation is shown as a volume integral to indicate that it is an internal process within the control volume.

Rearranging and dividing by the elapsed time:

$$\frac{S_{\mathrm{CV}@t+\Delta t} - S_{\mathrm{CV}@t}}{\Delta t} = \frac{1}{\Delta t}\iint\limits_{\mathrm{CS}} \frac{\delta Q_{\mathrm{IN}}}{T_B} + \frac{1}{\Delta t}\iiint\limits_{\mathrm{CV}} \Sigma - \frac{m_e}{\Delta t} s_e + \frac{m_i}{\Delta t} s_i$$

The left-hand side is the first derivative of the total energy stored within the control volume with respect to time in the limit of elapsed time shrinking to zero. Note that the entropy produced in the inlet and outlet masses need not be included in the derivation as this effect vanishes in the limit of elapsed time.

The heat transfer carries entropy with it (entropy transfer), with the boundary temperature where the heat crosses the control surface in the denominator (in absolute units). For devices that experience stray heat, the boundary is generally placed where the temperature equals the local ambient (often but not always the dead state), and there is entropy generation within the boundary layer as well as within the device itself. Dividing by the elapsed time and taking the limit yields the total heat transfer rate (a surface integral):

$$\frac{\dot{Q}_{\mathrm{in}}}{\overline{T}_B} = \lim_{\Delta t -> 0} \frac{1}{\Delta t} \iint\limits_{\mathrm{CS}} \frac{\delta Q_{\mathrm{IN}}}{T_B}$$

The instantaneous rate of entropy production is:

$$\underset{\mathrm{CV}}{\dddot{\Sigma}} = \lim_{\Delta t -> 0} \frac{1}{\Delta t} \iiint\limits_{\mathrm{CV}} \delta\Sigma$$

The entropy balance is thereby expressed as:

$$\frac{dS_{\mathrm{CV}}}{dt} = \frac{\dot{Q}_{\mathrm{IN}}}{\overline{T}_B} + \underset{\mathrm{CV}}{\dddot{\Sigma}} + \dot{m}_i s_i - \dot{m}_e s_e$$

The 2nd Law of Thermodynamics states that:

$$\overset{\cdots}{\underset{CV}{\Sigma}} \geq 0$$

If there are more than one inlet and/or outlet streams, the entropy balance equation is:

$$\frac{dS_{\mathrm{CV}}}{dt} = \frac{\dot{Q}_{\mathrm{IN}}}{\overline{T}_B} + \overset{\cdots}{\underset{\mathrm{CV}}{\Sigma}} + \sum_{\mathrm{inlets}} \dot{m}_i s_i - \sum_{\mathrm{outlets}} \dot{m}_e s_e$$

A more general expression of the entropy balance equation is:

$$\frac{\partial}{\partial t} \iiint_{\mathrm{CV}} \rho s d\forall = \frac{\dot{Q}_{\mathrm{IN}}}{\overline{T}_B} + \overset{\cdots}{\underset{\mathrm{CV}}{\Sigma}} - \iint_{\mathrm{CS}} \rho s \ \vec{V} \cdot d\vec{A}$$

The left-hand side is a volume integral over the control volume (the rate of change of entropy stored) and on the right-hand side is a surface integral over the entire control surface and represents the net rate of influx of entropy.

The eXergy destruction rate is the entropy production rate multiplied by the dead state temperature. That is,

$$\begin{pmatrix} \text{eXergy} \\ \text{Destruction} \\ \text{Rate} \end{pmatrix} = \underset{D}{e\overset{\cdots}{X}} = T_\infty \overset{\cdots}{\Sigma}$$

Momentum Equation for Open Systems

A derivation of the momentum equation (Newton's Law of Motion) has some subtleties that are not fully detailed here as the focus is on energy concepts. More details can be found in fluid mechanics textbooks. For a control volume that is linearly accelerating with respect to a fixed coordinate frame (not rotating), the result is

$$M_{\mathrm{CV}} \frac{d\vec{V}_{\mathrm{CM}}}{dt} = \vec{F}_{\mathrm{surface}} + \vec{F}_{\mathrm{body}} + \sum_{\mathrm{inlets}} \dot{m}_i \vec{V}_i - \sum_{\mathrm{outlets}} m_e \vec{V}_e$$

where

$\vec{V}_{\mathrm{CM}}$ is the velocity of the center of mass of the control volume

$\vec{F}_{\mathrm{surface}}$ is the net surface force (pressure, fluid friction and reaction forces, generally)

$\vec{F}_{body}$ is the net body force acting (gravity)

$\sum_{inlets} \dot{m}_i \vec{V}_i$ is the momentum flux into the control volume relative to the control volume.

$\sum_{outlets} m_e \vec{V}_e$ is the momentum flux out of the control volume relative to the control volume

The momentum flux terms are considered to be virtual forces. They are momentum fluxes, not actual forces, but they have the equivalent effect of a force on a structure. They are put to use as the propulsive effect of jet engines, for example.

Open System Devices: A Toolbox

In Chap. 3, heat engines and heat pumps were defined based on their overall thermodynamic objective (converting a thermal input to mechanical output or vice versa) without specifying a working fluid or the type of equipment used. Actual engineering systems designed to convert energy from one form to another do so through a series of processes that act on a working fluid. Each process has a different thermodynamic objective and is of very different physical design. This section introduces a toolbox of individual devices, each with a very specific thermodynamic purpose. Power plants (mobile or stationary), refrigeration, and other thermodynamic engineering systems are constructed out of several of these organized in a way to meet an overall objective. Part 2 of this book investigates alternative ways these individual components are combined into systems.

Table 5.2 summarizes the objectives of important engineering devices from a thermodynamic point of view. Most of these devices are open systems (i.e., those through which mass flows). Most are single-inlet, single-outlet devices that operate at steady state, shown schematically in Fig. 5.9. The total power (energy flux) of a flowing stream at a given inlet or outlet (state i) is given by the sum of the thermal (u), flow work (Pv), kinetic ($V^2/2$), and potential (gz). For thermodynamic analysis, the thermal and flow work is conveniently grouped using the enthalpy ($h = u + Pv$), but "u" and "Pv" play a very different role from a 2nd Law perspective.

For a given stream ("i"), the energy flux is given by Energy Flux$_i$ = $\dot{m}\left(h_i + \frac{V_i^2}{2} + gz_i\right)$

The general form of the energy equation for the single-inlet, single-outlet device (with a profile shape factor α set to unity) is

$$\frac{dE_{cv}}{dt} = \dot{Q} - \dot{W}_{shaft} + \dot{m}\left(h_1 + \frac{V_1^2}{2} + gz_1\right) - \dot{m}\left(h_2 + \frac{V_2^2}{2} + gz_2\right)$$

Table 5.2 Engineering devices and their thermodynamic objectives

Device	Thermodynamic objective	Process idealization
Nozzle	Convert enthalpy to kinetic energy	Isentropic expansion
Diffuser	Convert kinetic energy to enthalpy	Isentropic compression
Compressors	Convert shaft energy to enthalpy of a gas (with substantial pressure rise)	Isentropic compression
Fans	Convert shaft energy to enthalpy of a gas (with minimal pressure rise)	Isentropic compression
Steam or gas turbine	Convert enthalpy of a gas to shaft energy	Isentropic expansion
Hydroelectric turbine	Convert potential energy of a liquid to shaft energy	Isentropic expansion
Wind turbine	Convert kinetic energy of air to shaft energy	Isentropic deceleration
Pump	Convert shaft energy to enthalpy of a liquid	Isentropic compression
Heat exchangers	Transfer enthalpy from one fluid to another	Isobaric heating/cooling (for each stream)
Heaters/coolers	Change enthalpy of a single stream	Isobaric heating/cooling
Throttle	Drop pressure of a stream	Isenthalpic expansion
Combustor	Convert chemical enthalpy into thermal enthalpy	Adiabatic (chemical reaction)
Motor	Convert electrical energy to shaft energy	Adiabatic (closed system)
Generator	Convert shaft energy to electrical energy	Adiabatic (closed system)

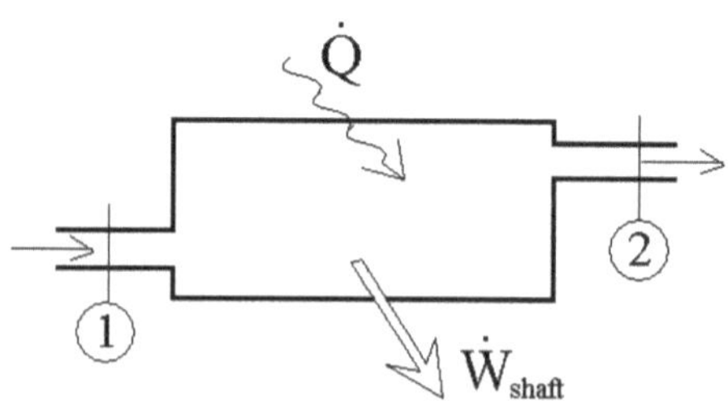

Fig. 5.9 General schematic of a single-inlet, single-outlet open system device

Table 5.3 Primary and secondary energy flux terms

Device	q	w_{shaft}	h_1	$V_1^2/2$	gz_1	h_2	$V_2^2/2$	gz_2
Nozzle	~0	=0	✓	✓	~0	✓	✓	~0
Diffuser	~0	=0	✓	✓	~0	✓	✓	~0
Compressors	~0	✓	✓	~0	~0	✓	~0	~0
Steam or gas turbine	~0	✓	✓	~0	~0	✓	~0	~0
hydroelectric turbine	~0	✓	✓	~0	✓	✓	✓	✓
Fans	~0	✓	✓	~0	~0	✓	~0	~0
Pump	~0	✓	✓	~0	~0	✓	~0	~0
Throttle	~0	=0	✓	~0	~0	✓	~0	~0
Heat exchangers (multiple inlet, exit)	~0	=0	✓	~0	~0	✓	~0	~0
Heaters	✓	=0	✓	~0	~0	✓	~0	~0

At steady state and expressed on a per-unit mass flow basis (divided by mass flow rate):

$$0 = q - w_{\text{shaft}} + \left(h_1 + \frac{V_1^2}{2} + gz_1\right) - \left(h_2 + \frac{V_2^2}{2} + gz_2\right)$$

Table 5.3 shows which terms are associated with the primary objective of the device (✓), those that are not identically zero and not part of the objective and usually (but not always) neglected in engineering calculations (~0), and those that are identically zero (=0).

Mass conservation applied to the control volume of a single-inlet, single-outlet is

$$\frac{dM_{\text{CV}}}{dt} = \dot{m}_1 - \dot{m}_2$$

In terms of the specific volume of the fluid:

$$\frac{d}{dt}\left(\frac{V\!\!\!/_{\text{CV}}}{\overline{v}}\right) = \frac{(\text{AV})_1}{v_1} - \frac{(\text{AV})_2}{v_2}$$

where $\overline{v}$ is the average specific volume of the fluid in a control volume of volume $V\!\!\!/_{CV}$

In terms of the density of the fluid:

$$\frac{d}{dt}(\overline{\rho}V\!\!\!/) = \rho_1(\text{AV})_1 - \rho_2(\text{AV})_2$$

where $\overline{\rho}$ is the average density of the fluid in the control volume.

Mass Implications for Steady-State Flows and Compressibility

If the flow is steady in nature, three equivalent forms of the mass conservation equation for single-inlet, single-outlet devices are

$$\dot{m}_1 = \dot{m}_2$$

$$\frac{(\mathrm{AV})_1}{v_1} = \frac{(\mathrm{AV})_2}{v_2}$$

$$\rho_1(\mathrm{AV})_1 = \rho_2(\mathrm{AV})_2$$

Note that if the inlet and exit specific volumes (or densities) are different, which is common for engineering devices like compressors, turbines, or heaters, the mass flow rate is conserved, but the volumetric flow rate is not.

The ratio of the exit velocity to inlet velocity is related to ratios of density (or specific volume) and cross-sectional area:

$$\frac{V_2}{V_1} = \left(\frac{\rho_1}{\rho_2}\right)\left(\frac{A_1}{A_2}\right) = \left(\frac{v_2}{v_1}\right)\left(\frac{A_1}{A_2}\right)$$

Nozzles and diffusers are flow devices whose thermodynamic objective is to convert flow work into kinetic energy (for nozzles) or vice versa (for diffusers). This change is accomplished by a simple flow channel whose exit area differs from the inlet. Typically, engineering students will first study incompressible flows, where the density ratio is unity. A nozzle will become associated with a converging flow channel (i.e., whose exit area is smaller than the inlet) and a diffuser will be associated with a diverging channel. This association is erroneous, however, and it can be difficult to unlearn when compressible flows are studied, where just the opposite trend can occur. Formal treatment of this behavior is the topic of a more advanced study of compressible flow phenomena, where it is shown that the qualitative behavior of nozzles and diffusers depends on the local Mach number (the ratio of the flow velocity to the local speed of sound). A converging channel will act as a nozzle for a subsonic flow (Mach number less than unity), but as a diffuser for a supersonic flow. A diverging channel will act as a subsonic diffuser, but a supersonic nozzle. The definition of a nozzle and diffuser is based on its function (accelerate or decelerate a flow, respectively), and not on its shape.

Having said that, it is reasonable to treat a large class of flows of air (or other gas) as incompressible. Compressible flow theory shows that when the Mach number is less than approximately 0.3, the flow can be reasonably treated as being incompressible. The speed of sound of air at nominal room conditions is 346 m/s (774 mph). Therefore, the flow can be treated as incompressible for air velocities below approximately 115 m/s (258 mph). That includes the aerodynamic flow around cars and trucks, the flow of air over a wind turbine (but maybe not near the tip of the rotating blade), or through any fan. On the other hand, it is not difficult to develop high-enough flow speeds for internal flows of air in variable area channels for compressible flow phenomena to be important.

Thus, there is a subtle distinction between the term "incompressible flow" and "incompressible fluid." The speed of sound of a truly incompressible fluid is infinite. Liquid water has a finite speed of sound, and the important technology of SONAR depends on it. Therefore, liquid water can give rise to compressible flow phenomena. On the other hand, air would seem to be inherently "compressible," but there are many applications where air can be safely treated as incompressible. A more precise definition of incompressible behavior is that kinetic energy changes of the fluid are much less than the changes in enthalpy (which could be due to the difference in flow work at inlet and exit, pressure times specific volume, or due to differences in internal energy).

An incompressible flow is one where the specific volume is assumed to be a constant, a reasonable approximation for most (not all) flows of a liquid and for many flow applications of a gas. In that case, the total mass stored within the control volume (or average specific volume or density) does not change, even for transient processes for which both mass flow rates change with time together. That is,

$$\dot{m}_1 = \dot{m}_2 \quad \text{(for flows of incompressible fluids, even transient flows)}$$

The volumetric flow rate is also conserved:

$$(\mathrm{AV})_1 = (\mathrm{AV})_2$$

The average flow velocities are related by the ratio of the cross-sectional area:

$$V_2 = V_1 \left(\frac{A_1}{A_2} \right)$$

That is, if the exit area is larger than the inlet (diverging channel), the exit velocity will be lower than the inlet (a diffuser), for incompressible flows. This relationship does not hold for rocket nozzles, however. Careful study of incompressible flows develops an intuition that proves to be limited and must be unlearned. Therefore, a general definition of the colloquial term "rocket science" could be "the study of phenomena for which acquired intuition fails."

Ideal Pump Analysis

The objective of a pump, from a thermodynamic point of view, is to convert mechanical shaft power into mechanical fluid power. Since this conversion is from one mechanical form to another, it should be theoretically possible to do so completely, with no conversion into thermal forms of power (internal energy or heat transfer). The ideal pump is therefore defined accordingly, and the minimum power input required to achieve the objective is the main desired quantity.

Recall the general forms of the mass, energy balance (1st Law) and entropy balance (2nd Law) for a single-inlet, single-outlet device (the pump of Fig. 5.7):

$$\frac{dM_{\mathrm{CV}}}{dt} = \dot{m}_i - \dot{m}_e$$

$$\frac{dE_{\mathrm{CV}}}{dt} = \dot{Q}_{\mathrm{IN}} + \dot{W}_{\mathrm{SHAFT,IN}} + \dot{m}_i\left(e_i + \frac{P_i}{\rho_i}\right) - \dot{m}_e\left(e_e + \frac{P_e}{\rho_e}\right)$$

$$\frac{dS_{\mathrm{CV}}}{dt} = \frac{\dot{Q}_{\mathrm{IN}}}{T_B} + \underset{\mathrm{CV}}{\dddot{\Sigma}} + \dot{m}_i s_i - \dot{m}_e s_e$$

The following assumptions define an ideal pump:

- Steady state operation: $dM_{\mathrm{CV}}/dt = dE_{\mathrm{CV}}/dt = dS_{\mathrm{CV}}/dt = 0$
- Negligible change in kinetic and potential energy: $e = u$
- Adiabatic (i.e. no stray heat): $\dot{Q}_{\mathrm{IN}} = 0$
- Reversible: $\underset{\mathrm{CV}}{\dddot{\Sigma}} = 0$
- Incompressible fluid: $\rho_i = \rho_e = \rho$ and u, s are function of temperature only

These assumptions result in the simplified 1st and 2nd laws, written first in the same form (with zeroes for neglected quantities):

$$0 = \dot{m}_i - \dot{m}_e$$

$$0 = 0 + \dot{W}_{\mathrm{SHAFT,IN}} + \dot{m}_i\left(u_i + \frac{P_i}{\rho_i}\right) - \dot{m}_e\left(u_e + \frac{P_e}{\rho_e}\right)$$

$$0 = 0 + 0 + \dot{m}_i s_i - \dot{m}_e s_e$$

The mass balance simplifies to

$$m_i = \dot{m}_e = \dot{m}$$

The incompressible approximation results in equal volumetric flow rates as well. The entropy balance simply states that the inlet and exit entropies are equal:

$$s_i = s_e$$

A further consequence of the incompressible fluid approximation is that the temperature is the same at inlet and exit, and therefore, the internal energy is the same.

Notice that the two assumptions adiabatic and reversible result in an isentropic flow. It is common to state these assumptions as simply isentropic, which implies adiabatic and reversible. However, it is theoretically possible for there to be irreversibilities (entropy generation) and simultaneous heat loss so that the entropy of the fluid is the same entering and exiting from a device with internal irreversibilities. Invoking both adiabatic and reversible assumptions rather than invoking an isentropic assumption eliminates that possibility.

The resulting input shaft power needed to pressurize the fluid is, in several equivalent forms,

$$\dot{W}_{\text{SHAFT,IN}} = \frac{\dot{m}}{\rho}(P_e - P_i)$$

$$\dot{W}_{\text{SHAFT,IN}} = \dot{m}\ v(P_e - P_i)$$

$$\dot{W}_{\text{SHAFT,IN}} = (\text{AV})(P_e - P_i)$$

$$\dot{W}_{\text{SHAFT,IN}} = \dot{m}\ (h_e - h_i)$$

The work of the pump is often called "Vee dee Pee" work (*vdP*), as indicated by the second form. Integrated between two pressure levels, *vdP* represents the minimum work required to change the pressure of a fluid in an open system. In closed systems, by contrast, work of expansion (i.e., from a moving piston in a cylinder) is "Pee dee Vee" work (*PdV*), which, when integrated, is the work required to change the volume of a closed system. Do not confuse open and closed system analysis.

A similar analysis for hydroelectric turbines holds, but the maximum power is extracted as the incompressible fluid undergoes a pressure decrease.

The last form, which involves enthalpy directly, is commonly obtained when the general energy equation is reduced from the form:

$$\frac{d(U_{\text{CV}} + KE_{\text{CV}} + PE_{\text{CV}})}{dt} = \dot{Q}_{\text{IN}} - \dot{W}_{\text{SHAFT}} + \dot{m}_i\left(h_i + \alpha_i\frac{V_i^2}{2} + gz_i\right) - \dot{m}_e\left(h_e + \alpha_e\frac{V_e^2}{2} + gz_{e3}\right)$$

Most traditional thermodynamics textbooks work with this form, which combines the internal energy and the flow work terms together into enthalpy. This form is convenient for many devices, particularly those that are designed to convert thermal energy forms, and the flow work is considered to be irrelevant. However, a common error is to then treat the enthalpy like the specific volume and internal energy, namely, as functions of temperature only. Recall that the change in enthalpy between two states that are both compressed liquids is

$$h_2 - h_1 = u_2 - u_1 + v(P_2 - P_1) \quad \text{for incompressible fluids.}$$

The pressure effect of the enthalpy for compressed liquids is very important for the pump.

For many thermodynamic systems, pumps are necessary in order to move fluid through various devices. However, their energy consumption is actually quite small compared to that in other devices. It is therefore reasonable to consider the pump to be ideal, or reversible. For situations where the performance of a real pump (with irreversibilities) is sought, it is common to express them in terms of a 2nd Law efficiency, called isentropic efficiency, defined next.

Isentropic Pump Efficiency

Determination of irreversibilities from first principles is usually not possible. It is common engineering practice, then, to define appropriate measures that compare the performance of actual devices to ideal (reversible) devices, whose performance can often be determined from first principles. Therefore, some form of a 2nd Law efficiency is defined. The value of the 2nd Law efficiency can be determined from measurements on an actual device. Furthermore, even if 2nd Law efficiencies cannot be determined, they can be introduced into theoretical models, and their effect on the performance of systems evaluated, a kind of sensitivity analysis.

In the case of devices that convert energy from one mechanical form to another (like pumps, compressors, and turbines), efficiencies called "isentropic efficiencies" are defined. For work-absorbing devices (conversion of shaft power into fluid power, such is in pumps, fans, and compressors), the isentropic efficiency is defined as the minimum work (or power) required divided by the actual work:

$$\eta_{II} = \frac{\text{Minimum Work Required}}{\text{Actual Work Required}} \quad \text{for work-absorbing devices (pumps, fans, compressors)}$$

For work-producing devices (conversion of fluid power into shaft power, such as in turbines):

$$\eta_{II} = \frac{\text{Actual Work Output}}{\text{Maximum Possible Work Output}} \quad \text{for work-producing devices (turbines)}$$

Nozzles and diffusers have similar 2nd Law efficiencies, except that the objective of those devices is to convert mechanical fluid energy into kinetic energy (nozzles) or vice versa (diffusers). Also, the objective of wind turbines is to convert kinetic energy of an air stream into output mechanical power.

The comparison between the actual device and the hypothetical reversible device is generally defined such that the inlet state and flow rates are the same, and there is something common with the output, usually pressure. The fluid state at the exit, which is defined by two independent thermodynamic properties, is different in the two cases. The ideal device is uniquely defined by the pressure and the entropy (which is the same as the inlet). This hypothetical state is given the subscript "2s," while the actual output state is given the subscript "2." The real device has a different entropy value and, therefore, a different temperature and internal energy.

Applied to work devices, the following basic assumptions are retained for both devices:

- Steady-state operation
- Adiabatic (i.e., neglect stray heat)
- Negligible kinetic and potential energy changes

The 1st Law balance under these conditions becomes for the shaft input power (negative for work-producing devices):

$$\dot{W}_{\text{SHAFT,IN,ACTUAL}} = \dot{m}(h_2 - h_1)$$

$$\dot{W}_{\text{SHAFT,IN,REVERSIBLE}} = \dot{m}(h_{2s} - h_1)$$

Therefore, the isentropic efficiencies are, noting that the mass flow rate cancels:

$$\eta_{II} = \frac{\text{Minimum}}{\text{Actual}} = \frac{h_{2s} - h_1}{h_2 - h_1} \quad \text{for work absorbing devices (pumps, compressors)}$$

$$\eta_{II} = \frac{\text{Actual}}{\text{Maximum}} = \frac{h_1 - h_2}{h_1 - h_{2s}} \quad \text{for work producing devices (turbines)}$$

Example: A fire pump is designed to deliver a flow rate of (AV) = 60 gallons per minute of seawater (density $\rho = 1050$ kg/m^3, specific heat $c = 4200$ J/kg/K) with $H = 100$ feet of head. Determine the input shaft power required and the temperature rise of the water if the isentropic efficiency (η_{pump}) of the pump is 60%.

The minimum possible enthalpy change is

$$h_{2S} - h_1 = u_{2s} - u_1 + v(P_2 - P_1)$$

Since this ideal is isentropic, it is also isothermal (for incompressible fluids) and the pressure change equals the head (in pressure units):

$$h_{2S} - h_1 = \frac{1}{\rho}(\rho_{H2O}gH)$$

$$= \frac{1}{1050kg/m^3}\left(\left(1000\frac{kg}{m^3}\right)\left(9.81\frac{m}{\sec^2}\right)(100ft) \cdot \left|\frac{0.3048m}{ft}\right|\right)$$

$$h_{2S} - h_1 = 285\frac{J}{kg}$$

Using the definition of isentropic efficiency to solve for the actual power:

$$\dot{W}_{\text{actual}} = \frac{\dot{W}_{\text{minimum}}}{\eta_{\text{pump}}} = \frac{\rho(AV)(h_{2s} - h_1)}{0.6}$$

$$= \frac{1050 kg/m^3(60\text{gal}/\min)(285J/kg)}{0.6} \cdot \left|\frac{m^3}{264\text{gal}}\right| \cdot \left|\frac{\min}{60\text{sec}}\right|$$

$$\dot{W}_{\text{actual}} = 1892 \text{ Watts}$$

To estimate the temperature rise, consider the actual change in enthalpy:

$$h_2 - h_1 = \frac{h_{2s} - h_1}{\eta_{\text{pump}}} = \frac{285J/kg}{0.6} = 475J/kg = u_2 - u_1 + v(P_2 - P_1)$$

Since $v(P_2 - P_1) = 285J/kg$, the change in specific internal energy is

$$u_2 - u_1 = 190J/kg$$
$$u_2 - u_1 = c_v(T_2 - T_1) = 285J/kg$$

The change in temperature is

$$T_2 - T_1 = \frac{190J/kg}{4200J/kg/\Delta^o C} = 0.045^o C$$

Notice that the temperature rise associated with the eXergy destruction is rather small, essentially unmeasurable, and certainly impractical to put to use, even as a water heater.

Deeper Dives: Flow System Analysis

Note: The remainder of this chapter involves analysis closer to traditional courses in fluid mechanics than thermodynamics, and can be skipped without significant loss of continuity. A basic understanding of matching pumps to flow systems is methodically developed. The thermodynamic possibilities of extracting useful energy from the wind are also developed.

Before analyzing the pump, the basics of a flow system is considered from a thermodynamic point of view. More thorough treatments are developed in textbooks on fluid mechanics, generally from a momentum point of view. A pump is usually required to move a liquid through a flow system. For example, the right side of the heart pumps blood to the lungs, through a network of capillaries that allow dissolved CO_2 to pass into the breathed air, and for O_2 to be absorbed into the blood, which returns to the left side of the heart, which pumps oxygenated blood to the body. In that complex network, O_2 is used by the body for metabolism and CO_2, a waste product, is carried back to the heart with the blood. Two values of blood pressure are commonly measured, low and high. The difference in pressure between the systolic (high number) and diastolic (low number) blood pressure is the pressure difference required to move the blood across the complex capillary network and therefore through the circulatory system. The heart maintains that pressure difference by repetitive beating.

The circulatory system has an interesting energy storage characteristic that helps to reduce the pulsing nature of the heartbeat. The arteries and veins (the large vessels) have a natural compliance, which is to say they act like balloons. Their volume increases with pressure. During active heart contractions, energy is stored in the spring-like nature of the vessels, and as the heart relaxes, the vessels release that energy and continue to maintain pressure as blood flows through the capillaries to the veins. The common "hardening of the arteries" as people age represents a reduction in the effectiveness of this remarkable characteristic. It will not, however, be investigated further here, although the effect is often used in engineering practice, not by pipes whose inner diameter varies with pressure, but by devices called accumulators.

The engineering design of flow systems involves the selection of a pump that is suitable for a specific flow application. An undersized pump will not be able to provide sufficient flow rate and/or pressure (head) to the intended target, and an oversized pump will incur unnecessary capital expenditures (cost too much) and will operate below its capacity, which generally will result in a low 1st Law efficiency and unnecessarily high operating costs (too much input power required).

Generally, liquids are delivered through pipes and fittings. Pipes are straight sections of constant cross-sectional area, and the flow is opposed by a fluid friction force along the inner wall. A gravitational effect is superimposed if the pipe is not horizontal and can assist the flow (downward flow) or oppose it (upward flow). The frictional effect gives rise to eXergy destruction in what is often called a "flow loss," a misleading term. It is not a loss of flow, but a loss of eXergy. These losses are classified, by convention, as major losses (even if they turn out to be minor in magnitude), and they involve an irreversible conversion of mechanical to thermal energy. The fluid gets immeasurably hotter. On the other hand, it is generally impossible to configure a flow delivery system with a completely straight run. The flow path must inevitably be turned. There are also good reasons to change the cross-sectional area, or split or merge flows. Irreversibilities are associated with each of these components, classified as minor losses (even though they may turn out to be major in magnitude).

Liquid Flow in Straight Pipes: Major Flow Losses

A schematic of the flow of a fluid through a straight section of pipe at an angle θ to the horizontal is shown in Fig. 5.10 in the form of a free-body diagram. The control volume (dashed line) is defined to be the fluid within the pipe. Fluid enters across location 1 and exits at 2. A pressure force acts in the direction of flow at 1 and

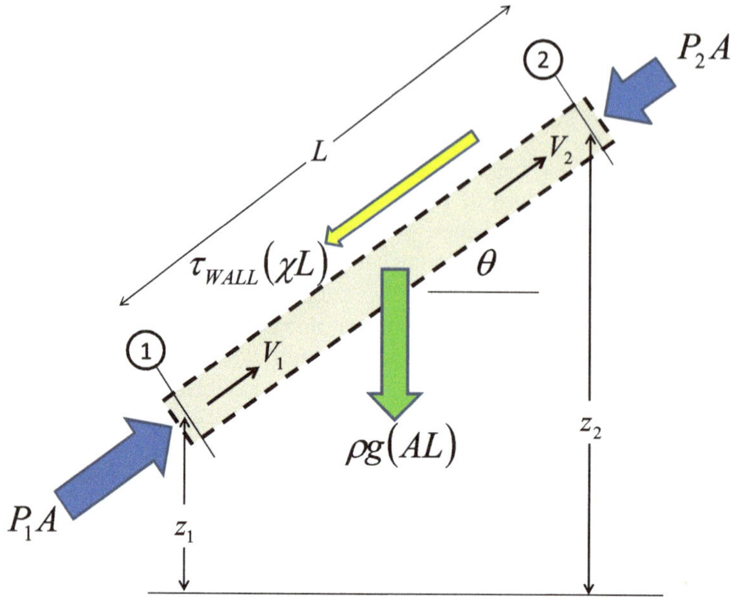

Fig. 5.10 Free-body diagram of fluid in a straight section of pipe of length L, cross-sectional area A, perimeter χ at an angle θ from the horizontal

opposes the flow at 2. The boundary in contact with the pipe inner wall experiences a frictional force that opposes the flow. The shear stress, τ_{wall}, is the frictional force per unit area. The weight of the fluid acts in a vertical direction. The pipe need not be circular in cross-section, and is defined by a hydraulic diameter, defined as $D_H = 4A/\chi$, where A is the cross-sectional area and χ the perimeter (which equals πD for a circular cross-section). The hydraulic diameter gives the same area to perimeter ratio as a circle.

Mass conservation applied to the control volume is

$$\frac{dM_{\text{CV}}}{dt} = \dot{m}_1 - \dot{m}_2$$

Assuming the flow is incompressible, the mass flow rate and volumetric flow rate are equal at the inlet and outlet. Since the cross-sectional area is also constant, the average inlet velocity and exit velocity are equal. That is,

$$V_1 = V_2 \quad \text{(for a constant area incompressible flow)}$$

The general momentum conservation $\left(M\vec{a} = \sum \vec{F}\right)$ applied to the control volume in the direction of flow is

$$\rho\left(\frac{\pi D_H^2}{4}\right)L\frac{dV_{x,\mathrm{CM}}}{dt}=(P_1-P_2)\frac{\pi D_H^2}{4}-\tau_{\mathrm{wall}}\pi D_H L-\rho g\sin\theta\left(\frac{\pi D_H^2}{4}\right)L$$
$$+\dot{m}(V_1-V_2)$$

The left side is the mass times acceleration, often called an inertial force (a virtual force). The first term on the right is the net pressure force, the second term is the net friction force, the third term is the gravity force, and the last term represents a difference in momentum flux between inlet and exit. Notice from the geometry that $L\sin\theta = z_2 - z_1$, the difference in vertical height between inlet and exit, which can be negative for a downward slope. The difference in momentum flux (the last term) is zero for incompressible flows in a constant area pipe since the velocities are the same. It can be important for compressible flows or when the cross-sectional area changes in the direction of flow. For steady-state flow, the transient term is identically zero. For transient flows, there can be an imbalance in the forces that gives rise to acceleration. However, even if the flow is transient (like in the pulsating reciprocating pump), it may be reasonable to neglect the inertial effect in comparison to at least two other terms that balance themselves in the momentum equation. That is, a quasi-equilibrium may be reasonable. In that case, rearrangement of the momentum equation results in

$$\left(\frac{P_1}{\rho}+gz_1\right)-\left(\frac{P_2}{\rho}+gz_2\right)=4\left(\frac{\tau_{wall}}{\rho}\right)\left(\frac{L}{D}\right)$$

for quasi-steady flow of an incompressible fluid in a constant area duct.

The left-hand side contains the mechanical quantities (pressure and gravity). The right-hand side contains the wall friction effect. The magnitude of the mechanical quantities is reduced by friction, which a formal 2nd Law analysis will show is inherently positive. The shear stress increases with flow velocity and increases strongly with a decrease in diameter, and its formal evaluation depends on the flow regime (i.e., Reynolds number). These dissipative effects are termed "head loss" in practice.

The 1st Law applied to the control volume is

$$\frac{dE_{\mathrm{CV}}}{dt}=\dot{Q}_{\mathrm{IN}}-\dot{W}_{\mathrm{SHAFT,OUT}}+\dot{m}\left(u_1+\frac{P_1}{\rho_1}+\alpha_1\frac{V_1^2}{2}+gz_1\right)$$
$$-\dot{m}\left(u_2+\frac{P_2}{\rho_2}+\alpha_2\frac{V_2^2}{2}+gz_2\right)$$

The shaft power is identically zero as there are no devices extracting work or any boundary work associated with a changing control volume. Rearranging, for the case of a quasi-steady flow, with the heat transfer per unit mass flow defined as $q_{\mathrm{IN}} = \dot{Q}_{\mathrm{IN}}/\dot{m}$.

$$\left(\frac{P_1}{\rho_1}+gz_1\right)-\left(\frac{P_2}{\rho_2}+gz_2\right)=[(u_2-u_1)-q_{IN}] \text{ for quasi-steady flow of an incompressible fluid.}$$

It is amazing how similar the momentum equation and energy equation look. The mechanical terms are exactly the same. Forces in the momentum equation become forces acting through distances, or work, in the energy equation. The head loss (momentum) and thermal effects (energy), however, are expressed differently. Combining momentum and energy, it is clear that

$$4\left(\frac{\tau_{wall}}{\rho}\right)\left(\frac{L}{D}\right)=[(u_2-u_1)-q_{IN}]$$

The friction gives rise to a combination of an internal energy rise (i.e., temperature) and heat transfer out of the system. The magnitude of the temperature rise would, in practice, be a mild warming, except for extreme cases.

The 2nd Law gives a more formal view of the dissipation associated with friction. Applied to the control volume:

$$\frac{dS_{CV}}{dt}=\frac{\dot{Q}_{IN}}{T_B}+\underset{CV}{\dddot{\Sigma}}+\dot{m}(s_1-s_2)$$

Assuming that the temperature of the wall is fixed at the dead state (by external means, not detailed), and with the entropy generation per unit mass flow defined as $\sigma_{CV}=\underset{CV}{\dddot{\Sigma}}/\dot{m}$, and rearranging for the heat transfer (per unit mass flow):

$$q_{IN}=T_\infty(s_2-s_1)-T_\infty\sigma_{CV}$$

Combining the 1st and 2nd Laws, and using the specific eXergy, $x=u-T_\infty s$

$$\left(\frac{P_1}{\rho_1}+gz_1\right)-\left(\frac{P_2}{\rho_2}+gz_2\right)=[(x_2-x_1)+T_\infty\sigma_{CV}]$$

Even for cases where the eXergy change is not zero, there is no device in place to extract useful work, and the entire right-hand side represents an inherently positive quantity, a specific energy dissipation (conversion to internal energy). That is,

$${}_ex_{D,EFF}=[(x_2-x_1)+T_\infty\sigma_{CV}]\geq 0$$

The nomenclature ${}_ex_{D,\,EFF}$ is used to indicate an "effective" eXergy destruction. Mechanical energy forms have been converted to a combination of a mild warming (x_2-x_1) with no physical means to subsequently convert into useful work and heat loss to the walls.

The relationship between the flow velocity and the frictional shear stress is established in fluid mechanics textbooks, and the results are presented in the form of an empirically derived dimensionless friction factor, *f*, defined by

$$\frac{\tau_{\text{wall}}}{\rho} = f\frac{V^2}{2}$$

The friction factor, in general, depends on the flow velocity and the roughness of the wall as the latter influences the intensity of the turbulence associated with the flow. The momentum equation becomes

$$\left(\frac{P_1}{\rho} + gz_1\right) - \left(\frac{P_2}{\rho} + gz_2\right) = 4f\left(\frac{V^2}{2}\right)\left(\frac{L}{D}\right)$$

The flow velocity can always be related to the volumetric flow rate. $V = (\text{AV})/A = 4(\text{AV})/(\pi D^2)$..

A little algebra yields

$$\left(\frac{P_1}{\rho} + gz_1\right) - \left(\frac{P_2}{\rho} + gz_2\right) = \frac{32f(\text{AV})^2 L}{\pi^2 D^5}$$

Note that for a given flow rate and length of pipe, the head loss has an inverse fifth-order dependence on diameter (but, depending on flow regime, the friction factor has at most an inverse dependence on D). That is, increasing the diameter by a factor of 2 decreases the head loss by a factor of between $2^5 = 32$ and $2^6 = 64$. In design, there is an economic trade-off between capital cost (size of pipe) and operating cost (pumping losses). The diameter of pipes plays a crucial role.

Liquid Flow Through Fittings, Turns, and Restrictions: Minor Flow Losses

In practice, fluid flow systems rarely involve only straight sections of pipe. Flows must invariably be turned, or transition from a pipe to one of different diameter, or be split or merged. Every change in a flow path results in additional eXergy destruction, or head loss. Consider again the steady-state energy equation for flow through a device with no moving parts (so shaft work is impossible) with a single inlet (state 1) and single outlet (state 2), and with an assumed uniform flow profile at inlet and exit (so the profile shape factor $\alpha = 1$):

$$\left(\frac{P_1}{\rho_1} + \frac{V_1^2}{2} + gz_1\right) - \left(\frac{P_2}{\rho_2} + \frac{V_2^2}{2} + gz_2\right) = -[(u_2 - u_1) + q_{\text{in}}]$$

In this form, the Bernoulli “constant” is changed by thermal effects. A 2nd Law analysis would show that the right-hand side must be negative (i.e., the exit temperature is higher and/or heat is lost to the environment).

In fluid mechanics, each component device that involves a change in the flow path can be expressed in a form that uses an empirically determined dimensionless loss coefficient (K), defined by

$$\left(\frac{P_1}{\rho}+\frac{V_1^2}{2}+gz_1\right)-\left(\frac{P_2}{\rho}+\frac{V_2^2}{2}+gz_2\right)=K_m\left(\frac{V^2}{2}\right)$$

The values of the loss coefficient can be obtained from a variety of sources. Care must be taken in interpreting whether the velocity, V, is based on the inlet or exit velocity. Any change in elevation usually has a negligible effect in these devices.

The total head loss for a system that consists of multiple straight sections and multiple components with minor losses can be expressed as

$$\left(\frac{P_1}{\rho}+\frac{V_1^2}{2}+gz_1\right)-\left(\frac{P_2}{\rho}+\frac{V_2^2}{2}+gz_2\right)=\sum_{\text{pipe i}}4f_i\left(\frac{V_i^2}{2}\right)\left(\frac{L_i}{D_i}\right)+\sum_{\text{minor } j}K_{m,j}\left(\frac{V_j^2}{2}\right)$$

Care must be taken in determining the system loss if the flow splits into multiple parallel streams, as is common.

By suitably multiplying and dividing by some specific area, such as that of the discharge of the pump driving the flow, the entire flow system can be expressed into a form that groups all the head loss effects into a single head loss coefficient and expressed in terms of the volumetric flow rate:

$$\left(\frac{P_1}{\rho}+\frac{V_1^2}{2}+gz_1\right)-\left(\frac{P_2}{\rho}+\frac{V_2^2}{2}+gz_2\right)=K_{\text{System}}\frac{(AV)^2}{2A^2}$$

Test stands designed to measure the performance of pumps are constructed with a regulating valve downstream of the pump that can be adjusted to simulate any system load. Opening the valve completely allows the pump to operate with minimal load, and closing the valve maximizes the load. A pump, combined with a flow system, will find a unique operating point.

A steady-state entropy balance applied to the single-inlet, single-outlet device, in which any heat transfer takes place between the fluid in the device and the dead state, is

$$s_1+\frac{q_{\text{in}}}{T_\infty}+\sigma=s_2$$

Combining the energy and entropy balances (by eliminating the heat transfer in):

$$\left(\frac{P_1}{\rho_1}+\frac{V_1^2}{2}+gz_1\right)-\left(\frac{P_2}{\rho_2}+\frac{V_2^2}{2}+gz_2\right)=-[(u_2-T_\infty s_2)-(u_1-T_\infty s_1)+T_\infty\sigma]$$

In terms of the eXergy of each state (x) and the eXergy destruction that takes place within the device (eX_D):

$$\left(\frac{P_1}{\rho_1}+\frac{V_1^2}{2}+gz_1\right)-\left(\frac{P_2}{\rho_2}+\frac{V_2^2}{2}+gz_2\right)=-\left[(x_2-x_1)+{}_ex_D\right]=-{}_ex_{D,\mathrm{EFF}}\leq 0$$

The eXergy destruction is inherently positive. Also, while the eXergy change of the fluid can be negative, the right-hand side will always be negative (or zero), and the Bernoulli "constant" will be degraded by internal irreversibilities.

Example of a Minor Loss: Flow Through a Sharp Orifice

An example of a simple inline flow device for which a minor loss coefficient can be approximated from first principles is an orifice flow meter, shown schematically in Fig. 5.11. A simple plate with a hole in it creates a restriction to the flow. Measurement of the pressure before and after the orifice (indicated by the manometer with taps before and after the orifice plate) can be related to the flow rate through the meter, as will be demonstrated. The light lines suggest streamlines, and the dashed lines show two control volumes that will be used for the analysis, which share the plane at the back of the orifice itself. The analysis that follows is restricted to incompressible flows. Compressible flows and the phenomenon of flow choking occur for flows of gases if the overall absolute pressure ratio P_1/P_2 is greater than approximately 2.

The cross-sectional areas at the entrance (location 1) and exit (location 2) are equal ($A_1 = A_2$), and the cross-sectional area at the orifice plane is A_{orifice}.

The mass conservation equations applied to the upstream and downstream control volumes (CV1 and CV2, respectively) are

$$\frac{dM_{\mathrm{CV1}}}{dt}=\dot{m}_1-\dot{m}_i$$

$$\frac{dM_{\mathrm{CV2}}}{dt}=\dot{m}_i-\dot{m}_2$$

$\dot{m}_1$ is the mass flow rate entering the meter.
$\dot{m}_i$ is the mass flow rate that crosses the plane at the orifice
$\dot{m}_2$ is the mass flow rate that exits the meter

Both control volumes are completely filled with an incompressible fluid at all times, regardless of the magnitude of the flow rates, and therefore the rate of change of mass stored within the control volumes is zero. Therefore, all mass flow rates are equal.

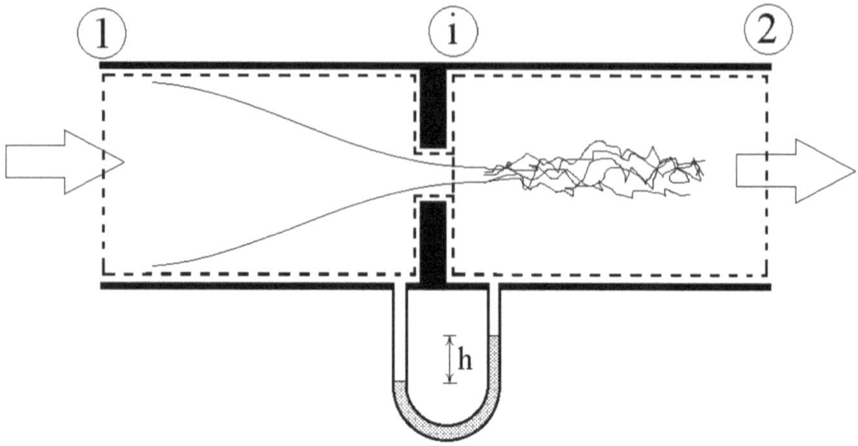

Fig. 5.11 Schematic of an orifice meter

$$\dot{m}_1 = \dot{m}_i = \dot{m}_2 = \dot{m}$$

The velocities at each section can be related through the general flow rate relation

$$\dot{m} = \rho VA$$

For incompressible flows, the density cancels and mass conservation results in simple relationships between the average velocities at each section:

$$V_1 = V_2 = \left(\frac{A_{\text{orifice}}}{A_1}\right) V_i$$

The velocity accelerates in the upstream control volume and decelerates from the orifice to the exit. However, the physical mechanism upstream is completely different from that downstream, as suggested by the streamline sketches. Flow is smoothly directed toward the orifice plate (with a dead zone of mild recirculation, not shown), and the upstream section behaves like a reversible nozzle. A high-velocity jet of fluid is created that enters the downstream region. The high shear forces generate turbulence and internal processes that destroy eXergy within the downstream control volume.

A momentum balance in the x direction applied to the downstream control volume is

$$M_{\text{CV}} \frac{dV_{\text{CM}}}{dt} = -\tau_{\text{wall}} \pi D L_{i2} + (P_i - P_2) A_2 + \rho g_x A L_{i2} + \dot{m}(V_i - V_2)$$

This result is based on the assumption that the pressure is uniform across the orifice area and the back wall of the orifice plate. Note that this is not a good assumption for the front of the orifice plate, where high pressure in the corners accelerates fluid into the orifice. At quasi-steady state, the transient term is neglected, and using mass relations, the pressure change from the orifice plate to the exit is

$$(P_i - P_2) = \rho V_1^2\left(1 - \frac{A_2}{A_{\text{orifice}}}\right) + 4\tau_{\text{wall}}\frac{L_{i2}}{D} - \rho g_x L_{i2}$$

Note that in the absence of friction and for horizontal flows ($g_x = 0$) there is some pressure recovery. This momentum result will be combined with the energy balance.

Energy balances applied to both of the control volumes are

$$\frac{dE_{\text{CV1}}}{dt} = \dot{m}\left(\frac{P}{\rho} + u + \frac{V^2}{2} + gz\right)_1 - \dot{m}\left(\frac{P}{\rho} + u + \frac{V^2}{2} + gz\right)_i + \dot{Q}_{\text{IN,CS1}} - \dot{W}_{\text{SHAFT,OUT,CV1}}$$

$$\frac{dE_{CV2}}{dt} = \dot{m}\left(\frac{P}{\rho} + u + \frac{V^2}{2} + gz\right)_i - \dot{m}\left(\frac{P}{\rho} + u + \frac{V^2}{2} + gz\right)_2 + \dot{Q}_{IN,CS2} - \dot{W}_{SHAFT,OUT,CV2}$$

The following approximations are made, with explanations to follow:

$$\rho V_1\left(1 - \frac{A_2}{A_{\text{orifice}}}\right) >> 4\tau_{\text{wall}}\frac{L_{i2}}{D}$$

that is, wall friction is negligible compared to the change in momentum flux

$$\frac{dE_{\text{CV1}}}{dt} \approx \frac{dE_{\text{CV1}}}{dt} \approx 0 \quad \text{quasi-steady state approximation}$$

$$\dot{W}_{\text{SHAFT,OUT,CV1}} = \dot{W}_{\text{SHAFT,OUT,CV2}} = 0 \quad \text{no non-flow work}$$

$$z_1 \approx z_i \approx z_2 \quad \text{no significant elevation change}$$

$$(u_i - u_1) - \dot{Q}_{\text{IN,CS1}}/\dot{m} = {}_e x_{D,\text{eff,CV1}} \approx 0$$

This approximation is informed by a 2nd Law analysis of upstream, and assumption of minimal entropy generation in the upstream control volume

$$(u_2 - u_i) - \dot{Q}_{\text{IN,CS2}}/\dot{m} = {}_e x_{D,\text{eff,CV2}} > 0$$

This expression is a consequence of the kinetic energy of the jet being dissipated in the downstream control volume.

The upstream control volume behaves like a nozzle, in which flow work is converted into kinetic energy in a nearly reversible process. The downstream control volume, on the other hand, is highly irreversible as it is difficult to decelerate a flow by recovering pressure without a well-designed and gradual flow contour. The downstream control volume decelerates the flow, but not as an effective diffuser. It could be called a "dissipater."

The net result, with the assumptions made, is

$$P_1 - P_2 = \frac{\rho V_i^2}{2}\left(1 - \frac{A_{\text{orifice}}^2}{A_1^2}\right) \approx \rho_e x_{D,\text{CV2}}$$

This form is useful in relating the velocity at the restriction to the overall pressure drop across the orifice. On the other hand, use of mass conservation can be used to relate the pressure drop to the velocity entering the device, expressed as a minor loss coefficient, K_m:

$$P_1 - P_2 = \frac{\rho V_1^2}{2}\left(\frac{A_1}{A_{\text{orifice}}} - 1\right)^2 = K_m \frac{\rho V_1^2}{2} = K_m \frac{\rho (\text{AV})^2}{2A_1^2} \approx \rho\left({}_e x_{D,\text{CV2}}\right)$$

$$K_{m,\text{orifice}} = \left(\frac{A_1}{A_{\text{orifice}}} - 1\right)^2$$

Rearranging for the flow rate:

$$(\text{AV}) = A_1 \sqrt{\frac{2(P_1 - P_2)}{K_m}}$$

Measurement of the pressure drop across the orifice plate results in an indirect measurement of the flow rate. In practice, meters are calibrated to account for real effects such as irreversibilities that were neglected in the analysis. These effects give rise to an effective diameter of the jet that narrows downstream of the orifice (the vena contracta effect). In the analysis of the reciprocating pump to follow, the intake and exhaust check valves will be assumed to behave like an orifice meter.

Check Valve

Check valves are common devices used to prevent backward flow. Their function is analogous to that of diodes in electronics that allow electricity to flow in one direction only. The main objective of this analysis is to estimate a minor loss coefficient for a check valve. There are many different designs, but a simple idea is shown schematically in Fig. 5.12. The valve consists of an orifice plate with an orifice diameter d inside a pipe of inner diameter $D_1 = D_2$ (the same diameter as the pipe before and after the valve). A sphere of diameter d_b (larger than d) is placed between the plate and a grating that keeps the sphere in place with minimal blockage. If the downstream pressure is greater than the upstream pressure, the sphere is pushed to the left, and blocks the orifice, preventing back flow. If the pressure downstream is less than the upstream pressure, the sphere is pushed against the

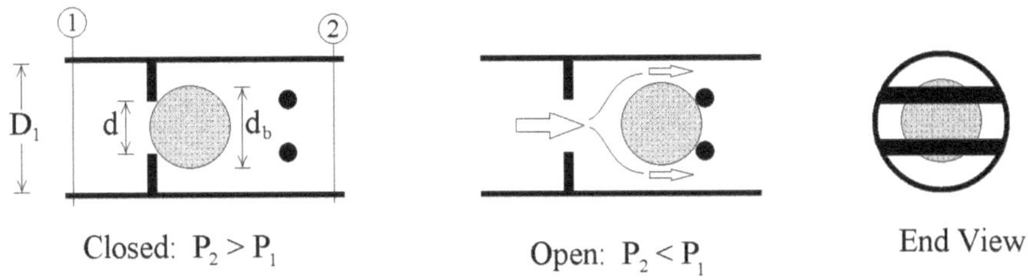

Fig. 5.12 Schematic of a basic check valve

grating, and fluid can flow through the orifice and around the sphere. For a given pipe diameter, the main design parameters are the diameters of the orifice and sphere. By intuition, the maximum area will be attained when the effective area through the orifice equals that around the sphere. That is, neglecting the area occupied by the grating:

$$\pi d^2/4 = \pi\left(D^2 - d_b^2\right)/4 \quad \text{for minimal flow losses}$$

$$d^2 + d_b^2 = D^2$$

Since $d < d_b$ this constraint can be written as

$$2d^2 < D^2$$

Or

$$d < D/\sqrt{2} = 0.707D$$

A good design might be to set $d_b = 0.71\text{D}$ and $d = 0.6D$. The minimum loss coefficient would be associated with that of the orifice plate:

$$K_{m,\text{orifice}} > \left(\frac{A_1}{A_{\text{orifice}}} - 1\right)^2 = \left(\frac{D^2}{d^2} - 1\right)^2 = (2-1)^2 = 1$$

The flow around the sphere creates a second orifice plate-type restriction (through an annular region) and, therefore, it is reasonable to model two minor losses in series, of approximately equal loss coefficient, and therefore $K_{m,\text{ check valve}} > 2$

Two check valves are integral parts of the reciprocating pump, and this result will be used when that specific design is reconsidered in the next section. It is possible to decrease the flow losses associated with a check valve by using a larger check valve of this design. For example, a diffuser (a gradual area expansion designed to reduce the flow velocity and increase the pressure) would be followed by an orifice plate with orifice diameter equal to the inlet pipe area. The sphere would be housed in a larger diameter section, followed by a nozzle (a gradual area contraction) after the grating back to the original pipe diameter. In practice, the losses could theoretically be reduced, but not eliminated, as there would still remain sudden changes in area.

A general treatment of pump behavior is investigated before a more detailed analysis of the reciprocating pump is conducted.

General Pump Characteristic Curve

Mechanical work involves force acting through a distance. Mechanical power involves force acting through a distance per unit time (a velocity), that is, effort and flow. If a large force is applied to an object (large effort) but the object does not move (no flow), then no work is done. At the other extreme, no work is performed if a large displacement (large flow) is accomplished without a force (no effort). In a similar way, the objective of a pump is to deliver mechanical power from a shaft to mechanical power of a fluid. The effort is the increase in flow energy per unit mass, while the throughput of fluid is the flow.

There are three components to the mechanical energy of a fluid stream: flow work (sometimes called pressure energy), kinetic energy, and potential energy. For pumps, the objective is generally to develop high pressure, and the flow work term generally dominates. The mechanical power of the fluid requires a simultaneous specific energy and a flow rate (force and velocity, effort and flow). Mechanical power delivered by the shaft that does not go into mechanical power of the fluid goes into thermal energy, a dissipative process.

In practice, sound engineering design that involves the selection of a pump for a given application requires an understanding of a pump characteristic curve, a plot of the pressure rise across the pump versus the flow rate through the pump, shown schematically in Fig. 5.13 (left axes). The mechanical power (pressure rise multiplied by flow rate) is shown schematically on the right axes. A given pump is capable of operating anywhere along that curve, depending on the flow system it drives. The pump characteristic curve generally involves a maximum pressure rise delivered (with no flow through the pump), a decreasing pressure rise across the pump with flow rate, and a maximum flow rate (with no pressure rise). In contrast, the pressure drop across a flow system increases with flow rate (a parabolic curve if the effective loss parameter is constant). The operating point of a pump coupled with a system occurs where the pump characteristic curve intersects the system curve. The set of axes on the right is the pump characteristic curve multiplied by the flow rate, that is, the flow power delivered to the fluid. It is typically shaped as shown, with a definite peak toward the higher flow rate end.

Three cases with different systems driven by the same pump are shown. For case A, there is minimal flow restriction in the system (losses), resulting in the pump's maximum possible flow rate delivered with minimal pressure rise (effort), and therefore a low power delivered to the fluid. A pump designed to move a large quantity of liquid might be designed to operate under these conditions, such as in a

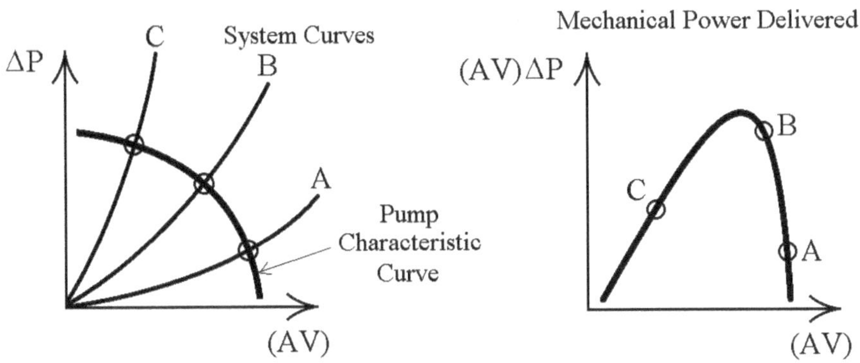

Fig. 5.13 Pump characteristic curve and three system curves, with operating points at the intersections (left). The right curve shows the mechanical power delivered to the fluid (pressure rise times volumetric flow rate) and the corresponding operating points

fire-fighting application. Case B shows a system with a higher flow loss. The operating point of this system results in a mechanical power delivered that is near the peak of the mechanical power that could be delivered by this pump. This case represents a good thermodynamic match between pump and system. Pumps in engineering systems would typically be sized to operate near the peak of the power curve. This behavior is analogous to impedance matching in electrical systems. Case C is one for which the pump develops a pressure near its maximum, but because the flow rate is low, there is a low delivery of mechanical power. Operation in this situation is usually to be avoided in a good design, but there can be applications that require a high pressure and a low flow rate to be satisfied by a pump that is capable of delivering higher flows (and lower pressures) when required.

A general treatment of a pump from a thermodynamic point of view will be used to construct similar curves. Consider an application, shown in Fig. 5.14, in which a liquid is drawn from a source plenum through an intake system to a pump at a different elevation, and from the pump to a discharge system whose exit is at another elevation. The system could be a closed loop (the discharge returns the flow to the intake plenum, like the human circulatory system).

Intake System

The intake is drawn from a source at low velocity and a zero elevation (reference). The flow analysis of the intake yields

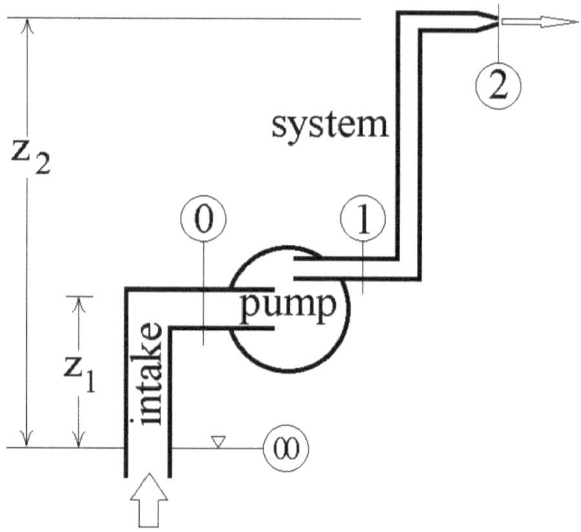

Fig. 5.14 Schematic of a general pump system with an intake drawn from a source at one elevation, and delivering flow at another, with the pump at an intermediate elevation

$$\left(\frac{P_\infty}{\rho}\right) - \left(\frac{P_0}{\rho} + \frac{D_1^4}{D_0^4}\frac{V_1^2}{2} + gz_1\right) = K_{IN}\frac{D_1^4}{D_0^4}\frac{V_1^2}{2}$$

The intake loss is based on the area that enters the pump, but continuity is used to express the loss in terms of the pump exit velocity. Rearranging for the pressure at the pump intake (expressing velocities in terms of the pump discharge):

$$P_0 = P_\infty - (K_{\mathrm{IN}} + 1)\frac{D_1^4}{D_0^4}\frac{\rho V_1^2}{2} - \rho g z_1$$

The pressure at the intake to the pump is a partial vacuum (relative to the source, which may be pressurized above the dead state). The intake side of a pump is therefore often called the "suction" side. The intake system must develop kinetic energy at the pump intake, overcome head losses, and raise the liquid above the source level. Leaks in such a system can be detrimental as ambient air will be sucked into the intake, which could stall the pump. It can also be difficult to start these pumps, which often must be primed to fill the intake system with liquid. Finally, cavitation (the boiling of liquid due to the low pressure) can be problematic if the pump intake pressure is below the saturation pressure that corresponds to the liquid temperature. On the other hand, if the elevation at the intake is lower than the source, this gravitational head can be of assistance in raising the pressure at the pump inlet. In general, pumps should be placed as low in elevation as possible and as close to the liquid source as possible, with large inlet diameters to minimize intake losses.

Discharge System

The flow analysis of the discharge system is

$$\left(\frac{P_1}{\rho}+\frac{V_1^2}{2}+gz_1\right)-\left(\frac{P_\infty}{\rho}+\frac{D_{\text{exit}}^4}{D_1^4}\frac{V_1^2}{2}+gz_2\right)=K_{\text{sys}}\frac{V_1^2}{2}$$

The system loss coefficient is based on the area exiting the pipe (A_1). Rearranging for the pressure at the pump exit:

$$P_1=P_\infty+\left(K_{\text{sys}}+\frac{D_{\text{exit}}^4}{D_1^4}-1\right)\frac{\rho V_1^2}{2}+\rho g(z_2-z_1)$$

The pump exit pressure is higher than the ambient due to head losses in the system, any difference in kinetic energy between the pump exit and the ultimate discharge and to elevation.

The system curve, including intake and discharge, rearranged for the required pressure difference across the pump is

$$\Delta P_{\text{pump}}=(P_1-P_0)=\left(K_{\text{sys}}+K_{\text{in}}\left(\frac{D_1}{D_0}\right)^4+\left(\frac{D_{\text{exit}}}{D_1}\right)^4-1\right)\frac{8\rho(AV)^2}{\pi^2D_1^4}+\rho gz_2$$

An effective head loss coefficient can be defined as:

$$K_{\text{EFF}}=K_{\text{sys}}+K_{\text{in}}\left(\frac{D_1}{D_0}\right)^4+\left(\frac{D_{\text{exit}}}{D_1}\right)^4-1$$

The pressure rise that will be required by the pump to push fluid through the system is then given by:

$$\Delta P_{\text{pump}}=(P_1-P_0)=K_{\text{EFF}}\frac{8\rho(\text{AV})^2}{\pi^2D_1^4}+\rho gz_2$$

This requirement includes all irreversibility associated with the intake process (the K_{eff} term), the development of kinetic energy of an exit jet, and any rise in head. The result was obtained by analyzing before and after the pump. The next section analyzes the pump directly.

Pump

A 1st Law balance applied to the pump that has an inlet pipe (state 0) and outlet pipe (state 1) is

$$\frac{dE_{CV,PUMP}}{dt} = \dot{Q}_{IN} + \dot{W}_{SHAFT,IN} + \dot{m}\left(u_0 + \frac{P_0}{\rho} + \frac{V_0^2}{2} + gz_0\right) - \dot{m}\left(u_1 + \frac{P_1}{\rho} + \frac{V_1^2}{2} + gz_1\right)$$

Assuming a quasi-steady state, and that the inlet and outlet are at the same elevation (although it would be easy to include the elevation change, if needed), and rearranging for the net flow power delivered to the fluid:

$$(\mathrm{AV})(P_1 - P_0) = \dot{W}_{SHAFT} - \rho(\mathrm{AV})\left(\frac{V_1^2}{2} - \frac{V_0^2}{2}\right) - \left[\dot{m}(u_1 - u_0) - \dot{Q}_{IN}\right]$$

A 2nd Law balance is

$$\frac{dS}{dt} = \frac{\dot{Q}_{IN}}{T_B} + \underset{CV}{\dddot{\Sigma}} + \dot{m}(s_0 - s_1)$$

Assuming that heat transfer takes place to the dead state (and a quasi-steady state):

$$\dot{Q}_{IN} = T_\infty \dot{m}(s_1 - s_0) - T_\infty \dddot{\Sigma}$$

Combining the 1st and 2nd Law balances, using mass conservation ($\dot{m} = \rho V_1 A_1 = \rho V_2 A_2$), to express the velocities in terms of the exit pump velocity, and introducing the eXergy ($x = u - T_\infty s$):

$$(\mathrm{AV})(P_1 - P_0) = \dot{W}_{SHAFT} - \frac{\rho(\mathrm{AV})^3}{2A_1^2}\left(1 - \left(\frac{A_1}{A_0}\right)^2\right) - \left[\dot{m}(x_1 - x_0) + T_\infty \underset{PUMP}{\dddot{\Sigma}}\right]$$

The term in brackets contains dissipative effects, and while the pump may increase the eXergy, it does so by merely warming it. It is unreasonable in practice to consider adding a heat engine to extract work from this source of eXergy. The second term in brackets is the actual eXergy destruction. Combining these two effects gives an effective eXergy destruction:

$$\left[\dot{m}(x_1 - x_0) + T_\infty \underset{PUMP}{\dddot{\Sigma}}\right] \equiv e\,\underset{D,eff}{\dddot{X}}$$

Defining a dimensionless diameter ratio (outlet to inlet) as $\widehat{D} = D_1/D_0$, the flow power delivered to the fluid is

$$(\mathrm{AV}) \cdot \Delta P_{\mathrm{PUMP}} = \dot{W}_{\mathrm{SHAFT}} - e\dot{X}_{D,\mathrm{eff}} - \frac{8\rho(\mathrm{AV})^3}{\pi^2 D_1^4}\left(1 - \widehat{D}^4\right)$$

The system curve is introduced to find the operating point (eliminating the pressure drop):

$$K_{\mathrm{EFF}} \frac{8\rho(\mathrm{AV})^3}{\pi^2 D_1^4} = \dot{W}_{\mathrm{SHAFT}} - e\dddot{X}_{D,\mathrm{eff}} - \frac{8\rho(\mathrm{AV})^3}{\pi^2 D_1^4}\left(1 - \widehat{D}^4\right)$$

Rearranging

$$\dot{W}_{\mathrm{SHAFT}} - e\dddot{X}_{D,\mathrm{eff}} = \left(K_{\mathrm{EFF}} + 1 - \widehat{D}^4\right) \frac{8\rho(\mathrm{AV})^3}{\pi^2 D_1^4}$$

The shaft power input to the fluid (less eXergy destruction from internal sources) develops kinetic energy power (velocity squared times flow rate) and overcomes the flow losses in the system (which include elevation changes). Rearranging for the flow rate delivered:

$$(\mathrm{AV}) = \left[\frac{\pi^2 D_1^4\left(\dot{W}_{\mathrm{SHAFT}} - e\dddot{X}_{D,\mathrm{eff}}\right)}{8\rho\left(K_{\mathrm{EFF}} + 1 - \widehat{D}^4\right)}\right]^{1/3}$$

For a given input shaft power input, the maximum possible flow rate that could be delivered by the pump is that for which there is no eXergy destruction, and no system flow losses. That is,

$$(AV)_{\mathrm{MAX}} = \left[\frac{\pi^2 D_1^4 \dot{W}_{\mathrm{SHAFT}}}{8\rho\left(1 - \widehat{D}^4\right)}\right]^{1/3}$$

Under these conditions, the power input to the pump changes the kinetic energy of the fluid. Notice that the exit diameter must be less than the inlet diameter. A dimensionless flow rate is the flow rate normalized by this maximum:

$$\hat{\dot{V}} = \frac{(\mathrm{AV})}{(\mathrm{AV})_{\mathrm{MAX}}} = \left[\frac{\left(1 - \hat{D}^4\right)\left(1 - e\overset{\cdots}{X}_{D,\mathrm{eff}} / \dot{W}_{\mathrm{SHAFT}}\right)}{\left(K_{\mathrm{EFF}} + 1 - \hat{D}^4\right)} \right]^{1/3}$$

A 2nd Law efficiency can be defined as the useful power that is actually delivered divided by the total power delivered:

$$\eta_{II} = \frac{\dot{W}_{\mathrm{SHAFT}} - e\overset{\cdots}{X}_{D,\mathrm{eff}}}{\dot{W}_{\mathrm{SHAFT}}} = 1 - \frac{e\overset{\cdots}{X}_{D,\mathrm{eff}}}{\dot{W}_{\mathrm{SHAFT}}}$$

The dimensionless flow rate is therefore:

$$\hat{\dot{V}} = \frac{(\mathrm{AV})}{(\mathrm{AV})_{\mathrm{MAX}}} = \left[\frac{\left(1 - \hat{D}^4\right)\eta_{II}}{\left(K_{\mathrm{EFF}} + 1 - \hat{D}^4\right)} \right]^{1/3}$$

The pressure drop is given by back-substitution into the system equation:

$$\Delta P_{\mathrm{pump}} = \left(\frac{8\rho}{\pi^2 D_1^4}\right)^{1/3} \left[\frac{K_{\mathrm{EFF}}}{\left(K_{\mathrm{EFF}} + 1 - \hat{D}^4\right)^{2/3}} \right] \left(\dot{W}_{\mathrm{SHAFT}} - e\overset{\cdots}{X}_{D,\mathrm{eff}}\right)^{2/3}$$

A pressure rise across the pump requires that there be a system through which the flow is driven. Therefore, the maximum pressure developed by the pump is considered to be that which occurs without eXergy destruction and a large K_{EFF}. In the limit, this pressure is termed the "reversible pressure":

$$\Delta P_{\mathrm{pump,REV}} = \left(\frac{8\rho}{\pi^2 D_1^4}\right)^{1/3} K_{\mathrm{EFF}}^{1/3} \dot{W}_{\mathrm{SHAFT}}^{2/3}$$

This pressure grows to infinity with K_{EFF}, and in practice, there will be a mechanical limit to the maximum pressure that the liquid can attain. A dimensionless pressure rise is

$$\Delta\hat{P} = \frac{\Delta P_{\mathrm{pump}}}{\Delta P_{\mathrm{pump,REV}}} = \left[\frac{\eta_{II} K_{\mathrm{EFF}}}{\left(K_{\mathrm{EFF}} + 1 - \hat{D}^4\right)} \right]^{2/3}$$

It is common for the intake diameter to the pump to be larger than the discharge and, therefore, with the fourth power of the ratio, for that ratio to be much less than unity. The dimensionless flow rate and pressure in those cases are

$$\hat{V} = \frac{(\mathrm{AV})}{(\mathrm{AV})_{\mathrm{MAX}}} = \left[\frac{\eta_{II}}{K_{\mathrm{EFF}} + 1}\right]^{1/3} \quad \text{for} \quad \hat{D}^4 << 1$$

$$\Delta\hat{P} = \frac{\Delta P_{\mathrm{pump}}}{\Delta P_{\mathrm{pump,REV}}} = \left[\frac{\eta_{II} K_{\mathrm{EFF}}}{K_{\mathrm{EFF}} + 1}\right]^{2/3} \quad \text{for} \quad \hat{D}^4 << 1$$

Eliminating the system loss parameter yields the pump characteristic curve, a relationship between the pressure and flow rate delivered by the pump. After some manipulation,

$$K_{\mathrm{EFF}} = \frac{\eta_{II}}{\hat{V}^3} - 1 \quad \text{from the flow rate relation}$$

$$\Delta\hat{P} = \left[\eta_{II} - \hat{V}^3\right]^{2/3}$$

A family of curves of the dimensionless pressure as a function of dimensionless flow rate is plotted in Fig. 5.15 for fixed values of the 2nd Law efficiency. However, it is not expected that the 2nd Law efficiency will be a constant for a given pump as that would imply that, as $K_{\mathrm{EFF}} \rightarrow \infty$, the actual pressure would also go to infinity (while the dimensionless pressure goes to unity). In reality, there must be a limit to the pressure that an actual pump would be able to deliver, set by mechanical issues.

The pump is capable of operating at any point on its characteristic curve, in principle. It can either deliver fluid at its highest pressure (with a high K_{eff}), but with a low flow rate, or near its highest flow rate (a low K_{eff}), but with a low-pressure rise. On the other hand, the delivery of power to the fluid requires a force (pressure) acting through a distance (flow). The mechanical power that is delivered to the fluid is the force times the velocity or pressure change times flow rate:

$$\text{Power Delivered} = (\mathrm{AV})\Delta P = (AV)_{\max} \Delta P_{\mathrm{pump,REV}} \hat{V} \Delta\hat{P}$$

A dimensionless power delivered is defined by

$$\frac{\text{Power Delivered}}{(\mathrm{AV})_{\max} \Delta P_{\mathrm{pump,REV}}} = \hat{V}\Delta\hat{P} = \hat{V}\left[\eta_{II} - \hat{V}^3\right]^{2/3}$$

A family of curves of the dimensionless power delivered as a function of dimensionless flow rate is plotted in Fig. 5.16 for fixed values of the 2nd Law efficiency. It is clear that there is a sweet spot for the delivery of power. If the system coefficient is too high, a high pressure will result with a low flow and low power

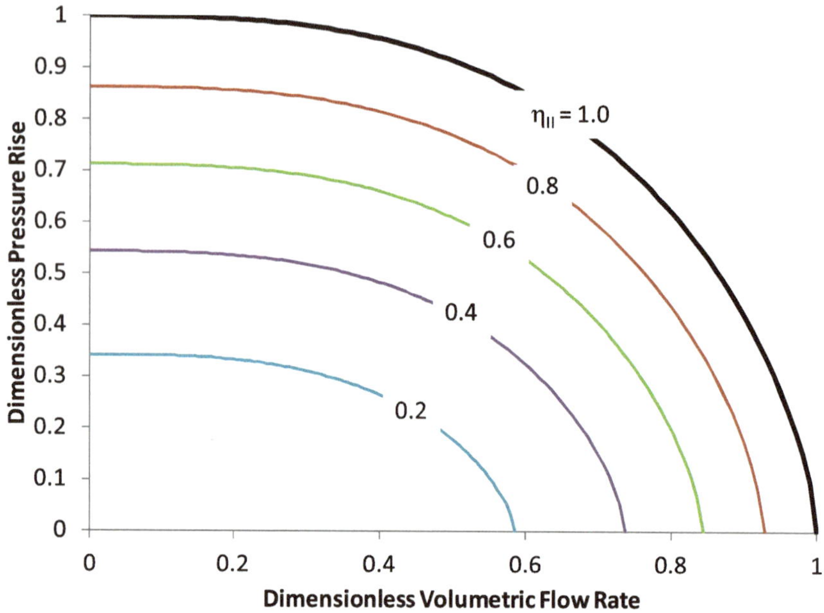

Fig. 5.15 Pump characteristic curves (pressure rise vs. volumetric flow rate) for fixed values of the 2nd Law efficiency

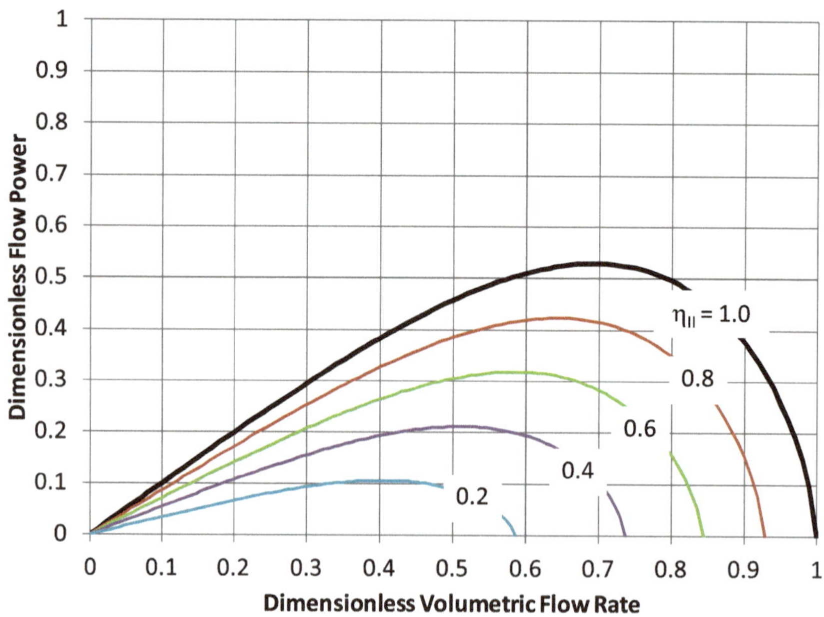

Fig. 5.16 Dimensionless flow power delivered to fluid

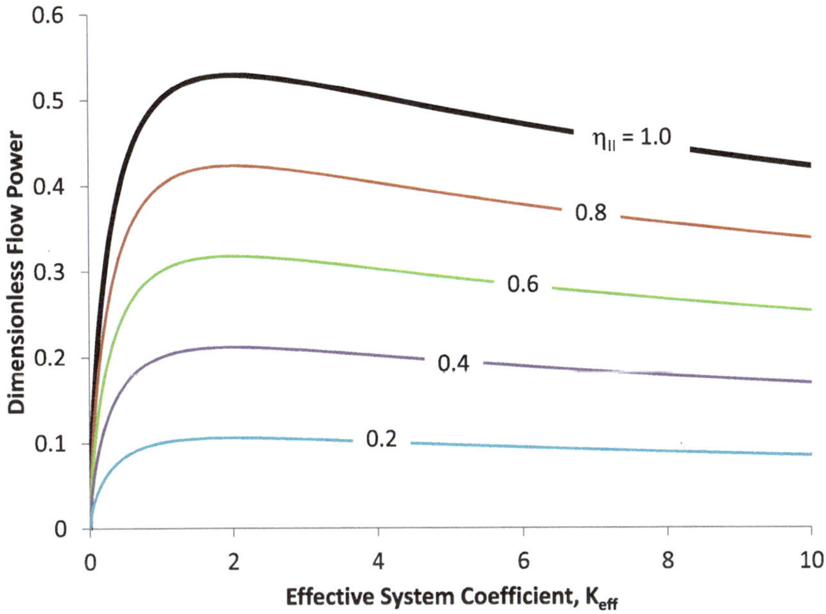

Fig. 5.17 Dimensionless flow power plotted against effective system coefficient

delivery. If there is no load on the system (low K_{eff}), there will be a high flow, but little energy added to it.

Fig. 5.17 shows how the effective system coefficient affects the dimensionless flow power delivered. The power peaks at a K_{eff} of approximately 2 and falls slowly afterward. This behavior would be helpful in choosing parameters like pipe diameters and the like for engineers charged with designing flow systems.

Reciprocating Pump Revisited

A more detailed analysis of a reciprocating pump is conducted in this section. The flow is inherently unsteady, and an open system, so this case is a somewhat general situation with the simplest type of fluid for thermodynamic analysis, namely, an incompressible fluid. However, it will serve as a good lead into the study of the gas exchange process in reciprocating engines that use air as the working fluid (i.e., automobile engines). However, the analysis of a reciprocating pump is simpler because the flow rate through the valves is directly related to the motion of the piston, a result of the constant density approximation.

The basic design and operation of a piston/cylinder reciprocating pump were introduced with Fig. 5.3. The valves are modeled as check valves, with an effective loss coefficient K_{valve}. The source and discharge pressures are assumed to be constant in time, and the intake and exit system curves modeled as in the previous section with effective loss coefficients. The control volume is defined to include the fluid within the cylinder plus the void volumes between the cylinder and the two valves. Mass enters the cylinder across the intake valve during the intake stroke and exits across the exhaust valve during the exhaust stroke. The analysis will be done for a constant rotational speed. In practice, even if the rotational rate is constant, there may be accelerations and decelerations within the cycle, depending on the inertia of the mechanical load. These mechanical effects are not considered and do not substantially change the basic thermodynamics of the pump. The analysis will also be conducted for a long connecting rod length compared to crank angle to simplify the analysis.

Cylinder Geometry

The geometry of the piston/cylinder arrangement is shown in Fig. 5.17. The crank angle (θ), shown as positive in the counterclockwise direction from the vertical, is considered to be a key independent variable. The angle between the connecting rod and the vertical (α) is geometrically related to the crank angle, connecting rod length (L), and crank arm (a). The crank angle is related to the angular velocity and rotational speed (N), which are positive in the counterclockwise direction by

$$\omega = 2\pi N = \frac{d\theta}{dt}$$

Integrating for constant angular speed:

$\theta = \omega t$ for an initial angle of 0 [called top dead center (TDC) with the volume at a minimum] (Fig. 5.18)

The volume of fluid within the control volume is

$$\forall = \forall_C + A_{cyl} x$$

where $\forall_C$ is the clearance volume (the minimum volume of fluid in the control volume) and $A_{cyl} = \pi B^2/4$ is the cross-sectional area of the cylinder and B is its diameter (bore).

The distance "x" varies with the crank angle (θ) and connecting rod angle (α):

$$x = L + a - (L\cos\alpha + a\cos\theta) = L(1 - \cos\alpha) + a(1 - \cos\theta)$$

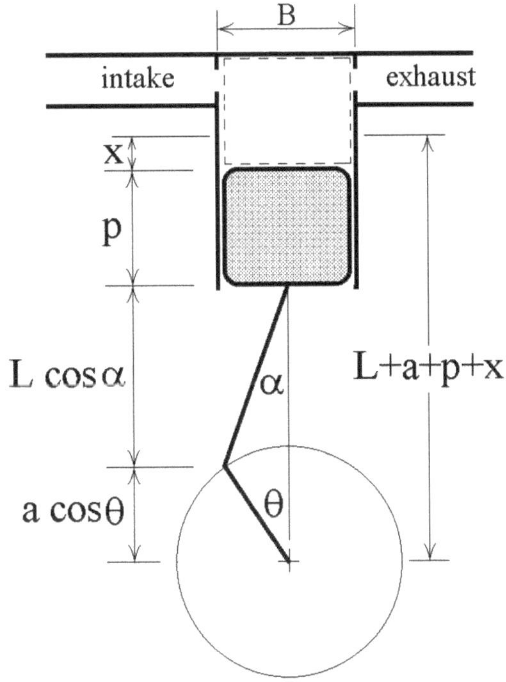

Fig. 5.18 Geometry of reciprocating pump

The crank angle and connecting rod angle are related through the horizontal distance from the centerline to the pin that connects the connecting rod and crank arm (or the Law of Sines):

$$L \sin \alpha = a \sin \theta$$

Squaring both sides and rearranging (and using the trigonometric identity $1 = \cos^2\alpha + \sin^2\alpha$):

$$1 - \cos^2\alpha = \left(\frac{a}{L}\right)^2 \sin^2\theta$$

Rearranging further:

$$\cos \alpha = \sqrt{1 - \left(\frac{a}{L}\right)^2 \sin^2\theta}$$

The distance 'x' is now related to the crank angle and lengths:

$$x = L\left(1 - \sqrt{1 - (a/L)^2 \sin^2\theta}\right) + a(1 - \cos\theta)$$

This expression simplifies for the case of a connecting rod that is long compared to the crank arm. That is,

$$x = a(1 - \cos\theta) \quad \text{for} \quad L >> a$$

The cylinder volume is

$$\forall = \forall_C + A_{\text{cyl}}a(1 - \cos\theta) \quad \text{for} \quad L >> a$$

The displacement volume is the volume swept during a stroke $\left(\forall_D = 2aA_{\text{cyl}}\right)$. Therefore, the volume is

$$\forall = \forall_C + \frac{\forall_D}{2}(1 - \cos\theta)$$

The rate of change of volume with respect to time is:

$$\frac{d\forall}{dt} = \frac{\forall_D}{2}\sin\theta\frac{d\theta}{dt}$$

In terms of the angular speed:

$$\frac{d\forall}{dt} = \frac{\omega\forall_D}{2}\sin(\omega t) \quad \text{for a constant angular speed and} \quad L >> a$$

The relationship between the bore and the crank arm can be an important design parameter, defined by the "squareness" of the piston/cylinder arrangement, measured by the bore to stroke ratio (*B*/*S*, where *S*=2a). A square pump has a bore to stroke ratio of unity.

$$\forall_D = 2a\frac{\pi B^2}{4} = 2a\frac{\pi}{4}\left(\frac{B}{2a}\right)^2(2a)^2 = 2\pi(B/S)^2a^3$$

These geometric relations and nomenclature are similar to that in reciprocating engines that use air as the working fluid.

Intake Stroke

During the intake stroke, as the piston withdraws, liquid is pulled through the intake valve, while the exhaust valve is closed. The mass conservation equation, with the subscript "IN" to indicate the inlet to the control volume (the minimum area of the intake valve):

$$\frac{dM_{\mathrm{CV}}}{dt} = \dot{m}_{\mathrm{IN}} \quad \text{during the intake stroke}$$

$$\frac{d(\rho \mathcal{V})}{dt} = \rho A_0 V_0$$

The density cancels, and the velocity of the liquid as it enters is therefore directly related to the change in volume of the cylinder:

$$V_0 - \frac{1}{A_0}\frac{d\mathcal{V}}{dt} \quad \text{(during intake stroke)}$$

In chapter 7, in which a similar problem but with a gas as the fluid, the density in the cylinder will be different than that at the entrance and this direct relationship between the flow velocity through the valve and the motion of the piston does not generally hold.

The previous analysis of the check valve relates the required pressure drop across the valve to accelerate the liquid from the source pressure (at a low kinetic energy) to the minimum area of the valve, corresponding to the inlet plane of the control volume, where the pressure is modeled to be equal to the cylinder pressure. The energy equation across the intake valve (treated as a sharp orifice with an effective minimum cross-sectional area $A_i << A_1$) yields an expression for the pressure in the cylinder as a function of the velocity at the orifice:

$$P_0 - P = K_{\mathrm{valve}} \frac{1}{2} \rho V_0^2$$

The energy and entropy equations are not formally analyzed at this point as it is not necessary to do so to determine the work (or power) associated with the entire intake stroke, and the eXergy destruction associated with the irreversible flow across the valve. In practice, other irreversibilities would also exist (piston friction and leakage, turbulent dissipation within the cylinder, heat transfer).

The total work done by the liquid on the piston during the entire intake stroke is

$$W_{\mathrm{INTAKE}} = \int_{V_C}^{V_C+V_D} P\,d\mathcal{V}$$

The pressure in the control volume is expressed and split into two integrals:

$$W_{\text{INTAKE}} = \int\limits_{V_C}^{V_C+V_D} P_0 d\forall - \int\limits_{t=0}^{\frac{1}{2N}=\frac{\pi}{\omega}} \left(\frac{K_{\text{valve}}\rho V_0^2}{2}\right)\left(\frac{d\forall}{dt}\right)dt$$

The first integral evaluates easily and represents the work that would be done by the fluid on the piston if there were no valve losses. That is,

$$W_{\text{INTAKE, REVERSIBLE}} = \int\limits_{V_C}^{V_C+V_D} P_0 d\forall = P_0 V_D$$

The second integral (converted to time, by multiplying and dividing by *dt*) is due to the pressure in the cylinder being below the source pressure, and the magnitude of that partial vacuum depends on the rate at which the piston withdraws. The vacuum is a maximum halfway through the stroke, when the piston is withdrawing at its greatest rate. Near the beginning and end of the stroke, the flow through the valve is able to maintain the pressure close to the source pressure.

Combining mass conservation relation between intake velocity and change in volume:

$$\int\limits_{0}^{\frac{1}{2N}=\frac{\pi}{\omega}} \left(\frac{K_{\text{valve}}\rho V_0^2}{2}\right)\left(\frac{d\forall}{dt}\right)dt = \int\limits_{0}^{\frac{\pi}{\omega}} \left(\frac{\rho K_{\text{valve}}}{2A_0^2}\right)\left(\frac{d\forall}{dt}\right)^3 dt$$

$$= \frac{\rho K_{\text{valve}}\omega^3 \forall_D^3}{16A_0^2} \int\limits_{0}^{\frac{\pi}{\omega}} \sin^3(\omega t)dt$$

If the effective intake area is a variable (e.g., for poppet valves), the valve area must remain inside the integral. It is treated as a constant here, however, because the intake valve is a check valve that is either fully open or fully closed. Converting the integral to angle (in favor of time) by substituting $\theta = \omega t$ and $d\theta = \omega dt$. That is,

$$\int\limits_{0}^{\frac{1}{2N}=\frac{\pi}{\omega}} \left(\frac{K_{\text{valve}}\rho V_0^2}{2}\right)\left(\frac{d\forall}{dt}\right)dt = \frac{\rho K_{\text{valve}}\omega^2 \forall_D^3}{16A_{IN}^2} \int\limits_{\theta=0}^{\pi} \sin^3\theta d\theta$$

A little integration by parts (or visit to a table of integrals…):

$$\int_{\theta=0}^{\pi} \sin^3\theta d\theta = \left[\cos\theta\left(\frac{1}{3}\cos^2\theta - 1\right)\right]_0^{\pi}$$

$$= \left[\left(-1\left(\frac{1}{3}(-1)^2 - 1\right)\right) - \left(1\left(\frac{1}{3}(1)^2 - 1\right)\right)\right] = \frac{4}{3}$$

Finally, the total work done by the fluid during the intake stroke is:

$$W_{\text{INTAKE}} = P_0 V_D - \frac{\rho K_{\text{valve}} \omega^2 \mathcal{V}_D^3}{12 A_0^2}$$

Creating a dimensionless parameter for the second term:

$$W_{\text{INTAKE}} = P_0 V_D \left(1 - \frac{\rho K_{\text{valve}} \omega^2 \mathcal{V}_D^2}{12 A_0^2 P_0}\right)$$

The fluid is at a lower pressure than the intake plenum pressure (which is lower than the supply pressure due to losses in the intake) and is a minimum halfway through the stroke. The assumption of a compressed liquid provides a hard limit on the design of the pump. If the pressure drops to the saturation pressure corresponding to the liquid temperature during the stroke, the liquid will tend to change phase (boil), a form of cavitation. This limitation puts a restriction on the rotational speed, displacement volume, and effective intake area. A criterion, then, is that the pressure must always be greater than the saturation pressure. The flow through the orifice yields the lowest pressure:

$$P = P_0 - \frac{1}{2}\rho V_{\text{orifice}}^2 > P_{\text{SAT}}(T)$$

$$\frac{\rho}{2A_{\text{orifice}}^2}\left(\frac{d\mathcal{V}}{dt}\right)^2 = \frac{\rho\omega^2 V_D^2 \sin^2(\omega t)}{8A_{\text{orifice}}^2} < (P_0 - P_{SAT}(T))$$

The rate of change of volume (and therefore flow velocity through the intake valve) is maximum halfway through the stroke (when $\sin(\omega t) = 1$). Therefore, the criterion to avoid cavitation is

$$\frac{\rho\omega^2 V_D^2}{8A_{orifice}^2} < (P_0 - P_{SAT}(T))$$

The intake plenum pressure (P_0) will be lower than the supply pressure due to flow losses (or lifting the liquid before the pump) in the intake system. Note that it is more difficult to meet the cavitation criterion the larger these losses.

Exhaust Stroke

The analysis of the exhaust stroke is similar to the intake stroke, except that flow exits through the exhaust valve through the exhaust check valve and the intake valve is closed. The pressure in the cylinder is greater than the pressure at the start of the discharge flow system. At the end of the intake stroke, when the volume is at a maximum, there is a sudden and instantaneous change in the cylinder pressure as the exhaust stroke commences. Since the liquid is incompressible, as soon as the cylinder volume decreases, the pressure rises to the exit pressure and the exhaust valve opens as the intake valve shuts. For this reason, the magnitude of the clearance volume does not play a role in reciprocating pumps. For air-based reciprocating compressors, a definite volume decrease must take place to first raise the pressure from intake to exit before mass can be moved across the exhaust valve.

The details of the calculations are the same as the intake, with some signs reversed. The exit velocity, from the mass conservation (with mass exiting), is

$$V_1 = \frac{-1}{A_1}\frac{d\forall}{dt} \quad \text{(during exhaust stroke)}$$

The pressure during the exhaust stroke is:

$$P - P_1 = K_{\text{valve}}\frac{1}{2}\rho V_1^2$$

The work done by the liquid is

$$W_{\text{EXHAUST}} = \int_{\forall_C+\forall_D}^{\forall_C} P d\forall = -P_1 V_D - \frac{\rho K_{\text{valve}}\omega^2 \forall_D^3}{12 A_1^2}$$

Notice how the sign of the first term (the reversible work) is opposite to the intake stroke, but the sign of the second term is the same. During intake, the partial vacuum caused by the intake valve means the liquid does less work on the piston as it moves (in the direction of the pressure force) than it would without valve losses. During exhaust, when the pressure force by the liquid and displacement is in opposite directions, the pressure is higher than the exit pressure, and more work is done.

This work represents the work done by the fluid on the piston. The piston also experiences a pressure (typically atmospheric) on the back side. However, that work is equal and opposite during the exhaust stroke and cancels. There is, in general, a friction force between the piston and cylinder wall that affects the transmission of work from the crankshaft to the piston and to the fluid in a more complete analysis.

Pump Efficiency

The 1st Law efficiency of the pump is defined as the work (or power) delivered to the fluid (the uphill pressure climb from intake to exhaust) divided by the total work (or power) done by the piston on the fluid in the cylinder.

$$\eta_{\text{pump}} = \frac{\text{PowerDelivered}}{\text{PowerInput}} = \frac{(AV)(P_1 - P_0)}{W_{\text{cycle}}N}$$

The work done by the piston for the cycle is the negative sum of the work done by the fluid during both strokes:

$$W_{\text{CYCLE}} = -W_{\text{INTAKE}} - W_{\text{EXHAUST}} = (P_1 - P_0)V_D + \frac{\rho K_{\text{valve}}\omega^2 \forall_D^3}{12}\left(\frac{1}{A_0^2} + \frac{1}{A_1^2}\right)$$

The total power input required is the total work divided by the total time (1/N = 2π/ω). The power input to the fluid is

$$\dot{W}_{\text{IN,CYCLE}} = \tau_{\text{CYCLE}}\ \omega = \frac{(P_1 - P_0)V_D\omega}{2\pi} + \frac{\rho\omega^3\forall_D^3}{24\pi A_0^2}\left(1 + \frac{A_0^2}{A_1^2}\right)$$

The expression $\tau_{\text{CYCLE}}\ \omega$ assumes the piston and all mechanisms are frictionless. Dividing through by the angular speed and rearranging for the pressure rise across the pump:

$$(P_1 - P_0) = \frac{2\pi\tau_{\text{CYCLE}}}{\forall_D} - \frac{\rho K_{\text{valves}}\forall_D^2\omega^2}{24A_0^2}\left(1 + \frac{A_0^2}{A_1^2}\right)$$

The volumetric flow rate for the cycle is the volume moved per cycle divided by the cycle time:

$$(AV)_{\text{CYCLE}} = V_D N$$

The efficiency of this reciprocating pump (with only flow losses in the intake and exhaust valves) is then

$$\eta_{\text{pump}} = \frac{1}{1 + K_{\text{valve}}\left(\frac{\rho\omega^2\forall_D^2}{12\Delta P A_0^2}\right)\left(1 + \frac{A_0^2}{A_1^2}\right)}$$

Instantaneous Versus Cycle Torque and Power

The relationships obtained above are integrations over the entire cycle. It is instructive to consider how all variables vary with time during the stroke. For example, the flow rate of the cycle $(AV)_{CYCLE}$ represents the total volume moved through the pump divided by the total time. However, the flow rate is pulsed. Similarly, the instantaneous power of the cycle and torque varies with time through the cycle.

The instantaneous torque depends on both the pressure in the cylinder and the moment arm. For the case of a long connecting rod ($L>>a$) and frictionless piston, the force in the connecting rod is approximately equal to the in-cylinder gage pressure times the area. The gage pressure is used because the backside of the piston is assumed to be open to atmospheric pressure. The moment arm is $a\ \sin\theta$. That is,

$$\tau = \left((P - P_\infty)\pi B^2/4\right)(a \sin\theta)$$

$$\tau = \frac{V_D}{2}(P - P_\infty)\sin\theta$$

The instantaneous power is

$$\dot{W}_{IN} = \tau\omega = -P\frac{d\forall}{dt}$$

Wind Turbine Fundamentals: Betz's Law

Wind power has been harnessed to meet human needs since early recorded history, but has only served as a minor niche energy source in the modern era. That is changing, and wind power is growing as an economically viable renewable energy resource with minimal greenhouse gas emissions (generated during production and disposal of equipment). This section is designed to demonstrate the key thermodynamic principles that can be used to estimate the available power in the wind. The thermodynamic objective is to convert kinetic energy in the wind to output shaft power. Figure 5.19 defines the kinetic energy associated with an undisturbed air stream (with ambient pressure and temperature) that passes through an imaginary

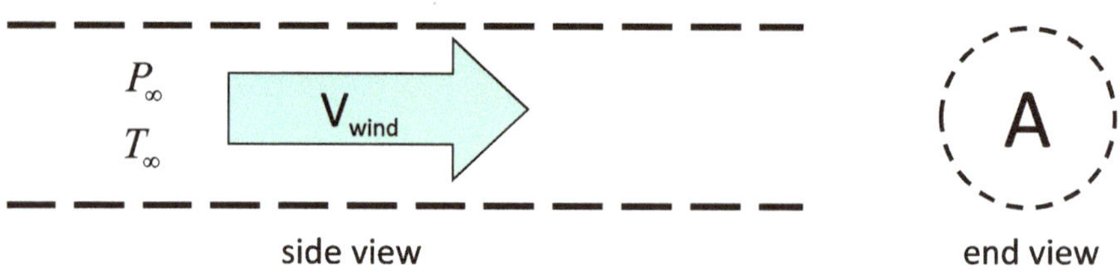

Fig. 5.19 Side and end views of an undisturbed column of air into which a wind turbine is to be placed

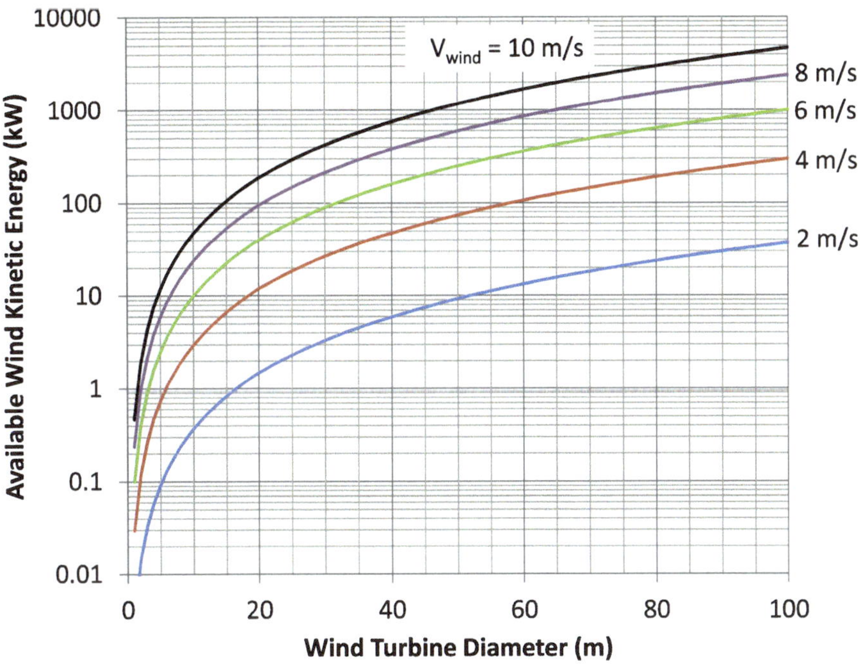

Fig. 5.20 Kinetic energy rate of flux as a function of wind turbine diameter for several wind speeds that span average values. Ambient conditions of 25°C, 1 atm, which results in ambient density of 1.18 kg/m^3. A velocity unit conversion is 1 m/s = 2.24 mph

circle into which a wind turbine will be placed. This streamtube defines the kinetic energy of the wind from which power can be extracted. That power, the denominator of the 1st Law efficiency of wind turbine, is

$$\text{Wind Power} = \dot{m}\left(\frac{V_{\text{wind}}^2}{2}\right) = \frac{1}{2}\rho_\infty V_{\text{wind}}^3 A$$

The air density can be calculated from the ideal gas law (discussed in depth in future chapters) from the ambient pressure and temperature and is generally of the order of 1.0 kg/m^3. The wind power is very sensitive to the wind speed, as might be expected. A doubling of the wind speed increases the wind power by a factor of 8. Figure 5.20 shows the wind power as a function of streamtube diameter for several typical wind speeds on a semi-logy scale.

Betz's law is derived in the following, which will demonstrate that only up to 59% of the power in the wind can be converted to useful power. A turbine will disturb an air stream into which it is placed. Since the objective is to convert kinetic energy of the flow into shaft power, the velocity behind the wind turbine will be lower than the incoming wind speed. A control volume for thermodynamic analysis is shown in Fig. 5.21 that defines a modified streamtube that contains air that will

Fig. 5.21 Process diagram of a wind turbine, with a streamtube defining a single-Sinlet, single-outlet flow path

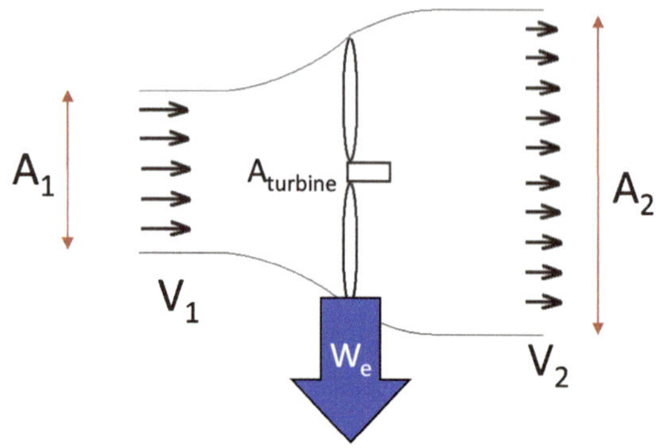

pass through the turbine plane. The entering section (state 1) is sufficiently far upstream that the velocity is the undisturbed wind speed and at ambient pressure and temperature. The exit plane is sufficiently far downstream that the pressure is ambient (although the temperature will be slightly higher due to irreversibilities).

At steady state, the mass flow rate is the same at the inlet, where it crosses the turbine plane and at the exit, that is,

$$\dot{m} = \rho_1 V_1 A_1 = \rho_{\text{rotor}} V_{\text{@turbine}} A_{\text{turbine}} = \rho_2 V_2 A_2$$

The flow in this unconfined streamtube can safely be treated as being incompressible (discussed further in future chapters) or constant density. The local flow near rotating blades can be compressible, but not at the planes considered in this analysis. With that, the mass balance is given by

$V_1 A_1 = V_{\text{rotor}} A_{\text{rotor}} = V_2 A_2$ assuming density changes are negligible (incompressible flow)

The flow decelerates before and after the turbine. To conserve mass, the cross-sectional area of the streamtube must increase to the area of the wind turbine, defined as A_{turbine}. The streamtube area must continue to increase to the exit plane.

Steady-state energy and entropy balances applied to the control volume, with a horizontal flow, are

$$\frac{dE}{dt} = 0 = \dot{Q}_{\text{in}} - \dot{W}_{\text{shaft}} + \dot{m}_1 \left(h_1 + \frac{V_1^2}{2} \right) - \dot{m}_2 \left(h_2 + \frac{V_2^2}{2} \right)$$

$$\frac{dS}{dt} = 0 = \frac{\dot{Q}_{in}}{T_B} + \dot{m}_1 (s_1) - \dot{m}_2 (s_2) + \dddot{\Sigma}$$

The enthalpy at the entrance and exit consists of a mechanical term (flow work, P/ρ) and a thermal term (internal energy, u), and it is useful to separate them in this application. Doing so and rearranging for the desired output shaft power:

$$\dot{W}_{\text{shaft}} = \dot{m}\left(\frac{P_1}{\rho_1} - \frac{P_2}{\rho_2}\right) + \dot{m}\left(\frac{V_1^2}{2} - \frac{V_2^2}{2}\right) + \left[\dot{Q}_{\text{in}} - \dot{m}(u_2 - u_1)\right]$$

Setting the pressures equal to ambient (this device is not pressure driven) and manipulating the kinetic energy term:

$$\dot{W}_{\text{shaft}} = \dot{m}\frac{V_1^2}{2}\left(1 - \frac{V_2^2}{V_1^2}\right) + \left[\dot{Q}_{\text{in}} - \dot{m}(u_2 - u_1)\right]$$

Invoking the mass conservation for the velocity ratios and eliminating the heat transfer with the entropy balance (with the boundary temperature for entropy transfer equal to the dead state temperature, $T_B - T_\infty$) and manipulating

$$\dot{W}_{\text{shaft}} = \dot{m}\frac{V_1^2}{2}\left(1 - \frac{A_1^2}{A_2^2}\right) - \left[T_\infty \dddot{\Sigma} + \dot{m}(x_2 - x_1)\right]$$

The term in brackets represents eXergy destruction and a rise in eXergy of the exiting stream, $x_2 - x_1 = (u_2 - T_\infty s_2) - (u_1 - T_\infty s_1)$. The latter does not represent useful energy (it is just warmed air) and is defined as an effective eXergy destruction rate:

$$e\dddot{X}_{D,\text{eff}} = \left[T_\infty \dddot{\Sigma} + \dot{m}(x_2 - x_1)\right]$$

The goal at this point of the analysis is to obtain an expression for the shaft power in terms of the wind turbine area and ratio of the inlet and exit areas. To do so, the mass flow rate is expressed at the turbine plane. However, the velocity there is taken to be the geometric average of the inlet and outlet. That is not a rigorous application of the conservation laws, but it is a reasonable engineering-type approximation to make. That is,

$$\dot{m} = \rho V_{@\text{turbine}} A_{\text{turbine}} \approx \rho\frac{V_1 + V_2}{2}A_{\text{turbine}} = \rho\frac{V_1}{2}\left(1 + \frac{V_2}{V_1}\right)A_{\text{turbine}}$$
$$= \rho\frac{V_1}{2}\left(1 + \frac{A_1}{A_2}\right)A_{turbine}$$

The final result for the output shaft power is

$$\dot{W}_{shaft} \approx \frac{1}{2}\rho_\infty V_1^3 A_{turbine}\left[\frac{1}{2}\left(1 + \frac{A_1}{A_2}\right)\left(1 - \frac{A_1^2}{A_2^2}\right)\right] - e\dddot{X}_{D,eff}$$

The 1st Law efficiency is defined as the actual output shaft power divided by the available wind power (the kinetic energy flux in the undisturbed column). That is,

$$\eta_I = \frac{\dot{W}_{\text{shaft}}}{\frac{1}{2}\rho_\infty V_1^3 A_{\text{turbine}}}$$

The 2nd Law efficiency is defined as the ratio of the actual shaft power (with irreversibilities) to the shaft power that would be obtained without irreversibilities (i.e., eXergy destruction set to zero). That is,

$$\eta_{II} = \frac{\frac{1}{2}\rho_\infty V_1^3 A_{\text{turbine}}\left[\frac{1}{2}\left(1+\frac{A_1}{A_2}\right)\left(1-\frac{A_1^2}{A_2^2}\right)\right] - e\,\dddot{X}_{D,\text{eff}}}{\frac{1}{2}\rho_\infty V_1^3 A_{\text{turbine}}\left[\frac{1}{2}\left(1+\frac{A_1}{A_2}\right)\left(1-\frac{A_1^2}{A_2^2}\right)\right]}$$

With a little manipulation, the 1st Law efficiency of the wind turbine is

$$\eta_I = \eta_{II}\frac{1}{2}\left(1+\frac{A_1}{A_2}\right)\left(1-\frac{A_1^2}{A_2^2}\right)$$

A family of curves of 1st Law efficiency as a function of streamtube area ratio (A_2/A_1) for several fixed 2nd Law efficiency values is shown in Fig. 5.22. This result shows the trade-off between flow throughput (the first term in parentheses) and the energy extracted (the second term). If $A_1 = A_2$, the maximum possible flow through is attained, but no kinetic energy is converted to shaft energy. As A_1 shrinks toward zero, all the kinetic energy is converted to shaft power, but there is little mass throughput. The approximate nature of the analysis (setting the velocity through the turbine plane equal to the average of the inlet and outlet) is inconsistent in this low limit. If the fluid is brought to completely to rest, the streamtube is not well represented by the schematic of Fig. 5.21.

The curve with a 2nd Law efficiency of 1.0 represents the maximum possible efficiency that can be obtained. A little basic calculus (find the maximum of a curve) shows that a streamtube area ratio of 1/3 results in the maximum value, and the corresponding maximum is

$$\eta_{I,\max} = \left[\frac{1}{2}\left(1+\frac{1}{3}\right)\left(1-\frac{1}{9}\right)\right] = 0.593$$

This basic understanding of the relationship between power extracted and the disruption of the air stream is an important tool for design of high-efficiency wind turbines. It is also important to design wind farms. A turbine placed in the wake of another will experience reduced output compared to an isolated turbine.

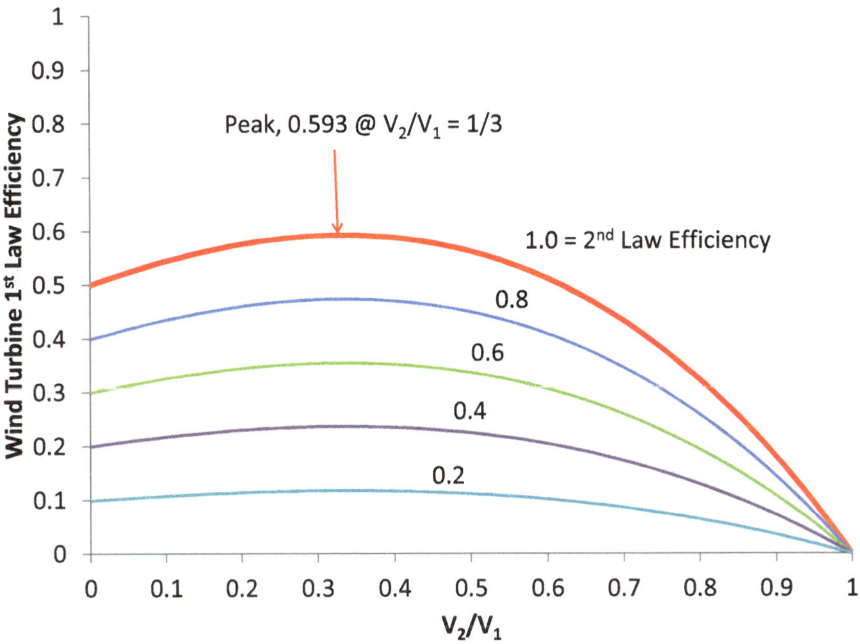

Fig. 5.22 First Law efficiency of wind turbine as a function of streamtube area ratio for fixed values of 2nd Law efficiency

Workshop 5.1: Hydroelectric Power Generation

Part 1: A dam creates a lake whose height is H above the river below it. A hydroelectric turbine is placed within the dam, whose objective is to convert the potential energy available in the water to useful work. The turbine discharges from piping of equivalent diameter d. The turbine can be bypassed through the same diameter pipe. Derive expressions for the

- Maximum possible mass flow rate and corresponding power of exit jet if the turbine is bypassed
- Mass flow rate and power that can be generated if the turbine is engaged (at steady state) as a function of the elevation difference (H)
- Maximum power that can be generated and corresponding flow rate
- Maximum theoretical efficiency of the system, defined as power generated over maximum power of discharge jet (with turbine bypassed)

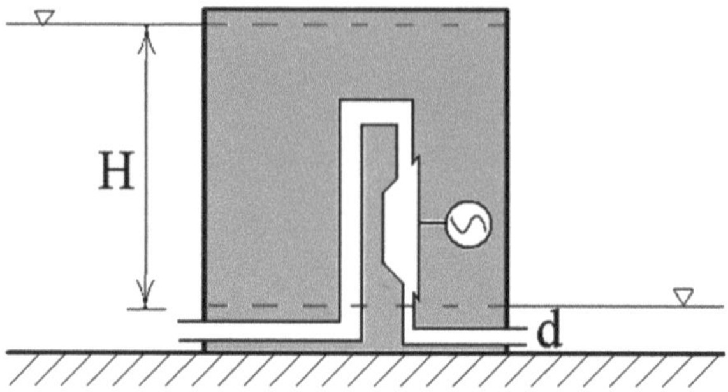

Part II: Hydroelectric Power Generation in a Canal Lock

It is proposed that useful power be generated by placing a hydroelectric turbine at the inlet to a canal lock. The lock has an area A_L, and the height difference between the up- and downstreams is H. The turbine discharges from piping of equivalent diameter d. The turbine can be bypassed through the same diameter pipe to reduce the time to fill or empty the lock. Derive expressions for the

- Maximum possible mass flow rate and corresponding depth in the lock (h) as functions of time and the time to fill the lock if the turbine is bypassed
- Maximum power that can be generated by a turbine operating at steady state and corresponding mass flow rate as a function of the instantaneous elevation difference (H–h)
- Mass flow rate and depth in the lock (h) as functions of time and the time to fill the lock if the turbine is engaged to maximize output power
- Total maximum energy that can be extracted and the total time to fill the lock
- Maximum theoretical first law efficiency of the system

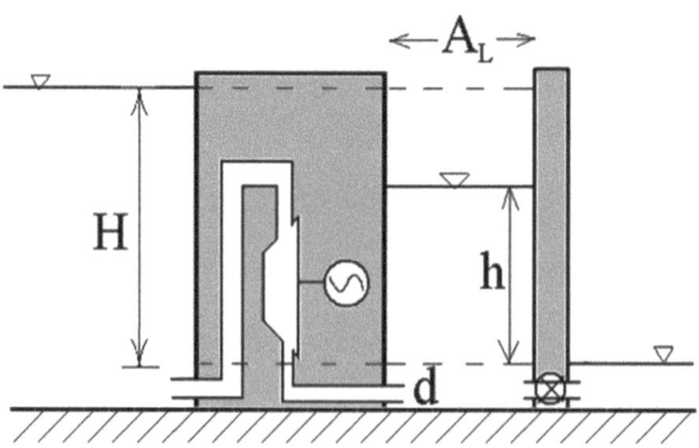

Chapter 6
Compressors and the Ideal Gas

The main goal of this chapter is to develop working relationships for ideal gases. The example of a reciprocating compressor serves to introduce operating principles of engineering devices that have closed system compression and expansion of a trapped gas as well as open system processes that involve gas exchange (intake and exhaust). Many engineering devices use working fluids that obey the ideal gas law and property relations can be obtained from simple algebraic expressions. The thermodynamic objective of a compressor is the same as the pump, namely, to convert shaft power into fluid power. However, the density of the fluid on which it acts changes appreciably. That is, the fluid is compressible. In contrast to liquids, a gas cannot easily be pressurized without performing work on it, which also increases its internal energy, a thermal form of energy. Like pumps, compressor designs come in two basic categories: positive displacement and dynamic. The reciprocating compressor (a positive displacement type) will be analyzed in some detail from a thermodynamic viewpoint as it serves as a case study for many other applications that involve gases, such as reciprocating engines. Reciprocating compressors are used in many vapor/compression refrigeration cycles.

The *Ts* diagram is shown to be a useful construct and an appreciation for the statement "gas turbines work because isobars diverge on a *Ts* diagram" that I remember hearing Professor Irvin Glassman (my PhD thesis advisor) say in a class on propulsion. It intrigued me at the time, but it took years of iterative learning to appreciate its significance.

Reciprocating Compressor

The schematic of a basic reciprocating compressor is exactly the same as that used for the pump of Chap. 5 (Fig. 6.1, repeated). The compressor operates on a mechanical cycle as an open system with two strokes, intake and exhaust. However, the crossover between intake and exhaust flows does not take place instantaneously

G. Sidebotham, *An Inductive Approach to Engineering Thermodynamics*,
Mechanical Engineering Series, https://doi.org/10.1007/978-3-030-20430-3_6

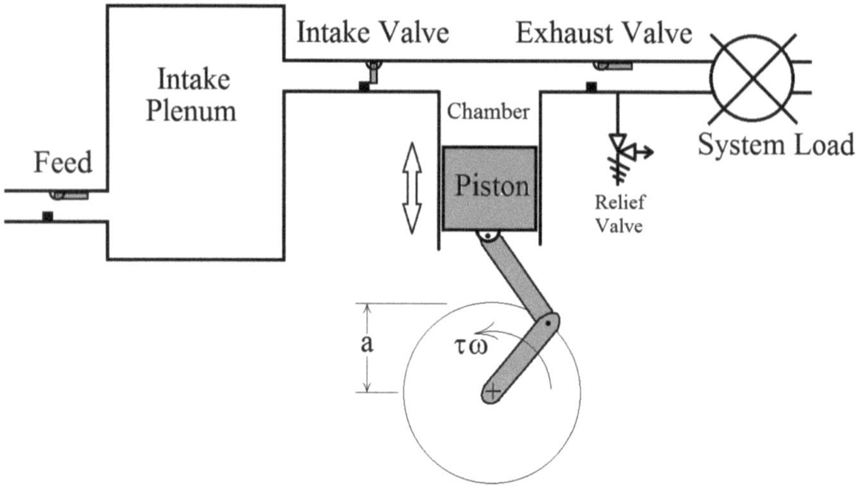

Fig. 6.1 Schematic of a reciprocating compressor

Table 6.1 Ideal reciprocating compressor processes

Process	Process name	Type of system	Ideal cycle approximations
1–2	Expansion	Closed	Adiabatic, reversible (hence, isentropic)
2–3	Intake	Open	Adiabatic, isobaric
3–4	Compression	Closed	Adiabatic, reversible (hence, isentropic)
4–1	Exhaust	Open	Adiabatic, isobaric

as it does with incompressible liquids. Therefore, two different types of process occur during each stroke, as summarized in Table 6.1. Two of the processes (expansion and compression) take place with both valves closed, and therefore are treated as closed systems.

The volume for a cylinder with a long connecting rod compared to the crank arm ($L \gg a$) is shown versus number of strokes in Fig. 6.2 for a compressor with a displacement volume $V_D = 1$ L, and a compression ratio $r = 10$ defined by the ratio of the maximum volume to the minimum volume, where V_C is the minimum volume, called the clearance volume:

$$r = \frac{V_C + V_D}{V_C}$$

If the working fluid is a liquid (i.e., a reciprocating pump), the clearance volume (and therefore compression ratio) plays no significant role because the mass contained within the clearance volume is always the same. On the other hand, if the working fluid is a gas, the compression ratio is very important, as will be apparent. Note that the unit for the x axis is number of strokes, which is equivalent to angle or time (for a constant rotational speed):

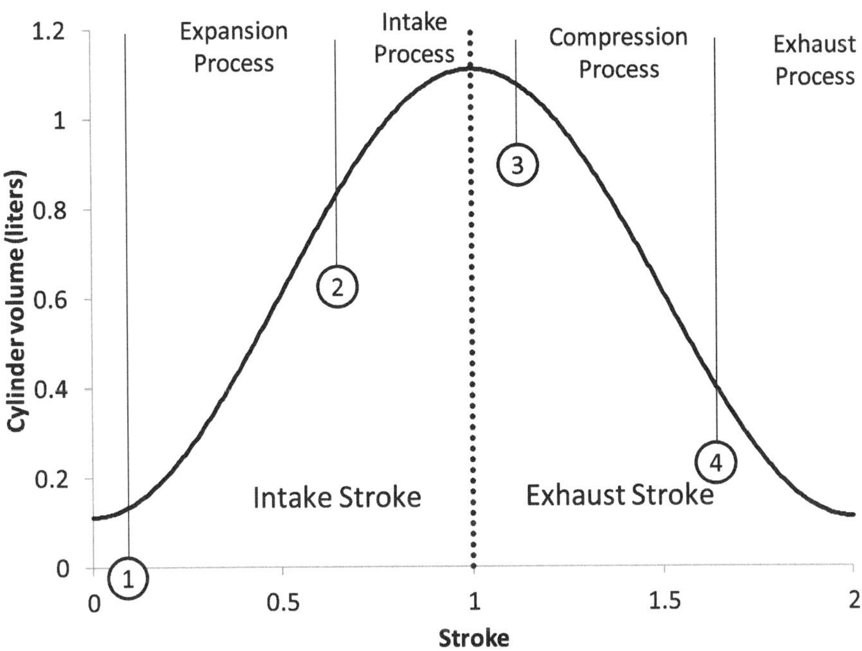

Fig. 6.2 Volume vs. crank angle, reported as number of strokes for a compression ratio of 10, and a displacement volume of 1 L. The connecting rod is much longer than crank arm. The locations of the four states are representative and depend on other operating parameters, such as overall pressure ratio, fluid type, and rotational rate. For the ideal cycle (no pressure drops across the valves), states 1 and 3 would correspond to the stroke values 0 and 1, respectively

$$1\ \text{stroke} = 180\ \text{degree} = \pi\ \text{rad} = 0.5\ \text{rev}$$
$$= 1/(2N)\ \text{s}\quad \text{for a rotational speed of N rev/s.}$$

In the ideal cycle, at the start of the cycle (state 1, the end of the exhaust stroke), the cylinder is filled with gas at the exhaust plenum pressure (P_1, the high pressure entering the flow system). As the piston begins to withdraw, the pressure does not immediately drop to the intake plenum pressure (P_0) as it would with a liquid. Rather, a fixed mass of gas is trapped between the intake valve and the exhaust valve, and the pressure gradually falls as the volume increases (and density decreases) until the pressure drops to the intake plenum pressure (state 2). This process is an expansion process, and it will be shown that the temperature decreases as well as the pressure. The intake valve opens (naturally), and further movement of the piston causes fluid (in the gaseous state) to flow across the intake valve. The cylinder pressure equals the intake plenum pressure during the intake stroke (for the ideal cycle, neglecting pressure drop across the intake valve). At the start of the exhaust stroke (state 3), a fixed mass of gas is trapped, greater in magnitude than the mass trapped during the start of the intake stroke. As the piston moves to decrease the volume, the pressure will gradually rise during a compression process until the pressure reaches the exhaust plenum pressure. The exhaust valve opens at state 4, and fluid flows into the exhaust plenum until the end of the exhaust stroke.

Traditional thermodynamics textbooks would treat the intake and exhaust processes as closed systems and introduce a fictitious heat transfer that would have the same effect (from a 1st Law point of view) as the gas exchange processes. That approach is avoided here in order to model the process more closely to the actual behavior and to consider the 2nd Law consequences.

The ideal model results in closed-form solutions and is developed in detail as it consists of several important types of processes found in other applications. The real case requires a numerical simulation and some basic heat transfer theory and is beyond the scope here. Some real effects are discussed qualitatively. The term "ideal" means two very different things in two different contexts, whether applied to the fluid or to the process. For both models, the fluid will be assumed to be an ideal gas, meaning that the relationship between pressure, volume, and temperature is governed by the ideal gas law $\left(PV = n\hat{R}T\right)$, discussed more fully shortly. However, the term "ideal" as applied to the processes means that side, unintended effects are neglected. The intention of the compressor is to convert shaft power into fluid power, and heat transfer has no part of that. Therefore, all processes are treated as adiabatic in the ideal cycle (Table 6.1). Also, the two gas exchange processes (intake and exhaust) are intended to simply move fluid across the valves. The ideal cycle neglects pressure drops across the valves (which would introduce irreversibilities), making the intake and exhaust processes isobaric (constant pressure).

The location of state 1 is indicated in Fig. 6.2 to be slightly after top dead center (TDC) when the volume is at its minimum. In the ideal cycle, this state would correspond exactly to TDC as the pressure in the cylinder during the exhaust stroke would be equal to the exhaust plenum pressure, and as soon as the piston begins its withdrawal, the pressure would drop below the exhaust plenum and the valve would shut. In the real cycle, the pressure in the cylinder is above the exhaust plenum pressure during the exhaust stroke due to a pressure drop across the exhaust valve. This overpressure can linger into the intake stroke and is more likely at higher rotational speeds. Similarly, the intake process creates a pressure below the intake plenum pressure, and this underpressure can linger into the exhaust stroke.

Properties of Ideal Gases

Building on the introduction in Chap. 4, a summary of the property relations for ideal gases is given, followed by a detailed analysis of the ideal reciprocating compressor cycle.

Recall from Chap. 4 that simple compressible substances are defined as those for which surface tension, magnetic, and electrical effects are insignificant compared to the effect of changing volume. That is, the only form of work on them is expansion (PdV) work. The important takeaway is the state postulate, namely, that the specification of two independent thermodynamic properties of a substance uniquely defines its state. For example, if the pressure and temperature of a substance are known and the substance is not undergoing a phase or chemical change, then the density (or specific volume), internal energy, enthalpy, and entropy can all be determined, in principle, either from tables of properties, or from appropriate formulas. It is important, in practice, to emphasize that pressure and temperature are not independent if the substance is undergoing a change of phase. For example, H_2O at 100 °C, 1 atm could be all saturated liquid, all saturated vapor, or a mixture of saturated liquid and vapor of any quality (mass fraction of vapor). However, if the temperature and specific volume are specified, and the specific volume is a value between that of a saturated liquid and vapor for the corresponding temperature, then all other properties can be determined (using quality calculations).

A table of properties (to be discussed in this chapter) of a few substances that are gases at nominal room temperature is provided in Table 6.2. Phase change effects were addressed in Chap. 4 and will return in future chapters when needed. This chapter focuses on ideal gases, which, by definition, are far removed from conditions that would cause phase change.

Ideal Gas Law and Generalized Compressibility Chart

The ideal gas law is used frequently in thermodynamic analysis, particularly as it relates to mass conservation principles. It is a relationship between the pressure, temperature, and density of a gas and can be expressed in several forms such as molar, mass, density, and specific volume. Historically, it was first discovered empirically based on macroscopic behavior (e.g., in the laws of Boyle and Charles). Later, it was derived from an atomic point of view (kinetic theory), in which a gas consists of many individual molecules that interact through elastic collisions.

Most students are first exposed in high school chemistry to the ideal gas law in its molar-based form. A mole of gas (n) in a volume (V) is related to the absolute temperature (T) and pressure (P) by

$$PV = n\widehat{R}T$$

where the universal gas constant is given in strict SI units by

$$\widehat{R} = 8314\frac{\text{J}}{\text{kmol}\cdot\text{K}}$$

The pressure and temperature must be in absolute units (not gage pressure and temperature in K, not °C, a point often missed in exam situations). A mole (whose official unit is spelled mol, not mole) is a number of molecules, given by Avogadro's number (to four significant figures):

Table 6.2 Properties of selected substances that are gases at nominal room conditions (25 °C, 1 atm)

Substance	Chemical formula	Molecular weight	R (J/kg/K)	c_V (J/kg/ΔK) @ room conditions	Normal boiling point (°C)	T_{CRIT} (K)	P_{CRIT} (bar)	T/T_{CRIT} @ room conditions	P/P_{CRIT} @ room conditions	Classification
Helium	He	4	2079	3120	−269.0	5.3	2.3	56.2	0.440	Gas
Air	Mixture	29.0	287	718	−215.0	132.5	37.7	2.25	0.027	Gas
Carbon monoxide	CO	28	297	720	−191.5	133.0	35.0	2.24	0.029	Gas
Carbon dioxide	CO2	33	252	655	−78.5	304.2	73.9	0.980	0.014	Vapor
Ethane	C2H6	30	277	1480	−88.0	305.5	44.8	0.975	0.023	Vapor
Propane	C3H8	44	189	1480	−43.0	370.0	42.6	0.805	0.024	Vapor
R-134a	CF3CH2F	102	81.5	2040	−26.3	374.2	40.6	0.796	0.025	Vapor
Ammonia	NH3	17	489	1660	−35.5	405.5	112.8	0.735	0.009	Vapor

Note: The compressibility factor (Z) is all 1.0 (by the low pressure limit)

$$1 \text{ mol} = 6.022(10)^{23} \text{ molecules}$$

That is, 1 mol = 602, 200, 000, 000, 000, 000, 000, 000 molecules.

The molecular weight (MW) of a chemical species relates the mass and moles. For a gas with mass *m* and number of moles *n*:

$$\text{MW} = \frac{m}{n}$$

The molecular weight (to 2 significant figures) is an integer for most elements. At higher precision, noninteger values of molecular weights can be used to reflect the ratio of naturally occurring isotopes.

The molar basis is most appropriate in chemistry where chemical changes occur. However, in a first course in engineering thermodynamics, where chemical changes are not considered, but conservation of mass, momentum, and energy is the principal effect, it is more appropriate to work on a mass basis. Simply multiplying and dividing the molar-based ideal gas law by the molecular weight yields a mass-based form for the ideal gas law:

$$PV = (n \cdot \text{MW})\left(\frac{\hat{R}}{\text{MW}}\right)T$$

The mass-based form of the ideal gas law is therefore:

$$PV = mRT$$

To reiterate, the pressure (P) and temperature (T) must be in absolute units. Alternative forms are obtained by dividing both sides by the mass:

$$Pv = RT \quad \text{and} \quad P = \rho RT$$

where the specific volume and density ($v = V/m = 1/\rho$) are used.

There are a few alternative forms of the ideal gas law, summarized in Table 6.3.

There are two common conditions of pressure and temperature that are considered to be "standard." In chemistry, *standard temperature and pressure* (STP) is used, and it is common for people to think that it is the same as an average room temperature. It is not. STP is defined as 0 °C, 1 atm. Students learn (and memorize) that a mole of gas occupies 22.4 L at STP. That is,

Table 6.3 Alternate forms of the ideal gas law

Form	Molar	Mass
Fixed mole or mass	$PV = n\hat{R}T$	$PV = mRT$
Mol or mass per unit volume	$P = c\hat{R}T$	$P = \rho RT$
Volume per unit mol or mass	$P\hat{v} = \hat{R}T$	$Pv = RT$

Note: $c = n/V$ is the molar concentration, while $\rho = m/V$ is the mass concentration

$$\begin{aligned} V_{@\mathrm{STP}} &= \frac{n\widehat{R}T}{P} \\ &= \frac{(1\ \mathrm{mol})\left(8314\frac{\mathrm{N}\times\mathrm{m}}{\mathrm{kmol}\times\mathrm{K}}\right)(273\ \mathrm{K})}{1\ \mathrm{atm}} \cdot \left|\frac{1\ \mathrm{kmol}}{1000\ \mathrm{mol}}\right| \cdot \left|\frac{1\ \mathrm{atm}}{1.013(10)^5\ \mathrm{N/m^2}}\right| \\ &\quad \cdot \left|\frac{1000\ \mathrm{L}}{\mathrm{m}^3}\right| \\ &= 22.4\ \mathrm{L} \end{aligned}$$

Please unlearn that. Instead, if the volume occupied by a gas at some pressure and temperature is needed, use the ideal gas law. Also, do not "look up" or search for the density of gases, like you would for a solid or liquid. Look up a molecular weight if needed instead. For example, an Internet search for "density of air" returned 1.225 kg/m^3. On closer inspection of the website, that value is for 15 °C, 1 atm. In engineering applications, the density (or volume) of a gas can change appreciably.

The other set of standard conditions is nominal room conditions. In this textbook, *nominal room conditions* are defined to be 25 °C, 1 atm. The volume occupied by that gas can be determined from an ideal gas relation applied for a fixed mass of an ideal gas:

$$mR = \left(\frac{PV}{T}\right)_{@\mathrm{Room}} = \left(\frac{PV}{T}\right)_{@\mathrm{STP}}$$

Or

$$V_{@\mathrm{Room}} = V_{@\mathrm{STP}}\left(\frac{T_{\mathrm{Room}}}{T_{\mathrm{STP}}}\right)\left(\frac{P_{\mathrm{STP}}}{P_{\mathrm{Room}}}\right) = 22.4\ \mathrm{L}\left(\frac{298\ \mathrm{K}}{273\ \mathrm{K}}\right)\left(\frac{1\ \mathrm{atm}}{1\ \mathrm{atm}}\right) = 24.5\ \mathrm{L}$$

Generalized Compressibility Chart and the Compressibility Factor, Z

A test for whether a gas behaves ideally can be obtained by comparing the actual pressure and temperature to the critical pressure and temperature. Then, a compressibility factor (Z) can be used to quantify deviations from ideal gas behavior, when needed. Consider again a vessel of volume "V" that contains a mass of gas "m" with molecular weight "MW" at a uniform absolute pressure "P" and absolute temperature "T." The ideal gas law (with real effects included) can be expressed in its mass form as

$$Pv = ZRT$$

In this expression,

- Pressure and temperature must be in absolute units.
- Z is the compressibility factor and is equal to 1 for ideal gases.
- $R = \widehat{R}/\mathrm{MW}$ is the gas constant.

Note that the gas could be a mixture of chemical species (such as air, which is composed of N_2, O_2, argon, CO_2, water vapor, and trace species) and the molecular weight of the gas as a whole is a number (molar) weighted average molecular weight (29.0 kg/kmol). The composition of the gas changes in chemical reactions and applications that involve changes in humidity levels, which will get a closer look in future chapters.

It is common to consider differential changes, and a useful way to express the ideal gas law is in terms of fractional changes. First, express the ideal gas law in the following form:

$$RZ = \frac{Pv}{T}$$

Second, take the derivative, using the chain rule:

$$RdZ = \frac{v}{T}dP + \frac{P}{T}dv - \frac{Pv}{m^2}dT$$

Finally, divide by RZ:

$$\frac{dZ}{Z} = \frac{dP}{P} + \frac{dv}{v} - \frac{dT}{T} = 0 \quad \text{for ideal gases}$$

This form of the ideal gas law will be useful going forward for certain processes to eliminate one property in terms of others.

Another useful form for finite changes in state (with specific volume expressed as total volume divided by mass) is

$$R = \frac{P_1 V_1}{Z_1 m_1 T_1} = \frac{P_2 V_2}{Z_2 m_2 T_2}$$

The compressibility factor (Z) can be estimated from a *generalized compressibility chart*, reproduced in Fig. 6.3, with regions indicated that correspond to the phase diagram sketch of Fig. 6.4. The conditions (temperature and pressure) are compared to the critical properties of the fluid in question. The x-axis is the reduced pressure, defined as the system absolute pressure normalized by the critical pressure. The solid lines are lines of constant reduced temperature, defined as the system absolute temperature normalized by the reduced temperature. The dotted lines are lines of constant reduced volume, defined as $v_R = RT_R/P_R$, which is useful for certain types of calculations. Critical property values and gas constants for common substances that are in the gas phase at nominal room temperature are given in Table 6.2, sorted by critical temperature. Values of the reduced pressure and temperature are listed for nominal room conditions (25 °C, 1 atm).

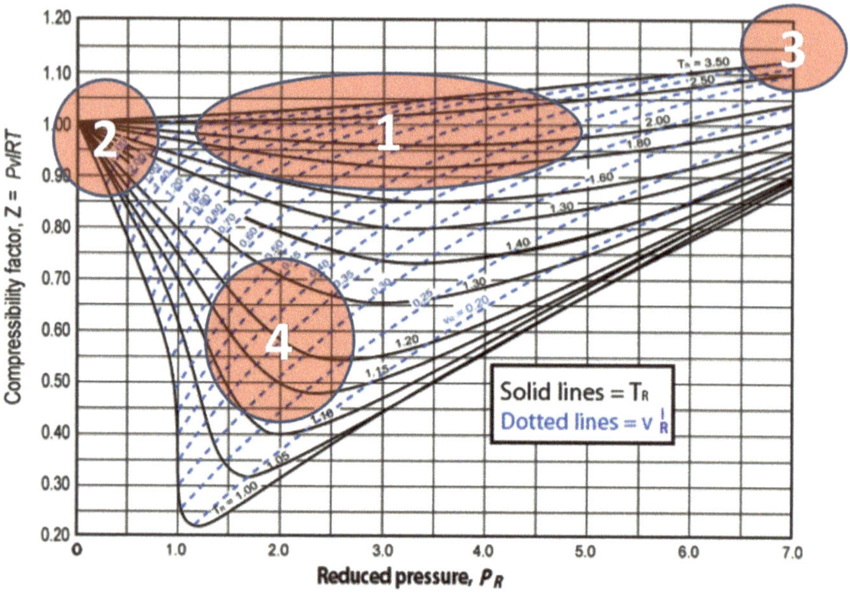

Fig. 6.3 Generalized compressibility chart. (Main chart downloaded 8/17/2015 from http://www.wiley.com/college/moran/CL_0471465704_S/user/tutorials/tutorial2/igl8/index.html with regions indicated that correspond to the phase diagram to follow)

Fig. 6.4 Phase diagram indicating regions of ideal gas behavior. The triple point (TP) and critical point (CP) are shown separating liquid from vapor (condensation), and the boundary between solid and liquid (freezing) and between solid and vapor (sublimation) is shown

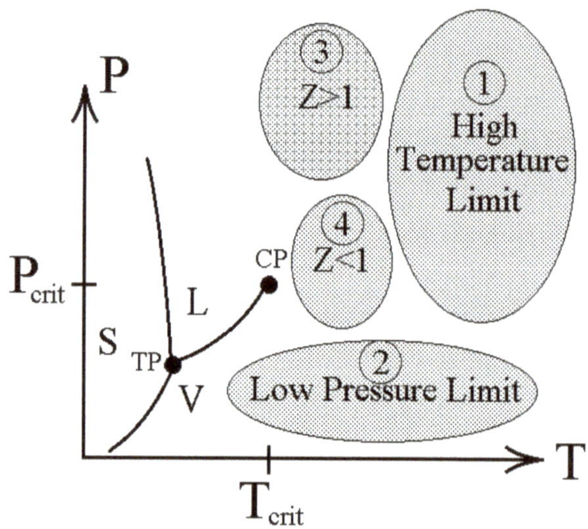

The isotherm corresponding to the critical temperature is the lowest isotherm in the particular chart shown in Fig. 6.3. Many charts show several isotherms for temperatures below the critical temperature, with upper limits of pressure equal to the corresponding saturation pressure. Such curves would correspond to vapors, not gases. Several of the entries of Table 6.2 are thus classified as vapors at nominal room conditions. There are two important limits for which $Z = 1$, corresponding to the high-temperature and low-pressure limits (regions 1 and 2 of Fig. 6.3, respectively). Stated mathematically:

$$\text{High-temperature limit}: \quad Z \approx 1 \quad \text{if } T > 2T_{\mathrm{R}}$$

$$\text{Low-pressure limit}: \quad Z \approx 1 \quad \text{if } P < 0.2P_{\mathrm{R}}$$

Many gases meet both requirements. For example, air at nominal room temperature has a reduced temperature of 2.25 and a reduced pressure of 0.027. On the other hand, the reduced temperature of propane at room conditions is less than 1 (0.805), but its reduced pressure is 0.024 and is well within the low-pressure limit. Propane at room conditions is therefore classified as a vapor, not a gas. It would condense if pressurized isothermally.

At very high reduced pressures, particularly those above the limit of this particular chart, the compressibility factor can be greater than 1 (region 3), and generalized compressibility charts at those high pressures are available. A physical interpretation in this regime is of molecular crowding. The volume occupied by the molecules themselves becomes significant compared to the total volume occupied by the gas.

Significant deviations from ideal gas behavior occur if the temperature is closer to the critical point temperature, and the pressure is on the order of the critical pressure (region 4). In this region, the weak intermolecular forces of attraction (Van der Waals forces) have a significant influence on the collisions between molecules.

Figure 6.4 shows a sketch of a phase diagram that separates the three phases (solid, liquid, and vapor). For pressures below the triple point, the sublimation line separates solid from vapor.

There is a subtle difference between a vapor and a gas. A vapor is a substance in a gaseous state, but would condense if pressurized isothermally. That is, a vapor is under the two evaporation lines (solid/vapor and liquid/vapor) on a PT phase diagram. Gases are to the right of that.

Many conscientious students exhibit a reluctance to use the ideal gas law, thinking that it represents a compromise of some sort, as it is an approximation. On the other hand, less conscientious students apply it haphazardly, for example, to devices that involve liquids or a change in the phase of its working fluid. When in doubt, the validity of the ideal gas law can usually be confirmed by a comparison of the conditions of a particular situation to the critical temperature and pressure of the fluid. The ideal gas law is perfectly reasonable if the working fluid of a device is air unless the device is designed to separate air into its components cryogenically (i.e., by making it very cold). On the other hand, there is an important difference between types of perfect gases (calorically or thermally perfect), the subject of the next topic.

In short, if the working fluid is air, use $PV = mRT$ religiously and unapologetically (unless it is really cold or under enormous pressure), and, as discussed next, assume that the specific heat is constant unless its minor variation with temperature is of interest. Just remember to use absolute temperature and pressure, and know the difference between the universal gas constant (molar based) and the gas constant for a particular gas (mass based).

Perfect Gases and Energy Changes

The historical nomenclature regarding gases is somewhat illogical, yet it is ingrained. Furthermore, there are inconsistencies in its usage in the literature. For example, what is the difference between an ideal gas and a perfect gas? Well, there are two kinds of perfect gas and both obey the ideal gas law. Huh? It is important for the responsible engineer to know the context and what the various implications are of the different terms. In a summary form, as explained in more detail in the following paragraphs:

> Perfect gases obey the ideal gas law (with $Z = 1$) and the specific internal energy is a function of temperature only. The specific heat (internal energy or enthalpy) varies with temperature for thermally perfect gases and is a constant for calorically perfect gases.

So there are two kinds of perfect gases, thermally perfect and calorically perfect. The calorically perfect gas approximation is more restrictive than the thermally perfect approximation. It has enormous advantages in modeling, however, and is fully justified when properly used, as described in Appendix 1. A summary of relations for calorically perfect gases is shown in Fig. 6.5.

Consider for the moment that the specific internal energy of a gas is a function of temperature and specific volume. That is,

$$u = u(T, v)$$

This choice of variables corresponds with the natural way that the 1st Law for closed systems relates to the various properties, with temperature related to heat transfer and volume associated with work. Now consider a differential change in the internal energy for a process where both the temperature and specific volume change. The total derivative of u is:

$$du = \left.\frac{\partial u}{\partial T}\right|_{\mathrm{v}} dT + \left.\frac{\partial u}{\partial v}\right|_{\mathrm{T}} dv$$

The first partial derivative that is often called (unfortunately) “the specific heat at constant volume” (c_{v}) is a thermodynamic variable, defined as the first partial derivative:

For any finite process (1-2) on a closed system (of fixed mass, m) of a calorically perfect gas:

Mass Relation: $$mR = \frac{PV}{T} = \frac{P_1V_1}{T_1} = \frac{P_2V_2}{T_2}$$

Energy Relation: $$\Delta U = mc_v\Delta T$$

$$from\ which\ \Delta H = mc_p\Delta T$$
$$and\ c_p = c_v + R$$

For any differential process on a closed system of a thermally perfect gas (c_v may vary with temperature):

Mass Relation: $$\frac{dP}{P} + \frac{dv}{v} - \frac{dT}{T} = 0$$

Energy Relation: $$du = c_v dT$$

Fig. 6.5 Summary of calorically perfect gas relations. For air at nominal room temperature, $c_v = 718$ J/kg/ΔK and $R = 287$ J/kg/K

$$c_v = \left.\frac{\partial u}{\partial T}\right|_v$$

The name "specific heat at constant volume" is often interpreted as being related to constant volume processes. It is nothing of the sort. Rather, it is a material property that relates internal energy changes to temperature changes. A better (but still flawed) name that will be used in this text would be the "internal energy specific heat." This name is flawed because it has nothing directly to do with heat transfer (as shown shortly), and the term "specific" does not refer to per unit mass like it does for other properties. It does show that the property relates internal energy changes to the temperature change (as opposed to the enthalpy specific heat to be discussed shortly).

The internal energy is a function of temperature only for a perfect gas (both thermally and calorically perfect). Therefore, the 2nd partial derivative is zero. That is,

$$\left.\frac{\partial u}{\partial v}\right|_T = 0 \quad \text{for perfect gases}$$

Therefore, the partial derivatives are unnecessary, and a change in internal energy is related to a change in temperature through the internal energy specific heat.

$$du = c_v dT \quad \text{for perfect gases}$$

For a finite change, this differential expression can be integrated between the states:

$$u_2 - u_1 = \int_{T_1}^{T_2} c_v dT \quad \text{for thermally perfect gases (variable specific heat)}$$

$$u_2 - u_1 = c_v (T_2 - T_1) \quad \text{for calorically perfect gases (constant specific heat)}$$

Appendix 1 compares the two models directly in a case study. The variation of specific heat with temperature is due to quantum mechanical effects, discussed qualitatively in the deeper dive of Appendix 2. In practice, tables or polynomial curve fits (valid over specific temperature ranges) can be obtained in the literature when needed. However, a general guideline is to assume suitably chosen constant specific heat values for simplified models aimed at capturing the key controlling physics, and variable specific heats for more detailed models that require numerical solutions (which are not developed in this text).

This relationship between internal energy and temperature is the same as for compressed liquids. However, it is important to differentiate the internal energy specific heat versus the enthalpy specific heat, a thermodynamic property defined as

$$c_p \equiv \left.\frac{\partial h}{\partial T}\right|_P$$

The relationship between the two specific heats (internal energy and enthalpy) is obtained from the definition of the combination thermodynamic property specific enthalpy, h:

$$h = u + Pv$$

For ideal gases: $h = u + RT$. Taking the derivative with respect to temperature:

$$\frac{dh}{dT} = \frac{du}{dT} + R$$

Recognizing the two specific heats:

$$c_p = c_v + R$$

This expression is valid for both thermally perfect (specific heats vary with temperature) and calorically perfect (constant specific heats) gases.

The ratio of specific heats is commonly introduced, given the symbol "k" in some publications, and "γ" in others. k will be used here, and the ratio of specific heats is defined as:

$$k = \frac{c_p}{c_v} = \frac{R + c_v}{c_v}$$

This ratio is always greater than unity ($c_p > c_v$) but less than 1.667 (the value for monatomic gases).

There are thus five related thermodynamic properties that commonly arise: c_v, c_p, k, R, and MW. Given any two of the five as inputs, the other three can be derived.

The enthalpy specific heat is always greater than the internal energy specific heat, that is, $c_p > c_v$. This result can be demonstrated on a physical basis by analyzing the following problem.

Example: Constant Volume (Isochoric), Constant Pressure (Isobaric), and Adiabatic, Reversible (Isentropic) Processes

Consider a well-insulated piston/cylinder of negligible thermal mass that is filled with $M = 1$ kg of air initially at nominal room conditions. The cylinder is fitted with an electrical wire placed inside. Determine the input electrical work needed to increase the temperature by a fixed amount $\Delta T = 100\,°C$ and the eXergy destruction associated with the process. Choose a control volume that includes the gas plus the electrical wire and neglect the thermal mass of the electrical wire in comparison with that of the gas. Consider three processes, one in which the piston is restrained from moving (a constant volume process), one in which the piston moves freely (a constant pressure process), and one for which no electrical work is added but the pressure is increased gradually. Sketch the processes on a PV diagram and a TS diagram (Figs. 6.6, 6.7, and 6.8).

Defining the system to be the gas and the electrical wire combined, energy and entropy balances for all three processes are

$$\Delta E = Q_{IN} - W_{OUT}$$

$$\Delta S = \frac{Q_{IN}}{T_B} + \Sigma$$

Neglecting kinetic and potential energy and the thermal mass of the wire (so $E = U = Mu$, where u is the specific internal energy of the gas), assuming adiabatic processes, and two forms of work (expansion and electrical), the 1st and 2nd Law balances are:

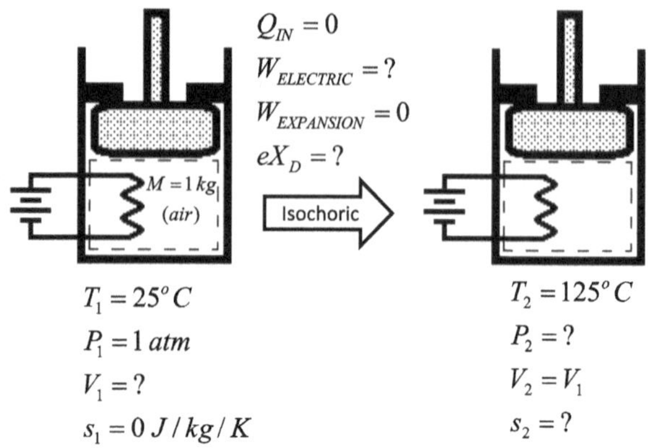

Fig. 6.6 Sketch of isochoric process

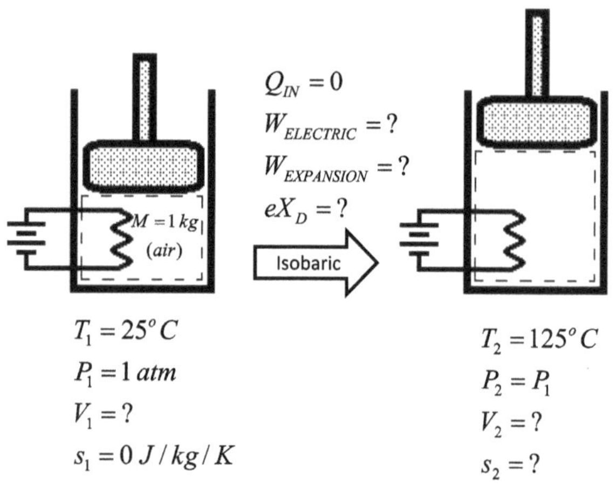

Fig. 6.7 Sketch of isobaric process

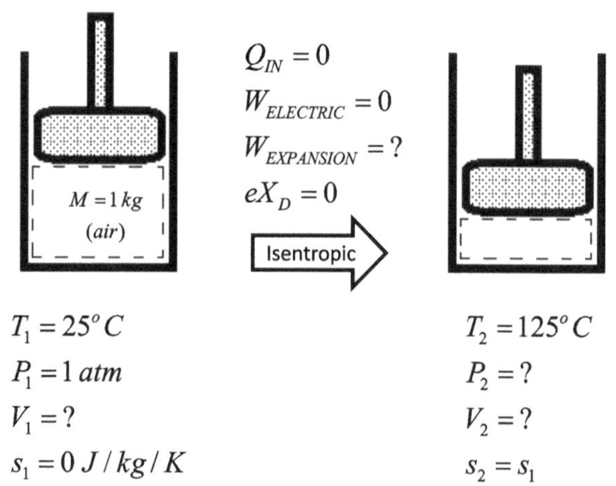

Fig. 6.8 Sketch of isentropic compression process

$$M\Delta u = -\int_{V_1}^{V_2} PdV + W_E$$

$$eX_D = T_\infty \Sigma = T_\infty \Delta S$$

Rearranging for the input electrical work:

$$W_E = M\Delta u + \int_{V_1}^{V_2} PdV$$

This expression shows that input work is required to raise the internal energy and to expand the gases against the opposing pressure.

$$W_E = M\Delta u \quad \text{for the constant volume process}$$

$$\begin{aligned} W_E &= M\Delta u + P\Delta V = M\Delta u + M\Delta(Pv) \\ &= M\Delta h \quad \text{for the constant pressure process} \end{aligned}$$

$$\begin{aligned} W_E = 0 \quad \text{so} \quad M\Delta u \\ = -\int_{V_1}^{V_2} PdV \quad \text{for the adiabatic, reversible (isentropic) process} \end{aligned}$$

These results are valid for any simple compressible substance. The constant volume and constant pressure processes can be analyzed further from the 1st Law. The isentropic process requires development of entropy relations and will be continued in the next section.

For ideal gases:

$$W_E = Mc_v\Delta T \quad \text{for the constant volume process}$$

$$\begin{aligned} W_E &= Mc_v\Delta T + P\Delta V = Mc_v\Delta T + MR\Delta T \\ &= Mc_p\Delta T \quad \text{for the constant pressure process} \end{aligned}$$

Rearranging these equations:

$$c_v = \frac{W_E}{M\Delta T} \quad \text{for the constant volume process}$$

$$c_p = \frac{W_E}{M\Delta T} \quad \text{for the constant pressure process}$$

The specific heats can therefore be interpreted as the input work required to change the temperature of a mass of gas by a fixed amount. This example is a deliberately misleading way to interpret the specific heat, but is equivalent to the historical record, which used a different particular case, namely, one in which heat was added instead of work. If the system had been defined as just the gas (without the wire), then the agent for change would have been heat, not work.

It is unfortunate, for three reasons, that the language "specific heat at constant volume" has become so ingrained in the language of thermodynamics. First, the word "specific" means "per unit mass" in other contexts (specific volume, specific internal energy, specific entropy). Second, specific heats have nothing directly to do with heat transfer. The internal energy or enthalpy can be changed by heat or work as this example shows. The phrase that defines specific heat as the "heat required to raise the temperature of a substance by one degree" refers to a specific type of process. It could equally have been defined as the "work required to raise the temperature of a substance by one degree" and then called a specific work. Third, the phrase "at constant volume" or "at constant pressure" confuses many students, who interpret that the use of these thermodynamic properties restricts their use to constant volume or constant pressure processes.

This discussion is simply a reminder to regard the specific heat as a thermodynamic property of a substance (like internal energy or specific volume), and not restricted to any type of process. The term "internal energy specific heat" will be used for c_v and "enthalpy specific heat" for c_p, which is better, but not perfect (or rather, ideal?). It can usually simply be referred to as c_v or c_p, depending on the context.

The two processes are shown (to scale) on a *PV* diagram in Fig. 6.9 (and the third case is discussed in the next section). Two isotherms for air are shown, as are the input parameters and derived parameters used to construct the plot. The constant volume process is a vertical line, with the final pressure being greater than the initial. The constant pressure process is a horizontal line, with the final volume being greater than the initial.

To calculate the eXergy destruction, to evaluate the integral for the isentropic case, and to plot the processes on *Ts* coordinates, the change in entropy must be calculated, the subject of the next section. To anticipate the result, for these adiabatic processes, the eXergy destruction is tied directly to the change in entropy, so the process with the larger entropy change will experience a larger eXergy destruction. All three processes experience the same temperature rise, but the pressure increases for the isochoric process and the volume increases for the isobaric process. However, it seems likely that the isobaric process will result in a greater eXergy destruction because it requires more input electrical work converted to thermal energy, which is a highly dissipative process.

This example could be redone entirely by choosing the system to consist of just the gas (not the wire). The 1st Law analysis would be similar, except that there would be heat exchange between the wire and the gas. For the wire, the electrical input would equal the heat transfer to the gas. For the gas, heat would be added from the wire and expansion work would be done. The advantage of choosing the combined wire and gas system is not apparent from a 1st Law analysis alone. In fact, it might seem overly complicated or misleading because the mechanism for raising the

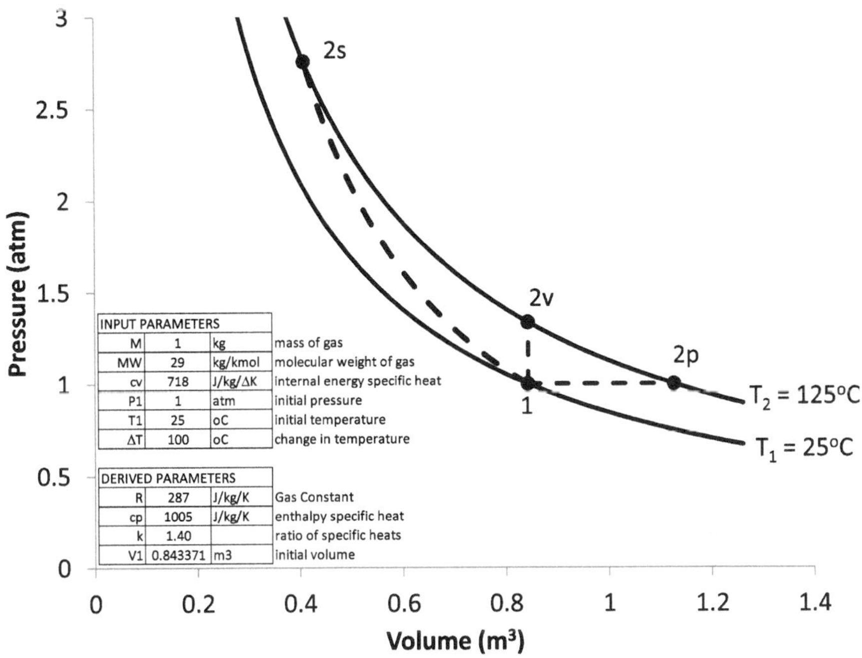

Fig. 6.9 *PV* diagram for constant volume, constant pressure, and constant entropy example problem

internal energy and performing work of expansion is a heat transfer process (internal to the combined wire/gas system). The advantage comes from a 2nd Law perspective since the entropy transfer between the wire and gas is not well characterized directly.

Entropy Changes of Perfect Gases

The goal of this section is to develop expressions for the entropy changes associated with pressure, temperature, and/or volume changes for perfect gases, then apply it to the example problem in question.

Consider the 1st and 2nd Laws for a differential change of a substance of mass M at temperature T, pressure P, volume V, and heat exchange with a surface at T_B with negligible change in kinetic or potential energy:

$$dU = \delta Q - PdV$$

$$dS = \frac{\delta Q}{T_B} + \delta\Sigma$$

Combining them, by eliminating the heat exchange:

$$dU = T_B dS - PdV - T_B \delta\Sigma$$

Consider a reversible process, where the boundary temperature equals the system temperature and there is no entropy generation:

$$dU = TdS - PdV$$

This combination of 1st and 2nd Law expression can be rearranged to determine the differential change in entropy:

$$dS = \frac{dU}{T} + \frac{PdV}{T}$$

Integrating between two states:

$$S_2 - S_1 = \int_1^2 \frac{dU}{T} + \int_1^2 \frac{PdV}{T}$$

Another useful form for the entropy change is obtained by introducing the enthalpy:

$$dS = \frac{d(H - PV)}{T} + \frac{PdV}{T} = \frac{dH}{T} - P\frac{dV}{T} - V\frac{dP}{T} + \frac{PdV}{T}$$

$$dS = \frac{dH}{T} - \frac{VdP}{T}$$

$$S_2 - S_1 = \int_1^2 \frac{dH}{T} - \int_1^2 \frac{VdP}{T}$$

It may seem like this result only applies to reversible processes. However, it is generally true, as the following argument addresses. Most actual processes have some degree of irreversibility, but a hypothetic reversible process path can be constructed, in theory, that depends only on total differentials of state variables. Once established, the change in entropy (a state variable) between two states is independent of the path taken. So, with that, the important expression just derived is valid for all quasi-equilibrium differential changes of simple compressible substances. Now, it can be applied to perfect gases, and integrated, as follows:

$$S_2 - S_1 = \int_1^2 Mc_v \frac{dT}{T} + \int_1^2 MR\frac{dV}{V}$$

Dividing through by the mass and evaluating the 2nd integral (since MR is constant) but not the 1st (since the internal energy specific heat may vary with temperature):

$$s_2 - s_1 = \int_{T_1}^{T_2} c_v \frac{dT}{T} + R\ln\left(\frac{V_2}{V_1}\right) \quad \text{for thermally perfect gases}$$

Most thermodynamic textbooks provide "ideal gas tables" that list the numerical results of the temperature integral in this expression. If the internal energy specific heat is treated as a suitably chosen constant value, the first integral can be integrated:

$$s_2 - s_1 = c_v \ln\left(\frac{T_2}{T_1}\right) + R\ln\left(\frac{V_2}{V_1}\right) \quad \text{for calorically perfect gases}$$

The first term is the same type of term for compressed liquids (with c_v) and shows that the entropy is an increasing function with temperature. The 2nd term shows that the entropy also increases with the volume. Liquids do not experience a significant change in volume, and this effect does not occur.

Students are more likely to misuse property relations if given the option to use ideal gas tables. A general "rule of thumb" is, for devices whose working fluids undergo phase change and/or are near the saturation curve, use the saturation and/or superheat tables. For devices whose working fluids can be treated as ideal gases, assume that it is calorically perfect and use the handful of algebraic expressions that ensue, summarized in Fig. 6.5. Panicking students tend to start using the ideal gas laws for phase-changing processes and ideal gas tables when constant specific heats are much simpler to use.

Similar treatment of the enthalpy form yields an entropy change expressed in terms of temperature and pressure ratios. Alternatively, the effect of pressure can be obtained by eliminating the temperature or volume in using the ideal gas law in the form for a fixed mass:

$$MR = \frac{P_1V_1}{T_1} = \frac{P_2V_2}{T_2}.$$

Eliminating the temperature:

$$s_2 - s_1 = c_v \ln\left(\frac{P_2 V_2}{P_1 V_1}\right) + R \ln\left(\frac{V_2}{V_1}\right) = c_v \ln\left(\frac{P_2}{P_1}\right) + (R + c_v) \ln\left(\frac{V_2}{V_1}\right)$$
$$= c_v \ln\left(\frac{P_2}{P_1}\right) + c_p \ln\left(\frac{V_2}{V_1}\right)$$

Eliminating the volume:

$$s_2 - s_1 = c_v \ln\left(\frac{T_2}{T_1}\right) + R \ln\left(\frac{T_2 P_1}{T_1 P_2}\right) = (c_v + R) \ln\left(\frac{T_2}{T_1}\right) - R \ln\left(\frac{P_2}{P_1}\right)$$
$$= c_p \ln\left(\frac{T_2}{T_1}\right) - R \ln\left(\frac{P_2}{P_1}\right)$$

It can be useful to nondimensionalize the entropy change by normalizing it by the gas constant, and then expressing the specific heats in terms of their ratio $k = c_p/c_v$. For the first case:

$$\frac{s_2 - s_1}{R} = \frac{c_v}{R} \ln\left(\frac{T_2}{T_1}\right) + \ln\left(\frac{V_2}{V_1}\right) = \frac{1}{k-1} \ln\left(\frac{T_2}{T_1}\right) + \ln\left(\frac{V_2}{V_1}\right)$$

A summary of all the entropy change relations is given in Table 6.4.

Isentropic processes for calorically perfect gases result in simple power law relationships between two thermodynamic variables. They can be obtained by setting the entropy change to zero and rearranging and are summarized in Table 6.5.

In the example problem, the isentropic compression case follows these relationships. The eXergy destruction is zero. The eXergy destruction for the other two processes is

Table 6.4 Summary of entropy change relations for calorically perfect gases (constant specific heats)

	In terms of T and V	In terms of P and V	In terms of T and P
$s_2 - s_1$	$c_v \ln\left(\frac{T_2}{T_1}\right) + R \ln\left(\frac{V_2}{V_1}\right)$	$c_v \ln\left(\frac{P_2}{P_1}\right) + c_p \ln\left(\frac{V_2}{V_1}\right)$	$c_p \ln\left(\frac{T_2}{T_1}\right) - R \ln\left(\frac{P_2}{P_1}\right)$
$\frac{s_2 - s_1}{R}$	$\frac{1}{k-1} \ln\left(\frac{T_2}{T_1}\right) + \ln\left(\frac{V_2}{V_1}\right)$	$\frac{1}{k-1} \ln\left(\frac{P_2}{P_1}\right) + \frac{k}{k-1} \ln\left(\frac{V_2}{V_1}\right)$	$\frac{k}{k-1} \ln\left(\frac{T_2}{T_1}\right) - \ln\left(\frac{P_2}{P_1}\right)$

Note: The first row is the entropy change. The second row is normalized by the gas constant

Table 6.5 Isentropic relations for calorically perfect gases

	In terms of T and V	In terms of V and P	In terms of T and P
R, c_p,c_v	$\left(\frac{T_2}{T_1}\right) = \left(\frac{V_2}{V_1}\right)^{\frac{-R}{c_v}}$	$\left(\frac{P_2}{P_1}\right) = \left(\frac{V_2}{V_1}\right)^{\frac{-c_p}{c_v}}$	$\left(\frac{T_2}{T_1}\right) = \left(\frac{P_2}{P_1}\right)^{\frac{R}{c_p}}$
k	$\left(\frac{T_2}{T_1}\right) = \left(\frac{V_2}{V_1}\right)^{\frac{-1}{k-1}}$	$\left(\frac{P_2}{P_1}\right) = \left(\frac{V_2}{V_1}\right)^{-k}$	$\left(\frac{T_2}{T_1}\right) = \left(\frac{P_2}{P_1}\right)^{\frac{k-1}{k}}$

$$eX_D = T_\infty \Delta S = T_\infty \left[c_p \ln\left(\frac{T_2}{T_1}\right) - R \ln\left(\frac{P_2}{P_1}\right) \right] \quad \text{for the constant volume process}$$

$$eX_D = T_\infty \Delta S$$

$$= T_\infty \left[c_v \ln\left(\frac{T_2}{T_1}\right) + R \ln\left(\frac{V_2}{V_1}\right) \right] \quad \text{for the constant pressure process}$$

It is not obvious, yet, which is larger. The temperature effect on entropy is larger for the constant volume process ($c_p > c_v$), but the pressure effect decreases entropy, while the volume effect increases the entropy for constant pressure process.

A temperature versus entropy plot is shown in Fig. 6.10, which shows the initial state (with entropy set to zero), an isochoric line with the corresponding path of the constant volume process, and an isobaric line with its constant volume process. Note that a plot of the isochoric and isobaric lines is obtained by expressing the appropriate form of the entropy change with temperature. The initial condition can be set to any value to obtain

$$T = T_1 \left(\frac{V_1}{V_2}\right)^{k-1} e^{\frac{s-s_1}{c_v}} \quad \text{for isochores}$$

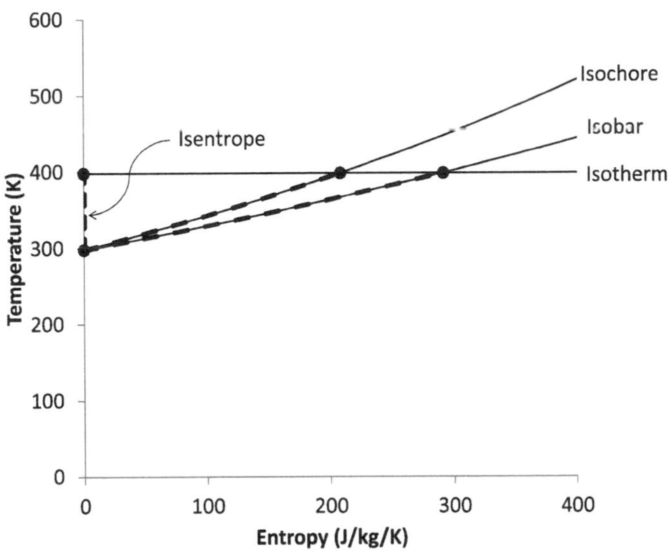

Fig. 6.10 Temperature entropy diagram for constant volume versus constant pressure example problem. The entropy at the initial state is set equal to zero. Air is treated as a calorically perfect gas, with the same parameters as in Fig. 6.9

$$T = T_1 \left(\frac{P_2}{P_1}\right)^{\frac{k-1}{k}} e^{\frac{s-s_1}{c_p}} \quad \text{for isobars}$$

In this case, the lines are drawn from the initial point forward in entropy, so the ratio of the variable held constant is unity. These equations can be used in other cases to plot different isochores or isobars. In both cases, however, it is clear that the temperature increases with entropy exponentially. The slope is higher for isochores than for isobars. Therefore, it is clear from inspection of Fig. 6.10 that the entropy increases more for the constant pressure case than the constant volume case, and therefore there is a greater eXergy destruction for the constant pressure case, as was anticipated.

Deeper Dive: Ideal Reciprocating Compressor

Application is made of these perfect gas relations for the ideal reciprocating compressor as a case study (Fig. 6.1). 1st and 2nd Law principles are applied with an ideal gas as a working fluid. This analysis serves to introduce the principles of reciprocating engines, the subject of the first section of Part 2 of this textbook.

Expansion Process

The expansion process takes place with both valves closed, and, assuming no leakage past the piston (or elsewhere), the system is defined to be the trapped mass at the instant the intake valve closes. In the ideal cycle, the start of the expansion process coincides with the piston at its topmost point (top dead center), when the volume equals its minimum (the clearance volume). The pressure is also well characterized, equal to the exhaust plenum pressure. However, the temperature is less well characterized because it depends on the history of the cycle. The mass of gas during the ideal expansion process is given by the ideal gas law (with pressure and temperature in absolute units):

$$M_1 = M_2 = \frac{P_1 V_1}{RT_1} = \frac{P_{\text{exhaust}} V_C}{RT_1}$$

The temperature is left unspecified for the moment. In the real cycle, the precise timing of state 1, (when the exhaust valve shuts) is not well defined. It will be slightly delayed, an effect that will not be considered in detail here.

This mass is called the "residual gases" as they represent the gases that remain at the end of the exhaust process. Subsequently, gases will enter from the intake manifold and mix with these gases, so the properties of the gases at the end of the intake will reflect a mixture of fluids that in general have different properties.

The expansion process is the same as the isentropic compression of the previous example, but with a volume increase instead of decrease. The process will be analyzed here, and while redundant, the isentropic process is important in so many applications that seeing it several times in slightly different ways is a worthwhile iterative reinforcement of fundamental thermodynamic principles.

The 1st Law applied to the trapped gases for the entire process is:

$$\Delta E = Q_{12,\mathrm{IN}} - W_{12,\mathrm{OUT}}$$

The following assumptions are made:

- Kinetic and potential energy changes are negligible
- Quasi-equilibrium (pressure is uniform in the cylinder)
- Adiabatic

With these approximations, the 1st Law becomes:

$$M_1(u_2 - u_1) = -W_{12,\mathrm{OUT}}$$

That is, any work done by the gases decreases the internal energy stored in the gases. Expressed as an integral across the entire process:

$$M_1 \int_{T_1}^{T_2} c_v dT = -\int_{V_1}^{V_2} PdV \quad \text{for a thermally perfect gas (variable specific heat)}$$

Integration of the left-hand side requires the specific heat to be explicitly expressed in terms of temperature. For a calorically perfect gas, the specific heat is treated as a constant, and the left-hand side is integrated as follows:

$$M_1 c_v (T_2 - T_1) = -\int_{V_1}^{V_2} PdV$$

$$= -W_{12,\mathrm{OUT}} \quad \text{for a calorically perfect gas (constant specific heat)}$$

Solving for the temperature at state 2:

$$T_2 = T_1 - \frac{W_{12,\mathrm{OUT}}}{M_1 c_\mathrm{v}}$$

This important result shows that if the gases do work on the environment (i.e., expand), their temperature decreases. That is, a gas can be made colder than its environment if it does work. This effect can be the basis for creating cold spaces (i.e., refrigeration). Conversely, if the work is done on the gases by the environment (negative work out), the temperature of the gases will increase. This effect will occur during the compression process (3-4).

The work done by the gases cannot be determined (integrate *PdV*) until the pressure is related to volume (or vice versa). Since temperature also changes, introducing the ideal gas law introduces another variable. The problem will be approached from a differential point of view.

The 1st Law applied to the trapped gas on a differential basis is:

$$dE = \delta Q_\mathrm{IN} - \delta W_\mathrm{OUT}$$

With the assumptions made previously (adiabatic, ideal gas):

$$M_1 c_\mathrm{v} dT = -PdV$$

It is reiterated that the left-hand side does not yet imply that the specific heat is constant, and clearly the volume changes during the process, so use of c_v does not imply a constant volume process. Dividing both sides by the temperature (in absolute units), for reasons that will be apparent shortly,

$$M_1 c_\mathrm{v} \frac{dT}{T} = -\frac{P}{T} dV$$

This 1st Law expression involves simultaneous changes in pressure, temperature, and volume. In the present problem, the constraint that will dictate the end of the expansion process is defined by the pressure. The pressure starts at the exhaust plenum pressure and reaches the intake plenum pressure at some unknown volume. Therefore, the ideal gas law can be used to eliminate the temperature. Introducing the differential expression of the ideal gas law for a fixed mass on the left side and the ideal gas law on the right side:

$$M_1 c_\mathrm{v} \left(\frac{dP}{P} + \frac{dV}{V} \right) = -M_1 R \frac{dV}{V}$$

This expression is a combined mass (ideal gas law) and energy equation. Rearranging and defining the ratio of specific heats:

$$k = \frac{c_p}{c_v}$$

and using the ideal gas relation $R = c_p - c_v$:

$$\frac{dP}{P} = -k\frac{dV}{V}$$

The equation represents a 1st Law balance applied to a closed mass for an adiabatic expansion (or compression) of a thermally perfect gas. It can be integrated between the initial and final states:

$$\int_{P_1}^{P_{22}} \frac{dP}{P} = -\int_{V_1}^{V_2} k\frac{dV}{V}$$

The left-hand side can be integrated right away as it only involves pressure in the integrand. On the other hand, the ratio of specific heats is a function of temperature, and therefore cannot be integrated until a relationship between them is made. If the ratio of specific heats is a constant, however,

$$\ln\left(\frac{P_2}{P_1}\right) = -k \ln\left(\frac{V_2}{V_1}\right) \quad \text{for a calorically perfect gas}$$

Using those curious properties of natural logarithms that $a \ln(x) = \ln(x^a)$ and $-\ln(x) = \ln\left(\frac{1}{x}\right)$ rearranging for the unknown volume at which the expansion process ends and the intake begins:

$$V_2 = V_1\left(\frac{P_1}{P_2}\right)^{1/k}$$

$$V_2 = V_C\left(\frac{P_{\text{exhaust}}}{P_{\text{intake}}}\right)^{1/k}$$

Notice that the volume at the end of the stroke is independent of the temperature of the gases. However, the work associated with the process does depend on the mass, which in turn depends on the starting temperature. Now that the volume is known, the final temperature of the gases at the intake can be determined, at least in terms of the initial temperature:

$$T_2 = \left(\frac{P_2}{P_1}\right)\left(\frac{V_2}{V_1}\right)T_1 = \left(\frac{P_2}{P_1}\right)\left(\frac{P_1}{P_2}\right)^{1/k} T_1 = \left(\frac{P_2}{P_1}\right)^{1-1/k} T_1$$

$$T_2 = \left(\frac{P_{\text{intake}}}{P_{\text{exhaust}}}\right)^{\frac{k-1}{k}} T_1$$

The work of the expansion process can be obtained either by direct integration of PdV or more simply from the overall energy balance:

$$\Delta U = M_1 c_v (T_2 - T_1) = -W_{12,\text{OUT}}$$

Rearranging and using the result for T_2:

$$W_{12,\text{OUT}} = M_1 c_v T_1 \left(1 - \left(\frac{P_{\text{intake}}}{P_{\text{exhaust}}}\right)^{\frac{k-1}{k}}\right)$$

Intake Process

The intake stroke consists of two processes: expansion (just analyzed) followed by intake. The intake process involves an open system. As the piston withdraws, gases from the intake plenum enter the cylinder and mix with the residual gases. The residual gases are at temperature T_2 at the start of the intake process, and the gases that enter are at the intake plenum temperature, T_{intake}. The temperature at the end of the intake process (T_3) will be between them. If the intake process begins near the start of the intake stroke, the temperature will be closer to the intake plenum temperature. If the intake process begins near the end of the intake stroke, the temperature will be closer to T_2. Note that the gases do work on the piston, which tends to decrease the temperature of the gases, but simultaneously, flow work is done on the control volume. For the ideal cycle, these two work terms are equal and opposite, as will be shown, so that there is no tendency for the temperature to change during intake. The temperature change is due to a pure mixing process. Since two gases of different temperatures are being mixed, there is an irreversibility associated with the intake process (eXergy is destroyed). The main purpose of the detailed analysis to follow is to demonstrate how to apply mass and energy balances to an inherently transient control volume. Mass and energy stored change with time, and the inlets and outlets do not balance.

A differential mass balance during the intake process is

$$dM = \delta m_{\text{in}}$$

where δm_{in} is a differential mass element that enters, changing the mass stored in the control volume (M).

The 1st Law applied to a differential process on the control volume is

$$dE = \delta Q_{\text{IN}} - \delta W_{\text{OUT}} + \delta m_{\text{in}}\left(\frac{P_{\text{intake}}}{\rho_{\text{intake}}} + e_{\text{intake}}\right)$$

The mass term represents the flow work and the internal energy that enters the control volume. The properties of the mass of the fluid that enter are the intake plenum properties. Neglecting kinetic and potential energy:

$$d(Mu) = Mdu + udM = \delta Q_{\text{IN}} - PdV + \delta m_{\text{in}}(P/\rho + u)_{\text{intake}}$$

Introducing perfect gas approximations, the mass relation ($dM = \delta m_{\text{in}}$) and an adiabatic process:

$$Mc_{\text{v}}dT = -PdV + dM(RT + u)_{\text{intake}} - udM$$

Since the pressure is fixed and the volume sweeps to a known value, the temperature is eliminated using the differential mass equation (noting that $dv/v = dV/V - dM/M$):

$$Mc_{\text{v}}T\left(\frac{dP}{P} + \frac{dV}{V} - \frac{dM}{M}\right) = -PdV + dM(RT + u)_{\text{intake}} - udM$$

The intake process is isobaric for the ideal cycle ($dP = 0$), and rearranging:

$$\left(\frac{c_{\text{v}}MT}{V} + P\right)dV = dM(RT_{\text{intake}} + u_{\text{intake}} - u + c_{\text{v}}T)$$

Since $MT/V = P/R$ and $P = P_{\text{intake}}$ during the intake stroke (for the ideal cycle):

$$\frac{P_{\text{intake}}}{R}(c_{\text{v}} + R)dV = dM(RT_{\text{intake}} + c_{\text{v}}(T_{\text{intake}} - T) + c_{\text{v}}T) = dM(c_{\text{v}} + R)T_{\text{intake}}$$

Simplifying:

$$dV = \frac{RT_{\text{intake}}}{P_{\text{intake}}}dM = \frac{dM}{\rho_{\text{intake}}}$$

This result, which in retrospect may seem it should be obvious, indicates that the change in volume of the cylinder equals the volume of the mass that is inducted. However, in the more general case (with a pressure change across the valve), that will not be precisely true. The equation can be integrated to yield the total mass that is inducted during the intake process:

$$M_3 - M_2 = \rho_{\text{intake}}(V_3 - V_2) = \frac{P_{\text{intake}}}{RT_{\text{intake}}}(V_3 - V_2)$$

The work done by the gases on the piston is

$$W_{23,\mathrm{OUT}} = P_{\mathrm{intake}}(V_3 - V_2)$$

The temperature of the gases at the end of the intake process is obtained from the ideal gas law:

$$T_3 = \frac{P_3 V_3}{RM_3} = \frac{P_{\mathrm{intake}}(V_{\mathrm{C}} + V_{\mathrm{D}})}{RM_3}$$

It is not apparent that this temperature is intermediate between intake plenum pressure and the temperature of the residual gases. To show that it is, return to the 1st Law, but eliminate *dV*:

$$Mc_{\mathrm{v}}dT = -PV\left(\frac{dM}{M} + \frac{dT}{T} - \frac{dP}{P}\right) + dM(h_{\mathrm{intake}} - u)$$

The intake process is isobaric for the ideal cycle ($dP = 0$), and

$$\left(Mc_{\mathrm{v}} + \frac{PV}{T}\right)dT = dM\left(h_{\mathrm{intake}} - u - \frac{PV}{M}\right)$$
$$M(c_{\mathrm{v}} + R)dT = (h_{\mathrm{intake}} - h)dM$$

$$Mc_{\mathrm{p}}dT = c_{\mathrm{p}}(T_{\mathrm{intake}} - T)dM$$
$$\frac{dT}{T_{\mathrm{intake}} - T} = \frac{dM}{M}$$

This differential expression for the intake process shows how the sign of the temperature change depends on the relative temperature between the intake and in-cylinder gases. If the intake temperature is lower than the in-cylinder, then the temperature decreases as the colder gases mix with the hotter residual gases, and vice versa. Integrating:

$$\int_{T_2}^{T_3} \frac{-d(T_{\mathrm{intake}} - T)}{T_{\mathrm{intake}} - T} = \int_{M_2}^{M_3} \frac{dM}{M}$$
$$-\ln\left(\frac{T_{\mathrm{intake}} - T_3}{T_{\mathrm{intake}} - T_2}\right) = \ln\left(\frac{M_3}{M_2}\right)$$
$$\left(\frac{T_{\mathrm{intake}} - T_3}{T_{\mathrm{intake}} - T_2}\right) = \left(\frac{M_2}{M_3}\right) = \left(\frac{P_2}{P_3}\right)\left(\frac{V_2}{V_3}\right)\left(\frac{T_3}{T_2}\right)$$

Solving for T_3:

$$T_3 = \frac{T_{\text{intake}}}{1 + \left(\frac{V_2}{V_3}\right)\left(\frac{T_{\text{intake}}}{T_2} - 1\right)}$$

Consider the limits:

if $V_3 \gg V_2$, $T_3 = T_{\text{intake}}$
if $V_3 \sim V_2$, $T_3 = T_2$

However, it will be demonstrated that the temperature at the start of the intake process (T_2) equals the intake temperature, and therefore, the temperature of the residual gases equals the temperature of the gases being inducted. That will not generally be true for the real cycle, however, or for reciprocating engines, as studied in Chap. 7.

Compression Process: 3–4

The compression process is similar to the expansion process. It just takes place with different mass and starting conditions, and the volume decreases. The relationship between pressure and volume is

$$V_4 = V_3\left(\frac{P_3}{P_4}\right)^{1/k} = (V_C + V_D)\left(\frac{P_{\text{intake}}}{P_{\text{exhaust}}}\right)^{1/k}$$

The exhaust valve will naturally open when this volume is reached. The work of the compression process raises the internal energy, hence temperature. The temperature at state 4 is given by

$$T_4 = T_3\left(\frac{P_4}{P_3}\right)^{\frac{k-1}{k}} = T_3\left(\frac{P_{\text{exhaust}}}{P_{\text{intake}}}\right)^{\frac{k-1}{k}}$$

The work done by the gases is

$$\Delta E = M_3 c_v (T_4 - T_3) = -W_{34,\text{OUT}}$$

Exhaust Process: 4–1

The exhaust process has an important difference compared to the intake. For the intake process, the mass that enters has fixed properties (for a fixed inlet plenum), and they mix with fluid of different properties. For the exhaust process, other than the pressure, the properties of the mass that exit define the properties of the exhaust plenum. For the ideal cycle, the work done on the gases by the piston is equal and opposite to the flow work done by the gases as they exit, and there is no tendency for the temperature to change. This result will be formally shown.

The differential mass balance is

$$dM = -\delta m_e$$

The 1st Law is

$$dE = \delta Q_{\text{IN}} - \delta W_{\text{OUT}} - \delta m_{\text{in}}\left(\frac{P}{\rho} + e\right)$$

The properties of the mass flow terms are that of the cylinder (whereas it was that of the intake plenum for the intake process). For the ideal cycle approximations:

$$Mdu + udM = -PdV + dM\left(\frac{P}{\rho} + u\right)$$

Multiplying PdV by V/V and eliminating dV/V:

$$Mdu = -PV\left(\frac{dM}{M} + \frac{dT}{T} - \frac{dP}{P}\right) + dM\left(\frac{PV}{M} + u\right) - udM$$

Simplifying:

$$\left(M_4 c_v + \frac{PV}{T}\right)dT = 0$$

Or simply:

$$dT = 0$$

The temperature does not change during the exhaust process, that is,

$$T_4 = T_1$$

The mass exhausted is

$$M_4 - M_1 = \frac{P_4 V_4}{RT_4} - \frac{P_1 V_1}{RT_1} = \frac{P_{\text{exhaust}}}{RT_1}(V_4 - V_1)$$

The work done is

$$W_{41,\text{OUT}} = P_{\text{exhaust}}(V_1 - V_4)$$

Balancing the Cycle

All four processes of the ideal reciprocating compressor cycle have been analyzed, but the analysis is not complete. The starting temperature, T_1, is not known, and this value affects all processes indirectly. In practice, if the compressor were started cold (and at the start of the intake stroke), the initial cycle temperature for the very first cycle would be ambient temperature (which is likely to also be the intake temperature). The temperature would drop during the expansion, then warm to a temperature somewhat below ambient during intake, then rise during the compression stroke to a temperature above ambient, which would remain the same during the exhaust process. When the piston returned to top dead center to start the next cycle, the fluid would be at a different temperature than the first cycle. The next cycle would see a change also, but eventually, and in practice after only a few cycles, the temperature at the start would equal the temperature at the end. Prior to that, the compressor operates with a mechanical cycle (the piston returning to its initial state), but not a thermodynamic cycle.

What is not readily apparent is that the mass inducted during the intake process would be different than the mass exhausted. The cycle is unbalanced. To balance the cycle, the mass inducted can be set equal to the mass exhausted, and solving for the resultant temperature. That is,

$$\text{Mass inducted} = \text{Mass exhausted}$$

$$M_3 - M_2 = M_4 - M_1$$

$$\frac{P_{\text{intake}}}{RT_{\text{intake}}}(V_3 - V_2) = \frac{P_{\text{exhaust}}}{RT_1}(V_4 - V_1)$$

$$\frac{T_1}{T_{\text{intake}}} = \frac{P_{\text{exhaust}}}{P_{\text{intake}}}\frac{(V_4 - V_1)}{(V_3 - V_2)}$$

Define the pressure ratio:

$$\widehat{P} = \left(\frac{P_{\text{exhaust}}}{P_{\text{intake}}}\right)$$

$$\frac{T_1}{T_{\text{intake}}} = \widehat{P}\frac{\left(V_3\widehat{P}^{-1/k} - V_1\right)}{\left(V_3 - V_1P^{1/k}\right)} = \widehat{P}^{1-\frac{1}{k}}\frac{\left(V_3 - V_1P^{1/k}\right)}{\left(V_3 - V_1P^{1/k}\right)}$$

$$\frac{T_1}{T_{\text{intake}}} = \widehat{P}^{\frac{k-1}{k}}$$

This really elegant result shows that the temperature of the exhaust ($T_4 = T_1$) equals the temperature that would be attained if the intake were compressed adiabatically directly from the intake plenum to the exhaust plenum. That also implies that the temperature at the end of the intake stroke equals that of intake plenum. The

Table 6.6 State properties for ideal reciprocating compressor

State	P	V	T	M
1	P_{exhaust}	V_{C}	$T_{\text{intake}}\widehat{P}^{\frac{k-1}{k}}$	$\frac{P_{\text{exhaust}}V_{\text{C}}}{RT_{\text{intake}}\widehat{P}^{\frac{k-1}{k}}}$
2	P_{intake}	$V_{\text{C}}\widehat{P}^{1/k}$	T_{intake}	$\frac{P_{\text{exhaust}}V_{\text{C}}}{RT_{\text{intake}}\widehat{P}^{\frac{k-1}{k}}}$
3	P_{intake}	$(V_{\text{C}} + V_{\text{D}})$	T_{intake}	$\frac{P_{\text{intake}}(V_{\text{C}}+V_{\text{D}})}{RT_{\text{intake}}}$
4	P_{exhaust}	$(V_{\text{C}} + V_{\text{D}})\widehat{P}^{-1/k}$	$T_{\text{intake}}\widehat{P}^{\frac{k-1}{k}}$	$\frac{P_{\text{intake}}(V_{\text{C}}+V_{\text{D}})}{RT_{\text{intake}}}$

residual gases (at state 1) are hot, and they mix with fluid from the intake plenum. The work of expansion tends to lower the temperature of this mixture, and is of just the right magnitude to set the temperature at the end of the intake stroke to equal that of the intake plenum. Again, a real cycle would experience heat transfer along the way and would deviate from this behavior somewhat. A summary of all the state properties is shown in Table 6.6.

The mass exchanged during cycle is

$$\Delta M = M_3 - M_2 = \frac{P_{\text{intake}}}{RT_{\text{intake}}}(V_{\text{C}} + V_{\text{D}}) - \frac{P_{\text{exhaust}}V_{\text{C}}}{RT_{\text{intake}}}\widehat{P}^{\frac{1-k}{k}}$$

$$\Delta M = \frac{P_{\text{intake}}(V_{\text{C}} + V_{\text{D}})}{RT_{\text{intake}}}\left(1 - \left(\frac{V_{\text{C}}}{V_{\text{C}} + V_{\text{D}}}\right)\widehat{P}\cdot\widehat{P}^{\frac{1-k}{k}}\right)$$

Introducing the compression ratio ($r = (V_{\text{C}} + V_{\text{D}})/V_{\text{C}}$) and rearranging:

$$\Delta M = \rho_{\text{intake}}V_{\text{D}}\left(\frac{r - \widehat{P}^{1/k}}{r - 1}\right)$$

Consider the limits. For an infinite compression ratio (attained with a finite displacement volume and a clearance volume that vanishes), the mass exchanged is

$$\Delta M \rightarrow \rho_{\text{intake}}V_D \quad r \rightarrow \infty \quad \text{as} \quad r \rightarrow \infty$$

This mass represents the ideal mass that would be inducted if a complete intake stroke were to fill fluid with the density in the intake plenum. This behavior would be similar to what would occur with a liquid, in which the intake valve opens immediately during the intake stroke. There would be no expansion process (as there is no gas to expand). Note, however, that the exhaust stroke would still have a compression process over a finite volume until the pressure reached the exhaust plenum pressure.

The other limit occurs when the mass exchanged becomes zero

$$\Delta M \rightarrow 0 \quad \text{as} \quad r \rightarrow \widehat{P}^{1/k}$$

Table 6.7 Summary of work expressions for ideal reciprocating compressor

Process	Description	Work in (direct form)	Alternate form
1–2	Expansion	$M_1 c_v(T_2 - T_1)$	$\frac{-P_{\text{exhaust}} V_D}{(r-1)(k-1)} \left(\frac{\widehat{P}^{\left(\frac{k-1}{k}\right)} - 1}{\widehat{P}^{\left(\frac{k-1}{k}\right)}} \right)$
2–3	Intake	$P_{\text{intake}}(V_2 - V_3)$	$\frac{-P_{\text{intake}} V_D}{r-1} \left(r - \widehat{P}^{1/k} \right)$
3–4	Compression	$M_1 c_v(T_4 - T_3)$	$\frac{r \ P_{\text{intake}} V_D}{(r-1)(k-1)} \left(\widehat{P}^{\left(\frac{k-1}{k}\right)} - 1 \right)$
4–1	Exhaust	$P_{\text{exhaust}}(V_4 - V_1)$	$\frac{P_{\text{exhaust}} V_D}{r-1} \left(\frac{r - \widehat{P}^{1/k}}{\widehat{P}^{1/k}} \right)$

For compression ratios below this critical value, compression of the gases from the intake would never reach the pressure of the exhaust plenum needed to open the exhaust valve. There is therefore a restriction on the compression ratio for a given pressure ratio, namely,

$$r \geq \widehat{P}^{1/k}$$

In reciprocating engines, the volumetric efficiency is an important measure of the gas exchange process and is defined as the ratio of the actual mass exchanged normalized by the ideal mass exchanged defined as what would be exchanged if a complete intake stroke (V_D) took in gas with the density of the intake plenum. For the ideal reciprocating compressor:

$$\eta_{\text{volumetric}} = \text{Volumetric efficiency} = \frac{\text{Mass exchanged}}{\text{Ideal mass exchanged}} = \frac{r - \widehat{P}^{1/k}}{r - 1}$$

The net work of the cycle is the sum of the work of all four processes, summarized in Table 6.7.

Case Study

The ideal reciprocating compressor analysis yields closed-form expressions for all variables and can be implemented into a general spreadsheet solution. Results from a particular case study are reported in this section. Figure 6.11 shows an indicator diagram (a plot of pressure vs. volume) for a reciprocating compressor with the specific input parameters shown as an inset. This compressor draws in air at nominal room conditions and delivers them to an exhaust plenum with a compression ratio (ratio of maximum to minimum volume) of 10. Various derived parameters are reported in Fig. 6.12. It is notable that the mass exchange processes take place over a small portion of the corresponding stroke.

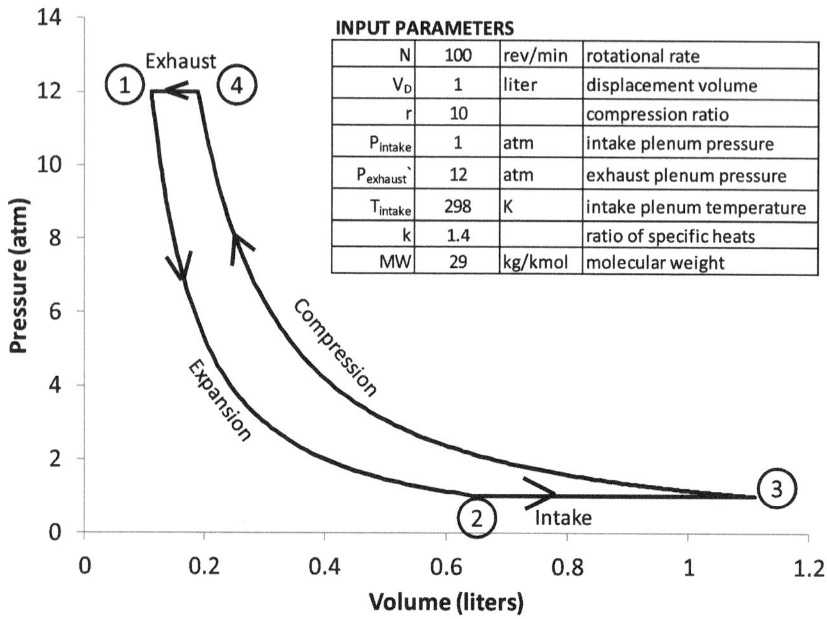

N	100	rev/min	rotational rate
V_D	1	liter	displacement volume
r	10		compression ratio
P_{intake}	1	atm	intake plenum pressure
$P_{exhaust}$`	12	atm	exhaust plenum pressure
T_{intake}	298	K	intake plenum temperature
k	1.4		ratio of specific heats
MW	29	kg/kmol	molecular weight

Fig. 6.11 Indicator diagram (pressure vs. volume) for an ideal reciprocating compressor. Input parameters are shown in the inset

DERIVED PARAMETERS

Phat,MAX	25.12	atm	maximum pressure ratio
Pratio	12		pressure ratio
Vc	0.111	liter	clearance volume
V2	0.656	liter	volume at start of intake
V4	0.188	liter	volume at start of exhaust
R	286.7	J/kg/K	gas constant
rho,intake	1.186	kg/m^3	intake density
w	10.47	rad/sec	angular velocity
(k-1)/k	0.286		adiabatic exponent
W12	-171.7	J	expansion
W23	-46.1	J	intake work
W34	290.9	J	compression
W41	93.9	J	exhaust work
Wcycle	167.0	J	Cycle Work

Fig. 6.12 Derived parameters for the reciprocating compressor case study

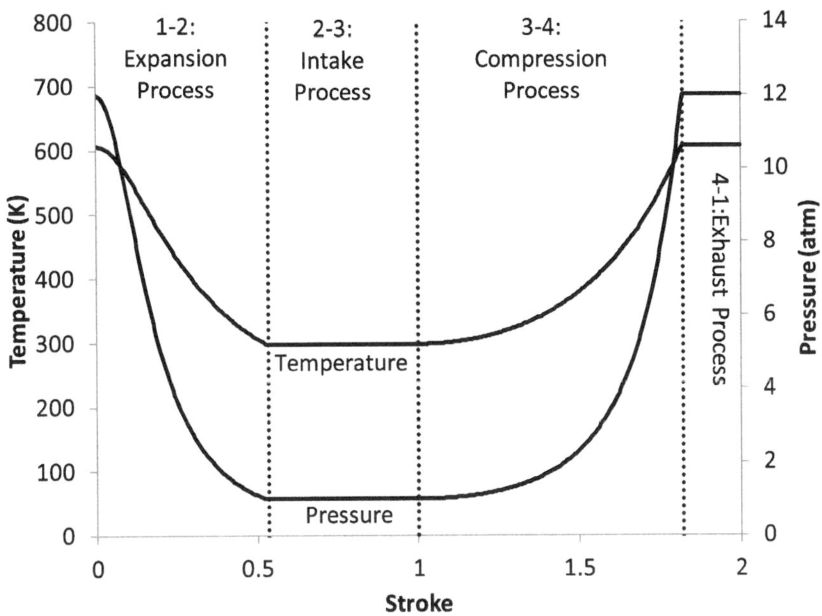

Fig. 6.13 Pressure and temperature as a function of crank angle, reported in units of strokes

The pressure and temperature as functions of the stroke are shown in Fig. 6.13. The pressure starts at the high pressure and decreases during the expansion process until it equals the intake manifold pressure. In a real cycle, the cylinder pressure would be below the intake manifold pressure due to pressure drops across the intake valve. The temperature starts high (over 600 K) and decreases to the intake plenum pressure, and remains constant during the intake process. The first quarter of the compression process results in very little pressure or temperature rise, and then increases dramatically during the second half until the pressure reaches the exhaust plenum pressure when the exhaust value opens. The pressure would be above that in a real cycle due to exhaust valve pressure drops.

The mass contained in the cylinder is shown in Fig. 6.14 and the mass flow rate (derivative of mass) in Fig. 6.15. It is notable that the mass contained during the expansion is more than half of the mass contained during compression despite the factor of 10 difference in volume at the start of each of these processes. The gases start at a higher density than the gases at the start of compression. The mass flow rates are highest at the beginning of the gas exchange steps and tail off as the change in piston volume diminishes as the end of the stroke is approached. The duration of the exhaust process is much shorter than that of the intake process, but the mass flow rate is higher. The total mass exchanged is the same during the intake as the exhaust.

The torque on the crank during each process is shown in Fig. 6.16, as is the work done by the gases for each process. This model is for the limiting case of connecting rod that is much longer than the crank arm. The torque changes sign with direction of

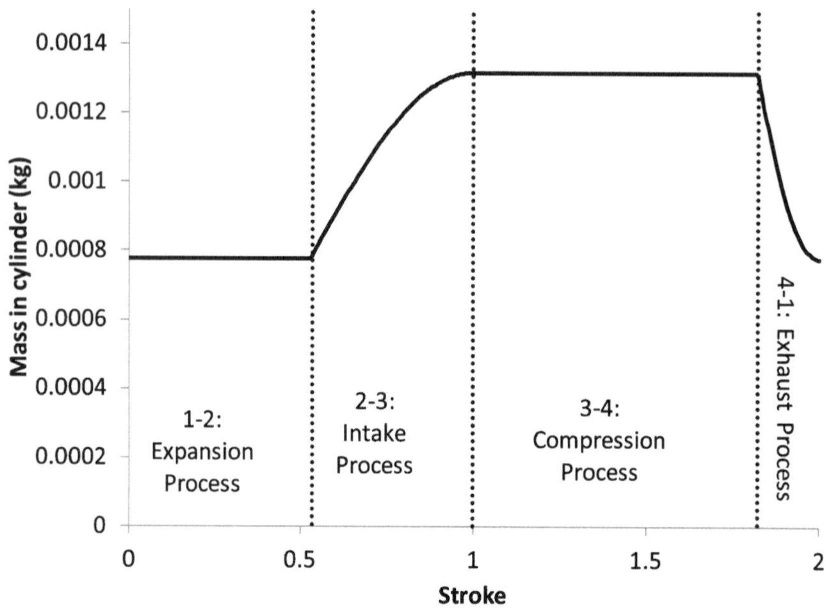

Fig. 6.14 Mass in cylinder as a function of crank angle

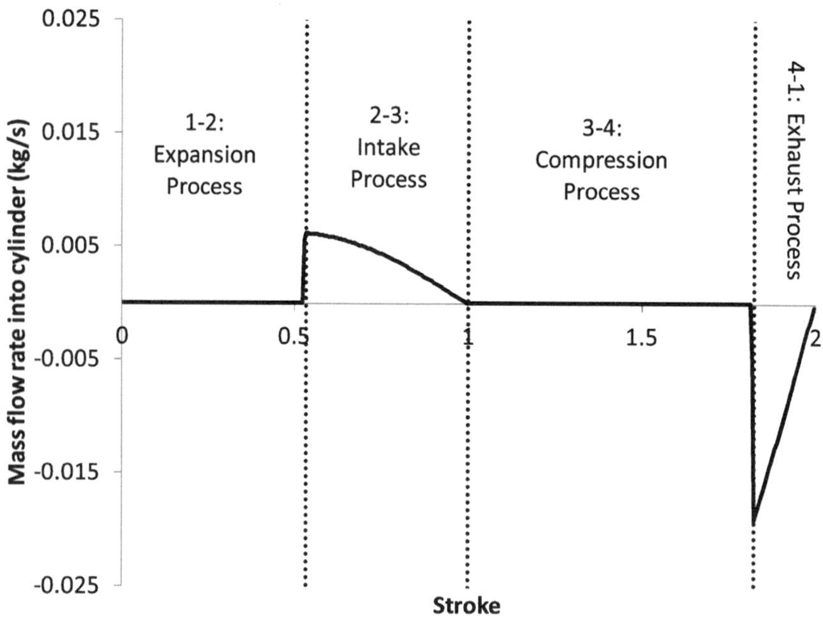

Fig. 6.15 Mass flow rate into the cylinder as a function of crank angle

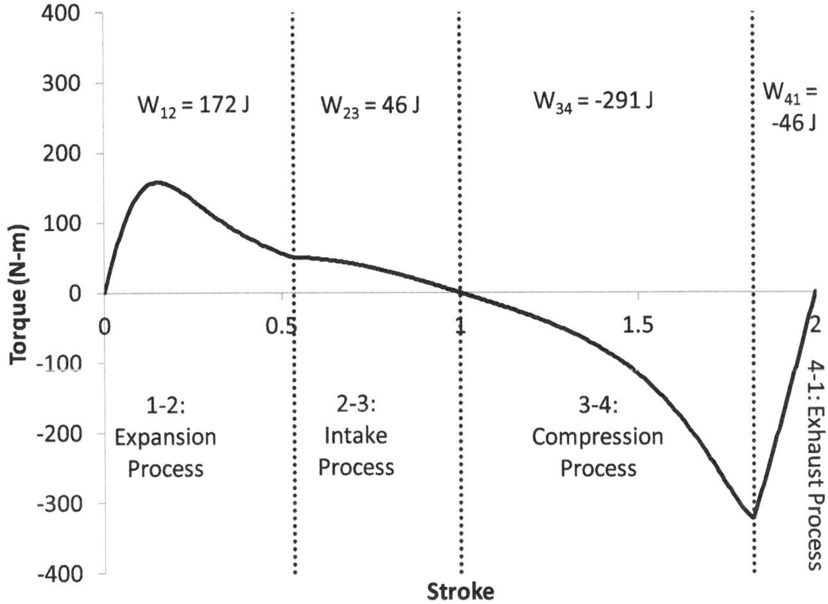

Fig. 6.16 Torque applied to crank as a function of crank angle. The work done by the gases during each process is shown

motion and is larger in magnitude during the exhaust stroke. The average torque equals the net work times the total angular displacement (2π). It is intriguing that the magnitudes of the work during the gas exchange processes are equal.

These base case results demonstrate how the key variables change with position during the cycle. The effects of compression ratio and the exhaust plenum pressure on the overall cycle requirements are shown in Figs. 6.17, 6.18, and 6.19. The net cycle work (input work required by the crankshaft) is shown as a function of overall pressure ratio (exhaust pressure to intake pressure) for three values of compression ratio (maximum to minimum volume). For a given compression ratio, there is a clear peak for input work required per cycle.

Reciprocating Compressor as a Steady Flow Device

A reciprocating compressor draws pulses of mass at low pressure and delivers them at high pressure. However, they are generally viewed in thermodynamic systems (such as vapor/compression refrigeration cycles) as quasi-steady-state devices that supply a continuous stream of high-pressure gas from a low-pressure source. Reciprocating engines are viewed similarly. The general mass relation is

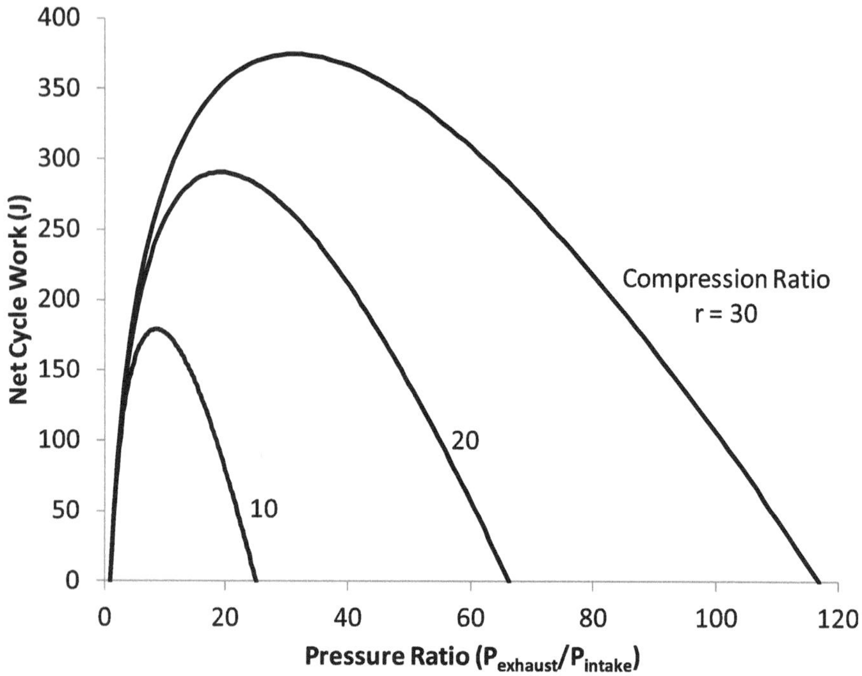

Fig. 6.17 Net cycle work as a function of pressure ratio for several compression ratios

$$\frac{dM}{dt} = \dot{m}_{\text{in}} - \dot{m}_{\text{out}}$$

Integrated over a cycle, recognizing no net change in mass stored, the mass flow rate is the mass of the intake (or exhaust) per cycle multiplied by the number of cycles per unit time (the rotational speed, N):

$$\dot{m}_{\text{in}} = \dot{m}_{\text{out}} = \dot{m} = \left(\frac{\text{Mass}}{\text{Cycle}}\right)\left(\frac{\text{Cycle}}{\text{Time}}\right) = \Delta M_{\text{intake}} N$$

The energy equation is

$$\frac{dE}{dt} = \dot{Q}_{\text{in}} + \dot{W}_{\text{in}} + \dot{m}(h_{\text{in}} - h_{\text{out}}) + \dot{m}\left(V_{\text{in}}^2 - V_{\text{out}}^2\right) + \dot{m}(gz_{\text{in}} - gz_{\text{out}})$$

Assuming no net change in energy stored, adiabatic, negligible kinetic and potential energy changes, the energy equation simplifies to

$$\dot{W}_{\text{in}} = \dot{m}(h_{\text{out}} - h_{\text{in}})$$

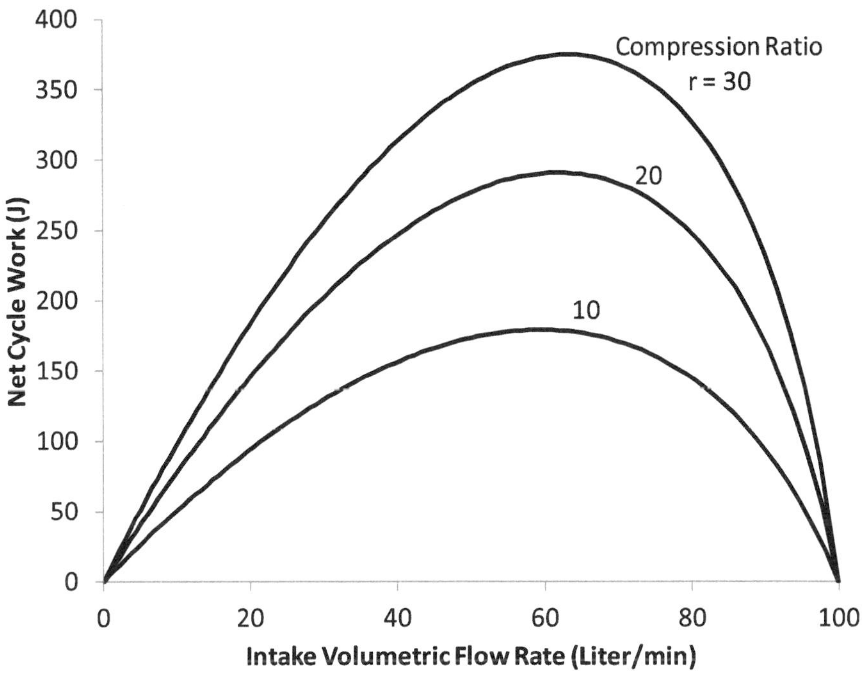

Fig. 6.18 Net cycle work as a function of volumetric flow rate for several compression ratios

Appendix 1: Perfect Gas Approximations

The purpose of this Appendix is to distinguish between thermally and calorically perfect gases (both of which obey the ideal gas law, $Pv = RT$) and demonstrate (by example) that the use of the more restrictive calorically perfect gas assumption to estimate the change in internal energy is usually completely justified. The calorically perfect gas approximation allows for simple closed-form solutions to be developed for various quantities (such as the thermal efficiency of an Otto, Diesel, or Brayton cycle).

First, some definitions, summarized in Table 6.8, and recalling that the internal energy specific heat is a thermodynamic property defined by $c_\mathrm{v} = \frac{\partial u}{\partial T}\big|_\mathrm{v}$.

Some comments:

- *The ideal gas law ($Pv = RT$) should not be applied to liquids.*
- $Z \approx 1$ if the pressure is much less than the critical pressure and/or if the absolute temperature is more than twice the critical temperature.
- A vapor is in the gas phase below the critical temperature and would therefore condense if compressed isothermally.
- The nomenclature of c_v often confuses people into thinking that its use involves processes that take place at constant volume. It merely relates changes in internal energy to changes in temperature.

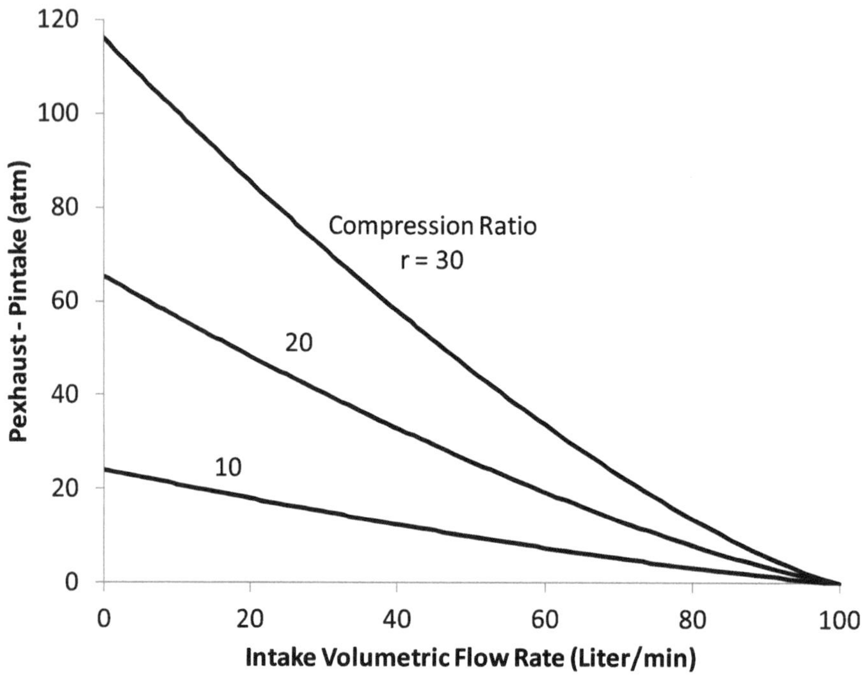

Fig. 6.19 Characteristic curve (pressure change vs. volumetric flow rate) for several compression ratios

Table 6.8 Classifications and gas property relations

Classification(s)	*PVT* state relation	Internal energy relation
Thermally perfect gas	$Pv = RT$	$c_v = c_v(T)$
Calorically perfect gas	$Pv = RT$	c_v = constant
Real gas	$Pv = ZRT$	$u = u(T, V)$
Vapor	$Pv = ZRT$	$u = u(T, V)$

- A thermally perfect gas is sometimes referred to as an ideal gas, while a calorically perfect gas is referred to as a perfect gas.
- Standard textbooks typically provide c_v data in a variety of forms (i.e., values at room temperature, mass basis specific heat values as a function of temperature, polynomial fits on a molar basis, and internal energy as a function of temperature). The main objective of this Appendix is to convince the reader that use of a single value for c_v is the easiest to use and often completely justified.

Problem Statement

Heat is added to 1 kg of air in a rigid vessel at STP (1 atm, 0 °C). Compare the required heat addition as a function of final temperature from the calorically perfect and thermally perfect models.

Table 6.9 Models considered

Model	Description	Formula
1	Thermally perfect gas	$\frac{\widehat{c}_p}{\widehat{R}} = \alpha + \beta T^1 + \gamma T^2 + \delta T^3 + \varepsilon T^4$
2	Calorically perfect gas, specific heat evaluated at initial temperature	$c_v = c_v(T_1)$
3	Calorically perfect gas, specific heat evaluated at average temperature	$c_v = c_v\left(\frac{T_1+T_2}{2}\right) = c_v\left(\overline{T}\right)$
4	Calorically perfect gas, average specific heat	$c_v = \frac{1}{2}\left[c_v(T_1) + c_v(T_2)\right] = \overline{c}_v$

Notes: $\widehat{c}_p/\widehat{R}$ is listed for model 1 as often published, where the "carat (^)" refers to molar quantities. c_v is obtained using the ideal gas relation $c_p - c_v = R$ and converting the basis from molar to mass, namely, $c_v = \widehat{c}_v/MW$ where MW is the molecular weight. The overbars refer to average quantities

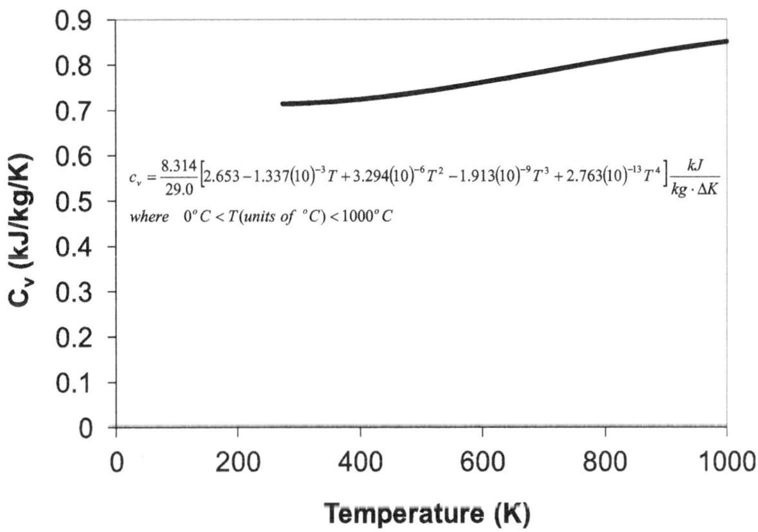

Fig. 6.20 c_v for air as a function of temperature

Solution Methodology

The 1st Law for this process, $\Delta E = Q - W$, reduces to $Q = m(u_2 - u_1)$ since there is no work (rigid vessel) or kinetic or potential energy changes. That is, the problem reduces to determining the change in internal energy. This calculation will be conducted for several models and then compared. Table 6.9 lists the models considered.

Figure 6.20 shows c_v for air plotted as a function of temperature (a plot of the fourth-order polynomial from the literature). The specific heat is a mildly increasing function of temperature between 273 K and 1000 K. The change in internal energy between any two temperatures is represented by the area under the curve between those two temperatures.

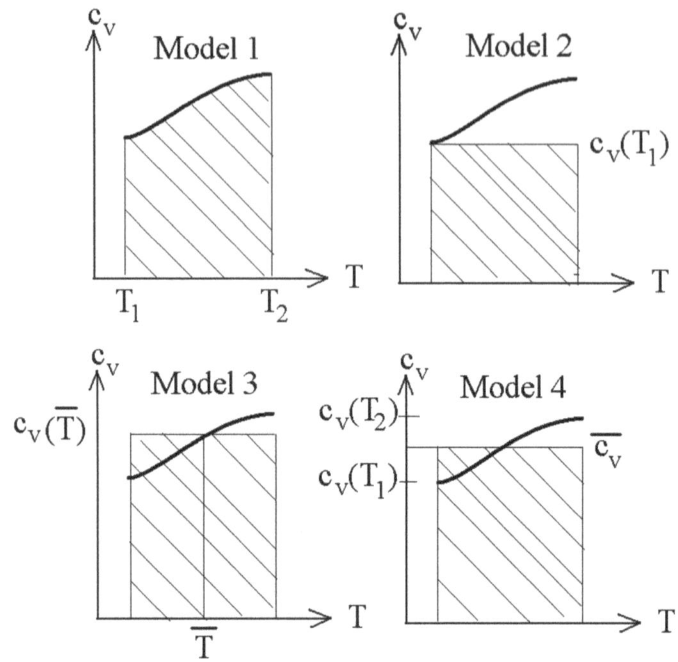

Fig. 6.21 Schematic representation of the area under the curve for the four models considered. The overbars here represent average quantities (not molar quantities)

$$\frac{\widehat{c}_P}{\widehat{R}} = \frac{c_P}{R} = \left[3.653 - 1.337(10)^{-3}T + 3.294(10)^{-6}T^2 - 1.913(10)^{-9}T^3 \right.$$
$$\left. +2.763(10)^{-13}T^4\right]$$

The units of T are °C. Converting to mass-based specific heat at constant volume:

$$c_v = \frac{8.314}{29.0}\left[2.653 - 1.337(10)^{-3}T + 3.294(10)^{-6}T^2 - 1.913(10)^{-9}T^3 \right.$$
$$\left. +2.763(10)^{-13}T^4\right] \frac{\text{kJ}}{\text{kg} \cdot \Delta\text{K}}$$

where $0\ ^\circ\text{C} < T(\text{units of}\ ^\circ\text{C}) < 1000\ ^\circ\text{C}$.

Figure 6.21 represents the four models schematically. For model 1, the thermally perfect gas (or ideal gas model), the area under the c_v curve is shown and is the most accurate model. The other three models are different versions of the calorically perfect gas model, with different values assumed for the (constant) specific heat. In all three cases, the change in internal energy is given by $\Delta u = c_v(T_2 - T_1)$, but a different value is used for c_v. For model 2, the specific heat is evaluated at the initial

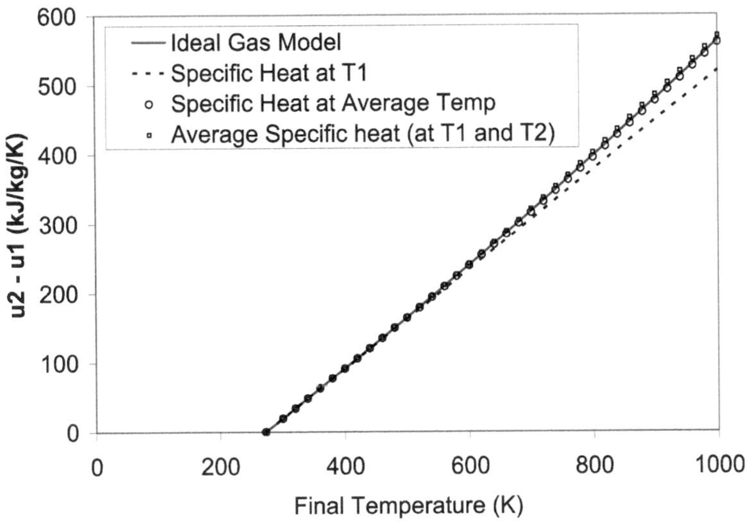

Fig. 6.22 Change in internal energy as a function of final temperature (T_2) for four models

temperature, T_1. This model is commonly used in the analysis of engine cycles (both reciprocating and gas turbine), and is a major assumption called the "cold air-standard analysis." This model underestimates the change in internal energy as it misses the area between the c_v curve and the rectangular area. Model 3 uses a specific heat value evaluated at the geometric average temperature while model 4 uses the average of the specific heat evaluated at the two temperature extremes. These latter two cases generally yield similar (but not identical) results to each other, with less error than model 2 (as they overcompensate at low temperatures and undercompensate at high temperatures).

Figure 6.22 shows the change in internal energy (equal to the heat added for the problem in question) as a function of final temperature. The ideal gas model is shown as a solid line, and the cold air standard model (model 2, initial specific heat) is shown as a dotted line. Symbols are used to show the results for models 3 and 4 (averaged specific heat). It is apparent that, in this view, all four models are indistinguishable for final temperatures below 600 K (a change in 300 °C), and models 3 and 4 continue to perform well to the upper limit of 1000 K, the highest temperature for the particular polynomial fit used.

Figure 6.23 shows these same results as the error in each of the three perfect gas models, treating model 1 as exact. The error in using model 2 is less than 10% at the maximum temperature considered and is expected to continue to increase (in absolute value) if higher final temperatures were considered. Model 2 underestimates and model 3 overestimates by less than 2%. They are considered to be comparable, although model 2 is more accurate for this case study (for this substance over this temperature range).

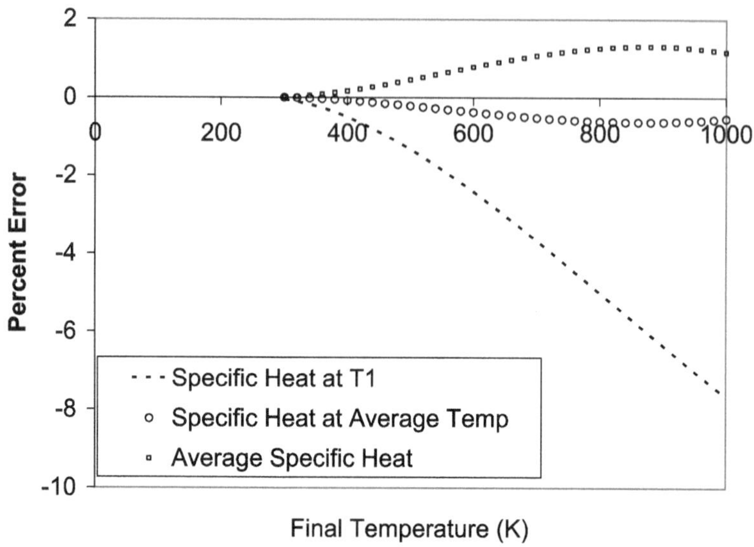

Fig. 6.23 Comparison of three perfect gas models with the ideal gas model

For problems for which either initial or final temperature is one of the unknowns, making the choice for a constant specific heat problematic, an iterative solution can be used by making a preliminary guess to the final temperature (using engineering judgment), and recalculating with a new value of specific heat if warranted. This is a form of numerical iterative solution.

The real power of the calorically perfect gas assumption is not in the analysis of isolated processes, but for the analysis of cycles (multiple processes in series). It allows for elegant closed-form solutions for quantities such as the net work or thermal efficiency. The calorically perfect gas assumption (constant specific heats) is very powerful and easier to implement than the thermally perfect model. Use it (whenever $Z \approx 1$) unless there is a compelling argument to do otherwise.

Appendix 2: Ideal Gases and Polytropic Processes

A deeper dive into the underlying physics associated with ideal gases is made in this Appendix. The development is qualitative and mathematical results are merely stated, rather than rigorously derived from kinetic theory, which studies the behavior of gases from an atomic view. The kinetic theory results in the two macroscopic mathematical relations needed for thermodynamic analysis (ideal gas law and property relation between internal energy and temperature).

From kinetic theory, an ideal gas consists of molecules that occupy a volume much less than the volume of their container. They travel in straight paths in random directions and only experience intermolecular forces during collisions that abruptly change their speed and direction. At any instant in time, each molecule contains stored internal energy in the form of the kinetic energy of its center of mass (translational kinetic energy), rotational kinetic energy, and vibrational energy (both potential and kinetic) associated with molecular bonds. The total internal energy of a gas is the sum of the internal energy of all its molecules. Momentum is conserved in every collision. For ideal gases, the collisions are elastic, with energy being redistributed between these modes of energy storage.

The principle of "equipartition of energy" states that the total energy inherent in the gas is distributed equally between translational, rotational, and the two modes of vibrational energy. Each fully excited energy mode stores energy equal to $\widehat{R}T/2$. However, quantum mechanics comes into play in determining the degree to which a given mode is fully excited. Translational energy quantum levels are sufficiently low that the translational modes are fully excited at temperatures above approximately 100 K (−173 °C). Rotational modes are fully excited near nominal room temperature. However, vibrational quantum energy levels are higher and not fully excited at room temperature. As the temperature increases, more and more higher quantum levels become populated. This behavior is the reason that specific heats of gases increase mildly with temperature.

A key result of kinetic theory is that the pressure, volume, and temperature are related by the ideal gas law ($Pv = RT$). Deviations from ideal gas behavior occur when the intermolecular forces and/or the volume occupied by the molecules themselves become appreciable.

Energy Property Relation

The other key result of kinetic theory is that the total internal energy is a function of temperature only (not pressure or volume). There are two corollary models that relate to the internal energy, the thermally perfect gas model and the more restrictive calorically perfect model. The ideal gas law ($Pv = RT$) is obeyed in both models. In the calorically perfect gas model, the specific heat (c_v) is constant, while it is a mild function of temperature in the thermally perfect model. With that, all property relations can be determined from simple algebraic relationships, with values for a selection of common gases at nominal room temperature listed in Table 6.10. The second to last column shows the molar-based specific heat normalized by the universal gas constant ($\widehat{R}$). Notice that the specific heat of the monatomic gases helium and argon is $3\widehat{R}/2$. Atoms themselves cannot store significant rotational energy, so monatomic gases can only store energy in three translational modes, or degree of freedom (corresponding to three dimensions of space). Each mode stores $\widehat{R}/2$. For diatomic gases (H_2, N_2, O_2), there are two rotational degrees of freedom (not three since there is minimum moment of inertia with respect to the axis along the two atoms) as well as the three translational. Therefore, the specific heat is approximately $5\widehat{R}/2$. For triatomic and polyatomic gases (CH_4, CO_2, and H_2O), there are three rotational modes, resulting in six total degrees of freedom, and specific heat of approximately $3\widehat{R}$.

Table 6.10 Ideal gas properties for a selection of common gases (specific heat values at nominal room temperature)

Gas	Molecular weight (kg/kmol)	R (J/kg/K)	$C_{v,mass}$ @ 298 K (J/kg/K)	$C_{v,molar}/R_u$ (@ 298 K)	$k = C_p/C_v$ (@298 K)
Hydrogen (H_2)	2	4157	10,160	2.44	1.41
Helium	4	2079	3120	1.50	1.67
Methane (CH_4)	16	520	1700	3.27	1.31
H_2O (vapor)	18	462	1460	2.81	1.36
Nitrogen (N_2)	28	297	743	2.50	1.40
Two component dry air	28.84	288	718	2.49	1.40
Four component dry air	28.95	287	718	2.50	1.40
Oxygen (O_2)	32	260	659	2.54	1.39
Argon	40	208	312	1.50	1.67
CO_2	44	189	655	3.47	1.29

Note: Two different models for dry air are shown, a two-component (21 mol% O_2 in N_2) and a four-component (78.09 mol% N_2, 20.95% O_2, 0.93% argon, 0.03% CO_2). Water vapor is for pressures below the corresponding saturation pressure

Deviations from these specific heat values at room temperature are due to the quantum nature of the vibrational modes. There are two vibrational degrees of freedom (stored kinetic and potential energy) for each molecular bond (1 for diatomic molecules, 2 for triatomic, and $n - 1$ for polyatomic molecules with n atoms), but the available energy levels are quantized. At room temperature, the energy levels are not significantly populated. As the temperature increases, a higher and higher percentage of vibrational modes becomes available, and is reflected in a temperature dependence of the specific heat.

In the thermally perfect gas model, the ideal gas law is obeyed but the specific heat depends on the temperature. Figure 6.24 shows the mass-based specific heat (molar value divided by molecular weight, which is more useful for engineering thermodynamic calculations than the molar value) for the five largest components of air (N_2, O_2, argon, CO_2, and water vapor) as a function of temperature. The top graph shows the specific heats over the full range of temperature available in the JANAF tables (for Joint Army, Navy, Air Force, a publicly available source of thermodynamic property data). The bottom graph shows this data set over a 300 °C temperature range around room temperature, designed to convince the reader that the calorically perfect gas model is quite reasonable for air in this range.

The specific heat of argon (a monatomic gas) is truly independent of temperature above 100 K because it only has three translational degrees of freedom and is fully excited. The diatomic gases (O_2 and N_2) exhibit a mild temperature dependency as a higher percentage of the two vibrational modes of energy become excited (and

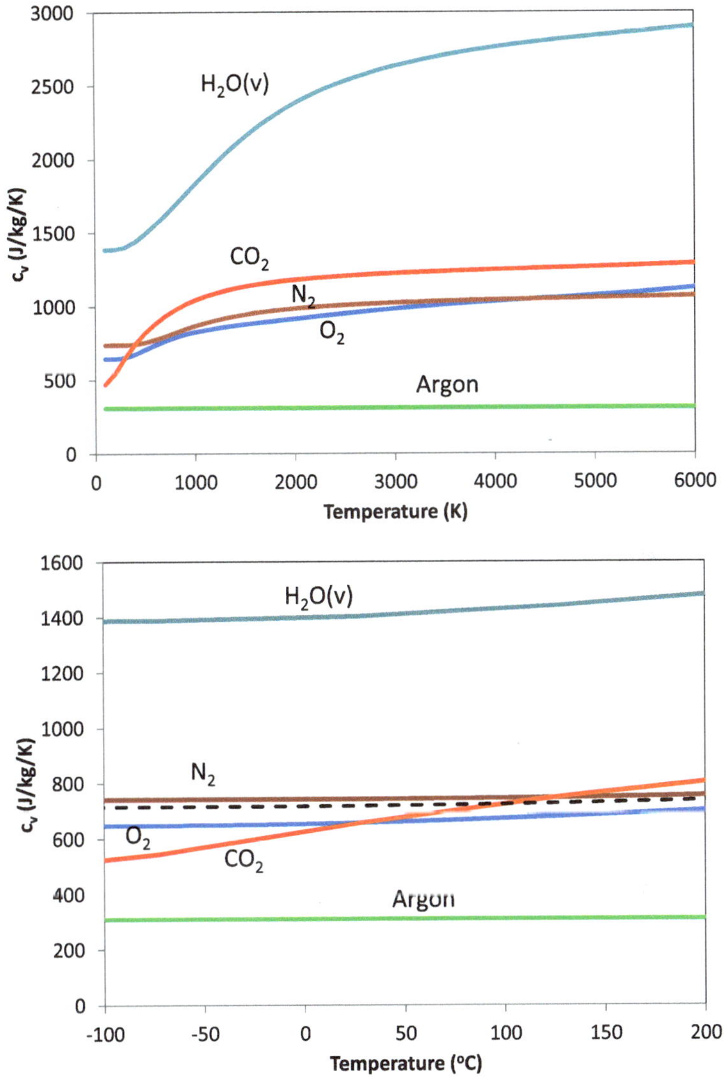

Fig. 6.24 Mass-based specific heat (c_v) as a function of temperature for the chemical components of air. Data adapted from the JANAF tables, available at https://janaf.nist.gov/. The full temperature range is shown in the top graph, and a 300 °C range around nominal room temperature is shown in the bottom. Air is shown as a dashed line, which follows N_2 closely

essentially complete above approximately 2000 K). CO_2 and H_2O vapor have a stronger temperature dependency, reflecting an additional molecular bond with two vibrational degrees of freedom. The mass-based specific heat is lower for CO_2 than H_2O vapor because of its larger molecular weight.

Polytropic Relationships

Several process types exhibit a power law relationship between pressure and volume, which can be expressed in a general form called a polytropic relationship, namely,

$$Pv^n = \text{constant}$$

where "n" is called a polytropic exponent. I am curious as to the origin of the term "polytropic" process and how it stuck. A cursory internet search was not helpful.

Figure 6.25 summarizes the process types, their fluid restrictions, and the general path on a Pv diagram. Figure 6.26 is a pressure/volume plot for compressions of air (initially at 1 L, 0.8 atm, 25 °C) for three cases, corresponding to adiabatic ($n = k = 1.4$ for air), isothermal ($n = 1$), and a case in between ($n = 1.2$).

Polytropic processes can be used in practice to relax some of the assumptions of the ideal model, yet retain the simple closed-form mathematical relationships that quantify process-dependent quantities (work and heat). For example, the compression and power strokes of an actual reciprocating engine are not perfectly adiabatic because the gas temperature is different than the wall temperature (colder during intake and the first part of compression, and hotter after combustion). Also, the specific heat varies with temperature (thermally perfect gases). Lastly, some mass might be exchanged near the start of compression (late intake valve closing) or the end of the power stroke (early exhaust valve opening). Deviations from the ideal gas law can also be important for a refrigerant, or historically, a steam engine, where superheated vapor is near the vapor dome. Including these effects in a theoretical model from first principles requires a numerical simulation (one well worth doing,

polytropic exponent, n	***Process Type***	***Constant Property***	***Fluid Type***
0	Isobaric	pressure	any
1	isothermal	temperature	thermally perfect gas
k = R/cv+1	adiabatic	entropy	calorically perfect gas
∞	isochoric	volume	any

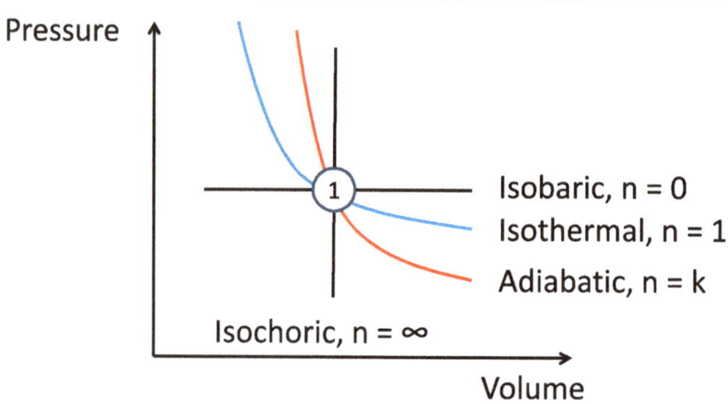

Fig. 6.25 Summary of polytropic relationships for specific process types

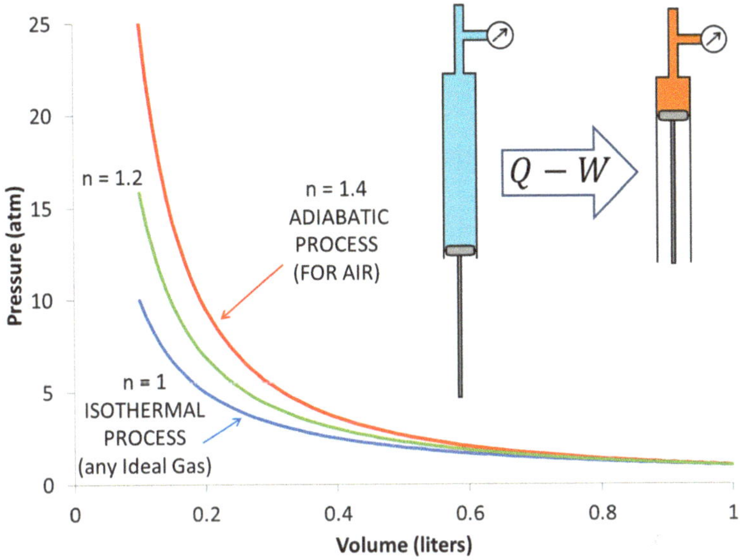

Fig. 6.26 Polytropic relationship for compressions of air (initially at 1 L volume at 0.8 atm, 25 °C) in a piston/cylinder for three values of the polytropic exponent

but not here). However, in a laboratory setting, the pressure within a cylinder might be measurable, and if the pressure is measured at two points (1 and 2) during a stroke that correspond to known total volumes, a polytropic exponent can be determined empirically as follows:

$$P_1V_1^n = P_2V_2^n = PV^n$$

Taking the natural log of both sides:

$$\ln(P_1) + n\ln(V_1) = \ln(P_2) + n\ln(V_2)$$

Rearranging for "n" and using $\ln(A) - \ln(B) = \ln(A/B)$:

$$n = \frac{\ln(P_1/P_2)}{\ln(V_2/V_1)}$$

The work associated with a polytropic process is obtained as follows:

$$W_{12} = \int_1^2 PdV = P_1V_1^n \int_{V_1}^{V_2} V^{-n}dV = \frac{P_1V_1^n}{1-n}\left[V_2^{1-n} - V_1^{1-n}\right] \quad \text{if } n \neq 1$$

Using the polytropic relationship for the first term:

$$W_{12} = \frac{P_2V_2^n}{1-n}\left[V_2^{1-n}\right] - \frac{P_1V_1^n}{1-n}\left[V_1^{1-n}\right]$$

Recognizing that $V^n \cdot V^{1-n} = V$, the final relationship for the work associated with a polytropic process is

$$W_{12} = \frac{P_2V_2 - P_1V_1}{1-n} \quad \text{for a polytropic process with exponent } n \neq 1$$

If the polytropic exponent is unity (corresponding to an isothermal process of a thermally perfect gas):

$$W_{12} = \int_1^2 PdV = P_1V_1 \int_{V_1}^{V_2} V^{-1}dV$$

$$W_{12} = P_1V_1 \ln\left(\frac{V_2}{V_1}\right) \quad \text{for a polytropic process with exponent } n = 1$$

Calorically Perfect Gas in a Closed System

The work of a polytropic process applies to any process in a piston/cylinder for which the pressure and total volume at two points can be measured even if some mass is exchanged through valves. However, if the mass is fixed and known (or can be reasonably estimated), then the total volume can be replaced with the specific volume, and both initial and final states are uniquely determined. Therefore, the 1st Law ($\Delta U = Q_{12} - W_{12}$) can be used to determine the heat transfer associated with the process.

For gases of fixed mass m that obey the ideal gas law, the work of a polytropic process is

$$W_{12} = \frac{mR}{1-n}(T_2 - T_1)$$

If the gas is also modeled as calorically perfect, then the internal energy change and work can be expressed in terms of the temperature change, and the heat transfer rate is

$$Q_{12} = mc_v(T_2 - T_1) + \frac{mR}{1-n}(T_2 - T_1)$$

$$Q_{12} = m\left(\frac{c_p - nc_v}{1-n}\right)(T_2 - T_1) \quad \text{for } n \neq 1$$

Expressed in terms of the ratio of specific heats ($k = c_p/c_v$) that must be greater than unity and in various forms:

$$Q_{12} = mc_vT_1\left(\frac{k-n}{1-n}\right)\left(\frac{T_2}{T_1} - 1\right)$$

$$Q_{12} = mc_vT_1\left(\frac{k-n}{1-n}\right)\left(\left(\frac{P_2}{P_1}\right)^{\frac{n-1}{n}} - 1\right)$$

$$Q_{12} = mc_vT_1\left(\frac{k-n}{1-n}\right)\left(\left(\frac{V_1}{V_2}\right)^{n-1} - 1\right)$$

Note that for $n = k$ the heat transfer rate reduces to zero as it must for an adiabatic process. Other cases for compressions ($V_2 < V_1$) are

$$\text{for} \quad 1 < n < k \quad Q_{12} < 0$$

$$\text{for} \quad n > k \quad Q_{12} > 0$$

$$\text{for} \quad n < 1 \quad Q_{12} > 0$$

A formal 2nd Law analysis could be done to demonstrate how the wall temperature must be above the gas temperature for cases in which heat is added to the gas.

Part II
Applications of Fundamental Thermodynamic Principles to Engineering Systems

Chapter 7
An Idealized Model of a Throttled Four-Stroke Reciprocating Spark Ignition Engine

Four-stroke spark ignition (SI) engines are the workhorses of the automotive industry and several other niche markets. Traditional thermodynamic textbooks develop closed-form working equations to approximate the performance of these engines based on the Otto cycle using the cold air standard model, which treats the cycle as a closed thermodynamic system (i.e., no mass crosses the boundary) with the combustion and blowdown processes as equivalent heat additions and the gas as a calorically perfect gas (constant specific heats). Actual engines operate on a mechanical cycle over four linear strokes of a piston (over two revolutions), but the working fluid does not undergo a cyclic process. Rather, the analysis of an idealized throttled engine recognizes the engine as an open thermodynamic system. Each mechanical cycle requires fresh fuel and air to be drawn in (intake) and burned gases to be expelled (exhaust). A basic understanding of the gas exchange process (the intake, blowdown, and exhaust processes) is thereby attained. There is no external heat addition (as modeled traditionally), but rather a rapid conversion of one form of internal energy (chemical) into another (thermal) within the working fluid itself, that is, an internal combustion process, given an introductory treatment. The blowdown process (venting of high-pressure gases after the power stroke) is accomplished by driving mass out of the cylinder by the pressurized gases after the power stroke, not by a fictitious heat transfer process to the environment. Finally, the model includes the throttling process as the means to vary the mechanical load (torque and speed) of the cycle. Throttling reduces the mass throughput in the cycle at the expense of thermal efficiency. The calorically perfect gas model is retained, allowing for closed-form expressions for all processes. This model is still idealized and therefore will provide an upper limit on the performance of actual engines to which the performance of real engines could be compared, but it more closely resembles the nature of actual engines and is therefore a launching point for more thorough analyses of reciprocating internal combustion engines.

G. Sidebotham, *An Inductive Approach to Engineering Thermodynamics*, Mechanical Engineering Series, https://doi.org/10.1007/978-3-030-20430-3_7

Ideal Throttled Spark Ignition Engine

A schematic of a throttled SI engine with a carburetor to mix fuel and air is shown in Fig. 7.1. The engine consists of an intake system, a piston/cylinder arrangement, and an exhaust system. Poppet valves that are designed to open (lift) and close at specific points of the cycle control the flow of gases. The schematic shows a single cylinder, but generally multiple cylinders are used (and their cycles are staggered in time). Each cylinder has its own intake and exhaust runner. An intake manifold feeds the intake runners, and an exhaust manifold collects flow from the exhaust runners. The engine is shown operating at a rotational rate N, generally specified in revolutions per minute (RPM). The four-stroke cycle undergoes two full revolutions per cycle. In this model, the engine speed is specified as a constant input, and one of the many quantities calculated is the applied torque required to maintain that speed. The volume inside the cylinder varies from a maximum value in the bottom dead center (BDC) position to a minimum (called the clearance volume, V_c) at top dead center (TDC). The difference is the volume swept by a stroke, called the displacement volume (V_D). The ratio of the maximum total volume to the minimum volume is an important parameter called the compression ratio, $r = (V_D + V_C)/V_C$

The function of the intake system is to draw in filtered ambient air, mix it with fuel (and evaporate a liquid fuel) in the proper proportion for combustion (near stoichiometric), and deliver the fuel/air mixture to the cylinder through an intake valve that separates the intake runner from the cylinder. Carburetors are replaced with fuel injectors in modern automobiles, but the basic concepts are the same.

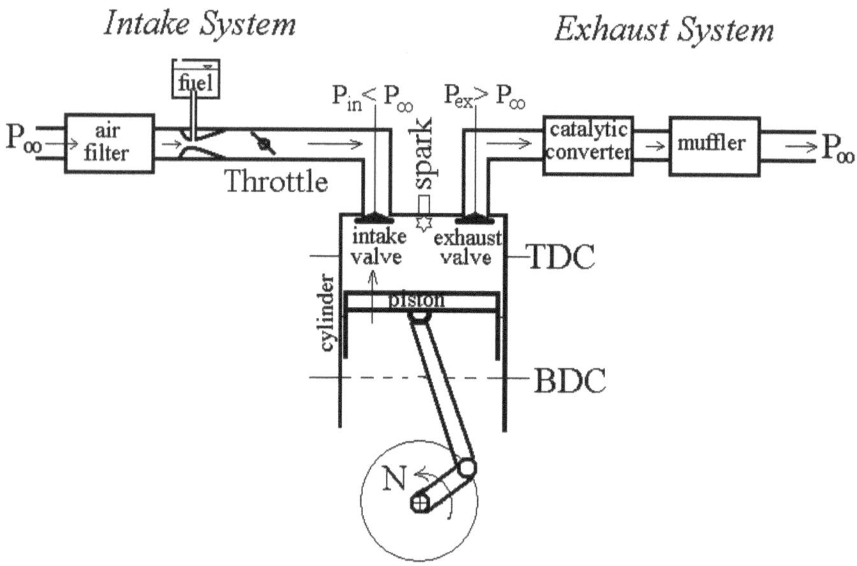

Fig. 7.1 Schematic of a throttled four-stroke spark ignition (SI) reciprocating engine, shown during the compression stroke

Carburetors are still used in niche markets like lawn mowers, jet skis, and marine engines. The throttle is generally a simple plate that can be opened and closed (controlled by a foot pedal) to introduce a restriction that deliberately introduces a pressure drop from ambient, resulting in a partial vacuum behind it. In a real engine, every component introduces pressure losses, either by design (the throttle) or due to frictional flow losses. In this idealized cycle, all pressure losses are neglected except by the throttle. The pressure inside the intake runner is specified as a user-defined input parameter. In practice, the pressure drop across the intake valve can be large, especially at high engine speeds, and more complete models would introduce that effect.

The piston/cylinder assembly converts the net mechanical work of the cycle (in the form of a pressure force acting through a distance) into the linear motion of a reciprocating piston that in turn is converted into rotational motion of an output shaft through a connecting rod and crank arm (Fig. 7.2). With both valves closed, the gases are contained within a volume that changes with crank angle (θ) and therefore time. Mechanical work must be done on the gases when the piston is moving upward (decreasing volume), and mechanical work is done by the gases when moving downward. The pressure of the gases is high after combustion, and more work is extracted during the expansion than the input work required to compress the gases prior to combustion.

The function of the exhaust system is to expel the hot burned gases into the environment, cleaning them of pollutants prior to their passing through a muffler (for noise reduction) and then exiting through the tailpipe. There are pressure drops across each component, and in this model, the pressure of the gases in the exhaust

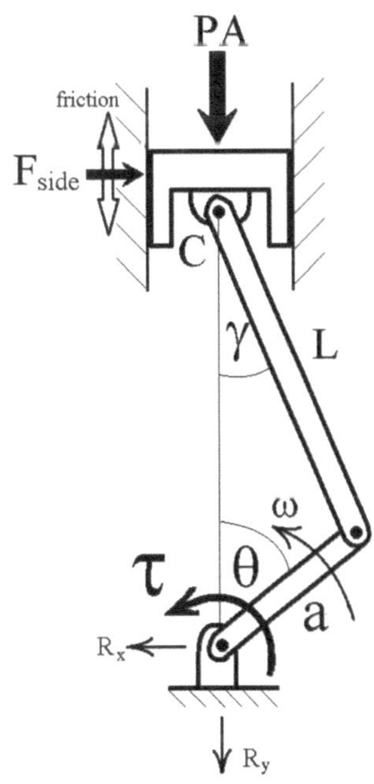

Fig. 7.2 Connecting rod geometry and free-body diagram of piston/connecting rod/crank assembly

runner is set as a user-defined fixed input value (above ambient). Some engines use some of the available thermal energy in the exhaust to extract additional mechanical power by placing a turbine in the exhaust stream. A turbo-compounded engine adds this energy directly to the output shaft. A turbo-charged engine uses the turbine to drive a compressor on the intake side to increase the mass of air and fuel inducted (the "charge"). These improvements in power and efficiency come at the cost of increased complexity and are the exception, not the rule in the market. Workshop 7.2 guides an analysis that compares the performance of several combinations of components added to the intake and/or exhaust systems.

Model Development

The seven processes that make up the ideal throttled Otto cycle are shown schematically on an indicator diagram (P-V) in Fig. 7.3. The pressures in the intake and exhaust runners are shown relative to ambient pressure. A brief description and the major assumptions of the idealized model for each process are summarized in Table 7.1. An analysis of each step follows, and relationships between all variables as a function of crank angle are developed. Results of case studies are then presented, implemented in a spreadsheet. A listing of input and derived parameters (with the values considered as default values) is shown in Fig. 7.4. If a spreadsheet is designed properly, a user can change any of the input parameters, and all calculations for the cycle are applied to that set. *Note: readers can skip directly to the next section, "Results and Discussion," if not interested in these details, much of which was developed in previous chapters.*

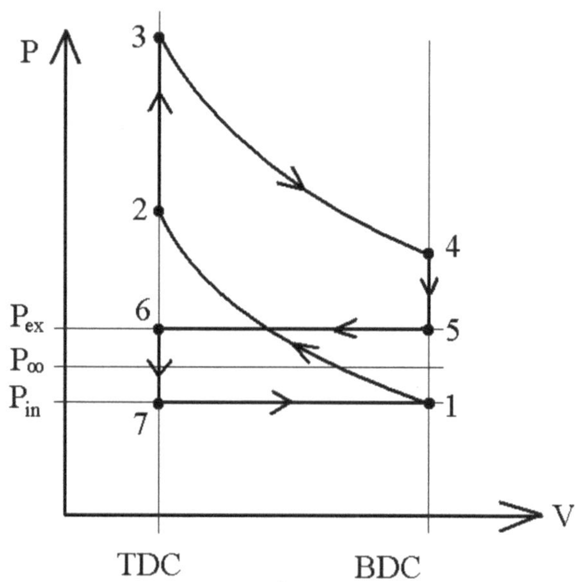

Fig. 7.3 Indicator diagram (pressure vs. volume) of the ideal throttled Otto cycle

Table 7.1 Summary of processes in throttled Otto cycle

Process	Name	Description of actual process	Model assumptions
1–2	Compression stroke	Fresh fuel/air mixture is compressed from BDC to TDC. In practice, the intake valve closes late, and there is heat exchange with the walls. Work must be put into system	Adiabatic, reversible compression
2–3	Combustion	A spark initiates combustion as TDC is approached. A turbulent premixed flame propagates at a finite rate. High heat loss rates to walls begin and cooling is needed in practice to prevent overheating	Adiabatic and instantaneous at TDC. Chemical energy converted to thermal energy
3–4	Power stroke	High-pressure and -temperature gases expand, and work is delivered to the output shaft. Heat loss rate is high	Adiabatic, reversible expansion
4–5	Blowdown	Exhaust valve begins to open as BDC is approached. High-pressure gases vent into exhaust manifold. Advanced engines can add a turbine downstream to extract more power	Instantaneous at BDC. Residual gases undergo adiabatic and reversible expansion to the exhaust manifold pressure
5–6	Exhaust stroke	With exhaust valve open, piston compresses gases, which flow across the valve with a pressure drop that is highest near the middle of the stroke (fastest piston speed)	No pressure drop across valve. Adiabatic
6–7	Valve overlap	As TDC is approached, intake valve begins to open and exhaust valve begins to close. Some backflow into the intake runner may occur at low speeds as gases continue to exhaust. After TDC, exhaust valve fully closes	Instantaneous at TDC. Remaining gases undergo adiabatic expansion to intake manifold pressure, with back flow into the intake runner
7–1	Intake stroke	With intake valve open, piston withdraws and fresh gases flow into the cylinder across the intake valve with a pressure drop	No pressure drop across valve. Adiabatic mixing of residual gases and fresh charge

Thermodynamic Relationships for Ideal Gases

A summary review of thermodynamic relationships for ideal gases is in order before diving fully into the analysis of the cycle. The calorically perfect gas model will be used to analyze the cycle, which means the ideal gas law ($Pv = RT$) is valid, and the

INPUT PARAMS

Vd	700	cc, Single Cylinder Displacement Volume
r	10	compression ratio
B/S	1.1	bore to stroke ratio
L/a	3	connecting rod length to crank radius ratio
n	6	# of Cylinders
N	2000	RPM, engine speed
Cv	0.718	kJ/kg/C, internal energy specific heat
R	0.287	kJ/kg/C, gas constant
Tin	298	K, Intake Manifold Temperature
Pin	0.7	atm, intake runner pressure
Pex	1.2	atm, Exhaust Manifold Pressure
LHV/(1+AFR)	2730	kJ/kg, chemical energy released per kg mixture
Tinf	298	K, Ambient Temperature
Pinf	1	atm, Ambient Pressure
mu	0.4	friction coefficient

DERIVED QUANTITIES

Vc	77.8	cc, Clearance Volume
B	9.9	cm, bore
a	4.5	cm, crank radius
Lcr	13.5	cm, connecting rod length
k	1.40	ratio of specific heats
Cp	1.005	kJ/kg/C, enthalpy specific heat
rhoinf	1.186	kg/m^3, ambient density
rhointake	0.830	kg/m^3, intake density
η_{OTTO}	0.601	Ideal Otto cycle efficiency (unthrottled)

Fig. 7.4 Input parameters and derived fixed quantities in a spreadsheet implementation of the ideal throttled Otto cycle

specific heat (c_v) is constant. With that, all property relations can be determined from the simple algebraic relationships summarized in Fig. 7.5. There is much redundancy with Chap. 6, by design, to reinforce key concepts.

For a gas with a molecular weight MW (=29.0 kg/kmol for dry air, based on its natural composition, modeled as 21 mol% O_2, 79 mol%N2), the gas constant is given by the universal (molar) gas constant divided by the molecular weight:

$$R = \frac{\widehat{R}}{\text{MW}}$$

where the universal gas constant (in strict SI units, with two different nomenclatures found in the literature) is:

Fig. 7.5 Summary of calorically perfect gas property relations and air values at room temperature. The specific heat is a mild function of temperature, and in applications, a suitable chosen constant value can almost always be justified

For any finite process (1-2) on a closed system (of fixed mass, m) of a calorically perfect gas:

Mass Relation: $mR = \frac{PV}{T} = \frac{P_1V_1}{T_1} = \frac{P_2V_2}{T_2}$

Energy Relation: $\Delta U = mc_v\Delta T$

Air Property Values
(near room temperature)

$$R = 287\frac{J}{kg.K}$$

$$c_v = 718\frac{J}{kg.K}$$

$$R_\mathrm{u} = \widehat{R} = 8314\frac{\mathrm{J}}{\mathrm{kmol}\cdot\mathrm{K}}$$

Specific heat values can be obtained from numerous sources, with a comprehensive list publicly available in the JANAF tables (for Joint Army Navy Air Force, found at https://janaf.nist.gov/). The specific heat is a mild monotonically increasing function of temperature. Use of the room temperature value for c_v for all processes in the analysis of the reciprocating engine cycle is called the "cold air standard model." More accuracy could be obtained by using a value based on an average cycle temperature, and a case study will be done to show this sensitivity. It is not too much harder to use different values based on the average temperature for each process. On the other hand, using a thermally perfect gas model (specific heats vary with temperature) requires a numerical solution and is difficult to generalize. There are so many other model assumptions (such as heat transfer to the walls) whose relaxation would require a much more involved analysis. Use of constant specific heat values yields closed-form solutions and obtains the correct qualitative trends and accuracy sufficient for a first formal study of engine cycles.

The relationship between the two specific heats (one based on internal energy and the other based on enthalpy) is obtained from the definition of the combination thermodynamic property specific enthalpy, h:

$$h = u + Pv$$

With the ideal gas law: $h = u + RT$. Taking the derivative with respect to temperature:

$$\frac{dh}{dT} = \frac{du}{dT} + R$$

The two specific heat properties are defined as:

$$c_{\mathrm{v}} \equiv \left.\frac{\partial u}{\partial T}\right|_{\mathrm{v}} \quad \text{and} \quad c_{\mathrm{p}} \equiv \left.\frac{\partial h}{\partial T}\right|_{\mathrm{P}} \quad \text{for any fluid}$$

For thermally perfect gases, the internal energy (u) and therefore enthalpy (h) are functions of temperature only, and the partial derivatives can be replaced by total derivatives:

$$c_{\mathrm{v}} = \frac{du}{dT} \quad \text{and} \quad c_{\mathrm{p}} = \frac{dh}{dT} \quad \text{for ideal gases}$$

Inserting these relations into the derivative of the definition of enthalpy yields the important relationship between specific heats for ideal gases:

$$c_{\mathrm{p}} = c_{\mathrm{v}} + R$$

This expression is valid for both thermally perfect (specific heats vary with temperature) and calorically perfect (constant specific heats) gases, the latter being used in the present analysis.

The ratio of specific heats is commonly used, given the symbol "k" in some publications, and "γ" in others. Since the latter is used for the connecting rod angle here, k will be used, and the ratio of specific heats is defined as follows:

$$k = \frac{c_{\mathrm{p}}}{c_{\mathrm{v}}} = \frac{R + c_{\mathrm{v}}}{c_{\mathrm{v}}}$$

In summary, there are five related thermodynamic properties that commonly arise: c_{v}, c_{p}, k, R, and MW. Given any two of the five as inputs, the other three can be derived.

Adiabatic Compression or Expansion

Isentropic (i.e., adiabatic and reversible) compression and expansion processes are important in many cycles, including the compression stroke and expansion (power) stroke of the reciprocating engine. The relationships between temperature, pressure, and volume are derived from the 1st Law for closed systems applied to a differential change on a control mass of a simple compressible substance, m, neglecting kinetic and potential energy, namely,

$$dU = \delta Q_{\mathrm{in}} - PdV$$

For simple compressible gases, the only form of work is expansion work (PdV, as opposed to surface tension, magnetic or electric effects).

For adiabatic processes, the heat added is by definition zero, and the energy equation reduces to

$$dU = -PdV \quad \text{for adiabatic processes}$$

For ideal gases, $dU = m\, c_v dT$:

$$mc_v dT = -PdV$$

Eliminating the pressure with the ideal gas law ($P = mRT/V$) and dividing through by temperature:

$$c_v \frac{dT}{T} = -R\frac{dV}{V}$$

The left-hand side is a function of temperature only, and the right-hand side is a function of volume only and can be integrated. Integrating from an initial state (1) to any other state (no subscript), and assuming the specific heat is a constant (calorically perfect):

$$c_v \ln\left(\frac{T}{T_1}\right) = -R\ln\left(\frac{V}{V_1}\right) \quad \text{for adiabatic processes on calorically perfect gases}$$

Invoking that odd feature of the natural logarithm that $A\ln(B) = \ln(B^A)$ and taking the inverse ln

$$\left(\frac{T}{T_1}\right)^{c_v} = \left(\frac{V}{V_1}\right)^{-R} \quad \text{for adiabatic processes on calorically perfect gases}$$

Rearranging for the temperature ratio or volume ratio:

$$\left(\frac{T}{T_1}\right) = \left(\frac{V}{V_1}\right)^{\frac{-R}{c_v}} \quad \text{or equivalently} \quad \left(\frac{T}{T_1}\right)^{\frac{-c_v}{R}} = \left(\frac{V}{V_1}\right)$$

The exponent is commonly expressed in terms of the single fluid property, k:

$$\frac{R}{c_v} = \frac{c_p - c_v}{c_v} = k - 1$$

Notice that an increase in volume (expansion) yields a decrease in temperature (and pressure, as will be shown shortly). The gases perform expansion work at the expense of internal energy, which gives rise to a temperature drop. Conversely, a decrease in volume (compression) yields an increase in temperature (and pressure). In the ideal Otto cycle, the compression stroke is modeled as an adiabatic compression and the power stroke as an adiabatic expansion. The blowdown and valve overlap processes are both adiabatic expansion processes as well, in which the control mass must be taken to be only that portion of the cylinder gases that remain in the cylinder at the end of the process.

A relationship between temperature and pressure can be obtained by eliminating the volume using the ideal gas law, resulting in a relationship between pressure and temperature:

$$\left(\frac{T}{T_1}\right)^{\frac{c_v+R}{R}} = \left(\frac{P}{P_1}\right) \quad \text{or equivalently} \quad \left(\frac{T}{T_1}\right) = \left(\frac{P}{P_1}\right)^{\frac{R}{c_v+R}}$$

The exponent is commonly expressed in terms of the single fluid property, k:

$$\frac{c_v+R}{R} = \frac{c_v + (c_p - c_v)}{(c_p - c_v)} = \frac{c_p}{c_p - c_v} = \frac{c_p/c_v}{c_p/c_v - 1} = \frac{k}{k-1}$$

A relationship between pressure and volume can be obtained by eliminating the temperature using the ideal gas law:

$$\left(\frac{V}{V_1}\right)^{\frac{-(c_v+R)}{c_v}} = \left(\frac{P}{P_1}\right) \quad \text{or equivalently} \quad \left(\frac{V}{V_1}\right) = \left(\frac{P}{P_1}\right)^{\frac{-c_v}{c_v+R}}$$

The exponent is commonly expressed in terms of the single fluid property, k:

$$\frac{c_v+R}{c_v} = \frac{c_v + (c_p - c_v)}{c_v} = \frac{c_p}{c_v} = k$$

This final form of the relationship between pressure and volume can be expressed as

$$Pv^k = \text{constant (valid for adiabatic processes of a calorically perfect gas).}$$

One way that deviations from an ideal cycle (real effects) are incorporated into models is the so-called polytropic model, in which pressure and volume are related by a power law, with a polytropic exponent (n) applied to the volume:

$$Pv^n = \text{constant for a polytropic process}$$

Details of polytropic processes were discussed in the previous chapter.

Otto Cycle Analysis

Without throttling, that is, if the intake and exhaust manifold pressures are equal to ambient pressure, the cycle reduces to a classic idealized cycle, namely, the Otto cycle. The analysis of the Otto cycle results in a simple closed-form expression for the 1st Law efficiency that depends only on the compression ratio (r) and the ratio of specific heats ($k = c_p/c_v$), as follows:

$$\eta_{\mathrm{OTTO}} = \frac{\text{Net Work Out}}{\text{"Heat" In}} = \frac{mc_v(T_3 - T_4) - mc_v(T_2 - T_1)}{mc_v(T_3 - T_2)} = 1 - \frac{(T_4 - T_1)}{(T_3 - T_2)}$$

The net work out follows from energy balances applied to the compression and power strokes, and the "heat" in follows from an energy balance on the combustion process. The relation can be manipulated as follows:

$$\eta_{\mathrm{OTTO}} = 1 - \frac{T_4}{T_3}\frac{\left(1 - \frac{T_1}{T_4}\right)}{\left(1 - \frac{T_2}{T_3}\right)} = 1 - \frac{T_4}{T_3}\frac{\left(1 - \frac{T_2}{T_4}\frac{T_1}{T_2}\right)}{\left(1 - \frac{T_2}{T_4}\frac{T_4}{T_3}\right)}$$

Recognizing that the compression and power strokes are adiabatic processes with the same volume ratios and assuming a calorically perfect gas:

$$\eta_{\mathrm{OTTO}} = 1 - \left(\frac{V_4}{V_3}\right)^{-R/c_v} \frac{\left(1 - \frac{T_2}{T_4}\left(\frac{V_1}{V_2}\right)^{-R/c_v}\right)}{\left(1 - \frac{T_2}{T_4}\left(\frac{V_4}{V_3}\right)^{-R/c_v}\right)}$$

Since $V_3/V_4 = V_2/V_1 = 1/r$, and $R/c_v = (c_p - c_v)/c_v = k - 1$, the final elegant result is:

$$\eta_{\mathrm{OTTO}} = 1 - \frac{1}{r^{k-1}}$$

The ideal Otto cycle efficiency is plotted versus compression ratio in Fig. 7.6 for several fixed values of the ratio of specific heats (k) that correspond to various gases and temperatures. For any gas, the efficiency increases sharply at low compression ratio and is a monotonically increasing function, but levels off for compressions ratios above approximately 10, which coincidentally is approximately the maximum compression ratio that can be used in an SI engine without experiencing premature combustion.

In practice, spark ignition reciprocating four-stroke engines (by far the dominant automobile engine) operate on a cycle closely modeled by the Otto cycle.

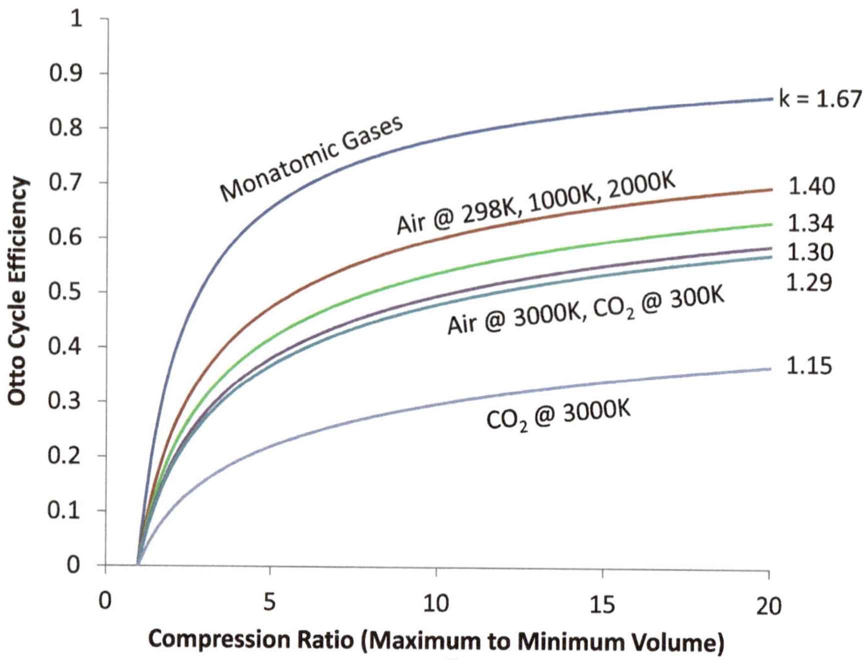

Fig. 7.6 Otto cycle efficiency as a function of compression ratio for values of the ratio of specific heats (k) corresponding to various gases and temperatures

The intake and exhaust manifold pressures are assumed to be equal to ambient pressure in the ideal unthrottled cycle, resulting in no net work associated with the gas exchange, and this clean expression for the 1st Law efficiency.

A formal 2nd Law analysis (i.e., identifying eXergy destruction) is not possible without consideration of the chemical changes. The "Heat In" in the idealized cycle is not a heat transfer (hence entropy transfer, with a known boundary temperature), it is a conversion of one form of internal energy (chemical) into another (thermal). Also, the blowdown process is not accomplished by a heat exchange from the still hot gases to ambient, but is done by an open system venting of the pressurized gases.

Geometric Relationships

The displacement volume (V_D), compression ratio (r), bore to stroke ratio (B/S), and ratio of connecting rod to crank radius (L/a) are key input geometric parameters in this model. The crank angle (θ) is an independent variable directly related to time (a consequence of the assumption of constant engine speed). Several other geometric quantities derive from these. The angular speed (consistent with θ defined as in Fig. 7.2, and positive angular velocity ω in the counterclockwise direction) is

$$\frac{d\theta}{dt} = -\omega = -2\pi N$$

With an imposed constant engine speed, N, the crank angle can be integrated to

$$\theta = \theta_0 - 2\pi N t$$

where θ_0 is an initial crank angle (π at the start of compression).

From the definition of compression ratio, the clearance volume is derived as follows:

$$r = \frac{V_C + V_D}{V_C}$$

Rearranging for the clearance volume:

$$V_C = \frac{V_D}{r - 1}$$

The cylinder diameter (bore, B) and crank radius ($a = S/2$) are derived from bore to stroke ratio:

$$V_D = \frac{\pi}{4} B^2 S = \frac{\pi}{4} B^2 \frac{B}{B/S} = \frac{\pi B^3}{4B/S}$$

Solving for the bore:

$$B = \left(\frac{4(B/S) V_D}{\pi} \right)^{1/3}$$

The crank radius is then

$$a = \frac{B}{2(B/S)}$$

The connecting rod length follows from the input ratio L/a.

A relationship between crank angle (θ, the angle from TDC) and connecting rod angle (γ, the angle from the vertical) is obtained by equating the lateral dimension of the crank to that of the connecting rod (Fig. 7.7):

$$a \sin\theta = L \sin\gamma$$

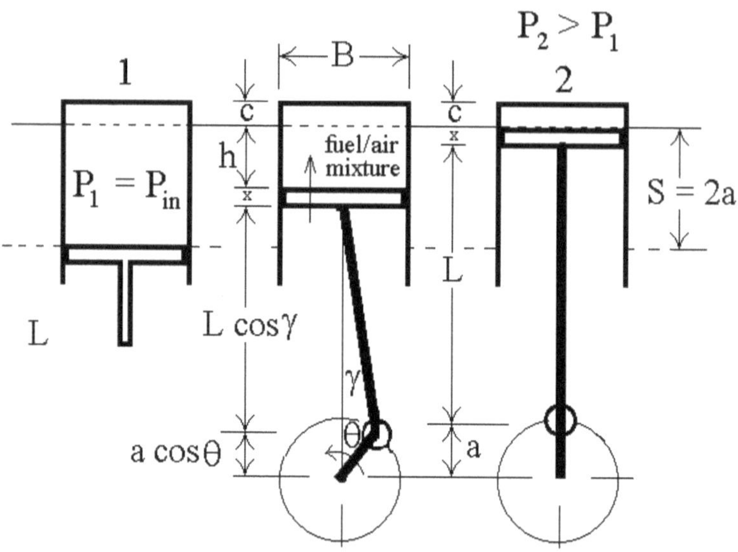

Fig. 7.7 Schematic of the compression process and definition of various dimensions used in the analysis

This expression is recognized as the law of sines. Solving for the connecting rod angle:

$$\gamma = \sin^{-1}\left[\frac{a}{L}\sin\theta\right]$$

The cylinder volume (V), from the center schematic of Fig. 7.7, is

$$V = V_C + \frac{\pi B^2 h}{4}$$

A relationship between the cylinder volume (V) and crank angle (θ) is seen in Fig. 7.7:

$$a\cos\theta + L\cos\gamma + x + h + c = a + L + x + c$$

Rearranging for h:

$$h = a(1 - \cos\theta) + L(1 - \cos\gamma)$$

Substituting for volume:

$$V = V_C + \left(\pi B^2/4\right)[a(1 - \cos\theta) + L(1 - \cos\gamma)]$$

Substituting for the displacement volume $V_D = \pi B^2(2a)/4$:

$$V = V_C + \frac{V_D}{2}\left[(1 - \cos\theta) + \frac{L}{a}(1 - \cos\gamma)\right]$$

It can be shown that the 2nd term reduces to zero in the limit of a long connecting rod ($L \gg a$) as $\cos\gamma$ goes to unity, and the volume is a pure sinusoidal, with an amplitude of $V_D/2$ and an average of $V_C + V_D/2$. The volume deviates from being a pure sinusoid due to the second term.

The mechanical power (rate of work done by the gases in the cylinder on the piston) is the pressure times the rate of change of volume (PdV/dt). The derivative of volume with respect to time is:

$$\frac{dV}{dt} = \frac{V_D}{2}\left[\left(\sin\theta\frac{d\theta}{dt}\right) + \frac{L}{a}\left(\sin\gamma\frac{d\gamma}{d\theta}\frac{d\theta}{dt}\right)\right]$$

To obtain the derivative relationship between γ and θ, take the derivative of the law of sines expression:

$$\frac{d}{dt}a\sin\theta = \frac{d}{dt}L\sin\gamma$$

$$a\cos\theta\frac{d\theta}{dt} = L\cos\gamma\frac{d\gamma}{dt}$$

Rearranging and eliminating the differential time, dt:

$$\frac{d\gamma}{d\theta} = \left(\frac{a}{L}\right)\left(\frac{\cos\theta}{\cos\gamma}\right)$$

Substituting back:

$$\frac{dV}{dt} = \frac{V_D}{2}[\sin\theta + \tan\gamma\cos\theta]\frac{d\theta}{dt}$$

Relating crank angle to engine speed:

$$\frac{dV}{dt} = -\pi V_D N[\sin\theta + \tan\gamma\cos\theta]$$

The velocity of the piston can be determined from

$$v_{piston} = -\frac{dh}{dt} = -\frac{4}{\pi B^2}\frac{dV}{dt} = \frac{4V_D N}{B^2}[\sin\theta + \tan\gamma\cos\theta]$$

A key underlying assumption of the reciprocating engine models is that the pressure remains uniform in the cylinder. A deeper dive topic, derived in Appendix

1, places a restriction on the engine speed to be much less than the speed of sound, or that the Mach number of the piston must be less than approximately 0.3.

Mechanics: Forces and Torques

A free-body diagram of a partially dismembered assembly is shown in Fig. 7.8. This analysis neglects the inertial effects (ma and $I\omega$) for all components, in which case the connecting rod is a two-force member and the direction of the force F_{ROD} (shown in compression) acts along the member. The force balances are as follows:

- x force balance: $\sum F_x = M_{\text{piston}} a_x = 0 = F_{\text{sides}} - F_{\mathrm{ROD}} \sin \gamma$
- y force balance: $\sum F_y = M_{\text{piston}} \frac{dv_{\mathrm{P}}}{dt} \approx 0 = -\mathrm{PA} \mp \mu |F_{\text{side}}| + F_{\mathrm{ROD}} \cos \gamma$

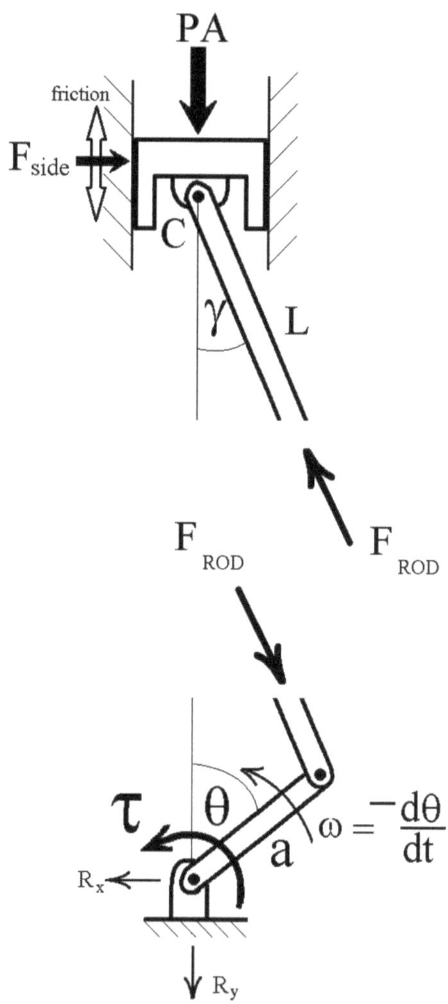

Fig. 7.8 Free-body diagram of dismembered assembly

The friction has been modeled as a simple kinetic coefficient model. The piston is pushed to the side, and the friction force is assumed to be proportional to the normal force (F_{side}) with a friction coefficient, μ. The sign depends on the direction the piston moves. If the piston is moving up ($v_p > 0$), the friction acts downward (in the same direction as the pressure force, PA). The plus or minus sign can be eliminated by multiplying the friction term by the piston speed normalized by its absolute value: the absolute value of the side force is needed because the force can act in the negative x direction.

It would not be difficult to include the inertial effects of the piston (its velocity is known and acceleration from taking its derivative). On the other hand, including the inertial effects on the connecting rod would be somewhat more involved. The thermodynamics are of primary interest here, not the mechanics, so these inertial effects are neglected. They would be more important at higher engine speeds.

Solving these two equations for the rod force in terms of the pressure force and geometry:

$$F_{ROD} = \frac{PA}{\cos\gamma - \mu|\sin\gamma|(v_P/|v_P|)}$$

The side force is obtained from the x force balance equation.

The instantaneous torque acting on the crank shaft is obtained from a vector-based moment balance on the bottom FBD of the dismembered assembly (Fig. 7.8):

$$\sum M = I\omega \approx 0 = \tau\vec{k} + \vec{r} \times \vec{F}_{rod}$$
$$= \tau\vec{k} + a\left(\sin\theta\ \vec{i} + \cos\theta\ \vec{j}\right) \times F_{ROD}\left(\sin\gamma\ \vec{i} - \cos\gamma\ \vec{j}\right)$$
$$\tau = -\vec{r} \times \vec{F}_{rod} = aF_{ROD}(\sin\theta\ \ \cos\gamma + \cos\theta\sin\gamma)\vec{k}$$

Note that this expression represents the torque applied to the crank shaft in the counterclockwise direction. When torque is negative (power and intake strokes), the torque is being supplied from the system to the mechanical load at the shaft. Positive torque means the external system must put torque in to maintain the speed (i.e., during the compression and exhaust strokes).

When the crank arm radius (a) is much less than the connecting rod length (L), the instantaneous torque simplifies to

$$\tau = aF_{ROD}\sin\theta \quad \text{for } a \ll L$$

The torque of an engine in a manufacturer specification is not the instantaneous torque that changes for each crank angle position, it is an integrated torque over the cycle, and is related to the overall engine power:

$$\dot{W}_{cycle} = 2\pi N\tau_{cycle}$$

Analysis of Cycle Processes

Recall the major assumptions of the model:

- Calorically perfect gas (constant specific heats) with properties of air ($R = 287$ J/kg, $c_v = 0.718$ kJ/kg/K for cold air approximation) during each process. A different single value could be used based on a cycle average temperature, or a different value for each process, but is not done so here. A separate sensitivity study of overall effects will be done.
- Every process is adiabatic, with complete combustion as an instantaneous internal energy conversion process.
- No pressure drops across intake and exhaust valves.
- Well-mixed model for blowdown and exhaust, meaning that the properties of the fluid as it exits the valve equal the average in-cylinder property value.
- Neglect kinetic and potential energy changes of the gas.
- Neglect inertial effects in the crank mechanism.

The mass of the charge in the cylinder can be evaluated at the start of compression:

$$m_1 = m_2 = m_3 = m_4 = \frac{P_1(V_C + V_D)}{RT_1}$$

However, as will be discussed, the temperature at the start of compression is not known precisely a priori (as in the reciprocating compressor example). Some hot gases from the previous cycle remain in the cylinder, and these "residual gases" are mixed with fresh intake. An iterative procedure will be used to determine the initial temperature, as explained shortly.

Process 1→2: Compression Stroke

The compression process (1–2 shown in Fig. 7.9) is assumed to be an adiabatic, frictionless compression of a calorically perfect gas (i.e., constant specific heats). The temperature (absolute units) is a function of cylinder volume:

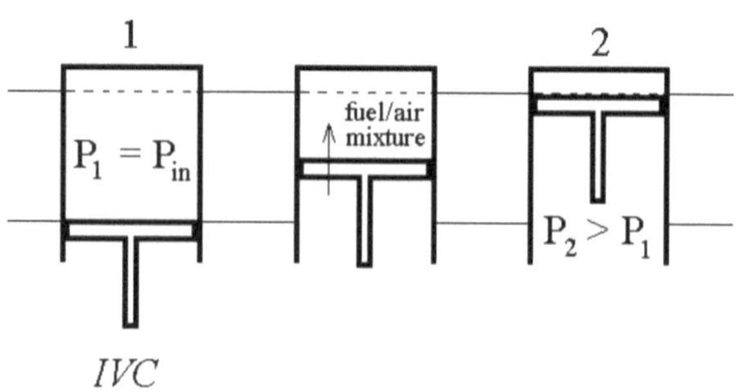

Fig. 7.9 Cylinder before, during, and after compression stroke

$$T = \left(\frac{V_1}{V}\right)^{\frac{R}{c_V}} T_1$$

The final state is related to the compression ratio, $r = (V_c + V_d)/V_c = V_1/V_2$:

$$T_2 = (r)^{\frac{R}{c_V}} T_1$$

The final pressure is obtained from the ideal gas law: $P_2 = m_2 R T_2 / V_2$.

The work done by the gases during compression is obtained from a 1st Law balance:

$$E_2 - E_1 = Q_{12} - W_{12}$$

The process is adiabatic ($Q_{12} = 0$) and with KE and PE neglected and the internal energy related to temperature through the definition of constant volume specific heat. Therefore, the work of compression is $W_{12} = -m_1 c_v (T_2 - T_1)$.

Process 2→3: Combustion Process

Combustion is assumed to take place completely and instantaneously at TDC (Fig. 7.10). A spark ignites the fuel/air mixture, which reacts chemically to form products of combustion. While the chemical composition changes, the nitrogen in the air is the dominant species, and the model assumes the gas has properties of air both before and after combustion. In practice, the combustion process takes place through the propagation of a premixed flame into the unburned gases, a process that takes time to complete and therefore not entirely at TDC. A deeper dive into combustion is taken in Appendix 2).

For the idealized model, the energy balance is

$$U_3 - U_2 = m_1 c_v (T_3 - T_2) + \Delta U_{\text{chemical}} = Q_{23} - W_{23}$$

The work is identically zero (constant volume process), and the heat addition is assumed to be zero (adiabatic). The chemical energy change is a conversion of

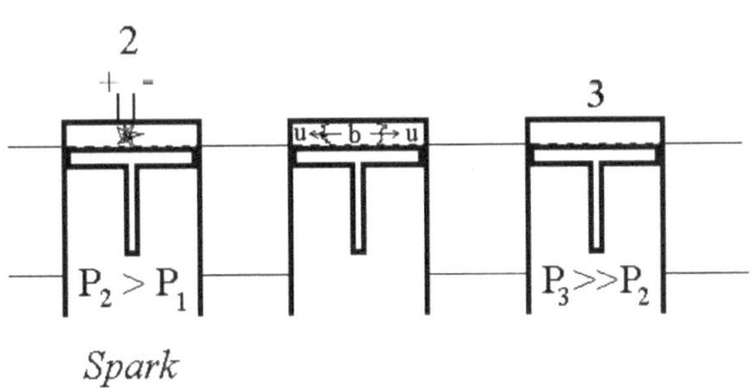

Fig. 7.10 Piston/cylinder before, during, and after combustion process

chemical into thermal energy. Solving for the temperature after combustion (the adiabatic flame temperature):

$$T_3 = T_2 + \frac{\Delta U_{\text{chemical}}}{m_2 c_{\text{v}}}$$

The heat of combustion is the chemical energy converted to thermal energy and is given by the word equation:

$$\Delta U_{\text{chemical}} = \left(\frac{\text{Energy Released}}{\text{kg Fuel Burned}}\right) \times \left(\frac{\text{kg Fuel Burned}}{\text{kg Fuel Fed}}\right) \times \left(\frac{\text{kg Fuel Fed}}{\text{kg Mixture Fed}}\right) \times (\text{kg Mixture Fed})$$

$$\Delta U_{\text{chemical}} = (\text{LHV})(\eta_{\text{c}})\left(\frac{1}{1+\text{AFR}_{\text{mass}}}\right) m_1$$

where *LHV* is the lower heating value of the fuel (chemical energy release per mass of fuel burned); η_C is the combustion efficiency (the fraction of fuel that actually burns); and AFR_{mass} is the air to fuel mass ratio.

In the ideal model, combustion is assumed to be complete ($\eta_C = 1$). For lean or stoichiometric mixtures, there is sufficient oxygen to burn all the fuel supplied. For rich mixtures, there is insufficient oxygen and the ideal combustion efficiency is a function of the equivalence ratio (ϕ).

$$\eta_{\text{c}} = 1 \quad \text{for} \quad \phi < 1$$

$$\eta_{\text{c}} = \frac{1}{\phi} \quad \text{for} \quad \phi > 1$$

The equivalence ratio is related to the air/fuel ratio by the following relation:

$$\phi = \frac{\text{AFR}_{\text{stoich}}}{\text{AFR}}$$

For gasoline, from Appendix 2,

$$\text{AFR}_{\text{stoich}} = 14.6 \quad \text{and} \quad \text{LHV} = 43{,}000 \frac{\text{kJ}}{\text{kg}}$$

The pressure after combustion is obtained from the ideal gas law:

$$P_3 = \left(\frac{T_3}{T_2}\right)\left(\frac{R_3}{R_2}\right)\left(\frac{V_3}{V_2}\right) P_2$$

Since the molecular weights are assumed constant (therefore R is constant) and the process is assumed to take place at TDC:

$$P_3 = \left(\frac{T_3}{T_2}\right) P_2$$

Process 3→4: Power Stroke

The power stroke (Fig. 7.11) is similar to the compression stroke, with different sign and starting conditions. In this idealized model, it is assumed to be an isentropic expansion:

$$T = \left(\frac{V_3}{V}\right)^{\frac{R}{c_{v,34}}} T_3$$

The subscript (34) on the specific heat indicates that a different value for the specific heat could be then used during the compression stroke, although that is not done here.

The final states are related to the compression ratio, $r = V_d/V_c$:

$$\frac{V_3}{V_4} = \frac{V_C}{V_C + V_D} = \frac{1}{r}$$

$$T_4 = \left(\frac{1}{r}\right)^{\frac{R}{c_{v,34}}} T_3$$

The pressure is obtained from the ideal gas law. The work done by the gases during the power stroke is given by

$$W_{34} = m_3 c_v (T_3 - T_4)$$

In a real engine, heat losses to the walls from the extremely high gas temperatures during the power stroke would reduce this work significantly, a major source of performance deviation between the ideal and real engines.

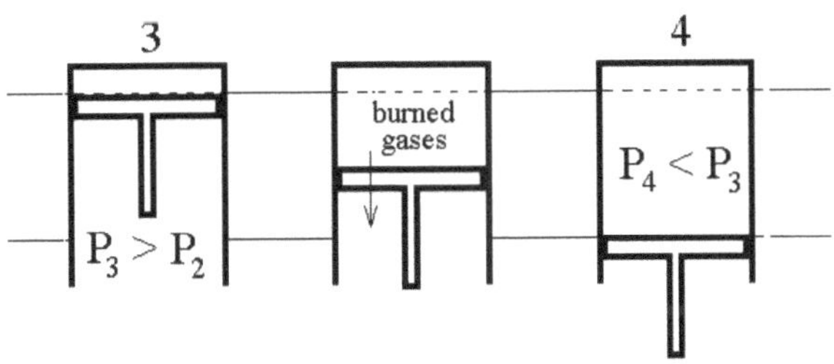

Fig. 7.11 Piston/cylinder before, during, and after power stroke

Process 4→5: Blow Down

In a real engine, the exhaust valve begins to open near the end of the power stroke, exposing the cylinder contents to the exhaust manifold, and gradually opens for mechanical reasons (inertial effects). Fluid will flow through the exhaust valve as long as the pressure in the cylinder is greater than that in the exhaust runner. The blowdown process (Fig. 7.12) refers to the initial venting of pressurized gases, to be followed by an exhaust stroke in which the movement of the piston is the key driver of the exhaust flow. In this idealized model, the pressure in the exhaust manifold is assumed to be a fixed input somewhat above atmospheric to account for the pressure drops across the pollution control systems, muffler, and tailpipe. In the idealized model, the exhaust valve opens fully and instantaneously at TDC and the blowdown process is assumed to be instantaneous.

A deeper dive in Appendix 3 shows that the remaining gases in the cylinder undergo an isentropic expansion process. There is no heat transfer and the cylinder volume remains constant, which would suggest to some (incorrectly) that the temperature remains constant. However, this process is a transient open thermodynamic system, and flow work is required to push fluid across the valve, even though no work is done at the boundary of the piston. The final temperature of the gases is therefore determined from the following:

$$T_5 = \left(\frac{P_5}{P_4}\right)^{\frac{R}{c_p}} T_4$$

There is no expansion work or heat exchange in the idealized model. The mass remaining after blowdown is obtained from the ideal gas law:

$$m_5 = \frac{P_5 V_5}{R T_5}$$

The mass flow rate in the idealized cycle during blowdown is infinite, but lasts for an infinitesimal amount of time. It is a Dirac delta function.

Fig. 7.12 Piston/cylinder before, during, and after blowdown

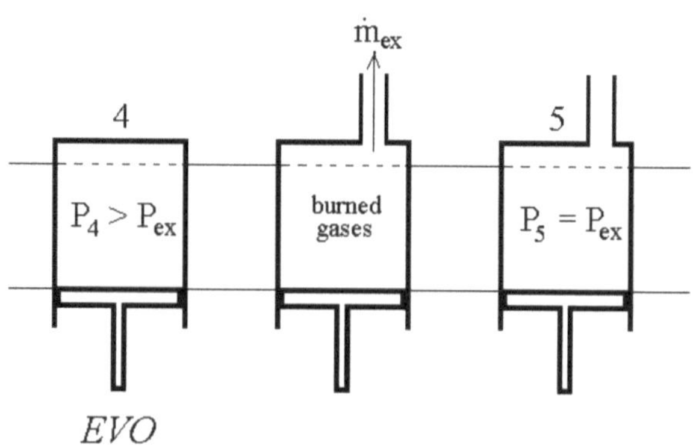

A turbine can be placed in the exhaust system aimed at producing power from the blowdown process, rather than simply venting this pressurized gas, explored in Workshop 7.2. The turbine can be used to drive a compressor on the intake side, which will increase the pressure of the intake air (and thereby the mass). This application is called turbo-charging, which gives a modest increase in both power and thermal efficiency over the basic engine. Alternatively, the turbine output can be tied directly to the output shaft, called turbo-compounding. Supercharging involves adding just a compressor to the intake, which increases the power output, but not the thermal efficiency because power is not extracted from the blowdown process.

The maximum theoretical work that could be extracted from a turbine can be calculated by assuming all the mass in the cylinder at the end of the power stroke (m_4) undergoes an adiabatic expansion and work is extracted. Then, an energy balance on this adiabatic, closed system yields

$$W_{\text{turbine}} = -\Delta U_{45} = m_4 c_v (T_4 - T_5)$$

For the case where there is no turbine to extract useful power from blowdown, this theoretical work represents the lost eXergy associated with the blowdown process.

Process 5→6: Exhaust Stroke

With the exhaust valve open, the piston moves from BDC to TDC during the exhaust stroke (Fig. 7.13). The pressure drop across the exhaust valve is neglected in the idealized model, so that the pressure in the cylinder is at the exhaust manifold pressure (fixed at a value somewhat above atmospheric). The total work done by the gases on the piston is:

$$W_{56} = -P_5(V_D)$$

The negative sign shows that work must be done by the piston on the gases to accomplish exhaust. A formal analysis of the exhaust stroke (an open system) in

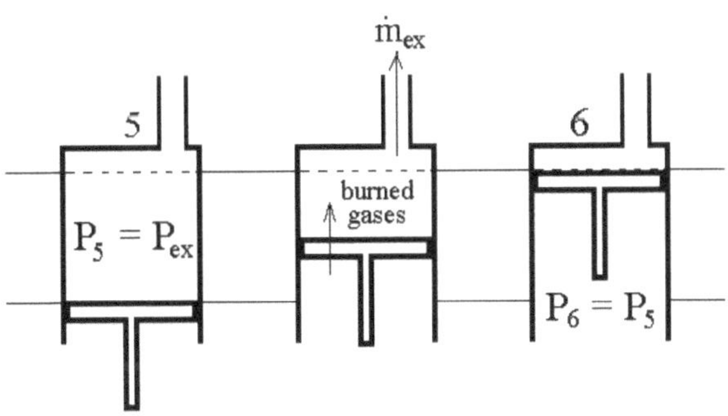

Fig. 7.13 Piston/cylinder before, during, and after exhaust stroke

Appendix 4 shows that the temperature remains constant at its value after blowdown. The compression work (PdV) is equal and opposite to the flow work ($m_e P/\rho$) and there is no change in the internal energy.

The mass at any time during the exhaust stroke is given by the ideal gas law:

$$m = \frac{P_5 V}{R T_5}$$

The mass flow rate can be obtained by differentiation or a numerical integration:

$$\dot{m} = \frac{dM}{dt} = \left(\frac{P_5}{R T_5}\right) \frac{dV}{dt}$$

The volume and rate of change of volume were derived previously, expressed either in terms of the crank angle or time and engine speed.

Process 6→7: Valve Overlap

When the piston reaches TDC (in the idealized model), the exhaust valve closes and the intake valve opens (Fig. 7.14). The intake manifold is generally at a lower pressure than atmospheric (by design, due to throttling), creating a tendency for backflow into the intake manifold. Similar to blowdown, a venting process occurs, but into the intake manifold (which can cause a form of backfire, where the hot gases ignite the fuel/air mixture in the intake runner). The temperature after valve overlap is given by

$$T_7 = \left(\frac{P_7}{P_6}\right)^{\frac{R}{c_P}} T_6$$

The mass remaining after blowdown is obtained from the ideal gas law:

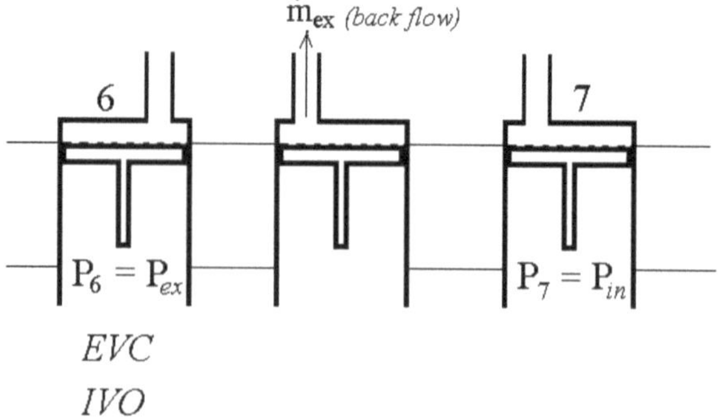

Fig. 7.14 Piston/cylinder before, during, and after valve overlap process

$$m_7 = \frac{P_7 V_7}{RT_7}$$

The mass flow rate in the idealized cycle is infinite for an infinitesimal amount of time.

Process 7→1: Intake Stroke

As the piston withdraws with the intake valve open, the cylinder pressure is maintained at the intake manifold pressure (neglecting the valve pressure drop) by the flow of fuel/air mixture (Fig. 7.15). The work done by the gases on the piston is:

$$W_{71} = P_7 V_{\mathrm{D}}$$

The difference between the intake and exhaust work is called "pumping work." In an unthrottled idealized engine (Otto cycle), $P_{\mathrm{in}} = P_{\mathrm{ex}} = P_{\infty}$ and the pumping work is zero.

The mass at any time during the intake stroke is given by the sum of the residual mass (M_7) and the mass of air that has been inducted with intake manifold properties, obtained from the ideal gas law:

$$m = m_7 + \frac{P_{\mathrm{in}}(V - V_{\mathrm{C}})}{RT_{\mathrm{in}}}$$

The mass flow rate can be obtained by differentiation:

$$\dot{m} = \frac{dm}{dt} = \left(\frac{P_{\mathrm{in}}}{RT_{\mathrm{in}}}\right)\frac{dV}{dt}$$

Note that in this case of a transient, single-inlet process, the rate of mass flow in ($\dot{m}$), a flux, is equal to the rate of change of the mass stored (dm/dt) both of which change with time, but together for this case only.

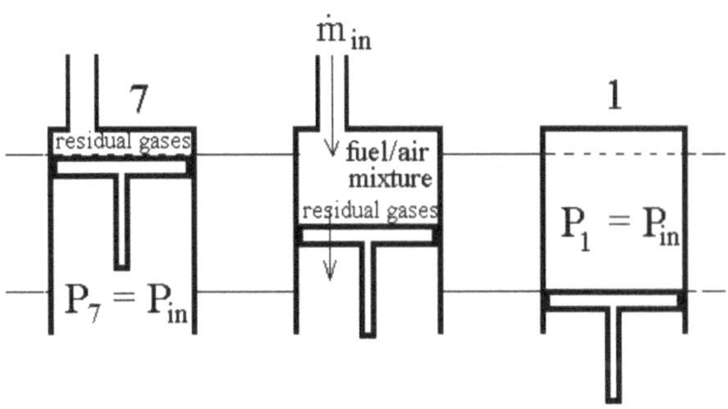

Fig. 7.15 Piston/cylinder before, during, and after intake stroke

Adiabatic Mixing of Fresh and Residual Gases

The pressure and volume at the start of the cycle (state 1, start of compression) are well defined, but the temperature is not. In this model, the temperature and pressure of the air/fuel mixture are set as inputs. However, these gases mix with residual gases from the previous cycle, whose temperature depends ultimately on the starting temperature. To get around this circular argument, an iterative scheme is conducted in which the temperature at the start of the cycle (T_1) is set to ambient as a first approximation. Then, the cycle is analyzed, resulting in a warmer gas temperature at state 7. These gases are assumed to mix adiabatically with inducted fuel/air mixture: an energy balance for this mixing process is:

$$m_7 c_v T_7 + (m_1 - m_7) c_v T_{in} = m_1 c_v T_1$$

$$T_1 = \frac{m_7 T_7 + (m_1 - m_7) T_{in}}{m_1}$$

Note that the value for m_7 depends on T_1, so this is not a closed-form solution. However, in the iterative procedure, after several cycle iterations, the temperature at state 1 approaches that of the previous cycle. This step represents the only instance of numerical analysis in the execution of this model.

Engine Performance

Considering all the results of the cycle performance, none of the state properties depend on the engine speed, so that, for example, the net work per cycle is independent of the engine speed for the ideal cycle. In a real cycle, pressure drops across the valves, intake and exhaust components, effects of heat transfer to the walls, and noninstantaneous and incomplete combustion are examples of effects that are dependent on engine speed. These effects are not considered here, but are good candidates for extended study.

On the other hand, the net power (work per unit time) for the ideal cycle depends directly on the engine speed. The higher the engine speed, the shorter time required for a cycle, and the greater the power.

While reciprocating engines are inherently unsteady, the performance of an engine can be modeled as an open system at steady state. For example, while the mass intake process is pulsed in time, the average mass flow rate into the cylinder during the cycle can be determined by the mass inducted divided by the time for a complete cycle:

$$\dot{m} = \frac{m_1 - m_7}{t_{cycle}} = \frac{\rho_1 V_1 - \rho_7 V_7}{2/N} = \frac{P_{in} N}{2R} \left(\frac{V_d + V_c}{T_1} - \frac{V_c}{T_7} \right)$$

The factor "2" arises because there are two revolutions per cycle. For multicylinder engines, the staggering of the cycles tends to smooth out pulses.

Another measure of flow rate is the volumetric flow rate of air inducted:

$$(\mathrm{AV})_{\mathrm{air}} = \frac{\dot{m}}{\rho_\infty} = \frac{RT_\infty \dot{m}}{P_\infty}$$

There are two expressions of work output, called the indicated and brake work. The indicated work is that associated with the net work done by the working fluid over a cycle. It is the work that would be "indicated" by a direct measurement of the pressure versus volume during the cylinder. The brake work is what would be obtained by a measurement of the total work applied to crankshaft restrained by a brake. The difference between the indicated and brake work is due to the work of friction in the crank mechanisms.

The indicated work is the net area under the p-V curve (the indicator diagram), which equals the sum of the work of all four strokes of the cycle:

$$W_{\mathrm{indicated}} = \int_0^{4\pi} PdV$$

The indicated power of the cycle is obtained by dividing the total work per cycle by the cycle time:

$$\dot{W}_{\mathrm{indicated}} = \frac{W_{\mathrm{indicated}}}{2/N}$$

The brake work is obtained by integrating the torque over all four strokes and is equal to the average torque over the cycle (called the brake torque):

$$W_{\mathrm{brake}} = 4\pi\tau_{\mathrm{cycle}} = \int_0^{4\pi} \tau d\theta$$

A common equivalent metric that eliminates scale is the mean effective pressure (MEP), which is the work per cycle divided by the displacement volume:

$$\mathrm{mep} = \frac{W_{\mathrm{brake}}}{V_{\mathrm{D}}}$$

The brake power is obtained by dividing the total work per cycle by the cycle time:

$$\dot{W}_{\mathrm{brake}} = \frac{W_{\mathrm{brake}}}{2/N}$$

The combustion process is classically treated as an "equivalent heat addition," which is appropriate for a 1st Law analysis, but not for a 2nd Law analysis. The rate of chemical energy addition (equivalent heat addition) is as follows:

$$\dot{H}_{\text{combustion,in}} = \frac{\Delta U_{\text{chemical}}}{2/N}$$

The 1st Law efficiency is defined as the net power output over the net rate of fuel energy (equivalent heat) addition:

$$\eta_1 = \frac{\dot{W}_{\text{out}}}{\dot{H}_{\text{combustion,in}}}$$

The 1st Law efficiency could be defined based on either the indicated or the brake efficiency (with friction losses included). A common metric is the specific fuel consumption, which is the inverse of the 1st Law efficiency and is generally given in dimensional units such as lb. per horsepower hour. It is not formally calculated here.

Results and Discussion

With the exception of the iterative procedure needed to determine the initial temperature of the cycle, all the above formulas can be entered directly into a spreadsheet format to conduct parametric analyses. The fixed parameter values (input and derived) for a case study are presented in Fig. 7.4. The iteration procedure for determining the initial state is shown in Fig. 7.16. In the first cycle, an initial guess for the gas temperature at the start of compression is taken to be the ambient temperature. Then, all other state variables and the mass can be calculated for the remaining six states of the first cycle. In particular, the mass and temperature of the residual gases after valve overlap (state 7) are known. These gases are mixed adiabatically with the fresh inducted gas, resulting in a warmer initial temperature ($T_1 = 327.8$ K for the 2nd cycle). This process is continued until the initial temperature is the same (within a convergence tolerance) for subsequent cycles (331.1 K). As is apparent, this temperature is the same to within four significant figures after four complete cycles. That is, the iterative process converges quickly.

Cycle #	Cycle State	Cycle Angle	V (cc)	P (atm)	T (K)	M (kg)
1	1	0	777.8	0.70	298.0	6.46E-04
	2	90	77.8	17.55	747.3	6.46E-04
	3	90	77.8	106.87	4549.5	6.46E-04
	4	180	777.8	4.26	1814.2	6.46E-04
	5	180	777.8	1.20	1263.7	2.61E-04
	6	270	77.8	1.20	1263.7	2.61E-05
	7	270	77.8	0.70	1083.5	1.78E-05
2	1	0	777.8	0.70	327.8	5.87E-04
	2	90	77.8	17.55	822.0	5.87E-04
	3	90	77.8	98.75	4624.3	5.87E-04
	4	180	777.8	3.94	1844.0	5.87E-04
	5	180	777.8	1.20	1313.7	2.51E-04
	6	270	77.8	1.20	1313.7	2.51E-05
	7	270	77.8	0.70	1126.4	1.71E-05
3	1	0	777.8	0.70	330.8	5.82E-04
	2	90	77.8	17.55	829.5	5.82E-04
	3	90	77.8	98.02	4631.8	5.82E-04
	4	180	777.8	3.91	1847.0	5.82E-04
	5	180	777.8	1.20	1318.6	2.50E-04
	6	270	77.8	1.20	1318.6	2.50E-05
	7	270	77.8	0.70	1130.6	1.70E-05
4	1	0	777.8	0.70	331.1	5.81E-04
	2	90	77.8	17.55	830.3	5.81E-04
	3	90	77.8	97.95	4632.5	5.81E-04
	4	180	777.8	3.91	1847.3	5.81E-04
	5	180	777.8	1.20	1319.1	2.50E-04
	6	270	77.8	1.20	1319.1	2.50E-05
	7	270	77.8	0.70	1131.1	1.70E-05
5	1	0	777.8	0.70	331.1	5.81E-04
	2	90	77.8	17.55	830.3	5.81E-04
	3	90	77.8	97.94	4632.6	5.81E-04
	4	180	777.8	3.91	1847.3	5.81E-04
	5	180	777.8	1.20	2586.8	1.27E-04
	6	270	77.8	1.20	2586.8	1.27E-05
	7	270	77.8	0.70	2218.1	8.67E-06

Initial Guess

This temperature is the basis for the iterative calculation. The gases at 7 have been mixed with the inducted gases: T1 = (M7*T7 + (M1-M7)*Tin)/M1

Fig. 7.16 Iteration table for five cycles

Base Case Results

A table of states is reported (after 10 cycles) in Fig. 7.17, a process table of energy and mass transfers in Fig. 7.18, and a table of engine performance measures in Fig. 7.19. Notice that the peak temperature after combustion is a staggering 4633 K and the peak pressure is nearly 100 atm. In a real engine, these values would be substantially reduced due to the dissociation of combustion products (resulting in lower chemical energy release), finite rate combustion, higher specific heat values than the cold air assumption, and heat loss to the walls. Nevertheless, the peak temperatures in a real engine would approach 3000 K. It is noteworthy that the melting temperature of cast iron is approximately 1500 K, and therefore, the walls of the cylinder must be cooled in practice.

STATE TABLE

Cycle State	Cycle Angle	V (cc)	P (atm)	T (K)	M (kg)
1	0	777.8	0.70	331.1	5.81E-04
2	90	77.8	17.55	830.4	5.81E-04
3	90	77.8	97.94	4632.6	5.81E-04
4	180	777.8	3.91	1847.3	5.81E-04
5	180	777.8	1.20	1319.2	2.50E-04
6	270	77.8	1.20	1319.2	2.50E-05
7	270	77.8	0.70	1131.1	1.70E-05

Fig. 7.17 Results of state property values after 10 cycles

PROCESS TABLE

Process	Description	Q (kJ)	W (kJ)	DM (g)
1-2	Compression	0	-0.208	0
2-3	Combustion	1.586	0	0
3-4	Power Stroke	0	1.162	0
4-5	BlowDown	0	0	-0.331
5-6	Exhaust Stroke	0	-0.085	-0.225
6-7	Valve Overlap	0	0	-0.008
7-1	Intake Stroke	0	0.050	0.564
sum		***1.586***	***0.918***	***0***

Fig. 7.18 Results of process values after convergence

The temperature of the gases remaining in the cylinder decreases during the blowdown (4–5) and valve-overlap processes (6–7). There is no piston work, but there is flow work associated with these open system processes, as derived formally in Appendix 3. This effect can be understood from a closed system point of view by considering the mass to be split into a portion that exits, and a portion that remains. The mass that remains occupies the full cylinder volume, but these gases occupied a lower volume before the venting occurred. They therefore undergo a rapid expansion process that is modeled as adiabatic (and justifiably so, due to the short time involved) between two known pressure values. The temperature drops due to flow work.

ENGINE PERFORMANCE

Chemical Energy Input	158.6	kW
Enthalpy Flux In	0.0	kW
Enthalpy Flux Out	57.9	kW
Indicated Power	89.7	kW
Indicated Torque	0.428	kN-m
Brake Power	79.0	kW
Friction Power	10.7	kW
Brake Torque	0.377	kN-m
eXergy Destruction at Intake	1.7	kW
eXergy Destruction at exhaust	0.9	kW
eXergy Destruction during blowdown	22.0	kW
Thermal Efficiency (indicated)	0.565	
Maximum Possible Efficiency	0.929	
2nd Law Efficiency	0.609	
Mass Flow Rate	0.0564	kg/s
Volumetric Flow Rate	2853.9	Liters/min
Texit	1319	K

Fig. 7.19 Engine performance

Indicator diagrams (absolute pressure vs. volume) are shown in Fig. 7.20. The top plot is on linear axes, which shows the basic shape of the cycle, to scale. The gas exchange processes (exhaust and intake) cannot be distinguished in this view. The semi-logy scale of the bottom plot reveals the gas exchange more clearly. The area under a given process curve represents the net pdV work exerted by the gas on the piston. The sign of the work depends on the direction (shown by arrows). The power stroke (3–4) and intake stroke (7–1) result in work being delivered to drive the crank shaft. The compression stroke (1–2) and exhaust stroke (5–6) require the crank to put work into the gases. There are two loops. The top loop (the power loop) acts in a clockwise direction and results in net positive work done by the gases. The bottom loop (the pumping work) is counterclockwise and is the net work required for the gas exchange process (fresh fuel/air in, spent products out). For a nonthrottled ideal cycle, the intake and exhaust strokes occur at the same pressure. Work is associated with both of them, but they are equal and opposite, and therefore the gas exchange contributes nothing to the net work of the cycle. On the other hand, the greater the throttling, the greater the pumping work. In turbo-charged and supercharged applications, the intake pressure is above ambient, and a net positive pumping work is common.

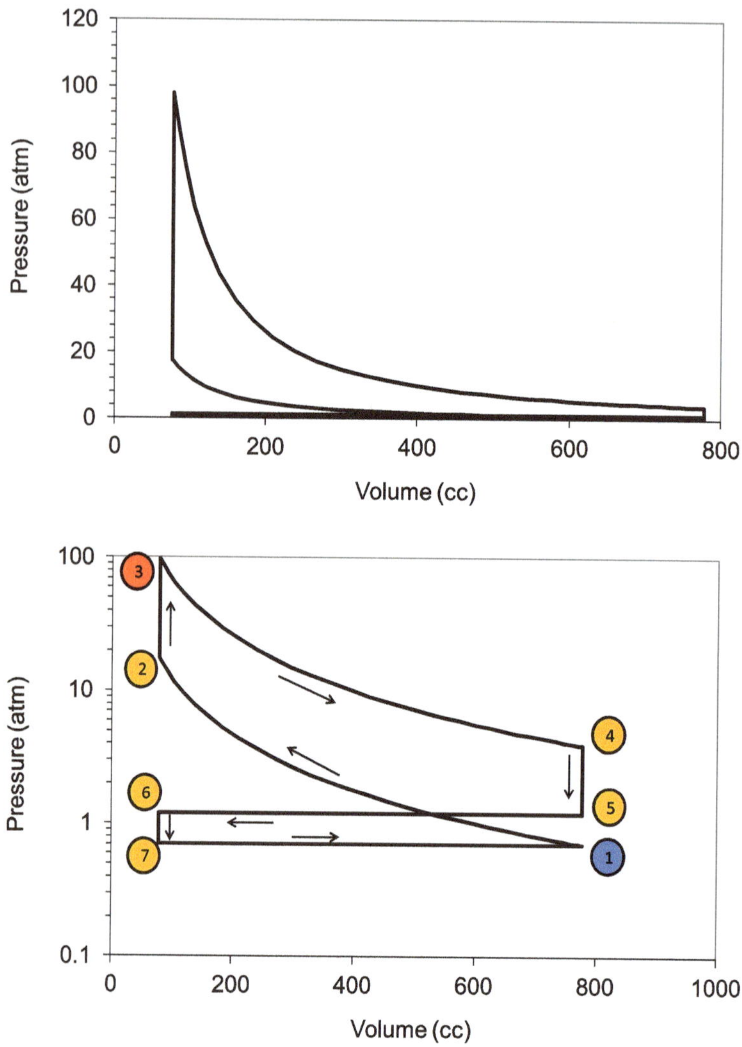

Fig. 7.20 Indicator diagrams (p-V). Top plot is on linear axes, bottom plot on semi-logy axes that show the throttling effect more clearly

Effect of Throttling

The effect of throttling is investigated by changing the pressure of the intake. The results are shown in Fig. 7.21 (mass flow rate), Fig. 7.22 (exit temperature), Fig. 7.23 (energy fluxes), Fig. 7.24 (eXergy destruction), and Fig. 7.25 (efficiencies). There is a minimum intake runner pressure (0.22 atm) below which blowdown would result

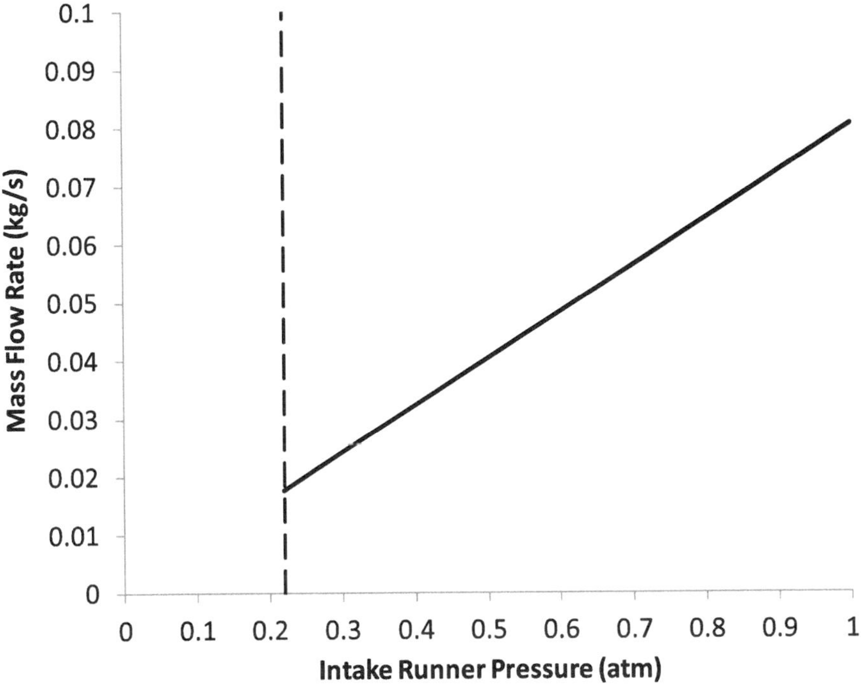

Fig. 7.21 Mass flow rate as a function of intake runner pressure. The lowest intake pressure (0.22 atm) corresponds to a condition below which the pressure at the end of the power stroke is less than the exhaust manifold pressure

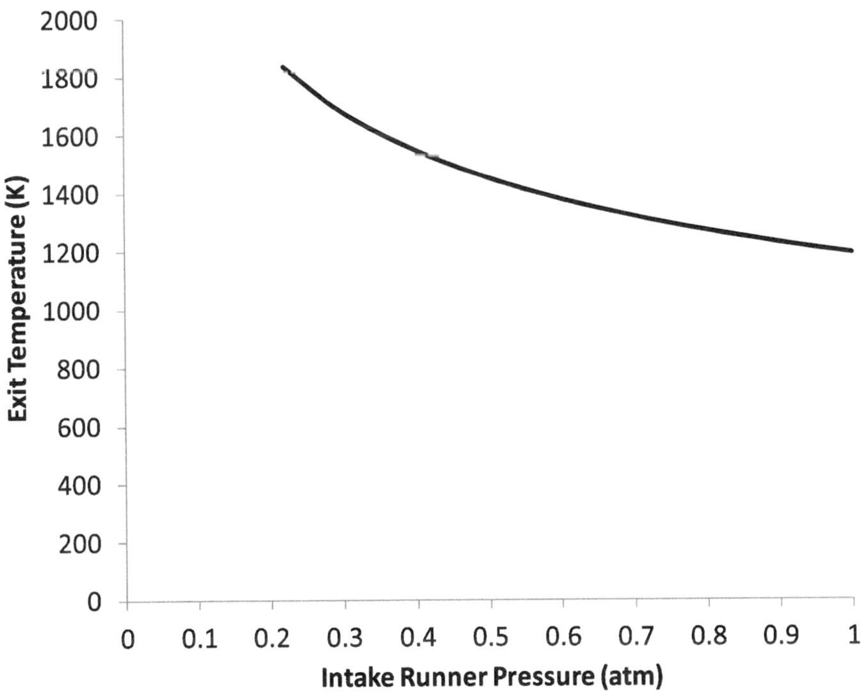

Fig. 7.22 Exit temperature as a function of intake pressure

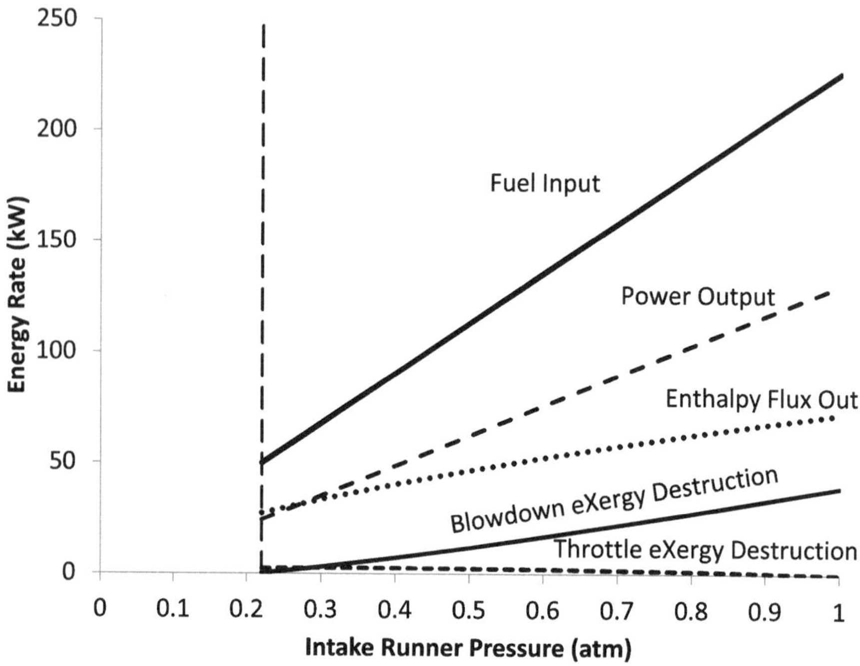

Fig. 7.23 Energy fluxes as a function of intake pressure

in backflow from the exhaust manifold into the cylinder because the pressure at the end of the power stroke would be below the exhaust manifold pressure.

The primary effect of throttling is to reduce the mass of the charge inducted per cycle, and therefore the mass flow rate through the engine (Fig. 7.21). The temperature of the exhaust stream (Fig. 7.22) increases with throttling (decreasing intake pressure) and is an indication of increased discarded thermal energy. All of the cycle averaged engine output parameters are proportional to the mass flow rate and some other quantity (Fig. 7.23). The only energy input to the engine is fuel input (the thermal energy of the air flow in is set to zero as a reference). All of this input is accounted for in an energy output, which includes the desired mechanical power, the undesired enthalpy flux of the exhaust stream (relative to the inlet enthalpy), and the eXergy destroyed in the intake (throttle) and exhaust. The power output is greater than the enthalpy flux of the exhaust stream at wide open throttle ($P_{in} = 1$ atm), but these outputs are approximately equal at high levels of throttling (low P_{in}), which is a reflection of the increasing exhaust temperatures with throttling. The eXergy destruction rate is much less than all the other energy streams, but becomes proportionally larger with throttling as the pressure drop across the throttling valve increases.

A consequence of the 2nd Law of Thermodynamics is that a thermal energy input (chemical energy of the fuel in this case) cannot be converted completely to a mechanical energy output (power output). A thermal energy rejection (hot exhaust)

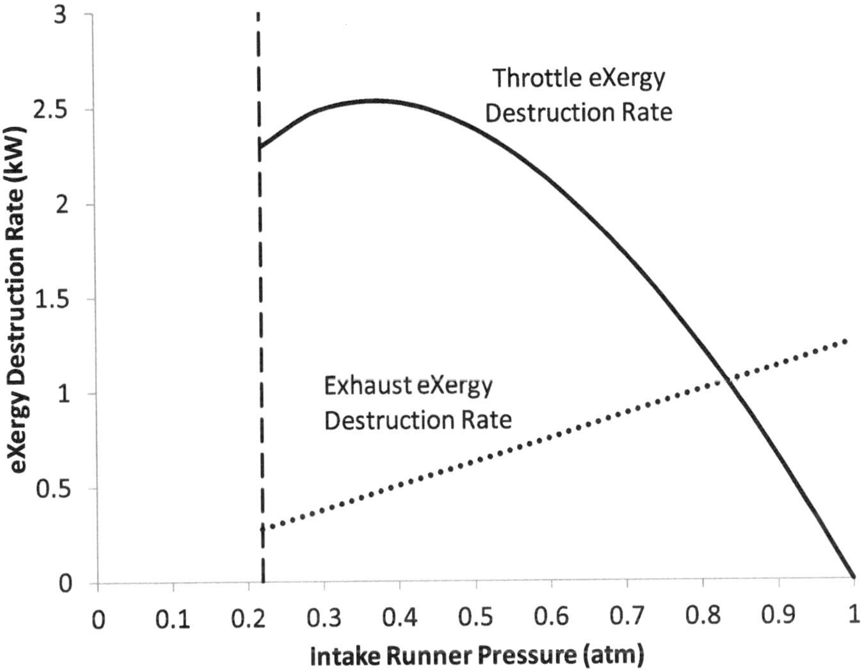

Fig. 7.24 eXergy destruction rates associated with intake and exhaust

is necessary. The 1st Law efficiency is the ratio of the mechanical output to the thermal input and is shown in Fig. 7.25 to be approximately 0.57 for an atmospheric pressure intake. This value should be close to, but not equal to, the simple closed-form Otto cycle result ($\eta_{\mathrm{I,\ OTTO}} = 1 - r^{-(k\ -\ 1)}$). The ideal Otto cycle efficiency is 0.601 for a compression ratio of 10 and ratio of specific heats of 1.4. The difference is due to the elevated exhaust pressure in this model. The 1st Law efficiency decreases with throttling, reflected in both higher exit temperatures and greater eXergy destruction associated with the throttle (although that output peaks at an intake pressure of approximately 0.4 atm).

The maximum theoretical efficiency (based on the high and low temperatures) in this case is 0.929, approaching unity because the temperature at the end of combustion is so high, and is independent of intake pressure, as the change in temperature due to compression and combustion are independent of pressure. The 2nd Law efficiency is the ratio of the 1st Law efficiency to the maximum possible efficiency, and is approximately 60% at wide open throttle and follows the 1st Law efficiency with throttling.

For all cases, there is considerable thermal energy leaving with the exhaust, but it cannot be used directly to produce mechanical power as it is at a nominal atmospheric pressure. The only reasonable addition that could be made is to add a turbine to capture a portion of the available mechanical energy (pressure) from the blowdown process (Workshop 7.2). For mobile applications, there is not much that can be

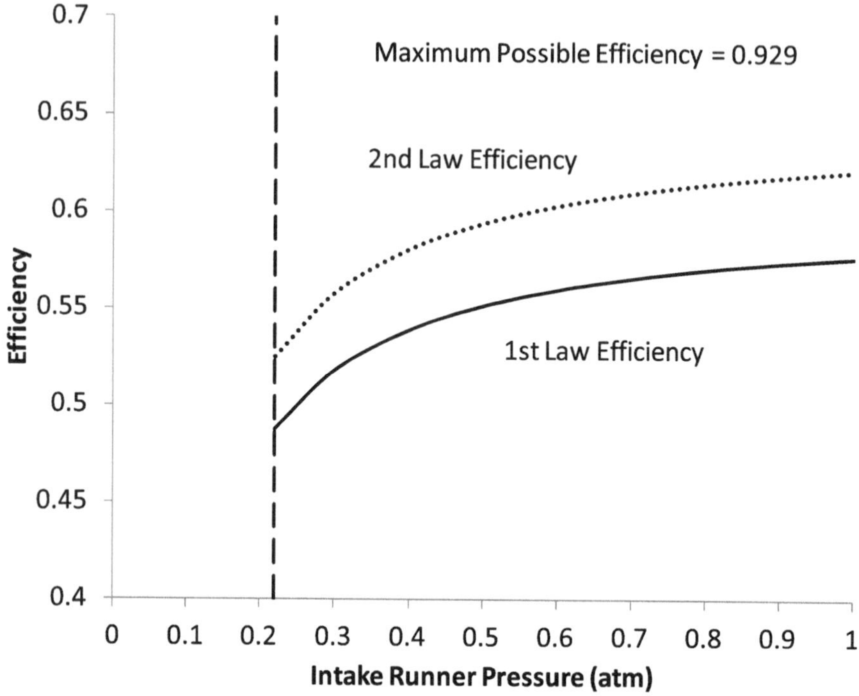

Fig. 7.25 Efficiencies as a function of intake pressure

done beyond that practically or economically to recover that energy, and it is simply vented to the atmosphere. For stationary applications, however, the exhaust can be fed to a heat exchanger to be used as either a heating source (in a cogeneration application) or to generate high-pressure steam to drive a steam turbine (a combined-cycle application, Chap. 11).

In practice, throttling is a simple and effective means to change the output power of a reciprocating engine. However, it does so at a penalty in thermal efficiency. Since, in practice, the vast majority of the time driving a car is spent in a throttled position, there is an incentive for automobile manufacturers to develop alternate means of varying the output power, such as delaying the intake valve closing or cutting off fuel to one or more cylinders. Any such scheme is more difficult and costly to implement than throttling.

Effect of Compression Ratio

When you pull up to a gas station to fill your car's gas tank, you are confronted by a choice of regular, supreme, or ultra (or some such descriptors), which have a listed "octane rating" of 87, 89, or 91, respectively (based on the $(R + M)/2$ method). This

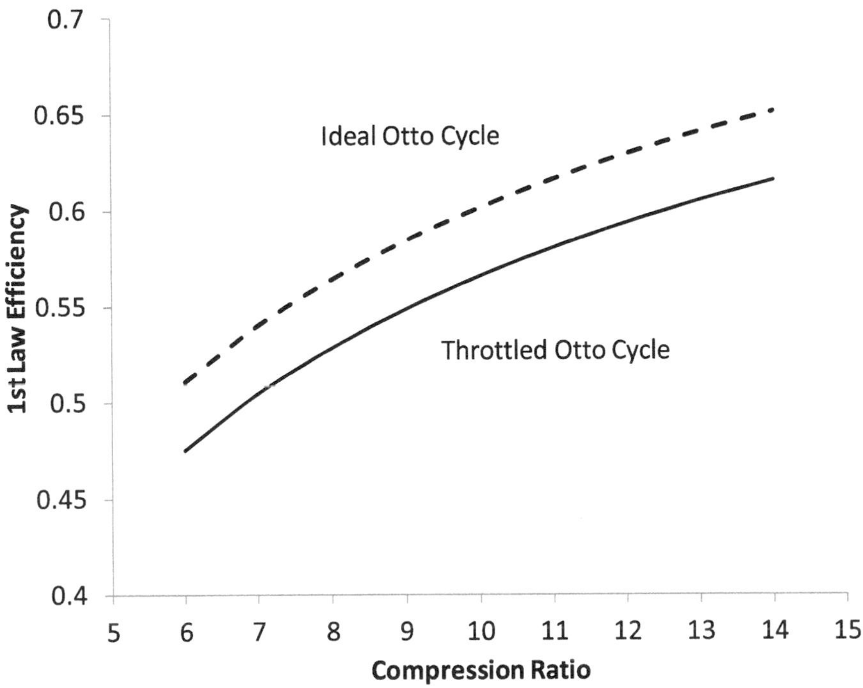

Fig. 7.26 Effect of compression ratio on First Law efficiency for the ideal Otto cycle and the base case throttled cycle

difference is primarily related to the compression ratio of the engine, discussed in Appendix 2.

The effect of compression ratio on the 1st Law efficiency is shown in Fig. 7.26. The base case throttled engine follows the trend of the ideal (unthrottled) engine at a lower efficiency. The energy effects are shown in Fig. 7.27. The fuel input is not affected (the same mass of fuel/air mixture is inducted), but the net power increases and exiting enthalpy decreases with compression ratio.

There is thermodynamic incentive, then, to design automobile engines with the highest possible compression ratio. In practice, however, the compression ratio of a spark ignition (SI) engine is limited by the chemical nature of the fuel/air mixture. At nominal room temperature, the fuel/air mixture will not react chemically without an ignition source. At elevated temperature, spontaneous combustion or autoignition (explosive chemical reaction without an ignition source) will occur. During the compression stroke, the temperature and pressure of the fuel/air mixture increases, and if it is too high, some form of spontaneous combustion will occur. For automotive fuels, the limiting compression ratio is approximately 10. Knocking and the octane rating is discussed a bit more in the deeper dive of Appendix 2.

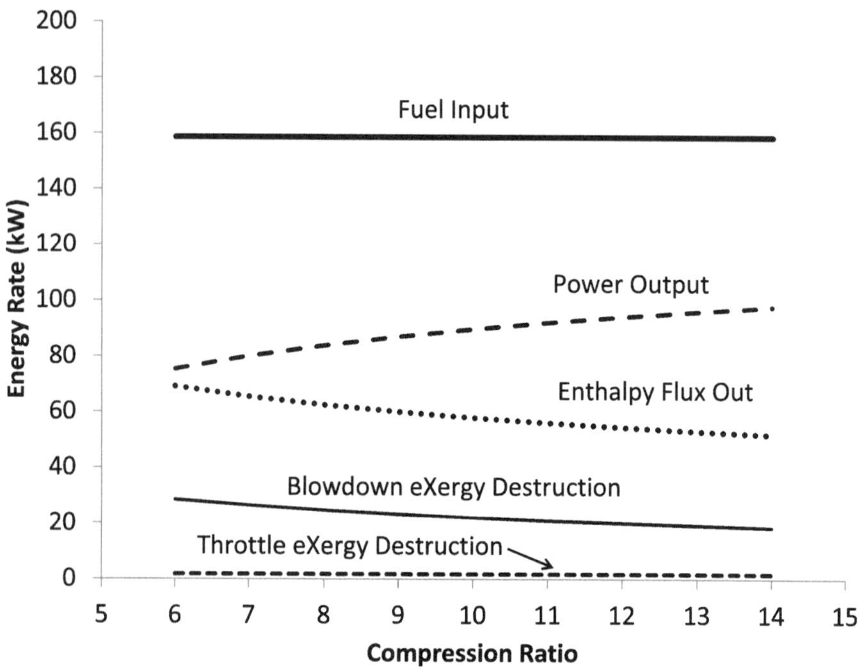

Fig. 7.27 Effect of compression ratio on energy rates for the base case throttled Otto cycle

Effect of Specific Heat Value

Almost all processes of the cycle depend on the value chosen for the specific heat. Traditionally, the cold air approximation is made, for which the value of the specific heat is that of air at room temperature. The base case results reported thus far correspond to that choice. Improved accuracy can be obtained by choosing a value that corresponds to an average temperature in the cycle. Therefore, the results of key performance measures are shown in Table 7.2 for the base case ($c_v = 0.718$ kJ/kg/K) and a value of specific heat evaluated at 2000 K ($c_v = 0.963$ kJ/kg/K, a 34.1% increase). The percentage change in each measure is calculated, as is the sensitivity to specific heat, defined as the ratio of the percentage change normalized by the percentage change of the specific heat. The chemical energy input depends only on the mass of fuel/air mixture inducted and is independent of specific heat. The work of the cycle and therefore power output and First Law efficiency are shown to decrease by 18.6%, with a sensitivity near 0.5. The decrease in Otto cycle efficiency is nearly the same. The 25% decrease in peak temperature is notable from the point of view of the more advanced combustion concepts. The higher specific heat values at higher temperatures result in significantly lower flame temperature, but still sufficiently high that chemical dissociation and heat loss to the walls would be significant in a real engine.

Table 7.2 Effect of the specific heat value chosen on key cycle performance measures

	Cv@300K	Cv@2000K	%change	Sensitivity
Cv, specific heat (kJ/kg/K)	0.718	0.963	34.1%	1.000
Chemical energy input (kW)	158.6	158.6	0.0%	0.000
Power output (kW)	89.7	73.0	−18.6%	0.545
Thermal efficiency	0.565	0.460	−18.6%	0.545
Ideal Otto cycle efficiency	0.601	0.496	−17.5%	0.512
T3 (K)	4633	3492	−24.6%	0.722

Note: The sensitivity is defined as the fractional change in a quantity normalized by the fractional change in specific heat value

Use of the same specific heat value for all processes (calorically perfect gas model) makes a simple spreadsheet solution possible. All output measures adjust by simply changing one input parameter. Different values of specific heat could be used for each process, increasing the complexity of the model somewhat. On the other hand, a thermally perfect gas model would require a numerical solution, and at that point, it would be advisable to relax many other assumptions, of most importance would be to model heat transfer from the hot gases to the cylinder walls.

Variations with Crank Angle

A series of plots of various quantities versus crank angle are shown in Figs. 7.28, 7.29, 7.30, 7.31, 7.32, 7.33, 7.34, and 7.35. The volume is shown on each graph, plotted as a red curve on a secondary *y* axis. The pressure and temperature are shown in Figs. 7.28 and 7.29. Their shapes are similar with the following observation of each process:

- 1–2: Compression stroke: With both valves closed, the pressure and temperature gradually rise during compression, but experience most of the change late in the stroke.
- 2–3: Combustion: A spark at TDC initiates a very large and sudden rise due to internal combustion.
- 3–4: Power stroke: A gradual fall during the power stroke (with most of the change early in the stroke).
- 4–5: Blowdown: The exhaust valve opens and the in-cylinder pressure falls suddenly to the exhaust runner pressure. The temperature falls according to the adiabatic expansion of the remaining gases. The temperature after blowdown will be the temperature of the exhaust gases.
- 5–6: Exhaust stroke: As the piston moves upward, mass flows through the exhaust valve and the pressure remains at the exhaust runner pressure. The work done on the gases by the piston is equal and opposite to the flow work of the exiting gases, so that the temperature remains constant.

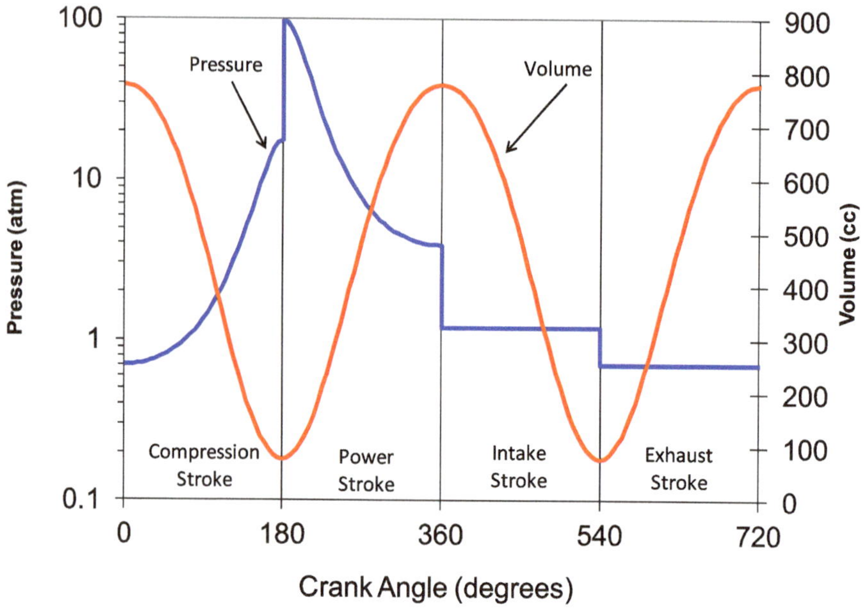

Fig. 7.28 Pressure versus crank angle

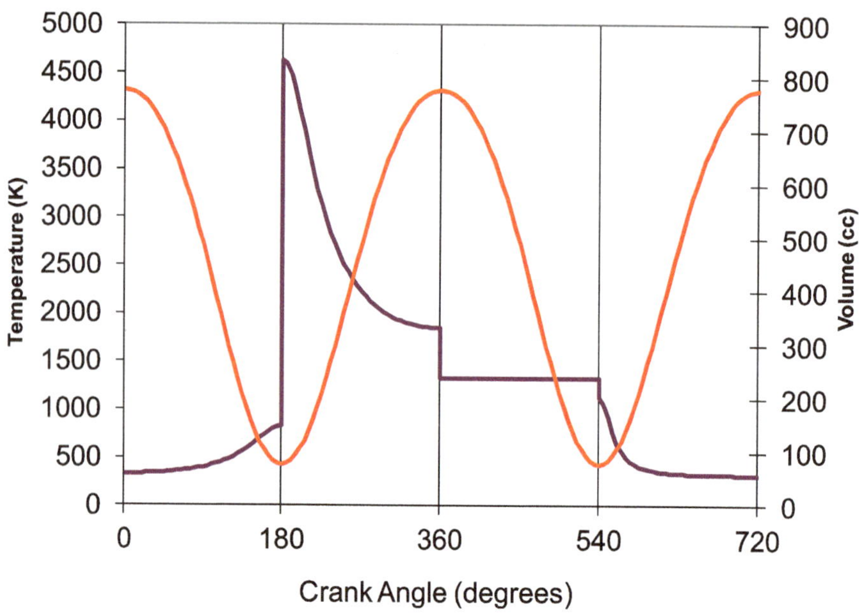

Fig. 7.29 Temperature versus crank angle

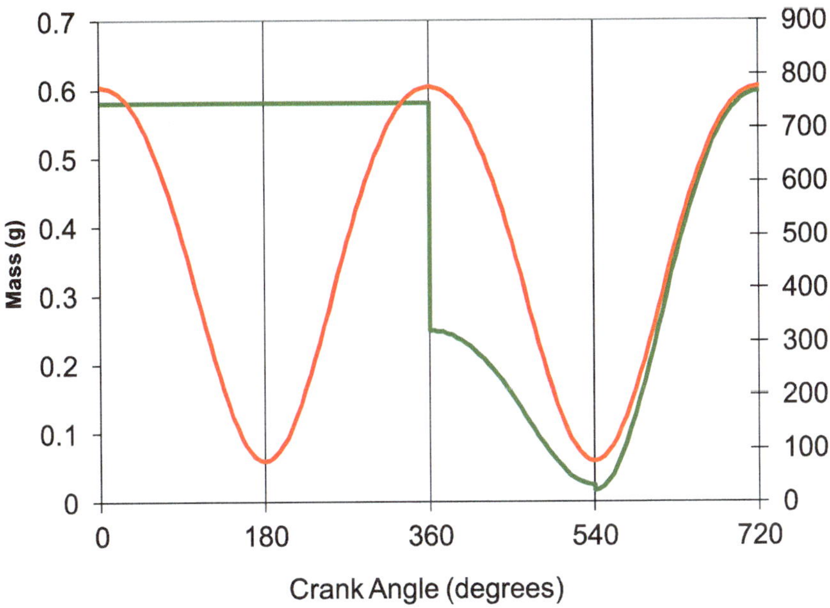

Fig. 7.30 Cylinder mass versus crank angle

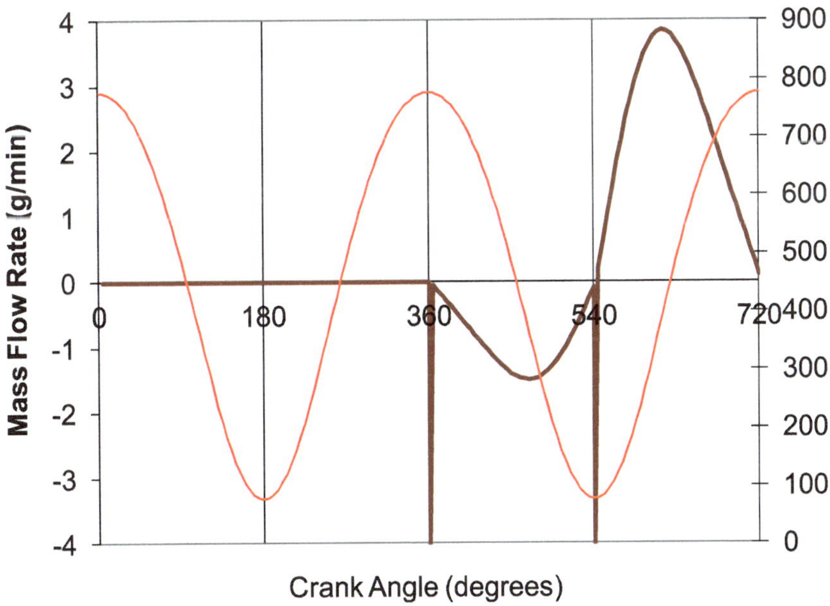

Fig. 7.31 Instantaneous mass flow rate versus crank angle for an engine speed of 2000 RPM

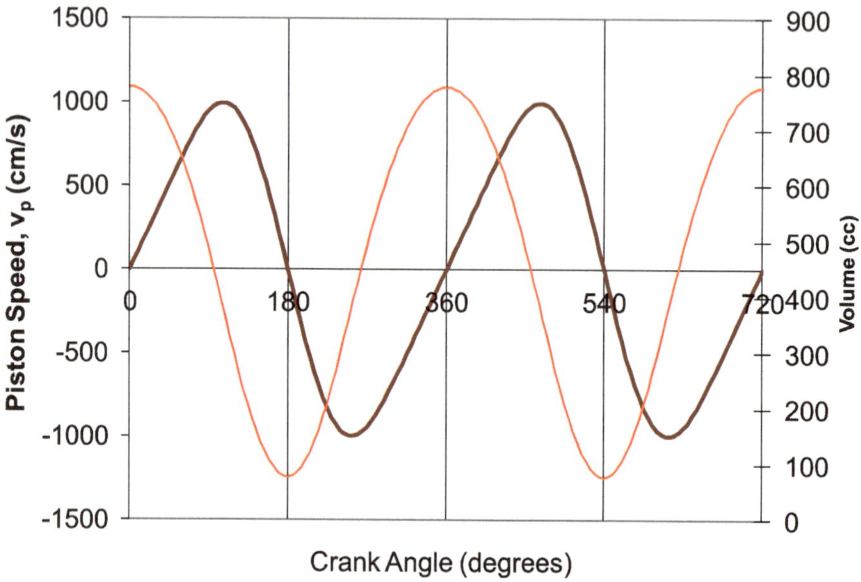

Fig. 7.32 Piston speed versus crank angle for an engine speed of 2000 RPM

- 6–7: Valve overlap: The exhaust valve closes and the intake valve opens. There is backflow into the intake manifold runner, and the pressure in the cylinder falls to the intake runner pressure. The temperature falls according to the adiabatic expansion of the residual gases.
- 7–1: Intake stroke: With the intake valve open, the piston moves downward, and fresh fuel and air flow through the valve, maintaining a constant pressure. The fresh gases mix with the residual gases, and as a result, the in-cylinder temperature decreases during the stroke, with most of the changes occurring early. The final temperature is warmer than ambient.

The in-cylinder mass and the instantaneous mass flow rate are shown in Figs. 7.30 and 7.31 for the base case as a function of crank angle. The piston speed is shown in Fig. 7.32. The mass in the cylinder during the compression and power stroke is constant (and the mass flow rate is zero). At the end of the power stroke, the exhaust valve opens, and more than half of the mass exits instantaneously (in the ideal model). The mass flow rate is a Dirac-delta function at that time. The cylinder mass decreases (mass flow rate is negative) during the exhaust stroke and peaks (in absolute value) approximately half way through the stroke. The instantaneous mass flow rate is directly related to the piston speed. The higher the piston speed, the greater the mass flow rate needed to maintain the pressure. In a real engine, there would be a pressure drop across the valve that would closely follow this mass flow rate curve. Note that the piston speed is not a pure sinusoid, owing to the nature of the conversion of the sinusoidal rotary motion into the linear motion of the piston through the crank arm and connecting rod mechanism.

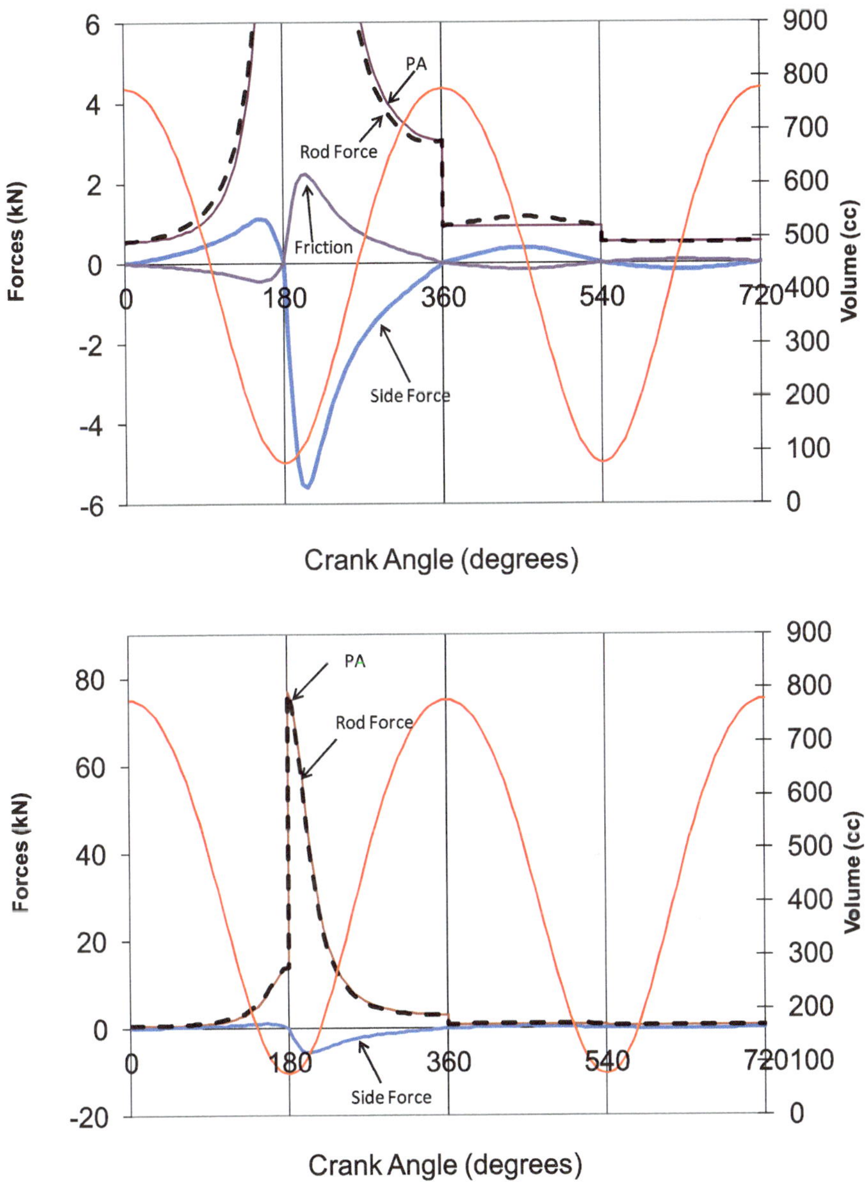

Fig. 7.33 Forces acting on the piston versus crank angle for the base case. Top graph is a at a smaller *y* scale

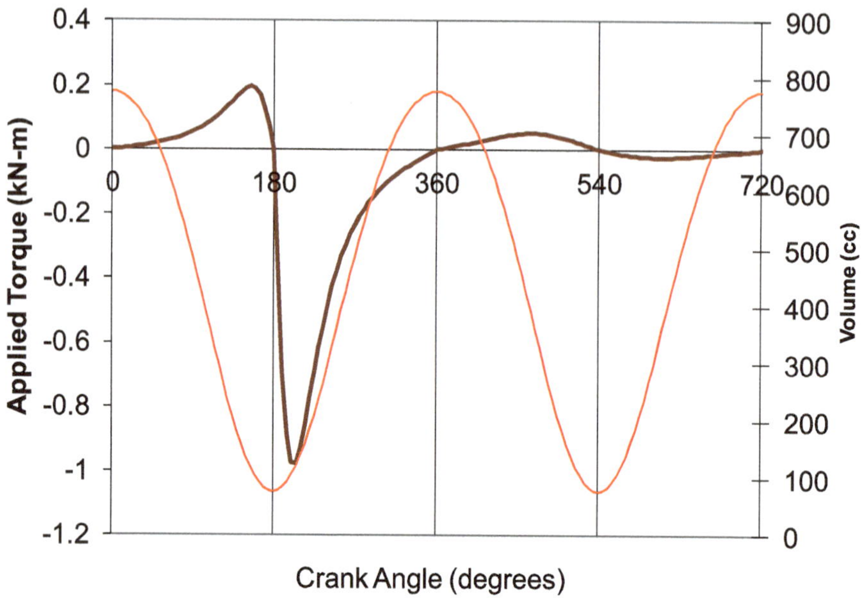

Fig. 7.34 Applied torque versus crank angle. A negative torque represents torque delivered to a mechanical load on the crank shaft

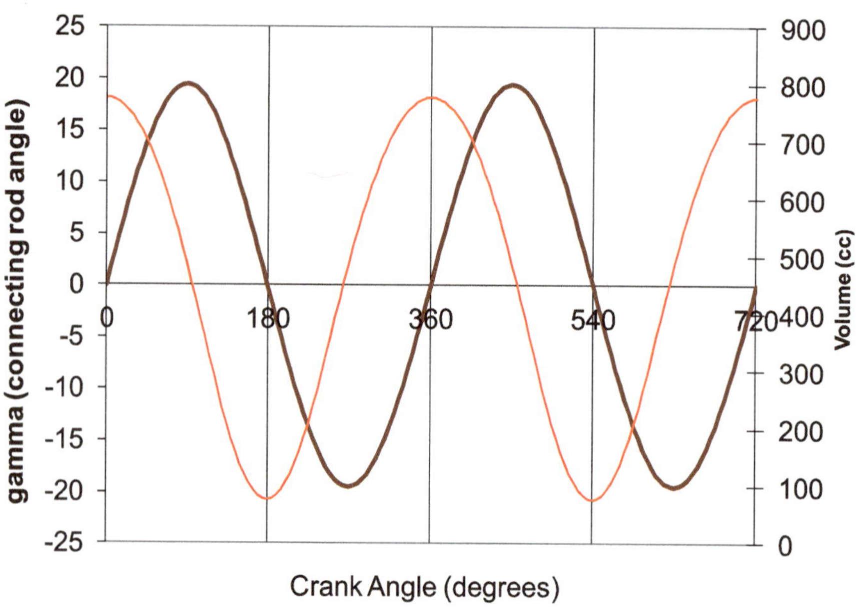

Fig. 7.35 Angle between the connecting rod and vertical versus crank angle

The valve overlap and intake stroke processes are similar to blow down and power, except that there is backflow during overlap (a negative flow rate since mass exits the cylinder) and a positive mass flow rate during intake. The total mass inducted during the intake stroke precisely equals the sum of the mass that exits during the other three gas exchange processes. The peak flow rate during intake is higher than the peak during exhaust because the density of the much colder gases is higher. In practice, for engines with a single exhaust valve and single intake valve, the size of the intake valve is larger than the exhaust since approximately half of the exhaust process occurs during blowdown, which is a pressure-driven flow event.

The mechanics of the cycle are demonstrated in Fig. 7.33, which show the forces acting on the piston (on two different y-scales), and Fig. 7.34 which shows the instantaneous applied torque (as opposed to the cycle average torque). The angle between the connecting rod and the vertical is shown in Fig. 7.35. The pressure force and rod force are nearly equal in magnitude, differing by the friction force that acts in the same direction and opposing the motion. This difference would be larger if inertial effects of the piston and rods were included in the mechanics analysis. The side force is essentially proportional to the pressure force, but is influenced heavily by the geometry. The higher the angle of the connecting rod (maximum at the midpoint of a stroke), the higher the component of the rod force that acts in the side direction. For the compression and power strokes, the side force is highest in magnitude near TCD, when the pressures are highest, but rapidly declines as TCD is approached, and the component of the rod force is nearly vertical. The friction force, proportional in magnitude to the side force, changes direction with the direction of the piston. Friction acts downward during compression and exhaust and upward during power and intake. The side force acts in the positive direction when the crank arm is on the right side and reverses direction when the crank arm moves to the other side. A piston ring and cylinder wall will exhibit greater wear on the side opposite that of the power stroke.

The torque is influenced by both the force in the connecting rod and the moment arm of that force about the central axis of the crankshaft. The shape of the curve therefore follows that of the side force closely. A positive torque on the crankshaft reflects negative work done by the gases in the cylinder.

The design of piston/cylinder arrangements that meet the contradictory aims of exhibiting low friction and not leaking past the piston is a considerable challenge to the engine manufacturer, a challenge that has been addressed in part by the action of piston rings. The piston itself has a considerable gap between it and the cylinder wall. Generally, two rings that fit into grooves in the piston are built in to create a seal against gases leaking past and to spread a thin layer of lubricating oil on the cylinder wall.

In practice, part of the noise of a reciprocating engine and challenge of manufacturing an engine is associated with the piston being pushed sideways back and forth and striking the side wall. Also, the piston rings moving back and forth vertically in their grooves would contribute to the audible noise of an engine.

Concluding Remarks

Think about how you drive. How much of the time do you spend with the "pedal to the metal?" In that condition, the engine is unthrottled (WOT, wide open throttle). The vast majority of the time your engine is operating under a throttled condition, even while cruising on a highway on long trips. You only need maximum torque applied when you want the maximum acceleration from your engine. You rarely want that. So, historically, the throttle was an ingenious invention whose primary purpose is to allow a single engine to operate under a wide range of operating output conditions. The main idea is to deliberately introduce a loss mechanism, an eXergy destruction device, to restrict the output. It works great (i.e., is effective), but the trade-off is a decrease in the thermodynamic efficiency of the engine.

There is an analogous history with braking, the other foot pedal in a conventional car with automatic transmission. Friction brakes have proven to be an effective way to slow a vehicle and are, or course, indispensable to its safe functioning. Like throttling, brakes work by introducing a loss mechanism. The kinetic energy of the vehicle (acquired from the engine, of course) is dissipated into waste heat during a braking event. An alternative would be any form of regenerative braking, that is, some way to convert the kinetic energy of the vehicle into another form of stored mechanical or electrical energy during deceleration, and to reclaim that energy during the next acceleration.

The relatively recent advent of hybrid engines addresses both of these deficiencies. A vehicle has two engines (a gasoline-powered engine and an electric motor), and a bank of rechargeable batteries for energy storage. The system can be designed with the gasoline engine operating near a single design point (at wide open throttle) and the varying net load supplied by the electric motor, which experiences less of a performance penalty off its design point than the gasoline engine. The gasoline engine can drive an electric generator to recharge the batteries. In addition, regenerative braking can be implemented in which application of the brakes engages an electric generator that converts a fraction of the kinetic energy of the vehicle into stored electrical energy. The result of this complex system is that there are two energy inputs to the system, fuel and regenerative braking (which is unclaimed in single engine vehicles). While active, the gasoline engine operates almost always at WOT, making hybrids an attractive and rapidly evolving technology.

Appendix 1: Compressible Flow Restriction on Piston Speed

One of the underlying assumptions made in the thermodynamic analysis of reciprocating engines is that the pressure is uniform within the cylinder at all times. In practice, there are two main effects that give rise to nonuniform pressures: the effect of the moving piston and undesired nonhomogeneous explosive combustion (called engine knock). This Appendix considers the first effect.

It might be expected that when the piston moves toward the gas (compression and exhaust strokes) the pressure on the piston face is somewhat higher than the average in cylinder pressure. When the piston moves away (power and intake), the piston face experiences a lower pressure than the cylinder average. These nonideal effects would increase the work required for gas exchange (pumping work) and reduce the work produced by the compression/expansion strokes.

Quantifying the effect requires application of compressible flow theory, in which dynamics of flows are influenced by temporal and/or spatial density variations. The physical mechanism by which the pressure tends toward uniformity is the speed at which pressure/density disturbances propagate, or the speed of sound, a thermodynamic property given by the relationship $c = \sqrt{kRT}$, not derived here. The Mach number of the piston is the ratio of the piston speed to the speed of sound in the gases. Therefore, a criterion for maintaining a uniform pressure is that the Mach number of the piston must be much less than some critical value (0.3 can be shown to be a reasonable criterion). If this criterion is met, any pressure disturbance is able to maintain a uniform pressure within the cylinder as the piston advances or withdraws. The pressure changes with time, but is the same at all locations.

To quantify the criterion, recall the expression for the piston velocity

$$v_{\text{piston}} = \frac{4V_{\text{D}}N}{B^2}[\sin\theta + \tan\gamma\cos\theta]$$

The instantaneous Mach number of the piston is as follows:

$$\text{Mach}_{\text{piston}} = \frac{|v_{\text{piston}}|}{c} = \frac{4V_{\text{D}}N|(\sin\theta + \tan\gamma\cos\theta)|}{B^2\sqrt{kRT}}$$

The Mach number is plotted as a function of crank angle for the base case in this chapter, with a very high engine speed of 10,000 RPM in Fig. 7.36. This result shows that the piston Mach number is highest during intake (because the temperature, hence speed of sound, is lowest), but that its maximum value midway during the stroke is approximately 0.13, which easily meets the criterion for uniform pressure.

This base engine is a typical automobile engine, and this analysis demonstrates that pressure uniformity during the entire four-stroke cycle is reasonable for engines of this scale. However, for larger engines, it might be expected that the piston speeds are higher. A more general scale-dependent restriction can be met by rearranging the Mach number expression for the engine speed at its maximum (angle term set to unity) and with the temperature set at ambient (approximating its intake stroke value):

$$N_{\text{crit}} < \text{Mach}_{\text{crit}}\frac{B^2\sqrt{kRT_\infty}}{4V_{\text{D}}}$$

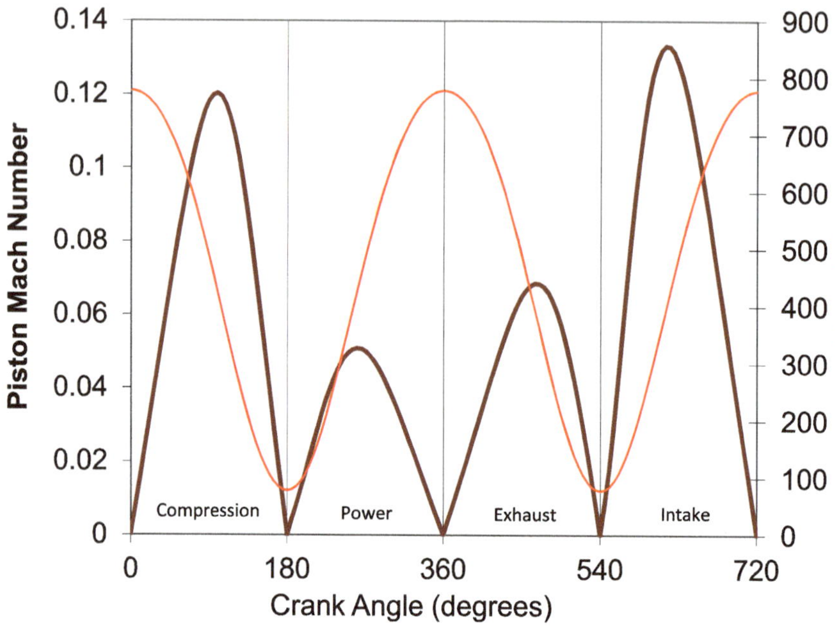

Fig. 7.36 Piston Mach number versus crank angle for the default engine parameters, with an engine speed of 10,000 RPM

This expression contains the geometric size parameters bore and displacement volume. Expressing the geometry in terms of the bore to stroke ratio (*B*/*S*), the critical engine speed above which nonuniform pressures would be experienced is

$$N_{\mathrm{crit}} < \mathrm{Mach}_{\mathrm{crit}}\sqrt{kRT_{\infty}}\left(\frac{(B/s)^2}{4\pi^2 V_{\mathrm{D}}}\right)^{\frac{1}{3}}$$

This function is plotted in Fig. 7.37 for a wide range of engine displacement volumes and bore to stroke ratios of 0.1, 1, and 10. A "square engine" is one for which the bore to stroke ratio is unity. A long narrow cylinder is one for which $B/s = 0.1$. The maximum speed for uniform pressure automobile engines (with displacement volumes of order 1 liter and nearly square) is approximately 20,000 RPM. The largest reciprocating engines (used in container ships) have per-cylinder displacement volumes on the order of 2000 liters, and their maximum engine speeds are approximately 1500 RPM. In practice, other mechanically related criterion (i.e., maximum dynamic stresses) places tighter restrictions on the engine speed.

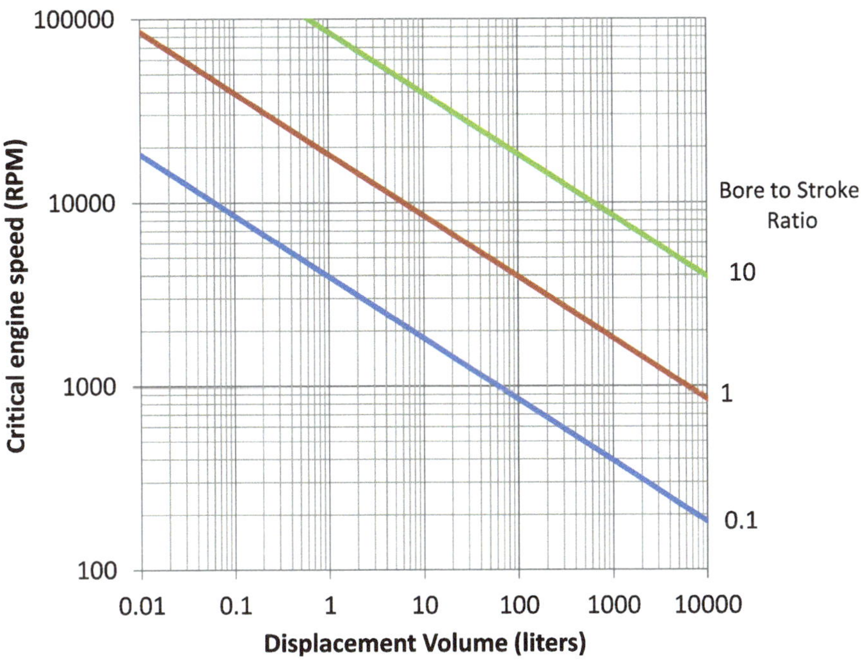

Fig. 7.37 Critical reciprocating engine speed required for uniform in-cylinder pressure as a function of displacement volume for three bore to stroke ratios

Appendix 2: Combustion Processes

Traditionally, the combustion process in reciprocating engines is treated as an "equivalent heat addition" to a working fluid. However, the actual process does not involve heat, the transfer of energy from a high to low temperature. Rather, it is an internal combustion process that involves a chemical reaction and the associated conversion of one form of internal energy to another, from chemical energy to thermal energy. Therefore, the working fluid is not strictly a pure substance. However, the properties of the working fluid are closely approximated by that of nitrogen (N_2), the dominant chemical species in air-breathing combustion devices. Table 7.3 lists several chemical properties of interest for several representative fuels. Brief derivations and their explanations are given in this Appendix, with an eye toward their application to engineering systems that involve combustion processes, such as reciprocating engines considered in this chapter.

Table 7.3 Chemical properties of typical fuels used in engineering systems

Fuel	Primary chemical species	Chemical formula	Stoichiometric AFR_{mass}	LHV (MJ/kg_{FUEL})	Δu_{comb} $(MJ/kg_{MIX})/\phi$	Density at room conditions (kg/m^3)	Volumetric energy density (MJ/liter)	kg_{CO2}/GJ
Hydrogen	Hydrogen	H_2	34.4	120.9	3.42	0.082	0.010	0
"Coal"	Carbon	CH	13.2	35.2	2.86	800	28.2	85.5
"Natural Gas"	Methane	CH_4	17.2	50.1	2.76	0.654	0.033	54.8
"Gasoline"	Octane	C_8H_{18}	15.1	42.4	2.73	750	31.8	74.7
"Diesel/Jet Fuel"	Dodecane	$C_{12}H_{26}$	14.7	41.6	2.69	800	33.3	76.6
"Wood"	Cellulose	$C_6H_{10}O_5$	4.9	16.1	2.64	380	6.1	101.3

Note: Fuels are listed in parentheses because actual fuels are generally complex and not well-defined mixtures of many chemical components. The chemical formula given is an approximate equivalent pure species to represent the fuel

Mass Considerations: Combustion Stoichiometry

The chemical reaction for complete combustion of a lean mixture of 1 mol of an oxygenated hydrocarbon (with chemical formula $C_nH_mO_p$) in air is as follows:

$$C_nH_mO_p + \left(\frac{i_{st}}{\phi}\right)(O_2 + 3.76N_2) \overset{\phi<1}{\rightarrow} nCO_2 + \frac{m}{2}H_2O + i_{st}\left(\frac{1}{\phi} - 1\right)O_2 + \left(\frac{3.76 i_{st}}{\phi}\right)N_2$$

where the stoichiometric coefficient (i_{st}) represents the moles of O_2 in the air required to completely burn the carbon in the fuel to CO_2 and the hydrogen to H_2O and is given by

$$i_{st} = n + \frac{m}{4} - \frac{p}{2}$$

The equivalence ratio, ϕ, is the actual fuel to air ratio of the mixture to the stoichiometric fuel to air ratio. If ϕ <1, there is excess air and not all of the O_2 can possibly react chemically. If ϕ >1, the mixture is rich and there is not sufficient O_2 to burn all the fuel. The rich case is not considered in this introductory treatment of combustion.

The air to fuel mass ratio is calculated (using the molecular weights of air and fuel) as

$$\text{AFR}_{\text{mass}} = \frac{\text{Moles of air (mass of air/moles of air)}}{\text{Moles of fuel (mass of fuel/moles of fuel)}} = \frac{4.76 i_{st}/\phi \, \text{MW}_{\text{air}}}{\text{MW}_{\text{fuel}}}$$

For a stoichiometric mixture:

$$\text{AFR}_{\text{mass,stoich}} = \frac{4.76\,(n + m/4 - p/2)\,(28.9)}{(12n + m + 16p)}$$

The results of this calculation, shown in Table 7.3, make it clear that the mass of air is generally an order of magnitude greater than the mass of fuel, approximately 15 for almost all hydrocarbon fuels. For that reason, the mass of the working fluid is dominated by air, and the air is dominated by N_2. The added fuel is considered to contribute chemical energy with little mass. It is reasonable therefore to model the properties of the gas as that of air as a first (and maybe last) approximation in thermodynamic analysis of engineering systems.

Chemical Energy

A bomb calorimeter is a device designed to measure the conversion of chemical energy available in fuels into thermal energy. It consists of a constant volume vessel (capable of withstanding high pressures) into which fuel and air (or pure O_2) are added, ignited, and then cooled (typically by a water bath). A schematic of the process is shown in Fig. 7.38. The calorimeter is initially charged with a lean fuel/air mixture. A spark or other ignition source initiates combustion that generally takes place quickly (so there is little time for heat transfer during the initial process), resulting in an intermediate state of high temperature (called the adiabatic flame temperature). Since the volume is constant, the pressure also rises dramatically according to the ideal gas law. The hot product gases are subsequently cooled (the heat transfer is negative in sign for an exothermic chemical reaction) until the product gases return to room temperature. The pressure will differ slightly from ambient due to the change in moles associated with the chemical reaction (Δn) from the original number of moles (n, which is generally much larger than Δn). For a 1 kg initial fuel charge, the heat transferred is equal to the negative of the lower heating value (LHV) of the fuel, which assumes the water vapor in the products is not condensed. The higher heating value (HHV) would be measured if all the water in the products were condensed (which does not generally occur during actual combustion processes). The LHV of fuels is approximately 40–50 MJ/kg$_{\text{fuel}}$ for pure hydrocarbon fuels (Table 7.3). It is very high for pure H2 (due largely to its low molecular weight) and low for wood (due to the oxygen in the fuel).

The internal energy consists of thermal energy and chemical energy. An energy balance for a closed system (with negligible kinetic and potential energy) can therefore be written as

$$\Delta U_{\text{thermal}} + \Delta U_{\text{chemical}} = Q - W$$

For a calorically perfect gas mixture of total mass m with a specific heat c_v,

$$\Delta U_{\text{thermal}} = mc_v \Delta T$$

and the temperature after a chemical reaction is

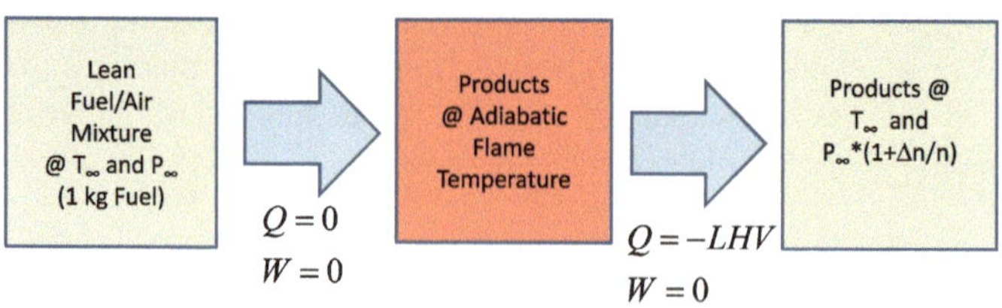

Fig. 7.38 Schematic of bomb calorimeter, scaled to measure lower heating value (LHV)

$$T_2 = T_1 + \frac{(-\Delta U_{\text{chemical}} + Q) - W}{mc_v}$$

The chemical energy change is grouped with the heat transfer to simply indicate how it is generally treated as being "equivalent" to a heat addition process, at least from a 1st Law perspective.

For an adiabatic, constant pressure process, the work of expansion can be grouped with the thermal internal energy as follows:

$$\Delta U_{\text{thermal}} - P\Delta V = \Delta(U_{\text{thermal}} - PV) = \Delta H_{\text{thermal}} = mc_p\Delta T$$

If heat transfer is negligible ($Q = 0$), the final temperature is called the adiabatic flame temperature, an important metric in combustion analysis. The final temperature of an adiabatic, constant volume process is as follows:

$$T_2 = T_1 + \frac{(-\Delta U_{\text{chemical}})}{mc_v} \quad \text{constant volume, adiabatic process}$$

The final temperature of an adiabatic constant pressure process is

$$T_2 = T_1 + \frac{(-\Delta U_{\text{chemical}})}{mc_p} \quad \text{constant pressure, adiabatic process}$$

Since the enthalpy specific heat is larger than the internal energy specific heat ($c_p = c_v + R$), the adiabatic flame temperature of a constant pressure process is lower than that of a constant volume process due to the work of expansion done by the gases required to maintain pressure constant.

The change in chemical energy (related to the heating value) is given by the difference in the inherent energy of the products and reactants.

$$\Delta U_{\text{chemical}} = \sum_{j,\text{products}} n_j \Delta u^0_{f,j} - \sum_{i,\text{reactant}} n_i \Delta u^0_{f,i}$$

Where each chemical species has an inherent chemical energy, called the "heat" of formation given by Δu^0_f. A better name would be internal energy of formation and is closely related to the enthalpy of formation Δh^0_f, which is typically listed in published data sets (such as the JANAF tables). If the chemical composition of a fuel is well defined, then the chemical energy change (and LHV or HHV) can be calculated from this relationship and from the stoichiometric chemical reaction (which defines the moles of fuel, components of air and products).

The chemical energy change divided by the total mass is an unconventional measure of the chemical energy, but one that is most relevant to thermodynamic analysis of combustion systems. It can be defined as follows:

$$\frac{(-\Delta U_{\text{chemical}})_{\text{stoich}}}{m} = \Delta u_{\text{comb}} = \frac{\text{LHV}}{1+\text{AFR}} [=] \frac{\text{J}}{\text{kg}_{\text{mix}}}$$

It is listed in Table 7.3 and is remarkably similar for hydrocarbons and approximately equal to 2.7 MJ/kg_{mix} and a bit higher for H2. In the table, it is listed as being divided by the equivalence ratio, ϕ, as it represents a maximum possible per-unit total mass chemical energy release of a stoichiometric mixture. Actual combustion processes can be overall lean (for diesel engines, gas turbines, boilers) and the chemical energy release can be adjusted accordingly.

The heating value is a mass-based chemical energy density of a fuel. However, an important practical consideration of a fuel is the space (volume) it occupies. Therefore, a volume-based energy density is calculated (simply the heating value multiplied by the density of the fuel) in Table 7.3, which shows that gaseous fuels have a low volume energy density, at least when stored at atmospheric pressure. Gaseous fuels would be pressurized, which would increase their volumetric energy density, but the increased mass of the storage vessel should be factored into their mass-based energy density.

A final metric associated with fuels is related to the controversial social issue of global warming due to the emission of greenhouse gases. The combustion of hydrocarbon fuels produces carbon dioxide (CO_2), which is not considered to cause direct health effects, but its gradual build-up in the atmosphere is cause for concern as a gas that traps radiant energy that would otherwise be emitted to outer space. A good metric for comparing the contribution of greenhouse gases of various fuels is the mass of CO_2 released normalized by the chemical energy release (Table 7.3). Of the fossil fuels, coal is approximately 50% higher than natural gas and oil-based fuels are in between. Interestingly, wood has the highest value (due to its low LHV), but wood is not a fossil fuel and trees absorb CO_2 from the atmosphere, making no net contribution to greenhouse gas emissions if the total biomass remains constant.

Knocking in Spark Ignition Engines

Gasoline is a complex blend of a mixture of hydrocarbons (carbon numbers ranging between 4 and 12, roughly) generally obtained from the distillation of crude oil. It is a liquid fuel, which has the advantage over a gaseous fuel in that it can be stored in an ambient pressure vessel. It also has a higher chemical energy volume density (heating value per unit volume). A liquid fuel has the advantage over a solid fuel in that it can be pumped, and it is volatile. The fuel is injected as a liquid, but evaporated prior to induction into the cylinder.

The octane rating refers to an industry standardized method for determining the tendency of a fuel/air mixture to knock or autoignite (react chemically without a spark) relative to two standard fuels: n-heptane (C_7H_{16}) and an isomer of octane

(2-2-4- trimethyl pentane, which has the chemical formula C_8H_{18}). Heptane has somewhat weaker bonds than iso-octane and autoignites at a lower compression ratio. The critical compression ratio above which the engine knocks is measured for both fuels in a laboratory engine under one of two standard protocols (research method or motor method). A scale is established with 0 being the compression ratio for n-heptane and 100 for the iso-octane. Then, the fuel whose octane rating is sought is tested for its critical compression ratio. The octane rating is determined by its location on the scale. The $(R + M)/2$ method refers to the test done on two different engine types (research and motor) with different operating conditions and the results averaged.

The octane rating of natural gas (primarily methane in composition) is 120. That means that automobiles designed to run on natural gas can be designed with significantly higher compression ratios than those designed for gasoline, and therefore will be more thermodynamically efficient. Also, the greenhouse gas emissions per mass of fuel burned is lower. Natural gas is an attractive automotive fuel for those reasons as a transition fuel to a hydrogen economy.

The octane rating of hydrogen (H2) is greater than 130. Ultimately, it would seem that a complete shift to an energy economy based on H2 is the logical direction for the evolution of man. The mass of the fuel would come from water, and the energy to convert would come from renewable sources (solar, wind, hydroelectric, tidal, etc.).

One of the arguments against H2 is its safety because it is "too explosive." This perception may be largely based on a single historical event, namely, the Hindenburg disaster. However, that disaster was due to the use of H2 as a buoyant medium, not as a fuel.

It would be easy to create an industry standard hydrogen-based fuel blend that is less explosive than pure H2 by simply adding an inert component. While N2 would be abundant, a better thermodynamic choice would be argon, which is plentiful (1 mol% of the Earth's enormous atmosphere) and has a specific heat ratio of 1.667, and that is independent of temperature. A fuel/air mixture of a H2/argon fuel would have a higher k and, therefore, more efficient reciprocating engine. Sounds like a good deeper dive workshop in the making . . . .

Appendix 3: Blowdown Process

Traditionally, the blowdown process is modeled as being due to a heat rejection to the environment, which is nothing of the sort. In practice, it is the first stage of the gas exchange process needed to expel burned gases and induct a fresh fuel/air mixture. As the piston approaches BDC during the power stroke, the pressure is above ambient. When the exhaust valve opens, mass is driven out of the cylinder in an open system venting, with a single exit stream. In the ideal cycle, the process is modeled to take place instantaneously at BDC.

To analyze this open system process, the mass, energy, and entropy balance equations are written in a general differential form with an incremental amount of mass exiting (m_e) as the independent variable:

$$dM_{CV} = -m_e$$

$$dE_{CV} = \delta q - (PdV + m_e P_e v_e) - m_e e_e$$

$$dS_{CV} = \delta q / T_B + \delta \Sigma - m_e s_e$$

The subscript CV on the left side refers to the average values within the control volume (not internal energy specific heat). Note that there are two forms of work: expansion work, which is zero during ideal blowdown, and flow work associated with pushing fluid out. With the model assumptions of no control volume change, negligible KE and PE and no heat transfer to the walls, and expressing the total energy and entropy in terms of their specific (per unit mass) properties, the energy equation reduces to

$$d(M_{CV} u_{CV}) = -m_e (u_e + P_e v_e)$$

Using the chain rule, $d(M_{CV}u_{CV}) = M_{CV}d(u_{CV}) + u_{CV}d(M_{CV})$.

The internal energy within the (u_{CV}) represents the average specific internal energy within the cylinder, while the exiting internal energy (u_e) represents that of the mass as it exits, which in general could be quite different. However, in this case, the conditions within the cylinder are considered to be uniform, and they are equal, that is, $u_{CV} = u_e$. The assumption that the exit conditions are equal to the average conditions within a control volume is called the "well-mixed model" approximation. Along with the mass relation ($dM_{CV} = - m_e$), the energy equation reduces to

$$M_{CV} d(u_{CV}) = -m_e P_{CV} v_{CV}$$

This result shows that as mass exits the cylinder, flow work decreases the specific internal energy stored within the cylinder. Rearranging,

$$\frac{d(u_{CV})}{P_{CV} v_{CV}} = \frac{dM_{CV}}{M_{CV}}$$

Invoking ideal gas relations and using the mass relation:

$$M_{CV} = \frac{P_{CV} V_{CV}}{R T_{CV}}$$

For a differential change:

$$\frac{dM_{CV}}{M_{CV}} = \frac{dP_{CV}}{P_{CV}} + \frac{dV_{CV}}{V_{CV}} - \frac{dT_{CV}}{T_{CV}}$$

Substituting, with the volume constant during blowdown:

$$\frac{c_V}{R}\frac{dT_{CV}}{T_{CV}} = \frac{dP_{CV}}{P_{CV}} - \frac{dT_{CV}}{T_{CV}}$$

Rearranging:

$$\frac{dP_{CV}}{P_{CV}} = \left(\frac{c_V}{R} + 1\right)\frac{dT_{CV}}{T_{CV}}$$

This relation is recognized as that of an adiabatic expansion of an ideal gas. Assuming constant specific heat and integrating from state 4 (the end of the power stroke) until the pressure decreases to the exhaust manifold pressure (state 5):

$$\ln\left(\frac{P_{\text{exhaust}}}{P_4}\right) = \left(\frac{c_V}{R} + 1\right)\ln\left(\frac{T_5}{T_4}\right)$$

The temperature at the end of the blowdown process is

$$T_5 = T_4\left(\frac{P_{\text{exhaust}}}{P_4}\right)^{\frac{R}{c_V+R}}$$

The mass of gas that remains after blowdown can be calculated with the ideal gas law.

With the key assumptions that blowdown takes place instantaneously at BDC (no time for heat transfer or volume change), this analysis demonstrates formally that the gases that remain inside the cylinder after blowdown have undergone an adiabatic expansion.

To demonstrate formally that the adiabatic, reversible process on the cylinder gases is isentropic, consider the entropy balance:

$$dS_{CV} = \delta q/T_B + \delta\Sigma - m_e s_e$$

If the process is considered to be adiabatic and reversible:

$$d(M_{CV}s_{CV}) = M_{CV}ds_{CV} + s_{CV}dM_{CV} = -m_e s_e$$

Using mass ($dM_{CV} = -m_e$) and the well-mixed model ($s_{CV} = s_e$) results in

$$ds_{CV} = 0$$

That is, the specific entropy within the cylinder remains constant. The gases that remain in the cylinder undergo an isentropic expansion process.

A good exercise (left to the reader) is to calculate the theoretical power that could be extracted by placing a turbine between the exit and ambient. The gases that exit during blowdown experience eXergy destruction as they are throttled across the valve (an isenthalpic process, with entropy generation). Note that the gases left behind in the cylinder undergo an isentropic process, while the gases that exit undergo an isenthalpic process. If a reversible turbine were placed there, the gases would expand isentropically as well as they extract useful power. The simpler way to calculate the total work extracted is to analyze a control mass.

Appendix 4: Exhaust Stroke

This Appendix formally shows the conditions for which the temperature remains constant during the exhaust and intake strokes (if the pressure drop across the valves is neglected). That is, the piston work is equal and opposite to the flow work.

With the control volume taken to be the in-cylinder gases, the differential mass, energy, and entropy balances are

$$dM_{\mathrm{CV}} = -m_{\mathrm{e}}$$

$$dE_{\mathrm{CV}} = \delta q - (PdV + m_{\mathrm{e}}P_{\mathrm{e}}v_{\mathrm{e}}) - m_{\mathrm{e}}e_{\mathrm{e}}$$

$$dS_{\mathrm{CV}} = \delta q/T_{\mathrm{B}} + \delta\Sigma - m_{\mathrm{e}}s_{\mathrm{e}}$$

Assuming adiabatic, and neglecting kinetic and potential energy changes, the energy equation becomes

$$d(M_{\mathrm{CV}}u_{\mathrm{CV}}) = M_{\mathrm{CV}}d(u_{\mathrm{CV}}) + u_{\mathrm{CV}}d(M_{\mathrm{CV}}) = -(PdV) - m_{\mathrm{e}}(P_{\mathrm{e}}v_{\mathrm{e}}) - m_{\mathrm{e}}u_{\mathrm{e}}$$

Inserting the mass balance and rearranging:

$$M_{\mathrm{CV}}d(u_{\mathrm{CV}}) = -(PdV) - m_{\mathrm{e}}(P_{\mathrm{e}}v_{\mathrm{e}}) + m_{\mathrm{e}}(u_{\mathrm{CV}} - u_{\mathrm{e}})$$

The first term on the right side represents the work on the piston on the gases (positive since the volume decreases). The second term represents the flow work of the gases as they exit the cylinder. The third term represents a difference between the average properties in the cylinder and the gases as they exit. This term is zero for the well-mixed model approximation.

The entropy balance is as follows:

$$d(M_{\mathrm{CV}}s_{\mathrm{CV}}) = M_{\mathrm{CV}}d(s_{\mathrm{CV}}) + s_{\mathrm{CV}}d(M_{\mathrm{CV}}) = \delta\Sigma - m_{\mathrm{e}}s_{\mathrm{e}}$$

$$M_{\mathrm{CV}}d(s_{\mathrm{CV}}) = \delta\Sigma + m_{\mathrm{e}}(s_{\mathrm{CV}} - s_{\mathrm{e}})$$

If the in-cylinder process is assumed to be reversible, and the well-mixed model is invoked ($s_{CV} = s_e$),then the entropy balance becomes as follows:

$$d(s_{CV}) = 0$$

That is, the specific entropy of the in-cylinder gases remains constant. Since the pressure remains equal to the exhaust runner pressure, the temperature must remain constant. Then, from the energy equation, the specific internal energy remains constant and the flow work is equal and opposite to the piston work.

Appendix 5: Second Law-Related Analysis

Pinpointing all sources of eXergy destruction through a thorough 2nd Law analysis of the engine cycle requires the entropy change associated with the combustion process, which is not developed here. Treating the combustion process as an "equivalent heat addition" is reasonable from a 1st Law perspective, but less so from a 2nd Law one, where the entropy transfer requires a thermal reservoir from which heat is drawn. For internal combustion engines, there is no such heat transfer. Actually, in a real engine, there is considerable heat transfer from the gases to the cylinder wall. A more advanced study of chemical reactions than done here can add an inherent eXergy destruction associated with the combustion reaction, similar to an internal energy change.

Nevertheless, the overall 2nd Law efficiency of the cycle is defined as the 1st Law efficiency divided by the maximum possible thermal efficiency, namely, the Carnot efficiency for a heat engine operating between two thermal reservoirs (one at the dead state temperature, T_∞, one at the maximum experienced in the cycle, T_{max}):

$$\eta_{I,MAX} = 1 - \frac{T_\infty}{T_{max}}$$

The 2nd Law efficiency is:

$$\eta_{II} = \frac{\eta_I}{\eta_{I,mas}}$$

eXergy Destruction Associated with Throttling

Most of the processes in the ideal throttled Otto cycle as defined here are reversible, and a 2nd Law analysis of them is not particularly insightful. On the other hand, the pressure drop associated with throttling in the intake system results in eXergy destruction. The steady-state, open system entropy balance, multiplied by the dead state temperature (T_{inf}), is:

$$T_\infty\left[\frac{dS}{dt} = \dot{m}(s_{in} - s_{out}) + \frac{\dot{Q}}{T_b} + \dddot{\Sigma}\right]$$

where T_b is the temperature at the boundary of the control volume where heat is transferred and $\dddot{\Sigma}$ is the entropy generation rate. At steady state and rearranging for the eXergy destruction rate:

$$e\dddot{X}_D = \dddot{\Sigma}\, T_\infty = \dot{m}(s_{out} - s_{in})T_\infty - \dot{Q}\left(\frac{T_\infty}{T_b}\right)$$

For ideal gases, the entropy change is related to temperature and pressure changes by:

$$s_{out} - s_{in} = c_p \ln\left(\frac{T_{out}}{T_{in}}\right) - R\ln\left(\frac{P_{out}}{P_{in}}\right)$$

The intake and exhaust processes are considered to be adiabatic, and an energy balance confirms that they are isenthalpic, and therefore isothermal (for ideal gases):

$$\frac{dE}{dt} = 0 = \dot{Q}_{in} + \dot{m}(h_{in} - h_{out})$$

Therefore, the eXergy destruction rate is given by

$$e\dddot{X}_D = \dddot{\Sigma}\, T_\infty = -\dot{m}RT_\infty \ln\left(\frac{P_{intake}}{P_\infty}\right) \quad \text{for the intake process}$$

If pressure losses occur in the exhaust system, the eXergy destruction rate is:

$$e\dddot{X}_D = \dddot{\Sigma}\, T_\infty = -\dot{m}RT_\infty \ln\left(\frac{P_{i\infty}}{P_{exhaust}}\right) \quad \text{for the exhaust process}$$

Workshop 7.1: Otto and Diesel Cycle Comparison

The spark ignition (SI) engine is considered to be most closely modeled by the Otto cycle for first-order thermodynamic analysis. A smaller, but very large, market exists for the compression ignition (CI) engine, which is modeled by the Diesel cycle. This workshop compares the performance of these two idealized cycles using a value of the compression ratio that is typical of their practical use.

In an SI engine, a fuel/air mixture is inducted during intake, and the combustion process takes place by the propagation of a turbulent premixed flame. The compression ratio of the engine is limited by the chemical nature of the fuel/air mixture. Combustion will initiate prematurely (during the compression stroke) and give rise to violent nonuniform pressure waves if the compression ratio is too high. This complex phenomenon (sometimes called knocking or other descriptive names) is problematic from both a thermal efficiency and mechanical view. Also, the fuel/air ratio must be maintained near a constant value to produce a reliable propagating flame. Therefore, power output is varied by throttling in practice.

In a CI engine, only air is inducted during intake, and fuel is injected directly into the cylinder under high pressure during the power stroke. The temperature after compression is sufficiently high that combustion initiates without a spark and the combustion process occurs in a turbulent non-premixed flame, which has very different nature than the premixed combustion in an SI engine. In the Diesel cycle, a model for this practical engine, the combustion process is treated as being isobaric (which is difficult and not necessary to really accomplish in practice, requiring a very precise and variable rate of fuel injection related to the rate of volume change). The compression ratio is not limited by preignition, but rather by more mechanical restrictions such as high instantaneous torque needed to compress the gases into a small clearance volume (not investigated further here). The gas exchange process is the same as for the SI engine, in that there is blowdown followed by exhaust and intake strokes. However, there is no need for throttling a Diesel engine to vary the power output. Rather, the power output of a CI engine is varied by changing the total amount of fuel injected.

The gas exchange process involves two strokes (exhaust followed by intake), but in the ideal cycle, the cylinder pressure equals ambient and there is no associated net work. Therefore, both ideal cycles consist of four internally reversible processes (in two strokes) in a piston/cylinder arrangement (Figs. 7.39 and 7.40). The only difference is process 2–3, the internal combustion process, which is constant volume (isochoric) in the Otto and constant pressure (isobaric) in the Diesel. They will be compared with the same displacement volume of 1.0 liters, but different

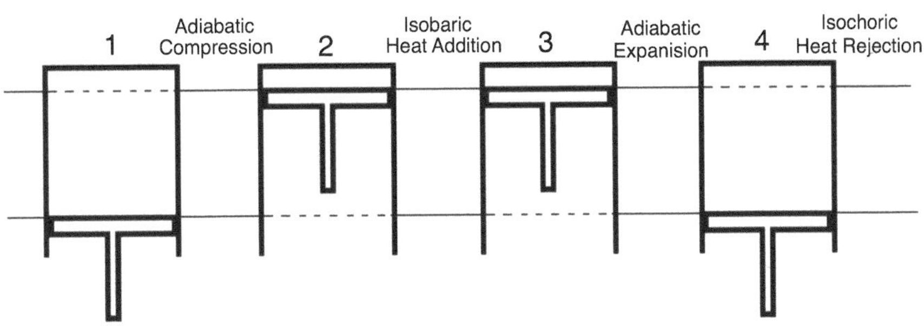

Fig. 7.39 Schematic of ideal Otto cycle

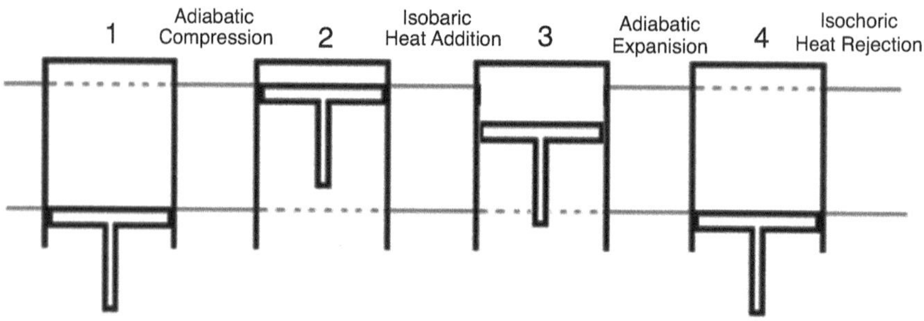

Fig. 7.40 Schematic of ideal diesel cycle

OTTO CYCLE

INPUT PARAMS

Vd	1	liter
Rotto	10	
cv	0.72	kJ/kg/K
R	0.287	kJ/kg/K
Q/M	2730	
Ncyl	8	
Nengine	2000	RPM

STATES	V(liters)	P (atm)	T (K)
1		1	298
2			
3			
4			

DIESEL CYCLE

INPUT PARAMS

Vd	1	liter
r	20	
cv	0.72	kJ/kg/K
R	0.287	kJ/kg/K
Q/M	583	kJ/kg
Ncyl	8	
Nengine	2000	RPM

STATES	V(liters)	P (atm)	T (K)
1		1	298
2			
3			
4			

Fig. 7.41 State tables

compression ratios, the ratio of maximum to minimum volume (10 for Otto, 20 for Diesel, considered to be representative values). Also, the conversion of chemical energy to thermal energy per unit mass during combustion will differ (2730 kJ/kg for Otto and 583 kJ/kg for Diesel, which runs fuel lean). For both cycles, invoke the cold air standard approximation, that is, assume the properties of the working fluid are that of air as a calorically perfect gas (with $c_v = 0.720$ kJ/kg/ΔK). The initial state is at 1.0 atm, 298 K. Thermal efficiency is defined as the net work divided by chemical energy added during combustion (process 2–3). For power output, assume an eight-cylinder engine operating at 2000 revolutions per minute (RPM), recognizing that two full revolutions are required to complete the cycle (one for the intake and exhaust, one for the two strokes analyzed) but there is no net work associated with the gas exchange (wide open throttle for Otto engine). Complete the state and results tables, and sketch both cycles on the PV scale provided (Figs. 7.41, 7.42, and 7.43).

	OTTO	DIESEL
Mass (kg)		
W12 (kJ)		
W23 (kJ)		
W34 (kJ)		
W41 (kJ)		
NET WORK (kJ)		
Q23 (kJ)		
1st Law Efficiency		
Power Output (kW)		
mep (atm)		
STEADY-STATE ENGINE		
Torque (kN-m)		
mdot (kg/sec)		
Heat Rate (kW)		
Exit Enthalpy Rate (kW)		
Exit Temperature (K)		

Fig. 7.42 Results table. The sign of the work is that done by the working fluid

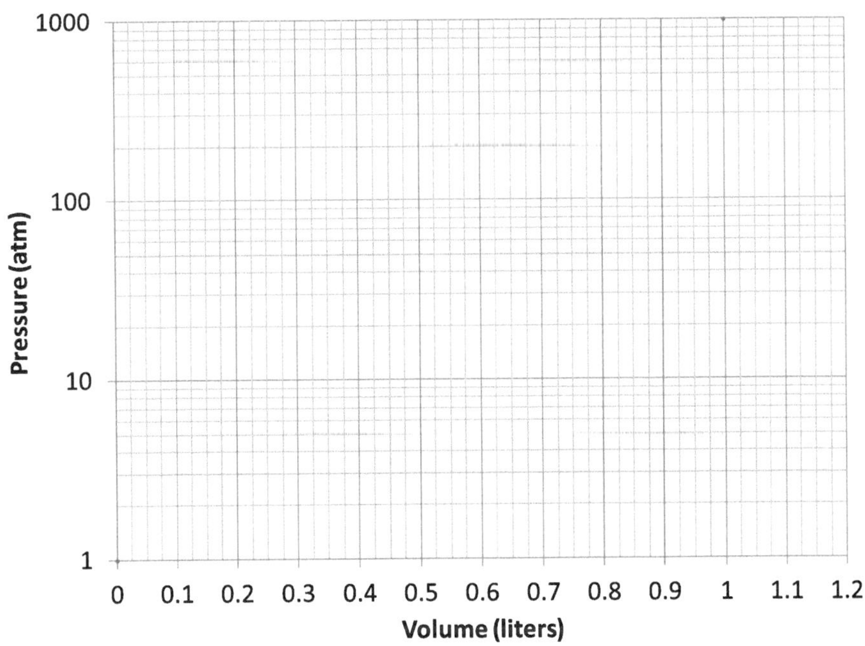

Fig. 7.43 Indicator diagram. Place the state properties according to the calculations for both cycles and sketch the processes in between. (You could calculate a few intermediate points for curved processes)

Workshop 7.2: Adding a Compressor and/or Turbine

This exercise builds on Workshop 7.1 (Otto and Diesel cycles) and is a good segue to the Brayton cycle, (Chap. 9). The thermodynamic performance of a reciprocating engine can be improved by adding a compressor (and possibly an intercooler) to the intake and/or a turbine to the exhaust. There are therefore several possible combinations. The schematic of Fig. 7.44 shows all components. The goal will be to determine how each combination affects the mass flow rate, power output, and First Law efficiency of the engine by completing the following table:

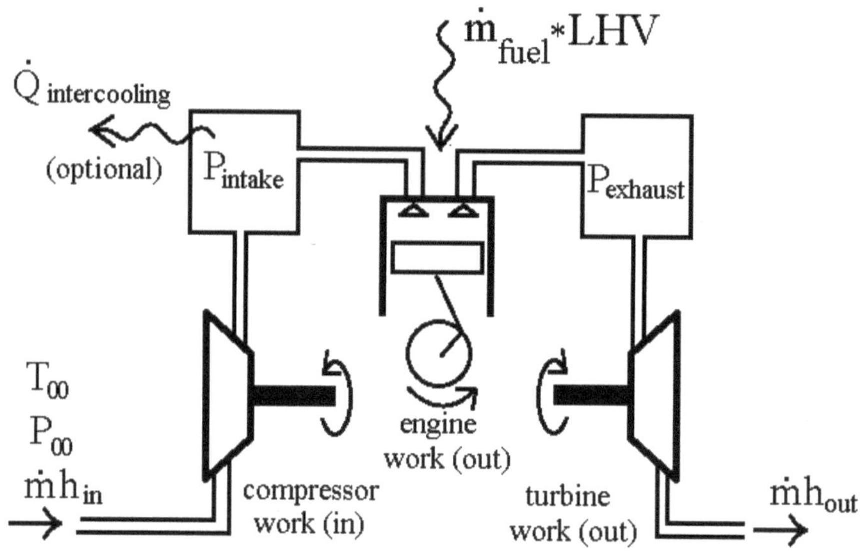

Fig. 7.44 Schematic of reciprocating engine turbo-charging and intercooling

ENGINE TYPE	Added Component	Mass Flow Rate (kg/sec)	Total Power (kW)	1st Law Efficiency	Exit Temperature (K)	COMMENTS?
Ideal 4 stoke SI Engine (OTTO)	NONE					
SI Engine w/SuperCharging	Compressor					
SI Engine w/SuperCharging and Intercooling	Compressor and Intercooler					
SI Engine w/Turbocompounding	Turbine					
SI Engine w/TURBO-CHARGING	Turbine and Compressor					
SI Engine w/TURBO-CHARGING and INTERCOOLING	Turbine, Compressor and Intercooler					

Consider the ideal four-stroke Otto cycle from Workshop 7.1 to be the base case (compression ratio of 10, 8 cylinders, each with a displacement volume of 1 liter operating unthrottled at 2000 RPM, and a calorically perfect gas, with $c_p = 1.0$ kJ/kg/ΔK). The key concept is that while the cycle is executed as a series of processes, the engine as a whole operates as a quasi-steady state open system, with a steady mass flow rate equal to the total mass exchanged during the intake and exhaust strokes divided by the cycle time (with four strokes per cycle). Note that the mass exchanged is the inlet density times the displacement volume, which is slightly lower than the mass contained in the cylinder. The difference is the "residual gases."

With supercharging, a portion of the output shaft power is used to drive a compressor (assumed to be isentropic) that maintains a pressurized intake plenum, as shown. The net power output to the shaft is the power produced by the four strokes of cycle minus that used to drive the compressor. The temperature after the compressor is elevated, and an option is to cool these gases prior to entry into the engine cycle (intercooling). For this exercise, choose an intake plenum pressure of 2 atm for all cases that involve a compressor.

At the end of the power stroke, the exhaust valve opens and the gases are hot and above ambient pressure. This available energy is dissipated in the engine cycle without a turbine. However, placement of a turbine in the exhaust stream is a means to extract a portion of that available energy. The maximum possible power that can be extracted from the turbine is obtained by setting the exhaust plenum pressure to the pressure at the end of the power stroke. The net power output to the shaft is the power produced by the four strokes of cycle and the turbine, less that used to drive the compressor.

The turbo-compounded case is when only a turbine is used (no compressor), and its mechanical output is tied directly into the output shaft. Turbo-charging involves both a compressor and a turbine. In practice, all the mechanical output of the turbine is used to drive the compressor, and the intake plenum pressure would be whatever this coupling yields. In this exercise, the compressor and turbine are considered to be uncoupled, so that the compressor work does not equal the turbine work.

This workshop is a little tricky to get precisely right because of the subtle ways that the added components alter the performance of the main engine (i.e., pumping work and mass undergoing the cycle). Assume that the temperature at the start of compression equals the intake manifold pressure, which neglects the effect of mixing the fresh intake with the hot residual gases (and requires a more complex iteration calculation). The main points of this exercise are as follows:

- To observe how a reciprocating engine can be viewed as a steady-state engine
- To observe how much additional output power and/or increase in efficiency are theoretically possible with these modifications to the basic cycles

Chapter 8
Vapor/Compression Refrigeration

The rate at which food spoils due to bacterial action increases dramatically with temperature. Therefore, nearly every household in the developed world has at least one refrigerator: an electrically driven appliance whose primary function is to maintain a space at a temperature suitable for the storage of perishable food items, although they are often simultaneously used to keep nonperishables such as bottled beverages cold. The vast majority of these systems use a variation of the vapor/compression refrigeration cycle, the main subject of this chapter. There are niche markets, not considered here, that use other means to create and maintain a cold space (absorption chillers, thermoelectric modules). The first part of the chapter focuses on the thermal behavior of a typical refrigerator in actual use, and the second part discusses the ideal vapor/compression cycle and deviations from ideal behavior in actual practice.

Basic Performance of a Household Refrigerator

Typically, the refrigerator appliance used in households is designed with two compartments: a freezer and a refrigerator space. The temperature of the freezer is below the freezing point of water, while the refrigerator space is as cold as possible without freezing. That is, a delicate balancing act must be met for the refrigerator space. If maintained too cold, some or all of the contents will freeze (as there is not perfect uniformity of temperature within the space), and if maintained too warm, the rate of spoilage will be excessive. The US Food and Drug Administration (FDA) recommends a refrigerator temperature below 4 °C (40 °F).

An experiment was conducted, with an actual refrigerator (model Amana ARS8265BW) that was instrumented by Mike Gutierrez (a Cooper Union master's student at the time) with thermocouples to measure the temperature of the air within a 8.0 cubic foot freezer space and a 13.6 cubic foot refrigerator space, and a power meter to measure the instantaneous electrical power drawn from a standard 120 VAC

G. Sidebotham, *An Inductive Approach to Engineering Thermodynamics*, Mechanical Engineering Series, https://doi.org/10.1007/978-3-030-20430-3_8

wall socket. This appliance has a side-by-side design, with the freezer compartment accessed by a door hinged at the left and the refrigerator compartment accessed by a door hinged at the right. Starting with the unit at room temperature, and empty, the unit was plugged into the electrical outlet and the temperatures and input electrical power were measured over a 12 h period. Results are shown in Fig. 8.1 (for the full 12 h) and Fig. 8.2 (detail of a 1 h period).

Initially, the freezer and refrigerator temperatures fall steadily over approximately the first 2 h. The electrical power is nearly constant at approximately 170 W during this time. Afterward, during normal operation, a feedback control system cycles the refrigeration system on and off in order to maintain the compartment temperatures within desirable ranges. If left on indefinitely, the compartment temperatures would be too low. Therefore, for this particular unit, when the temperature in the freezer compartment reaches a temperature of approximately −22 °C, the system turns off. The appliance then warms naturally toward ambient temperature until the freezer temperature reaches approximately −8 °C at which point the refrigeration system turns on. While specific designs can vary, a single heat exchanger (called an evaporator in the vapor/compression refrigeration cycle) cools the freezer compartment, and the refrigerator compartment is cooled indirectly by the freezer. The magnitude of the temperature range of the refrigerator is much less than that of the freezer. It is noteworthy that there is no true steady-state behavior of the main compartments. The temperature changes constantly about an average temperature.

Heat gain from the ambient (called the passive thermal load) is constantly present. For this unit, the on time is approximately 10 minutes and is nearly equal to the off time. That is, the total refrigeration cycle time is approximately 20 minutes with a duty cycle (ratio of on time to cycle time) of 0.5 when the unit operates with just the passive thermal load. Additional thermal loads, discussed in the next section, would alter the duty cycle.

An interesting feature of most modern refrigerators is the defrost cycle, which prevents the gradual buildup of ice in the freezer compartment from the freezing of water vapor in air (humidity) that enters during normal operation. In the sixth hour, the power drawn by the refrigerator surges to approximately 500 W, and the temperature of the freezer compartment increases above the freezing point. This power surge is due to a heating element strategically placed to melt any ice that has formed, and the liquid is directed (by a natural gravity flow) to an open pan underneath the main unit where it can evaporate into the room.

While the refrigerator is generally thought to operate continuously, it is clear that in a household refrigerator the compressor actually turns on and off several times an hour (approximately 4.5 times per hour for this refrigerator while it is simply operating empty, with the doors closed). There is a power surge each time the compressor turns on. The duty cycle would naturally adjust to accommodate a different thermal load if the refrigerator were being more actively used (opening and closing the door, adding and removing food).

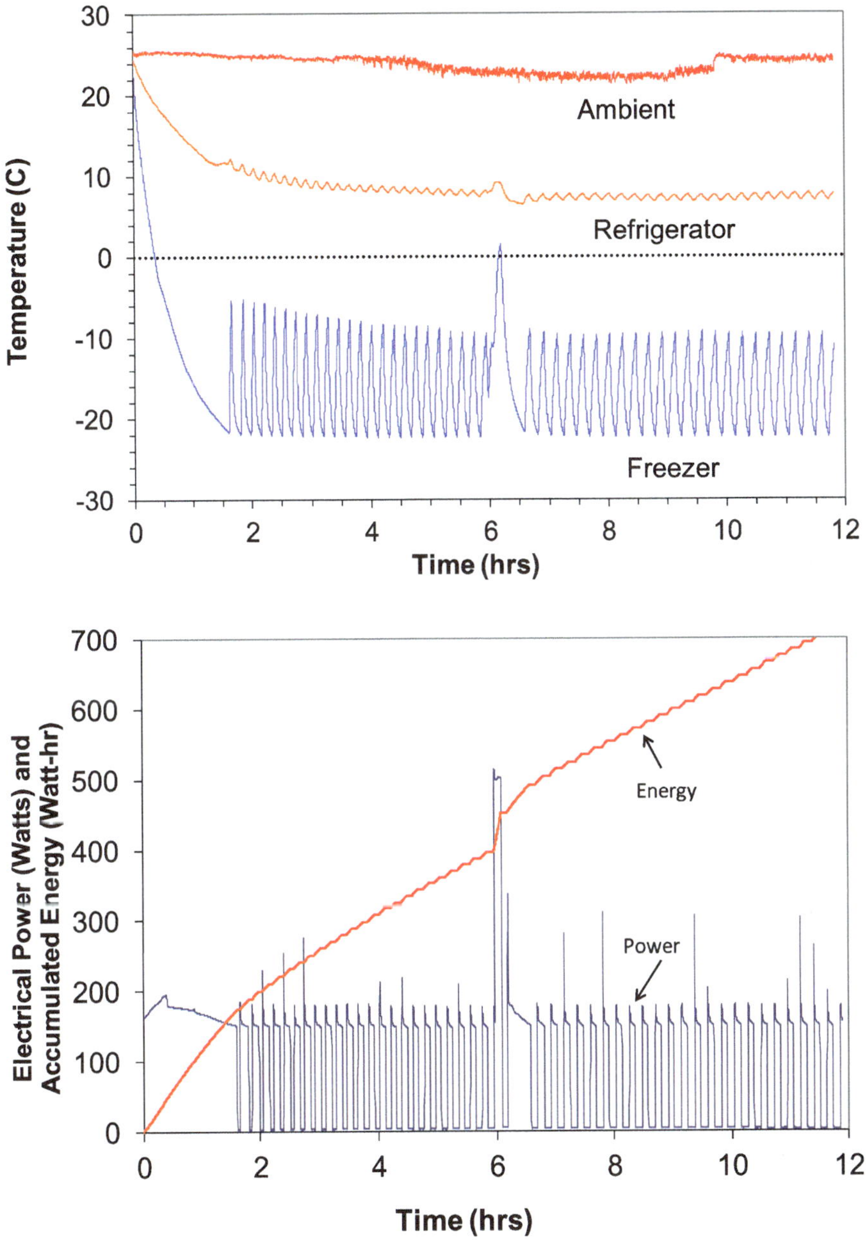

Fig. 8.1 Compartment and ambient temperatures (top) and electrical power and accumulated electrical energy consumption (bottom) of a typical household refrigerator (Amana model ARS8265BW) over a 12 h period. The unit was initially at room temperature and was empty and untouched for the entire test. The sampling rate was 0.25 Hz (that is, samples every 4 s)

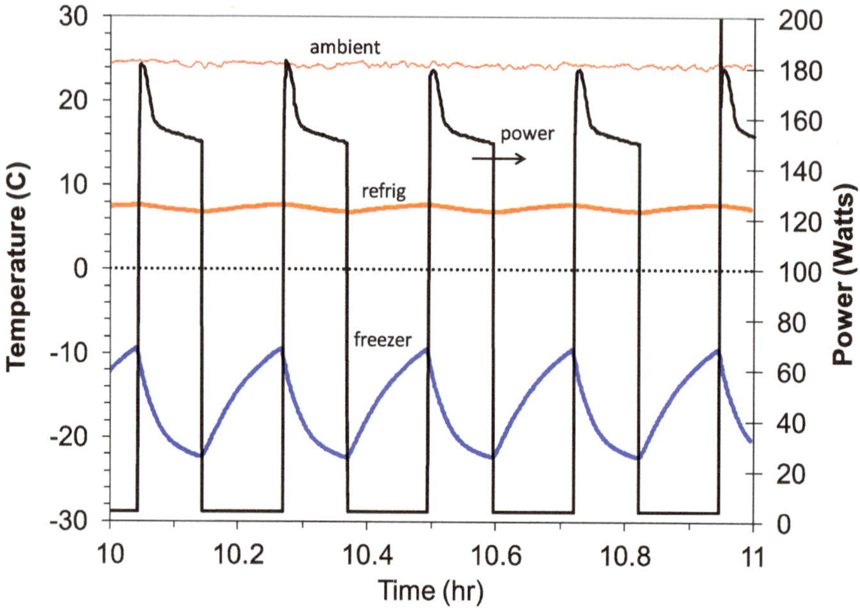

Fig. 8.2 Detail of temperatures (left axis) and electrical input power (right axis) for a 1 h period

Thermal Loads in Household Refrigeration

For the purpose of an overall thermodynamic analysis, the appliance is considered to be broken into two subsystems (Fig. 8.3): a refrigerated space (both compartments) and a refrigeration system (a heat pump cycle that is hidden from view in a consumer appliance, located at the bottom but shown on top to better reveal its thermodynamic purpose). The refrigerated space is maintained at an average cold temperature $T_{C\infty}$ in a room maintained at an average temperature of $T_{H\infty}$. The cold temperature is considered to be the average temperature in the freezer compartment, and the warmer refrigerated space is maintained at an intermediate temperature between the freezer and ambient by processes internal to the refrigerated space. Contents of the refrigerated space (called "food," although nonfood items are included) are considered as part of that control volume. A more detailed analysis would consider separate freezer, refrigerator spaces, and contents. The required heat transfer from the refrigerated space to the refrigeration system, Q_C, is called the "thermal load." The refrigeration system, hidden from view, draws electrical power to run a compressor and auxiliaries and must reject heat from a condenser to ambient. More details of these components will be considered in the next section. A general thermodynamic treatment is conducted first.

A consideration of the refrigerated space suggests three different classifications of thermal loads: active, passive, and parasitic. Consider an event that takes place over a time period Δt. During that time, an item of food that is at a different temperature than the cold space experiences a change in stored energy (ΔE_{food}). The heat transfer

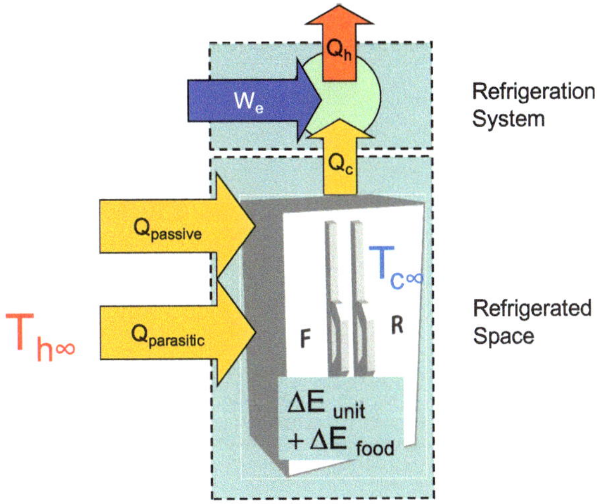

Fig. 8.3 Schematic energy diagram of household refrigerator

between the food and the refrigerated space is an internal process to the defined control volume. This change in energy is considered a desired effect and is classified as an "active" thermal load, which makes warm food cold. The metabolic action in foods that breathe, such as fruits and vegetables, is an additional active load associated with the contents.

Simultaneously, heat flows from ambient through the walls to the refrigerated space. This load is classified as the "passive" thermal load required to maintain the space cold. Passive loads can be reduced by insulating the box, but can never be eliminated entirely. The thermal load on the refrigerator considered in the previous section is entirely passive. While the electricity demand cycles between on and off, the passive load is a steady, continuous thermal load.

A third and final load is classified as a parasitic load. This load is due to operation of the refrigerator whose effect requires removal of heat by the refrigeration system, but is not part of the desired effects of making and maintaining cold food items. A major parasitic load occurs when the door is opened to add or remove a food item. The cold air in the space is replaced by warm ambient air, which must be cooled after the door is shut. Also, the defrosting systems and lights that automatically turn on when the door is opened contribute to the parasitic load as well as add to the electrical input. A leaking gasket would be considered a parasitic load, although it would act continuously (like a passive load). Also, any cooling of the unit itself (ΔE_{unit}), such as the walls, shelves, and components of the refrigeration system, is considered a parasitic load.

It is a common misconception that the total electricity demanded of a household refrigerator depends strongly on how it is actively used. Many people keep second or third refrigerators or freezers in their homes that operate 24/7 and justify it because they do not "use" it much. They open the door maybe once a week, if that. However, while an active thermal load *makes* food cold, a passive load is what *keeps* it cold. A rigorous analysis would likely show that passive thermal loads dwarf active loads under typical use (unless your teenager likes to stand there with the door open looking for something to eat).

First Law Quasi-Steady-State Analysis

A quasi-steady-state analysis is conducted to establish the relationship between the thermal loads and the cooling capacity required of the refrigeration system, and to define its coefficient of performance. Even though the temperature fluctuates over the course of a compressor on/off cycle, a long-term average over several cycles is considered.

Consider first the control volume that encompasses just the refrigerated space. An energy balance over a fixed period of time is as follows:

$$\Delta(E_{\mathrm{unit}} + \Delta E_{\mathrm{food}}) = Q_{\mathrm{passive}} + Q_{\mathrm{parasitic}} - Q_{\mathrm{C}} \quad \text{for refrigerated space}$$

$$Q_{\mathrm{C}} = (-\Delta E_{\mathrm{food}}) + Q_{\mathrm{passive}} + (-\Delta E_{\mathrm{unit}} + Q_{\mathrm{parasitic}})$$

The change in energy of food (temperature and metabolism) is considered to be the active load, and the change in energy of the unit is a form of parasitic load. Therefore, the thermal load required of the refrigeration system is the sum of the active, passive, and parasitic thermal loads:

$$Q_{\mathrm{C}} = Q'_{\mathrm{active}} + Q_{\mathrm{passive}} + Q'_{\mathrm{parasitic}}$$

Primes have been introduced to indicate that some of the terms are changes in energy, not heat transfer, a distinction that would be important for a 2nd Law analysis.

It is instructive next to analyze the refrigerator as a combined system (cold space and refrigeration system), again averaged over multiple compressor cycles. In that case, the heat transfer absorbed by the refrigeration system from the refrigerated space (Q_{C}) is internal. An energy balance is as follows:

$$\Delta(E_{\mathrm{unit}} + \Delta E_{\mathrm{food}}) = Q_{\mathrm{passive}} + Q_{\mathrm{parasitic}} - Q_{\mathrm{H}} + W_{\mathrm{e}}$$

Rearranging for the required input electricity with no change in energy stored:

$$W_{\mathrm{e}} = Q_{\mathrm{H}} - (Q'_{\mathrm{active}} + Q_{\mathrm{passive}} + Q'_{\mathrm{parasitic}}) = Q_{\mathrm{net}}$$

The electrical work into the refrigeration system equals the *net* heat transfer from the combined system to the ambient, the difference between the heat rejected by the refrigeration system and the thermal load of the refrigeration. In other words, the refrigerator acts like a 100% efficient electrical heater of the kitchen. This result answers the question, "If I hold open the door of a refrigerator, does it make the room cold?" Answer, "No. Refrigerators heat the room they are in. Keeping the door open increases the parasitic load and makes it heat the room even more. However, it might make you feel cooler since you are standing there right in front of it. Of course, your teenager might simply be trying to decide what to eat."

Finally, the coefficient of performance of the refrigeration system is defined by the desired output over the required input:

$$\mathrm{COP_R} = \frac{Q_\mathrm{C}}{W_\mathrm{e}}$$

All these quantities require more detailed analysis to conduct a quantitative assessment, the objective of the rest of this chapter. First, a classic treatment of the ideal vapor/compression refrigeration cycle will be presented followed by a deeper dive into real effects and off-design behavior associated with refrigeration system design. These topics are more appropriate for fluid mechanics and/or heat transfer courss than a first semester of thermodynamics.

Ideal Vapor/Compression Refrigeration Cycle

A refrigerated space, that is, one that is colder than the ambient in which it is surrounded, can be maintained if an engineered system creates and maintains a working fluid that is colder than the refrigerated space. The most common way in practice is to circulate a working fluid (refrigerant) through four open system components, shown in Fig. 8.4, namely, compressor, condenser, throttling valve, and evaporator. There are other components, considered to be auxiliary (such as an accumulator), that facilitate the cycling on and off of the system, and help to smooth out the inherent pulses associated with the reciprocating compressor, but do not play a significant thermodynamic role during the quasi-steady operation.

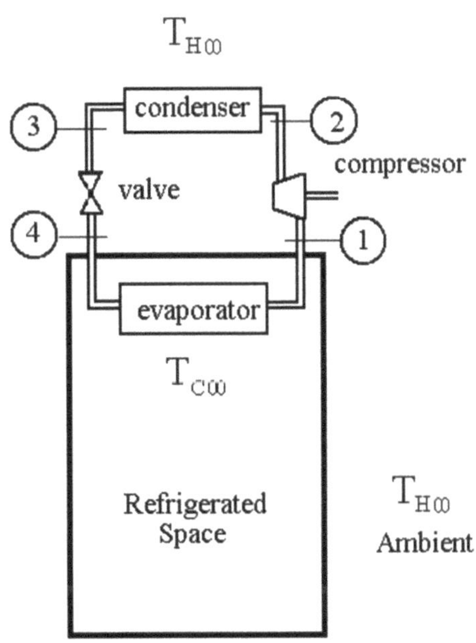

Fig. 8.4 Schematic of vapor/compression refrigeration cycle used to create a refrigerated space

A contrived refrigeration cycle with a piston/cylinder was studied in Chap. 4, which required physically moving the piston/cylinder from the cold space to ambient. Circulating a fluid through sequential open system components represents a more practical means to move the fluid in and out of the cold space.

The compressor and valve have a primarily mechanical function, namely, to change the pressure of the refrigerant, which changes the corresponding saturation temperature. The condenser and evaporator have a thermal function, namely, to transfer heat between the refrigerant and either the ambient (condenser) or cold space (evaporator).

A key design objective of a refrigeration system is to produce a working fluid that is at a lower temperature than the refrigerated space so that heat can be removed from it (without violating the 2nd Law). The temperature of a working fluid can be reduced below ambient if it does work on the environment (but that is not is done here). For example, the temperature of a gas in a piston/cylinder will decrease if it undergoes an expansion. This effect can form the basis for a gas refrigerator. This example of a closed device has an open system counterpart in a steady flow turbine in which pressurized gas at ambient temperature drives an adiabatic turbine and exits at a lower pressure, resulting in a temperature below ambient. Workshop 7.1 is written in the form of an in-class exam and shows how a gas refrigerator can be designed. It is really a closed Brayton cycle (Chap. 9) operating in reverse.

There are other ways to accomplish refrigeration, such as the use of thermoelectric modules. However, gas refrigerators, thermoelectric modules, or other methods have not found a significant market, and the vapor/compression refrigeration cycle is the dominant method.

In Workshop 4.3, the free expansion of a liquid into an evacuated space was shown to result in temperatures below ambient without extracting work (or heat). The total energy of the working fluid (liquid plus vapor) does not change, but phase change is needed to fill the larger volume. Since the liquid that evaporated moves to a higher internal energy state, the temperature of both remaining and evaporated liquid must decrease. This phenomenon forms the basis for the vapor/compression refrigeration cycle. First, the free expansion of a pressurized liquid in an open system device is studied. Such a process is often called throttling, and the device is generally called a throttling valve (or just valve). There are different fundamental designs, depending on the extent of control required for the throttling effect. The simplest valve is a capillary tube, investigated in a deeper dive, using a case study that corresponds to experimental data from a manufacturer. The free expansion process, followed by heat absorption from a cold space in an evaporator, is the key to producing a refrigeration effect. However, a cycle is required to continuously feed the high-pressure liquid to the capillary tube (throttling valve), accomplished by compressing vapor to high pressure, and then cooling to condense to a liquid.

The vapor/compression refrigeration cycle operates between two pressure levels, and therefore, two corresponding saturation temperature levels. On the high side, the saturation temperature must be higher than ambient in order for heat to flow. The refrigerant is condensed during this process, and the term condensing temperature refers to the saturation temperature on the high side (between the compressor and the valve). On the low side, the saturation temperature must be below the cold space temperature. The refrigerant is evaporated, so the term evaporating temperature is used.

In principle, any substance that has a critical temperature above the ambient and a triple-point temperature below the cold space can be used as a refrigerant. However, there is much more that goes into the thermal, chemical, and material properties of a suitable refrigerant. While the refrigerant is sealed in a refrigeration system, there are many physical connections where leaks can develop over time, and disposal of refrigeration systems can release most if not all of the refrigerant contained in a system.

Historically, propane and ammonia were popular substances that satisfied design objectives of refrigeration systems. However, they are both flammable and ammonia is poisonous, and they were phased out. A class of nonflammable, nontoxic substances called Freons (a trade name) or hydro-chloro-fluoro carbons (HCFCs) were developed as replacements. Some of these were proven to cause damage to the fragile ozone layer of Earth's upper atmosphere and were phased out. Some commercial refrigerants are blends of HCFCs, which experience phase change over a range of temperatures (between the bubble and dew points) at constant pressure. These mixtures are not investigated formally here as they involve changes in chemical composition. A common pure substance refrigerant is R134a (Fig. 8.5), which will be studied here as a case study, even though it has come under scrutiny as a greenhouse gas, and will likely be phased out as well. The numbering system is from the ANSI/ASHRAE system, with the "R" referring to refrigerant, the "1" meaning carbon number minus 1, the "3" means hydrogen number plus 1, and the "4" means number of fluorine atoms. The "a" refers to a specific isomer. I'm not sure why it is done that way, but I'm sure there is a good reason. A coarse listing of saturation property data is shown in Fig. 8.6, and a plot of the saturation pressure/temperature curve in a form that yields a Clausius–Clapeyron curve fit is shown in Fig. 8.7. Superheated property data at four pressures typical of applications is listed in Fig. 8.8. The enthalpy specific heat (c_p) is shown, calculated from the table as $c_{p\ @T_i} \approx (h_{i+1} - h_i)/(T_{i+1} - T_i)$ for temperature T_i. More detailed property relations are easily found in multiple sources, for example, https://webbook.nist.gov/chemistry/.

As a whole, the refrigeration system is a closed cycle, as shown in Fig. 8.9. Heat is added to the refrigerant in the evaporator (heat removed from the cold space), work is added to the compressor (to increase the pressure, and therefore corresponding saturation temperature), and heat is rejected to ambient in the condenser. There is no heat or work associated with the valve, but the throttling of the high-pressure liquid is a key step required to create a working fluid at a temperature below the cold space.

The ideal vapor/compression refrigeration cycle is defined by the following restrictions in states and processes.

- Saturated vapor entering compressor (state 1) at a temperature below the cold space ($T_{c\infty}$)
- Saturated liquid entering valve (state 3) at a temperature above ambient ($T_{H\infty}$)

Fig. 8.5 Chemical structure of R134a

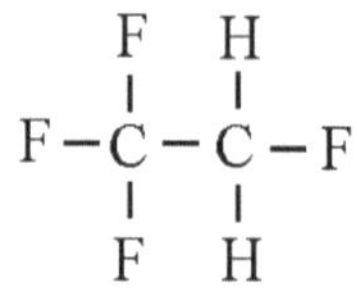

MW	102.02	kg/kmol
R	81.49	J/kg/K

R134a Saturation Properties

T (oC)	P (bar)	v_f(m^3/kg)	v_g(m^3/kg)	h_f (kJ/kg)	h_g (kJ/kg)	s_f (kJ/kg/K)	s_g (kJ/kg/K)
-40	0.5164	0.0007054	0.360810	0.00	222.88	0.0000	0.9560
-28	0.9305	0.0007234	0.206660	14.37	230.38	0.0600	0.9411
-20	1.330	0.0007362	0.147290	24.26	235.31	0.0996	0.9332
0	2.928	0.0007723	0.069255	50.02	247.23	0.1970	0.9190
20	5.716	0.0008161	0.035969	77.26	258.36	0.2924	0.9102
40	10.16	0.0008720	0.019952	106.19	268.24	0.3866	0.9041
60	16.81	0.0009498	0.011434	137.42	275.99	0.4814	0.8973
80	26.32	0.0010772	0.006436	172.71	279.12	0.5814	0.8827
90	32.44	0.0011932	0.004599	193.69	276.32	0.6380	0.8655
100	39.74	0.0015269	0.002630	224.74	259.13	0.7196	0.8117

Fig. 8.6 Coarse listing of saturation properties. A plot and Clausius–Clapeyron fit of the saturation pressure/temperature curve is given in Appendix 5

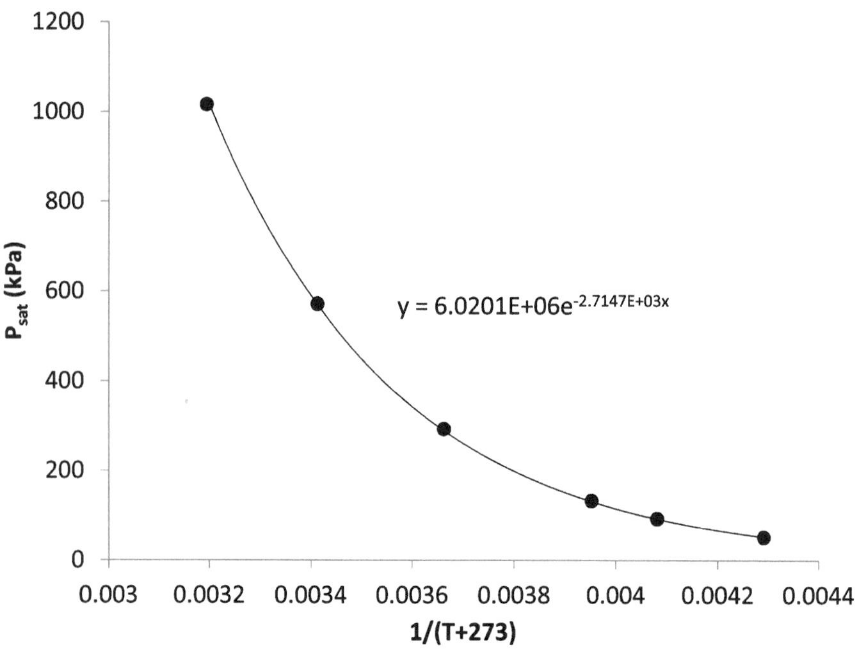

Fig. 8.7 Saturation pressure versus inverse absolute temperature for R134a over a temperature range from −40 to 40 °C. Markers are from tabulated data with a curve fit to the form of a Clausius/Clapeyron equation

R134a Superheated Vapor Properties

P = 100 kPa (Tsat = -26.37°C)

T (°C)	v (m³/kg)	h (kJ/kg)	s (kJ/kg/K)	c_p(kJ/kg/K
-26.37	0.19254	234.44	0.9518	0.794
-20	0.19841	239.50	0.9721	0.799
-10	0.20743	247.49	1.0030	0.809
0	0.2163	255.58	1.0332	0.823
10	0.22506	263.81	1.0628	0.836
20	0.23373	272.17	1.0918	0.851
30	0.24233	280.68	1.1203	0.866
40	0.25088	289.34	1.1484	0.882
50	0.25937	298.16	1.1762	0.897
60	0.26783	307.13	1.2035	0.913
70	0.27626	316.26	1.2305	0.929
80	0.28465	325.55	1.2572	0.944
90	0.29303	334.99	1.2836	0.961
100	0.30138	344.60	1.3096	

note: c_p based on first order forward difference approximation

R134a Superheated Vapor Properties

P = 320 kPa (Tsat = 2.46°C)

T (°C)	v (m³/kg)	h (kJ/kg)	s (kJ/kg/K)	c_p(kJ/kg/K
2.46	0.06360	251.88	0.9301	0.903
10	0.06609	258.69	0.9544	0.897
20	0.06925	267.66	0.9856	0.899
30	0.07231	276.65	1.0157	0.905
40	0.07530	285.70	1.0451	0.915
50	0.07823	294.85	1.0739	0.926
60	0.08111	304.11	1.1021	0.937
70	0.08395	313.48	1.1298	0.950
80	0.08675	322.98	1.1571	0.964
90	0.08953	332.62	1.1840	0.977
100	0.09229	342.39	1.2105	0.991
110	0.09503	352.30	1.2367	1.005
120	0.09775	362.35	1.2626	

R134a Superheated Vapor Properties

P = 800 kPa (Tsat = 31.31°C)

T (°C)	v (m³/kg)	h (kJ/kg)	s (kJ/kg/K)	c_p(kJ/kg/K
31.31	0.025621	267.29	0.9183	1.054
40	0.027035	276.45	0.9480	1.024
50	0.028547	286.69	0.9802	1.012
60	0.029973	296.81	1.0110	1.007
70	0.031340	306.88	1.0408	1.009
80	0.032659	316.97	1.0698	1.013
90	0.033941	327.10	1.0981	1.020
100	0.035193	337.30	1.1258	1.029
110	0.036420	347.59	1.1530	1.038
120	0.037625	357.97	1.1798	1.048
130	0.038813	368.45	1.2061	1.060
140	0.039985	379.05	1.2321	1.071
150	0.041143	389.76	1.2577	1.083
160	0.042290	400.59	1.2830	

note: c_p based on first order forward difference approximation

R134a Superheated Vapor Properties

P = 1000 kPa (Tsat = 39.37°C)

T (°C)	v (m³/kg)	h (kJ/kg)	s (kJ/kg/K)	c_p(kJ/kg/K
39.37	0.020313	270.99	0.9156	1.143
40	0.020406	271.71	0.9179	1.103
50	0.021796	282.74	0.9525	1.064
60	0.023068	293.38	0.9850	1.047
70	0.024261	303.85	1.0160	1.040
80	0.025398	314.25	1.0458	1.039
90	0.026492	324.64	1.0748	1.042
100	0.027552	335.06	1.1031	1.047
110	0.028584	345.53	1.1308	1.053
120	0.029592	356.06	1.1580	1.063
130	0.030581	366.69	1.1846	1.071
140	0.031554	377.40	1.2109	1.082
150	0.032512	388.22	1.2368	1.093
160	0.033457	399.15	1.2623	

Fig. 8.8 Superheated vapor properties (at four pressures that correspond to typical condensing and evaporating temperatures)

Fig. 8.9 Thermodynamic process diagram of refrigeration cycle

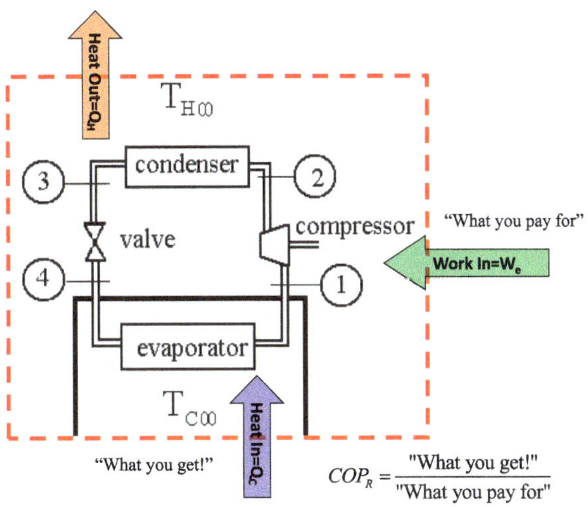

- Steady-state operation
- Negligible changes in kinetic and potential energy in all components
- Negligible stray heat in compressor and valve
- Reversible compressor (hence isentropic, since also adiabatic)

Mass, Energy, and Entropy Balances

Each component is a single-inlet, single-outlet device operating at steady state. The general mass balance for a single-inlet, single-outlet device is:

$$\frac{dM}{dt} = \dot{m}_{\text{in}} - \dot{m}_{\text{out}}$$

and therefore, at steady state, the mass flow rate at each station is the same:

$$\dot{m}_1 = \dot{m}_2 = \dot{m}_3 = \dot{m}_4 = \dot{m}$$

The general energy balance for single-inlet, single-outlet devices is:

$$\frac{dE}{dt} = \dot{Q}_{\text{in}} - \dot{W}_{\text{shaft}} + \dot{m}_{\text{in}}\left(h_{\text{in}} + \frac{V_{\text{in}}^2}{2} + gz_{\text{in}}\right) - \dot{m}_{\text{out}}\left(h_{\text{out}} + \frac{V_{\text{out}}^2}{2} + gz_{\text{out}}\right)$$

With the assumptions made for the ideal cycle (steady-state, negligible kinetic and potential energy changes), the energy balance simplifies to

$$0 = \dot{Q}_{\text{in}} - \dot{W}_{\text{shaft}} + \dot{m}(h_{\text{in}} - h_{\text{out}})$$

The general entropy balance is

$$\frac{dS}{dt} = \frac{\dot{Q}_{\text{in}}}{T_{\text{b}}} + \dot{m}_{\text{in}}s_{\text{in}} - \dot{m}_{\text{out}}s_{\text{out}} + \dddot{\Sigma}$$

With the assumptions made (steady state):

$$0 = \frac{\dot{Q}_{\text{in}}}{T_{\text{b}}} + \dot{m}(s_{\text{in}} - s_{\text{out}}) + \dddot{\Sigma}$$

Multiplying by the dead state temperature (which gives the entropy balance an energy interpretation):

$$0 = T_\infty \frac{\dot{Q}_{\text{in}}}{T_{\text{b}}} + \dot{m}_{\text{in}} T_\infty (s_{\text{in}} - s_{\text{out}}) + e\dddot{X}_{\text{D}}$$

where the eXergy destruction rate $\left(e\dddot{X}_{\text{D}} = T_\infty \dddot{\Sigma}\right)$ has been introduced.

Table 8.1 Governing equations for components of an ideal vapor/compression refrigeration cycle

Process	Component	Process specification	Energy balance	Entropy balance
1–2	Compressor	Adiabatic, reversible	$\dot{W}_{12} = \dot{m}(h_2 - h_1)$	$s_2 = s_1$
2–3	Condenser	Isobaric	$\dot{Q}_{23} = \dot{m}(h_3 - h_2)$	$\dot{\Sigma}_{23} = \frac{-\dot{Q}_{23}}{T_{H\infty}} + \dot{m}(s_3 - s_2)$
3–4	Valve	Adiabatic	$h_3 = h_4$	$\dot{\Sigma}_{34} = \dot{m}(s_4 - s_3)$
4–1	Evaporator	Isobaric	$\dot{Q}_{41} = \dot{m}(h_4 - h_1)$	$\dot{\Sigma}_{41} = \frac{-\dot{Q}_{41}}{T_{C\infty}} + \dot{m}(s_1 - s_4)$

The balance equations for the ideal vapor/compression cycle are further simplified for each component, which is defined by the process specifications, and simplified energy and entropy balances of Table 8.1. The sign of the heat transfer terms is positive into the device (e.g., so the heat transfer rate into the condenser will evaluate to a negative number).

Case Study

Analyze an ideal vapor/compression refrigeration cycle that is designed to maintain a cold space at $T_{C\infty} = -10\ ^\circ\mathrm{C}$ in an ambient environment at $T_{H\infty} = 25\ ^\circ\mathrm{C}$. A thermal load analysis of the cold space results in a required cooling rate of $\dot{Q}_C = 1000\ \mathrm{W} = 1.0\ \mathrm{kW}$. Set the condensing temperature to be $T_{COND} = 40\ ^\circ\mathrm{C}$ (must exceed $T_{H\infty}$) and the evaporating temperature to be $T_{EVAP} = -28\ ^\circ\mathrm{C}$ (must be below $T_{C\infty}$). Complete the state and process tables (paying attention to signs and magnitudes) and calculate relevant results (mass flow rate, 1st and 2nd Law efficiencies). Plot the cycle on *Ts* and *Ph* diagrams.

A template for the analysis of an ideal vapor/compression cycle is shown in Fig. 8.10. The analysis will be conducted for R134a, but the template could be used for other substances. Input values could be changed also. Completed thermodynamic diagrams are shown in Fig. 8.11 (*Ts*) and Fig. 8.12 (*Ph*). The completed spreadsheet is presented in Fig. 8.13 and will be discussed in detail. Figure 8.14 shows the results of a refrigeration cycle in which the valve is replaced by a reversible turbine, to be discussed.

Details of Analysis

Figure 8.13 shows a completed color-coded spreadsheet template. Each cell contains either an input value, when available, or an appropriate formula that references other cells. These calculations are detailed in the following, starting with the compressor.

The compressor is modeled in the ideal cycle as being adiabatic $\left(\dot{Q}_{in} = 0\right)$ and reversible $\left(\dddot{\Sigma} = 0\right)$. Therefore, the energy and entropy balance equations are as follows:

$$\dot{W}_{12} = \dot{m}(h_2 - h_1)$$

$$s_2 = s_1$$

INPUTS

$T_{C\infty}$	-10	oC	cold space temperature
$T_{H\infty}$	25	oC	ambient temperature
Qc	1	kW	Cooling capacity

STATE TABLE

state	P(bar)	T(°C)	h(kJ/kg)	s(kJ/kg/K)	quality
1		-28			1
2					
3		40			0
4					

PROCESS TABLE (all units in kW)

		1st Law Quantities			2nd Law Quantities		
process	**device**	***m'(h out - h in)***	***Q in***	***W out***	***T∞m'(s out - s in)***	***T∞*Q in/T b***	***X D***
1-2	compressor		0				0
2-3	condenser			0			
3-4	valve		0	0			
4-1	evaporator			0			
cycle							

RESULTS

mdot		kg/sec, mass flow rate
COP		$Q_c/W_{net,in}$, coefficient of performance
COP_{CARNOT}		$T_{C\infty}/(T_{H\infty} - T_{C\infty})$
η_{II}		$COP/COP_{CARNOT} = 1 - X_D/W_{in}$

Fig. 8.10 Spreadsheet template for analysis of an ideal vapor/compression refrigeration cycle

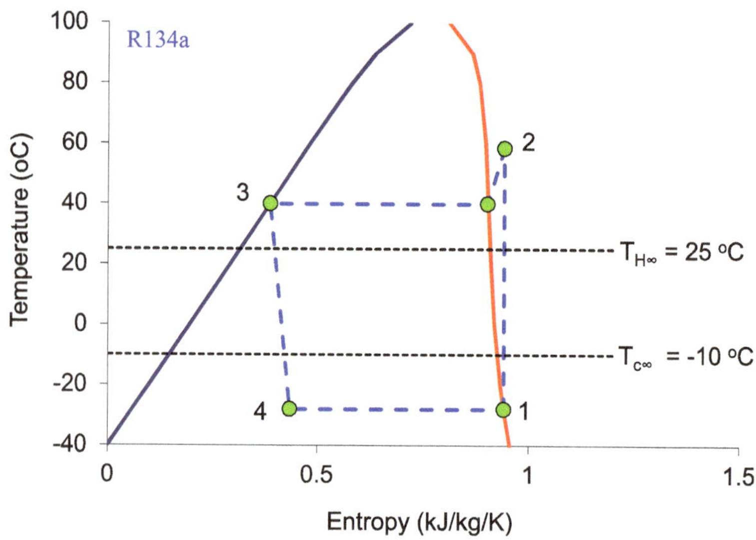

Fig. 8.11 Temperature/entropy diagram of ideal vapor/compression refrigeration cycle case study. The solid blue line is the saturated liquid line. The solid red line is the saturated vapor line. The dashed line is the cycle path (note that process 3–4 is not a well-defined path, and that for process 2–3 an intermediate state distinguishing cooling of the superheated vapor from the condensation process is shown)

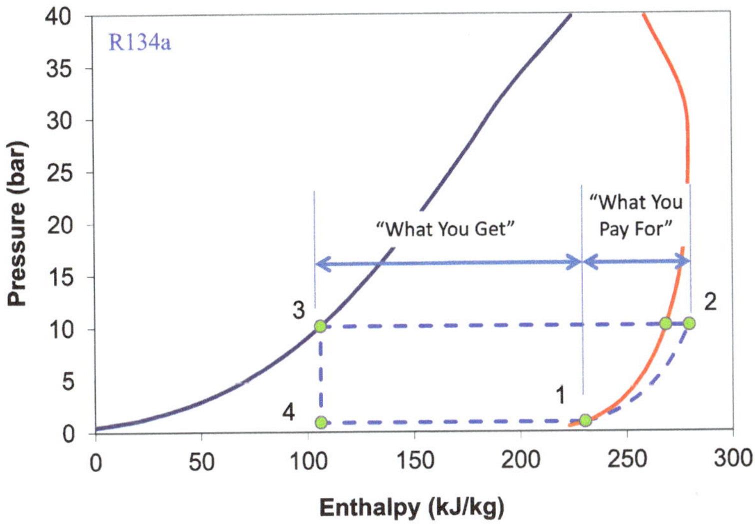

Fig. 8.12 Pressure/enthalpy diagram of ideal vapor/compression refrigeration cycle case study. The solid blue line is the saturated liquid line. The solid red line is the saturated vapor line. The dashed line is the cycle path. The curved path shown for Process 1–2 uses ideal gas behavior for an isentropic process rather than a straight line which would cross into, then out, of the vapor dome)

INPUTS

$T_{C\infty}$	-10	oC	cold space temperature
$T_{H\infty}$	25	oC	ambient temperature
Qc	1	kW	Cooling capacity

STATE TABLE

state	P(bar)	T(° C)	v(m³/kg)	h(kJ/kg)	s(kJ/kg/K)	quality
1	0.9305	-28	0.20666	230.38	0.9411	1
2	10.164	47.21		279.27	0.9411	NA
3	10.164	40	0.0008421	106.19	0.3866	0
4	0.9305	-28	0.0883	106.19	0.4345	0.4251

COLOR CODE

INPUT PARAMETERS
Process Specification
Direct entry from tables
quality calculation
2 point interpolation

PROCESS TABLE (all units in kW)

		1st Law Quantities			2nd Law Quantities		
process	device	$m'(h_{out} - h_{in})$	Q'_{in}	W'_{out}	$T_{\infty}m'(s_{out} - s_{in})$	$T_{\infty}{*}Q_{in}/T_b$	X_D
1-2	compressor	0.394	0	-0.394	0	0	0
2-3	condenser	-1.394	-1.394	0	-1.331	-1.394	0.063
3-4	valve	0	0	0	0.115	0	0.115
4-1	evaporator	1	1	0	1.216	1.133	0.082
cycle		0	-0.394	-0.394	0	-0.261	0.261

COLOR CODE

INPUT PARAMETERS
From State Table
From Energy Balance
From Entropy Balance

RESULTS

mdot	0.00805	kg/sec, mass flow rate
COP	2.540	$Q_c/W_{net,in}$, coefficient of performance
COP_{CARNOT}	7.514	$T_{C\infty}/(T_{H\infty} - T_{C\infty})$
η_{II}	0.338	$COP/COP_{CARNOT} = 1 - X_D/W_{in}$

Fig. 8.13 Completed template for case study. Specific volume of state 2 is intentionally left blank as it does not enter into energy calculations, but would be a good exercise in two-point interpolation for the interested reader

That is, the process is isentropic. For the case study, two independent thermodynamic properties are specified for state 1 (temperature and quality), and all other properties can be determined directly from the saturation tables (color-coded green).

INPUTS

$T_{C\infty}$	-10	oC	cold space temperature
$T_{H\infty}$	25	oC	ambient temperature
Qc	1	kW	Cooling capacity

STATE TABLE

state	P(bar)	T(°C)	v(m³/kg)	h(kJ/kg)	s(kJ/kg/K)	quality
1	0.9305	-28	0.20666	230.38	0.9411	1
2	10.164	47.21		279.27	0.9411	NA
3	10.164	40	0.0008421	106.19	0.3866	0
4	0.9305	-28	0.0771	94.4391	0.3866	0.3707

COLOR CODE
INPUT PARAMETERS
Process Specification
Direct entry from tables
quality calculation
2 point interpolation

PROCESS TABLE (all units in kW)

		1st Law Quantities			2nd Law Quantities		
process	device	$m'(h_{out} - h_{in})$	Q'_{in}	W'_{out}	$T_{\infty}m'(s_{out} - s_{in})$	$T_{\infty}{}^*Q_{in}/T_b$	X_D
1-2	compressor	0.360	0	-0.360	0	0	0
2-3	condenser	-1.273	-1.273	0	-1.216	-1.273	0.058
3-4	valve	-0.086441	0	0.086	0.000	0	0.000
4-1	evaporator	1	1	0	1.216	1.133	0.082
cycle		0	-0.273	-0.273	0	-0.140	0.140

COLOR CODE
INPUT PARAMETERS
From State Table
From Energy Balance
From Entropy Balance

RESULTS

mdot	0.00736	kg/sec, mass flow rate
COP	3.660	$Q_c/W_{net,in}$, coefficient of performance
COP_{CARNOT}	7.514	$T_{C\infty}/(T_{H\infty} - T_{C\infty})$
η_{II}	0.487	$COP/COP_{CARNOT} = 1-X_D/W_{in}$

Fig. 8.14 Effect of replacing the valve with a reversible turbine. Completed spreadsheet template for that modified case study

Process 1–2 is an isentropic compression and, therefore, the entropy at state 2 is determined from a process specification (color-coded red).

Similarly, state 3 is uniquely determined from the input temperature and quality, and all other properties at state 3 are obtained from saturation tables. Since process 2–3 is isobaric, the pressure at state 2 is determined from the process specification.

Now that two independent thermodynamic property values are known (pressure and entropy), all the other properties can be determined. First, from Fig. 8.8, the entropy value at state 2 (0.9411 kJ/kg/K) exceeds the saturated vapor value for the corresponding pressure (0.9041 kJ/kg/K) and, therefore, state 2 is a superheated vapor, and the quality is listed as not applicable (NA). Some students will enter a value of 1 here, a conceptual error.

A two-point interpolation is needed to precisely determine state 2 since the pressure (10.164 bar) does not correspond to one of the pressure tables available. A detail of such an interpolation conducted in a separate spreadsheet is shown in Fig. 8.15, and these results are entered into the template, and color-coded as a two-point interpolation result. An alternative interpolation method in such a case (called an ideal gas interpolation method) is introduced in Chap. 10.

Process 1–2 plots as a vertical line on a *Ts* diagram and as a curved line on a *Ph* diagram. The precise shape shown was obtained from a power law fit (a polytropic process) suggested by ideal gas behavior (not detailed here) rather than a straight line (which would cross into, then out of the liquid/vapor region).

Turning to the valve (process 3–4), the properties at state 3 have been determined, and the pressure at state 4 is equal to that at state 1 since the process specification for process 4–1 is isobaric. The valve process is isenthalpic and, therefore, two independent thermodynamic properties at state 4 (P_4 = 0.9305 bar and

P = 10.164 bar, specified by state 3 and process 2-3
S = 0.9411 kJ/kg/K, specified by state 1 and process 1-2

R134a Superheat Data (w/interpolations)
ENTROPY (kJ/kg/K)

P (bar) =	10	10.164	12.5
T (oC)			
40	0.9179		
46.29	0.939663	0.937914281	0.913
47.20668		0.9411	
50	0.9525	0.95080752	0.9267

R134a Superheat Data (w/interpolations)
ENTHALPY (kJ/kg/K)

P (bar) =	10	10.164	12.5
T (oC)			
40	271.71		
46.29			273.87
47.20668	279.659	279.27	274.9572
50	282.74		278.27

ORDER
1
2
3
4
5
6
7

Fig. 8.15 Detail of two-point interpolation needed to determine the properties of state 2. Unshaded values are entries from superheat tables. Multiple interpolations are made in the order indicated by the color coding

$h_4 = h_3 = 106.19$ kJ/kg) are determined. The enthalpy value is between the saturated liquid and vapor values for that pressure, and therefore state 4 is a liquid/vapor mixture. The temperature is the saturation temperature that corresponds to the pressure (and therefore is also the same as state 1 since process 4–1 is isobaric), and the quality is determined from a quality calculation, namely:

$$x_4 = \frac{h_4 - h_{f4}}{h_{g4} - h_{f4}} = \frac{106.19 - 14.37}{230.38 - 14.37} = 0.4251$$

This formula is entered directly into the template. Then, the entropy value is determined from

$$s_4 = (1 - x_4)s_{f4} + x_4 s_{g4} = (1 - 0.4251)0.06 + 0.4251{}^*0.9411$$
$$= 0.4345 \text{ kJ/kg/K}$$

Having completed the state table, appropriate formulas can be entered to complete the process and results tables. First, the mass flow rate can be determined from the energy balance on the evaporator (since the total cooling capacity is specified as an input). That is,

PRECISION DEMO ON EVAPORATOR EXERGY

SIG.FIG	$T_\infty m'(s_{out} - s_{in})$	$T_\infty * Q_{in}/T_b$	X_D
15	1.21553539188001	1.13307984790875	0.08245554397126
4	1.216	1.133	0.083
3	1.22	1.13	0.09
2	1.2	1.1	0.1
1	1	1	0

Fig. 8.16 Demonstration of effect of precision on eXergy calculations

$$\dot{m} = \frac{\dot{Q}_C}{h_4 - h_1}$$

Completion of the template is straightforward, given the simplified governing equations of Table 8.1, with careful attention to signs. Also, it is important to properly interpret what is meant by the dead state temperature and the boundary temperature. For the condenser, they are the same. For the evaporator, the boundary temperature is the cold space temperature. Of course, these temperatures must be in absolute temperature units, another common error. Celsius units are generally preferred over Kelvin for reporting, so unit conversions can be made simply in the formulas entered into the cells.

For eXergy calculations, it is important to use a high precision in performing calculations because the eXergy destruction is often a small number obtained from the difference of two large numbers (entropy transfer and eXergy transfer). This precision is automatic when conducting calculations in a spreadsheet (15 significant figures) even if fewer decimals are reported, as is appropriate. However, care must be taken if calculations are conducted with a hand calculator and reported on paper, as traditionally done. For example, consider the eXergy destruction in the evaporator. The result reported (0.082) seems inconsistent with the difference between entropy change (1.216) and eXergy transfer (1.133), which should return a value of 0.083. However, the underlying calculations carry 15 significant figures, and the reported result has been rounded to 4 significant figures (the choice of the format of the cells in question). A detailed example of this effect is shown in Fig. 8.16, which shows the reported results for different levels of precision.

Once the state and process tables are complete, the results are straightforward to calculate and worth commenting on. It is noteworthy that the mass flow rate, in strict SI units, has a relatively low value of 0.008 kg/s needed to deliver 1000 W of cooling. There are two reasons why this characteristic of vapor/compression cycles is advantageous over, say, a gas refrigeration system; constant temperature and high heat of vaporization. As heat is added to the evaporator, the temperature of the working fluid remains constant (or drifts for refrigerant blends) as the quality of the liquid/vapor mixture gradually increases. The refrigerant itself thereby acts as a thermal reservoir. The higher the heat of vaporization, the lower the mass flow rate required of the mechanical components (compressor and valve) to maintain the pressure levels.

The coefficient of performance (COP) of 2.54 seems attractive, meaning that the desired cooling effect is 2.5 times greater than the required work input. However, the Carnot COP for a refrigerant operating between the cold space and ambient is 7.51. The actual COP is about three times lower than the Carnot COP, making the 2nd Law efficiency 0.338. The actual performance of this ideal cycle is considerably lower than the theoretical ideal cycle (recall the analysis of a reversible heat pump with irreversible heat transfer in Chap. 3). A real refrigeration system would have additional irreversibility, some of which are discussed in the next sections, and even lower COP performance.

The source of the deviation of the ideal vapor/compression refrigeration cycle from the theoretical ideal is identified by the 2nd Law analysis, namely, from the eXergy destruction of individual components. The compressor is reversible, with no eXergy destruction. A real compressor would have some irreversibility associated with it. The other three components have irreversibilities that contribute to a reduction in the 2nd Law efficiency.

The largest irreversibility is associated with the valve. As discussed in detail in the next section, the throttling process that takes place in the valve is effective at dropping the high-pressure liquid (at a temperature above ambient) to a low-pressure liquid/vapor mixture at a temperature below the cold space, thereby preparing the refrigerant to receive heat from the cold space. It is effective, and simple to design and build, but it comes at an eXergy price. However, there has been insufficient market incentive to improve on its thermodynamic performance to date. In principle, the valve could be replaced by a turbine that would improve the thermodynamics in two ways. First, a turbine would extract useful work as the fluid pressure drops, rather than simply dissipating the eXergy associated with the high-pressure fluid. The net work of the cycle would decrease accordingly. Second, the quality of the liquid/vapor mixture exiting the turbine would be lower than that of a valve, and additional heat could be added to the fluid in the evaporator, increasing the cooling capacity. Both these effects would increase the COP of the cycle.

While difficult to design and build, replacing the valve with a reversible turbine can be readily done in a theoretical calculation. It requires only a few changes to the spreadsheet, and the completed spreadsheet for this case is shown in Fig. 8.14. The effects on performance are compiled in Fig. 8.17, where the cooling capacity is fixed at 1 kW. There is an 8% reduction in mass flow rate required (because the quality exiting the turbine is lower than that exiting the valve). The net input power is reduced by 30.7% because of the lower mass flow rate required in addition to the useful power extracted from the turbine. The COP, and therefore 2nd Law efficiency, increased by 44.1%, a rather large improvement. Considering that virtually every household in the developed world has at least one refrigerator running continuously and some form of air conditioning during the summer, not to mention commercial applications, there is enormous potential to reduce electrical power demand by the development of the technology to tap into the loss of useful energy associated with the throttling process in vapor/compression refrigeration systems. What is perceived as a technical challenge could be viewed more as a research and development opportunity that would pay dividends for eternity.

quantity	Ideal Vapor/Compression Cycle	Replace Valve with Turbine	%Change
mass flow rate (kg/s)	0.0081	0.0074	-8.6%
Net Cycle Input Power (kW)	0.394	0.273	-30.7%
COP	2.54	3.66	44.1%
2nd Law Efficiency	0.338	0.487	44.1%

Fig. 8.17 Effect of replacing the valve of an ideal vapor/compression cycle with a reversible turbine

The other source of eXergy destruction is in the heat transfer across a finite temperature difference in the two heat exchangers (condenser and evaporator). While the temperature hill to climb from the cold space to ambient is 35 °C, the refrigerant on the high side is 15 °C above ambient and is 18 °C below the cold space for an effective temperature hill of 63 °C. As will be shown, these differences are attained in practice by heat exchangers with a precise effectiveness to meet these design conditions. Improvement in the thermodynamic performance could be attained in principle by the specification of a lower condensing temperature and a higher evaporating temperature, which would require more effective heat exchangers.

Deviations from the Ideal Cycle

Study of the ideal vapor/compression cycle provides a first overview of how refrigeration is achieved in practice. As with many thermodynamic analyses, however, mass, energy, and entropy balances are useful theoretical tools for determining what is possible, but they do not give an indication of the principles for how those devices are designed. This section expands on each component, giving an indication of how they can be constructed, as well as giving a clearer picture of how they deviate from the ideal cycle.

A key objective of this section is to demonstrate the interconnectedness between components. A change in one component will shift the operating conditions in all the others. Going forward, the case study for the ideal vapor/compression refrigeration study will be considered to be the base case, and a sensitivity analysis will be conducted by considering one effect at a time. It is understood that if there will be changes in any one component, then the performance of all the other components will be affected. The analyses are generally considered to elucidate key functional dependencies, but may not capture all nonideal effects quantitatively. Each change would require calibration with actual experimental systems in any manufacturing application. The deeper dives in this chapter go further in the details of each component.

Degree of Superheat and the Suction Line

From the point of view of the specifications of state processes, there are two important differences that are designed into refrigeration systems. The ideal cycle specifies that the heat added to the refrigerant in the evaporator is just sufficient to completely evaporate the refrigerant, leaving the compressor inlet condition a saturated vapor. It can be problematic mechanically if liquid droplets impinge on the high-speed piston face, like a sand blasting effect. In practice, systems are generally designed so that the evaporator causes an overshoot of the saturated vapor, and the exit attains a certain degree of superheat, usually listed in manufacturing specification sheets as degrees of superheat (in °F or °C). Furthermore, in practice, the physical location of the compressor is generally outside of the cold space, so there must be some length of piping, called the suction line, between the evaporator and the compressor that generally absorbs some stray heat from the ambient (not the cold space) and therefore does not contribute to the desired thermal load.

A form of sensitivity analysis can be conducted to quantify the effect of superheat on the thermodynamic performance of the cycle. The fractional change in the COP is given by

$$\frac{d\mathrm{COP}}{\mathrm{COP}} = \frac{d\dot{Q}_\mathrm{C}}{\dot{Q}_\mathrm{C}} - \frac{d\dot{W}_\mathrm{in}}{\dot{W}_\mathrm{in}}$$

That is, the fractional change in COP increases with the fractional rate of cooling in the evaporator and decreases with the fractional change in the input power required. Considering the ideal cycle to be the base, and considering a change in which the evaporator exits at a differential temperature higher (dT), the cooling rate is

$$\dot{Q}_\mathrm{C} = \dot{m}\left(h_1 - h_3 + c_{\mathrm{p,vapor}}(dT)\right)$$

The work required to compress a vapor from P_1 to P_2 is given by

$$w_{12} = \int_1^2 dh = \int_1^2 (Tds + vdP)$$

For an isentropic compression of an ideal gas, the work is given by

$$w_{12} = \int_1^2 vdP$$

That is, vdP work. The specific volume of the superheated vapor is higher than the saturated vapor, and therefore, it takes more work to compress a superheated vapor a given pressure rise than it does to compress a saturated vapor at the same pressure. For this calculation, it is more insightful to express the work in terms of the pressure ratio, assuming ideal gas behavior, as follows:

$$w_{12} = h_2 - h_1 = c_{p,\text{vapor}} T_1 \left(\frac{T_2}{T_1} - 1 \right) = c_{p,\text{vapor}} T_1 \left(\left(\frac{P_2}{P_1} \right)^{R/c_{p,\text{vapor}}} - 1 \right)$$

For a given pressure ratio, the work is proportional to the compressor inlet temperature. The fraction change in COP becomes as follows:

$$\frac{d\text{COP}}{\text{COP}} = \frac{c_{p,\text{vapor}} dT}{h_1 - h_3} - \frac{c_{p,\text{vapor}} dT \left((P_2/P_1)^{R/c_{p,\text{vapor}}} - 1 \right)}{h_2 - h_1}$$

For the base case conditions, with $c_{p,\text{ vapor}} = 0.79$ kJ/kg/K, $R = 0.0815$ J/kg/K and all enthalpies from the base case, the fractional change in COP, per degree Celsius change in superheat is

$$\frac{d\text{COP}/\text{COP}}{dT} = 0.00636 - 0.00452 = 0.00184 \text{ per } ^\circ\text{C}$$

That is, if the suction line is neglected, the COP actually increases with superheat, by 0.18% for each degree of superheat, the difference between the increase in cooling (0.64%), and the increase in input work (0.45%). There is an increase in the input work required, but a larger increase in the cooling rate, so the thermodynamic performance improves. It should be noted that the effectiveness of the condenser and the valve would change as well in order to maintain the state entering the evaporator and to allow this comparison with a shift in only state 1.

However, this sensitivity analysis is a bit misleading, in that the last heat transfer that takes place before the compressor may not remove heat from the cold space. There is generally some distance between the evaporator (that part in thermal contact with the cold space) and the compressor inlet, called the suction line, and stray heat occurs there. The superheating that occurs there should not be considered part of the cooling capacity, so the COP decreases due to the increased work required of the compressor. There is therefore an incentive to minimize heat transfer in the suction line, achieved in practice by adding insulation.

Degree of Subcooling and the Discharge Line

Similarly, the condenser is generally designed to overshoot the saturated liquid condition, and the exit of the condenser is generally a compressed liquid (not saturated), and listed in manufacturer specification sheets as a certain degree of subcooling. This change from the base case would result in a lower enthalpy entering the evaporator, and therefore an increase in the cooling capacity. From the point of view of the COP, there is no other change. The fractional change in COP is therefore

$$\frac{d\mathrm{COP}}{\mathrm{COP}} = \frac{d\dot{Q}_\mathrm{C}}{\dot{Q}_\mathrm{C}} = \frac{c_\mathrm{liquid} dT}{h_1 - h_3}$$

With $c_{p,\,\mathrm{liquid}} = 1.27\ \mathrm{kJ/kg/K}$

$$\frac{d\mathrm{COP}/\mathrm{COP}}{dT} = 0.0102\ \text{per}\ {}^\circ\mathrm{C}$$

That is, the COP increases by 1.02% for every degree of subcooling.

The piping between the compressor and the condenser is called the discharge line. Refrigerant enters the discharge line with a high degree of superheat and the temperature quickly falls as stray heat is transferred to the local ambient. This stray heat is not considered to degrade the thermodynamic performance as it is part of the heat rejection that must take place between the compressor and the valve. There really is no need to try to minimize it, in practice, unless it causes problems with local heating, for example.

In short, this sensitivity analysis demonstrates that subcooling the liquid in the condenser improves the thermodynamic performance compared to the ideal vapor compression refrigeration cycle. Stray heat from the discharge line is generally not problematic. On the other hand, superheating can improve the thermodynamic performance if stray heat to the suction line can be eliminated (by designing a short distance between evaporator and compressor, and by heavy insulation). Generally, however, superheating will degrade thermodynamic performance due to stray heat loss.

A good exercise is to modify the ideal cycle in these ways and observe the effect they have on the performance. The reader can design a WORKSHOP based on that idea.

Deeper Dives: Component Analyses

Fundamental design principles of each of the four main components of the vapor/compression refrigeration cycle are developed in this section. The analysis requires basic fluid mechanics and heat transfer theory and is therefore not typically a part of strict thermodynamic analysis, which treats each component as a black box that can achieve a thermodynamic objective, without regard to how it works at a more fundamental level.

Capillary Tube Throttling of a Liquid

A capillary tube (cap tube, for short) is simply a constant area tube with an inner diameter D and length L, shown schematically in Fig. 8.18 with a sudden contraction from the condenser at the inlet and a sudden expansion to the evaporator, which is shown with a larger diameter than the condenser (to accommodate the larger increase in specific volume). In practice, the cap tube is usually coiled for spatial reasons, but behaves as an equivalent straight pipe at an angle θ to the horizontal to include gravity effects. The cap tube system therefore consists of three sections: an entrance region from a condenser, the cap tube itself, and an exit region leading to an

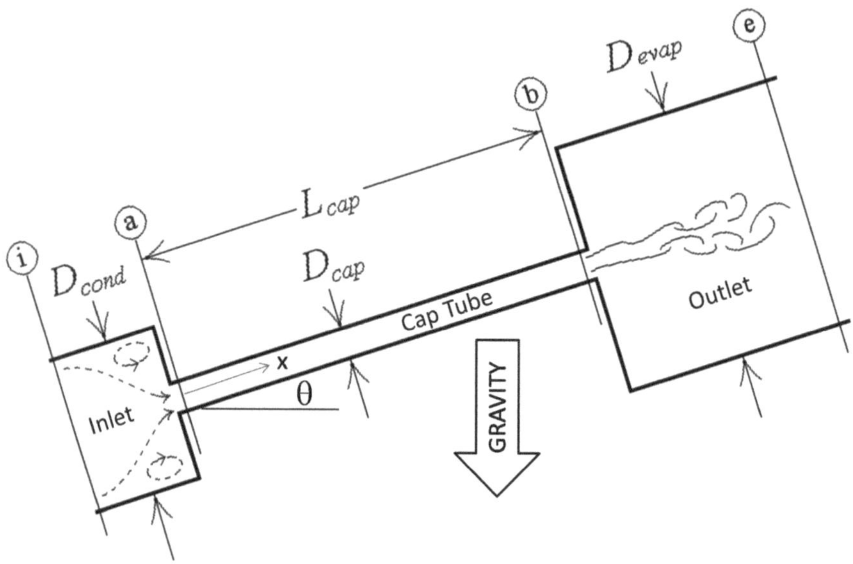

Fig. 8.18 Schematic of capillary tube system (cap tube), with a feed (inlet) from the condenser and flow delivered to the evaporator (outlet). The system is shown with a straight pipe at an angle θ to the horizontal to include effects of gravity. Suggestive streamlines are shown at the inlet and exit of the cap tube itself

evaporator. The three sections will be analyzed separately and then combined to determine the relationship between mass flow and overall pressure change as a function of cap tube length and diameter.

Fluid will flow through a cap tube when it is connected to a high-pressure source at the inlet and low-pressure source at the outlet. The relationship between the pressure drop and flow through a cap tube is very well characterized if there is no phase change. In refrigeration, however, a portion of the liquid tends to evaporate, creating a liquid/vapor mixture that is harder to analyze from a fundamental point of view, but nevertheless is expected to behave qualitatively the same as a nonphase-changing process. In the following analysis, two different limiting cases are considered: one in which no evaporation occurs (incompressible model), and one in which a quasi-equilibrium between the vapor and liquid is assumed (equilibrium model). The term "incompressible model" in the present context means the density is treated as a constant. It generally connotes a situation where any change in density has an insignificant effect on the particular dynamics of the flow in question, even if the density is not strictly constant. Any phase change will result in a significant change in density. The term "equilibrium model" allows for phase change to occur, and the flow is locally in thermodynamic equilibrium. It will be shown that the actual behavior of a cap tube falls in between these limits, and a semi-empirically obtained model will be used to demonstrate various trends.

Part of the objective of this analysis is to include effects of gravity and kinetic energy changes and determine criteria for which they can be neglected, so the cap tube is assumed to be straight and with an angle to the horizontal. The main goal is to determine what length and diameter are needed to obtain a desired mass flow rate, which will result in a cooling rate in the evaporator for specified condenser and evaporator temperatures (which defines the high- and low-side pressures). A more general objective is to conduct a more thorough thermal/fluid analysis than is typically done in a first look at vapor/compression refrigeration.

The source of the flow to the cap tube is liquid from the condenser (state "i" for inlet, which corresponds to state 3 of the ideal vapor/compression cycle of Fig. 8.4) from a tube with a larger diameter ($D_{\text{cond}} > D_{\text{cap}}$). Therefore, the flow experiences a sudden contraction prior to entering the cap tube. The flow must accelerate and therefore the pressure is favorable (i.e., it decreases in the direction of flow). Smooth streamlines are shown, with a mild recirculation in the corner of the contraction. The flow will eventually be modeled as reversible in the entrance section, which requires that the pressure on the wall of contraction be greater than the pressure of the fluid at the entrance to the cap tube.

On the other hand, the flow from the cap tube discharges to the evaporator, with a larger diameter ($D_{\text{evap}} > D_{\text{cap}}$). Therefore, the flow experiences a sudden expansion. In practice, as the analysis will show, there is some pressure recovery accompanied by internal dissipation (eXergy destruction) against an adverse pressure gradient. A jet of fluid enters the evaporator, and internal dissipative effects are felt as indicated by the suggestion of turbulence. The flow will be modeled such that the pressure on the wall of the expansion is equal to that of the fluid at the exit of the cap tube. The flow becomes fully developed in the first portion of the evaporator, which is likely to be several diameters downstream. State "e" (for exit), which corresponds to state 4 of the ideal vapor/compression cycle, is considered to be sufficiently downstream that the flow attains a nominal one-dimensional flow beyond any significant internal jet and/or recirculation regions.

Note that, in the limit of the length of the capillary tube reducing to zero, the fluid flows through a restriction from one pipe diameter to another. In principle, there is no real reason to use a capillary tube for throttling a liquid. In practice, the flow through the cap tube gives the fluid residence time for the required phase change to occur. The results of two models will suggest that equilibrium is not fully realized even with a cap tube (rather than a simple restriction). Consideration of nonequilibrium effects in a fundamental way would require an even deeper dive (with scuba gear!).

The following analysis is intended primarily to consider the controlling physics associated with throttling through a capillary tube. Ultimately, design of refrigeration systems requires the correct interpretation and use of manufacturer specification sheets. An example of experimental results adapted from a manufacturer is given in Fig. 8.19. The results of two model limiting cases (incompressible and equilibrium) will be compared to these results. The reported results considered the combined

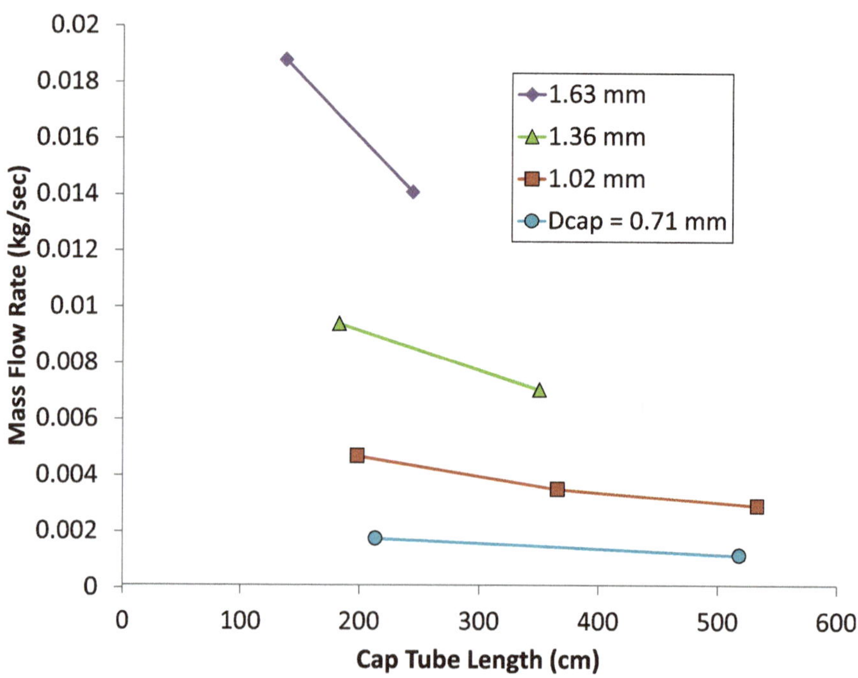

Fig. 8.19 Experimental cap tube data. (Adapted from Tecumseh (https://www.tecumseh.com/~/media/North-America/Files/Guidelines-and-Manuals/Engineering-Recommendations_North-America/Tecumseh-ER-3-Cap-Tube-Sizing.PDF)). The manufacturer lists cooling rates (combined effect of the valve plus the evaporator), but these are converted to refrigerant mass flow rates (assuming a change in enthalpy of 118.8 kJ/kg) to correspond to just the valve. Conditions are stated in the case study problem statement

effects of the cap tube and the subsequent evaporator, and are therefore reported as cooling load. They have been converted to mass flow rates assuming the evaporator behaves ideally, with negligible kinetic energy changes and a change in enthalpy corresponding to the stated inlet and exit conditions. It is clear from the data that the mass flow rate increases and is very sensitive to diameter, and decreases with length.

Case Study

The following problem statement is designed to correspond with the manufacturer data consulted, with a cap tube system of Fig. 8.18. Refrigerant R134a enters a capillary tube of inner diameter D_a and length L_{cap} from a condenser with pipe inner diameter $D_i = 5$ mm and a condensing temperature of 120 °F (49.2 °C) with 5 °F (2.78 °C) subcooling. This specification fixes the thermodynamic state of the inlet (i.e., pressure, enthalpy, specific volume). The capillary tube discharges to an evaporator with an inner diameter $D_e = 7$ mm with an evaporating temperature of −10 °F (−23.0 °C) with 10 °F (5.56 °C) superheat (which fixes the thermodynamic state of the exit). Determine the mass flow rate as a function of length for cap tube diameters that correspond to the manufacturer data ($D_a = D_b = 0.71$, 1.02, 1.36, 1.63 mm). Note that the condenser and evaporator diameters are not specified by the manufacturer.

Results

Detailed analysis of a general nature is conducted in Appendix 1 for three different models: incompressible, equilibrium without kinetic energy, and equilibrium with kinetic energy. The results of the model predictions are presented here. Spreadsheet results for a base case are shown in Fig. 8.20. The results of these models are compared with the incompressible model and the manufacturer data for a cap tube diameter of 1.02 mm in Fig. 8.21, for a constant value of the Fanning friction factor of 0.0025, chosen to correspond with the roughness of drawn copper tubing (1.28 $(10)^{-4}$ mm). Inclusion of kinetic energy changes requires a numerical (i.e., iterative) solution and is developed in Appendix 3.

The incompressible model predicts higher mass flow rates than the manufacturer data and the equilibrium model predicts lower mass flow rates. There is also a much weaker dependency on cap tube length for the equilibrium model compared to the incompressible model. Inclusion of kinetic energy effects plays a minor role. The experimental data is therefore between the incompressible and equilibrium models. A plausible explanation is that, in reality, some evaporation occurs, but there is insufficient time for full equilibrium to be maintained as the refrigerant flows along the cap tube and experiences a friction-induced pressure drop.

The incompressible model is easier to implement and produces similar shapes and trends to the data. It is possible, therefore, to introduce a constant factor K to scale the results into a semiempirical model. Using a value of $K = ½$, the results of the incompressible model with the manufacturer data are shown in Fig. 8.22. This calibrated incompressible model will be used to consider other quantities of interest.

Fig. 8.20 Spreadsheet solution for base case capillary tube

INPUT PARAMETERS

Di	5	mm, condenser diameter
D	1.02	mm, cap tube diameter
De	7	mm, evaporator diameter
L	366	cm, cap tube length
fFANNING	0.0025	friction factor
cliquid	1.3950	kJ/kg/K
Tin	46.4444	oC
x	0.4746	exit quality (equil. Model)
sf@Tevap	0.0982	kJ/kg/K, saturated liquid entropy
sfg@Tevap	0.8547	kJ/kg/K, entropy of vaporization

MOMENTUM EFFECTS

quantity	unit	INCOMPRESSIBLE	EQUIL. NO KE	EQUIL. w/ KE
rhoa	kg/m^3	1133.8	1133.8	1133.8
rhob	kg/m^3	1133.8	11.9	16.4
mdot	kg/s	6.92E-03	2.64E-03	2.90E-03
rho*gL	atm	0.4019	0.2030	0.2038
K	atm	0.6238	0.0912	0.1097
Pi	atm	13.03	13.03	13.03
Pa	atm	12.72	12.99	12.98
Pb	atm	1.13	0.96	1.41
Pe	atm	1.14	1.14	1.57

ENERGY EFFECTS (in kJ/kg)

quantity	INCOMPRESSIBLE	EQUIL. NO KE
h_i	120.5900	120.5900
$V_i^2/2$	0.0000483	0.0000071
h_a	120.5622	120.5859
$V_a^2/2$	0.0279	0.0041
h_b	120.5622	83.7175
$V_b^2/2$	0.0279	36.87
h_e	120.5900	120.5734
$V_e^2/2$	0.0000126	0.0166

ENTROPY (in kJ/kg/K)

state	INCOMPRESSIBLE	EQUIL. NO KE
i	0.44019	0.44019
a	0.44019	0.44019
b	0.44175	0.50391
e	0.44192	0.71333

eXergy EFFECTS (in kJ/kg)

T_∞ = 25 oC

quantity	INCOMPRESSIBLE	EQUIL. NO KE
$T_\infty\sigma_{ia}$	0	0
$T_\infty\sigma_{ab}$	0.47	18.99
$T_\infty\sigma_{be}$	0.05	62.41

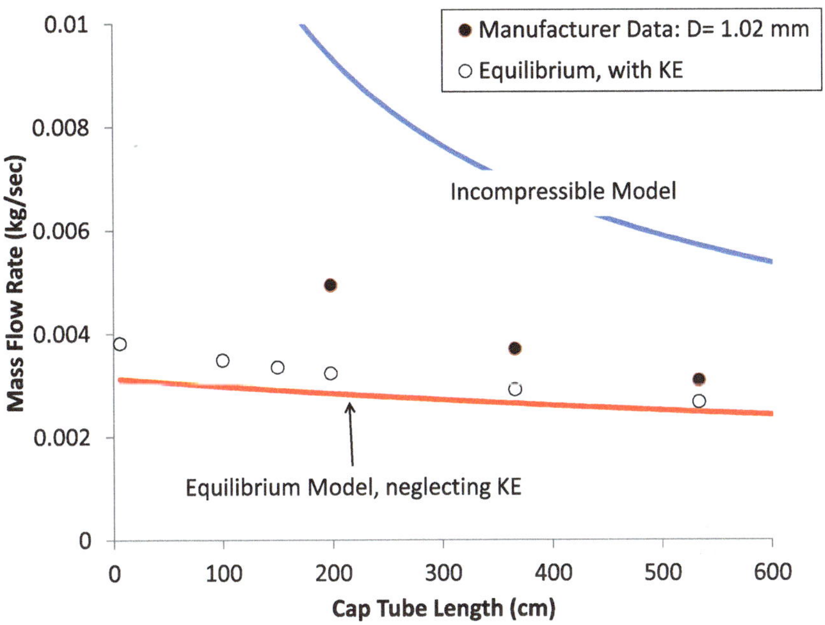

Fig. 8.21 Comparison of models for a cap tube diameter of 1.02 mm

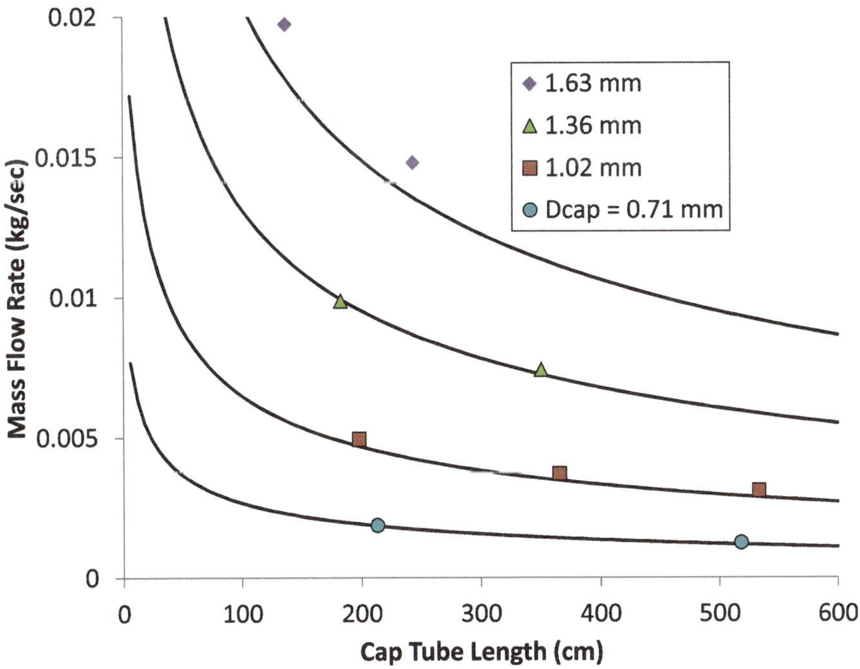

Fig. 8.22 Calibrated incompressible model. A simple factor of 1/2 is applied to the predicted mass flow rate. The solid lines are the model predictions, the symbols are the manufacturer data

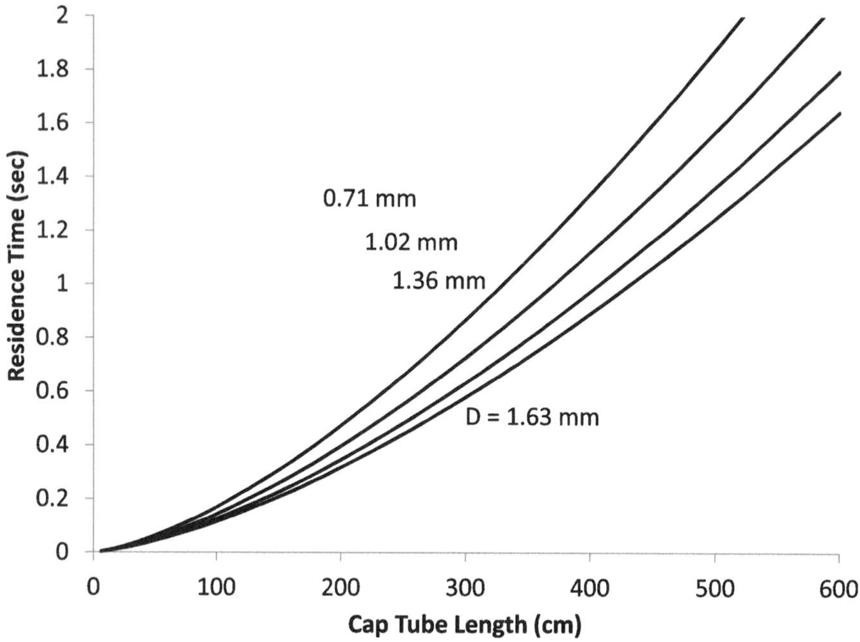

Fig. 8.23 Residence time in cap tube for incompressible model as a function of cap tube length for various diameters

Figure 8.23 shows an estimate of the residence time for the fluid in the cap tube defined as follows:

$$t_{\text{residence}} = \frac{L_{\text{cap}}}{\overline{V}} = \frac{\pi \overline{\rho} D_a^2 L_{\text{cap}}}{4\dot{m}}$$

The mass flow rate from the calibrated incompressible model is used for this plot, which is intended to result in an order of magnitude estimate and demonstration of trends. The residence time is less than 2 s, and weakly dependent on diameter, and strongly dependent on cap tube length. No further attempts are made to develop the fundamental processes needed to model the rate of evaporation. That deeper dive requires scuba gear.

Figure 8.24 shows the Reynolds number for the incompressible flow model. The Reynolds number is well above the critical value for laminar flow (2000) and supports the use of a constant friction factor (fully rough) model of the friction effects. Prior to conducting this analysis, I had expected the flow in the cap tube to be laminar and went ahead and derived the closed-form expressions (now included in Appendix 2 for generality and applications that are laminar, as would be found in even smaller scales, like blood vessels).

Figure 8.25 shows the intermediate pressures obtained for the case of a cap tube diameter of 1.02 mm and length of 366 cm for the incompressible and equilibrium

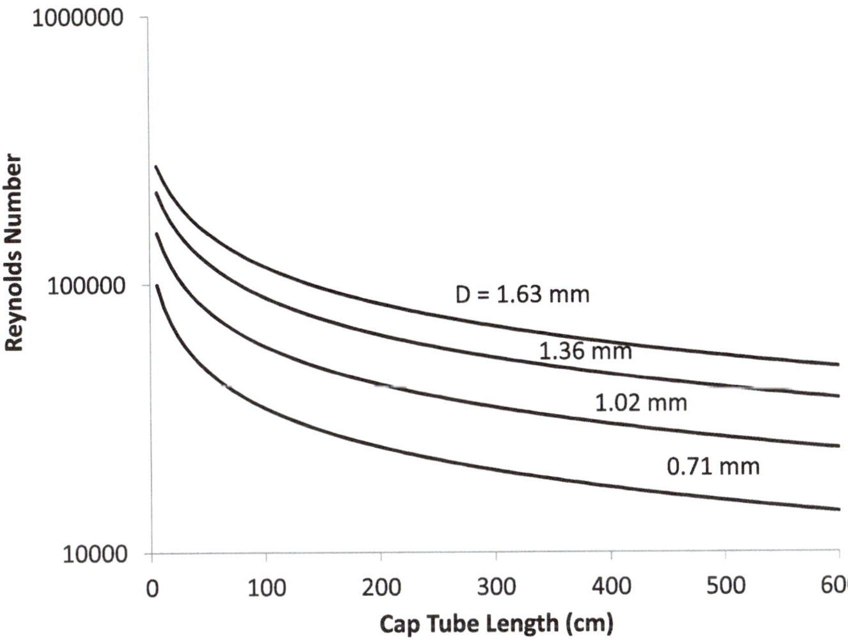

Fig. 8.24 Reynolds number for incompressible model as a function of cap tube length for various diameters

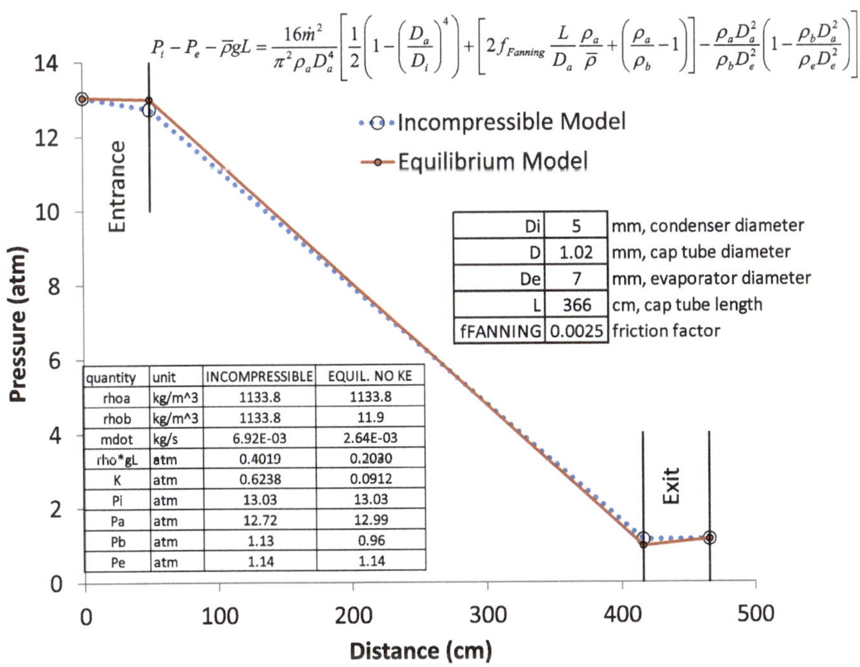

Fig. 8.25 Pressure versus position for incompressible and equilibrium models for a specific cap tube (1.02 mm diameter, 366 cm length)

models. It is clear that the pressure drop across the cap tube itself is dominant. There is a slight drop in pressure at the inlet (sudden contraction), more so for the incompressible model (due to the higher mass flow rate). There is a slight pressure recovery in the exit (sudden expansion), more so for the equilibrium model, due to the lower density.

Final Comments on Throttling

Throttling a liquid is an effective and easy way to achieve a refrigeration effect. It comes at an eXergy cost, however. In principle, it can be replaced by a turbine that would extract useful work as well as increase the cooling capacity by lowering the enthalpy entering evaporator. In the theoretical limit, the process could be reversible, while throttling is inherently irreversible.

There is an analogy with the use of regenerative brakes in automobiles and trucks. Friction brakes are very effective in stopping a vehicle, but they dissipate the kinetic energy stored in the vehicles motion (which was delivered by the engine, of course). It is becoming more common for systems that convert the kinetic energy of the vehicle into a storable mechanical or electrical form, a portion of which is reclaimed later.

Evaporator

The desired heat transfer from a cold space to a working fluid in a refrigeration system takes place in the evaporator, a heat exchanger so named because the entering liquid/vapor mixture from the throttling valve is evaporated. It practice, the exit of the evaporator must be of sufficiently high quality (mass fraction of vapor) or completely evaporated to avoid mechanical problems due to the impingement of droplets on the high-speed moving components in the compressor, the component that follows. Typically, refrigeration systems are designed with a certain amount of superheat, that is, a bit more heat is added than is needed to completely vaporize. In practice, significant superheat can take place by undesired stray heat in the suction line between the evaporator (the part in thermal contact with the cold space) and the compressor inlet.

The evaporator in a refrigeration application is frequently a single-inlet, single-outlet device that exchanges heat with a desired cold space. It is clear from Fig. 8.1 that in a consumer refrigerator the process is never at a true steady state and the cold space does not act like a thermal reservoir (because its temperature changes with time). The implications of this transient behavior will be discussed in this section, but the evaporator will be modeled as being in a quasi-steady state. The cold space temperature may change with time but that is controlled by the thermal balance of the cold space to be investigated in a subsequent section.

For the purposes of analysis, the evaporator will be modeled as consisting of a straight, constant area pipe of inner diameter D_{evap} and length L_{evap} to which heat is transferred from air in a cold space (Fig. 8.26). In practice, the main piping that

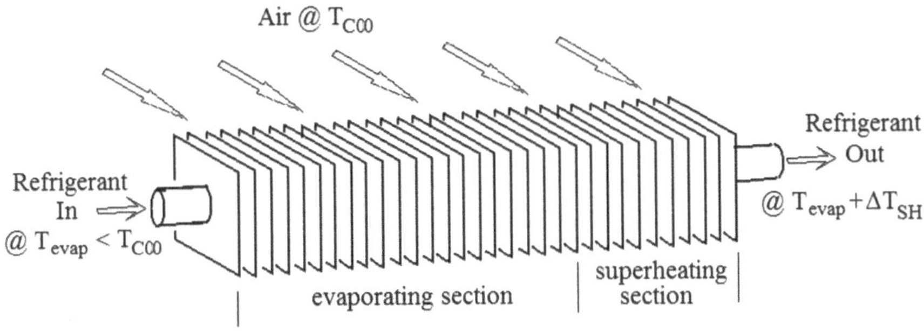

Fig. 8.26 Schematic of evaporator as a finned pipe

contains the fluid is not straight, but may consist of several 180° turns between straight sections, and continuous heat fins attached to the outer surfaces. Compared to the capillary tube, whose purpose is more mechanical than thermal, namely, to produce a pressure drop (throttling, or flashing), the diameter of the evaporator is intended to be large enough so that pressure is nearly constant from inlet to exit. A typical diameter expansion ratio from 1 to 5 (cap tube to evaporator) results in a cross-sectional area 25 times as large. On the other hand, the length of the evaporator is determined by the requirement of sufficient heat transfer from the cold space to completely evaporate the refrigerant. The primary function of the capillary tube is to attain the required pressure drop with minimal heat transfer, while the primary function of the evaporator is to achieve minimal pressure drop with the required heat transfer.

As shown, rectangular sheets of material (heat transfer fins) are attached to the outer wall of the pipe in order to increase the exposed surface area with the air, which flows across the fins. The design details of such a pipe are the subject of a more in-depth study of heat transfer, and not considered here. However, the net effect is to give an overall heat transfer coefficient (Ω) between the two fluids (refrigerant and air). The precise design of actual evaporators can be quite different, but the function is always to provide surface area between the evaporator and the cold space to facilitate heat transfer. The layout considered here allows a clearer visualization of the fundamental processes involved. For example, the quality of refrigerant increases monotonically in the direction of flow from the entrance (the exit of the throttling valve) until it reaches a value of unity (saturated vapor) and then the temperature rises into the superheated region.

Mass

Conservation of mass applied to the evaporator as a single-inlet and single-exit device with a constant cross-sectional area $A_{\text{evap}} = \pi D_{\text{evap}}^2/4$ is

$$\frac{dM}{dt} = \dot{m}_{\text{in}} - \dot{m}_{\text{out}}$$

As with the throttling valve, a quasi-steady-state behavior results in a balance between the inlet and outlet flows, even if the mass within the control volume is not strictly constant. That is,

$$\dot{m}_{\text{in}} \approx \dot{m}_{\text{out}} = \dot{m} \quad @ \text{ quasi-steady state}$$

The mass flow rate is also independent of position along the pipe. However, the specific volume changes dramatically with position, and therefore, the average flow velocity and volumetric flow rate vary considerably with position, unlike an incompressible flow model.

Momentum

The momentum equation is similar to that for the cap tube. Furthermore, the cap tube and evaporator tube represent two flow channels in series that therefore have the same mass flow rate, but through different tube diameters. The Reynolds number, expressed in terms of mass flow rate, is:

$$\mathrm{Re}_D = \frac{\rho V D}{\mu} = \frac{\rho(\dot{m}/\rho A)D}{\mu} = \frac{4\dot{m}}{\pi \mu D}$$

The Reynolds number tends to be lower in the evaporator than in the cap tube because the inner diameter is larger. On the other hand, the effective viscosity is expected to decrease as the quality of the liquid/vapor mixture increases since the viscosity of vapor is much lower than that of liquid.

The pressure drop relationships are the same as for the cap tube, but with the inlet being state 4 of the refrigeration system of Fig. 8.4 (exit of valve, after expansion) and the outlet being state 1 (inlet to compressor):

$$P_4 - P_1 = \Delta P_{\text{EVAP}} = \frac{16\dot{m}^2}{\pi^2 \rho_4 D_{\text{evap}}^4}\left[2 f_{\text{Fanning}}\left(\frac{L_{\text{evap}}}{D_{\text{evap}}}\right)\left(\frac{\rho_4}{\bar{\rho}_{41}}\right) + \left(\frac{\rho_4}{\rho_1} - 1\right)\right]$$

The density and viscosity are averages between the inlet and the exit. Figure 8.27 shows results that correspond to the cap tube case study using the mass flow rate results from the calibrated incompressible flow model. The pressure drop across the evaporator is shown normalized by the pressure drop across the cap tube. It is clear that the pressure drop across the evaporator is much lower than across the cap tube, as is intended. It is noted that the evaporator diameter is the same for all these cases, and that the evaporator diameter for a larger cap tube diameter would likely be larger in practice, resulting in even lower pressure drops than shown in this case study.

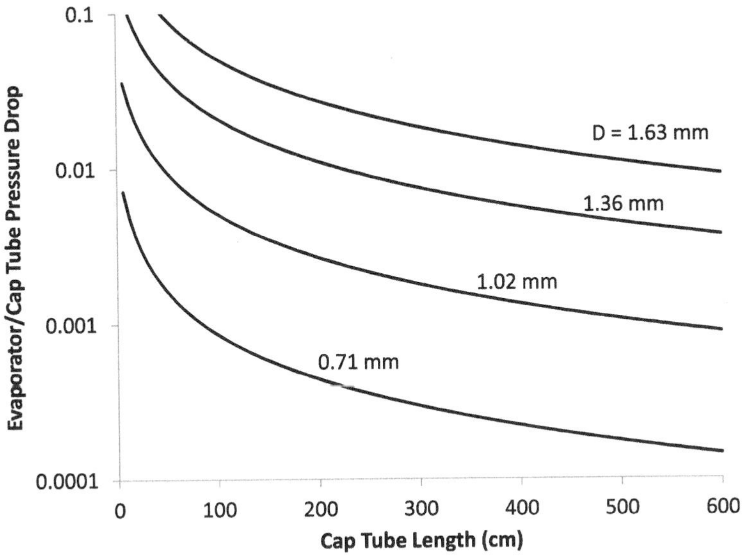

Fig. 8.27 Pressure drop across evaporator, normalized by pressure drop across cap tube. A 1.0-m-long evaporator with a 7 mm inner diameter is assumed for all cases using the calibrated incompressible cap tube model for mass flow rate

Energy

The general energy equation is

$$\frac{dE}{dt} = \dot{Q}_{\text{in}} - \dot{W}_{\text{shaft}} + \dot{m}_{\text{in}}\left(h_{\text{in}} + \frac{V_{\text{in}}^2}{2} + gz_{\text{in}}\right) - \dot{m}_{\text{out}}\left(h_{\text{out}} + \frac{V_{\text{out}}^2}{2} + gz_{\text{out}}\right)$$

For the evaporator, the heat transfer rate is the primary effect, unlike in the valve, where it is considered a secondary effect (stray heat) and normally neglected. It is expressed in terms of an overall heat transfer coefficient, surface area (inner perimeter times pipe length, not cross-sectional area), and temperature difference driving the heat. The shaft power is identically zero, and the gravity effect is neglected, while the change in kinetic energy is retained (at least for now), resulting in

$$0 = (\Omega \pi D L)_{\text{evap}}\left(T_{C\infty} - T_{\text{evap}}\right) + \dot{m}(h_{\text{in}} - h_{\text{out}}) + \dot{m}\frac{V_{\text{in}}^2}{2}\left(1 - \left(\frac{\rho_{\text{in}}}{\rho_{\text{out}}}\right)^2\right)$$

Rearranging in terms of the required evaporator length and expressing the inlet velocity in terms of the mass flow rate:

$$L_{\text{evap}} = \frac{\dot{m}\left[(h_4 - h_1) + \frac{8\dot{m}^2}{\pi^2 \rho_4^2 D_{\text{evap}}^4}\left(1 - \left(\frac{\rho_4}{\rho_1}\right)^2\right)\right]}{(\Omega \pi D)_{\text{evap}}\left(T_{\text{evap}} - T_{C\infty}\right)}$$

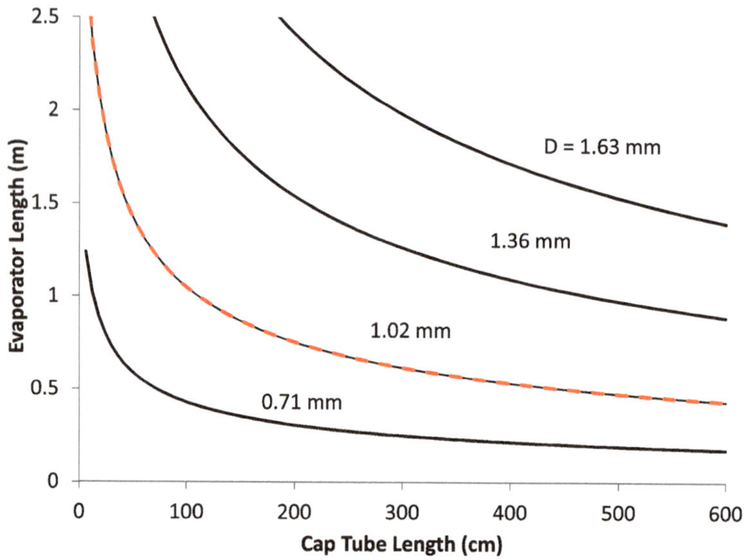

Fig. 8.28 Required evaporator length corresponding to the calibrated incompressible model. An overall heat transfer coefficient of 2500 W/m^2/K is used, based on a finned pipe heat transfer analysis conducted separately. The dashed red line for $D = 1.02$ mm neglects kinetic energy changes for that case

The first term, the change in enthalpy, is considered to be the primary effect, while the change in kinetic energy (the second term that is associated with density change) is considered to be secondary. The required evaporator length for the steady-state, incompressible model is plotted against cap tube length for various diameters in Fig. 8.28. The shape and trends follow those of the mass flow rate, namely, the evaporator length increases strongly with cap tube diameter and inversely with cap tube length.

An assumed value of overall heat transfer coefficient of $\Omega = 2500$ W/m^2/K is used based on a separate heat transfer analysis (not shown) of a finned pipe with a phase change fluid flowing inside. This value is highly uncertain and considered to be an order of magnitude estimate. The actual evaporator length would need to be determined empirically. It is also noted that, in practice, the overall heat transfer coefficient could change with time due to the buildup of ice around it, for example, if the evaporator were in direct contact with liquid water instead of air.

The evaporator length for the $D = 1.02$ mm case is shown with and without kinetic energy effects. These two curves are indistinguishable, even though the kinetic energy change is not identically zero due to the large density, hence velocity, difference between the inlet (state 4) and outlet (state 1). For practical purposes, then, kinetic energy effects in the evaporator of a refrigeration system can be safely neglected.

In a subsequent section, a transient analysis of the cold space will be conducted for which the temperature of the cold space, $T_{C\infty}$, changes with time, the exit enthalpy will change along with it. Expressing the energy equation in terms of the exit enthalpy:

$$h_1 = h_4 + \frac{(\Omega \pi DL)_{\text{evap}}}{\dot{m}}\left(T_{C\infty} - T_{\text{evap}}\right) + \frac{8\dot{m}^2}{\pi^2 \rho_4^2}\left(1 - \left(\frac{\rho_4}{\rho_1}\right)^2\right)$$

The energy equation is expressed this way to put it in the context of a transient refrigeration period, where the cold space temperature changes with time, or perhaps the overall heat transfer coefficient changes with time. The kinetic energy term is included but could be safely neglected in practice. During an active compressor cycle, the cold space temperature decreases, and therefore, the enthalpy of the refrigerant as it exits the evaporator decreases. The total heat transfer rate is less because the temperature difference driving the heat is less. Therefore, in designing refrigerators, care must be taken to ensure that the quality of the refrigerant that exits the evaporator is sufficiently high for the entire transient cycle to prevent damage to the compressor that follows.

Reciprocating Compressor

While the components of the refrigeration cycle are treated as if refrigerant flows steadily through them, a reciprocating compressor actually delivers pulses of flow, much like the way the heart delivers blood through arteries, across capillaries, and returns to the heart through the veins. An accumulator, an auxiliary component, is generally designed into the system to help smooth out the pulses in the other components. The stretching of arteries plays a similar role. A reciprocating compressor, commonly used in household refrigeration, is based around a piston/cylinder arrangement with intake and exhaust flapper valves that lead to inlet and exhaust ports. Figure 8.29 shows schematic diagrams of the compressor at different stages of its cycle. Figure 8.30 shows the cylinder volume and pressure as a function of time, and the corresponding pressure/volume curve for the idealized model considered here. The intake flapper valve opens naturally when the pressure in the cylinder is below the low-side pressure (the exit of the evaporator). The exhaust flapper valve opens when the cylinder pressure is higher than the high-side pressure (the feed to the condenser). The displacement volume (V_{disp}) is defined by the volume swept by the cylinder, and the clearance volume (V_{clear}) is the minimum volume remaining in the cylinder when the piston is at "top dead center." The compression ratio is the ratio of the maximum (V_{disp} + V_{clear}) to minimum (V_{clear}).

For modeling purposes, an ideal compressor cycle is defined as one that is adiabatic, with no pressure drops across the flapper valves and assuming the refrigerant behaves as a perfect gas (similar to the air compressor of Chap. 6).

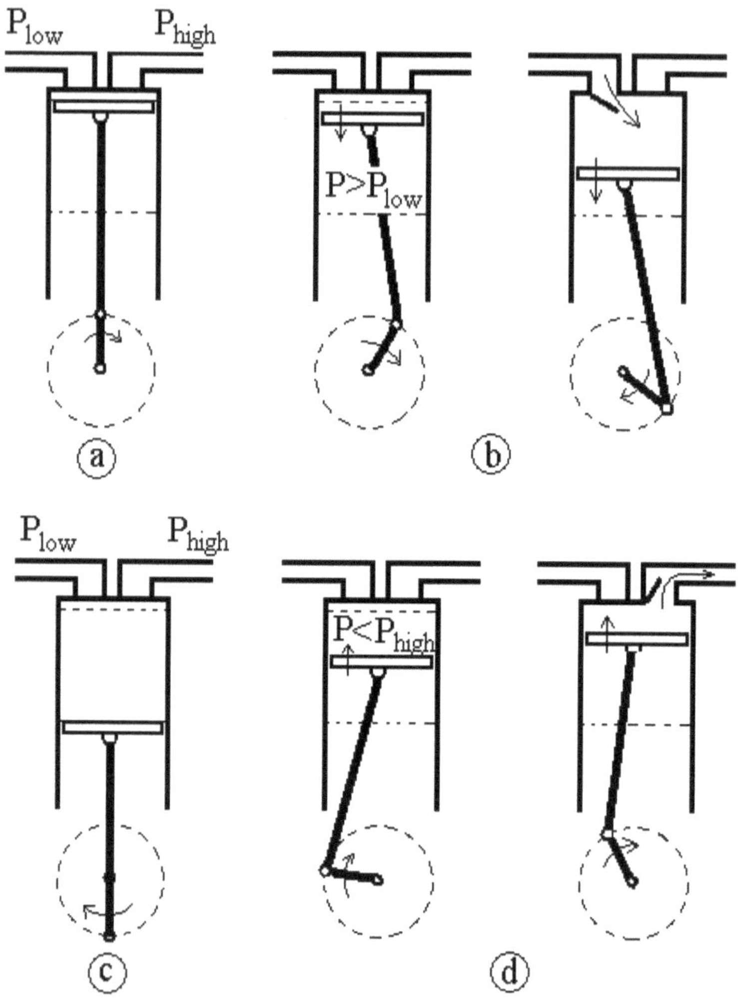

Fig. 8.29 Compressor cycle at various stages. The top row shows the intake stroke and the bottom row shows the exhaust stroke

Real effects are incorporated by using assumed isentropic and volumetric efficiencies as appropriate. Figure 8.30 shows results for a specific case, namely, with a displacement volume of 17.3 cc (1.06 in^3), a clearance volume of 0.583 cc, and a compressor speed of 1500 RPM. The high-side pressure is 1016.4 kPa (corresponding to a condensing temperature of 40 °C) and the low-side pressure is 93.05 kPa (corresponding to an evaporating temperature of −28 °C).

The cylinder cycle is shown starting at point "a," where the cylinder volume is at a minimum (the clearance volume), the pressure equals the high-side pressure (P_H), and the temperature equals that at state "d" when the exhaust valve opens

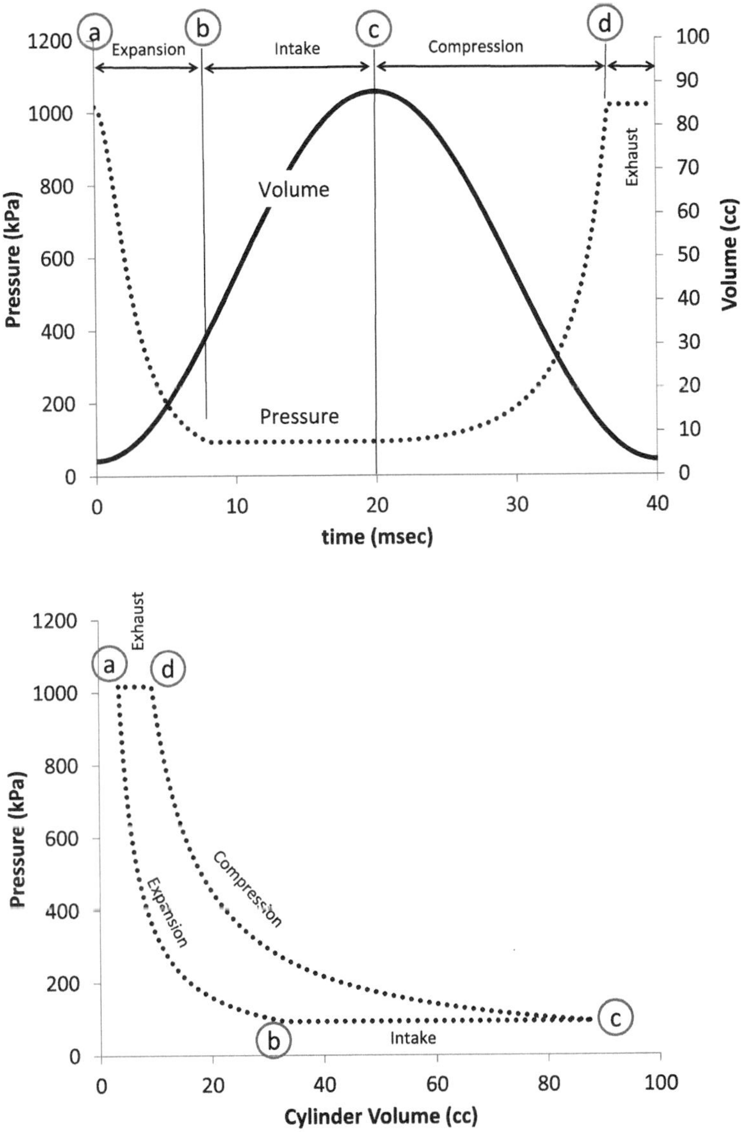

Fig. 8.30 Top graph shows cylinder pressure and volume versus time. For this case, the displacement volume is 84.5 cc, the clearance volume is 3.52 cc (for a compression ratio of 25), and the compressor speed is 1500 RPM. The bottom graph shows the corresponding PV diagram

(which equals the refrigerant at state 2 in the refrigeration cycle, T_2, unknown at this point).

During the intake stroke, the piston initially withdraws with both valves closed and the refrigerant undergoes an adiabatic expansion (with PV^k = constant, where

$k = c_p/(c_p - R_{gas})$ and c_p is the heat capacity of the vapor). When the volume expands to $V_c(P_{high}/P_{low})^{1/k}$, the pressure in the cylinder will equal the low-side pressure, defining state "*b*." Upon further volume increase, the flapper valve will open and allow refrigerant to flow into the cylinder from the low-pressure side. The total mass inducted (in the idealized cycle) is well characterized by the intake process b-c. This mass inducted is

$$M_{intake} = \rho_1(V_c - V_b) = \frac{P_L}{R_{gas}(T_1 + 273)}\left(V_{disp} - V_{clear}\left(\frac{P_H}{P_L}\right)^{\frac{1}{k}}\right)$$

The ideal gas law has been used to determine the refrigerant density, with $R_{gas} = 8.314/\text{MW}$ as the gas constant (units of kJ/kg/K) and MW is the molecular weight of the refrigerant (102 kg/kmol) and recognizing that the temperature at the intake equals the reservoir temperature (the thermally long evaporator assumption) and the pressure is the low-side pressure. In a real cycle, the cylinder pressure during intake will be below the low-side port pressure, and therefore, the mass inducted will be less than this expression suggests. This real effect is modeled using a volumetric efficiency (η_v), defined as the actual mass inducted divided by the ideal mass inducted. The mass flow rate of the refrigeration cycle is therefore determined by the mass inducted per cycle times the number of cycles per unit time (the compressor speed):

$$\dot{m} = \frac{\eta_v N P_L V_{disp}}{R_{gas}(T_1 + 273)}\left(1 - \frac{V_{clear}}{V_{disp}}\left(\frac{P_H}{P_L}\right)^{\frac{1}{k}}\right)$$

The exhaust stroke begins with an adiabatic compression of the refrigerant until the pressure in the cylinder equals the high-side pressure (state "d") at which point the exhaust flapper valve will open and refrigerant will flow into the exhaust port that leads to the condenser. The temperature of the refrigerant at the end of the intake (T_c) equals the evaporator exit temperature. The pressure equals the low-side pressure. As shown in Fig. 8.30, the exhaust process (d-a) occurs over a short sweep of the piston. The majority of the sweep during the exhaust stroke is required to compress the vapor to the point where it can exhaust. The mass exhausted, however, equals the mass inducted once a steady state has been reached (after a few cycles).

The power input required for the compressor can be calculated by analyzing the compressor as a steady flow device. The input power is obtained from an energy balance across the compressor:

$$\dot{W}_{elec} = \frac{\dot{W}_{shaft}}{\eta_m} = \frac{\dot{m}(h_2 - h_1)}{\eta_m} = \frac{\dot{m}(h_{2s} - h_1)}{\eta_m \eta_c}$$

In addition to the motor efficiency, η_m, a compressor efficiency, η_c, defined as the work required in an isentropic process divided by the actual, and a hypothetical state

2s (defined by an isentropic compression from the low pressure to the high pressure) has been introduced. A large contributor to this component efficiency being less than unity is the pressure drop across the flapper valves. Treating the vapor as a perfect gas (with $h_{2s} - h_1 = c_p(T_{2s} - T_1)$) and subsequent rearrangement:

$$\dot{W}_{\text{elec}} = \frac{\dot{m}c_{p,\text{vapor}}(T_{2s} - T_1)}{\eta_m \eta_c} = \frac{\dot{m}c_{p,\text{vapor}}T_1\left(\frac{T_{2s}}{T_1} - 1\right)}{\eta_m \eta_c}$$

The compression process (c–d in Fig. 8.29) is modeled as an isentropic compression of a perfect gas, $T_{2s}/T_1 = (P_H/P_L)^{(R/cp)}$, and the final expression for power delivered to the compressor is

$$\dot{W}_{\text{elec}} = \frac{\dot{m}c_{p,\text{vapor}}T_1\left(\left(\frac{P_H}{P_L}\right)^{\left(R/c_{p,\text{vapor}}\right)} - 1\right)}{\eta_m \eta_c}$$

The temperature of superheated vapor exiting the compressor (state 2) is determined from an energy balance across the compressor:

$$T_2 = T_1 + \frac{\dot{W}_{\text{shaft}}}{\dot{m}c_{p,\text{vapor}}}$$

Condenser

Once the evaporator picks up all the heat it can from the cold space while it is at a low pressure, the compressor is needed to raise the refrigerant to a pressure whose saturation temperature is above ambient. However, the refrigerant is left in a superheated vapor state. The condenser is needed to change the phase to a liquid, while maintaining the pressure sufficiently high, and thereby return the working fluid to the point where it can be flashed in the throttling valve. The function of the condenser is therefore opposite to that of the evaporator but accomplished by similar means. The direction of heat transfer is from the refrigerant to either a high-temperature reservoir (ambient air in a refrigeration application or the heated space in a heat pump application) or to another fluid in a heat exchanger with a secondary cooling loop. The reservoir case is considered here, while a secondary loop model is considered in Workshop 8.3. From an overall energy balance of the refrigeration system, the heat rate that must be rejected equals that picked up from the cold reservoir in the evaporator plus the mechanical power that is added to the working fluid in the compressor.

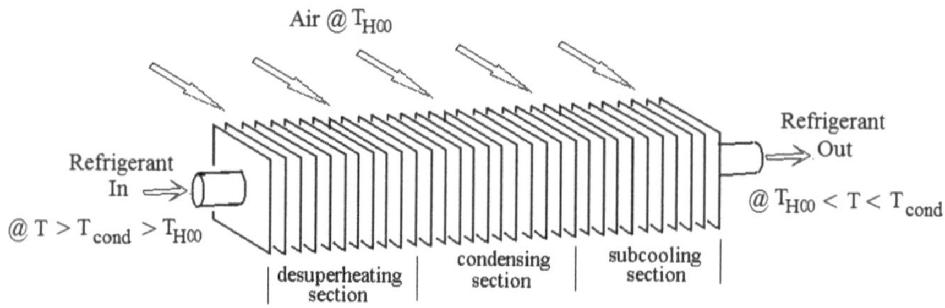

Fig. 8.31 Schematic of condenser

Figure 8.31 shows a schematic, similar to the evaporator, in which refrigerant enters a single-inlet, single-outlet heat transfer device (from the compressor, corresponding to state 2 of the refrigeration cycle of Fig. 8.4) as a superheated vapor. In practice, there would be a discharge line between the compressor and condenser proper where stray heat would reduce the temperature somewhat. This stray heat works in concert with the function of the condenser and need not be minimized in practice. While superheated, the temperature is very sensitive to heat transfer and at some point within the condenser the refrigerant becomes a saturated vapor (a quality of 1), at which point further heat transfer reduces the quality until the refrigerant becomes a saturated liquid. It is of note, from a fluid mechanical perspective, that the density increases dramatically and, therefore, for a constant area pipe, the velocity decreases in the direction of flow.

Momentum Effects

The pressure drop relationships are the same as for the cap tube and evaporator, but with the inlet being state 2 of the refrigeration system of Fig. 8.4 (exit of compressor) and the outlet being state 3 (inlet to throttling valve):

$$P_2 - P_3 = \Delta P_{\text{COND}} = \frac{16\dot{m}^2}{\pi^2 \rho_2 D_{\text{cond}}^4}\left[2f_{\text{Fanning}}\left(\frac{L_{\text{cond}}}{D_{\text{cond}}}\right)\left(\frac{\rho_2}{\overline{\rho}_{23}}\right) + \left(\frac{\rho_2}{\rho_3} - 1\right)\right]$$

The difference between the condenser and the evaporator is that since the density increases in the direction of flow, the pressure tends to rise in the direction of flow (an adverse pressure gradient). Friction will always tend to reduce the pressure, but the condensing to liquid can result in a pressure rise, which may seem counterintuitive. When applied to the case study using the calibrated incompressible model with a fixed condenser length of 1.5 m and diameter of 5 mm, there is a net pressure rise, plotted in Fig. 8.32 and normalized by the pressure drop across the cap tube. The magnitude of the pressure rise is small compared to the cap tube pressure drop. The pressure drop is considered to be negligible in most analyses, and the condenser process, like the evaporator, is considered to be isobaric.

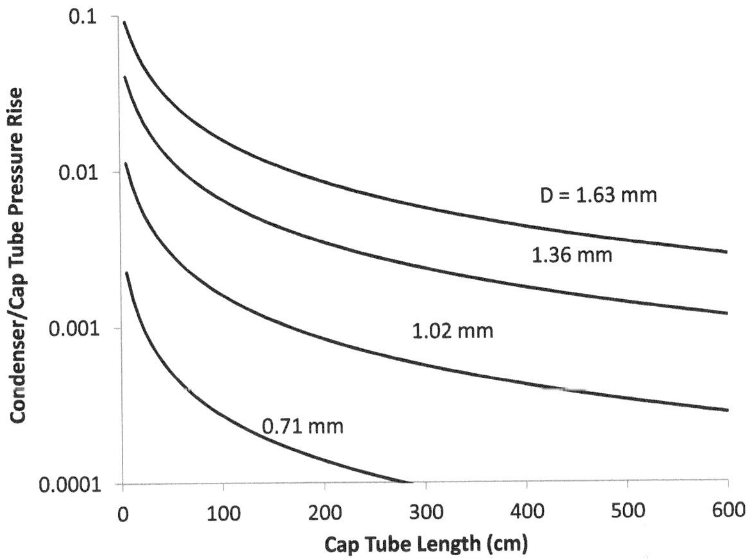

Fig. 8.32 Pressure rise in condenser normalized by pressure across cap tube. A 1.0-m-long condenser with a 5 mm inner diameter is assumed for all cases using the calibrated incompressible cap tube model for mass flow rate

Energy Effects

The energy equation is the same as for the evaporator, with a different inlet and exit. Arranged for the required length of condenser:

$$L_{\text{cond}} = \frac{\dot{m}\left[(h_2 - h_3) + \frac{8\dot{m}^2}{\pi^2 \rho_2^2 D_{\text{cond}}^4}\left(1 - \left(\frac{\rho_2}{\rho_3}\right)^2\right)\right]}{(\Omega \pi D)_{\text{cond}}(T_{\text{cond}} - T_{\text{H}\infty})}$$

Results for the calibrated equilibrium model are shown in Fig. 8.33. Kinetic energy effects are negligible, but not explicitly shown as it was for the evaporator. The condenser lengths have similar trends to evaporator lengths, but are longer to account for the greater total heat transfer rate required. Also, the temperature difference driving the majority of the heat is slightly different in the two cases.

Transient Behavior of Cold Space

There are two very different time scales associated with transient behavior for typical household refrigerator applications: reciprocating compressor transients (already discussed) and cold space temperature transients. A reciprocating pressure will give rise to pulses of flow in and out of a cylinder with a regular frequency, or

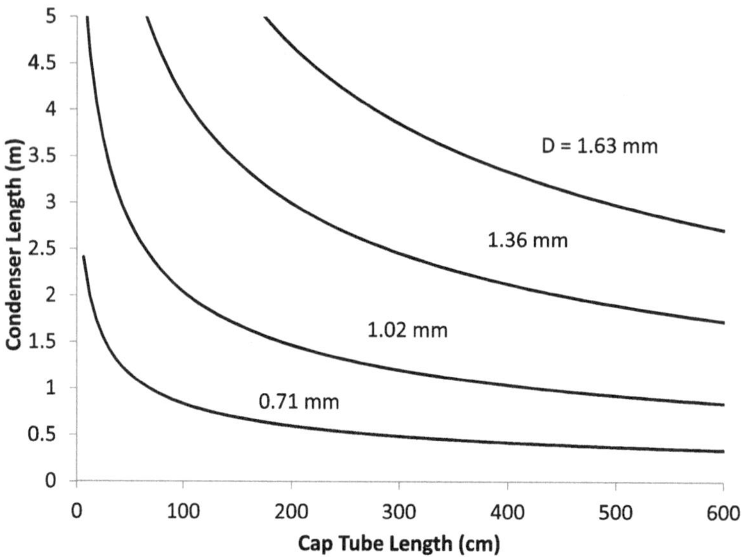

Fig. 8.33 Required condenser length corresponding to the calibrated incompressible model. An overall heat transfer coefficient of 2500 $W/m^2/K$ is used based on a finned pipe heat transfer analysis conducted separately

heartbeat. The other components (condenser, valve, and evaporator) will in principle experience a pulsed flow about a mean, but the amplitude of the pulses is generally considered minor, with an accumulator to assist. That is, a quasi-steady flow rate is assumed to take place in all other components (condenser, cap tube, and evaporator) while the compressor is operating. On a longer time scale, the cold space temperature changes during the compressor on and off cycles, with a resultant change in the heat transfer rate to the evaporator, which causes a change in the operating conditions of all other components. This section analyzes the transient behavior of the cold space.

The temperature of the cold space was treated as a thermal reservoir in the evaporator analysis, at temperature $T_{C\infty}$. However, it is clear from Fig. 8.1 that the cold space is not a strict thermal reservoir because its temperature changes with time. A feedback control system in a conventional refrigerator is designed to keep the cold space between two temperature extremes, T_{min} and T_{max}. In operation, if the compressor is active, the cold space will decrease in time until it reaches the minimum temperature, at which point the compressor will turn off. The temperature of the refrigerant flowing through the evaporator (T_{evap}) is below that of the cold space temperature during this time. While the compressor is dormant, the unit will warm toward ambient. When it reaches T_{max}, the compressor will kick on, completing a compressor cycle. Feedback control systems are not necessarily designed that way, but the behavior is similar. For example, there may be a temperature sensor to trigger when the compressor turns on (at T_{max}), and then the compressor is maintained on for a set time, rather than by a second temperature sensor set at T_{min}.

A First Law analysis of the cold space response to the on/off cycle of the compressor is conducted in this section. The transient energy balance, neglecting the active and parasitic thermal loads, is

$$\frac{dE_{\text{unit}}}{dt} = \dot{Q}_{\text{passive}} - \dot{Q}_{\text{C}} \quad \text{during active compressor stage (compressor on)}$$

$$\frac{dE_{\text{unit}}}{dt} = \dot{Q}_{\text{passive}} \quad \text{during dormant compressor stage (compressor off)}$$

As there is no phase change, the rate of change of energy stored within the unit (which includes the air, the cold food, the shelves unit, and an effective portion of the walls), an effective thermal capacitance can be identified. Therefore,

$$\frac{dE_{\text{unit}}}{dt} = C_{\text{EFF}} \frac{dT_{\text{C}\infty}}{dt}$$

The two heat transfer rates are not strict constants because they depend on the temperature difference that drives them. Therefore, using overall heat transfer coefficients Ω, multiplied by an effective surface area, A, the passive thermal load can be expressed as

$$\dot{Q}_{\text{passive}} = (\Omega A)_{\text{passive}} (T_{\text{H}\infty} - T_{\text{C}\infty})$$

The overall heat transfer coefficient depends on the precise construction of the unit, and heat transfer analysis would be needed to develop it further. However, increasing the insulation of the box would decrease it. Note, however, that the warmer the unit, the lower the passive thermal load (for a constant room temperature, $T_{\text{H}\infty}$).

On the other hand, as developed in this chapter, the cooling load is considered to take place between the cold space air and the temperature of a refrigerant flowing in an evaporator at a temperature T_{evap}. That is,

$$\dot{Q}_{\text{C}} = (\Omega A)_{\text{evap}} (T_{\text{C}\infty} - T_{\text{evap}})$$

Therefore, the energy equation for the cold space temperature becomes

$$C_{\text{EFF}} \frac{dT_{\text{C}\infty}}{dt} = (\Omega A)_{\text{passive}} (T_{\text{H}\infty} - T_{\text{C}\infty}) - (\Omega A)_{\text{evap}} (T_{\text{C}\infty} - T_{\text{evap}}) \quad \text{during active compressor stage}$$

$$C_{\text{EFF}} \frac{dT_{\text{C}\infty}}{dt} = (\Omega A)_{\text{passive}} (T_{\text{H}\infty} - T_{\text{C}\infty}) \quad \text{during dormant compressor stage}$$

Assuming constant coefficients (and ambient temperature), these first-order equations have a well-defined steady-state temperature (although the compressor state is changed long before equilibrium is reached) and a closed-form solution for the temperature as a function of time. The cold space temperature while the compressor is active is

$$T_{C\infty} = T_{SS,ON} + (T_{MAX} - T_{SS,ON})e^{\frac{-t}{\tau_{ON}}}$$

where the steady-state temperature is

$$T_{SS,ON} = \frac{(\Omega A)_{passive}T_{H\infty} + (\Omega A)_{evap}T_{evap}}{(\Omega A)_{passive} + (\Omega A)_{evap}}$$

and the time constant during the active compressor stage is

$$\tau_{ON} = \frac{C_{EFF}}{(\Omega A)_{passive} + (\Omega A)_{evap}}$$

During the dormant compressor stage, the cold space temperature is given by

$$T_{C\infty} = T_{H\infty} + (T_{MIN} - T_{H\infty})e^{\frac{-t}{\tau_{OFF}}}$$

where the steady-state temperature is ambient ($T_{H\infty}$), and the time constant during the dormant compressor stage is

$$\tau_{OFF} = \frac{C_{EFF}}{(\Omega A)_{passive}}$$

The duty cycle of the compressor is defined by the ratio of the active (ON) time to cycle time (ON + OFF):

$$\text{time}_{ON} = -\tau_{ON}\ln\left(\frac{T_{MIN} - T_{SS,ON}}{T_{MAX} - T_{SS,OFF}}\right)$$

$$\text{time}_{OFF} = -\tau_{OFF}\ln\left(\frac{T_{MAX} - T_{H\infty}}{T_{MIN} - T_{H\infty}}\right)$$

$$\text{Duty cycle} = \frac{\text{time}_{ON}}{\text{time}_{ON} + \text{time}_{OFF}}$$

This analysis serves to demonstrate that the design of a refrigerator involves a delicate balance between the thermal management of the cold space and the capability of the refrigeration system.

Appendix 1: Thermo/Fluid Analysis of Capillary Tube

The analysis of the cap tube with the sudden contraction from the condenser and sudden expansion to the evaporator (Fig. 8.18) is conducted in this Appendix. Compressibility effects associated with the large change in specific volume (and therefore flow velocity and kinetic energy) due to phase change are included.

Mass Flow Considerations

The general mass flow equation for a single-inlet, single-outlet control volume is

$$\frac{dM}{dt} = \dot{m}_{\text{in}} - \dot{m}_{\text{out}}$$

A quasi-steady state exists if the rate of change of mass stored is small compared to the magnitude of the inlet and outlet mass flow rates. For the cap tube system, at quasi-steady state, the mass flow rate at each station is equal:

$$\rho_i V_i A_i = \rho_a V_a A_a = \rho_b V_b A_b = \rho_e V_e A_e$$

For incompressible flows, the density is the same at each station, making the volumetric flow rate (AV) invariant, resulting in a simple relationship between flow velocity and cross-sectional area:

$$\begin{aligned} V_i A_i &= V_a A_a = V_b A_b \\ &= V_e A_e \quad \text{for constant density flows (incompressible approximation)} \end{aligned}$$

However, if any evaporation occurs, the specific volume and therefore density changes dramatically between inlet and outlet. For example, in the ideal vapor/compression case study in which the density of the liquid entering the valve (state 3) is $\rho_3 = 1/v_3 = 1187\ \text{kg/m}^3$, and that exiting is $\rho_4 = 1/v_4 = 11.32\ \text{kg/m}^3$, a factor of 100 lower density. These values correspond to the assumption of thermodynamic equilibrium at the inlet and exit. The diameter of the evaporator would have to be 10 times larger (which it typically is not) than the condenser in order to have equal velocities. Then, continued evaporation in the evaporator would give rise to a much larger density decrease. So, it is clear that there are significant changes in average flow velocities in the refrigeration cycle, and that kinetic energy changes are present. Whether they are important requires more careful consideration, part of the objective of the subsequent analyses.

Inlet Section, *i–a*

From Chap. 5, the general momentum equation is

$$M_{\text{CV}} \frac{d\vec{V}_{\text{CM}}}{dt} = \vec{F}_{\text{surface}} + \vec{F}_{\text{body}} + \sum_{\text{inlets}} \dot{m}_i \vec{V}_i - \sum_{\text{outlets}} m_e \vec{V}_e$$

At quasi-steady state, the momentum equation in the y direction applied to the fluid within the straight vertical section of pipe (of undefined length L_{ia}) with an upward flow immediately ahead of the contraction is

$$0 = \left[P_i A_i - P_a A_a - \overline{P}_{\text{wall},i}(A_i - A_a) - \overline{\tau}_w \pi D_i L_{ia}\right] - \overline{\rho}_{ia} g L_{ia} \pi D_i^2/4 + \dot{m}V_i - \dot{m}V_a$$

In this expression, there is a pressure force exerted on the fluid by the wall where the sudden contraction occurs. This pressure varies across the surface of the wall and is expressed as an average pressure. In the case of the inlet, there must be a transverse pressure gradient to accelerate the fluid from the outer edges to the restriction, so the average pressure is higher than the fluid pressure as it enters the cap tube proper, at state a. There is also a shear stress along the walls, which will be neglected in the inlet and exit, but is a primary effect in the cap tube itself. The undefined length can be shown to be of the order of the inlet (condenser) diameter, D_i. For the upward flow, gravity acts in the negative direction. The momentum equation is not particularly useful in establishing the relationship between flow rate and pressure change directly for the inlet section because of the unknown pressure on the wall. However, it will be of interest to solve for the average pressure on the upstream wall, so rearranging the momentum equation:

$$\overline{P}_{\text{wall},i} = \left(\frac{1}{A_i - A_a}\right) \times \left[P_i A_i - P_a A_a + \frac{\dot{m}^2}{\rho_i A_i}\left(1 - \frac{\rho_i A_i}{\rho_a A_a}\right) - \text{GRAVITY} - \text{FRICTION}\right]$$

The energy equation for the inlet section (sudden contraction), neglecting potential energy changes, is

$$\frac{dE}{dt} = \dot{Q}_{ia} + \dot{m}\left(h_i - h_a + \frac{V_i^2}{2} - \frac{V_a^2}{2}\right)$$

At quasi-steady state, assuming negligible heat flows into the fluid, the change in enthalpy is

$$h_i - h_a = \left(\frac{V_a^2}{2} - \frac{V_i^2}{2}\right) \quad \text{adiabatic and steady state}$$

There is a conversion of enthalpy to kinetic energy as the flow is accelerated. Introducing mass balance equations, rearranging, and expressing in different ways:

$$h_i - h_a = \frac{V_i^2}{2}\left(\frac{V_a^2}{V_i^2} - 1\right) = \frac{\dot{m}^2}{2\rho_i^2 A_i^2}\left(\frac{\rho_i^2 A_i^2}{\rho_a^2 A_a^2} - 1\right) = \frac{8\dot{m}^2}{\pi^2 \rho_i^2 D_i^4}\left(\frac{\rho_i^2 D_i^4}{\rho_a^2 D_a^4} - 1\right)$$

The entropy balance is

$$\frac{dS}{dt} = \frac{\dot{Q}_{ia}}{T_b} + \dot{m}(s_i - s_a) + \dddot{\Sigma}_{ia}$$

Assuming steady-state and adiabatic flow:

$$s_a - s_i = \frac{\dddot{\Sigma}_{ia}}{\dot{m}} = \sigma_{ia} \quad \text{adiabatic and steady state}$$

That is, any irreversibility gives rise to higher entropy exiting than entering.

Given state conditions at the inlet (e.g., pressure and temperature for compressed liquids) and assuming reversible flow in the inlet section, the energy and entropy balance equations represent two equations that can be solved in principle to relate the pressure at the inlet to the cap tube section and mass flow rate. However, if thermodynamic equilibrium is maintained (meaning that evaporation occurs as the pressure falls), an iterative method would be needed. For example, if the pressure were "guessed," then the quality at state "*a*" can be determined from the entropy balance, from which the enthalpy change and density can be determined, and the mass flow rate obtained from the energy equation.

On the other hand, a simple closed-form relationship between pressure and mass flow rate exists if the flow is assumed to be incompressible, that is, $\rho_i = \rho_a = \rho$, and reversible, $\sigma_{ia} = 0$. Breaking the enthalpy into its mechanical and thermal components $h = u + P/\rho$ and rearranging the energy equation:

$$P_i - P_a = \frac{8\dot{m}^2}{\pi^2 \rho D_a^4}\left(1 - \frac{D_a^4}{D_i^4}\right) + \rho(u_a - u_i) \quad \text{For incompressible adiabatic flow}$$

The internal energy change represents a thermal effect.

For reversible, adiabatic flow, $s_i = s_a$. The entropy of a compressed liquid is a function of temperature only, and therefore, an isentropic and incompressible flow is also isothermal. Furthermore, the internal energy is a function of temperature only, and therefore, the internal energy change is zero. Therefore, for steady-state, adiabatic, reversible, and incompressible flow in the inlet section, the pressure change expressed in several equivalent ways is

$$\begin{aligned} P_i - P_a &= \frac{8\dot{m}^2}{\pi^2 \rho D_i^4}\left(\left(\frac{D_i}{D_a}\right)^4 - 1\right) = \frac{1}{2}\frac{\dot{m}^2}{\rho A_i^2}\left(\left(\frac{A_i}{A_a}\right)^2 - 1\right) \\ &= \frac{8\dot{m}^2}{\pi^2 \rho D_a^4}\left(1 - \left(\frac{D_a}{D_i}\right)^4\right) = \frac{1}{2}\frac{\dot{m}^2}{\rho A_a^2}\left(1 - \left(\frac{A_a}{A_i}\right)^2\right) \end{aligned}$$

For the limiting case where the cap tube diameter is much less than the condenser diameter:

$$P_i - P_a = \left(\frac{8}{\pi^2 \rho D_a^4}\right)\dot{m}^2 \quad \text{if } D_a \ll D_i$$

This diameter restriction is not difficult to attain in practice due to the fourth power dependency of diameter ratio. For example, if $D_a = D_i/2$, there is a $1/2^4$ (approximately 6%) error associated with neglecting the diameter effect. A typical ratio is more like $D_a = D_i/5$, which results in a 0.16% error. The limiting case is retained going forward. It is just worth mentioning for other applications with more modest flow contractions.

The pressure at the cap tube entrance (state a), which decreases with flow rate squared, cannot be negative. Therefore, for a given inlet pressure and diameters, there is a maximum possible flow rate that would result in a perfect vacuum at the inlet. That is,

$$\dot{m}_{\max} = A_a \sqrt{\frac{2\rho P_i}{1 - A_a^2/A_i^2}}$$

There will clearly be a pressure drop in the inlet section and therefore a tendency for evaporation to begin, which would make the incompressible approximation invalid. However, it seems likely (but hard to justify rigorously) that there will be insufficient time for significant evaporation. Therefore, going forward, the flow in the inlet section will be considered to be reversible, adiabatic, and for the fluid to be incompressible (meaning that no evaporation occurs, even though the pressure is below the saturation pressure). The acceleration in this inlet section may be important, however, as the pressure drop across the cap tube itself (between states a and b) will be lower than the overall pressure drop between the inlet and exit.

Out of curiosity, as it is not necessary to proceed, the average pressure on the wall can now be obtained by back substitution into the momentum equation. The result, after some algebra, is

$$\overline{P}_{\text{wall},i} = P_i - \frac{1}{2}\rho V_i^2\left(\frac{A_i}{A_a} - 1\right) = P_a + \frac{1}{2}\rho V_i^2\left(\frac{A_i}{A_a} - 1\right)\left(\frac{A_i}{A_a}\right)$$

The factor $\rho V_1^2/2$ is the dynamic pressure or stagnation pressure associated with the upstream flow. This result shows that the pressure in the corner is higher than the inlet pressure (the flow accelerates to the restriction), but the pressure in the corner is higher than at the entrance to the cap tube. It is this relative high pressure that turns the flow from the larger area A_i to the restricted area A_a.

Exit Section *b–e*

The governing equations for the exit section are similar to the inlet section, with state "*i*" replaced by "*b*" and "*a*" replaced by "*e*." The difference in behavior is related to the mechanism by which the velocity change is achieved. The inlet experiences a favorable pressure gradient, while an adverse pressure gradient is associated with the exit (giving rise to flow separation). For the inlet, there is a high-pressure area in the corner that turns the approaching flow toward the cap tube entrance. For the exit (b to e), an analogous turning of the flow would require a lower pressure than at the exit, which is not possible. Rather, the high-velocity jet enters an effectively stagnant region, and the large-velocity gradients result in internal friction. Therefore, eXergy destruction is a primary effect of the exit, while it is considered secondary in the entrance, a result that is difficult to justify further without a much more rigorous thermo/fluid mechanics analysis. A portion, but not all, of the kinetic energy associated with the exit jet entering into the larger evaporator diameter is recovered as a pressure rise.

The steady-state momentum equation applied to the fluid within the straight section of pipe (of undefined length L_{be}) immediately after the expansion is

$$P_bA_b + \overline{P}_{\text{wall}}(A_e - A_b) + \dot{m}V_b = P_eA_e + \dot{m}V_e + \overline{\rho}_{be}gL_{be}\pi D_e^2/4 + \overline{\tau}_w\pi D_eL_{be}$$

The difference between inlet and outlet is that the pressure of the jet that exits the restriction into the larger diameter matches the pressure outside of it, and there is minimal transverse pressure gradient at the inlet section. That is, $P_b \approx \overline{P}_{\text{wall}}$. Neglecting friction and gravity, the pressure recovery in the exit section is

$$P_e - P_b = \frac{\dot{m}(V_b - V_e)}{A_e} = \frac{\dot{m}V_b}{A_e}\left(1 - \frac{V_e}{V_b}\right) = \frac{\dot{m}^2}{\rho_bA_eA_b}\left(1 - \left(\frac{\rho_bA_b}{\rho_eA_e}\right)\right)$$

Expressed in terms of diameters, and multiplying and dividing by the density at state a (to put in a form to combine with the cap tube analysis to come):

$$P_e - P_b = \frac{16\dot{m}^2}{\pi^2\rho_aD_a^4}\left(\frac{\rho_aD_a^2}{\rho_bD_e^2}\right)\left(1 - \frac{\rho_bD_b^2}{\rho_eD_e^2}\right)$$

For incompressible flows $\rho_b = \rho_e = \rho$

$$P_e - P_b = \frac{\dot{m}^2}{\rho A_eA_b}\left(1 - \left(\frac{A_b}{A_e}\right)\right) \quad \text{(incompressible flows)}$$

To demonstrate that there is eXergy destruction, consider the steady-state energy balance:

$$h_b + \frac{V_b^2}{2} + q_{be} = h_e + \frac{V_e^2}{2}$$

Breaking the enthalpy into its mechanical (P/ρ) and thermal (u) components, and grouping the thermal effects together in brackets:

$$\frac{P_e}{\rho_e} - \frac{P_b}{\rho_b} = \frac{V_b^2}{2}\left(1 - \frac{V_a^2}{V_b^2}\right) - [-q_{be} + (u_e - u_i)]$$

For incompressible flows, the energy equation is

$$P_e - P_b = \frac{1}{2}\frac{\dot{m}^2}{\rho A_b^2}\left(1 - \frac{A_b^2}{A_e^2}\right) - \rho[-q_{be} + (u_e - u_i)]$$

Combining the momentum and energy equations:

$$\rho[-q_{be} + u_e - u_b] = \frac{1}{2}\rho V_b^2\left(1 - \frac{A_b}{A_e}\right)^2 \quad \text{for incompressible flows}$$

Note that if there is no expansion ($A_b = A_e$), then there are no thermal effects (and no pressure rise either). On the other hand, if the expansion is large ($A_e \gg A_b$), then all the kinetic energy of the jet is dissipated into thermal effects.

The thermal effects can be related to the eXergy destruction by application of the 2nd Law, and with any heat transfer considered to take place ultimately to ambient. In that case, combining the entropy balance:

$$\rho[T_\infty \sigma_{be} + (u_e - T_\infty s_e) - (u_b - T_\infty s_b)] = \frac{1}{2}\rho V_b^2\left(1 - \frac{A_b}{A_e}\right)^2$$

The first term is the eXergy destruction, and the two other terms are the change in eXergy of the fluid. The latter eXergy is not useful, and the left-hand side represents an effective eXergy destruction.

A form of 2nd Law efficiency of the exit section can be defined as the ratio of the actual pressure rise to the pressure rise of a reversible deceleration (which could be approached in practice with a gradual, rather than a sudden, expansion). That is,

$$\eta_{\text{II,exit}} = \frac{\Delta P_{\text{exit}}}{\Delta P_{\text{exit,reversible}}} = \frac{\frac{\dot{m}^2}{\rho A_b A_e}\left(1 - \frac{A_b}{A_e}\right)}{\frac{1}{2}\frac{\dot{m}^2}{\rho A_b^2}\left(1 - \frac{A_b^2}{A_e^2}\right)} = \frac{2\frac{A_b}{A_e}\left(1 - \frac{A_b}{A_e}\right)}{\left(1 + \frac{A_b}{A_e}\right)\left(1 - \frac{A_b}{A_e}\right)}$$

$$\eta_{\text{II,exit}} = \frac{2}{1 + A_e/A_b}$$

The pressure recovery is nearly complete for a very small expansion and goes to zero for large expansions.

Capillary Tube Section *a–b*

Conservation of mass applied to the capillary tube (valve), a single-inlet–single-exit device with a constant cross-sectional area $A = \pi D_a^2/4$, is

$$\frac{dM}{dt} = \dot{m}_{\text{in}} - \dot{m}_{\text{out}} = \rho_a V_a A - \rho_b V_b A$$

At steady state, the inlet and exit mass flow rates are equal, and mass conservation reduces to

$$V_b = \left(\frac{\rho_a}{\rho_b}\right) V_a$$

If any evaporation occurs, the density entering the valve is greater than at the exit, and therefore, the exit velocity is greater than the inlet velocity. The inlet density is considered to be a known input, while the density of the exit requires a more detailed analysis. For the incompressible model, the inlet and exit densities (and therefore velocities) are equal.

Momentum

The momentum equation is needed to design a cap tube system, that is, to specify the length and diameter of a cap tube. This design is not typically presented in traditional thermodynamics treatments of refrigeration, where the overall energy consequence is sufficient. It is developed here in some detail as an important application of refrigeration, and a crossover between fluid mechanics and thermodynamics. An analysis of flow in a constant area duct was developed in Chap. 5 for incompressible flows (i.e., for constant density). Since the density can change dramatically, a more general treatment is given here. From Chap. 5, the general momentum equation in the direction of flow (x) is the dot product of the full vector momentum equation with the unit vector in the x direction:

$$\left[M_{\text{CV}} \frac{d\vec{V}_{\text{CM}}}{dt} = \vec{F}_{\text{surface}} + \vec{F}_{\text{body}} + \sum_{\text{inlets}} \dot{m}_i \vec{V}_i - \sum_{\text{outlets}} m_e \vec{V}_e\right] \cdot \hat{i}$$

With the control volume taken to consist of the fluid contained within the cap tube, the surface forces consist of the pressure forces at the inlet and outlet, and the fluid viscous (shear) force that acts along the inner wall of the tube. The body force is the weight of the fluid within the tube acting in the direction of flow. There is a momentum flux at the inlet and outlet. Applied to the capillary tube, at steady state, in the vertical (y) direction:

$$0 = -\pi D_a \int_{x=0}^{L} \tau_w dx + P_a A_a - P_b A_a - g \sin\theta A_a \int_{x=0}^{L} \rho dx + \dot{m}_a V_a - \dot{m}_b V_b$$

Here, the shear force is expressed as a shear stress (τ_w) times the differential inner surface area on which it acts (perimeter times dx). All quantities that vary with position within the tube are shown inside integrals. Using average values for these, that is,

$$\int_{x=0}^{L} \tau_w dx \equiv \overline{\tau}_w L \quad \text{and} \quad \int_{x=0}^{L} \rho dx \equiv \overline{\rho} L$$

The change in the momentum flux for a constant area duct at quasi-steady state can be expressed as

$$\dot{m}(V_b - V_a) = \dot{m}V_a\left(\frac{V_b}{V_a} - 1\right) = \frac{\dot{m}^2}{\rho_a A_a}\left(\frac{\rho_a}{\rho_b} - 1\right) = \frac{4\dot{m}^2}{\pi \rho_a D_a^2}\left(\frac{\rho_a}{\rho_b} - 1\right)$$

Dividing through by the area and rearranging the momentum equation, the pressure drop across the capillary tube is expressed as

$$P_a - P_b = 4\left(\frac{L_{cap}}{D_a}\right)\overline{\tau}_w + \overline{\rho} g L \sin\theta + \frac{16\dot{m}^2}{\pi^2 \rho_a D_a^4}\left(\frac{\rho_a}{\rho_b} - 1\right)$$

The pressure drop across the capillary tube must overcome the wall friction, gravity (which would be negative for downward flow), and the different momentum flux exiting than entering as the fluid must be either accelerated (if $\rho_a > \rho_b$) or decelerated (if $\rho_a > \rho_b$) as it flows through the pipe. The friction and gravity terms are both integrated averages along the length of the duct, while the momentum flux is related to inlet and exit conditions only, and does not directly involve the pipe length. The gravity term is a hydrostatic effect and can be grouped with the pressure. In practice, a cap tube is coiled and does not experience significant elevation changes, but is included here for generality for pipe flows. A similar expression will be used for the evaporator and condenser, whose momentum equation is similar (with different inlet and exit states).

For incompressible flows, there is no momentum flux effect for constant area pipe flows and the pressure drop is due entirely to wall friction and gravity effects. For the capillary tube, however, there is potentially a large change in density associated with the phase change.

A general treatment of the friction effect is to define a friction factor, f, which is the shear stress normalized by the dynamic pressure ($\rho V^2/2$).

$$\tau_w = f_{\text{Fanning}} \frac{\rho V^2}{2} = f_{\text{Fanning}} \frac{8\dot{m}^2}{\pi^2 \rho D^4}$$

The latter form introduces the mass flow rate (velocity divided by density and cross-sectional area), which is independent of position, in favor of the average flow velocity, V. The friction factor is not a known constant, in general, but depends on the Reynolds number of the flow $\text{Re}_D = \rho V D/\mu$, which is where viscosity effects come into play, and the roughness of the tube.

The flow is laminar for Reynolds numbers below a critical value (of approximately 2000), and an exact solution exists, one that is independent of the roughness of the tube but is inversely related to flow rate, as demonstrated in Appendix 2. For sufficiently large Reynolds numbers, the friction factor is a constant whose value increases with the roughness of the tube, the so-called "fully rough" behavior. The friction factor has a complex behavior for intermediate values of Reynolds number. The results of the case study produce sufficiently high Reynolds numbers to justify the use of a constant friction factor.

There are two versions of friction factor in the literature, one proposed by Fanning (defined here based on the shear stress) and one proposed by Darcy (defined based on pressure drop). They are closely related:

$$f_{\text{Darcy}} = 4f_{\text{Fanning}}$$

The factor 4 arises from the ratio of cross-sectional area to perimeter. It is important to know which friction factor is being used when estimating the value from the literature.

The density and therefore velocity changes along the duct, and the integrated average shear stress is approximated by using an average fluid density along the duct:

$$\overline{\tau}_w = f_{\text{Fanning}} \frac{8\dot{m}^2}{\pi^2 \overline{\rho} D^4} = f_{\text{Fanning}} \frac{8\dot{m}^2}{\pi^2 \rho_a D^4} \left(\frac{\rho_a}{\overline{\rho}} \right)$$

The last form is obtained by multiplying and dividing by the cap tube inlet density to give the friction term a similar functional dependency as the momentum flux term.

The momentum equation becomes

$$P_a - P_b - \overline{\rho} g L \sin\theta = \frac{16\dot{m}^2}{\pi^2 \rho_a D_a^4} \left[2f_{\text{Fanning}} \left(\frac{L}{D_a} \right) \left(\frac{\rho_a}{\overline{\rho}} \right) + \left(\frac{\rho_a}{\rho_b} - 1 \right) \right]$$

The laminar flow case is considered in Appendix 2. The gravity term is a hydrostatic effect that has been grouped with the pressure drop across the cap tube, and an average density is needed there. The first term in brackets represents the pressure drop due to wall friction, and the second term is due to the different momentum flux between inlet and exit. Assuming the friction factor to be independent of diameter (fully rough behavior), the friction term depends on the diameter to the inverse fifth power, while the momentum has a lower inverse fourth power dependency. For laminar flows, the friction is proportional to the mass flow rate (not the mass flow rate squared).

A formal demonstration of eXergy destruction and friction factor for incompressible flow is given in Appendix 4.

Cap Tube System

The inlet pressure to the cap tube (state a) does not equal the pressure before the sudden contraction. Recall, for incompressible entrance section, the pressure drop across the inlet section is

$$P_i - P_a = \frac{8\dot{m}^2}{\pi^2 \rho_a D_a^4}\left(1 - \left(\frac{D_a}{D_i}\right)^4\right)$$

At the exit of the cap tube, a jet of flow discharges into a sudden expansion. The pressure recovery after the expansion was shown to be

$$P_e - P_b = \frac{16\dot{m}^2}{\pi^2 \rho_a D_a^4}\left(\frac{\rho_a D_a^2}{\rho_b D_e^2}\right)\left(1 - \frac{\rho_b D_a^2}{\rho_e D_e^2}\right)$$

Therefore, the overall pressure drop across the cap tube system (contraction, cap tube, and expansion) is

$$P_i - P_e - \bar{\rho} g L = \frac{16\dot{m}^2}{\pi^2 \rho_a D_a^4} \times \left[\frac{1}{2}\left(1 - \left(\frac{D_a}{D_i}\right)^4\right) + \left[2 f_{\text{Fanning}} \frac{L}{D_a}\frac{\rho_a}{\bar{\rho}} + \left(\frac{\rho_a}{\rho_b} - 1\right)\right] - \frac{\rho_a D_a^2}{\rho_b D_e^2}\left(1 - \frac{\rho_b D_a^2}{\rho_e D_e^2}\right)\right]$$

The first term in brackets is due to the change in pressure at the inlet (contraction), the next two terms are across the cap tube itself, and the last term is the exit (expansion) effect. Rearranging for the mass flow rate:

$$\dot{m} = \frac{\pi D_a^2}{4}\sqrt{\frac{\rho_a\left[(P_i - P_e) - \rho_a g L_{\text{cap}} \sin\theta\right]}{\frac{1}{2}\left(1 - \frac{D_a}{D_i}\right)^4 + 2 f_{\text{fanning}} \frac{L_{\text{cap}}}{D_a}\frac{\rho_a}{\bar{\rho}} + \frac{\rho_a}{\rho_b} - 1 - \frac{\rho_a D_a^2}{\rho_b D_e^2}\left(1 - \frac{\rho_b D_a^2}{\rho_e D_e^2}\right)}}$$

The solution is not complete because the density of the exit gases is not known. However, if it is assumed that the density is constant throughout (incompressible model), the mass flow rate becomes the following closed-form expression:

$$\dot{m} = \frac{\pi D_a^2}{4}\sqrt{\frac{\rho_a\left[(P_i - P_e) - \rho_a g L_{\text{cap}} \sin\theta\right]}{\frac{1}{2}\left(1 - \frac{D_a}{D_i}\right)^4 + 2 f_{\text{fanning}} \frac{L_{\text{cap}}}{D_a} - \frac{D_a^2}{D_e^2}\left(1 - \frac{D_a^2}{D_e^2}\right)}} \quad \text{Incompressible model}$$

The equilibrium model assumes that evaporation occurs, but the extent of evaporation cannot be determined from mass and momentum considerations alone. Therefore, the energy balance on the cap tube is considered next.

The general energy equation for a single-inlet, single-outlet device is

$$\frac{dE}{dt} = \dot{Q}_{in} - \dot{W}_{\text{shaft}} + \dot{m}_{\text{in}}\left(h_{\text{in}} + \frac{V_{\text{in}}^2}{2} + gz_{\text{in}}\right) - \dot{m}_{\text{out}}\left(h_{\text{out}} + \frac{V_{\text{out}}^2}{2} + gz_{\text{out}}\right)$$

For the cap tube, the shaft power is identically zero, and at steady state, using mass conservation, applying the energy balance across both the contraction and the cap tube:

$$h_b = h_i - \frac{1}{2}V_i^2\left(\left(\frac{V_b^2}{V_i^2}\right) - 1\right) - gL\sin\theta + q_{ib}$$

$$h_b = h_i - \frac{8\dot{m}^2}{\pi^2\rho_i^2 D_i^4}\left(\left(\frac{\rho_i}{\rho_b}\right)^2\left(\frac{D_i}{D_b}\right)^4 - 1\right) - gL\sin\theta + q_{ib}$$

An equivalent expression is

$$h_b = h_i - \frac{8\dot{m}^2}{\pi^2\rho_i^2 D_b^4}\left(\left(\frac{\rho_i}{\rho_b}\right)^2 - \left(\frac{D_b}{D_i}\right)^4\right) - gL\sin\theta + q_{ib}$$

In principle, the enthalpy exiting a cap tube is different from that entering due to a change in kinetic energy, potential energy (gravity), and heat transfer. It is common (and generally justified) to consider these effects to be of secondary importance and therefore to neglect them for the cap tube system (including sudden contraction at the inlet and expansion at the outlet), and the basic relationship for a cap tube (or any other throttling valve design) is simply

$$h_i \approx h_a \approx h_b \approx h_e \quad \text{steady-state, adiabatic, neglect KE and PE}$$

That is, a throttling process (i.e., the adiabatic drop in pressure with no power extracted) is usually treated as isenthalpic. Notice that throttling of an ideal gas results in no temperature change, and therefore cannot be used for refrigeration. It is the energy consequence of the phase change required of a high-pressure liquid that results in the refrigeration effect when a liquid is throttled.

For isenthalpic cap tube approximation, two independent thermodynamic property values are known (evaporating temperature, T_c, and enthalpy $h_b = h_i$) and, if equilibrium is assumed, the quality exiting the cap tube is calculated from

$$x_b = \frac{h_b - h_{f@T_c}}{h_{fg@Tc}}$$

The specific volume (and therefore density) at state b is determined from

$$v_b = v_{f@Tc} + x_b v_{fg@Tc} = 1/\rho_b$$

An entropy balance will demonstrate that the throttling process involves eXergy destruction:

$$\frac{dS}{dt} = \frac{\dot{Q}_{\text{in}}}{T_b} + \dot{m}(s_{\text{in}} - s_{\text{out}}) + \dddot{\Sigma} = 0 \quad @SS$$

The eXergy destruction rate is

$$e\dddot{X}_{\text{D}} = T_{\infty}\dddot{\Sigma} = T_{\infty}(s_{\text{out}} - s_{\text{in}}) - \frac{T_{\infty}}{T_b}\dot{Q}_{\text{in}}$$

An adiabatic throttling process inherently involves an entropy increase, and therefore, eXergy destruction.

The entropy change in a straight pipe section (cap tube), obtained from combining mass, momentum, and energy, and using thermodynamic property relations, and assuming steady state, incompressible, and adiabatic flow is developed in detail in Appendix 4. The result is, with state variables a and b for inlet and exit of cap tube section, respectively,

$$s_b - s_a = c\ln\left(1 + \left(\frac{8\dot{m}^2}{\pi^2\rho_a D_a^4}\right)\left(f\frac{L}{D_a}\right)\left(\frac{1}{\rho_a c(T_{\text{in}} + 273)}\right)\right)$$

Since the pressure drop across the cap tube system is dominated by that across the cap tube section itself, it is useful to recall the relationship between pressure drops across cap tube, namely,

$$P_a - P_b = \left(\frac{8\dot{m}^2}{\pi^2\rho_a D_a^4}\right)4f_{\text{Fanning}}\frac{L}{D_a}$$

The entropy change across the cap tube in terms of the pressure drop across the cap tube becomes

$$s_b - s_a = c\ln\left(1 + \frac{(P_a - P_b)}{4\rho_a c(T_{\text{in}} + 273)}\right)$$

The results of this long Appendix are presented in the main text, itself a deeper dive.

Appendix 2: Shear Stress for Laminar Flow

A relationship between the shear stress and flow velocity, a topic of fluid mechanics, is needed to analyze the capillary tube. For Newtonian fluids, the shear stress within a fluid is proportional to the local velocity gradient (*dV/dr*). Applied at the wall of the tube:

$$\tau_w = -\mu \frac{dV}{dr}\bigg|_{\text{wall}}$$

where the dynamic viscosity, μ, a fluid property, has been introduced, with SI units of kg/s/m = Pa-s. For example, at nominal room temperature the dynamic viscosity of liquid H_2O is 0.001 kg/s/m, while the viscosity of air is some 50 times lower (0.00002 kg/s/m). Viscosity values of saturated liquids and vapors for R134a are shown in Fig. 8.34. The dynamic viscosity of saturated liquid is more than an order of magnitude higher than vapor, and is more temperature sensitive.

Laminar Flow Model

A relationship between the shear stress and flow velocity is needed. That topic of fluid mechanics is developed here for laminar flow (defined by the Reynolds number being less than a critical value of approximately 2000). The result for the shear stress at the wall at any location is expressed as

$$\tau_w = \frac{8\mu V}{D} = \frac{32\mu\ \dot{m}}{\pi\ \rho D_{\text{cap}}^3} \quad \text{valid for} \quad \text{Re}_D = \frac{\rho V D}{\mu} < \text{Re}_{D,\text{critical}} \approx 2000$$

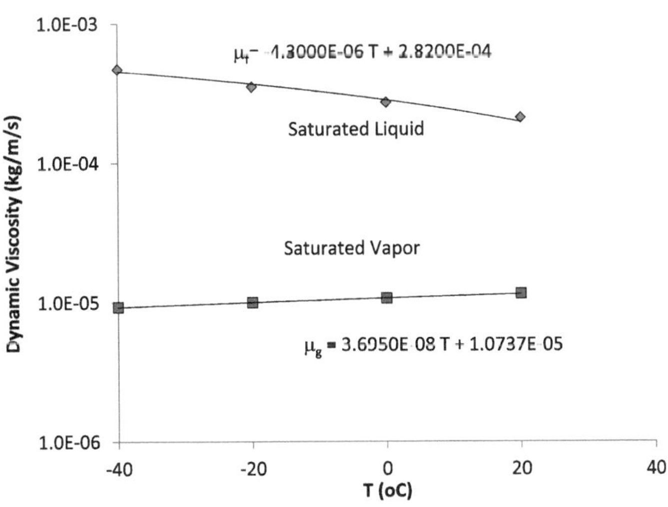

Fig. 8.34 Dynamic viscosity of R134a saturated liquid and vapor as a function of temperature. Linear fits are shown to be used for modeling

The latter form introduces the mass flow rate (divided by density and cross-sectional area), which is independent of position, in favor of the average flow velocity, V. The fluid properties (viscosity and density) change significantly between the inlet to exit, and therefore, so does the local shear stress.

To derive this result, recall that the fluid velocity used in mass conservation represents a mass-averaged flow velocity that crosses a plane. Determination of shear stress requires more detail about the shape of the velocity profile. While exact treatments are conducted in fluids textbooks, a technique used in many more advanced fluid mechanics studies (like boundary layer theory) is conducted here. It starts with assuming that a polynomial fit for the velocity profile can yield a suitable velocity profile, and then applying conditions or constraints to determine the polynomial coefficients. That is, showing only three terms:

$$V_{(r)} = a + br + cr^2 + \ldots$$

At the centerline ($r = 0$), due to symmetry, the velocity gradient must be zero:

$$\left.\frac{dV}{dr}\right|_0 = (b + 2cr)|_{r=0} = 0 \quad \text{by symmetry}$$

Therefore, the coefficient $b = 0$.

The no-slip condition applies, namely, that the fluid velocity at a fluid/solid interface equals the solid velocity, which is zero for a stationary tube. That is,

$$V_{(D/2)} = 0 = a + c\left(\frac{D}{2}\right)^2 \quad \text{and therefore} \quad c = -a\left(\frac{2}{D}\right)^2$$

A final condition is set by the definition of the mass average velocity, V. That is,

$$\dot{m} = \rho VA = \int_{r=0}^{D/2} \rho V_{(r)} 2\pi r dr = 2\pi\rho \int_{r=0}^{D/2} a\left(1 - \left(\frac{2}{D}\right)^2 r^2\right) r dr$$

$$= 2\pi\rho a\left[\frac{r^2}{2} - \left(\frac{2}{D}\right)^2 \frac{r^4}{4}\right]_0^{D/2} = \frac{\pi\rho a}{2}\left(\frac{D}{2}\right)^2$$

It has been assumed that the specific volume does not change in the radial direction (even as it changes in the direction of flow), and the velocity profile (in terms of a) has been introduced.

Evaluating the area results in $a = 2V$ and the velocity profile as a 2nd order polynomial is given by

$$V_{(r)} = 2V\left(1 - \left(\frac{r}{D/2}\right)^2\right)$$

A 2nd order polynomial shape to the velocity is obtained. This velocity profile is known to be an exact solution if the flow is laminar, which is found empirically to be valid if the Reynolds number $Re_D = \rho VD/\mu$ is below a critical value of approximately 2000. In this case, the velocity profile is

$$\left.\frac{dV}{dr}\right|_{r=D/2} = \frac{-8V}{D}$$

And the shear stress at the wall at any location is

$$\tau_w = \frac{8\mu V}{D} = \frac{32\mu\ \dot{m}}{\pi\ \rho D_{\text{cap}}^3}$$

The latter form introduces the mass flow rate (divided by density and cross-sectional area), which is independent of position, in favor of the average flow velocity, V. Since the viscosity and density of liquid and vapor are different, the viscosity and density of the fluid will change from the inlet to exit, and the total shear stress is expressed with an average viscosity.

Viscosity is a fluid mechanical property, not a thermodynamic property, and therefore the viscosity of a liquid/vapor mixture may not strictly follow the quality calculation. Nevertheless, for the purpose of demonstrating cap tube behavior, the quality formula $\mu = (1 - x)\mu_f + x\mu_g$ is used.

If the flow is assumed to be laminar in the duct, the momentum equation takes the following form:

$$P_a - P_b - \bar{\rho} gL = \left(\frac{128\bar{\mu}L}{\pi\bar{\rho}D^4}\right)m + \frac{16}{\pi^2\rho_{\text{in}}D^4}\left(\frac{\rho_{\text{in}}}{\rho_{\text{out}}} - 1\right)\dot{m}^2 \quad \text{For laminar flow}$$

Note that the pressure drop due to friction is proportional to the mass flow rate, while the momentum effect is proportional to the mass flow rate squared. Both the friction and momentum flux terms are inversely proportional to the diameter to the fourth power, a very strong effect.

Assuming that the flow in the inlet is incompressible, and that there is no pressure recovery in the exit section, an expression for the overall pressure drop across the total cap tube system is

$$\begin{aligned}\Delta P_{\text{ov}} &= P_i - P_e \\ &= \left(\frac{128\bar{\mu}L}{\pi\bar{\rho}D^4}\right)\dot{m} + g\bar{\rho}L + \frac{16}{\pi^2\rho_{\text{in}}D^4}\left(\left(\frac{\rho_{\text{in}}}{\rho_{\text{out}}} - 1\right) + \frac{1}{2}\left(1 - (D_a/D_i)^4\right)\right)\dot{m}^2\end{aligned}$$

Rearranging in the form of a quadratic equation for mass flow rate:

$$\frac{1}{2}\dot{m}^2 + \frac{4\pi\bar{\mu}L\rho_a}{\bar{\rho}\left(\frac{\rho_a}{\rho_b} - \frac{1}{2}\left(1 - \left(\frac{D_a}{D_i}\right)^4\right)\right)}\dot{m} - \frac{\pi^2(\Delta P_{\text{ov}} - \bar{\rho}gL)\rho_a D^4}{32\left(\frac{\rho_a}{\rho_b} - \frac{1}{2}\left(1 - \left(\frac{D_a}{D_i}\right)^4\right)\right)} = 0$$

The solution for the mass flow rate is therefore

$$\dot{m} = -b + \sqrt{b^2 + 2c}$$

where

$$b = \frac{4\pi\overline{\mu}L\rho_a}{\overline{\rho}\left(\frac{\rho_a}{\rho_b} - \frac{1}{2}\left(1 - \left(\frac{D_a}{D_i}\right)^4\right)\right)}$$

$$c = \frac{\pi^2(\Delta P_{\text{ov}} - \overline{\rho}gL)\rho_a D^4}{32\left(\frac{\rho_a}{\rho_b} - \frac{1}{2}\left(1 - \left(\frac{D_a}{D_i}\right)^4\right)\right)}$$

If the flow is considered to be incompressible through the cap tube, these expressions simplify to

$$b_{\text{incompressible}} = \frac{8\pi\overline{\mu}L}{1 - (D_a/D_i)^4}$$

$$c_{\text{incompressible}} = \left[\frac{\pi^2(\Delta P_{ov} - \overline{\rho}gL)\rho D_a^4}{16\left(1 - (D_a/D_i)^4\right)}\right]$$

Turbulent Flow Model

For higher Reynolds numbers, the flow is turbulent, and the parabolic profile shape does not adequately capture the shape, especially near the wall where fluid friction plays such an important role. A more general treatment is to define the friction factor, f, which is the shear stress normalized by the dynamic pressure ($\rho V^2/2$).

$$\tau_w = f\frac{\rho V^2}{2}$$

Appendix 3: Equilibrium Cap-Tube with Kinetic Energy Effects

The neglect of kinetic energy changes might be justified for the inlet and exit flows (states i and e), but just as the momentum change may not be negligible, the kinetic energy change across the cap tube itself may be significant due to the small diameter (and thus high velocity). In that case, the thermodynamic state at the exit to the cap tube (state b) is not uniquely specified from mass and momentum alone. The enthalpy at state b is lower than that entering since a portion of the inlet enthalpy is converted to kinetic energy, and therefore, there will be less evaporation (in the equilibrium model) than if kinetic energy were neglected.

Combined Mass, Momentum, and Energy

The momentum and energy equations represent two independent equations that allow for the full solution for the equilibrium model. However, the result cannot be expressed in closed form, and a numerical solution is required. The combined momentum and mass equations can be rearranged to yield the following unknown constant:

$$\frac{\dot{m}^2}{\rho_a A_a^2} = \frac{P_i - P_e - \bar{\rho} g L}{2 f_{\text{Fanning}} \left(\frac{L}{D_a}\right)\left(\frac{\rho_a}{\bar{\rho}}\right) + \left(\frac{\rho_a}{\rho_b} - 1\right) + \frac{1}{2}\left(1 - \left(\frac{D_a}{D_i}\right)^4\right) - \frac{\rho_a D_a^2}{\rho_b D_e^2}\left(1 - \frac{\rho_a D_a^2}{\rho_b D_e^2}\right)}$$

$$= K_{\text{mom}}$$

The combined mass and energy equations can be rearranged for the same constant:

$$\frac{\dot{m}^2}{\rho_a A_a^2} = \frac{2\rho_a((h_i - h_b) - gL + q_{ib})}{\left(\frac{\rho_i}{\rho_b}\right)^2 - \left(\frac{D_b}{D_i}\right)^4} = K_{\text{energy}}$$

The way the problem is posed, namely, specifying the evaporating and condensing temperatures, an iterative solution can be obtained by "guessing" a value for quality at state *b*, and then the density and enthalpy at state *b* can be determined, and then checking to see if $K_{\text{mom}} = K_{\text{energy}}$. If not, another guess for the quality.

A spreadsheet can be set up appropriately, and then the "goal seek" feature can be used for a given case. Alternatively, any number of iterative schemes can be implemented. The results of Fig. 8.21 were obtained using a Regula–Falsi algorithm, in which a known low value of quality (0) and a high value (predicted by the model neglecting kinetic energy) are selected to straddle the solution. The next guess is the average of the two, then either the high or low value is replaced. A sample spreadsheet that uses this algorithm is shown in Fig. 8.35.

INPUT PARAMETERS

Pi	1320391	N/m^2, inlet pressure
Pe	115794	N/m^2, inlet pressure
rhoa	1134	kg,m^3, inlet density
ha	120590	J/kg, inlet enthalpy
hf,b	23715	J/kg, sat'd liquid at Tc (NEGLECT PRESSURE EFFECT)
hfg@Tc	212218	J/kg, heat of vaporization at Tc
vf,b	7.33E-04	m^3/kg, sat'd liquid @Tc
vg,b	0.17598	m^3/kg, sat'd vapor @Tc
Di	5	mm, condenser diameter
De	7	mm, evaporator
q	0	J/kg, heat transfer
f	0.0025	Fanning friction factor
D	1.016	mm, cap tube diameter
L	366	cm, cap tube length

RESULTS

conv?	0.000000	
xb	0.344015	
K	11291	
mdot	0.002901	kg/sec

$$\frac{\dot{m}^2}{\rho_a A_a^2} = \frac{P_i - P_e - \bar{\rho} g L}{2 f_{Fanning} \left(\frac{L}{D_a}\right)\left(\frac{\rho_a}{\bar{\rho}}\right) + \left(\frac{\rho_a}{\rho_b} - 1\right) + \frac{1}{2}\left(1 - \left(\frac{D_a}{D_i}\right)^4\right) - \frac{\rho_a D_a^2}{\rho_b D_e^2}\left(1 - \frac{\rho_a D_a^2}{\rho_b D_e^2}\right)} = K_{mom}$$

$$\frac{\dot{m}^2}{\rho_a A_a^2} = \frac{2\rho_a((h_i - h_b) - gL + q_{ib})}{\left(\frac{\rho_i}{\rho_b}\right)^2 - \left(\frac{D_b}{D_i}\right)^4} = K_{energy}$$

ITERATION

xLOW	h (J/kg)	v (m^3/kg)	rho,b	rho,bar	Kmom	Kenergy
0	23715	0.00073	1364	1249	69593	318621379
0.237374	74090	0.04233	23.6	578.7	14298	45737
0.237374	74090	0.04233	23.6	578.7	14298	45737
0.296718	86684	0.05273	19.0	576.4	12456	21486
0.326389	92981	0.05793	17.3	575.5	11699	14493
0.341225	96129	0.06053	16.5	575.2	11353	11759
0.341225	96129	0.06053	16.5	575.2	11353	11759
0.341225	96129	0.06053	16.5	575.2	11353	11759
0.34308	96523	0.06086	16.4	575.1	11312	11446
0.344007	96719	0.06102	16.4	575.1	11291	11292

xHIGH	h	v	rho,b	rho,bar	Kmom	Kenergy
0.474748	124465	0.0839	11.9	572.9	8948	-979
0.474748	124465	0.0839	11.9	572.9	8948	-979
0.356061	99278	0.0631	15.8	574.8	11027	9417
0.356061	99278	0.0631	15.8	574.8	11027	9417
0.356061	99278	0.0631	15.8	574.8	11027	9417
0.356061	99278	0.0631	15.8	574.8	11027	9417
0.348643	97703	0.0618	16.2	575.0	11188	10543
0.344934	96916	0.0612	16.3	575.1	11270	11139
0.344934	96916	0.0612	16.3	575.1	11270	11139
0.344934	96916	0.0612	16.3	575.1	11270	11139

Xave	h	v	rho,b	rho,bar	Kmom	Kenergy
0.237374	74090	0.0423	23.6	578.7	14298	45737
0.356061	99278	0.0631	15.8	574.8	11027	9417
0.296718	86684	0.0527	19.0	576.4	12456	21486
0.326389	92981	0.0579	17.3	575.5	11699	14493
0.341225	96129	0.0605	16.5	575.2	11353	11759
0.348643	97703	0.0618	16.2	575.0	11188	10543
0.344934	96916	0.0612	16.3	575.1	11270	11139
0.34308	96523	0.0609	16.4	575.1	11312	11446
0.344007	96719	0.0610	16.4	575.1	11291	11292
0.344471	96818	0.0611	16.4	575.1	11280	11215

Fig. 8.35 Spreadsheet implementation of Regula–Falsi method to solve for the cap tube mass flow rate with kinetic energy effects included

Appendix 4: eXergy Destruction Due to Friction

The direct relationship between eXergy destruction and friction for the case of incompressible adiabatic steady flow in a straight pipe section is derived formally in this Appendix. States "in" and "out" represent inlet and outlet states (corresponding to states "*a*" and "*b*" for the cap tube application).

The momentum equation for incompressible fully developed steady flow, with a Fanning friction factor model for shear stress, is

$$P_{in} = P_{out} + f\frac{1}{2}\rho V^2\frac{L}{D} \quad \text{Momentum}$$

The momentum flux in equals the momentum flux out for a constant area flow and constant density (hence velocity), with an assumed fully developed velocity profile throughout. That is, entrance effects are neglected.

The energy equation for incompressible and adiabatic steady flow is

$$h_{in} = h_{out} \quad \text{Energy}$$

Breaking enthalpy into its mechanical (P/ρ) and thermal (u) components and combining momentum and energy:

$$P_{in} - P_{out} = f\frac{1}{2}\rho V^2\frac{L}{D} = \rho(u_{out} - u_{in}) \quad \text{Combined momentum and energy}$$

Converting velocity to mass flow rate, and relating internal energy to temperature with the specific heat, c:

$$\frac{f\dot{m}^2L}{2\rho^2A^2D} = c(T_{out} - T_{in})$$

The temperature change, expressing the area in terms of diameter, is therefore

$$T_{out} - T_{in} = \frac{8f\ \dot{m}^2L}{\pi^2\rho^2cD^5}$$

That is, friction gives rise to an increase in temperature for adiabatic, incompressible flows. It is to be noted that the temperature can decrease if the fluid is a gas (due to momentum flux differences) or if there is phase change. The result presented here is restricted to incompressible flows.

The entropy change for an incompressible fluid (with temperature in °C) is

$$s_{\text{out}} - s_{\text{in}} = c \ln \left(\frac{T_{\text{out}} + 273}{T_{\text{in}} + 273} \right)$$

Combining with the momentum and energy equations, the change in entropy associated with fully developed pipe flow is

$$s_{\text{out}} - s_{\text{in}} = c \ln \left(1 + \frac{8 f \dot{m}^2 L}{\pi^2 \rho^2 c (T_{\text{in}} + 273) D^5} \right)$$

Finally, the eXergy destruction is

$$eX_D = T_\infty (s_{\text{out}} - s_{\text{in}}) = c T_\infty \ln \left(1 + \frac{8 f \dot{m}^2 L}{\pi^2 \rho^2 c (T_{\text{in}} + 273) D^5} \right)$$

Since all the terms in the natural log expression are inherently positive, this result shows formally how fluid friction results in eXergy destruction.

Workshop 8.1: A Gas Refrigerator/Water Heater, a Refreater

A gas refrigerator simultaneously absorbs heat at the rate of $\dot{Q}_C = 0.10$ kW from a cold space (maintained at $T_C = 0$ °C in a room at $T_H = 25$ °C, the dead state) and produces hot water suitable for domestic heating in a heat exchanger. The cycle uses argon as the working fluid (which can be treated as a calorically perfect gas with $c_p = 0.5203$ kJ/kg/K and $R = 0.2081$ kJ/kg/K for all processes) in a reverse Brayton cycle, that is, adiabatic and reversible (hence isentropic) compression in a compressor (process 1–2), constant pressure heat rejection in the water heater (2–3), adiabatic and reversible (hence isentropic) expansion through a turbine (3–4), and constant pressure heat addition from the cold space (4–1) in the space chiller. Liquid water enters the water heater steadily at $T_a = 15$ °C and exits at $T_b = 45$ °C. Assume a specific heat of 4.2 kJ/kg/K and density of 1000 kg/m^3 for the water. Complete the state and results tables and sketch a *T-s* diagram, approximately to scale. Neglect stray heat loss, kinetic and potential energy changes, and assume steady-state operation.

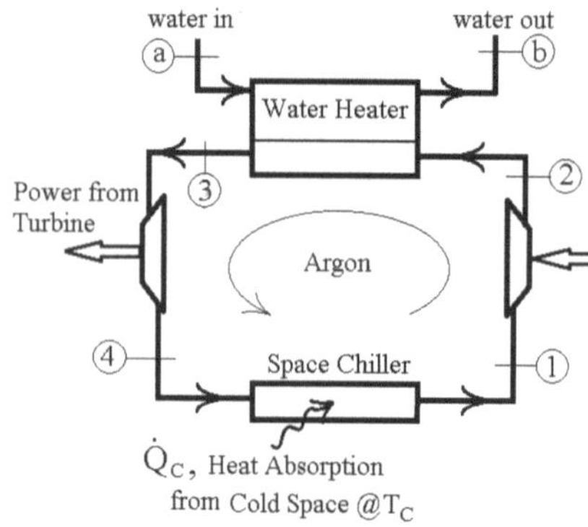

INPUT PARAMETERS

Qdot,C	**0.1**	kW
Ta	**15**	oC, inlet water temp
Tb	**45**	oC, outlet water temp
cp,water	**4.2**	kJ/kg/K
rho,water	**1000**	kg/m^3
Cp	**0.5203**	kJ/kg/K
R	**0.2081**	kJ/kg/K
Tcinf	**0**	oC
Thinf	**25**	oC

Argon State Table

STATE	T(oC)	P(atm)	h(kJ/kg)	s(kJ/kg/K)
1	**-10**	**10**		
2	**50**			
3	**30**			
4		**10**	**0**	**0**

(note: state 4 is a reference state for h and s)

Results Table

Mass flow rate of Argon in Brayton cycle		kg/sec
Mass flow rate of hot water		kg/sec
Volumetric flow rate of hot water		liters/day
Net Input Power to Cycle		Watts
eXergy Destruction rate in Water Heater		Watts
eXergy Destruction rate in Space Chiller		Watts
1st Law Efficiency, COP_R		
2nd Law Efficiency, COP_R/COP_{CARNOT}		

Extra Fun Repeat with isentropic efficiencies of 0.85 and 0.9 for the compressor and turbine, respectively (and comment on the effect).

Workshop 8.2: A Central Air Conditioning Application

Just about everyone knows that air conditioning makes air cold, but fewer people appreciate that it also lowers its humidity (i.e., removes water vapor from the air). Formal study of the mass, energy and eXergy consequences of humidity effects falls into the field of psychrometrics, with chemical composition changes in the working fluid. This workshop is constructed to avoid formal study of psychrometrics, but retains an important consequence, namely, that moisture is usually removed from air by lowering its temperature below the dew point (but above the freezing point). Therefore, even though the air temperature must be reduced by a relatively small temperature difference, say, from 25 °C to 20 °C (76.4 °F to 67.4 °F) when it is 35 °C (94.4 °F) outside ($T_{H\infty}$), the air temperature must be lowered to near the freezing point of water to remove water vapor and then reheated to its desired delivery temperature. Reheating the air seems ridiculous, but it is a comfort issue. The reheating can be accomplished (in principle) with a heat exchanger (called a recuperator) that also pre-cools the air that is sent to an HVAC (heating, ventilation and air conditioning) system.

The schematic shows a central air conditioning application that uses a standard vapor/compression refrigeration cycle and a recuperator to draw an air stream from an interior space, condition it, and return it at a lower temperature and humidity. The refrigeration system is modified from the ideal cycle by specifying a degree of subcooling in the condenser and an isentropic efficiency for the compressor. The heat exchangers (condenser and evaporator) are assumed to be isobaric on the refrigerant side. The dead state is taken to be the outside ambient, set at 35 °C (94.4 °F).

On the air side, a fan draws air (state a) steadily from the interior of a building maintained at a temperature $T_{c\infty}$ (and a relative humidity) and sends it to an HVAC system. The conditioned air is returned to the building at a lower temperature and relative humidity (state *e*). The cooling load is represented by the difference in enthalpy flux between the send and return air required to maintain the interior space temperature (which equals the send air temperature). This load would be somewhat higher than indicated by this workshop if the change in humidity were included. The send air first passes through a recuperator (a heat exchanger between the send and return air) that simultaneously lowers the send air temperature prior to it entering the evaporator (state *c*) and raises the temperature of the return air that exits the evaporator (state *d*).

Water vapor in the entering air can be condensed in both the recuperator and evaporator, but will not be formally studied in this workshop. A deeper dive (after studying psychrometrics) would be to redo this workshop with specified humidity levels for the send and return air, and compare the results to see how important the latent energy effects (those associated with phase change) are compared to sensible energy changes (lowering the temperature).

A spreadsheet template is shown with the systems sized to meet a cooling load of 18 kW (approximately 60,000 Btu/h, sized to meet a rule of thumb of 20 Btu/h per square foot of living space and a 3000 square foot living space), with cooling load defined as the difference in enthalpy flux of the send and return air streams. The pressures set for the air stream are highly uncertain in this workshop but are included for the purposes of allowing for a calculation of the power required by the fan and practice calculating 2nd Law quantities for ideal gases. More accurate representation of the air-side pressures (and therefore sizing of the fan) requires a fluid mechanical analysis of the duct network that distributes the conditioned air throughout the building. The air is treated as a calorically perfect gas, as usual. Conditions are set so that refrigerant properties can be obtained from the coarse listings of Figs. 8.6 and 8.8, but alternative published and/or online data sources can certainly be used.

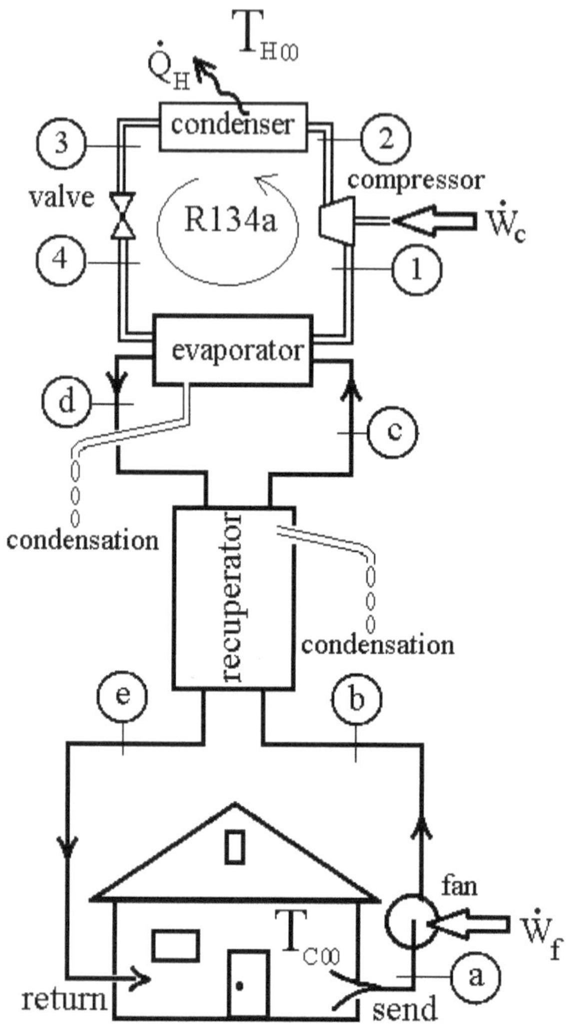

$T_{H\infty}$
$\dot{Q}_H$
condenser
3
2
compressor
valve
R134a
$\dot{W}_c$
4
1
evaporator
d
c
condensation
recuperator
condensation
e
b
fan
$T_{C\infty}$
$\dot{W}_f$
a
return
send

INPUT PARAMETERS

$m'_{air}c_{p,air}(T_{send} - T_{return})$	**18**	kW, cooling rate of air
ΔT_{sub}	**5**	oC, degrees of subcooling
η_c	**0.85**	isentropic efficiency of compressor
$T_{H\infty}$	**35**	oC, outdoor (dead state) temperature
cp,air	**1**	kJ/kg/K, specific heat of air
R,air	**0.287**	kJ/kg/K, gas constant for air

STATE TABLE FOR REFRIGERATION CYCLE

state	P (kPa)	T (oC)	h (kJ/kg)	s (kJ/kg/K)	x or phase
1		**0**			**1**
2s (isentropic state)	**1000**				
2	**1000**				
3					
4					

STATE TABLE FOR CONDITIONED AIR

state	P (atm)	T (oC)	h (kJ/kg)	s (kJ/kg/K)
a	**1**	**25**		
b	**1.003**			
c	**1.002**	**10**		
d	**1.001**			
e	**1**	**20**		

MASS FLOW RATE RESULTS (in kg/sec)

m',air		kg/s, mass flow rate of air
m',R		kg/s, mass flow rate of refrigerant

ENERGY RATE RESULTS (in kW)

W',c		power input to compressor
Q',H		heat transfer rate to outside ambient
W', f		power input to fan

EXERGY DESTRUCTION RATES (in kW)

Compressor	
Condenser	
Valve	
Evaporator	
Regenerator	

EFFICIENCIES

COP		Cooling Rate over total power in
COP,CARNOT		Carnot efficiency
$\eta 2$		2nd Law eff'y (COP/COPcarnot)

Workshop 8.3: Water Chiller

The objective of this workshop is to conduct 1st and 2nd Law analyses of a vapor/compression refrigeration cycle, a simplified model of typical systems in large buildings (Fig. 8.36). There are three main fluid loops in this model. The inner loop consists of refrigerant R134a operating in a vapor/compression refrigeration cycle. This system is located in the basement in an interior space (set at $T_C = 23\ °C$ for this workshop). The evaporator of the refrigeration cycle (bottom) is a heat exchanger that chills water to be distributed to a network of cooling panels throughout the building where heat is removed from interior spaces. In an actual system, there could be a secondary heat exchanger between the water chiller and the flow distribution to the building. The condenser of the refrigeration cycle (top) transfers heat (internally) from the refrigerant to a water stream that is sent to a remote cooling

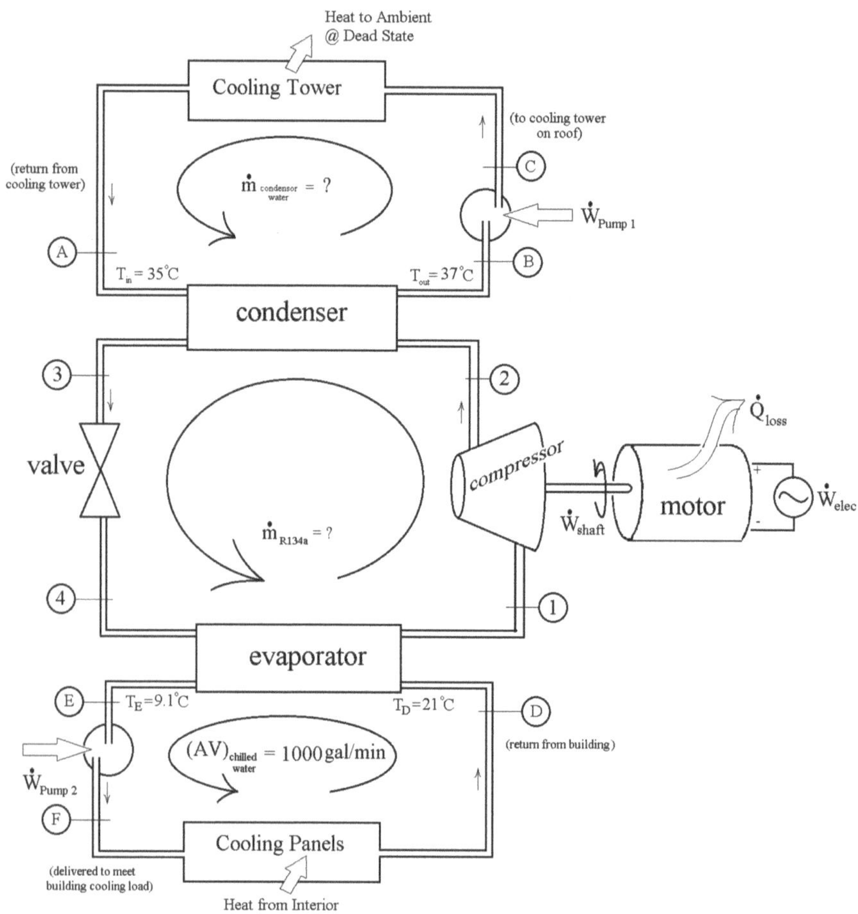

Fig. 8.36 Schematic of water chilling system

tower (e.g., on the roof of a 10-story building) where heat is exchanged (rejected heat) to the outside environment (set at $T_{\infty} = T_{\mathrm{H}} = 30\ °\mathrm{C}$, and considered to be the dead state temperature). The actual cooling tower is not generally a straight heat exchanger, but a type of open heat exchanger that mixes ambient air with the cooling water. Evaporation effects are a key element of this type of heat exchanger, and those topics require a study of psychrometrics.

In the refrigeration loop, a motor (with a default 1st Law efficiency of 0.85, defined as output shaft power/input electrical power) drives an isentropic compressor that delivers refrigerant R134a through a standard vapor/compression refrigeration cycle (compressor, condenser, valve, evaporator, with each process description given in Table 8.2). Each component operates at steady state, with negligible kinetic energy, potential energy, and stray heat (except the motor, which has stray heat equal to the difference in input and output power).

Both water loops (condensing and evaporating) are shown with a pump needed to drive the water through the system. Modeling of the required pressures (and therefore power required to drive pumps) is a topic for fluid mechanics, and not considered here. Treat the water in these loops as incompressible, with a specific volume of 0.001 $\mathrm{m^3/kg}$ and a specific heat of 4.2 kJ/kg/K (i.e., *use of* H_2O *tables is unnecessary in this workshop*, but R134a tables are needed). In practice, the power required to drive the pumps may not necessarily be insignificant compared to the power required by the compressor of the refrigeration system. It is just not analyzed in this workshop. Complete the state and results tables, and show the refrigeration cycle on *P-h* and *T-s* diagrams. Results are reported in Appendix 2 of the textbook.

Table 8.2 Process table for vapor/compression refrigeration cycle

Process	Device	Process type/idealization
1–2	Compressor	Adiabatic, reversible
2–3	Condenser	Isobaric in either stream
3–4	Throttling valve	Adiabatic
4–1	Evaporator	Isobaric is either stream

INPUT PARAMETERS

$(AV)_{CW}$	**1000**	gallons/minute, cooling water volumetric flow rate
T_C	**23**	oC, temperature of interior spaces
T_H	**30**	oC, temperature of outside ambient (dead state)
η_{motor}	**0.85**	motor efficiency (output shaft to input electric)
v_{H2O}	**0.001**	m3/kg, specific volume of all water streams
$c_{p,H2O}$	**4.2**	kJ/kg/K, specific heat of all water streams
TA	**35**	oC, water entering condenser
TB	**37**	oC, water exiting condenser
TD	**21**	oC, water entering evaporator
TE	**9.1**	oC, water exiting evaporator

R134a STATE TABLE

state	P (kPa)	T (oC)	h (kJ/kg)	s (kJ/kg/K)	quality
1	**320**	**10**			
2	**1000**				
3		**36**			
4					

RESULTS TABLES

m',CW		kg/sec, mass flow rate of chilled water
Q',C		kW, rate of heat transfer to the cooling panels
m',R		kg/sec, mass flow rate of R134a
m',cond		kg/sec, mass flow rate of condenser water
W',shaft		kW, shaft power to compressor
W',elec		kW, electric power to motor
Q'tower		kW, heat rejected by cooling tower
COP		Qdot,C/Wdot,elec: overall coefficient of performance
COP,CARNOT		maximum possible COP to pump heat from inside to out
2nd Law eff'y		COP/COP,CARNOT

eXergy Destruction rates:		Tinf=		K
cooling tower		kW		
condenser		kW		
valve		kW		
evaporator		kW		
cooling panels		kW		
motor		kW		
TOTAL				

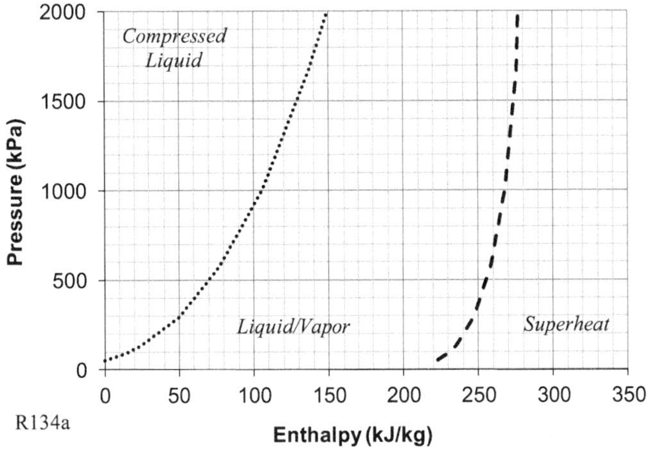
Compressed
Liquid
Liquid/Vapor
Superheat
Pressure (kPa)
2000
1500
1000
500
0
0
50
100
150
200
250
300
350
Enthalpy (kJ/kg)

R134a

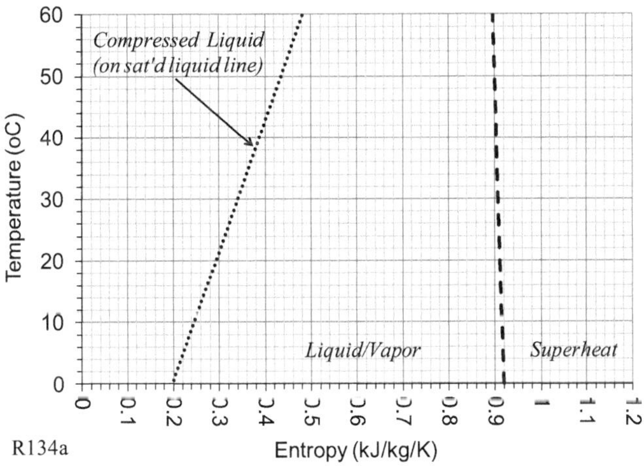
Compressed Liquid
(on sat'd liquid line)
Liquid/Vapor
Superheat
Temperature (oC)
60
50
40
30
20
10
0
0
0.1
0.2
0.3
0.4
0.5
0.6
0.7
0.8
0.9
1
1.1
1.2
Entropy (kJ/kg/K)

R134a

Chapter 9
Gas Turbines

This chapter focuses on the fundamentals of the gas turbine, an internal combustion engine used for propulsion (mobile power plant) and power generation (stationary power plant). The ideal Brayton cycle forms the basis for gas turbines, which has two main functions: to produce a high pressure gas stream and to do something with it. The high pressure stream is produced in a gas generator which consists of a compressor, combustor (or heater), and primary turbine. The high pressure stream delivered by the gas generator is either used to drive a secondary turbine stage (for mechanical power generation) or accelerated in a nozzle (for propulsion). Power generation is investigated first, as the kinetic energy entering and leaving all devices is safely neglected, and the functions of primary and secondary turbines can be considered as a single device. Real engine performance is modeled using isentropic efficiencies for the compressor and turbine. For propulsion, a diffuser is added before the compressor to convert kinetic energy of the inlet air to enthalpy, and the secondary turbine is replaced by a nozzle. The working fluid is justifiably treated as an ideal gas with constant specific heats (a calorically perfect gas) with properties of air for all applications. The *Ts* diagram is shown to be a useful construct, and an appreciation for the statement "gas turbines work because isobars diverge on a *Ts* diagram" is gained.

Gas turbines are the prime movers in propulsion (jet engines) where chemical energy is ultimately converted to kinetic energy of a jet whose momentum develops a thrust force. They are also important in stationary applications in utility-based electrical production where chemical energy is converted to shaft power, used to drive an electric generator. Gas turbines are thus used alone as peaking plants to meet rapidly changing demand. These applications result in a high temperature (but ambient pressure) exhaust that is simply vented to the atmosphere. Combined cycles (Chap. 11) are base-load power plants that use a gas turbine as a topping cycle and use its high temperature exhaust to drive a heat engine to extract as much additional work as possible. This bottoming cycle is typically a Rankine cycle, developed in the Chap. 10. The bottoming cycle, in practice, has a response that is too slow to adjust to changing demands, and is therefore not generally suited as peaking plants.

G. Sidebotham, *An Inductive Approach to Engineering Thermodynamics*, Mechanical Engineering Series, https://doi.org/10.1007/978-3-030-20430-3_9

The working fluid of gas turbines is generally air, with the intermediate addition of a fuel that adds chemical energy in an internal combustion process. The mass added by the fuel is of the order of a few percent, and therefore the working fluid is treated at this level as being dominated by air properties. Gas turbines are also ideal for cogeneration of work and heat, used in commercial and industrial applications.

Practical gas turbines can be categorized as either industrial or aero-derivative. Industrial gas turbines are designed specifically for stationary power applications. Size and weight are not a primary concern at the design stage. The pressure ratios tend to be in the range of 10 to 15. On the other hand, aero-derivative gas turbines are designed as mobile jet engines, for which size and weight are to be minimized and high performance and reliability are paramount. These engines can be modified for stationary power production, and their pressure ratios are in the range of 20 to 30. The aero-derivative tends to be more efficient as a straight peaking plant due to the higher pressure ratio and is designed to easily change power output. The industrial turbine then has a higher exit temperature, which can be used more favorably in a combined cycle in which the hot exhaust is used to generate steam for a Rankine cycle.

Stationary Gas Turbines and the Brayton Cycle

The objective of a stationary gas turbine (whether industrial or aero-derivative) is to generate mechanical shaft power (to drive an electric generator or a mechanical load directly, such as the propeller of a ship or grinding wheel in a factory) from a thermal source (such as combustion of a fuel). It is analyzed from a thermodynamic perspective in this section, first as an ideal cycle (the open Brayton cycle) and then with irreversibilities included to more closely simulate the performance of real gas turbines. That is, an overall balance of mass, energy, and entropy is conducted for each component treated as a "black box." The thermodynamic constraints provide hard upper limits that can be determined theoretically and approached in practice. Use of reasonable isentropic efficiencies can provide performance predictions that are quite close to actual devices. The design details and principles of operation associated with each component (i.e., how it works) are more advanced topics and would be developed from more of a fluid mechanics than thermodynamics perspective.

The basic gas turbine used for power production consists of a compressor, a combustor, and a turbine in series, shown schematically in Fig. 9.1. A single turbine is shown, yet it has two key functions: to drive the compressor (by the internal shaft) and to drive an output mechanical shaft (shown driving an electric generator). A control volume for the compressor cuts through the internal shaft, and mechanical power is transferred into it. A control volume for the turbine cuts through the output shaft as well and shares the internal shaft with the compressor (equal and opposite mechanical power). The net power of the engine is the difference between that generated by the turbine and that consumed by the compressor. The back work ratio,

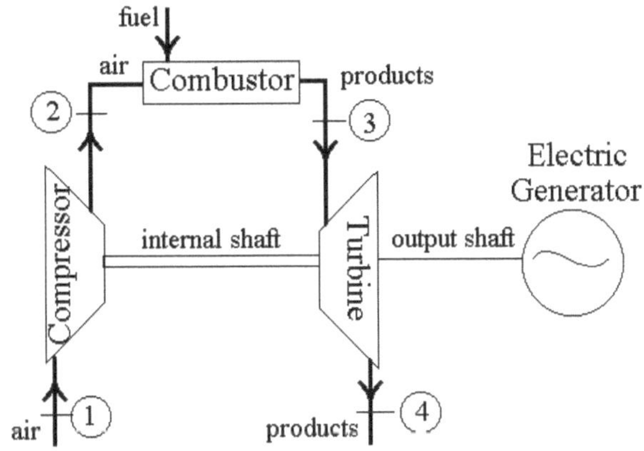

Fig. 9.1 Schematic of basic gas turbine for stationary power generation

defined as the ratio of the compressor power to the turbine power, is an important measure to be minimized. It differentiates the function of the gas generator from the desired output of the gas turbine.

In practice, a gas turbine is an internal combustion engine and is not operated as a cycle. It must be continuously fed with air at the intake (state 1) and fuel into the combustor (after state 2), which react chemically to form combustion products (states 3 and 4). The mass flow rate of the fuel is much less than that of air (on the order of a few percent by mass), but it supplies chemical thermal energy. Overall, air is drawn in at atmospheric conditions (the dead state), and hot combustion products are exhausted at (or near) ambient pressure to the environment (or to another device) in an overall open steady-state system. The exit stream (state 4) has eXergy associated with it, but it is thermal in nature (hot gases), not mechanical (pressure). Not much can be done with that eXergy in a propulsion application, but it can be used in stationary applications to produce a heating source (cogeneration) or as the heating source for another power cycle, making a combined cycle.

In traditional thermodynamic analysis, the combustion process is treated as an "equivalent heat addition," and a hypothetical process that returns the working fluid to the initial state by an equivalent heat transfer is included, creating a cycle. In Fig. 9.2, a closed system is produced by adding a hypothetical isobaric heat exchanger between states 4 and 1 that returns the working fluid to the initial state by the transfer of heat (q_C) to the dead state. The thermal energy added in the combustor is modeled as a heat addition (q_H) from a thermal reservoir maintained at a temperature above (or equal to) the maximum temperature of the working fluid (state 3).

The maximum possible efficiency of this power cycle is given by the Carnot efficiency, $\eta_{Carnot} = 1 - T_C/T_H$. For a dead state of 298 K and high temperature reservoir of 1800 K, the Carnot efficiency is 0.834, which is rather high. The performance of the various gas turbines will be compared with that standard.

There are niche markets and possible future novel applications where Brayton-type closed cycles can be used for either power generation or refrigeration in a closed cycle, with a circulating working fluid. The refrigeration cycle from Workshop 8.1 is

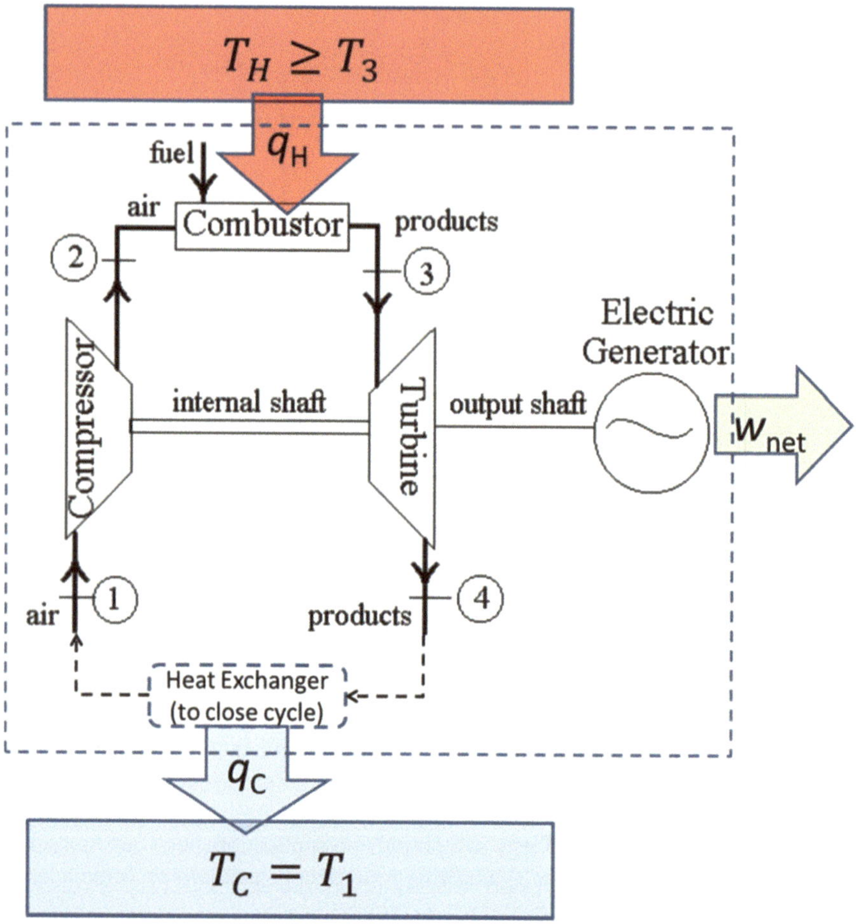

Fig. 9.2 Gas turbine modeled as a closed thermodynamic power cycle, with a heat exchanger to close the cycle and heat transfers (per unit mass) between hot and cold thermal reservoirs and net work (per unit mass) extracted

food for thought, using Argon as the working fluid because of its abundance (1% of Earth's atmosphere) and its high ratio of specific heats (it is a monatomic gas, with $k_{\text{Argon}} = 1.667$) which yields higher thermal efficiency (or coefficient of performance) than air (a diatomic gas, $k_{\text{Air @ 298K}} = 1.40$). For power cycles, the high side thermal source could be concentrated solar input and heat rejection to the environment (directly to air, or to cooling water), for example.

This section analyzes stationary gas turbines that take in air at a low velocity (kinetic energy), and whose objective is to produce as much output shaft power as possible from a chemical energy source. The following section, a deeper dive, considers mobile gas turbines, or jet engines, that take in air which may have a significant kinetic energy relative to the engine, and whose objective is to expel a high kinetic energy jet with momentum that provides propulsion. Higher flight speeds eliminate the need for a compressor (and therefore turbine), and RAM and SCRAM jets are introduced in an Appendix.

Problem statements are formulated in a Workshop format to investigate various permutations of Brayton-type cycles, with detailed solutions posted in Appendix 2 of the textbook. All cases are scaled to provide a specific power input to the compressor (10 MW) and the same primary turbine inlet temperature (1800 K, typical values attainable in the early twenty-first century). The problems are formulated so that each component can be analyzed in a spreadsheet format, which can be either filled out in hard copy, or a spreadsheet of similar structure can be built. With the assumption of a calorically perfect gas (constant specific heat), formulas can be placed in all cells outside the input parameter section, facilitating a parametric study. This flexibility is not generally possible when state properties are obtained manually from tables. State and process tables for the entire system are shown with input parameters specified. The state table shows actual and isentropic states, where appropriate. A single results table is included for the first two cases so that they can be directly compared.

The turbine can be viewed either as a single device with one output shaft (as in Fig. 9.1), or broken into two turbines (a primary and secondary) that serve very different functions (as in the example and workshops to follow). In the latter case, the primary turbine is coupled directly to the compressor. The combination of compressor, combustor, and primary turbine is called a "gas generator," and it delivers a gas stream at elevated pressure and temperature that can be used for either stationary power (a secondary turbine/generator combination) or for mobile power (sent through a nozzle). All shaft power developed by the primary turbine is used to drive the compressor. Example problem statements for an ideal Brayton cycle and a gas turbine with irreversibilities are introduced, followed by a thermodynamic analysis of each component and the entire system. Temperature/entropy (*Ts*) diagrams are useful constructs for visualizing these processes, as will be shown. I remember something my PhD thesis advisor (Prof. Irv Glassman) said when I was a graduate student, namely, "gas turbines work because isobars diverge on a *Ts* diagram." This adage intrigued me and stuck with me, but it was years until I understood or appreciated it.

The working fluid is treated in all cases as a calorically perfect gas, with a value chosen for the specific heat at an elevated temperature (usually 1000 K, somewhat arbitrarily but representative of a cycle average) and as having properties of air throughout. The chemical composition change that occurs during combustion is not directly considered, as the mass of fuel injected is a few percent of that of air, and the inert N_2 is the dominant component of air. In traditional thermodynamic analysis, an *air standard analysis* treats the gas as a thermally perfect gas (specific heat varies with temperature). A *cold air standard analysis* treats the working fluid as a calorically perfect gas with the specific heat evaluated at the inlet temperature, which underestimates its value for all processes. The approach taken here could be called a *hot air standard analysis*. Treating the specific heat as a constant allows for closed form expressions for all processes, a big advantage. Using different specific heat values for different processes is a viable approach, but not taken here.

Ideal Open Brayton Cycle

Air is drawn into a stationary gas turbine that operates with a pressure ratio of 25 (typical of jet engines). The gas generator (states 1–4) consists of an adiabatic, reversible compressor (isentropic efficiency of 1.0), isobaric combustor and adiabatic, reversible primary turbine. The compressor takes an input shaft power of 10 MW from the primary turbine. Chemical energy is added in the combustor from a fuel with a lower heating value (LHV) of 50.1 MJ/kg (natural gas value). The primary turbine inlet temperature is 1800 K. The gas enters an adiabatic, reversible secondary turbine that produces net power. The turbine exhaust enters a heat exchanger (HEX), the heat from which is used for space heating an interior space maintained at the dead state temperature in a cogeneration application. The exit is therefore returned to the same conditions as the inlet (giving the thermodynamic performance of a closed cycle).

Complete the state and process tables, and plot the processes on a Ts diagram (to scale). Assume that the mass flow rate of the fuel is much less than that of air, and the working fluid has properties of air throughout, treated as a calorically perfect gas with $c_p = 1.140$ *kJ/kg/ΔK (its value at 1000 K, an average cycle temperature).*

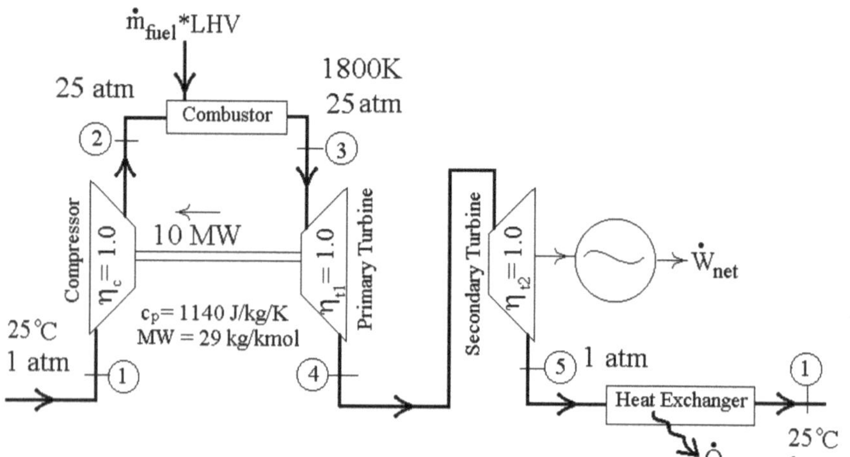

INPUT PARAMETERS

Wdot,c	10,000	kW	power into compressor
T1	25	oC	inlet temperature
P1 = P4	1	atm	inlet pressure
P2 = P3	25	atm	compressor outlet pressure
T3	1800	K	turbine inlet temperature
LHV	50,100	kJ/kg	lower heat value of fuel
cp	1.140	kJ/kg/K	specific heat at constant pressure
MW	29	kg/kmol	molecular weight

DERIVED PARAMETERS

R		kJ/kg/K	gas constant
k			ratio of specific heats

Gas Turbine with Irreversibilities

Analyze the gas turbine of the previous problem with the following irreversibilities:

- *The compressor has an isentropic efficiency* $\eta_C = 0.85$.
- *There is a pressure drop across the combustor such that the pressure entering the primary turbine is 24 atm.*
- *Both turbines have isentropic efficiencies of* $\eta_T = 0.90$.
- *There is a pressure drop across the HEX such that the turbine exit pressure is 1.2 atm.*
- *The HEX exit temperature is higher than the dead state (100 °C).*

Complete the state and process tables and add results to compare to ideal cycle. Add process sketches to Ts diagram of ideal cycle.

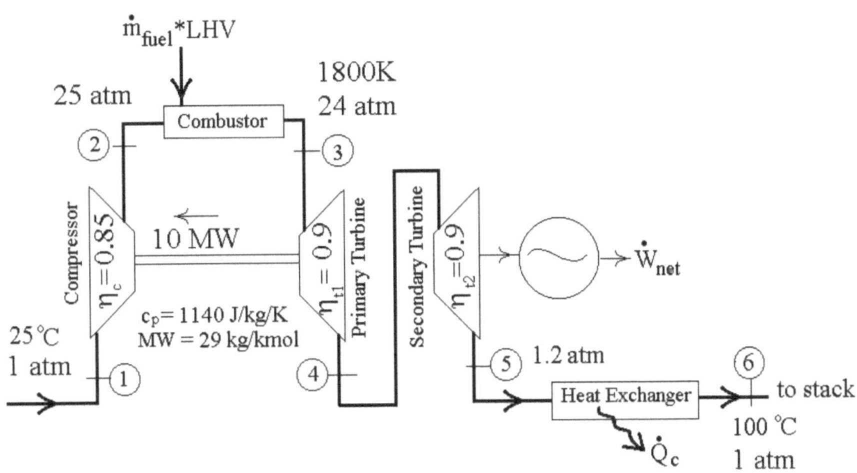

INPUT PARAMETERS

Wdot,c	10,000	kW	power into compressor
T1	25	oC	inlet temperature
P1	1	atm	inlet pressure
P2	25	atm	compressor outlet pressure
P3	24	atm	turbine inlet pressure
T3	1800	K	turbine inlet temperature
P5	1.2	atm	turbine exit pressure
T6	100	oC	Stack temperature
LHV	50,100	kJ/kg	lower heat value of fuel
cp	1.140	kJ/kg/K	specific heat at constant pressure
MW	29	kg/kmol	molecular weight
η_C	0.85		compressor isentropic efficiency
η_T	0.9		turbine isentropic efficiency
η_G	0.95		electric generator efficiency

DERIVED PARAMETERS

R		kJ/kg/K	gas constant
k			ratio of specific heats

STATE TABLE FOR IDEAL BRAYTON CYCLE

State	*P (atm)*	*T (K)*	*h (kJ/kg)*	*s (kJ/kg/K)*
1	**1**	**298**		
2	**25**			
3	**25**	**1800**		
4				
5	**1**			

STATE TABLE FOR BRAYTON CYCLE WITH IRREVERSIBILITIES

State	P (atm)	T (K)	h(kJ/kg/K)	s(kJ/kg/K)
1	**1**	**298**	**0**	**0**
2s	**25**			
2	**25**			
3	**24**	**1800**		
4s				
4				
5s	**1.2**			
5	**1.2**			
6	**1**	**373.0**		

PROCESS TABLE FOR IDEAL BRAYTON CYCLE

		1st Law Quantities			Combined 1st and 2nd Law Quantities		
Process	**Device**	$h_{out} - h_{in}$	q_{in} or m_{fuel}*LHV	w_{out}	$T_\infty(s_{out} - s_{in})$	$T_\infty * q_{in}/T_B$	eX_d
1-2	**Compressor**						
2-3	**Combustor***						
3-4	**Primary Turbine**						
4-5	**Secondary Turbine**						
5-1	**Heat Exchanger**						
	net						

* chemical energy treated as equivalent heat transfer from a thermal reservoir at T3

PROCESS TABLE FOR BRAYTON CYCLE WITH IRREVERSIBILITIES

		1st Law Quantities			Combined 1st and 2nd Law Quantities		
Process	**Device**	$h_{out} - h_{in}$	q_{in} or m_{fuel}*LHV	w_{out}	$T_\infty(s_{out} - s_{in})$	$T_\infty * q_{in}/T_B$	eX_d
1-2	**Compressor**						
2-3	**Combustor***						
3-4	**Primary Turbine**						
4-5	**Secondary Turbine**						
5-1	**Heat Exchanger**						
	net						

* chemical energy treated as equivalent heat transfer from a thermal reservoir at T3

RESULTS TABLE (IDEAL AND WITH IRREVERSIBILITIES)

Quantity	***Unit***	***Ideal Brayton Cycle***	***w/Irreversibilities***	***%Change***
inlet density	kg/m3			
air mass flow rate	kg/sec			
fuel mass flow rate	kg/sec			
Fuel to air mass flow ratio				
Compressor/Primary Turbine Power	***MW***	***10***	***10***	0.0%
Chemical Input Power	MW			
Secondary Turbine (output) Power	MW			
Process Heating Rate	MW			
eXergy Destruction Rate in HEX	MW			
Backwork Ratio				
1st Law Efficiency as Power Cycle				
Carnot Efficiency of Power Cycle*				
2nd Law Efficiency of Power Cycle				
1st Law Efficiency of Ideal Combined Cycle**				
1st Law Efficiency as Cogeneration				

* based on high temperature source at T3

**Assumes all eXergy destruction in HEX converted to power

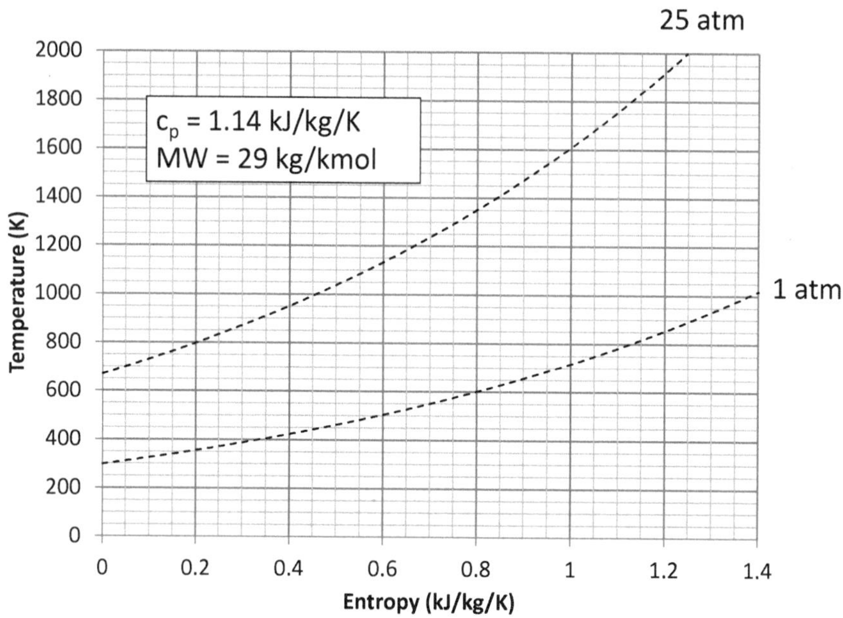

Compressor

Compressors, designed to convert input shaft power into fluid enthalpy, can usually be treated as steady-state open devices if they are run continuously with constant (or gradually varying) conditions, even if operating in a pulsed mode, like reciprocating compressors (which are not generally used in gas turbines). Turbines are treated similarly, the only difference being that the objective is to convert fluid enthalpy into useful shaft work.

For a single inlet, single outlet compressor, the mass, energy, and entropy (multiplied by the dead state temperature) balance equations are:

$$\frac{dM}{dt} = \dot{m}_1 - \dot{m}_2$$

$$\frac{dE}{dt} = \dot{Q}_{\mathrm{IN}} + \dot{W}_{\mathrm{IN}} + \dot{m}_1\left(h_1 + \frac{V_1^2}{2} + gz_1\right) - \dot{m}_2\left(h_2 + \frac{V_2^2}{2} + gz_2\right)$$

$$T_\infty\left[\frac{dS}{dt} = \frac{\dot{Q}_{\mathrm{IN}}}{T_{\mathrm{B}}} + \dot{m}_1 s_1 - \dot{m}_2 s_2 + \overset{\cdots}{\underset{12}{\Sigma}}\right]$$

The intention of the compressor is to steadily convert input shaft power into fluid enthalpy power (not kinetic or potential energy). Neglecting all unintended effects (stray heat, kinetic and potential energy), but allowing for irreversibilities to exist, the mass, energy and entropy balances at steady-state simplify to:

$$\dot{m}_1 = \dot{m}_2 = \dot{m}_{\text{air}}$$

$$\dot{W}_{\text{IN}} = \dot{m}(h_2 - h_1)$$

$$e\dddot{X}_{\text{D},12} = T_\infty \dddot{\Sigma}_{12} = \dot{m}T_\infty(s_2 - s_1)$$

Practical calculations are conducted by the use of isentropic efficiencies, as was developed for pumps. An ideal compressor is reversible (no entropy generation), and therefore isentropic. By definition, the ideal compressor and real compressor operate with the same pressure ratio, mass flow rate, and inlet conditions. The isentropic efficiency is:

$$\begin{pmatrix} \text{Isentropic} \\ \text{compressor} \\ \text{efficiency} \end{pmatrix} = \eta_{\text{C}} = \frac{\text{Minimum power required}}{\text{Actual power required}} = \frac{\dot{m}(h_{2\text{s}} - h_1)}{\dot{m}(h_2 - h_1)}$$

$$\eta_{\text{C}} = \frac{h_{2s} - h_1}{h_2 - h_1}$$

This efficiency is valid for any type of fluid. For calorically perfect gases (ideal gases with constant specific heats):

$$\eta_{\text{C}} = \frac{T_{2\text{s}} - T_1}{T_2 - T_1}$$

The isentropic process experienced by the fluid as it passes through the compressor is the same as would be obtained by an isentropic compression process in a piston/cylinder. Since the pressure ratio across the compressor is an input parameter, the temperature of the hypothetical isentropic compression can be expressed in terms of input parameters by the following type of manipulation:

$$T_{2\text{s}} = T_1\left(\frac{T_{2\text{s}}}{T_1}\right) = T_1\left(\frac{P_2}{P_1}\right)^{\frac{k-1}{k}} = 298\left(\frac{25}{1}\right)^{\frac{1.34-1}{1.34}} = 669.5 \text{ K}$$

This temperature is the hypothetical temperature that would be obtained in an isentropic compressor to the pressure of state 2. The actual temperature is obtained from the definition of the isentropic efficiency (set to a value of 0.85, typical of modern compressors):

$$T_2 = T_1 + \frac{T_{2\text{s}} - T_1}{\eta_{\text{C}}} = 298 + \frac{669.5 - 298}{0.85} = 735.1 \text{ K}$$

The exit temperature is rather hot. Isentropic compression of gases increases both the pressure and temperature (work is done on the gases), and decreases its density. On the other hand, isentropic compression of liquids (in a pump) results in a pressure rise but not a temperature rise or decrease in density.

There is incentive to cool the gases (a nonadiabatic compressor) as they are being compressed to reduce the input work required to raise the pressure. The residence time of the gases is too short to make cooling within the compressor practical, but it is possible to compress the gases part way in a first stage, then send the flow to a heat exchanger (called an intercooler) to cool the gases approaching ambient and then to an upper stage to compress to the combustor pressure. This effect is investigated in Workshop 9.2.

The way the problem statement is formulated, with input mechanical power to the compressor and the pressure ratio specified, the energy equation can be rearranged to solve for the mass flow rate:

$$\dot{m} = \frac{\dot{W}_C}{h_2 - h_1} = \frac{\dot{W}_C}{c_p(T_2 - T_1)}$$
$$= \frac{10\ \text{MW}}{1.14\ \text{kJ/kg/K}(735.1 - 298)\ \text{K}} \cdot \left|\frac{10^6\ \text{W}}{\text{MW}}\right| \cdot \left|\frac{\text{kJ}}{10^3\ \text{J}}\right| = 20.1\ \frac{\text{kg}}{\text{s}}$$

The volumetric flow rate is obtained from $\dot{m} = \rho(AV)$. The density of the inlet is obtained from the ideal gas law:

$$\rho_1 = \frac{P_1}{RT_1} = \frac{1\ \text{atm}}{(0.287\ \text{kJ/kg/K})298\ \text{K}} \cdot \left|\frac{101.3\ \text{kN/m}^2}{\text{atm}}\right| = 1.186\ \frac{\text{kg}}{\text{m}^3}$$

The volumetric flow of the inlet (not sought in the problem statement, but calculated to demonstrate a few points) is therefore:

$$(\text{AV})_1 = \frac{\dot{m}}{\rho_1} = \frac{20.1\ \text{kg/s}}{1.186\ \text{kg/m}^3} = 16.9\ \frac{\text{m}^3}{\text{s}}$$

The volumetric flow rate at the exit of the compressor can be simply calculated by the following string:

$$(\text{AV})_2 = (\text{AV})_1 \frac{\rho_1}{\rho_2} = (\text{AV})_1 \left(\frac{P_1}{P_2}\right)\left(\frac{T_2}{T_1}\right) = (AV)_1 \left(\frac{1\ \text{atm}}{25\ \text{atm}}\right)\left(\frac{735.1\ \text{K}}{298\ \text{K}}\right)$$
$$= (\text{AV})_1 \left(\frac{1}{10.13}\right) = 1.67\ \frac{\text{m}^3}{\text{s}}$$

The density ratio, which depends on both the temperature ratio and pressure ratio (with opposite trends) is 10.13. Therefore, the volumetric flow rate is an order of magnitude lower at the exit than the entrance. In practice, in order to keep the velocity of the gases relative to the equipment nominally constant, the cross-sectional area of the flow path decreases in the direction of flow. Therefore, a common symbol for a compressor is as shown in the schematics as a trapezoid. The opposite effect takes place in a turbine, which is generally designed with an increase in the flow path area.

The eXergy destruction rate is obtained from this string of calculations:

$$e\dddot{X}_{\mathrm{D},12} = T_\infty \dddot{\Sigma}_{12} = T_\infty \dot{m}(s_2 - s_1) = T_\infty \dot{m}\left(c_\mathrm{P} \ln\left(\frac{T_2}{T_1}\right) - R \ln\left(\frac{P_2}{P_1}\right)\right)$$

Inserting values:

$$e\dddot{X}_{\mathrm{D},12} = (298\ \mathrm{K})\left(20.12\frac{\mathrm{kg}}{\mathrm{s}}\right)\left(1.140 \ln\left(\frac{735.1\ \mathrm{K}}{298\ \mathrm{K}}\right) - 0.287 \ln\left(\frac{25\ \mathrm{atm}}{1\ \mathrm{atm}}\right)\right) \times \frac{\mathrm{kJ}}{\mathrm{kg}\cdot\mathrm{K}}$$

$$e\dddot{X}_{\mathrm{D},12} = (298\ \mathrm{K})\left(20.12\frac{\mathrm{kg}}{\mathrm{s}}\right)(1.029 - 0.924)\frac{\mathrm{kJ}}{\mathrm{kg}\cdot\mathrm{K}} \cdot \left|\frac{\mathrm{MW}}{10^6\ \mathrm{W}}\right| \cdot \left|\frac{10^3\ \mathrm{J}}{\mathrm{kJ}}\right|$$

$$= 0.637\ \mathrm{MW}$$

The eXergy destruction rate is 6.34% of the input shaft power. This calculation is done laboriously in order to demonstrate the separate effects of temperature and pressure change. The increasing temperature increases the entropy, while the increasing pressure decreases it. These effects exactly counteract each other for an isentropic process. For an irreversible process, the temperature effect is greater. However, the difference between the two effects is small (about 8%) compared to the magnitude of these two effects. It is therefore important when conducting entropy change calculations to carry extra significant figures in the course of calculations. Premature setting of final reported significant figures (3, say) can result in large calculation errors. Conducting the calculations in a spreadsheet guarantees that, while the output shown can be formatted to show an appropriate number of significant figures (3 or 4). To reiterate, when conducting calculations, carry extra significant figures and report 3 or 4.

The corresponding *Ts* diagram for this compressor is shown to scale in Fig. 9.3. Two isobars, corresponding to the inlet and exit pressures (1 and 25 atm, respectively) are shown as solid lines, as are 3 isotherms (dotted lines) that correspond to the inlet temperature, the temperature after an isentropic and irreversible compression (state 2s and 2). The dashed lines show the ideal and actual processes as labeled. The path of the irreversible process is not generally a straight line as drawn, but has a curvature that is not easy to predict. State 2s is the hypothetical state that would be obtained with an adiabatic reversible compression to the final pressure, and state 2 is the actual state (which could include stray heat exchange and internal irreversibilities). The entropy and temperature are both higher for the actual state. The block arrows show the actual work and the isentropic work.

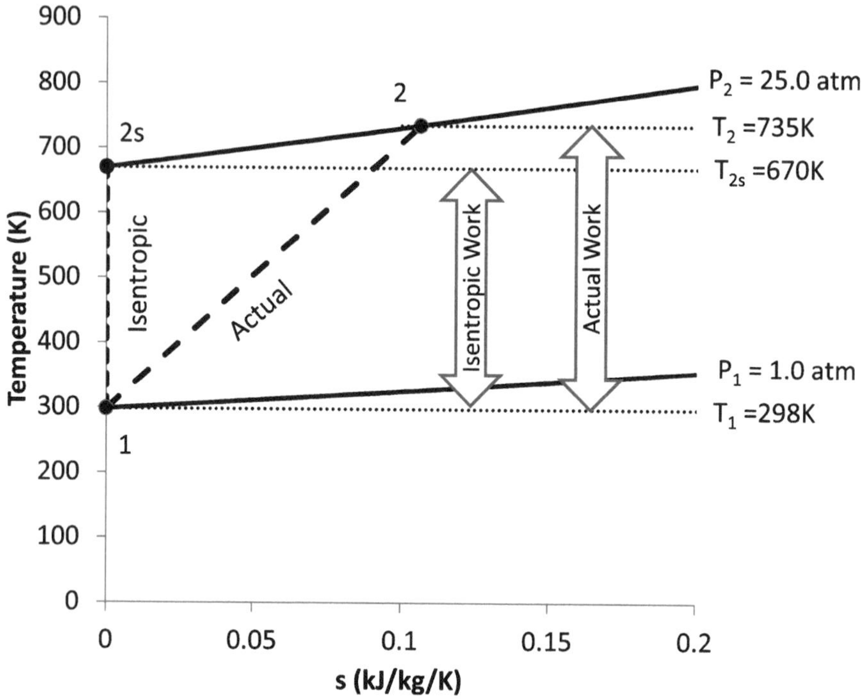

Fig. 9.3 *Ts* diagram for compressor. Isobars for initial and final states and isotherms for the initial and isentropic compression final state are shown. The block arrows indicate the temperature change, which, when divided by the specific heat at constant pressure, yields the associated work per unit mass

Combustor

A schematic of a gas turbine combustor is shown in Fig. 9.4. Air from the compressor (state 2s for the ideal Brayton cycle, state 2 for the gas turbine with irreversibility) is split into multiple streams that flow into a combustor. Primary air mixes and burns with fuel that is injected into a region of the combustor where the main exothermic chemical reactions occur. However, the temperatures required in this zone for rapid chemical reaction are too high for the turbine, and therefore secondary air is mixed with the burned gases in a dilution zone, which brings the temperature to a safe limit. These products (state 3) are delivered to the primary turbine. Much of the advance in gas turbine technology has been the development of turbine materials and that can handle higher temperatures as they simultaneously experience high mechanical stresses due to the high rotational rate. Active cooling techniques (injecting compressed air, water or steam out of the turbine blades) have also been developed. The turbine inlet temperature, state 3, which is the exit of the combustor, is a key parameter in gas turbine performance. There are secondary, or side, chemical reactions that result in pollutant formation, particularly nitrogen oxides from the oxidation of nitrogen in the air or soot when jet fuels are used, important practical matters that are not discussed here.

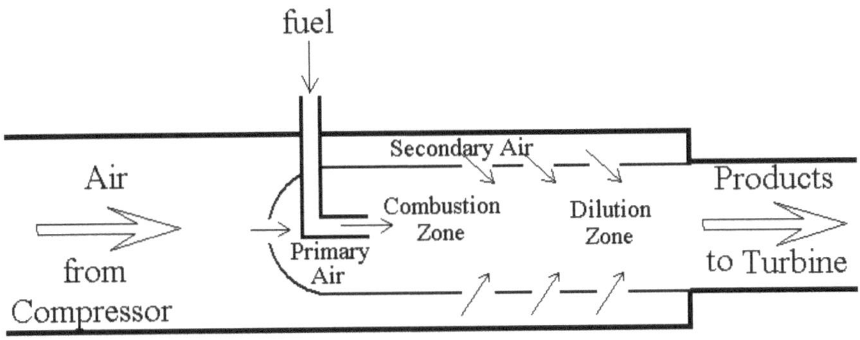

Fig. 9.4 Schematic of a gas turbine combustor

The combustor has two mass inputs, fuel and air, and one output (products of combustion). There are chemical reactions and a conversion of chemical energy into thermal energy, and the means to address those changes were introduced in Chap. 7. As will be calculated, however, the mass flow rate of fuel added is on the order of a few percent the total flow, and only the oxygen in the air reacts (less than half of it reacts in gas turbines). That is, the working fluid is dominated by the nitrogen in the air. Therefore, the combustion process will be treated as the addition of internal energy plus flow work (enthalpy) associated with the fuel, whose mass flow rate is neglected. Traditional thermodynamics textbooks would state that the chemical energy release is equivalent to an external heat transfer, and that view will be used here to estimate eXergy destruction.

The mass balance for the combustor is:

$$\frac{dM}{dt} = \dot{m}_2 + \dot{m}_{\text{fuel}} - \dot{m}_3$$

At steady state, upon rearrangement, the outlet mass flow rate out equals the inlet mass flow rate:

$$\dot{m}_3 = \dot{m}_{\text{products}} = \dot{m}_{\text{air}} + \dot{m}_{\text{fuel}} = \dot{m}_{\text{air}}\left(1 + {}^{1}\!/_{\text{AFR}}\right)$$

Here, *AFR* is the air to fuel mass ratio. Its calculation requires basic combustion stoichiometry calculations, as discussed in Chap. 7, and will be much larger than unity in practice.

The energy equation, neglecting kinetic and potential energy, is (keeping the exit mass flow equal to fuel plus air):

$$\frac{dE}{dt} = \dot{Q}_{23} + \dot{W}_{23} + \dot{m}_{\text{air}}h_2 + \dot{m}_{\text{fuel}}h_{\text{fuel}} - \left(\dot{m}_{\text{air}} + \dot{m}_{\text{fuel}}\right)h_3$$

At steady state, with the work identically zero, neglecting stray heat and rearranging:

$$\dot{m}_{\text{air}}(h_3 - h_2) = \dot{m}_{\text{fuel}}(h_{\text{fuel}} - h_3) = \dot{m}_{\text{fuel}}\ \text{LHV}$$

Most traditional thermodynamics textbooks would leave the reader thinking that the combustor is a heater, and that it is the heat transfer rate that causes the necessary rise in temperature. However, the process is nearly adiabatic (with stray heat loss possibly included), but there is a conversion of chemical energy to thermal energy (internal combustion). This conversion is expressed here as $\dot{m}_{\text{fuel}}$ LHV for the chemical energy conversion rate associated with combustion, where *LHV* is the lower heating value of the fuel, an input parameter. Better, but still awkward, names than "heat of combustion" would be "equivalent heat rate" or "chemical energy conversion rate" or "chemical input power" as used in the spreadsheet template.

The turbine inlet temperature is specified in the problem statement, however, so the chemical energy release rate can be calculated. For the irreversible case:

$$\begin{aligned}\dot{m}_{\text{fuel}}\ \text{LHV} &= \dot{m}_{\text{air}} c_{\text{p}}(T_3 - T_2) \\ &= \left(18.8\frac{\text{kg}}{\text{s}}\right)\left(1.14\frac{\text{kJ}}{\text{kg}\cdot\Delta\text{K}}\right)\left(1800 - 735.1\right)\ \text{K}\cdot\left|\frac{\text{MW}}{1000\ \text{kW}}\right|\end{aligned}$$

$$\dot{m}_{\text{fuel}}\ \text{LHV} = 24.4\ \text{MW}$$

With a lower heating value of natural gas, the mass flow rate of fuel is:

$$\dot{m}_{\text{fuel}} = \frac{24.4\ \text{MW}}{50.1\ \text{MJ/kg}} = 0.486\frac{\text{kg}}{\text{s}}$$

The air to fuel mass ratio is:

$$\text{AFR} = \frac{\dot{m}_{\text{air}}}{\dot{m}_{\text{fuel}}} = \frac{20.1\ \text{kg/s}}{0.486\ \text{kg/s}} = 41.4$$

This calculation confirms that the mass flow rate of fuel is on the order of a few percent of the total flow, while it contributes chemical energy, the nature of internal combustion.

The entropy balance equation for the combustor is:

$$\frac{dS}{dt} = \frac{\dot{Q}_{23}}{T_{\text{B}}} + \dot{m}_{\text{air}} s_2 + \dot{m}_{\text{fuel}} s_{\text{fuel}} - (\dot{m}_{\text{air}} + \dot{m}_{\text{fuel}}) s_3 + \dddot{\Sigma}_{23}$$

At steady state and neglecting stray heat:

$$\dddot{\Sigma}_{23} = \dot{m}_{\text{air}}(s_3 - s_2) + \dot{m}_{\text{fuel}}(s_3 - s_{\text{fuel}})$$

There is a significant entropy change associated with the chemical reaction $\left(\dot{S}_C\right)$ that is not developed here, and a formal calculation of the entropy production (and eXergy destruction) requires a much deeper dive not taken. However, a first approximation can be made by treating the chemical entropy change as an equivalent heat transfer with a thermal reservoir at the maximum temperature (T_3). The entropy production rate is therefore approximated as:

$$\dddot{\Sigma}_{23} \approx \dot{m}_{\text{air}}(s_3 - s_2) - \frac{\dot{m}_{\text{fuel}}\,\text{LHV}}{T_3}$$

The entropy change of the air is expressed in terms of temperatures and pressures, and multiplying by the dead state temperature to express as eXergy destruction rate:

$$e\dddot{X}_{\text{D},23} = T_\infty \dddot{\Sigma}_{23} \approx \dot{m}_{\text{air}} T_\infty\left(c_{\text{p}}\ln\left(\frac{T_3}{T_2}\right) - R\ln\left(\frac{P_3}{P_2}\right)\right) - \frac{T_\infty}{T_3}\left(\dot{m}_{\text{fuel}}\,\text{LHV}\right)$$

Evaluating for the irreversible gas turbine case:

$$\begin{aligned} e\dddot{X}_{D,23} \approx &\left(20.1\frac{kg}{s}\right)(298K) \\ &\times\left(1.14\frac{kJ}{kg\cdot\Delta K}\ln\left(\frac{1300K}{735.1K}\right) - 0.287R\frac{kJ}{kg\cdot\Delta K}\ln\left(\frac{25}{24}\right)\right) - \frac{298K}{1300K} \\ &\times\left(24,400\frac{kJ}{s}\right) \end{aligned}$$

$$e\dddot{X}_{\text{D},23} = 6115 + 70 - 4040 = 2145\ \text{kW} = 2.15\ \text{MW}$$

The first term (6115 kW) is due to the entropy change of the working fluid associated with the temperature rise. The second term (70 kW) is due to the pressure drop. The third term (−4040) is the eXergy transfer.

A *Ts* diagram is shown in Fig. 9.5 with the isobars corresponding to states 2 (25 atm) and 3 (24 atm). These two isobars are close together when viewed across the entire process. Therefore, Fig. 9.6 is a detail focused on the combustor exit state.

Some people think, erroneously, that there is a rise in pressure associated with combustion. That may be true for closed systems. However, for open systems, pressure drives the flow into the combustor, more of a momentum consequence than energy. The specific volume of the exit gases is higher than that of the inlet gases, so the gases expand as they pass through and flow work is done by the gases. The combustion process is modeled as isobaric in the ideal Brayton cycle, and there is a pressure drop due to flow friction in the real cycle. For high speed propulsion engines (RAM and SCRAM jets), the kinetic energy of the flow cannot be safely neglected and there is a greater pressure change in the combustors of those advanced engines, not treated here formally.

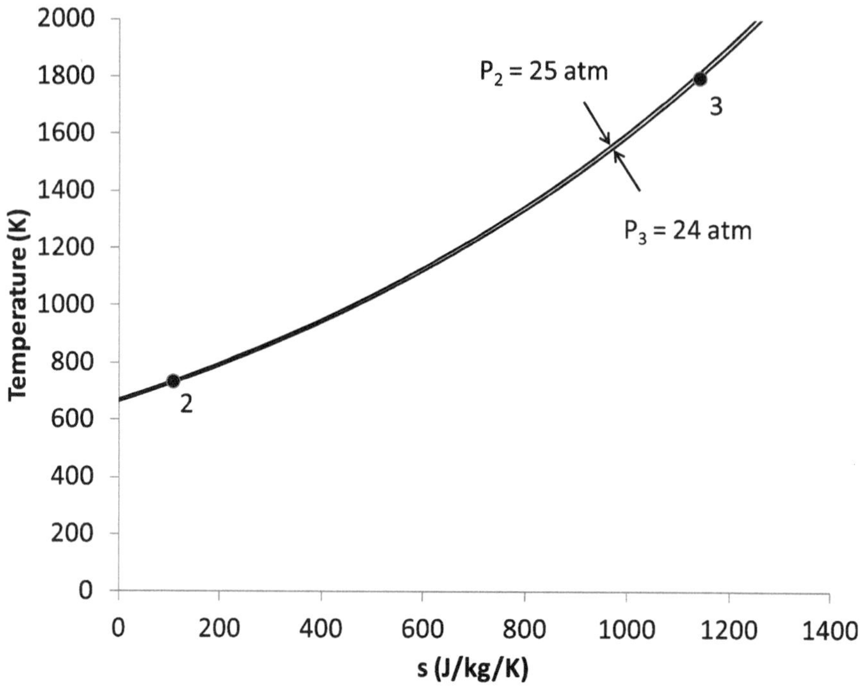

Fig. 9.5 *Ts* diagram for combustor. Isobars corresponding to state 2 (25 atm) and state 3 (24 atm) are shown

Turbines

Treatment of turbines is similar to that of compressors, except that the objective is reversed. That is, the goal of the turbine is to convert fluid enthalpy into output shaft work. The output of the primary turbine is used to drive the compressor. The Ts diagram for the turbines is shown in Fig. 9.7. The inlet mass flow rate equals the outlet mass flow rate, and the energy equation (at steady state, neglecting stray heat, kinetic and potential energy changes) is:

$$\dot{W}_{\text{OUT},34} = \dot{m}_{\text{air}}(h_3 - h_4)$$

The isentropic efficiency of a turbine is defined as the actual work divided by the maximum possible work (obtained from a hypothetical reversible turbine):

$$\eta_{T,\text{primary}} = \frac{\dot{W}_{\text{ACTUAL}}}{\dot{W}_{\text{IDEAL}}} = \frac{\dot{m}_{\text{air}}(h_3 - h_4)}{\dot{m}_{\text{air}}(h_3 - h_{4s})}$$

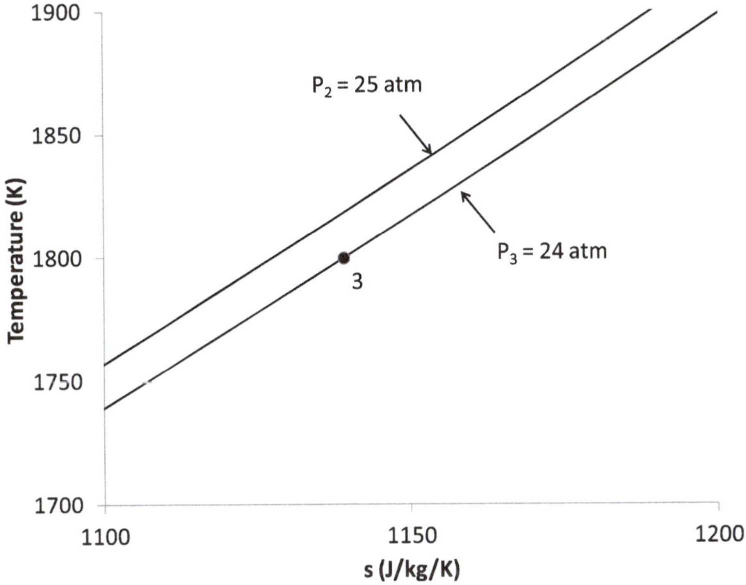

Fig. 9.6 *Ts* diagram of combustor. Detail near exit state 3

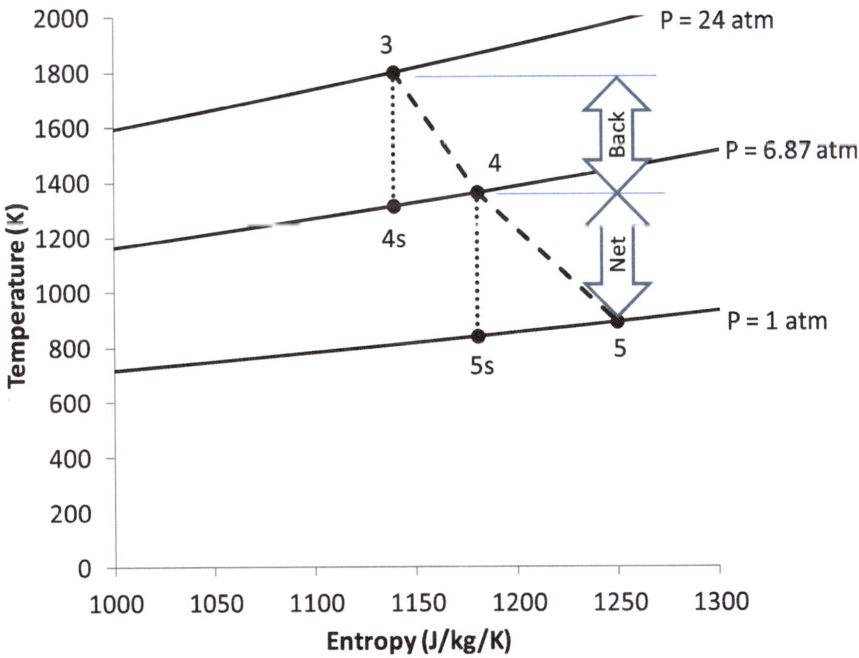

Fig. 9.7 *Ts* Diagram for turbines. Isobars are shown as solid lines. Dotted lines are for hypothetical isentropic process. Dashed lines are for actual processes. The net work out (from turbine 2 to output shaft power) and back work (from turbine 1 to compressor) are shown by block arrows

For calorically perfect gases:

$$\eta_{T,\text{primary}} = \frac{T_3 - T_4}{T_3 - T_{4s}}$$

Since the input power to the compressor is specified, the exit state of the primary turbine (state 4) is obtained in the following round-about way. For the problem as stated, the total power output is specified as an input, as are the inlet conditions. Therefore, the energy equation can be used to obtain the exit temperature. From there, the definition of isentropic efficiency yields the temperature of the hypothetic state 4s. Isentropic expansion yields the pressure at state 4, which must be greater than ambient in order to produce net power. That is:

$$T_4 = T_3 - \frac{\dot{W}_{\text{OUT},34}}{\dot{m}_{\text{air}} c_\text{p}} = 1800\ \text{K} - \frac{10(10)^6\ \text{W}}{(20.1\ \text{kg/s})(1140\ \text{J/kg/K})} = 1363\ \text{K}$$

From the isentropic efficiency definition:

$$T_{4\text{s}} = T_3 - \frac{T_3 - T_4}{\eta_{T,\text{primary}}} = 1800 - \frac{1800 - 1363}{0.9} = 1314\ \text{K}$$

For an isentropic expansion between known temperatures:

$$P_4 = P_3 \left(\frac{T_{4\text{s}}}{T_3}\right)^{\frac{k}{k-1}} = 24\ \text{atm} \left(\frac{1314}{1800}\right)^{\frac{1.34}{1.34-1}} = 6.87\ \text{atm}$$

After power is extracted from the primary turbine to drive the compressor, the pressure is still above ambient, and therefore the gases are capable of converting additional thermal energy into mechanical energy. The gases enter the secondary turbine with known conditions, and vent to ambient pressure. Therefore, the hypothetical isentropic process results in an exit temperature given by:

$$T_{5\text{s}} = T_4 \left(\frac{P_5}{P_4}\right)^{\frac{k-1}{k}} = 1363 \left(\frac{1}{6.87}\right)^{\frac{1.4-1}{1.4}} = 839\ \text{K}$$

The definition of the isentropic efficiency of the secondary turbine is used to determine the actual exit temperature:

$$T_5 = T_4 - \eta_{T,\text{primary}} (T_4 - T_{5\text{s}}) = 1363 - 0.9(1363 - 839) = 892\ \text{K}$$

Process Heat Exchanger

The last component is the process heater used for cogeneration. This component would not be used in a peaking power plant, where the hot exhaust (at ambient pressure) would be vented to the stack. The energy and entropy balances are (neglecting kinetic and potential energy, and a single inlet, single exit device at steady state):

$$\frac{dE}{dt} = 0 = -\dot{Q}_{\mathrm{C}} + \dot{m}_{\mathrm{air}}(h_5 - h_6)$$

$$T_\infty\left[\frac{dS}{dt} = 0 = \frac{-\dot{Q}_{\mathrm{C}}}{T_{\mathrm{B}}} + \dot{m}_{\mathrm{air}}(s_5 - s_6) + \dddot{\Sigma}_{56}\right]$$

Rearranging and evaluating for calorically perfect gases:

$$\dot{Q}_{\mathrm{C}} = \dot{m}_{\mathrm{air}} c_{\mathrm{p}}(T_5 - T_6)$$

$$e\dddot{X}_{\mathrm{D,56}} = T_\infty \dddot{\Sigma}_{56} = \dot{Q}_{\mathrm{C}}\frac{T_\infty}{T_{\mathrm{B}}} + \dot{m}_{\mathrm{air}} T_\infty\left(c_{\mathrm{p}}\ln\left(\frac{T_6}{T_5}\right) - R\ln\left(\frac{P_6}{P_5}\right)\right)$$

For the ideal cycles, state 6 corresponds to the inlet state 1 and the process is isobaric. For the real device, state 6 is warm, and there is a pressure drop across the heat exchanger.

Cycle Results

The stationary gas turbine results are compiled in Figs. 9.8, 9.9, 9.10, 9.11, and 9.12. The shaded and bolded entries show input parameters. Problems can be posed differently than done here, and it is easy to either over-specify or under-specify problem statements. It can be, however, a good exercise to do so, or to conduct parametric studies using a properly constructed spreadsheet by simply changing input parameter values. Rather than venting the exit gases directly to the environment, the heat exchanger was added in the problem statements to make a closed cycle for the Brayton cycle, and to demonstrate the inherent energy of the exit stream, while the eXergy destruction reveals how it is dissipated.

Looking at the big picture, the 1st Law efficiency as a power cycle, defined as the net mechanical power output divided by the chemical power input, is 0.555 for the ideal Brayton cycle, and 0.389 for the gas turbine with irreversibilities, a 30% decrease. The performance of the irreversible engine is consistent with that advertised by manufacturers of aero-derivative gas turbines.

State	P (atm)	T (K)	h (kJ/kg)	s (kJ/kg/K)
1	1	298	0	0
2	25	669.5	423.5	0
3	25	1800	1712.3	1.1274
4	9.970	1428	1288.7	1.1274
5	1	801.2	573.6	1.1274

State	P (atm)	T (K)	h(kJ/kg/K)	s(kJ/kg/K)
1	1	298	0	0
2s	25	669.5	423.5	0.000
2	25	735.1	498.3	0.107
3	24	1800	1712.3	1.139
4s	6.87	1314.33	1158.6	1.139
4	6.87	1362.9	1214.0	1.180
5s	1.2	878.7	662.0	1.180
5	1.2	927.1	717.2	1.242
6	1	373.0	85.5	0.256

Fig. 9.8 State tables for stationary gas turbines (ideal Brayton cycle on left, gas turbine with irreversibilities on right)

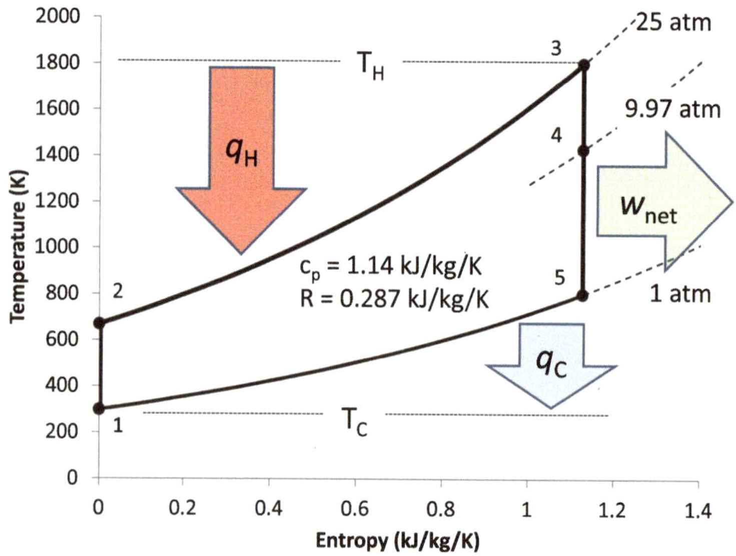

Fig. 9.9 *Ts* diagram (to scale) for an ideal Brayton cycle (with nominal room temperature and pressure inlet air, a compression ratio of 25, and turbine inlet temperature of 1800 K). High, intermediate, and low pressure isobars shown as dashed lines. Equivalent heat exchange is shown between thermal reservoirs at a high temperature and ambient

The performance of the aero-derivative could be compared to that of an industrial turbine by simply lowering the pressure ratio (to approximately 15 to be representative). In recognition of the higher development efforts for aero-derivatives, the component isentropic efficiencies (compressor and turbine) and the turbine inlet temperature could be lowered slightly. One would expect to find lower efficiencies as a power cycle and a shift to higher heat exchange from the exhaust due to primarily to the lower compression ratio. The lower component efficiencies and turbine inlet temperature would also impact the industrial gas turbine performance.

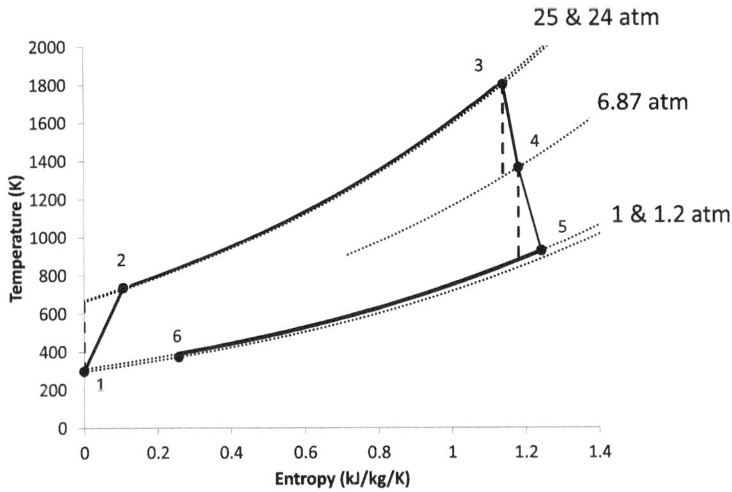

Fig. 9.10 *Ts* diagram for gas turbine with irreversibilities. The 5 main isobars are shown as dotted lines. Vertical dashed lines are hypothetical isentropic processes. The paths for the combustor and heat exchanger follow the higher isobar on this plot, but actually cross over from higher to lower pressure isobars

		1st Law Quantities			Combined 1st and 2nd Law Quantities		
Process	*Device*	$h_{out} - h_{in}$	q_{in} or m_{fuel}*LHV	w_{out}	$T_\infty(s_{out} - s_{in})$	$T_\infty * q_{in}/T_B$	eX_d
1-2	**Compressor**	423.5	0	-423.5	0	0	0
2-3	**Combustor***	1288.7	1288.7	0	336.0	213.4	122.6
3-4	**Primary Turbine**	-423.5	0	423.5	0	0	0
4-5	**Secondary Turbine**	-715.1	0	715.1	0	0	0
5-1	**Heat Exchanger**	-573.6	-573.6	0.0	-336.0	-573.6	237.6
	net	0.0	715.1	715.1	0.0	-360.2	360.2

* chemical energy treated as equivalent heat transfer from a thermal reservoir at T3

		1st Law Quantities			Combined 1st and 2nd Law Quantities		
Process	*Device*	$h_{out} - h_{in}$	q_{in} or m_{fuel}*LHV	w_{out}	$T_\infty(s_{out} - s_{in})$	$T_\infty * q_{in}/T_B$	eX_d
1-2	**Compressor**	498.3	0	-498.3	31.7	0.0	31.7
2-3	**Combustor***	1214.0	1214.0	0	307.7	201.0	106.7
3-4	**Primary Turbine**	-498.3	0	498.3	12.3	0.0	12.3
4-5	**Secondary Turbine**	-496.8	0	496.8	18.2	0.0	18.2
5-1	**Heat Exchanger**	-631.7	-631.7	0.0	-293.7	-631.7	338.0
	net	85.500	582.3	496.8	76.3	-430.7	507.0

* chemical energy treated as equivalent heat transfer from a thermal reservoir at T3

Fig. 9.11 Process tables for stationary gas turbines (ideal Brayton cycle on top, gas turbine with irreversibilities on bottom)

The complete *Ts* diagram is a helpful visual aid in understanding why gas turbines perform the way they do. The work into the compressor is the vertical jump at the low entropy level. The work out of the turbines is the vertical jump down

Quantity	Unit	Ideal Brayton Cycle	w/Irreversibilities	%Change
inlet density	kg/m3	1.186	1.186	0.0%
air mass flow rate	kg/sec	23.61	20.1	-15.0%
fuel mass flow rate	kg/sec	0.6073	0.4863	-19.9%
Fuel to air mass flow ratio		0.0257	0.0242	-5.8%
Compressor/Primary Turbine Power	**MW**	**10**	**10**	0.0%
Chemical Input Power	MW	30.43	24.4	-19.9%
Secondary Turbine (output) Power	MW	16.88	9.97	-41.0%
Process Heating Rate	MW	13.54	12.68	-6.4%
eXergy Destruction Rate in HEX	MW	5.61	6.78	20.9%
Backwork Ratio		0.372	0.501	34.6%
1st Law Efficiency as Electric Generator		0.555	0.389	-29.9%
Carnot Efficiency of Power Cycle*		0.834	0.834	0.0%
2nd Law Efficiency of Power Cycle		0.665	0.466	-29.9%
1st Law Efficiency as Cogeneration		1.000	0.909	-9.1%
1st Law Efficiency of Ideal Combined Cycle		0.739	0.688	-7.0%

Fig. 9.12 Results table for stationary gas turbines

at the higher entropy level, where entropy is added by the addition of chemical energy. The total turbine work is greater than the compressor work in the Brayton cycle because the isobars diverge on a *Ts* diagram, that old adage. For a real gas turbine, net work is possible provided the irreversibilities are not excessive. The backwork ratio (compressor work to net work) is nearly 1 to 1 for this irreversible gas turbine example. That is, the work required to compress the gas is of the same order as the net work out. The intermediate pressure (the pressure developed by the gas generator) for the ideal Brayton cycle is 9.97 atm and 6.87 for the gas turbine with irreversibilities, a significant decrease due to the higher backwork ratio.

For the Brayton cycle *Ts* diagram (Fig. 9.9), block arrows show the main energy flows, namely, heat in from a fictitious high temperature reservoir, net work out (secondary turbine) and heat out to a fictitious low temperature reservoir. In the combustor, the heat added raises the entropy of the working fluid, but the heat transfer takes place over a finite temperature difference, particularly during the early part of the process, and is the source of irreversibility. Similarly, the heat transfer across the exchanger takes place over a finite temperature difference. These two effects account for the 2nd Law efficiency (0.665), that is, the ratio of the ideal Brayton cycle efficiency and the maximum possible efficiency (Carnot efficiency) for a power cycle operating between the two temperature reservoirs.

Ideal Brayton Cycle

The Brayton cycle results of the previous section were obtained by analyzing the basic components of the gas turbine cycle (with both isentropic efficiencies set to unity and no pressure drop across the combustor). As with the Otto cycle for reciprocating engines, the 1st Law efficiency reduces to an elegant simple solution that depends only on the pressure ratio and the ratio of specific heats:

$$\eta_{\text{BRAYTON}} = \frac{\text{Net work}}{\text{Chemical energy in}} = \frac{h_4 - h_{5s}}{h_3 - h_{2s}} = \frac{(h_3 - h_{5s}) - (h_{2s} - h_1)}{h_3 - h_{2s}}$$

$$\eta_{\text{BRAYTON}} = \frac{(T_3 - T_{5s}) - (T_{2s} - T_1)}{T_3 - T_{2s}} \quad \text{for calorically perfect gases}$$

Rearranging, and manipulating,

$$\eta_{\text{BRAYTON}} = 1 - \frac{T_{5s} - T_1}{T_3 - T_{2s}} = 1 \quad \frac{\left(\frac{T_{5s}}{T_3}\right)T_3 - \left(\frac{T_1}{T_{2s}}\right)T_{2s}}{T_3 - T_{2s}}$$

$$= 1 - \frac{\left(\frac{P_{5s}}{P_3}\right)^{\frac{k-1}{k}} T_3 - \left(\frac{P_1}{P_{2s}}\right)^{\frac{k-1}{k}} T_{2s}}{T_3 - T_{2s}}$$

The pressure ratios are equal, and therefore:

$$\eta_{\text{BRAYTON}} = 1 - \left(\frac{P_1}{P_{2s}}\right)^{\frac{k-1}{k}}$$

The 1st Law efficiency of the Brayton cycle, plotted in Fig. 9.13, depends only on the pressure ratio and the ratio of specific heats. It is independent of the turbine inlet temperature. The efficiency increases monotonically with pressure ratio, very strongly below 10. The effect of the ratio of specific heats is fairly strong. The values chosen for Fig. 9.13 correspond to monotonic gases ($k = 1.67$), diatomic gases (O_2, N_2, CO, etc.) at nominal room temperature ($k = 1.4$), steam at moderate temperatures and pressures ($k = 1.25$), and mists or particle-laden gas streams ($k = 1.1$). The value of k decreases with increasing specific heat. Generally, the more bonds in a molecule, the higher the specific heat, as raising the temperature requires more energy to increase the vibrational and rotational kinetic energy of the molecule. Also, except for monotonic gases, raising the temperature raises the specific heat due to quantum mechanical effects (more energy levels are available).

The 1st Law efficiency of the ideal Brayton cycle is independent of the turbine inlet temperature. On the other hand, the total work per unit mass is a strong function of turbine inlet temperature, as shown in Fig. 9.14. For a given calorically perfect gas (air with $k = 1.34$), the net work increases strongly with pressure ratio below approximately 10, then levels off, and is nearly constant. However, the higher the turbine inlet temperature, the higher the net work output.

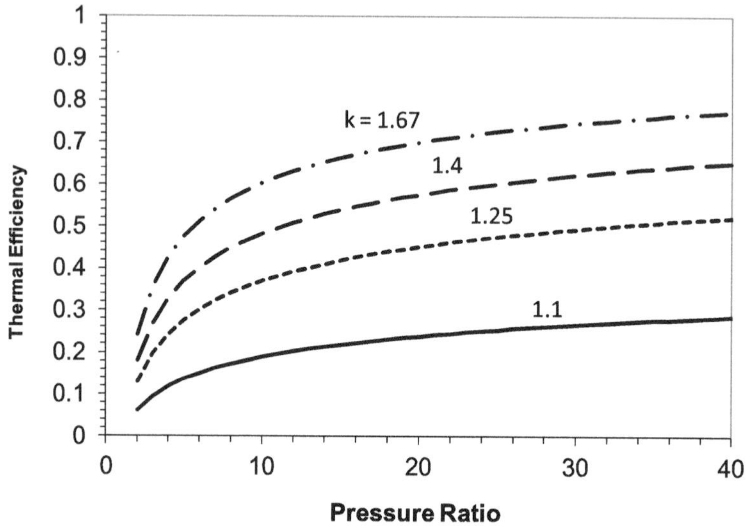

Fig. 9.13 Ideal Brayton cycle efficiency as a function of pressure ratio for several values of the ratio of specific heats

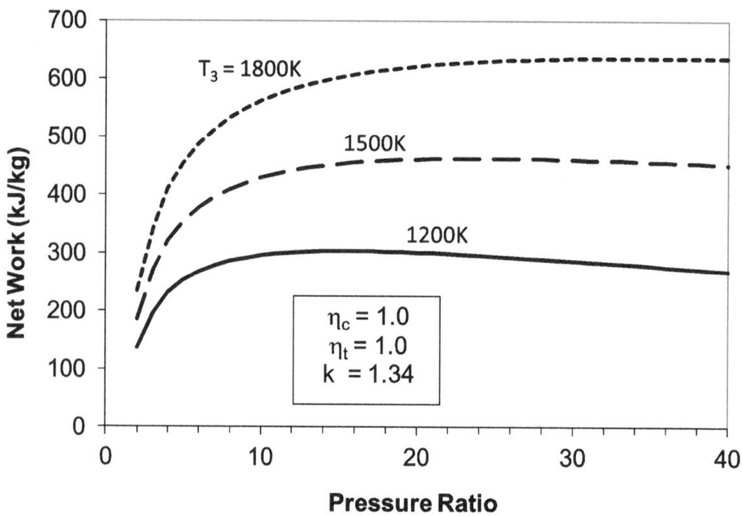

Fig. 9.14 Net work (per unit mass throughput) of the ideal Brayton cycle as a function of pressure ratio for several turbine inlet temperatures. The ratio of specific heats ($k = 1.34$) is based on air at 1000 K

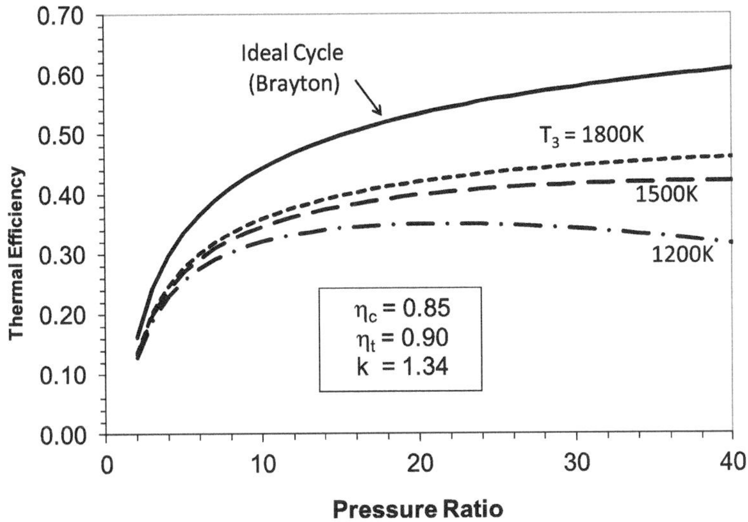

Fig. 9.15 1st Law efficiency of gas turbines as a function of pressure ratio for several turbine inlet temperatures for the ideal Brayton cycle and for cycles with component losses. The isentropic efficiencies for the compressor and turbine are 0.85 and 0.90, respectively, and ratio of specific heats is 1.34 (based on air at 1000 K). Pressure losses in the combustor are neglected

Effect of Component Irreversibilities

For gas turbines (with compressor and turbine irreversibilities), the 1st Law efficiency depends on turbine inlet temperature, as shown in Fig. 9.15, for fixed values of compressor and turbine efficiencies, considered to be state of the art values. Therefore, higher turbine inlet temperatures result in a dramatic increase in net work, which is very important for mobile applications, and a significant increase in 1st Law efficiency (for real gas turbines), which is very important for stationary applications. Increasing turbine inlet temperature has been a focus of research, an example of the dependency of thermodynamic application on material development, and vice versa.

The goal of the gas generator (compressor, combustor and primary turbine) is to produce a high pressure fluid that can drive the output compressor. That high pressure fluid can be used for a different purpose, namely, to generate a propulsive thrust in a mobile power plant.

Deeper Dive: Mobile Gas Turbines

The objective of a jet engine is to produce thrust, a virtual force that can propel an aircraft by Newton's 3rd Law (action/reaction), a momentum concept. For an air-breathing engine (as opposed to a rocket that carries all of its propellant), the

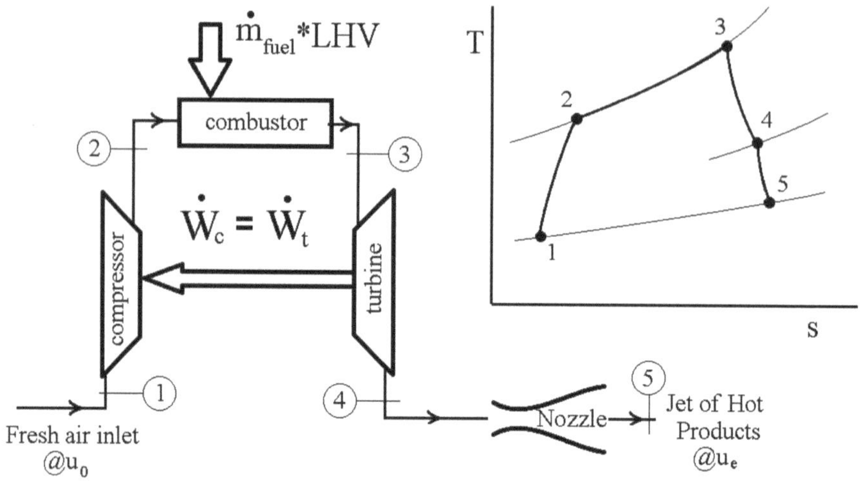

Fig. 9.16 Schematic and *Ts* diagram of a mobile gas turbine, which consists of a gas generator (states 1–4) followed by a nozzle designed to develop a high velocity jet that provides propulsion

air that is taken into an engine pushes it backward, while air sent out the back as a jet pushes it forward. Therefore, the main objective is to expel a jet out the back at a higher velocity than it takes in. Figure 9.16 is a schematic and *Ts* diagram of a basic jet engine which consists of a gas generator (states 1–4) that feeds a nozzle. Inlet air (state 1) enters the compressor at the flight speed (u_0) and is compressed to the high pressure isobar (state 2). Fuel is injected into the combustor which burns with the air, converting chemical energy into thermal energy. The high pressure, high temperature gas (state 3) enters the turbine and expands to an intermediate pressure. All the mechanical power developed by the turbine is used to drive the compressor. Combustion is generally treated as an isobaric process, but deviations from that model can easily be included in an analysis. Isentropic efficiencies can be applied to the compressor and turbine, and the schematic shows the entropy increase in these mechanical devices due to irreversibilities.

Kinetic energy was neglected in all the components of the stationary gas turbine. Kinetic energy is not identically zero, just like stray heat would rarely be zero, but the primary thermodynamic objective of each device is to convert energy between other forms, and in practice, that is usually a justifiable assumption, certainly for understanding the basic operation and gaining a first approximation of performance metrics.

However, for mobile jet engines, the kinetic energy of the intake (for sufficiently high speed vehicles) may not be negligible. Also, the primary thermodynamic objective of the nozzle is to convert enthalpy into kinetic energy so clearly the kinetic energy of the exit jet is not negligible in any analysis. Therefore, the first part of the compression process can be accomplished by an added component, namely a diffuser. The diffuser is located just before the compressor, and its goal is to convert kinetic energy of the inlet air (relative to the vehicle) to enthalpy. It involves no

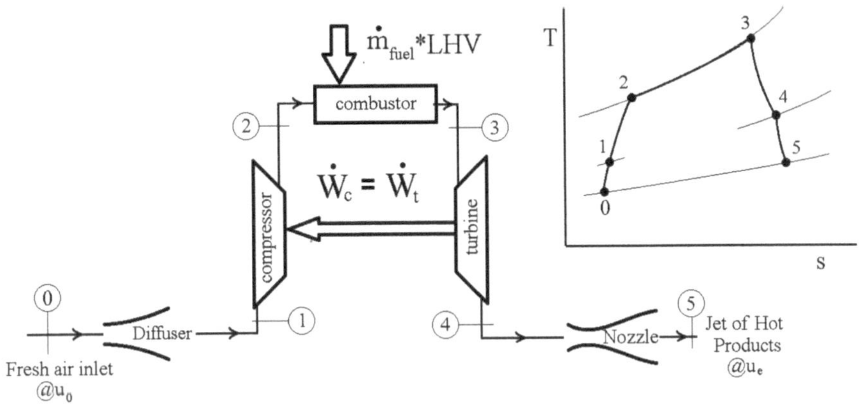

Fig. 9.17 Schematic and *Ts* sketch of a mobile gas turbine, with a diffuser placed before the compressor and the secondary turbine replaced by a nozzle. The two intermediate pressures (states 1 and 4) are not directly related

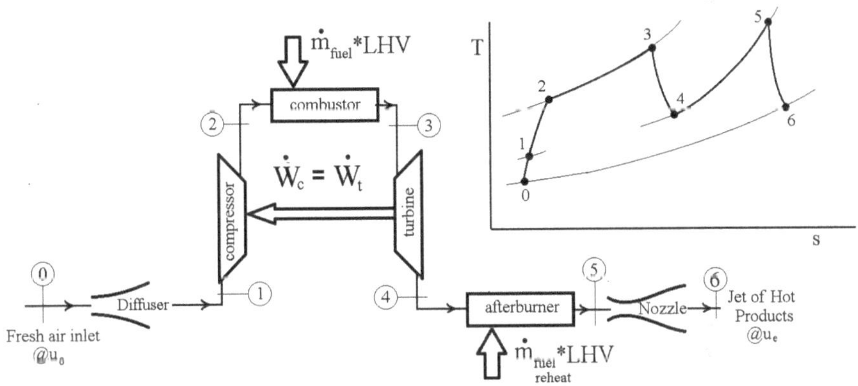

Fig. 9.18 Schematic and *Ts* sketch of a gas turbine with an afterburner (reheat) placed between the turbine and nozzle, with a diffuser placed before the compressor and the secondary turbine replaced by a nozzle

moving parts, and is simply a carefully designed flow channel that decelerates the fluid (relative to the engine). Figure 9.17 shows a modified schematic that includes a diffuser before the compressor. The *Ts* diagram shows an adiabatic compression (with irreversibilities) from the ambient pressure to an intermediate pressure. The higher the flight speed, the higher the pressure entering the compressor (state 1). The performance of both nozzles and diffusers can be modeled using isentropic efficiencies in a similar way as turbines and compressors.

Another device that can be added to the mix is an afterburner, or the use of reheat in the basic cycle, as shown in Fig. 9.18. In general, the thermodynamic performance of a gas turbine that includes reasonable irreversibilities improves with the turbine inlet temperature. However, in practice, the turbine inlet temperature is limited by material constraints to approximately 1800 K (early twenty-first century). From a combustion point of view, if not for material limits, the turbine inlet temperature would be limited by the fuel flow required to burn all the O_2 in the air (i.e., the fuel/air stoichiometry). The material limit temperature is met in practice while there is still plenty of O_2 available (typically 15 mol%), as indicated in Fig. 9.4 by the necessary addition of dilution air downstream of the main combustion zone. However, there is a considerable decrease in temperature across the turbine associated with the expansion of the working fluid. Therefore, additional fuel can be burned with the remaining O_2 in a second combustor (called an afterburner, process 4–5) that operates at the intermediate pressure delivered by the gas generator before delivery to the nozzle. Furthermore, since the nozzle contains no moving parts, the material constraints are not as severe as with the turbine, and a higher temperature can be attained (i.e., $T_5 > T_3$ is possible in practice).

To follow with the adage "gas turbines work because isobars diverge on a *Ts* diagram," notice that the vertical distance across the nozzle with the afterburner (process 5–6) is greater than it would be if the flow entered the nozzle at state 4 (see process 4–5 of Fig. 9.17). In addition, the temperature can be set higher than the turbine inlet temperature because the gases enter a nozzle with no moving parts. Therefore, the kinetic energy of the exit jet is higher with the afterburner. However, additional fuel is burned, and the *Ts* diagram shows that higher exit temperatures vented out at the back are associated with the afterburner. When the numbers are crunched, the thermal efficiency with the afterburner is lower than without it. Therefore, afterburners are generally used when high thrust is required (i.e., at take-off or to accelerate rapidly) and are generally turned off (by simply shutting down the fuel flow to the afterburner) during most of flight.

The compression of the working fluid in a diffuser raises both temperature and pressure. As flight speeds increase, the compression obtained in the diffuser can become sufficiently high to eliminate the need for an actual compressor, which eliminates the need for the primary turbine. That is, there is no need for a gas generator at all. Such engines, called RAM jets, are deployed in military aircraft and missiles (the SR-71 Blackbird being the most famous). Figure 9.19 shows a schematic of a RAM jet (and SCRAM jet). Understanding their design requires a study of compressible fluid mechanics, but their theoretical thermodynamic performance can be investigated by analysis of a diffuser/combustor/nozzle string, at least as a first-order calculation. Appendix 1 provides a deeper dive and workshop designed to investigate some fundamental trends of an ideal RAM jet cycle, defined with isentropic diffuser and nozzle and negligible kinetic energy in the combustor. A general treatment of propulsion performance parameters is included.

As flight speeds increase, the pressure/temperature rise in the diffuser becomes excessive if the working fluid is decelerated to negligible kinetic energy (typical of subsonic flow). Therefore, at flight speeds greater than approximately 5 times the speed of sound, (Mach 5, approximately 1500 m/s = 3350 miles/h, considered to be a rough boundary above which the term hypersonics applies) the compression process in the diffuser cannot bring the working fluid to a low kinetic energy. A

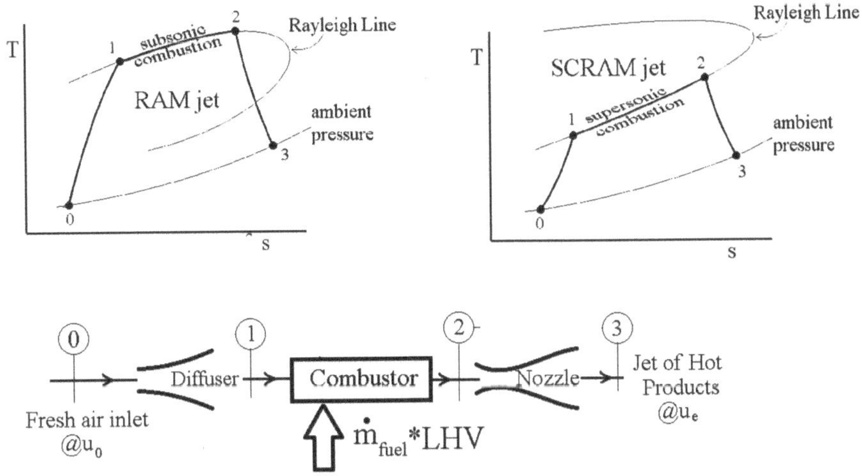

Fig. 9.19 Schematic and *Ts* diagrams for RAM and SCRAM jet engines

SCRAM jet (Supersonic Combustion RAM jet) is an engine for which the diffuser decelerates the flow but it is still supersonic when it enters the combustor. The combustion process can be modeled to take place along a thermodynamic line called the Rayleigh line, which is obtained by the thermo/fluid analysis of a constant area flow with heat addition (or internal combustion), but for the case where the kinetic energy is significant (a deeper dive with scuba gear). A limiting result of Rayleigh flow is that if the kinetic energy of the inlet and exit streams (states 1 and 2) are negligible, the Rayleigh line reduces to an isobar (the top left side of the Rayleigh line). This result is the fundamental reason why the combustion process is treated as isobaric in gas turbine applications in which kinetic energy effects in the combustor have a negligible effect.

The thermodynamic analysis of all these combinations involves the devices associated with the gas generator (with negligible kinetic energy), plus the addition of the diffuser and nozzle, analyzed next.

Diffusers

The diffuser is a single inlet, single outlet device and at steady state the inlet mass flow rate equals the outlet mass flow rate. Potential energy changes are neglected, but kinetic energy is included as a primary thermodynamic objective of the device. With the inlet defined as state 0 and the exit as state 1 (as in the schematics), the energy balance for a diffuser is:

$$\frac{dE}{dt} = \dot{Q} - \dot{W}_{\text{shaft}} + \dot{m}\left(h_0 + \frac{V_0^2}{2}\right) - \dot{m}\left(h_1 + \frac{V_1^2}{2}\right)$$

At steady state, neglecting stray heat, with the shaft work identically zero, the energy balance becomes:

$$h_1 - h_0 = \frac{V_0^2}{2} - \frac{V_1^2}{2} = \frac{V_0^2}{2}\left(1 - \frac{V_1^2}{V_0^2}\right)$$

That is, the change in enthalpy is equal and opposite to the change in kinetic energy. The best performance is obtained for a reversible diffusion process to a negligible exit kinetic energy. Real effects can be introduced with an isentropic efficiency, defined by comparing the actual diffuser to a hypothetical isentropic one that achieves the same pressure rise with the same flow rate. Therefore, similar to the compressor:

$$\eta_{\text{DIFFUSER}} = \frac{h_{1s} - h_0}{h_1 - h_0}$$

$$\eta_{\text{DIFFUSER}} = \frac{T_{1s} - T_0}{T_1 - T_0} \quad \text{for calorically perfect gases}$$

On a *Ts* diagram, the diffuser process looks the same as the compressor on a *Ts* diagram to an intermediate isobar, followed by a compressor rise to the pressure that enters the combustor.

For a calorically perfect gas, the enthalpy change can be related to temperature rise and pressure rise as follows:

$$h_1 - h_0 = c_p(T_1 - T_0) = c_p T_0\left(\frac{T_1}{T_0} - 1\right)$$

Solving for the temperature ratio:

$$\frac{T_1}{T_0} = 1 + \frac{h_1 - h_0}{c_p T_0}$$

This result is valid if the compressor is irreversible. The pressure rise is obtained for the hypothetical irreversible process:

$$\frac{P_1}{P_0} = \left(\frac{T_{1s}}{T_0}\right)^{\frac{c_p}{R}} = \left(\frac{T_{1s}}{T_0}\right)^{\frac{k}{k-1}}$$

A 2nd Law analysis (adiabatic and steady state) will show that the eXergy destruction rate is given by:

$$\dddot{eX}_D = \dot{m}T_\infty(s_1 - s_0)$$

For calorically perfect gases,

$$\dddot{eX}_D = \dot{m}T_\infty\left(c_p \ln\left(\frac{T_1}{T_0}\right) - R \ln\left(\frac{P_1}{P_0}\right)\right)$$

Nozzles

The nozzle is also a single inlet, single outlet device, and at steady state, the inlet mass flow rate equals the outlet mass flow rate. Like the diffuser, kinetic energy is included as a primary thermodynamic objective. The energy balance (with states labeled "in" and "out") is:

$$\frac{dE}{dt} = \dot{Q} - \dot{W}_{\text{shaft}} + \dot{m}\left(h_{\text{in}} + \frac{V_{\text{in}}^2}{2}\right) - \dot{m}\left(h_{\text{out}} + \frac{V_{\text{out}}^2}{2}\right)$$

At steady state, neglecting stray heat, with the shaft work identically zero, the energy balance becomes:

$$h_{\text{in}} - h_{\text{out}} = \frac{V_{\text{out}}^2}{2} - \frac{V_{\text{in}}^2}{2} = \frac{V_{\text{out}}^2}{2}\left(\frac{V_{\text{in}}^2}{V_{\text{out}}^2} - 1\right)$$

That is, the change in enthalpy is equal and opposite to the change in kinetic energy. Isentropic efficiency for a nozzle is defined by comparing the actual nozzle to a hypothetical isentropic one that achieves the same pressure drop with the same flow rate. Therefore, similar to the turbine:

$$\eta_{\text{NOZZLE}} = \frac{h_{\text{in}} - h_{\text{out}}}{h_{\text{in}} - h_{\text{out,s}}}$$

$$\eta_{\text{NOZZLE}} = \frac{T_{\text{in}} - T_{\text{out}}}{T_{\text{in}} - T_{\text{out,s}}} \quad \text{for calorically perfect gases}$$

The nozzle process looks the same as the secondary turbine of a stationary gas turbine on a *Ts* diagram.

Note that the exit pressure of a nozzle in practice may not match ambient pressure to which it is venting due to compressible flow phenomena.

A 2nd Law analysis (adiabatic and steady state) will show that the eXergy destruction rate is given by:

$$e\dddot{X}_{\mathrm{D}} = \dot{m}T_{\infty}(s_{\mathrm{out}} - s_{\mathrm{in}})$$

For calorically perfect gases,

$$e\dddot{X}_{\mathrm{D}} = \dot{m}T_{\infty}\left(c_{\mathrm{p}}\ln\left(\frac{T_{\mathrm{out}}}{T_{\mathrm{in}}}\right) - R\ln\left(\frac{P_{\mathrm{out}}}{P_{\mathrm{in}}}\right)\right)$$

Appendix: A First Look at RAM and SCRAM Jets

This deeper dive and workshop provide a simplified view of high speed aircraft performance, which requires that kinetic energy effects be considered. The gas turbine engine used for propulsion is based on the Brayton cycle which consists of a compressor, combustor and a turbine (to drive the compressor). The turbine exit is at an intermediate pressure and is followed by a nozzle to accelerate the working fluid to a high exit velocity jet, which gives a propulsive force. The thermodynamic goal is to convert enthalpy out of the gas generator to kinetic energy of the exit jet. Whether the kinetic energy of the inlet is important depends on the flight speed. The higher the speed of an aircraft, the more important kinetic energy effects at the inlet becomes.

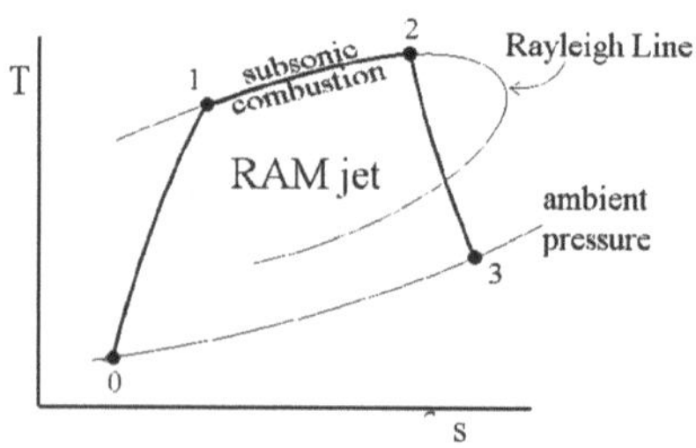

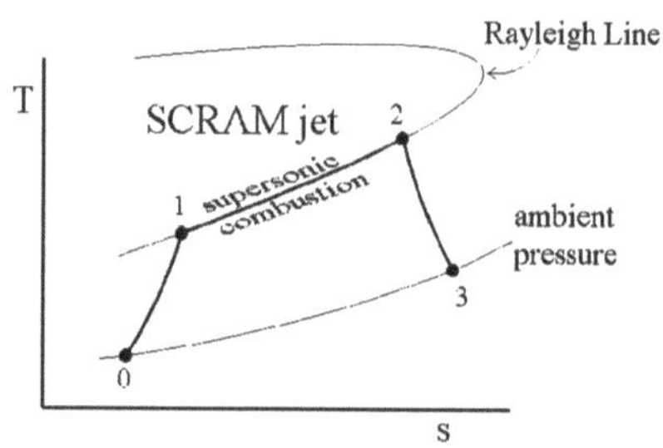

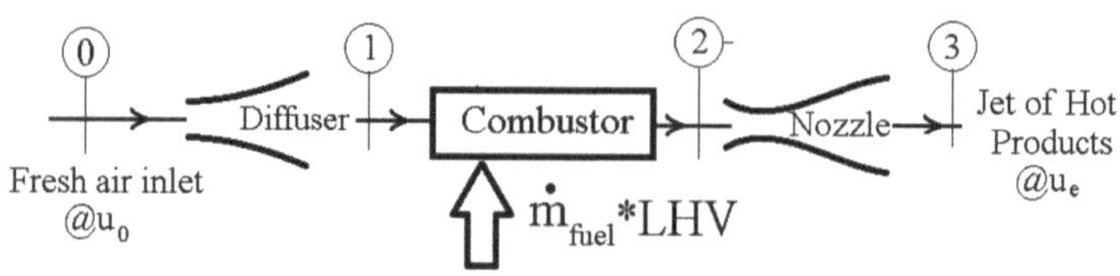

The speed of an aircraft is characterized by the flight Mach number (M_0), defined as the flight speed divided by the speed of sound of the local ambient. The speed of sound of a gas (symbol c or a in the literature) is a thermodynamic property, not derived here but defined by $c = \sqrt{kRT}$. At nominal room conditions (298 K), the speed of sound in air is 356 m/s = 774 mph.

If the flight speed is sufficiently high, adequate compression to operate a Brayton-type cycle can be achieved with a diffuser (with no moving parts) instead of a compressor. If the compressor is not needed, then neither is a turbine to drive it. Therefore, RAM jets and SCRAM jets, shown schematically in Fig. 9.19 and repeated here, operate on a Brayton-type string of components but with no moving parts: diffuser followed by a combustor followed by a nozzle. The overall goal is to produce a net thrust, a momentum concept (i.e., fluid mechanics), given by the relation:

$$\text{Thrust} = \dot{m}(u_{\text{exit}} - u_0) + \text{pressure effect}$$

where u_{exit} is the exit velocity (state 3) and u_0 is the inlet velocity (the flight speed, state 0). If the exit pressure matches ambient pressure at the nozzle design point, there are no pressure effects. The mass flow rate of fuel is also neglected in comparison to the air flow rate.

As shown on the *Ts* diagram sketches, the combustion process takes place along a thermodynamic process line called a Rayleigh line (an advanced thermo/fluid topic) which differs from the constant pressure combustion in the Brayton cycle because the kinetic energy is not negligible. However, the Rayleigh line reduces to an isobar when kinetic energy is neglected (at low Mach numbers, to the left of state 1 on the RAM jet *Ts* diagram). The diffuser of a RAM jet decelerates the flow to a subsonic speed, with low kinetic energy. An ideal RAM jet is defined here as one in which the diffuser isentropically decelerates the inlet stream to a negligible kinetic energy.

As the flight speed increases in RAM mode, the pressure and temperature after compression in the diffuser (process 0–1) increases as the inlet supersonic flow is decelerated to a subsonic velocity (at much lower kinetic energy than the inlet). Eventually, those pressures and temperatures are excessive and chemical energy addition will be impossible. Therefore, at sufficiently high flight speeds, the diffuser must be designed to only partially decelerate the flow, and the flow that enters the combustor is still supersonic, that is, it is a RAM jet, but the relative speed between the engine and the gases in the combustor is supersonic. It is a Supersonic Combustion RAM jet, or SCRAM jet.

Thermodynamic Performance Measures

The objective of a propulsion aircraft is more of a mechanics (forces) objective than a thermodynamic one (energy). Nevertheless, the propulsion engine is a thermodynamic device. The thrust force is obtained from a momentum balance in the direction of flight. The result (with +... to indicate that many assumptions are made) is:

$$\text{Thrust} = \dot{m}_{\text{air}}(V_{\text{exit}} - V_{\text{inlet}}) + \cdots$$

An energy balance of the overall aircraft with a chemical energy input is:

$$\frac{dE}{dt} = \dot{Q}_{\text{in}} + \dot{m}_{\text{fuel}}\,\text{LHV} - \dot{W}_{\text{out}} + \dot{m}_{\text{air}}\left(h_{\text{inlet}} + \frac{u_{\text{inlet}}^2}{2}\right) - \dot{m}_{\text{air}}\left(h_{\text{exit}} + \frac{u_{\text{exit}}^2}{2}\right)$$

This equation and all parameters in this introductory view neglect the mass flow rate of the fuel in comparison to that of the air. The main useful output as a propulsion device is the mechanical work associated with the thrust, and the associated power is given by the thrust force times the flight speed. That is:

$$\dot{W}_{\text{out}} = \text{Thrust} * \text{Flight speed} = \dot{m}_{\text{air}}(u_{\text{exit}} - u_0)u_0$$

At steady state, rearranging the energy equation for the input chemical energy rate:

$$\dot{m}_{\text{fuel}}\,LHV = \dot{m}_{\text{air}}(u_{\text{exit}} - u_0)u_0 + \dot{m}_{\text{air}}\left(\frac{u_{\text{exit}}^2}{2} - \frac{u_{\text{inlet}}^2}{2}\right) + \left[\dot{m}_{\text{air}}(h_{\text{exit}} - h_{\text{inlet}}) - \dot{Q}_{\text{IN}}\right]$$

The first term is called the thrust power, the desired quantity of a propulsion device. The 2nd term is the change in kinetic energy, which is a pure thermodynamic output. The last term represents thermal energy that is lost by the engine to the environment. That is, whatever input chemical energy that is not used to produce propulsive power or increase the kinetic energy of the jet is sent out the back with the hot exhaust (and/or as heat lost to the environment).

The overall efficiency of propulsion is defined by the ratio of the thrust power divided by the input chemical energy rate:

$$\text{Overall efficiency} = \eta_{\text{ov}} = \frac{\text{Thrust power}}{\text{Chemical input power}} = \frac{\dot{m}_{\text{air}}(u_{\text{exit}} - u_0)u_0}{\dot{m}_{\text{fuel}}\,\text{LHV}}$$

Traditionally, the conversion of chemical power into thrust power is considered to be built into two conceptual steps: the conversion of chemical input energy into kinetic energy of the working fluid, and conversion of that kinetic energy into thrust power. Therefore, two intermediate efficiencies are:

$$\text{Thermal efficiency} = \eta_{\text{th}} = \frac{\text{Kinetic energy change}}{\text{Chemicall input power}} = \frac{\dot{m}_{\text{air}}\left(u_{\text{exit}}^2/2 - u_0^2/2\right)}{\dot{m}_{\text{fuel}}\,\text{LHV}}$$

$$\text{Propulsion efficiency} = \eta_{\text{p}} = \frac{\text{Thrust power}}{\text{Kinetic energy change}} = \frac{\dot{m}_{\text{air}}(u_{\text{exit}} - u_0)u_0}{\dot{m}_{\text{air}}\left(u_{\text{exit}}^2/2 - u_0^2/2\right)} = \frac{2u_0}{u_{\text{exit}} + u_0}$$

Note that $\eta_{\text{ov}} = \eta_{\text{th}}\eta_{\text{p}}$.

A final measure of performance of aircraft is the specific impulse (I_{sp}), which is a measure of how effectively a propulsion engine (rocket or air breather) uses fuel. It is defined as the ratio of the total impulse (force multiplied by burn time) imparted to the vehicle normalized by the launch weight of the fuel (its initial mass, $M_{f,o}$, times gravity, g). That is:

$$I_{sp} = \frac{\text{Thrust} * \text{burntime}}{M_{f,0}g} = \frac{\text{Thrust} * (M_{f,0}/\dot{m}_{fuel})}{M_{f,0}g} = \frac{\text{Thrust}}{\dot{m}_{fuel}g}$$

An important difference between rockets (which carry all the propellant) and air-breathing engines is apparent. For rockets, there is no inlet velocity, and the thrust depends on the fuel flow rate. That is:

$$I_{sp,rockets} = \frac{\dot{m}_{fuel}u_{exit}}{\dot{m}_{fuel}g} = \frac{u_{exit}}{g}$$

For air-breathing engines:

$$I_{sp,air\text{-}breathing} = \frac{\dot{m}_{air}(u_{exit} - u_{inlet})}{\dot{m}_{fuel}g}$$

Introducing the air to fuel mass ratio,

$$I_{sp,air\text{-}breathers} = \text{AFR}\frac{(u_{exit} - u_{inlet})}{g}$$

The factor AFR in the specific impulse reveals the advantage of an air-breathing engines in terms of the effectiveness of using the fuel. Rockets do not have inlet air pushing them backward like air-breathers, and can generate much higher exit velocities (much higher chamber pressures), but the use of air as a propellant and the fuel to add chemical energy to it gives air-breathers higher specific impulse.

The following problem statement is designed to demonstrate the performance of an ideal RAM jet and serves as an upper performance limit to which actual RAM jets can be compared.

Problem Statement

An ideal RAM jet engine with a 1.3 m diameter cylindrical intake flies at Mach 3.2 at an altitude of 85,000 ft (25.9 km) where the ambient air conditions are 0.024 atm, 216 K (all these parameters are similar to the cruising SR71 Blackbird). The fuel is H_2 (while the SR71 used a JP-7, a kerosene-based fuel). The temperature entering the nozzle is set at $T_2 = 1800$ K, a material constraint. The assumptions that constitute an ideal RAM jet (similar to the ideal Brayton cycle) are as follows:

- Steady state, with negligible potential energy changes.
- The working fluid is a calorically perfect gas with properties of air (a value for specific heat at an elevated temperature can be chosen)
- Both diffuser and nozzle are isentropic
- Kinetic energy is negligible at the entrance and exit of combustor

INPUT PARAMETERS

M0	3.2	flight Mach number
Dintake	1.3	m, diameter of intake
Tinf	216	K, ambient temperature
Pinf	0.024	atm, ambient pressure
T2	1800	K, nozzle inlet temperature
cp	1.15	kJ/kg/K
R	0.287	kJ/kg/K
LHV	120900	kJ/kg, heating value of fuel (H2)

DERIVED PARAMETERS

k		ratio of specific heats
c		m/s, speed of sound of ambient
v1		m^3/kg, specific volume of ambient air

RAM jet

state	P (atm)	T (K)	h (kJ/kg)	V^2/2 (kJ/kg)	V (m/s)	c (m/s)	M
0							
1							
2							
3							

RESULTS

P1/P0		pressure ratio
m'air		kg/sec
m'fuel		kg/sec
thrust		kN
Thrust Power		MW
m'airΔKE		MW
m'fuel*LHV		MW
Thermal effy		
Propulsive effy		
Overall effy		
Specific Impulse		sec, specific impulse

Fig. 9.20 Blank spreadsheet template for the Ideal RAM jet problem

- The combustor is isobaric
- The diffuser inlet and nozzle exit are at ambient pressure.

Create a spreadsheet (like the blank template shown in Fig. 9.20) with an Input Parameter section, and sections for Derived Parameters, State and Results. Add cell references and formulas so that a parametric study can be conducted by changing input parameters. Once the spreadsheet is working, investigate the effect of the flight Mach number. Reason out why the RAM jet will not work if the flight Mach number is too low or too high. The assumption that the kinetic energy in the combustor is negligible would have to be relaxed to start studying higher speeds, and that engine would be a SCRAM jet. Development of the Rayleigh line would be needed for that, a deeper dive with scuba gear.

Results

The results of this default case are shown in Fig. 9.21 and discussed after which a parametric study in conducted over the full range of flight speeds for which the ideal

INPUT PARAMETERS

M0	3.2	flight Mach number
Dintake	1.3	m, diameter of intake
Tinf	216	K, ambient temperature
Pinf	0.024	atm, ambient pressure
T2	1800	K, nozzle inlet temperature
cp	1.15	kJ/kg/K
R	0.287	kJ/kg/K
LHV	120900	kJ/kg, heating value of fuel (H2)

DERIVED PARAMETERS

k	1.33	ratio of specific heats
c	287.42	m/s, speed of sound of ambient
v1	25.49852	m^3/kg, specific volume of ambient air

RAM jet

state	P (atm)	T (K)	h (kJ/kg)	V^2/2 (kJ/kg)	V (m/s)	c (m/s)	M
0	0.024	216	0.0	423.0	920	287	3.20
1	1.289	584	423.0	0.0	0	473	0.00
2	1.289	1800	1821.6	0.0	0	830	0.00
3	0.024	666	517.5	1304.1	1615	505	3.20

RESULTS

P1/P0	53.7	pressure ratio
m'air	47.9	kg/sec
m'fuel	0.554	kg/sec
thrust	33.3	kN
Thrust Power	30.6	MW
m'airΔKE	42.2	MW
m'fuel*LHV	67.0	MW
Thermal effy	0.630	
Propulsive effy	0.726	
Overall effy	0.457	
Specific Impulse	6126.3	sec, specific impulse

Fig. 9.21 Results of the ideal RAM jet with default parameter values

RAM jet can be operated. It is intriguing that the inlet and exit Mach numbers are identical, but the speed of sound (and therefore its velocity) is higher at the exit.

As with all idealized devices, the goal of the analysis is to provide an upper limit in performance, and to accurately reveal important trends. In practice, a diffuser in a RAM jet will involve decelerating the flow partly through shock waves, which generate entropy. Also, there can be significant kinetic energy entering the combustor and the chemical energy addition follows a Raleigh line in a region that is not isobaric, and the kinetic energy increases. Finally, it may be difficult in practice to design the nozzle to operate with its exit pressure matching ambient (what is called the design point of a nozzle).

The pressure ratio and diffuser exit temperature are shown in Fig. 9.22 which both increase strongly with flight Mach number. Pressure ratio is plotted on semilogy axes since the pressure ratio varies over 4 orders of magnitude from unity at low subsonic Mach numbers to approximately 5000 (for a pressure of 117 atm for ambient pressure of 0.023 atm). The diffuser exit temperature is shown on a linear scale.

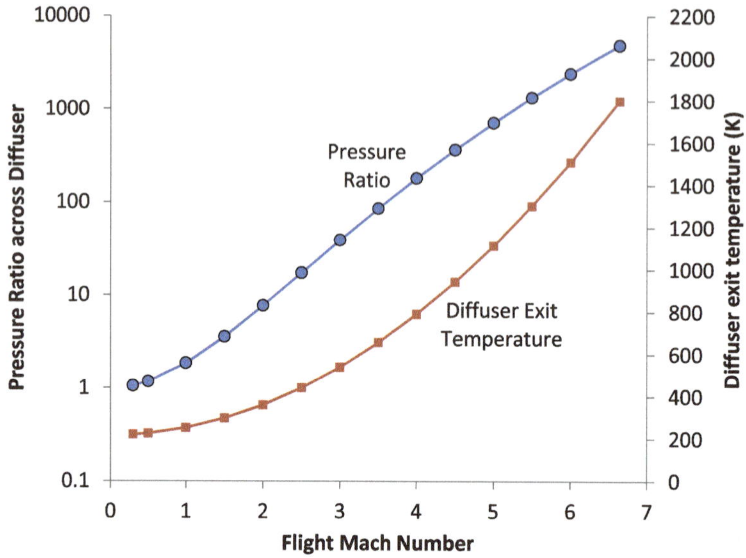

Fig. 9.22 Pressure ratio and diffuser exit temperature of ideal RAM jet over the full range of flight Mach numbers for which a net thrust is achieved

The thermodynamic efficiency of a Brayton-type engine increases with pressure ratio, hence flight speed, but there is a limit because chemical energy must be added in the combustor. The upper flight Mach number of 6.64 occurs when the diffuser exit temperature (combustor inlet temperature) equals the material limit (set to 1800 K in this study).

Ts diagrams are shown for 3 flight Mach number cases in Fig. 9.23. Recall the Glassman adage "gas turbines work because isobars diverge on a *Ts* diagram." That requires a combination of vertical movement (to higher pressure) and horizontal movement (to higher entropy level). At lower Mach numbers, there is very little pressure rise, so the isobars are close together and there is minimal vertical movement. At higher Mach numbers, the isobars are far apart (large vertical movement), but the chemical input is limited and there is minimal horizontal movement.

The inlet and exit velocities and their difference (necessary for thrust) are shown in Fig. 9.24. The velocity difference peaks at a flight Mach number of 2.5. The thrust and specific impulse are shown in Fig. 9.25. Fuel and air mass flow rates are shown in Fig. 9.26. The air mass flow rate increases linearly with flight speed for this fixed inlet area. Therefore, the thrust (air mass flow rate times velocity difference) peaks at a higher Mach number (approximately 4) than the velocity difference. The specific impulse peaks with Mach number, but remains high as the fuel flow rate (in the denominator) decreases. The peak value of approximately 6000 s is approximately three times that of the SR71 Blackbird (reported as 1900 s). This discrepancy seems reasonable and in line with the comparison of other ideal cycles with actual engines.

Energy rates and efficiencies are shown in Figs. 9.27 and 9.28, respectively. These results follow from the effect of flight Mach number on the pressure ratio,

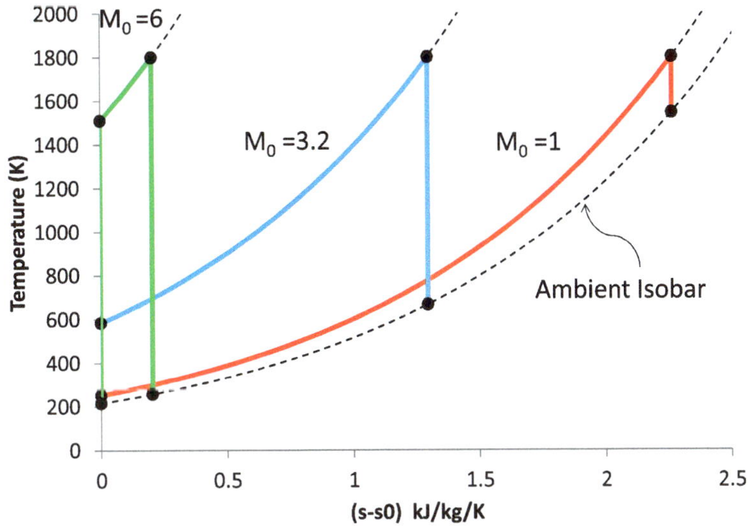

Fig. 9.23 *Ts* diagram for 3 flight Mach number cases: $M\ 0 = 1$, 3.2, and 6, corresponding to pressure ratios of 1.85, 53.7, and 2414, respectively

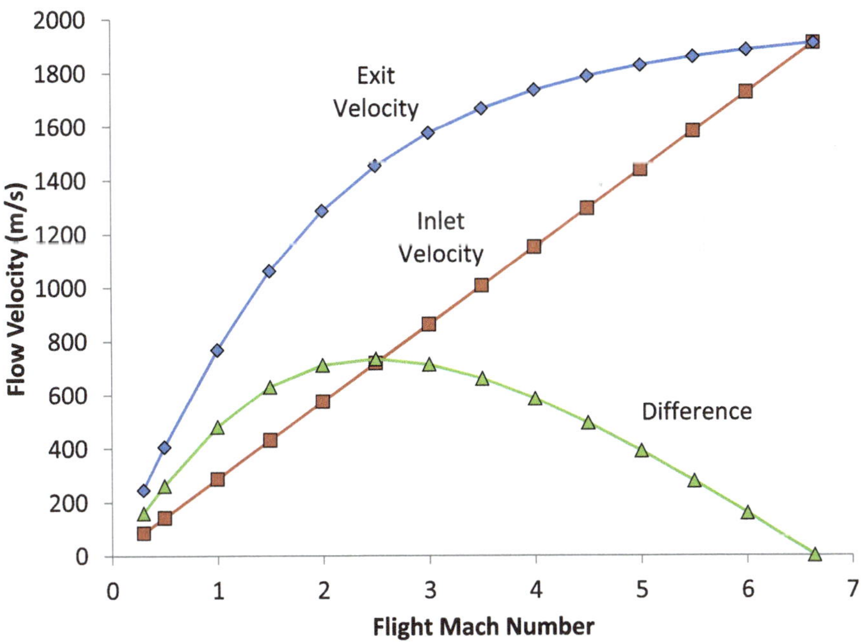

Fig. 9.24 Inlet and exit velocities as a function of flight Mach number

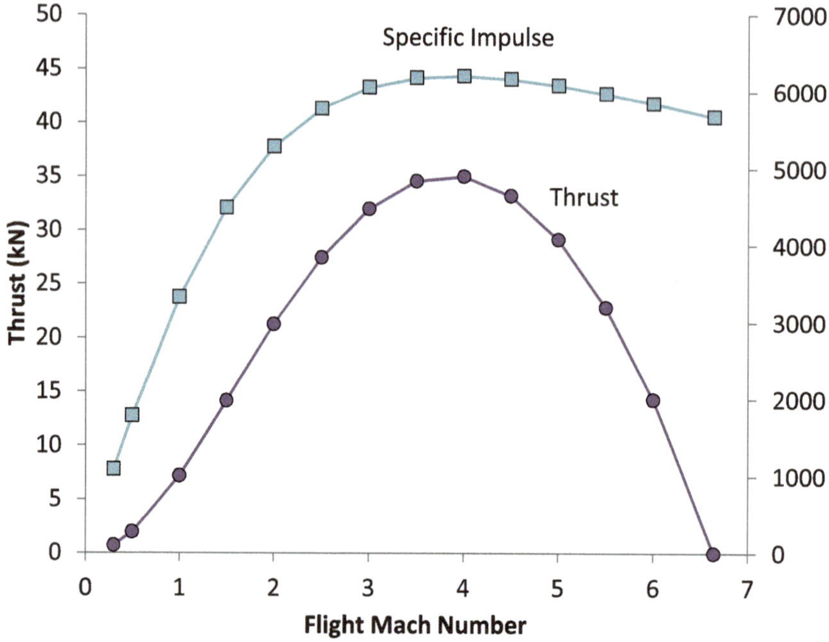

Fig. 9.25 Thrust and specific impulse as a function of flight Mach number

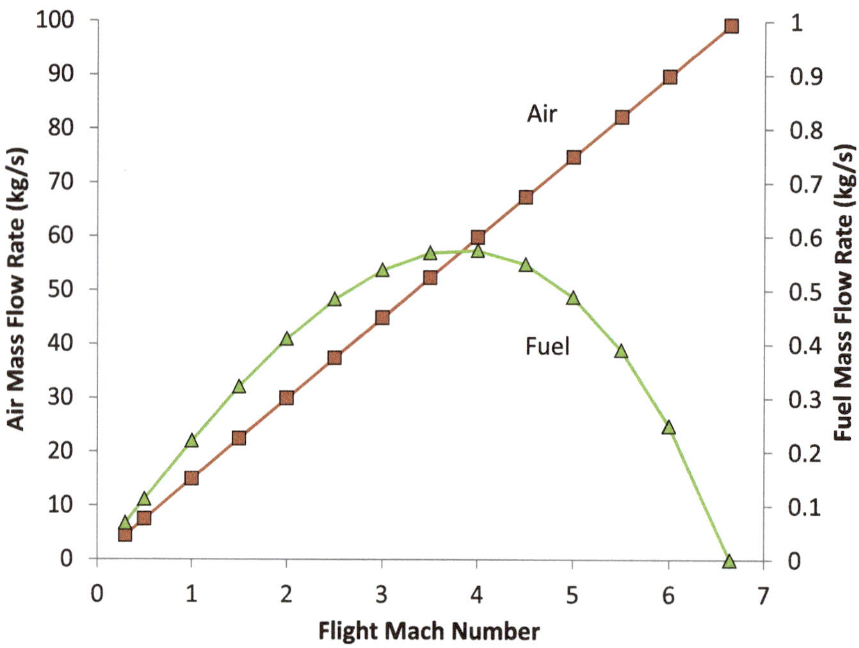

Fig. 9.26 Air and fuel flow rates as a function of flight Mach number

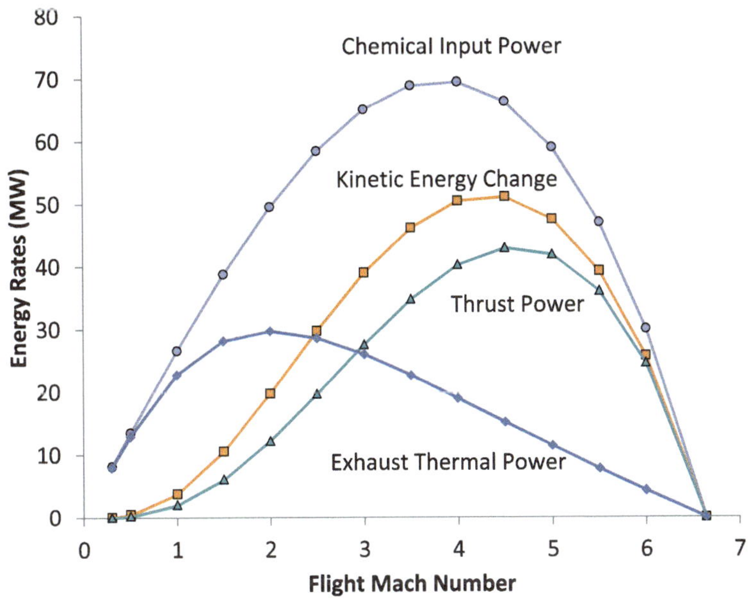

Fig. 9.27 Energy rates as function of flight Mach number

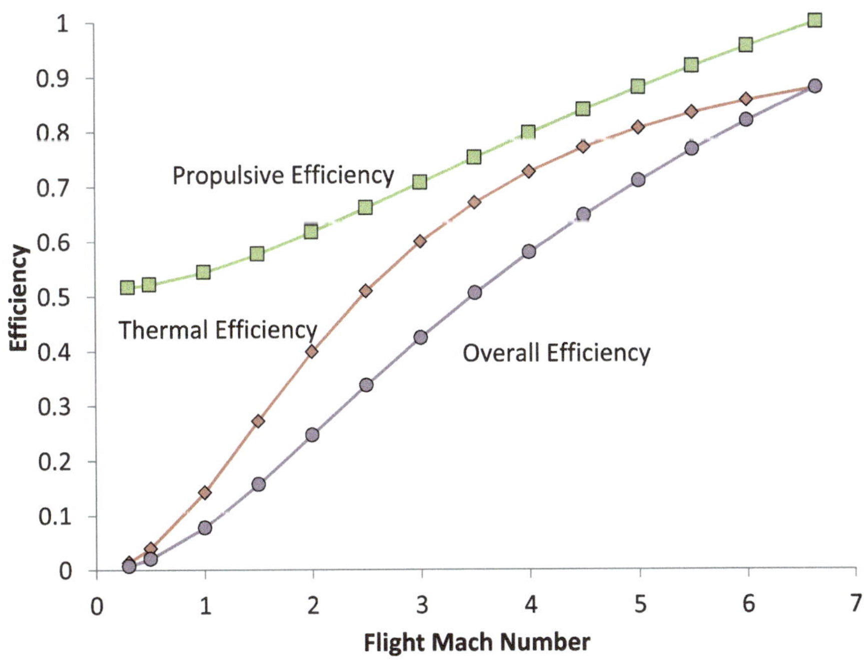

Fig. 9.28 Efficiencies as a function of flight Mach number

coupled with the temperature limit of the combustor exit. The input chemical energy power peaks at an astonishing 70 MW (the equivalent electrical consumption rate of approximately 130,000 people in the USA). At subsonic flight speeds, almost all the input chemical power is vented out the back as exhaust thermal power $\dot{m}_{\text{air}}(h_3 - h_0)$ and the thermal efficiency is very low. As flight speed (and therefore pressure ratio) increases, so does the chemical input power and the conversion efficiency to kinetic energy. Above a Mach number of approximately 4, the temperature limit becomes dominant and all energy rates drop (while the efficiencies continue to climb).

Workshop Variations

1. Include the effect of an isentropic efficiency for the diffuser
2. Relax the temperature limit, but determine chemical input limit (stoichiometric)

Workshop 9.1: Electric Hand Dryer

"(now wipe hands on pants)"

Room air at 1 atm, 25 °C enters an electric hand dryer at low velocity (state 0), and is compressed adiabatically and reversibly (i.e., no entropy generation) by a fan into a duct with diameter $D_1 = 0.05$ m. The air (state 1) enters a constant area ($D_2 = D_1$) heater section with an electric heating element, then enters an adiabatic and reversible nozzle (state 2), and exits at 1 atm, 57 °C through a $D_3 = 0.015$ m exit with a velocity of 95.0 m/s. Complete the state and results tables. Assume steady state and no pressure drop in the heater. Neglect stray heat and potential energy. For energy balances, neglect kinetic energy everywhere except at the nozzle exit. Use mass conservation for interior velocities, but neglect their kinetic energy in energy balances. Treat air as a calorically perfect gas with $c_p = 1005$ J/kg/K, $R = 287$ J/kg/K.

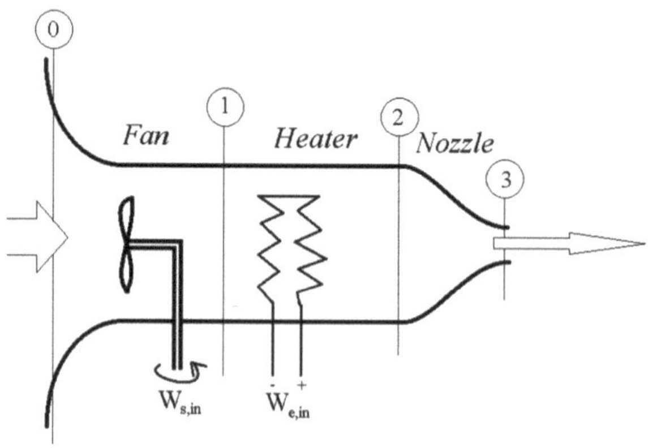

INPUT PARAMETERS

Cp=	**1005**	J/kg/K
R=	**287**	J/kg/K
D1=D2=	**0.05**	m
D3=	**0.015**	m

STATE TABLE (note: state 0 is set as a reference state)

state	T	P(gage)	h (J/kg)	V (m/s)	s(J/kg/K)	v(m^3/kg)
0	**25**	**0**	**0**	**0**	**0**	
1						
2						
3	**57**	**0**		**95**		

RESULTS

			hint
air mass flow rate		kg/s	analyze state 3
Total input power (Ws + We)		Watts	over-all energy balance from 0 to 3
eXergy destruction		Watts	over-all entropy balance
Power to Heater		Watts	
Power to Fan		Watts	

Workshop 9.2: Brayton Cycle with Intercooling, Reheat, and Regeneration

Three modifications to the basic Brayton cycle can be combined to increase the thermal efficiency: intercooling, reheat, and regeneration shown schematically in Fig. 9.29 with one stage of intercooling and one stage of reheat and regeneration. A process heat exchanger is included after the regenerator to provide for cogeneration and to return the working fluid to its initial state (for the ideal cycle). The effects on gas turbine performance are implemented and observed in this workshop. All components are treated as ideal in order to observe the thermodynamic objectives, shown in the color-coded *Ts* diagrams of Fig. 9.30, which shows 2 stages of intercooling and reheat, with and without regeneration, and a case of perfect intercooling and reheat (with an infinite number of stages). The 1st Law efficiency of the ideal cycle with all modifications equals the Carnot efficiency (treating the chemical energy addition as equivalent to a thermal reservoir at the turbine inlet temperature). Real effects are discussed briefly in this background and can be included quantitatively in a follow-up investigation (or take-home project).

The main thermodynamic goal of intercooling is to reduce the input power required to compress the gases to the combustor pressure. Intercooling shifts the

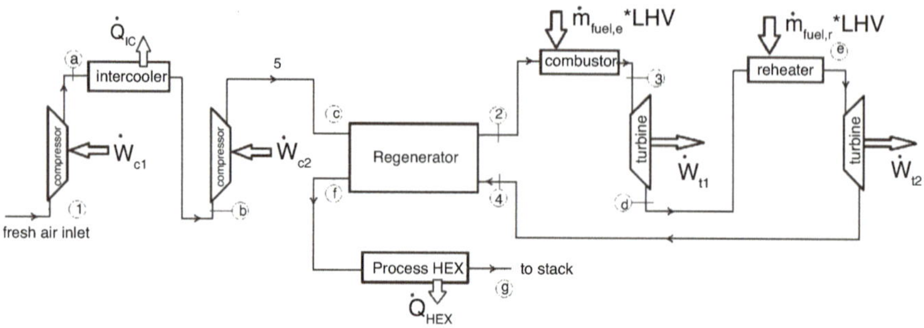

Fig. 9.29 Schematic of modified gas turbine cycle with one stage of intercooling and one stage of reheat and regeneration

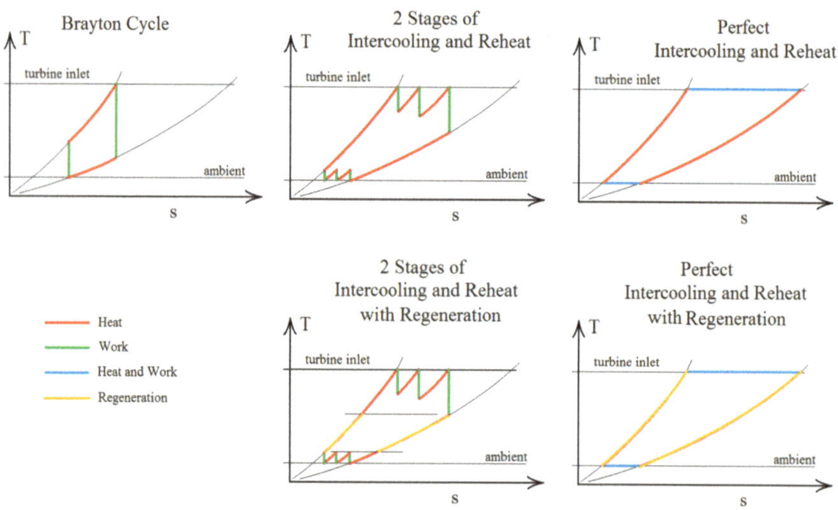

Fig. 9.30 *Ts* diagrams that show the effect of intercooling, reheat and regeneration. Each process is color-coded to indicate whether heat, work, both, or neither (regeneration) are exchanged with the surroundings

compression to lower entropy levels, where the vertical rise across isobars is lower, that is "gas turbines work because isobars diverge on a *Ts* diagram." Recall also "Vee dePee" work associated with compressors and turbines. As gases are compressed in an open system process, both the mechanical (Pv) and thermal energy (u) increase. The work required to compress gases to a target pressure is reduced if they are simultaneously cooled (which lowers the specific volume). In practice however, the residence time of gases in a compressor it too short to allow for heat transfer and therefore compression can be accomplished in alternating stages of adiabatic compression to an intermediate pressure (in a compressor) followed by a cooling toward ambient (in a heat exchanger). The gas through an ideal intercooler undergoes an isobaric process and exits in thermal equilibrium with ambient. A real intercooler will result in a pressure drop and exit temperature above ambient.

A single stage of intercooling is shown in Fig. 9.29 and two ideal stages are shown in the middle *Ts* diagrams of Fig. 9.30. The right side *Ts* diagrams show the result of perfect simultaneous intercooling and compression, accomplished by adding an infinite number of stages of compressor and intercooler. The heat transfer to ambient in the intercooler represents heat rejection to the dead state in the heat engine analysis. In the idealized limit of perfect intercooling, the working fluid is always equal to the dead state temperature, and there is no eXergy destruction associated with this heat transfer. The compression process is isothermal (not isentropic), with simultaneous heat and work (blue line in Fig. 9.30).

The thermodynamic downside of intercooling is that, used alone, it requires a higher chemical energy addition in the combustor to reach the turbine inlet temperature. This effect will be shown by example to result in a power plant with a higher power output (per unit mass of working fluid) but a lower 1st Law efficiency. The regenerator will address that shortcoming. With the addition of perfect regeneration, all the heat rejection of the cycle takes place during intercooling.

The thermodynamic effect of reheat is to increase the power output, at the expense of greater chemical energy addition. Similar arguments as for intercooling hold, with the goal to ultimately add "heat" (in the form of chemical energy) simultaneously with mechanical power extraction. In practice, a stage of reheat involves expansion to an intermediate pressure in an adiabatic turbine, followed by an isobaric chemical energy addition (reheat). Perfect reheat results in simultaneous heat and work, and an isothermal expansion to ambient pressure.

In practice, reheat can be used in mobile power plants (high performance jet engines) because the reheat process does not involve an actual heat exchanger, but rather internal combustion. Reheat alone increases power output (useful for taking-off a short runway like an aircraft carrier) or accelerating in a dog-fight, but reduces the 1st Law efficiency. Reheat is accomplished by internal combustion, not by actual heat exchange (like the intercooler) and is possible in practice only because the primary combustor is overall fuel lean. The O_2 for combustion in the reheater is limited by stoichiometry, but not considered here.

The combined use of intercooling and reheat, then, will result in a power plant with a higher power output (per unit mass of working fluid) but a lower 1st Law efficiency. Regeneration addresses that shortcoming. One effect of intercooling is to lower the temperature after compression and an effect of reheat is to raise the temperature after expansion. If the temperature of the turbine exit is higher than that of the compressor exit, then a heat exchanger (called the regenerator) can be added to simultaneously raise the temperature of the high pressure gases prior to the combustor (reducing the required chemical energy addition) and lower the temperature of the low pressure gases prior to exhaust (reducing the required heat rejection to the dead state, or reducing the available process heat in a cogeneration application).

Thermal equilibrium is achieved in an ideal regenerator, with the exit temperature of the high pressure side gas (state 2, combustor inlet) equal to the inlet temperature on the low side (state 4, turbine exit) and the exit temperature of the low pressure gas (state f, to process or stack) equal to the inlet temperature on the high side (state c, compressor exit). With perfect intercooling, reheat and regeneration, all of the heat exchange takes place during compression or expansion. Thermal equilibrium cannot be achieved in practice, and this real effect could be incorporated into follow-up workshops.

Workshop

The spreadsheet template provided is designed to allow analysis of any combination of modifications to the Brayton cycle with a single stage of intercooling and reheat, all on a per unit mass basis of the working fluid. Enter appropriate formulas in each blank cell, with states and processes that correspond to the schematic of Fig. 9.29. Design the spreadsheet to update completely to a change in any input parameter. The default values given are for an overall pressure ratio of 30 to 1, with both intercooler and reheat intermediate pressure at 5.5–1, making the compression and expansion ratios of all stages equal.

Combinations can be investigated by appropriate choice of intermediate pressure. If, for example, the Brayton cycle is to be investigated, the intermediate intercooler pressure can be set to ambient (i.e., $Pa = Pb = 1$ atm) and the intermediate reheater pressure can be set to the high pressure ($Pc = Pd = 30$ atm). Once a base spreadsheet is set up, it can be copied for each combination. Set up a compiled results table to allow side by side comparison of performance parameters. Plot each case on the *Ts* diagram (which is to scale for the default fluid properties c_p and R).

Each process as treated as ideal, with the following approximations:

- The working fluid is a calorically perfect gas with properties of air (and c_p evaluated at an intermediate temperature, like 1000 K).
- Both compressors and both turbines are adiabatic and reversible (hence isentropic).
- Each heat exchanger stream is isobaric (intercooler, regenerator, and reheater) and the exit temperature achieves thermal equilibrium with its heat exchange partner.
- The chemical energy addition in the combustor and reheater are treated as equivalent to the heat transfer from a reservoir at the turbine inlet temperature.
- The intercooler and process heat exchanger exchange heat with the dead state.

INPUT PARAMETERS

cp	**1.14**	kJ/kg/K, enthalpy specific heat
R	**0.287**	kJ/kg/K, gas constant
Tinf	**298**	K, dead state temperature
Pinf	**1**	atm, dead state pressure
T3 = Tc	**1800**	K, turbine inlet temperatures
Pa = Pb	**5.5**	atm, intercooler pressure
P2 = P3	**30**	atm, combustor pressure
Pd = Pe	**5.5**	atm, reheater pressure

STATE TABLE

STATE	P (atm)	T(K)	h(kJ/kg/K)	s(kJ/kg/K)
1				
a				
b				
c				
2				
3				
d				
e				
4				
f				
g				

PROCESS TABLE (all in kJ/kg)

device	process	"q"in	wout	eXD
compressor 1	1-a			
intercooler	a-b			
compressor 2	b-c			
combustor	2-3			
turbine 1	3-d			
reheater	'd-e			
turbine 2	e-4			
process HEX	f-g			
REGENERATOR	c-2/4-f			
	net			

RESULTS

1st Law efficiency	
Carnot efficiency	
2nd Law efficiency	

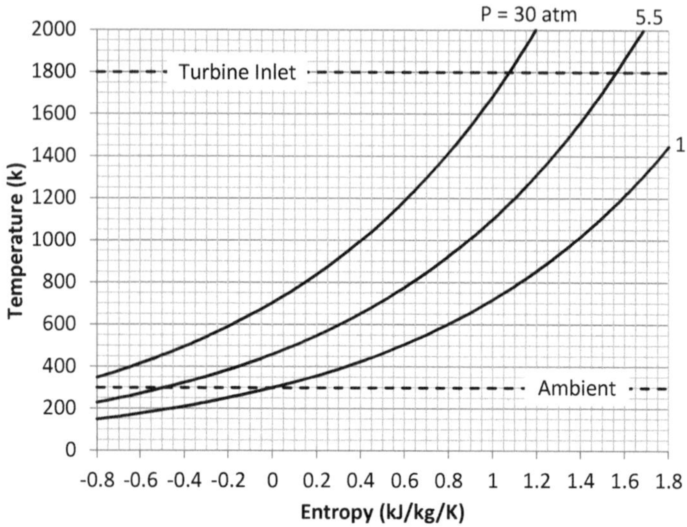

COMPILED RESULTS

Cycle	qin (kJ/kg)	wnet (kJ/kg)	qprocess (kJ/kg)	eXd (kJ/kg)	η_I	η_{II}
Brayton						
Intercooling						
Reheat						
Intercooling and Reheat						
Intercooling, reheat and regeneration						

Workshop 9.3: Jet Engine with Afterburner

The goal of this workshop is to observe the effect of an afterburner (or reheater) on the performance (thrust and cycle efficiency) by comparing the performance of a nozzle with a combination afterburner/nozzle using the same gas generator, shown schematically. Air enters at ambient conditions at the flight speed (u_0) and is first decelerated to a low relative velocity in a diffuser. The air enters a compressor (with a specified pressure ratio). Fuel is injected in a combustor and burned. The products enter a turbine and are expanded to an intermediate pressure (state 4) so that the output power of the turbine equals the input power required of the compressor. These basic processes constitute a "gas generator." The gases then either enter a nozzle directly, or an afterburner followed by a nozzle. Complete the state and results tables, with the following assumptions:

- Steady-state operation, with negligible potential energy changes
- The working fluid is a calorically perfect gas (i.e., constant specific heat)
- Neglect kinetic energy everywhere except at the entrance (state 0) and exit (state 5 or 6).
- Neglect stray heat in all devices (chemical energy is added in an internal combustion process in the combustor and afterburner).
- Adiabatic and reversible (i.e., isentropic) diffuser, compressor, turbine, and nozzle.
- Neglect pressure drops in the combustor and afterburner.
- Mass flow rate of fuel is negligible in comparison to air flow rate.
- Nozzle exit pressure is ambient.

- 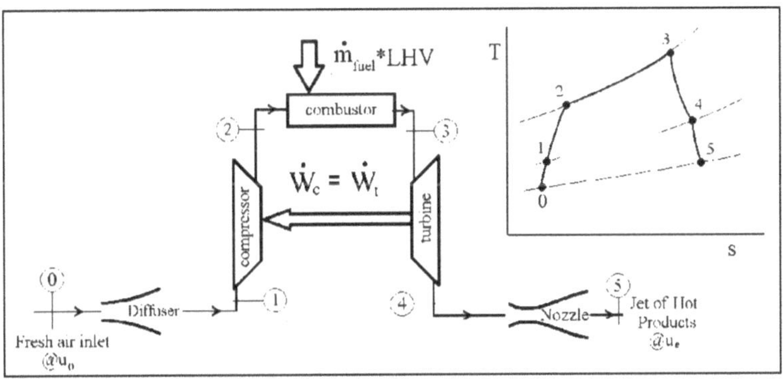

- 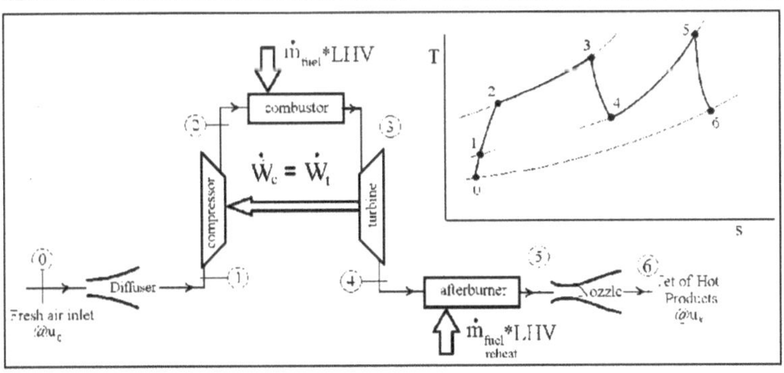

INPUT PARAMETERS

m',air	100	kg/sec, mass flow rate of air
u0	300	m/s, cruise speed
P2/P1	25	pressure ratio across compressor
cp	1.14	kJ/kg/K, specific of air at 1000K
R	0.287	kJ/kg/K, gas constant for air

STATE PROPERTIES OF GAS GENERATOR

state	T (K)	P (atm)	h (kJ/kg)	$V^2/2$ (kJ/kg)	s (kJ/kg/K)
0	**230**	**0.3**	**0**		**0**
1					
2					
3	**1800**				
4					

NOZZLE ONLY

state	T (K)	P (atm)	h (kJ/kg)	$V^2/2$ (kJ/kg)	s (kJ/kg/K)
5		**0.3**			

AFTERBURNER + NOZZLE

state	T (K)	P (atm)	h (kJ/kg)	$V^2/2$ (kJ/kg)	s (kJ/kg/K)
5	**2000**				
6		**0.3**			

RESULTS

	BASIC CYCLE	w/Afterburner
Exit Temperature (K)		
Exit Velocity (m/s)		
Thrust (kN)		
Compressor Power (kW)		
m'*ΔKE (kW)		
m',fuel*LHV (kW)		
Thermal Efficiency		

Workshop 9.4: Gas Turbine-Based Cogeneration with Supplemental Firing

The stoichiometry of the internal combustion process in the combustor of practical gas turbines is limited by the allowable turbine inlet temperature, not by the O_2 available in the air. It is possible, however, to burn more fuel after the turbine since the temperature of the gases has been lowered by the work extracted. In reheating, the goal is to produce more power, and therefore a primary turbine stage expands to an intermediate pressure. Chemical energy added at this intermediate pressure results in a greater power output than full expansion without reheat.

On the other hand, in a cogeneration application, where the extraction of heat from the engine is one of the desired outputs, the gases can be expanded to ambient in the turbine, and then additional fuel added at ambient pressure to raise the temperature before entering a process heat exchanger, a process called supplemental firing (shown schematically). Typically, all the O2 can be burned without the restriction of the turbine inlet because the gases enter a heat exchanger with no moving parts, not a turbine with highly stressed rotating blades. This workshop explores this modification.

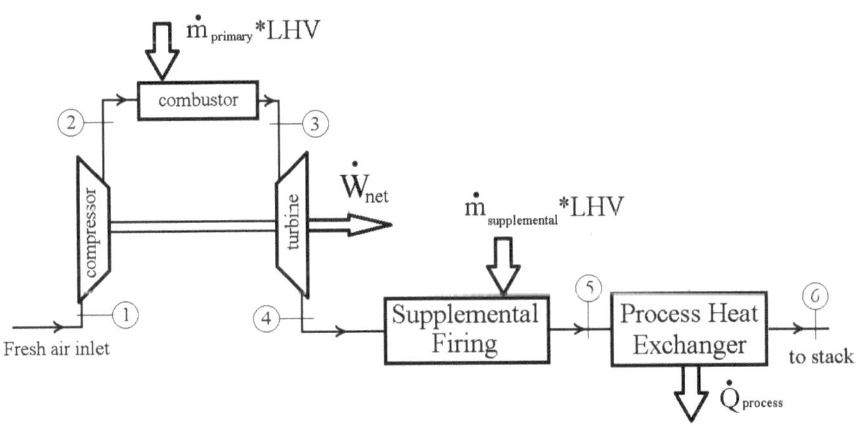

The rate of fuel addition is limited by the available O2 in the turbine exhaust, and that is related to the chemical energy change and stoichiometry, discussed in Chap. 7. The lower heating value (LHV) of a fuel is the chemical energy release associated with complete combustion per unit mass of fuel added. The heating value of the fuel/air mixture ($\Delta h_{\text{mix, comb}}$) is the chemical energy release per unit mass of a stoichiometric mixture. It incorporates the chemical stoichiometry and energy, and represents the maximum possible chemical energy release that can be achieved. Therefore, the chemical energy addition rate available for supplemental firing is given by:

$$\dot{m}_{\text{supplemental}}\,\text{LHV} = \dot{m}_{\text{air}}\Delta h_{\text{mix,comb}} - \dot{m}_{\text{primary}}\,LHV$$

The workshop template sets default values typical of an industrial gas turbine (15 to 1 pressure ratio) with compressor and turbine isentropic efficiencies as inputs. The stack temperature is set, and the process heat is assumed to take place with the dead state. In practice, a secondary loop might be used to extract heat from the engine and deliver it to a process. The combustion, supplemental firing, and process heating processes are all considered to be isobaric, but reasonable pressure drops in each could be easily added.

The 1st Law efficiency is defined as the ratio of the desired output (heat rate plus power) to the required input (total chemical energy addition rate). The 2nd Law efficiency is defined as the ratio of the desired output to the sum of the desired output and total eXergy destruction rate. There is considerable eXergy destruction associated with the process heating process, and better thermodynamic use could be made by the addition of a heat engine to extract more power and deliver the heat rejection of that to the process (an example of a combined cycle, the subject of Chap. 11).

Note that in this workshop, the state table cannot be completely filled without referencing results, specifically, in the determination of the available chemical energy for the supplemental firing.

INPUT PARAMETERS

m'air	**100**	kg/s, flow rate of working fluid
cp	**1.14**	kJ/kg/K, enthalpy specific heat
R	**0.287**	kJ/kg/K, gas constant
Tinf	**298**	K, dead state temperature
Pinf	**1**	atm, dead state pressure
T3	**1800**	K, turbine inlet temperatures
P2 = P3	**15**	atm, combustor pressure
T6	**100**	oC, stack temperature
ηc	**0.85**	compressor isentropic efficiency
ηt	**0.9**	turbine isentropic efficiency
LHV	**50100**	kJ/kgfuel, lower heat value of fuel
Δhmix,comb	**2760**	kJ/kgmix, stoich misture heating value

STATE TABLE

state	P (atm)	T (K)	h (kJ/kg)	s (kJ/kg/K)
1				
2				
3				
4				
5				
6				

PROCESS TABLE (kJ/kg of working fluid)

device	process	"qin"	wout	eXd
compressor	1-2			
combustor	2-3			
turbine	3-4			
supplemental firing	4-5			
Process Heater	5-6			
	NET			

RESULTS

m'primary		kg/s, primary fuel flow rate
m'supplemental		kg/s, secondary fuel flow rate
W'net		MW, output power
m'fuel*LHV		MW, total chemical energy input rate
Q'process		MW, total process heat addition
eX'D		MW, total eXergy destruction rate
Cogen 1st Law		(Q'process+W'net)/m'fuel*LHV
2nd Law eff'y		(Q'p + W'n)/(Q'p + W'n +eX'D)

Chapter 10
Steam Turbine Cycles

Steam turbines are used to meet a large percentage of the global electrical demand. In practice, a steam turbine is fed superheated H_2O (steam) as a working fluid at a high pressure. Enthalpy is converted into output shaft work, and the working fluid exits at a lower pressure as either a superheated vapor or a liquid/vapor mixture with a sufficiently high quality to avoid mechanical erosion of the final stages of the turbine. To operate continuously, the working fluid must be returned to the turbine inlet state, with the primary goal of extracting net power. A Rankine cycle with modifications satisfies that requirement and forms the basis for virtually all steam-based stationary power plants. The turbine exit is condensed at the low pressure to liquid, then pumped to the high pressure needed at the turbine inlet, and then preheated, boiled, and superheated to the turbine inlet conditions. The source of the heat addition can be from the combustion of fossil fuels, nuclear energy, concentrated solar energy, incineration of municipal solid waste or geothermal energy, to name a few. In this chapter, several Rankine-based cycles are investigated, all scaled in size to generate a net power output of 1.0 GW, typical of large stationary power plants. The scaling factor is the mass flow rate of working fluid common to each device. First, an ideal Rankine cycle with operating parameters set by material-driven practical constraints is defined and analyzed. A second study introduces another device (reheater) that allows the cycle to operate at an overall higher pressure ratio without violating the practical restrictions. In the third study, an open feedwater heater is introduced, a device that preheats liquid from the pump using a portion of the steam extracted from an intermediate pressure from the turbine in a process called regeneration. A final study introduces realistic deviations from ideal behavior to more closely model actual power plants. The effect of each of these changes on the overall performance (mass flow rate, 1st and 2nd Law efficiencies), and the eXergy destruction in each device is compared.

The basic Rankine cycle with its four components is shown schematically in Fig. 10.1 operating between a low and high temperature reservoir. This control surface defines a closed power cycle, with heat addition from a high temperature source (Q_{in}) in a boiler, net work (the difference between the work extracted by a

G. Sidebotham, *An Inductive Approach to Engineering Thermodynamics*,
Mechanical Engineering Series, https://doi.org/10.1007/978-3-030-20430-3_10

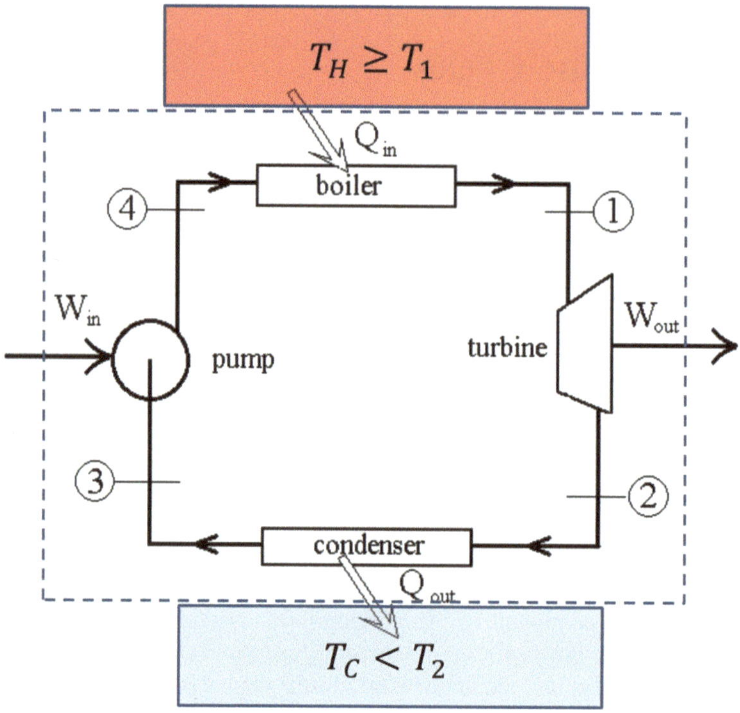

Fig. 10.1 Schematic and thermodynamic process diagram of a basic Rankine cycle with a control surface chosen to define a closed system

turbine, W_{out} and that put into a pump, W_{in}), and heat rejection to a low temperature source (Q_{out}) in the condenser, typically at the dead state temperature. Any stray heat loss within the cycle (from the pump, turbine, and piping, for example) could be considered heat transfer to the low temperature source. The working fluid undergoes a closed thermodynamic cycle from this view and will be analyzed in this introductory section that way (a review of Chap. 3). Then control volumes will be drawn about each component to be analyzed as open systems.

In practice, the heat addition in the boiler and heat rejection in the condenser takes place in heat exchangers, and not directly with a thermal reservoir. On the boiler side, the heat addition can come from a variety of sources, including, for example, from the hot gases created by a combustion process (fossil fuel, municipal solid waste, and biomass), a fluid heated by nuclear reactions, geothermal energy, or concentrated solar energy. In Chap. 11, the heating source from the hot exhaust of an internal combustion engine (reciprocating or gas turbine) will be used to create a combined cycle. On the condenser side, heat is transferred to a coolant, usually liquid water which is either drawn from a source like a river, or sent to a cooling tower where heat exchange between the coolant and ambient air occurs. These heat exchanger devices will be investigated in Deeper Dives and the next chapters.

Overall, then, the energy and entropy balances for the cycle are given by

$$\Delta E_{\text{cycle}} = 0 = Q_{\text{in}} - Q_{\text{out}} + W_{\text{in}} - W_{\text{out}}$$

$$\Delta S_{\text{cycle}} = 0 = \frac{Q_{\text{in}}}{T_{\text{H}}} - \frac{Q_{\text{out}}}{T_{\text{C}}} + \Sigma_{\text{cycle}}$$

The changes in state (energy and entropy) are zero for the cycle. Combining the energy and entropy balances by eliminating the heat rejected by the condenser and rearranging for the 1st Law thermal efficiency:

$$\eta_{\text{I}} = \frac{W_{\text{out}} - W_{\text{in}}}{Q_{\text{in}}} = \left(1 - \frac{T_{\text{C}}}{T_{\text{H}}}\right) - \left(\frac{T_{\text{C}}\Sigma_{\text{cycle}}}{Q_{\text{H}}}\right)$$

The first term in parentheses is the Carnot efficiency of a cycle operating between the two thermal reservoirs. With the low temperature as the dead state, the second term is the total eXergy destruction within the cycle (the sum of eXergy destruction in all component), normalized by the heat addition.

A 2nd Law efficiency can be defined as the 1st Law efficiency normalized by the maximum possible efficiency (the Carnot efficiency). That is:

$$\eta_{\text{II}} = 1 - \frac{T_{\text{H}}T_{\text{C}}\Sigma_{\text{cycle}}}{(T_{\text{H}} - T_{\text{C}})Q_{\text{H}}}$$

This global analysis, valid for any heat engine cycle, shows that any eXergy destruction within the cycle will reduce the 1st Law efficiency. Subsequent analysis of individual devices (open systems) will pinpoint sources of eXergy destruction, first in an idealized cycle, and then with component efficiencies included.

Both Rankine-based and Brayton-based power plants take advantage of the adage "isobars diverge on an h-s diagram." That is, they both operate across a pressure ratio which requires an input of mechanical power at a low entropy level, an increase in the entropy level (by entropy transfer in, not entropy generation) and an output power that exceeds the required input. The back work ratio (ratio of required input power to the output power) is therefore an important practical measure. Figure 10.2 shows to scale *T-s* and *h-s* property diagrams for H_2O with 4 isobars. *T-s* diagrams for gas turbines were discussed in Chap. 8, where there is no real distinction between *T-s* and *h-s* diagrams. For a working fluid that undergoes phase change, however, the *T-s* and *h-s* diagrams look different within the liquid/vapor zone, within which isobars are also isotherms, whereas the enthalpy increases continuously with entropy. These blank plots can be used to help locate states when specific problems are analyzed.

Workshop 10.1 shows a warm-up problem that focuses on a steam turbine that has a single inlet and multiple exits. The bulk of the working fluid undergoes an expansion, but a portion of the flow is extracted at intermediate points (used elsewhere in the power plant as a form of regeneration). This workshop is designed to include kinetic and potential energy effects. The thermodynamic objective of a

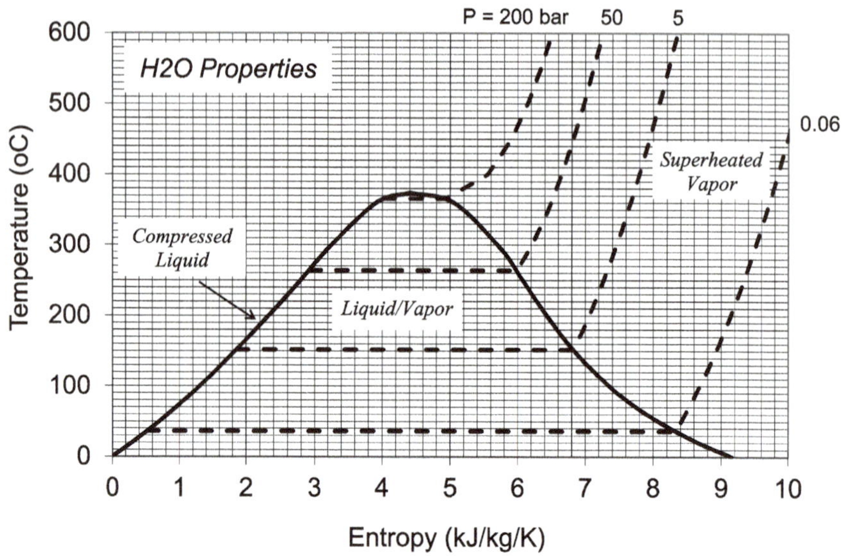

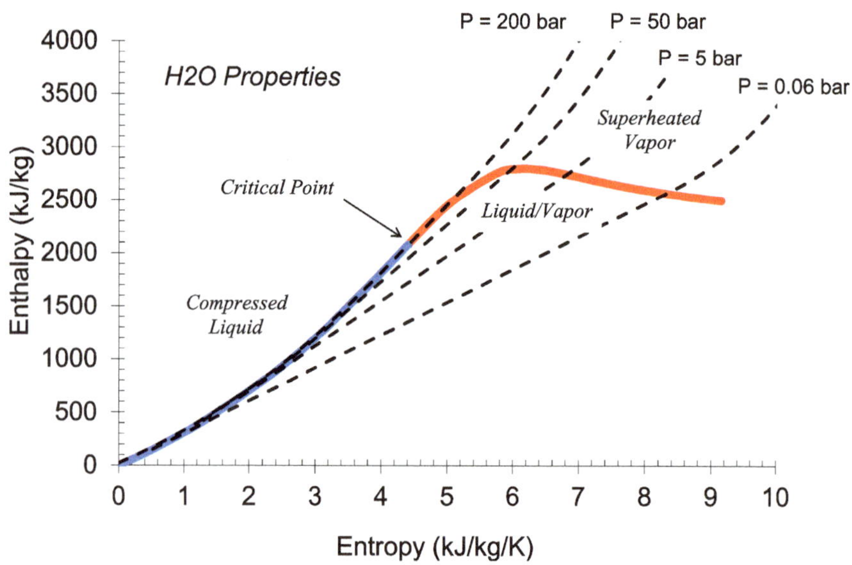

Fig. 10.2 *T-s* and *h-s* diagrams for H_2O for four isobars. The saturated liquid line is shown in blue and saturated vapor in red on the *h-s* diagram. The reference state for this table is the triple point (0.01 °C, 0.6117 kPa) where *u* and *s* are assigned values of 0

steam turbine is to convert fluid enthalpy into output shaft power. Kinetic and potential energy changes are side effects that are generally not identically zero. This workshop is designed as practice/review of the thermodynamic analysis of open system devices and to justify the neglect of kinetic and potential energy changes going forward.

Table 10.1 Practical constraints on steam turbine cycles

Thermodynamic property	Practical reason for constraint	Thermodynamic constraint used for analysis
Turbine inlet temperature	Metallurgic restriction on boiler containment	$T_1 \leq 600\ ^\circ\mathrm{C}$
Turbine exit quality	Erosion of turbine blades by condensed liquid mist	$x_2 \geq 0.9$
Condenser temperature	Adequate heat transfer rates	$T_2 \geq T_C + 10\ ^\circ\mathrm{C}$
Pump inlet	Inlet to pump must be all liquid	$x_3 = 0$ or $T_3 < T_{SAT\ @P_3}$

In industrial practice, elaborate chemical treatment of the working fluid is needed to maintain a high purity of H_2O that would otherwise cause problems, particularly in the boiler and turbines. These auxiliary components are not addressed further here.

There are a number of practical constraints that define upper limits to steam turbine cycle performance, summarized in Table 10.1 and discussed in some more detail in the Deeper Dive of Appendix 1. These constraints are placed on the steam turbine cycles in the following analyses to obtain the maximum possible thermodynamic performance of Rankine-based cycles.

The focus of this chapter is on stationary power plants in which the working fluid must be isolated from the source of heating, which requires the boiler to be a heat exchanger, called external combustion (as opposed to equivalent heating from internal combustion). An important difference between internal and external combustion is that the working fluid in internal combustion is higher than the temperature of the walls of its containment, and the temperature of the working fluid in external combustion is lower than the containment walls.

The trend in the industry is for new Rankine cycle-based power plants to employ either solid fuel combustion (coal, biomass, municipal solid waste) or require a secondary fluid to be recirculated (nuclear, geothermal). In contrast, gaseous or liquid fuels can be used directly in internal combustion engines (reciprocating or gas turbine). However, their hot exhaust gases can be used as the heating source for a Rankine cycle, creating a combined cycle with a topping cycle as an internal combustion engine and a Rankine bottoming cycle, treated directly in Chap. 11.

The Energy Information Agency (EIA) is a valuable resource for up-to-date and historical energy-related information. For example, the site https://www.eia.gov/analysis/studies/powerplants/capitalcost/pdf/capcost_assumption.pdf provides both technical background and economic perspectives on power plant development.

In general, the higher the pressure and temperature supplied to the turbine, the better the thermodynamic performance, but the higher the installed capital cost of a power plant, as more expensive metals are needed to contain the working fluid safely. As is typical, there is an economic trade-off between the capital costs and the operating costs associated with the heating source (fuel cost). In addition, environmental concerns can play an important economic role, which can depend on shifting governmental regulations. For example, coal is a relatively plentiful and inexpensive heating source, but it has a high carbon footprint compared to other sources. These and other important policy issues will be ongoing, and the more informed the policy-makers and influencers, the better. A good grounding in a fundamental understanding of thermodynamic principles will result in better and fairer decisions.

Ideal Rankine Cycle Case Study

The basic Rankine cycle consists of a turbine, condenser, pump, and boiler and was shown operating between two thermal reservoirs (a high temperature source and low temperature sink) in Fig. 10.1. Each device is assumed to operate as a single-inlet, single-outlet open system at steady state, with negligible stray heat, kinetic, and potential energy changes. The process specifications for the ideal Rankine cycle are listed in Table 10.2. A spreadsheet template for a base case Rankine cycle is shown in Fig. 10.3, and the *T-s* diagram of Fig. 10.2 can be used to trace the cycle. A detailed solution and discussion follows. The steps taken to complete the State Table are summarized in Table 10.3. The results are shown in Fig. 10.4, which is color-coded to indicate what type of calculation is needed for each entry.

Thermodynamic property data needed for the analysis are provided, readily available from online or other published sources.

The results of this case study represent a realistic upper bound on the performance of an actual basic Rankine cycle. Further improvements will then be made by adding additional components.

With the problem formulated this way, a natural place to begin the analysis is at state 3, because two independent thermodynamic properties (temperature and quality) are specified. Saturation data for this temperature (35 °C) and two states needed for the next case are listed in Table 10.4. The pressure at state 3 is the corresponding saturation pressure (5.6291 kPa = 0.056291 bar), with the unit conversion 1 bar = 100 kPa applied. All other properties (v, h, s) for state 3 are the saturated liquid values (since the steam quality $x_3 = 0$).

State 2 is tackled next. The process specification of an isobaric condenser (process 2–3) allows P_2 to be entered (same value as P_3). With the specified quality of 0.9, the temperature is the corresponding saturation temperature. The process is also isothermal, since the process is isobaric and both states are within the dome (the saturated liquid line being the end point). The other properties are obtained from a quality calculation. For example

Table 10.2 Ideal rankine cycle process specifications

Process	Device	Process specification
1–2	Turbine	Adiabatic and reversible expansion
2–3	Condenser	Isobaric heat rejection to dead state
3–4	Pump	Adiabatic and reversible compression
4–1	Boiler	Isobaric heat addition from thermal reservoir

INPUT PARAMETERS

W'_{net}	**1**	GW, net power output
T_C	**25**	oC, dead state temperature (sink)
T_H	**650**	oC, thermal reservoir temperature (source)

STATE TABLE

state	P (bar)	T (oC)	v (m^3/kg)	h (kJ/kg)	s (kJ/kg/K)	quality
1		**600**				
2						**0.9**
3		**35**				**0.0**
4						

PROCESS TABLE (all in MW)

process	$m'(h_{out}-h_{in})$	Q'_{in}	W'_{out}	$m'T_c(s_{out}-s_{in})$	$T_c(Q'_{in}/T_b)$	$T_c{*}\Sigma'$
1-2, turbine						
2-3, condenser						
3-4, pump						
4-1, boiler						
CYCLE						

RESULTS

m'		kg/sec, mass flow rate
Back Work Ratio		ratio of pump work to turbine work
η_1		1st Law Efficiency, defined as NET WORK over HEAT IN BOILER
η_{CARNOT}		Carnot efficiency, for heat engine operating between source and sink
η_{II}		2nd Law Efficiency, defined as ACTUAL over CARNOT 1st law efficency

Fig. 10.3 Spreadsheet template for ideal rankine cycle base case

Table 10.3 Solution steps for completing state table

Step	Objective	Explanation of state table calculation
1	State 3	Since 2 independent properties are specified (temperature and quality) all properties can be obtained directly from saturation tables
2	State 2	P_2 from process spec (2–3 is isobaric). Since quality is an input, T_2 is the corresponding saturation temperature and quality calculations for all other properties
3	State 1	With T_1 as input and $s_1 = s_2$ (process spec) ideal gas interpolation used to solve for P_1, v_1, h_1
4	State 4	With $P_4 = P_1$ and $s_3 = s_4$ (process specs), confirm compressed liquid ($s_{f@P4} < s_4$) and use compressed liquid approximations

INPUT PARAMETERS

W'net	1	GW, net power output
T_C	25	oC, dead state temperature (sink)
T_H	650	oC, thermal reservoir temperature (source)

STATE TABLE

state	P (bar)	T (oC)	v (m³/kg)	h (kJ/kg)	s (kJ/kg/K)	quality
1	26.727	600	0.1490	3686.8	7.56704	NA, SHV
2	0.056	35	22.6846	2322.8	7.56704	0.9
3	0.056	35	0.001006	146.64	0.50510	0.0
4	26.727	35	0.001006	149.32	0.50510	NA, CL

COLOR CODE
Input Parameters
Process Specification
Entry from tables
quality calculation
Compressed Liquid Approx

PROCESS TABLE (all in MW)

process	$m'(h_{out}-h_{in})$	Q'_{in}	W'_{out}	$m'T_c(s_{out}-s_{in})$	$T_c(Q'_{in}/T_b)$	$T_c{*}\Sigma'$
1-2, turbine	-1002.0	0	1002.0	0	0	0
2-3, condenser	-1598.6	-1598.6	0	-1545.9	-1598.6	52.7
3-4, pump	1.971	0	-1.971	0	0	0
4-1, boiler	2598.6	2598.6	0	1545.9	839.0	706.9
CYCLE	0	1000	1000	0	-759.6	759.6

COLOR CODE
From State Table
Process Specification
From Energy Balance
From Entropy Balance

RESULTS

m'	734.6	kg/sec, mass flow rate
Back Work Ratio	0.00197	ratio of pump work to turbine work
η_I	0.385	1st Law Efficiency, defined as NET WORK over HEAT IN BOILER
η_{CARNOT}	0.677	Carnot efficiency, for heat engine operating between source and sink
η_{II}	0.568	2nd Law Efficiency, defined as ACTUAL over CARNOT 1st law efficency

Fig. 10.4 Solution of ideal rankine cycle

Table 10.4 Saturation data used for ideal rankine cycle and ideal rankine cycle with reheat

Saturation property data used for analysis

T (°C)	P (kPa)	v_f (m³/kg)	v_E (m³/kg)	h_f (kJ/kg)	h_g (kJ/kg)	s_f (kJ/kg/K)	s_g (kJ/kg/K)
35	5.6291	0.001996	25.205	146.64	2564.6	0.5051	8.3517
223.95	2500	0.001197	0.079952	961.87	2801.9	2.5542	6.2558
233.85	3000	0.001217	0.066667	1008.3	2803.2	2.6454	6.1856

$$s_2 = s_{2f} + x_2(s_{2g} - s_{2f}) = 0.5051 + 0.9(8.3517 - 0.5051) = 7.56704 \ \text{kJ/kg/K}$$

In the worksheet file, the saturation property data is listed in a worksheet, and appropriate formulas that refer to the appropriate cells can be input. It is a good idea to pinpoint these states on a *T-s* diagram before moving on, and observe how the cycle is building up.

Ideal Gas Interpolation in Superheat Tables

Moving up the state table to state 1, the temperature is specified ($T_1 = 600$ °C, the turbine inlet temperature constraint), and the turbine process 1–2 is isentropic, as specified. Therefore, two independent thermodynamic properties are known (T_1 and

SUPERHEATED VAPOR DATA

BASIC CYCLE STATE 1

	R (kJ/kg/K)	0.462	gas constant (8.314/MW)		
	P_0 (MPa)	2.5			
	T (°C)	v (m³/kg)	h (kJ/kg)	s (kJ/kg/K)	
T_0	600	0.15931	3686.8	7.5979	
T_1	700	--	3915.2	--	

COLOR CODE

Superheat Table Data

Values from Problem

Ideal Gas Interpolation

IDEAL GAS INTERPOLATION

INPUT

s (kJ/kg/K)	7.5670
T (°C)	600

INTERPOLATED

cp (kJ/kg/K)	2.284
P (MPa)	2.673
v (m³/kg)	0.1490
h (kJ/kg)	3686.8

$$c_p = \frac{h_1 - h_0}{T_1 - T_0}$$

$$P = P_0 \left(\frac{T+273}{T_0+273}\right)^{\frac{c_p}{R}} e^{\frac{-(s-s_0)}{R}}$$

$$v = v_0 \left(\frac{T+273}{T_0+273}\right)\left(\frac{P_0}{P}\right)$$

$$h = h_0 + c_p (T - T_0)$$

Fig. 10.5 Spreadsheet template for ideal gas interpolation for state 1 (turbine inlet) of the basic ideal Rankine cycle case study

s_1). Locating the point on the Ts diagram, it is clearly in the superheated region, with a temperature well above the critical temperature (373.95 °C). Consulting superheated tables, it is clear that the pressure will not correspond exactly to any listed pressure, and some form of 2 point interpolation will be needed. This is a case where an "ideal gas interpolation" comes in handy, as detailed in Fig. 10.5, a form of spreadsheet template that can be adapted for interpolating within the superheat region, especially when the pressure does not correspond to a readily available value, as in this case. The top section shows superheat data (from the internet) for a pressure for which the entropy and temperature are close to their target values (600 °C and 7.567 kJ/kg/K). Specific volume, enthalpy, and entropy values that correspond to the known temperature are listed, as is just the enthalpy at the closest available temperature (700 °C). In this case, the known temperature corresponds to one of the temperatures listed, but the method works if an intermediate temperature (say, 613.2 °C from a measurement) were known.

The ideal gas interpolation method involves use of the mass, energy and entropy property relations for calorically perfect gases (constant specific heat) to determine properties at a nearby state. The enthalpy-specific heat (c_p) is needed in the ideal gas relations, and is calculated first (and why only the enthalpy is listed at 2 temperatures from the superheat tables). That is:

$c_p = \frac{h_1 - h_0}{T_1 - T_0}$ from two superheat table values (subscript 0 and 1) at a constant pressure.

The entropy relation is used first in this case as it involves 2 of the knowns, and one of the unknowns (P). That is, starting with the entropy change relation:

$$s - s_0 = c_p \ln\left(\frac{T + 273}{T_0 + 273}\right) - R \ln\left(\frac{P}{P_0}\right)$$

The temperature is listed in °C, and the pressure is in absolute units. Here, state 0 corresponds to the known value from the superheat tables, and the state that requires interpolation is shown without a subscript. This method will work for any specific entry, but is most accurate the closer the actual state is to a tabulated state. In this case, since entropy and temperature are specified, the entropy change relation can be rearranged for the pressure. That is:

$$P = P_0\left(\frac{T + 273}{T_0 + 273}\right)^{\frac{c_p}{R}} e^{\frac{-(s-s_0)}{R}}$$

Note that if T and P were known, the original entropy change equation could be used to solve for s. If s and P were known (as is common), then the entropy change could be rearranged for the temperature. That is:

$$T = (T_0 + 273)\left(\frac{P}{P_0}\right)^{\frac{R}{c_p}} e^{\frac{(s-s_0)}{c_p}}$$

The specific volume is determined from the ideal gas relation $Pv/T = P_0 v_0/T_0$

$$v = v_0\left(\frac{P_0}{P}\right)\left(\frac{T + 273}{T_0 + 273}\right)$$

Finally, the enthalpy is determined from:

$$h = h_o + c_p(T - T_0)$$

The state table of Fig. 10.4 can reference the appropriate cells for these values.

Another case study that details and compares a 2-point interpolation and the ideal gas interpolation method is conducted as part of the solution to Workshop 10.2.

Compressed Liquid Approximations for State 4

In the experience of this engineering Professor, proper analysis of the pumping process 3–4 has proven to be a difficult concept for many students, yet it involves a simple application of compressed liquid approximations which were discussed in

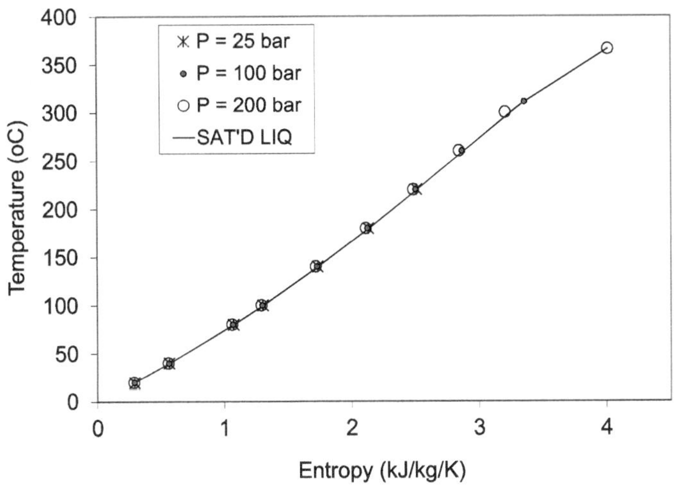

Fig. 10.6 Compressed liquid data plotted in *T-s* coordinates for 3 pressures and the saturated liquid line. Figure repeated from Chap. 4

detail in Chap. 4. The confusion may come in part from a misleading representation of the process on Ts diagrams reported in the literature. The following analysis highlights a few pitfalls. The *T-s* plot of data from compressed liquid tables from Chap. 4 is repeated in Fig. 10.6 to reinforce the point. Since there is a dramatic change in pressure of the working fluid as it passes through the pump, students expect a dramatic change in something else. In reality, the only property that changes dramatically is the pressure, and the enthalpy is the only other property affected by pressure. It is like pushing a solid wall. You can lean gently against it or put your full weight into it and it will not move. It is a very different state, but only the force applied changes, not the position of the wall so no work is done. In contrast, compression of a gas or vapor results in a significant temperature rise and decrease in specific volume due to the work done on the gas.

At this point in this case study, all property values for states 1 and 3 have been determined. Process 3–4 is isentropic and, therefore, $s_4 = s_3$ is known (0.5051 kJ/kg/K) from a process specification. Process 4–1 is isobaric, and therefore, $P_4 = P_1$ is known (26.73 bar = 2673 kPa). It is proper at that point to consult the saturation tables, Table 10.4, from which it is seen that the entropy is lower than the saturated liquid value at that pressure. It is then appropriate, but ill-advised, to seek out the property values from the compressed liquid tables. It is ill-advised because the results of Chap. 4 and Workshop 4.1, with the results of compressed liquid data on *T-s* coordinates repeated in Fig. 10.6 showed that the entropy of a compressed liquid is effectively independent of pressure. Alternatively stated, the entropy of a compressed liquid equals that of a saturated liquid at the same temperature. That is, $s(T, P) = s_{f\ @\ T}$ for compressed liquids. A student that attempts to determine the properties from readily available compressed liquid data will generally find the entries to be so coarse that linear interpolations would be cumbersome and give a false sense of accuracy. Alternatively, an interpolation methodology similar to what was done for superheated state 1, but with compressed liquid approximations, as detailed in the deeper dive of Appendix 2. But in practice, this should only be used for extremely high pressures and temperatures (where there is some small deviation from the compressed liquid approximations). It is poor engineering judgment to spend time going down that blind alley.

Another pitfall is an egregious conceptual error, which is to interpret the compressed liquid approximation as being $s(T, P) = s_{f\,@\,P}$. That would result in students entering, in this case study, erroneous and inconsistent property values, like a temperature of 225 (interpolating "by eye" from Table 10.4). This common error can be corrected by designing practice problems that specify a pressure and either an internal energy or entropy value lower than the saturated liquid value, and determining all other properties. This exercise forces the student to search the saturation tables for the temperature for which the saturated liquid value corresponds to the specified internal energy or entropy. The specific volume could be specified instead, but since the specific volume is a weak function of temperature, the effect is not as clear. Using enthalpy requires an iterative procedure since it depends on both temperature and pressure.

In short, with sound engineering judgment, the compressed liquid approximations can be used unapologetically, and the result is then very simple. First, since the entropy is a function of temperature only, and the entropy does not change in the pump, neither does the temperature. Furthermore, the specific volume is a function of temperature only and it does not change. The enthalpy, however, has both a temperature and pressure dependency, namely, $h(T,\ P) = h_{f\,@\,T} + v_{f\,@\,T}(P - P_{\mathrm{SAT}\,@\,T})$, taking care to conduct proper unit conversions. In this case study:

$$h_4 = h_{@35^\circ\mathrm{C},\ 26.73\mathrm{bar}} = \left[146.64\frac{\mathrm{kN}\cdot\mathrm{m}}{\mathrm{kg}}\right]$$

$$+\left[0.001006\frac{\mathrm{m}^3}{\mathrm{kg}}(26.73 - 0.056)\mathrm{bar}\cdot\left|\frac{100\ \mathrm{kPa}}{\mathrm{bar}}\right|\cdot\left|\frac{\mathrm{kN/m^2}}{\mathrm{kPa}}\right|\right]$$

$$h_4 = h_{@35\ ^\circ\mathrm{C},\ 26.73\mathrm{bar}} = 149.32\frac{\mathrm{kJ}}{\mathrm{kg}}$$

The T-s diagram can be completed, recognizing that the boiler process 4–1 is isobaric. This process path will follow the saturated liquid line (even though the working fluid is a compressed liquid) until it reaches the saturation temperature (227.4 °C by linear interpolation), then phase change will occur at constant temperature to the saturated vapor line and then follow a saturated vapor isobar to state 1.

Most thermodynamics textbooks and internet sources show sketches of T-s diagrams that are not to scale, and are misleading with regards to the pump process. The left sketch of Fig. 10.7 shows the misleading representation, in which the pumping process 3–4 is shown with an increase in temperature at constant entropy. An internet search (for example "Rankine cycle Ts diagram") will show that virtually

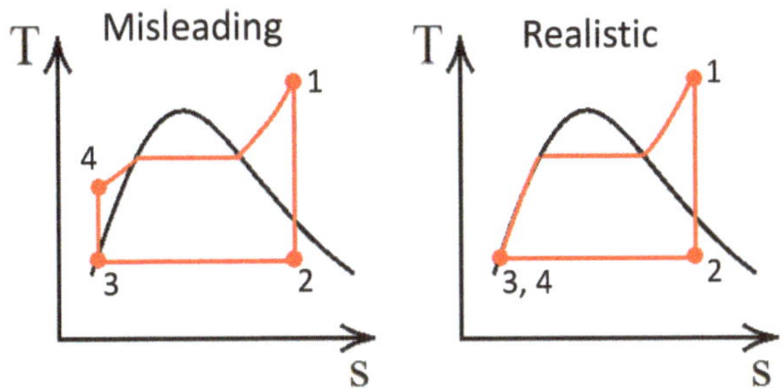

Fig. 10.7 Two representations of Ts diagrams of the ideal Rankine cycle. The left schematic is shown in most textbook and sources, and is misleading. The right sketch is more accurate, with states 3 and 4 coinciding (according to the compressed liquid approximation)

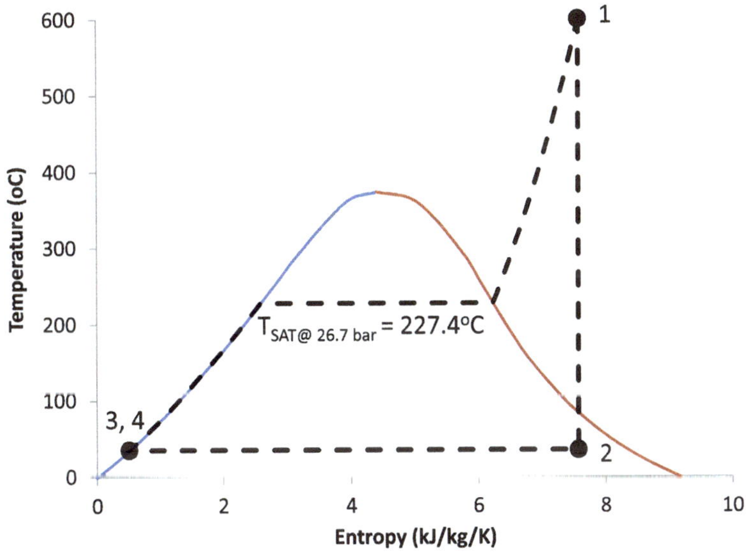

Fig. 10.8 *T-s* diagram of ideal Rankine cycle, to scale. Note that the temperature of state 4 is obtained from the compressed liquid tables in the plot, emphasizing that there is no effective change in temperature for an isentropic compression of a compressed liquid

all sources depict the pumping process with a significant temperature change. The right sketch shows the more realistic view, invoking the compressed liquid approximation. If the process is isentropic, then it is also isothermal. The actual Ts diagram, with numbers to scale for this case study, is shown in Fig. 10.8. Note that for this plot, the temperature of state 4 using data from compressed liquid table (35.13, calculated in Appendix 2 rather than 35.0 for the compressed liquid approximation) is used, emphasizing that there is no noticeable difference in temperature.

It should not be so difficult to accept that process 3 and 4 are indistinguishable on Ts coordinates. For example, consider a phase diagram (P-T space) where the saturation line that separates compressed liquid from superheated vapor. A saturated liquid and saturated vapor at the same temperature and pressure would be at the same point, even though the specific volume, enthalpy and entropy is very different.

Process Table: Energy and Entropy Balances

The process table is completed from state table data and from energy and entropy balances on each device. The general forms for single-inlet, single-exit devices are (with the entropy balance multiplied by T_c considered to be the dead state temperature):

$$\frac{dE}{dt} = \dot{Q}_{\text{in}} - \dot{W}_{\text{out}} + \dot{m}(h_{\text{in}} - h_{\text{out}}) + \text{KE} + \text{PE}$$

$$T_{\text{C}}\frac{dS}{dt} = \dot{Q}_{\text{in}}\frac{T_{\text{C}}}{T_{\text{B}}} + T_{\text{C}}\dddot{\Sigma} + \dot{m}T_{\text{C}}(s_{\text{in}} - s_{\text{out}})$$

For the ideal Rankine cycle, each device is operated at steady state ($dE/dt = dS/dt = 0$), and kinetic and potential energy changes are neglected. Table 10.5 summarizes the simplified energy and entropy balance equation for each device in the ideal Rankine cycle. Appropriate formulas can be quickly entered into the process table in a spreadsheet, and the results are shown in Fig. 10.4. The signs of these terms are consistent with the convention that heat into and work extracted from a device are positive, whereas the energy terms in Table 10.5 are consistent with the overall control volume.

Postprocessing calculations give the mass flow rate (input power of 1.0 GW divided by the net cycle work per unit mass, with appropriate unit conversions), back work ratio, 1st Law efficiency, Carnot efficiency, and 2nd Law efficiency, all shown in the Results block.

All the eXergy destruction in the ideal cycle is in the two thermal devices (condenser and boiler), ultimately due to heat transfer across a finite temperature difference. There is no eXergy destruction in the two mechanical devices (pump and turbine). The eXergy destruction in the condenser is relatively low, being the

Table 10.5 Energy terms for ideal Rankine cycle, with signs of energy flux terms consistent with Fig. 10.1

Device	Energy balance	T_{C} * entropy balance
Turbine	$w_{\text{turbine}} = w_{\text{out}} = (h_1 - h_2)$	$eX_{\text{D, turbine}} = 0$
Condenser	$q_{\text{C}} = q_{\text{out}} = (h_2 - h_3)$	$eX_{\text{D, condenser}} = q_{\text{C}} - T_{\text{C}}(s_2 - s_3)$
Pump	$w_{\text{pump}} = w_{\text{in}} = (h_4 - h_3)$	$eX_{\text{D, pump}} = 0$
Boiler	$q_{\text{H}} = q_{\text{in}} = (h_1 - h_4)$	$eX_{\text{D,condenser}} = -q_{\text{H}}\frac{T_{\text{C}}}{T_{\text{H}}} + T_{\text{C}}(s_1 - s_4)$

difference between two relatively large numbers of opposite sign, namely, the entropy transfer with heat and the difference in entropy of the working fluid between outlet and inlet. The process is isothermal, at a temperature designed to be as close to the cold reservoir (at the dead state temperature) as is practical. In practice, significant effort to reduce this eXergy destruction would not be warranted.

On the other hand, most of the eXergy destruction in the cycle occurs in the boiler. The working fluid undergoes a large temperature change, and the heat transfer occurs over a large temperature difference between the reservoir (at T_H) and working fluid. The choice of reservoir temperature does not affect any 1st law quantities. That is, the 1st Law efficiency of 38.5% for this ideal Rankine cycle subject to reasonable practical constraints is independent of either reservoir temperature. The Carnot efficiency and therefore 2nd Law efficiencies are dependent on choice of reservoir temperature, however. The main result of the 2nd Law analysis is that the primary source of irreversibility in an ideal Rankine cycle is in the boiler.

The boiler has three primary sections to meet different heat transfer objectives: preheater (sometimes called an economizer) where compressed liquid water is heated to the saturation temperature, a boiler section where the phase change occurs from which saturated vapor emerges, and a superheater section where the temperature of the vapor is raised. The nature of the working fluid is very different (liquid, liquid/vapor, and vapor) as are the heat transfer properties. The design of the boiler, not discussed in detail here, is a difficult task that has evolved significantly. The objective here is an overall balance and not in the design details, which are addressed in more detailed studies.

Three variations are considered in the following sections, considered to be deeper dives, with key results compiled in Table 10.6. The results of the basic ideal Rankine cycle are listed in the first column. The next two cases (reheat and reheat plus regeneration) include additional components aimed at improving thermodynamic performance. The last case includes component efficiencies for the pumps and turbines to more closely model the performance of actual power plants

Deeper Dive: Ideal Rankine Cycle with Reheat

The pressure ratio in the basic Rankine cycle subject to the constraints of a maximum allowable turbine inlet temperature and steam quality exiting the turbine results in its best possible thermodynamic performance. Improved performance without violating the constraints of Table 10.1 can be obtained by pumping to a higher pressure in the boiler, expanding partially to an intermediate pressure, then reheating the steam, shown schematically in Fig. 10.9. This approach involves adding a high pressure (HP) turbine, and another heating device called a reheater. The rest of the cycle remains the same, except the pump raises the pressure higher and the original turbine is called a low pressure (LP) turbine. A blank spreadsheet template is shown in Fig. 10.10, which can be used to conduct various studies. The case study conducted here will build upon the basic ideal Rankine cycle, subject to the constraints of Table 10.1, to demonstrate the potential of adding reheat.

Table 10.6 Compilation of results from case studies. All energy quantities are in MW

	Ideal Rankine cycle	Ideal Rankine cycle with Reheat	Ideal rankine cycle with Reheat and Regeneration	Rankine cycle with Reheat and Regeneration with component effy
Boiler mass flow rate (kg/s)	735	506	644	770
Total turbine power (MW)	1002	1015	1019	1046
eXd, turbine	0	0	0	148
Total pump power	2.0	15.4	19.1	45.8
eXd, pumps	0	0	0	14.1
Total heat addition rate	2599	2100	1964	2282
eXd, boiler and reheater	707	386	242	272
eXd, regenerator	NA	NA	56.1	68.3
Condenser heat rate	1599	1100	964	1282
eXd, condenser	52.7	36.3	31.8	42.2
1st law efficiency	38.5%	47.6%	50.9%	43.8%
2nd law efficiency	56.8%	70.3%	75.2%	64.7%

In practice, the reheating process is accomplished in the main boiler, with the high pressure turbine exit flow being ducted back into the superheater section of the boiler and then returned to the low pressure turbine. However, the reheater is shown schematically as a separate heat exchanger device, drawing heat from the same high temperature reservoir. Both turbines are generally also driven by the same shaft, in practice. The schematic shows the functional relations, not the lay-out of equipment.

In this case study, the maximum possible performance with a single reheater is obtained by defining an additional constraint, namely, specification of a saturated vapor at the exit of the high pressure turbine. Two additional states are added (a and b) to the base case and the numbering of all other states corresponds with the basic Rankine cycle. The high pressure turbine is specified as being adiabatic and reversible (hence isentropic, like the LP turbine), and the reheater process is isobaric, just like the boiler.

States 1, 2, 3, and 4 all have the same state as for the basic Rankine cycle. State b has a specified quality of 1.0, and the process specification of the reheater (isobaric) fixes the pressure at 26.7 bar $=$ 2670 kPa $=$ 2.67 MPa. The temperature is the corresponding saturation temperature and the other properties (specific volume, enthalpy, and entropy) are the corresponding saturated vapor values. All these

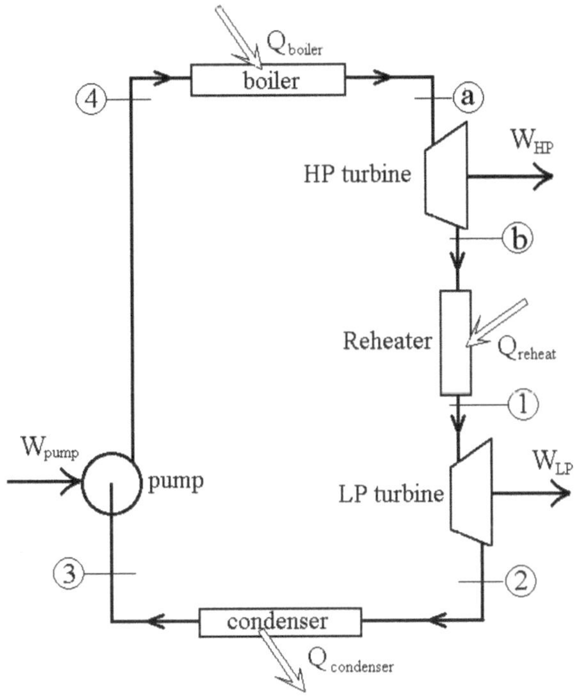

Fig. 10.9 Schematic of rankine cycle with reheat

require a linear interpolation from the saturation tables (entries at pressures of 25 bar and 30 bar are listed in Table 10.4).

State a, the high pressure (HP) turbine inlet, is fixed by the turbine inlet temperature and the isentropic process specification of the turbine. However, an interpolation similar to state 1 is needed to determine all other thermodynamic properties, but at a higher pressure level. The ideal gas interpolation template was modified with superheated data near the target value. These results are shown in Fig. 10.11. The resulting pressure is 30.38 MPa = 303.8 bar. This value exceeds the critical pressure of 220.6 bar for H_2O. Modern power plants are routinely designed with pressures higher than the critical pressure, and are called supercritical (or ultra-supercritical) power plants. Containing such high pressures at elevated temperatures requires advanced design and metallurgy. In this case study, the need for supercritical pressure is a natural result of imposing the turbine inlet and exit constraints, and represents the best possible thermodynamic performance with a single reheating stage.

Results are shown in Fig. 10.12 (spreadsheet) and Fig. 10.13 (*T-s* diagram). There is a significant improvement in overall performance associated with reheat. The mass flow rate required to produce 1 GW of net power decreases by 31.2%, the net heat addition rate decreases by 19.2%, and the 1st Law and 2nd Law efficiencies increase by 23.7% (the Carnot efficiency is unaffected by reheat). The back work ratio increases from 0.002 to 0.015, due to the much higher pump power required, but still is rather low, a major reason that the Rankine cycle is so practical in industry.

INPUT PARAMETERS

W'_{net}		GW, net power output
T_C		oC, dead state temperature (sink)
T_H		oC, thermal reservoir temperature (source)

STATE TABLE

state	P (bar)	T (oC)	v (m^3/kg)	h (kJ/kg)	s (kJ/kg/K)	quality
a						
b						
1						
2						
3						
4						

PROCESS TABLE (all in MW)

process	$m'(h_{out}-h_{in})$	Q'_{in}	W'_{out}	$m'T_C(s_{out}-s_{in})$	$T_C(Q'_{in}/T_b)$	$T_C*\Sigma'$
a-b, HP turbine						
b-1, reheater						
1-2, LP turbine						
2-3, condenser						
3-4, pump						
4-1, boiler						
CYCLE						

RESULTS

m'		kg/sec, mass flow rate
Back Work Ratio		ratio of pump work to turbine work
η_1		1st Law Efficiency, defined as NET WORK over HEAT IN BOILER
η_{CARNOT}		Carnot efficiency, for heat engine operating between source and sink
η_{II}		2nd Law Efficiency, defined as ACTUAL over CARNOT 1st law efficency

Fig. 10.10 Blank template for conducting a study of the ideal Rankine cycle with reheat. This template can be used with a basic Rankine cycle (without reheat) by ignoring states a and b

A good classroom exercise is to assign a different boiler pressure to different student teams, facilitate the completion of a blank template (either on paper, or preferably in a spreadsheet) and then compile the results. Care must be taken in formulating problems, as it is easy to inadvertently either over-specify or underspecify a problem. In this case, if the boiler pressure is specified, then either the turbine inlet temperature specification or the HP turbine exit state must be left unspecified. Many interesting studies can be conducted in this way, and the ideal gas interpolation method can greatly facilitate matters.

REHEAT STATE a

R (kJ/kg/K)	0.462		
P (MPa)	30		
T (°C)	v (m³/kg)	h (kJ/kg)	s (kJ/kg/K)
600	0.011445	3446.8	6.2373
650	--	3599.4	--

COLOR CODE

Superheat Table Data
Values from Problem
Ideal Gas Interpolation

IDEAL GAS INTERPOLATION

INPUT

s (kJ/kg/K)	6.2316
T (°C)	600

DERIVED

cp (kJ/kg/K)	3.052
P (MPa)	30.38
v (m³/kg)	0.01130
h (kJ/kg)	3446.8

Fig. 10.11 Ideal gas interpolation for state a of ideal Rankine cycle with reheat case study

INPUT PARAMETERS

W'_{net}	1	GW, net power output
T_C	25	oC, dead state temperature (sink)
T_H	650	oC, thermal reservoir temperature (source)

STATE TABLE

state	P (bar)	T (oC)	v (m³/kg)	h (kJ/kg)	s (kJ/kg/K)	quality
a	303.76	600	0.011303	3446.8	6.2316	NA, SHV
b	26.73	227.4	0.075363	2802.3	6.2316	1.0
1	26.73	600	0.149014	3686.8	7.5670	NA, SHV
2	0.06	35	22.684601	2322.8	7.5670	0.9
3	0.06	35	0.001006	146.6	0.5051	0.0
4	303.76	35	0.001006	177.2	0.5051	NA, CL

COLOR CODE

Input Parameters
Process Specification
Entry from tables
quality calculation
Compressed Liquid Approx

PROCESS TABLE (all in MW)

process	$m'(h_{out}-h_{in})$	Q'_{in}	W'_{out}	$m'T_C(s_{out}-s_{in})$	$T_C(Q'_{in}/T_b)$	$T_C*\Sigma'$
a-b, HP turbine	-325.8	0	325.8	0	0	0
b-1, reheater	447.2	447.2	0	201.2	144.4	56.8
1-2, LP turbine	-689.6	0	689.6	0	0	0
2-3, condenser	-1100.2	-1100.2	0	-1064.0	-1100.2	36.3
3-4, pump	15.4	0	-15.4	0	0	0
4-a, boiler	1653.1	1653.1	0	862.8	533.7	329.1
CYCLE	0.0	1000	1000	0.0	-422.2	422.2

COLOR CODE

From State Table
Process Specification
From Energy Balance
From Entropy Balance

RESULTS

m'	505.6	kg/sec, mass flow rate
Back Work Ratio	0.01521	ratio of pump work to turbine work
η_1	0.476	1st Law Efficiency, defined as NET WORK over HEAT IN BOILER + REHEATER
η_{CARNOT}	0.677	Carnot efficiency, for heat engine operating between source and sink
η_{II}	0.703	2nd Law Efficiency, defined as ACTUAL over CARNOT 1st law efficency

Fig. 10.12 Spreadsheet solution for ideal Rankine cycle with reheat case study

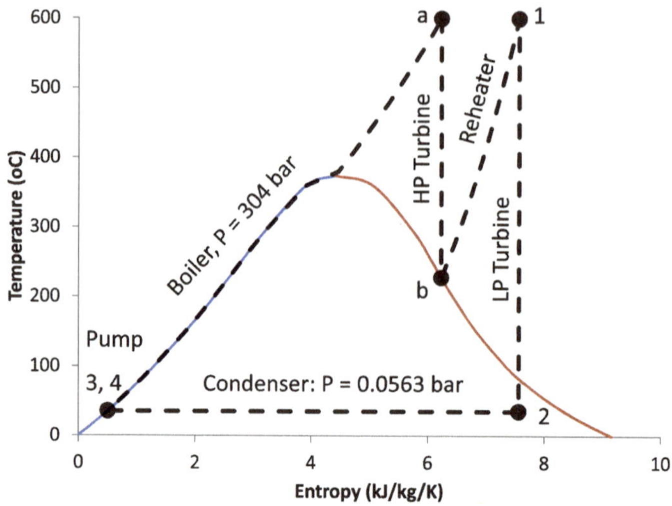

Fig. 10.13 *T-s* diagram for ideal Rankine cycle with reheat case study

Ideal Rankine Cycle with Reheat and Regeneration

The largest source of eXergy destruction in a Rankine cycle occurs in the boiler, due to the heat transfer across a finite temperature difference. The boiler has three requirements, or sections: preheating compressed liquid to the boiling point, phase change at the saturation temperature and superheating. The temperature difference between the high temperature reservoir and the working fluid is greatest in the preheater section. Improved thermodynamic performance can be achieved by reducing the external heat addition to the working fluid at its lowest temperature. Regeneration is the technique of partially preheating the pump exit (the feedwater) using a portion of the working fluid at some point after the boiler.

A schematic of a Rankine cycle modified to include both reheating and regeneration is shown in Fig. 10.14. A portion of the flow (fraction "f") that exits the high pressure turbine is extracted at the intermediate pressure of the reheater. This steam will be used to preheat the feedwater (compressed liquid leaving the pump). The thermodynamic properties of the extracted flow and the bulk flow that enters the reheater are the same (state b), the flow being simply split. The bulk flow goes through the same processes as the base Rankine cycle, but with a reduced mass flow rate. The value of the extracted fraction (f) is not arbitrary, but set by the specified conditions and appropriate energy balances.

After the LP pump, two streams at the same pressure are mixed adiabatically in a device called an open feedwater heater (OFWH); the compressed liquid water at the condenser temperature (state c), and the extracted steam from the HP turbine exit (state b). The properties of the resultant exit stream (state d) depend on the relative flow rates. The state will be specified as being a saturated liquid ($x_d = 0$), and that will uniquely define the fraction of steam that must be extracted from the turbine (f).

Fig. 10.14 Schematic of Rankine cycle with reheat and regeneration

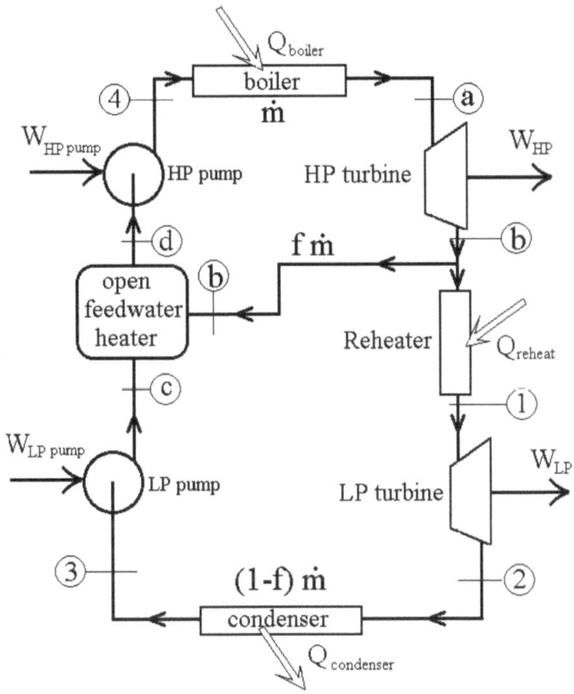

In effect, the feedwater (after the LP pump) has been partially preheated by an internal process (and not in the boiler). The combined saturated liquid stream is at the intermediate pressure and it enters the HP pump that brings it to the boiler pressure as a compressed liquid (but at an elevated temperature compared to the condenser temperature).

The pressure of all three streams in the open feedwater heater must be the same, a consequence of momentum concepts. The two entering streams are brought into direct contact and simply mixed together. Closed feedwater heaters are heat exchangers that accomplish regeneration and often used (but not analyzed here) which allow different pressures in the feedwater and extracted steam because the streams do not mix.

The spreadsheet results for this case are shown in Fig. 10.15 and the *T*-*s* diagram for the cycle is shown in Fig. 10.16. To analyze the cycle, the only new state added is the exit of the feedwater, which will be specified as a saturated liquid at the intermediate pressure defined by the requirements previously set by the reheater. The properties are obtained from a linear interpolation for saturated liquid values from Table 10.4.

Completion of the process table requires a calculation of the mass flow rates in each device. First, the fraction of steam extracted (f) can be determined from a 1st Law analysis of the feedwater heater. Assuming steady-state operation, neglecting stray heat, kinetic and potential energy changes, the energy balance for the feedwater heater is:

INPUT PARAMETERS

W'_{net}	1	GW, net power output
T_C	25	oC, dead state temperature (sink)
T_H	650	oC, thermal reservoir temperature (source)

STATE TABLE

state	P (bar)	T (oC)	v (m^3/kg)	h (kJ/kg)	s (kJ/kg/K)	quality
a	303.76	600	0.011303	3446.8	6.2316	NA, SHV
b	26.73	227.4	0.075363	2802.3	6.2316	1.0
c	26.73	35	0.001006	149.3	0.5051	NA, CL
d	26.73	227.4	0.001204	977.9	2.586	0.0
1	26.73	600	0.149014	3686.8	7.5670	NA, SHV
2	0.0563	35	22.684601	2322.8	7.5670	0.9
3	0.0563	35	0.001006	146.6	0.5051	0.0
4	303.76	227.4	0.001204	1005.8	2.5857	NA, CL

COLOR CODE
Input Parameters
Process Specification
quality calculation
Compressed Liquid Approx
Entry from tables

PROCESS TABLE (all in MW)

process	$m'(h_{out}-h_{in})$	Q'_{in}	W'_{out}	$m'T_C(s_{out}-s_{in})$	$T_C(Q'_{in}/T_b)$	$T_C*\Sigma'$
a-b, HP turbine	-415.0	0	415.0	0	0	0
b-1, reheater	391.7	391.7	0	176.3	126.5	49.8
1-2, LP turbine	-604.1	0	604.1	0	0	0
2-3, condenser	-963.8	-963.8	0.0000	-932.0	-963.8	31.8
3-c, LP pump	1.188	0	-1.188	0	0	0
d-4, HP pump	17.9	0	-17.95	0	0	0
b,c,d; FWH	NA	0	0	NA	0	56.1
4-a, boiler	1572.1	1572.1	0	699.7	507.6	192.1
CYCLE	NA	1000	1000	NA	-330	330

COLOR CODE
From State Table
Process Specification
From Energy Balance
From Entropy Balance

RESULTS

f_{ext}	0.3123	fraction of mass extracted after turbine 1
m'_{boiler}	644.0	kg/sec, mass flow rate through boiler
$m'_{condenser}$	442.9	kg/sec, mass flow rate through condenser
η_1	0.509	1st Law Efficiency, defined as NET WORK over HEAT IN BOILER + REHEATER
η_{CARNOT}	0.677	Carnot efficiency, for heat engine operating between source and sink
η_{II}	0.752	2nd Law Efficiency, defined as ACTUAL over CARNOT 1st law efficency

Fig. 10.15 Spreadsheet results for ideal Rankine cycle with reheat and regeneration

$$0 = (1-f)\dot{m}h_c + f\dot{m}h_b - \dot{m}h_d$$

where the mass flow rate is that through the boiler (equal to the HP pump and HP turbine flows). Solving for the fraction of mass extracted:

$$f = \frac{h_d - h_c}{h_b - h_c}$$

With all three states specified, the fraction of steam extracted evaluates as:

$$f = \frac{977.9 - 149.3}{2802.3 - 149.3} = 0.3123$$

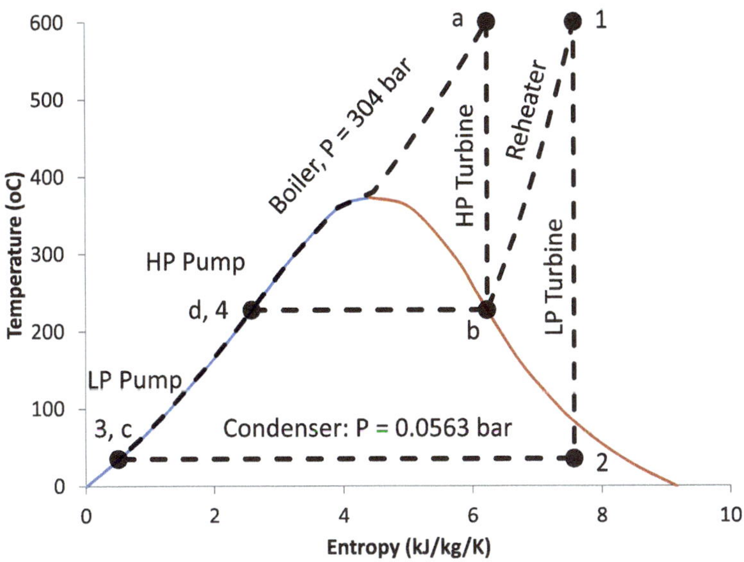

Fig. 10.16 *T-s* diagram for Ideal Rankine cycle with reheat and regeneration case study

That is, 31.23% of the flow is extracted. Note that if a higher fraction were extracted, the exit of the feedwater heater would not be completely condensed (it would have a quality greater than 0). On the other hand, a lower f would result in a compressed liquid.

The flow rate through the boiler is obtained from the scaling equation for the specified net power:

$$\dot{W}_{\text{net}} = \dot{W}_{\text{HPturbine}} - \dot{W}_{\text{HPpump}} + \dot{W}_{\text{LPturbine}} - \dot{W}_{\text{LPpump}}$$

$$\dot{W}_{\text{net}} = \dot{m}[(h_{\text{a}} - h_{\text{b}}) - (h_4 - h_{\text{d}}) + (1 - f)((h_1 - h_2) - (h_{\text{c}} - h_3))]$$

Solving for the mass flow rate (through the boiler)

$$\dot{m} = \frac{\dot{W}_{\text{net}}}{(h_{\text{a}} - h_{\text{b}}) - (h_4 - h_{\text{d}}) + (1 - f)((h_1 - h_2) - (h_{\text{c}} - h_3))}$$

The mass flow rate through all devices below the intermediate pressure (LP turbine, condenser, LP pump) is $(1 - f)\dot{m}$.

The boiler mass flow rate required to generate 1.0 GW is significantly higher for the cycle with regeneration (644 kg/s) than for the cycle with reheat only (506 kg/s). This trend is a result of the use of a portion of the flow rate from the boiler to preheat liquid internally at the expense of generating power in the LP turbine. Nevertheless, the net heat addition rate in the boiler plus reheater is lowered (from 2100 to 1964 MW) as a result of a higher enthalpy entering the boiler.

The mixing of two streams of very different properties in the regenerator comes at a thermodynamic price in the form of entropy generation (hence eXergy destruction). The entropy balance, multiplied by the dead state temperature (in absolute units), is:

$$0 = T_C[(1-f)\dot{m}s_c + f\dot{m}s_b - \dot{m}s_d + \dddot{\Sigma}]$$

The eXergy destruction rate is:

$$e\dddot{X}_D = T_C\dddot{\Sigma} = T_C\dot{m}[(s_d - s_c) + f(s_c - s_b)]$$

For the case study, the eXergy destruction rate in the feedwater heater (regenerator) evaluates to 56.1 MW. The mixing process is therefore dissipative. However, the eXergy destruction rate in the boiler plus reheater decreased from 386 MW for the cycle with the reheater only to 242 MW for the cycle with reheater and regeneration. The net effect is a lower significantly lower net eXergy destruction rate for the cycle with regeneration (330 MW vs. 422 MW).

The net effect of all the changes associated with regeneration results in an increase in 1st Law efficiency from 47.6% to 50.9%.

In this case study, regeneration and reheat take place at the same intermediate pressure stage. It is possible to employ regeneration without reheat and reheat without regeneration. It is also possible to use multiple intermediate pressures for reheat and regeneration. Figure 10.17 shows a schematic of a turbine that has a single reheat section (with extraction) and four extraction points after that. The physical size of succeeding sections is shown to increase dramatically, in response to the large increase in specific volume of the working fluid to keep the flow velocity from increasing too much, as discussed more fully in the next section. Liquid is shown being added to the reheater which gives plant operators the ability to control the returning temperature to adapt to changing load requirements.

Theoretically, use of reheat and regeneration at a large number of intermediate pressure stages results in a thermodynamic efficiency that can approach the Carnot efficiency (based on the high temperature source at the highest working fluid temperature). All the external heat addition takes place with the working fluid approaching the high temperature reservoir temperature. There is still eXergy destruction in the feedwater heaters, however.

Nonideal (Real) Rankine Cycle: Including Component Efficiencies

The three case studies conducted thus far have all been considered to be ideal Rankine cycle variations. The sources of eXergy destruction in these ideal cases stem from the external heat transfer across a finite temperature difference that takes

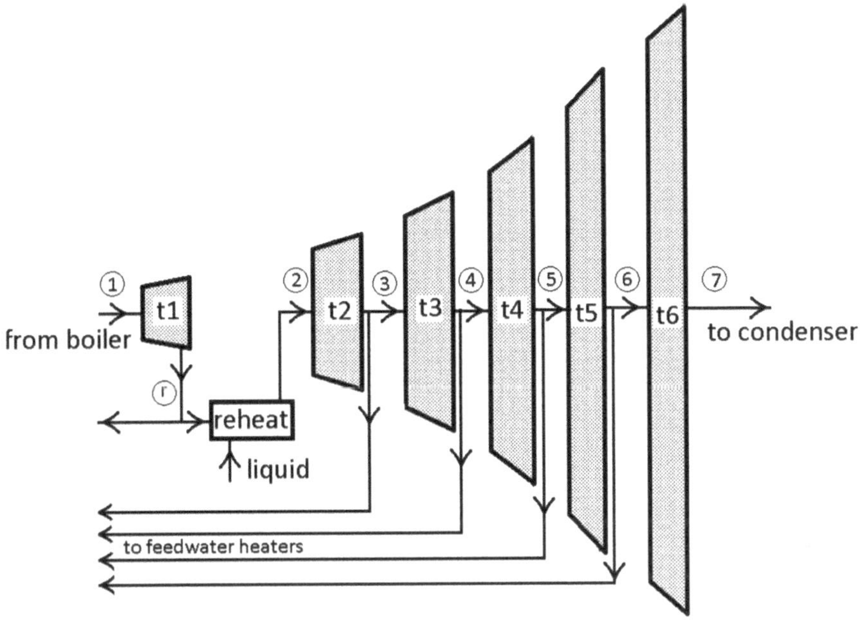

Fig. 10.17 Schematic of a steam turbine with a single reheat stage after a HP turbine followed by steam extraction at multiple stages used for regeneration (without reheating)

place between the thermal reservoirs and the working fluid. Technically, the regeneration process using an open feedwater heater involves a deviation from ideal behavior, but that eXergy destruction is considered to be part of the overall heating process.

Any real cycle will involve deviations from ideal behavior associated the following secondary effects that are neglected in the analysis of the ideal cycle. A brief description of these effects is given in Table 10.7.

Piping Design Criterion to Minimize Kinetic Energy Changes

A criterion to justify the neglect of kinetic energy changes for a single-inlet, single-outlet device can be expressed as:

$$\left|\frac{V_{\text{out}}^2}{2} - \frac{V_{\text{in}}^2}{2}\right| \ll |h_{\text{out}} - h_{\text{in}}|$$

Table 10.7 Brief description of real effects that cause a deviation from ideal Rankine cycle performance

Effect	Comments
Transient effects	Power plants are generally designed to operate at steady-state, but in practice, changes in operating conditions are frequently made in the control room to adjust to changing conditions and demand for load. All components of the cycle will respond differently to these changes. Power plants are equipped with sensors throughout and feedback/control systems are vital to the practical functioning of a power plant. No further attempts to analyze transient behavior are made here
Stray heat effects	The external housing of all components is generally higher than the local ambient, and stray heat loss is never identically zero. Means to insulate key components is often made. Stray heat is not considered to significantly impact performance of a real power plant, and no further analysis is conducted here
Changes in potential energy	The inlet and exit into each device is generally at a different elevation. As workshop 10.1 shows, while these effects can be easily included in an analysis (assuming the physical layout is known), potential energy changes are indeed negligible in Rankine cycle-based power plants
Changes in kinetic energy	The kinetic energy associated with an inlet is usually different than the exit. It can be considered negligible if the difference is small compared to the difference in enthalpy. Generally, the size of the piping is made to be sufficiently large that the kinetic energy is not important. An analysis similar to that conducted for the evaporator in a refrigeration cycle could be conducted, but not done so here. Instead, a simple criterion for the ratio of the inlet to outlet diameter that negates any kinetic energy change is derived
Pump and turbine irreversibility	Pumps and turbines cannot be made in practice to completely avoid irreversibilities. This effect is the major source of deviation between an ideal and real power plant performance. The effect is considered in the analysis of this section by introducing isentropic efficiencies
Pressure drops in heat exchangers	Internal friction in all devices, especially heat exchangers, tends to give rise to a decreasing pressure in the direction of flow. Therefore, the boiler and condenser are not isobaric. This effect could be included by specifying lower exit pressures, but is not done in this analysis

Using mass conservation, namely, $\dot{m} = \mathrm{VA}/v$ this expression could be rearranged for the ratio of outlet to inlet diameter. Kinetic energy effects can be reduced by increasing the piping diameter. The change in specific volume can be rather large, depending on the device, and the diameter of piping is generally designed to keep the flow velocities nominally the same.

A simpler way is to set a design goal for which the velocity of inlet and outlet are equal. That is:

$$V_{\mathrm{out}} = V_{\mathrm{in}}$$

Introducing the mass flow rate expression:

Table 10.8 Inlet-specific volume and inlet to outlet area and diameter ratio required to make kinetic energy changes identically zero for the ideal Rankine cycle with reheat case study

Device	v_{in} (m^3/kg)	$(A_{out}/A_{in})_{design}$	$(D_{out}/D_{in})_{design}$
HP Turbine	0.0113	6.7	2.58
Reheater	0.0754	2.0	1.41
LP turbine	0.1490	152.2	12.34
Condenser	22.6846	0.0000443	0.0067
Pump	0.0010	1.0	1.0
Boiler	0.0010	11.2	3.35

$$\frac{\dot{m}v_{out}}{\pi D_{out}^2/4} = \frac{\dot{m}v_{in}}{\pi D_{in}^2/4}$$

Rearranging for the ratio of outlet to inlet diameter:

$$\left(\frac{D_{out}}{D_{in}}\right)_{design} = \sqrt{\frac{v_{out}}{v_{in}}}$$

Consider the reheat case (where all devices are single inlet, single outlet). Table 10.8 shows the inlet-specific volume of each device and the ratio of outlet to inlet diameters that result in identically zero kinetic energy change. The inlet diameter to the HP turbine is not specified, but it would be designed with the known mass flow rate and specific volume to give a "manageable" inlet velocity. Then, an expansion in area of 6.7 (2.58 expansion in diameter) is required in the turbine. A moderate change in the reheater, and a very large change in the LP turbine. The diameter in the condenser requires a very large decrease in area.

In practice, it is not necessary to design the piping to give identical flow velocities, but the general trends of that design criterion are generally adhered to. Workshop 10.1 (with a turbine with 2 extraction points) showed a case where there is area expansion, but the exit velocity increased with each exit stream somewhat, but the kinetic energy change, while not identically zero, does not play a significant role thermodynamically.

Component Isentropic Efficiencies

The only real effect that is included in this case study is the irreversibilities associated with the pumps and turbines. While difficult to determine from first principles, the use of component efficiencies (a form of 2nd Law efficiency, as was done for the compressor in the refrigeration cycle and for the compressor and turbine in the gas turbines) allows real effects to be quantified. The performance of a

INPUT PARAMETERS

W'_{net}	**1**	GW, net power output
T_C	**25**	oC, dead state temperature (sink)
T_H	**650**	oC, thermal reservoir temperature (source)
η_T	**0.85**	isentropic turbine efficiencies
η_C	**0.5**	isentropic pump efficiencies
c	**4.2**	kJ/kg/kg, H2O liquid specific heat

STATE TABLE

state	P (bar)	T (oC)	v (m^3/kg)	h (kJ/kg)	s (kJ/kg/K)	quality
a	303.76	**600**	0.011303	3446.8	6.2316	NA, SHV
bs	26.73	227.4	0.075363	2802.3	6.2316	**1.0**
b	26.73	257.0	0.08252	2899.02	6.4142	NA, SHV
cs	26.73	35	0.001006	149.3	0.5051	NA, CL
c	26.73	35.64	0.001006	152.0	0.5138	NA, CL
d	26.73	227.4	0.001204	977.9	2.5857	**0.0**
1	26.73	**600**	0.149014	3686.8	7.5670	NA, SHV
2s	0.0563	35	22.684601	2322.8	7.5670	**0.9**
2	0.0563	35	24.817291	2527.40	8.2310	**0.984617**
3	0.0563	**35**	0.001006	146.6	0.5051	**0.0**
4s	303.76	227.4	0.001204	1005.8	2.5857	NA, CL
4	303.76	234.0	0.001204	1033.6	2.6410	NA, CL

COLOR CODE
Input Parameter
Process Specification
Component Efficiency
quality calculation
Compressed Liquid Approx
Entry from tables

PROCESS TABLE (all in MW)

process	$m'(h_{out}-h_{in})$	Q'_{in}	W'_{out}	$m'T_c(s_{out}-s_{in})$	$T_c(Q'_{in}/T_b)$	$T_c{*}\Sigma'$
a-b, HP turbine	-421.7	0	421.7	41.9	0	41.9
b-1, reheater	424.1	424.1	0	184.9	136.9	48.0
1-2, LP turbine	-624.1	0	624.1	106.5	0	106.5
2-3, condenser	-1281.6	-1281.6	0	-1239.4	-1281.6	42.2
3-c, LP pump	2.89	0	-2.89	1.3960	0	1.4
d-4, HP pump	42.90	0	-42.90	12.6920	0	12.7
a,c,d; FWH	NA	0	0	NA	0	68.3
4-a, boiler	1857.5	1857.5	0	823.6	599.7	223.9
CYCLE	NA	1000	1000	NA	-545.0	545.0

COLOR CODE
From State Table
Process Specification
From Energy Balance
From Entropy Balance

RESULTS

fext	0.3007	fraction of mass extracted after turbine 1
m'boiler	769.8	kg/sec, mass flow rate through boiler
m'condenser	538.3	kg/sec, mass flow rate through condenser
η_I	0.438	1st Law Efficiency, defined as NET WORK over HEAT IN BOILER
η_{CARNOT}	0.677	Carnot efficiency, for heat engine operating between source and sink
η_{II}	0.647	2nd Law Efficiency, defined as ACTUAL over CARNOT 1st law efficency

Fig. 10.18 Spreadsheet solution for the nonideal Rankine cycle with reheat and regeneration. Hypothetic states (subscripts "s") are introduced to facilitate the solution

real device is compared to the theoretical best performance of a device that has the same inlet conditions, and the same exit pressure. The theoretical best performance is determined by considering an adiabatic, reversible (hence isentropic) process, and a hypothetical state with subscript "s" is defined. In this case study, the isentropic states have already been determined.

The isentropic efficiency of a turbine is therefore the actual power extracted divided by the theoretical power from a reversible turbine:

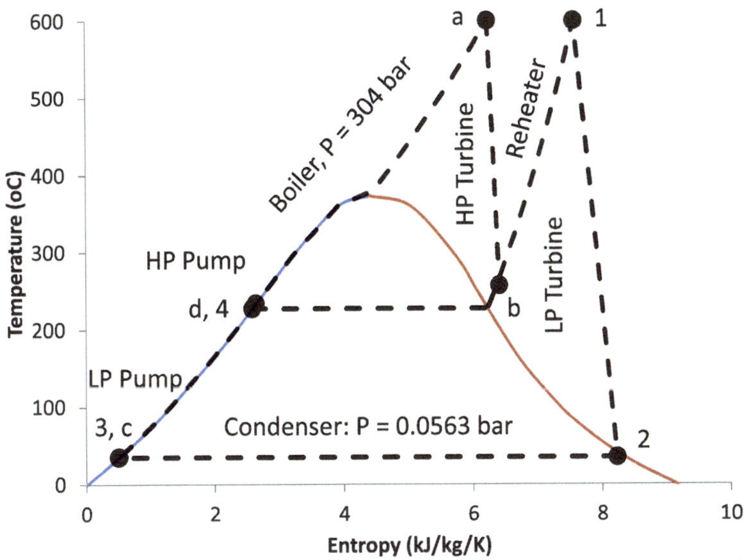

Fig. 10.19 Ts diagram of Rankine cycle with reheat and regeneration and component efficiencies

$$\eta_{\text{turbine}} = \frac{\dot{W}_{\text{actual}}}{\dot{W}_{\text{reversible}}} = \frac{h_{\text{in}} - h_{\text{out}}}{h_{\text{in}} - h_{\text{out,s}}}$$

The isentropic efficiency of a pump is the ratio of the theoretical work (the minimum input required to raise the pressure) to the actual work:

$$\eta_{\text{pump}} = \frac{\dot{W}_{\text{reversible}}}{\dot{W}_{\text{actual}}} = \frac{h_{\text{out,s}} - h_{\text{in}}}{h_{\text{out}} - h_{\text{in}}}$$

The results of this case study are shown in Fig. 10.18. The *T-s* diagram is shown in Fig. 10.19. An isentropic efficiency of 0.85 is assumed for both turbines (a typical value) and a much lower isentropic efficiency of 0.5 is assumed for the pumps. This low value is used to demonstrate the point that even a pump with a low 2nd Law efficiency has a minor impact on the overall thermodynamics of the power plant.

Another reason for using an unrealistically low value for the pump is to re-emphasize the point about the compressed liquid region on a Ts diagram (Fig. 10.7). Since the pumps are not isentropic, there is a difference on *T-s* coordinates for states 3 and c (LP pump) and a larger difference for states d and 4 (HP pump). On the Ts plot, there is no distinction between states 3 and c, and a small but noticeable difference in the points for states 3 and 4. Furthermore, these points are all on the saturated liquid line (although the outlet states are compressed liquids).

SUPERHEATED VAPOR DATA

HP Turbine Exit

	R (kJ/kg/K)	0.462	gas constant (8.314/MW)	
	P_0 (MPa)	2.5		
	T (°C)	v (m³/kg)	h (kJ/kg)	s (kJ/kg/K)
T_0	250	0.08705	2880.9	6.4107
T_1	300	--	3009.6	--

COLOR CODE

Superheat Table Data
Values from Problem
Ideal Gas Interpolation

IDEAL GAS INTERPOLATION

INPUT

h (kJ/kg)	2899.0167
P (MPa)	2.672738

INTERPOLATED

cp (kJ/kg/K)	2.574
T (oC)	257.038
v (m³/kg)	0.0825
s (kJ/kg/K)	6.4142

Fig. 10.20 Ideal gas interpolation for HP turbine exit with isentropic efficiency

Calculation of state b requires an interpolation in the superheat tables that is different than previously encountered. Figure 10.20 shows a spreadsheet template modified to conduct an ideal gas interpolation for the case where the pressure and enthalpy are specified.

Workshop 10.1: Steam Turbine with Extraction

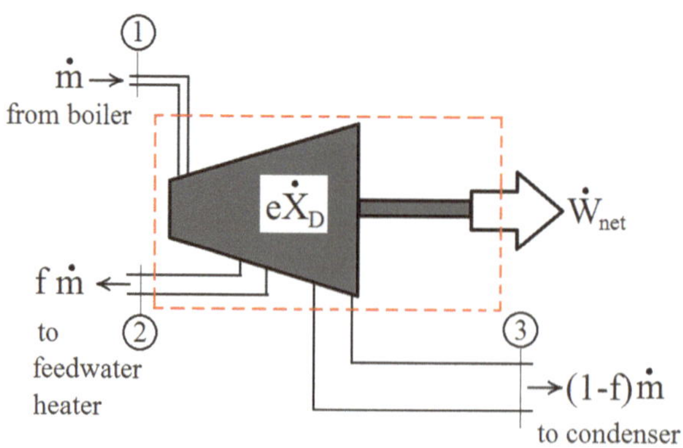

The main goals of this workshop are to review the analysis of open systems (with multiple streams) and property evaluation, and to demonstrate that the effects of kinetic and potential energy changes are generally of secondary importance in a device like a turbine. The problem is stated with unreasonably high values of elevation and reasonable values for velocity.

Steam steadily enters an adiabatic turbine (state 1) with a specified elevation, velocity, pipe diameter, temperature and pressure. A fraction (f) of the stream is extracted at an intermediate state and sent to a feedwater heater. The remainder of the stream continues to expand through the turbine and is sent to a condenser. Output power is extracted from a single shaft. Conditions are specified in the State table. Complete the State and Results tables and sketch the process on a *T-s* diagram (Fig. 10.2). Use the control volume indicated with the dashed line.

The following spreadsheet templates contain superheated vapor data (from an on-line source) for conditions near states 1 and 2 (which require 2-point interpolations). The ideal gas interpolation method can be used to quickly determine these states (Fig. 10.5). State 3 saturation property data can be found in Table 10.4. Color coding: yellow is superheat data, orange is values from problem statement, green is ideal gas interpolation.

SUPERHEATED VAPOR DATA

STATE 1

	R (kJ/kg/K)	0.462	gas constant	
	P_0 (MPa)	20		
	T (°C)	v (m³/kg)	h (kJ/kg)	s (kJ/kg/K)
T_0	550	0.016571	3396.2	6.339
T_1	600	--	3539.0	--

IDEAL GAS INTERPOLATION

INPUT

P (MPa)	
T (°C)	

INTERPOLATED

cp (kJ/kg/K)	
s(kJ/kg/K)	
v (m³/kg)	
h (kJ/kg)	

SUPERHEATED VAPOR DATA

STATE 2

	R (kJ/kg/K)	0.462	gas constant	
	P_0 (MPa)	3		
	T (°C)	v (m³/kg)	h (kJ/kg)	s (kJ/kg/K)
T_0	350	0.09056	3116.1	6.745
T_1	400	--	3231.7	--

IDEAL GAS INTERPOLATION

INPUT

T (oC)	
s (kJ/kg/K)	

INTERPOLATED

cp (kJ/kg/K)	
P (Mpa)	
v (m³/kg)	
h (kJ/kg)	

INPUT PARAMETERS

Tinf	**25**	oC, dead state temperature
f	**0.21**	fraction of mass extracted at state 2
D1	**1**	m, inlet pipe diameter
D2	**1.5**	m, pipe diameter of extracted stream
D3	**12**	m, exit pipe diameter

STATE TABLE

State	P (bar)	T (oC)	v (m^3/kg)	h (kJ/kg)	s(kJ/kg/K)	quality	V (m/s)	z (m)	V^2/2 (kJ/kg)	gz (kJ/kg)
1	**189**	**574**					**20**	**100**		
2		**342**			**6.8000**			**50**		
3		**35**				**0.9**		**0**		

RESULTS

m'1		kg/s, mass flow rate into turbine
W'net		MW, net power extracted from turbine
W'net, neglecting PE		MW, net power neglecting PE
%difference		Error in neglecting PE
W'net, neglecting KE		MW, net power neglecting KE and PE
%difference		Error in neglecting KE
eXd'		MW, Total eXergy destruction rate
2nd Law Efficiency		W'/(W'+eX'd)

Workshop 10.2: Effect of Boiler Pressure on Ideal Rankine Cycle

The four states and processes of the idealized Rankine cycle are shown schematically, with H_2O being the working fluid. Superheated vapor at elevated pressure (state 1) enters an adiabatic, reversible (hence isentropic) steam turbine. Each student or student team is assigned a different turbine inlet pressure. The turbine exit (state 2, fixed for all cases at 50 kPa with a quality of 0.95) enters a condenser where heat is rejected at constant pressure to a thermal reservoir at the dead state temperature (25 °C). Saturated liquid (state 3) enters an adiabatic, reversible pump and is compressed to high pressure (state 4). The working fluid enters a boiler and is heated at constant pressure from a thermal reservoir at temperature T_1 to complete the cycle.

Complete the property, state and results tables and sketch on the *T-s* diagram provided (use the one most suited to your case). Assume steady-state operation and neglect kinetic and potential energy changes and stray heat. The mass flow rate through the cycle is 200 kg/sec. Results can be compiled to observe the effect of boiler pressure on various quantities.

INPUT PARAMETERS

mdot	200	kg/sec, mass flow rate
T_{inf}	25	oC, dead state temperature

STATE TABLE

state	P (MPa)	T (oC)	h (kJ/kg)	s (kJ/kg/K)	quality
1	22.064	915.47	4358.21	7.268005	NA, SV
2	**0.05**	81.32	2530.01	7.268005	**0.95**
3	**0.05**	81.32	340.54	1.0912	**0**
4	22.064	81.32	363.21	1.0912	NA, CL

PROCESS TABLE (all in MW)

process	$m'*(h_{out}-h_{in})$	Q'_{in}	W'_{out}	$T_{inf}*m'*(s_{out}-s_{in})$	$T_{inf}*(Q'_{in}/T_b)$	$T_{inf}*\Sigma'$
1-2	-365.6	0.0	365.6	0.0	0.0	0.0
2-3	-437.9	-437.9	0.0	-368.3	-437.9	69.6
3-4	4.5	0.0	-4.5	0.0	0.0	0.0
4-1	799.0	799.0	0.0	368.3	200.4	167.9
net	0.0	361.1	361.1	0.0	-237.5	237.5

RESULTS

A	0.4519	1st Law Efficiency, defined as NET WORK over HEAT IN BOILER
η_{CARNOT}	0.7492	Carnot efficiency, operating between source and sink
η_{II}	0.6033	2nd Law Efficiency, ACTUAL over CARNOT 1st law efficency

Fig. 10.21 Results of Workshop 10.2 for an assigned pressure of 22.064 MPa (the critical pressure)

Results for a case study with the pressure set to the critical pressure are listed at the bottom of the pressure assignment worksheet and details are shown in Fig. 10.21. This case requires a 2-point interpolation or ideal gas interpolation for state 1. Figure 10.22 shows a spreadsheet detail for both methods. Figure 10.23 shows a *T-s* plot of both methods.

2 point interpolation for state 1

TABLE			INTERPOLATION			TABLE		
P	20	MPa	P	22.064	MPa	P	25	MPa
T	h	s	T	h	s	T	h	s
900	4325.4	7.2829	900	4317.84576	7.23497392	900	4307.1	7.1668
			915.4746188	4358.214188	7.268			
1000	4584.7	7.495	1000	4578.7144	7.44839488	1000	4570.2	7.3821

IDEAL GAS INTERPOLATION FOR STATE 1

	R (kJ/kg/K)	0.4619	gas constant (8.314/MW)	
	P_0 (MPa)	20		
	T (°C)	v (m^3/kg)	h (kJ/kg)	s (kJ/kg/K)
T_0	900	--	4325.4	7.2829
T_1	1000	--	4584.7	--

COLOR CODE

- SUPERHEAT TABLE
- INPUT
- LINEAR INTERPOLATION
- 2-POINT INTERPOLATION
- IDEAL GAS INTERPOLATION

IDEAL GAS INTERPOLATION

INPUT

s (kJ/kg/K)	7.2680
P (MPa)	22.064

INTERPOLATED

cp (kJ/kg/K)	2.593
T (oC)	913.863
v (m^3/kg)	--
h (kJ/kg)	4361.3

Fig. 10.22 Comparison of 2 point Interpolation and Ideal Gas interpolation for state 1 (turbine inlet) for an assigned pressure of 22.064 MPa (the critical pressure of H_2O). The top portion shows a spreadsheet approach to conducting 2 point interpolations. The bottom portion shows the ideal gas interpolation method

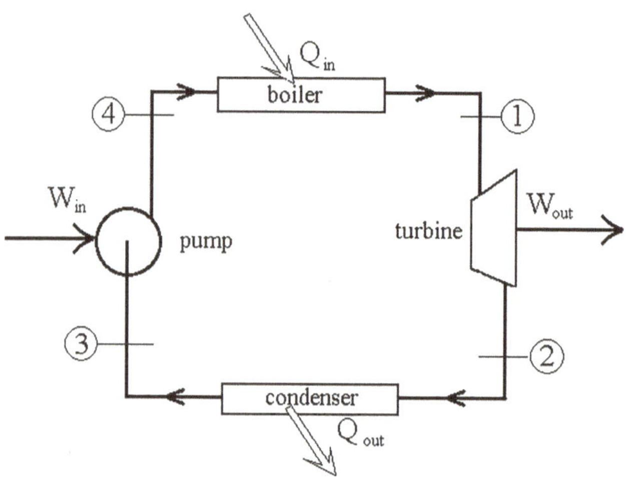

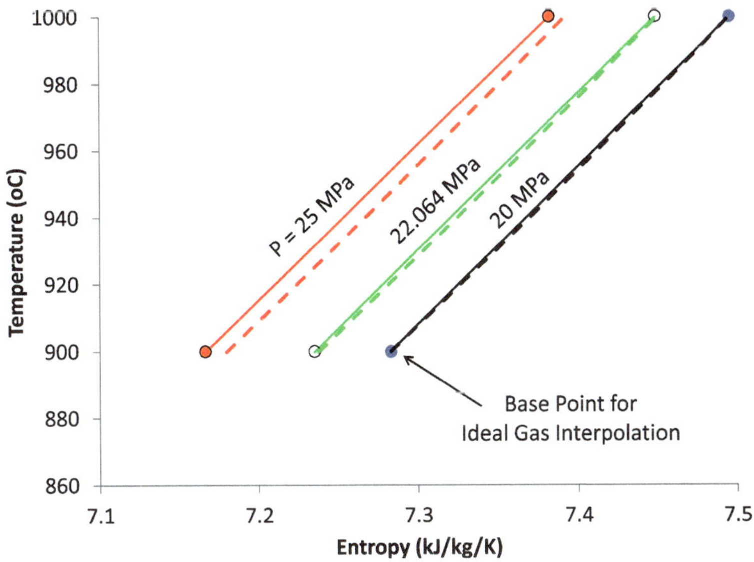

Fig. 10.23 *T-s* detail of interpolation methods near state 1. Filled symbols are superheated table data. Open symbols are for linear interpolation for the target pressure of 22.064 MPa. Solid lines represent linear interpolation. Dashed lines represent ideal gas interpolations with the base point indicated (20 MPa, 900 °C)

INPUT PARAMETERS

mdot	200	kg/sec, mass flow rate
T_{inf}	25	oC, dead state temperature

STATE TABLE

state	P (MPa)	T (oC)	h (kJ/kg)	s (kJ/kg/K)	quality
1					
2	**0.05**				**0.95**
3	**0.05**				**0**
4					

PROCESS TABLE (all in MW)

process	$m'*(h_{out}-h_{in})$	Q'_{in}	W'_{out}	$T_{inf}*m'*(s_{out}-s_{in})$	$T_{inf}*(Q'_{in}/T_b)$	$T_{inf}*\Sigma'$
1-2						
2-3						
3-4						
4-1						
net						

RESULTS

η_1		1st Law Efficiency, defined as NET WORK over HEAT IN BOILER
η_{CARNOT}		Carnot efficiency, for heat engine operating between source and sink
η_{II}		2nd Law Efficiency, defined as ACTUAL over CARNOT 1st law efficency

Case	High Pressure (MPa)	Turbine Power (MW)	Pump Power (MW)	Heat Addition (MW)	Heat Rejection (MW)	Boiler eXergy Destruction (MW)	Condenser eXergy Destruction (MW)	1st Law Efficiency	Max. Theoretical Efficiency	2nd Law Efficiency	student
1	0.1										
2	0.2										
3	0.3										
4	0.4										
5	0.5										
6	0.6										
7	0.8										
8	1										
9	1.2										
10	1.4										
11	1.6										
12	1.8										
13	2										
14	2.5										
15	3										
16	3.5										
17	4										
18	4.5										
19	5										
20	6										
21	7										
22	8										
23	9										
24	10										
25	12.5										
26	15										
27	17.5										
28	20										
	22.064	360.3	4.54	793.8	438.04	167.9	69.57	0.448	0.699	0.641	Professor

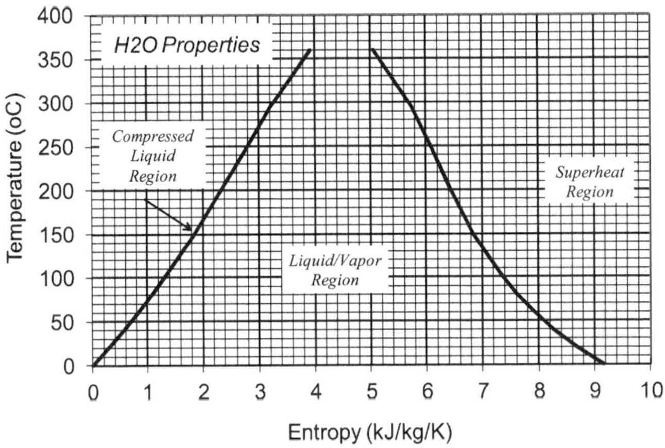

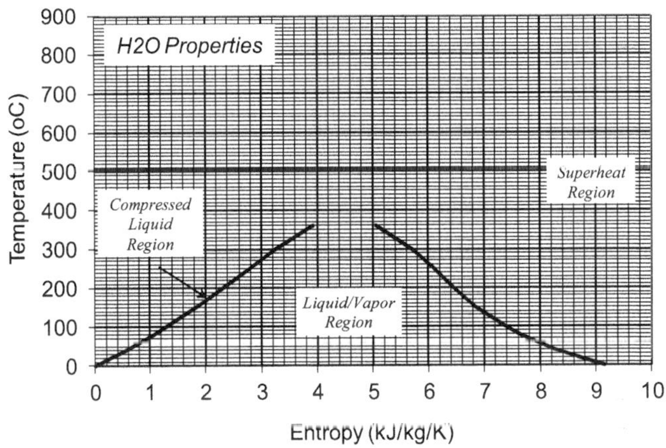

Appendix 1: Gas and Steam-Based Power Plants: Practical Considerations

Steam turbine cycles (Rankine-based power plant) and gas turbines (Brayton-based power plant) have much in common, and several important differences. The overall objective of both is to produce output mechanical power from a thermal source. The terminology is a little different, however. The term "steam turbine cycle" is used here, and is often just called a Rankine cycle. The term "steam turbine" would just refer to the turbine itself. The term "gas turbine" in practice is used to mean the entire Brayton-based engine (compressor, combustor, and turbine) and not just the turbine itself. The term "gas generator" applies to gas turbines designed for propulsion where the purpose of the turbine is to drive the compressor and deliver a high pressure gas to a nozzle in a mobile power plant. Aero-derivative gas turbines repurpose the engine to stationary power production by driving a secondary turbine to produce net power.

A mechanical device is used to raise the pressure in both gas turbines and steam turbine cycles. A compressor (which acts on a gas) is used in a gas turbine and a pump (which acts on a liquid) is used in a steam turbine cycle. The steady-state energy balance on such a device with an inlet and outlet (neglecting stray heat and kinetic and potential energy changes) is:

$$\frac{\dot{W}_{out}}{\dot{m}} = w_{out} = h_{inlet} - h_{outlet} = \int_{outlet}^{inlet} \mathrm{dh} = \int_{outlet}^{inlet} (\mathrm{Tds} + \mathrm{vdP})$$

The design goal of a pump, compressor or turbine is to approach adiabatic, reversible (hence isentropic) compression or expansion of the working fluid as it flows from inlet to outlet. That is, the isentropic work from an open-system device is given by:

$$w_{out,s} = \int_{outlet}^{inlet} vdP = \overline{v}(P_{inlet} - P_{outlet})$$

That is, the work for an open system device is *vdP* or "Vee Dee Pee" work (as opposed to *Pdv* or "Pee Dee Vee" work in a closed system, where the change in internal energy, not enthalpy, is the key thermodynamic property change).

The average specific volume is defined in this expression so show that the work required to compress or expand a working fluid across a given pressure ratio is directly related to the average specific volume of the working fluid as it passes from inlet to outlet. Rough estimates are:

$$\mathrm{BWR}_{\mathrm{RANKINE}} = \frac{\overline{v}_{compression}}{\overline{v}_{expansion}} = \frac{v_{liquid}}{v_{steam}} \frac{0.001 \ \tilde{}\ \mathrm{m^3/kg}}{1\mathrm{m^3/kg}} = 0.001$$

$$\mathrm{BWR}_{\mathrm{BRAYTON}} = \frac{\overline{v}_{compressor}}{\overline{v}_{turbine}} \frac{R\overline{T}_{compression}/\overline{P}\ \tilde{500}\,\mathrm{K}}{R\overline{T}_{compression}/\overline{P}1500\,\mathrm{K}} = 0.33$$

A main advantage, then, of the Rankine cycle is that the compression to high pressure is done on a liquid, with a low specific volume. Therefore, the required input work is very low compared to the work that is extracted by the turbine. For the Brayton cycle, a large fraction of the output work of the turbine is needed to drive the compression process.

Consider next the restrictions on the operating pressures. In both applications, the turbine, the key component, is designed to produce the desired mechanical power output. The objective of the other components is to continuously maintain a pressure difference across the turbine. That is, to supply the high pressure gases to it and maintain a low pressure to which it exits. The other components, then, are designed to maximize the turbine inlet pressure and minimize the turbine exit pressure. The high pressure of the steam turbine cycle is limited primarily by economic trade-off between the capital and operating and maintenance costs associated with containing high pressure fluid (thick-walled components throughout) and the operating cost associated with the heating source in the boiler. Both the power output (per unit mass flow) and thermal efficiency increase with pressure ratio. Workshop 10.2 outlines a means for seeing this trend as well as giving students practice in basic thermodynamic calculations. Historically, pressures exceeding the critical pressure (220.6 bar) were difficult to justify economically, but more and more power plants with higher pressures (called supercritical power plants) are being built with increased experience and improvements in material properties. In the analysis of this chapter, the high side pressure was fixed by other restrictions and not considered to be limited by any material constraint.

The low side pressure of a steam turbine cycle is limited by the operational requirement of the condenser. That is, the temperature of the working fluid in the condenser must be higher than the low temperature reservoir (the local ambient temperature) in order for heat to flow. The corresponding saturation pressure for nominal room temperature (25 °C) is 0.0317 bar, which is a rather high vacuum. In practice, the temperature must be somewhat higher (approximately 10 °C) and condenser pressure of approximately 0.05 bar is a reasonable lower limit. Such low pressures can be maintained because the Rankine cycle is closed, but requires careful design to prevent ambient air from leaking in.

In contrast, the low pressure of a gas turbine is fixed by the local ambient pressure from which intake air is drawn and to which the turbine must discharge. The high side pressure is limited in large part because of the temperature rise associated with compressing a gas. After taking into account reasonable isentropic efficiencies of the compressor and turbine and the limits on the turbine inlet temperature (discussed next), the performance of a gas turbine degrades for pressure ratios exceeding approximately 30.

There are practical limits to the turbine inlet temperature for both steam turbine cycles and gas turbines that result from material constraints on the containment vessel, but for very different reasons that stem from the difference between internal and external combustion processes. The temperature of the working fluid in an internal combustion engine (like a gas turbine or reciprocating engine) is higher than the physical vessel that contains it, and can be significantly higher as the containment vessel can be cooled from the outside. A practical upper limit to turbine inlet temperatures in gas turbines is approximately 1600 °C (1873 K). On the other hand, in external combustion engines, combustion creates high temperature gases and heat is transferred through the walls of the containment vessel and then to the working fluid. That is, the temperature of the working fluid must be lower than the temperature of its containment vessel in order for heat to flow into it. A practical limit on the temperature of the high pressure working fluid in a boiler is approximately 600 °C (873 K), much lower than that of gas turbines.

There is an important constraint on the turbine for the steam turbine cycle that does not exist for the gas turbine. As superheated steam undergoes the (nearly) isentropic expansion in the turbine, the temperature drops with the pressure, and the steam approaches the liquid/vapor dome. This trend is clear by following a vertical path from high to low pressure on a *T*-*s* or *h*-*s* diagram (Fig. 10.2), starting anywhere in the superheated region. This phenomenon has an important practical consequence. If the working fluid enters the liquid/vapor region, condensation can begin and result in a fine mist of liquid droplets. These droplets impinge on the high speed turbine blades and causes mechanical erosion. In practice, the turbine can tolerate a certain degree of condensation, and also, the working fluid may not maintain thermodynamic equilibrium and become supersaturated due to its rapid expansion. Therefore, a practical lower limit on the steam quality can be tolerated, with a value of approximately 0.9.

A final constraint on steam turbine cycles is the practical requirement that the pump requires fully condensed liquid to operate reliably.

Appendix 2: Compressed Liquid Interpolation

As discussed multiple times, saturation tables can be used to determine all properties of a compressed liquid using the compressed liquid approximations. That is, compressed liquid tables are usually unnecessary. Nevertheless, this Appendix demonstrates a simple way to conduct interpolations with compressed liquid tables to obtain more precise values when warranted. 2 point interpolation (or extrapolation) is often needed, so an interpolation method that uses compressed liquid approximation formulas from compressed liquid tables is described. The method is similar to the ideal gas interpolation method used for superheated vapors.

Several states from the ideal Rankine cycle case study are evaluated from compressed liquid data. First, for the basic ideal cycle, state 4 has a specified entropy (0.5051 kJ/kg/K) and pressure (26.73 bar). The closest pressure available in compressed liquid tables consulted is 50 bar = 5 MPa. A 2-point interpolation is not possible without data at a pressure below the actual pressure. However, using the nearest available data entry (5 MPa) and compressed liquid approximations expanded about that point as detailed in Fig. 10.24 yields accurate results. Notice that the temperature does indeed rise, but only by 0.122 °C (from 35 °C). A second case, state 4 for the reheat cycle, is shown in Fig. 10.25. This state has the same entropy but a much higher temperature (303 bar). The temperature rise in this case is 0.778 °C.

COMPRESSED LIQUID DATA

BASIC CYCLE STATE 4

	P_0 (MPa)	5		
	T (°C)	v (m³/kg)	h (kJ/kg)	s (kJ/kg/K)
T_0	20	0.0009996	88.61	0.2954
T_1	40	0.0010057	171.95	--

COLOR CODE

Compressed Liquid Table Data
Values from Problem
Compressed Liquid Interpolation

COMPRESSED LIQUID INTERPOLATION

INPUT

s (kJ/kg/K)	0.5051
P (MPa)	2.673

INTERPOLATED

c (kJ/kg/K)	4.167
T (oC)	35.122
v (m³/kg)	0.0010103
h (kJ/kg)	149.27

$$c = \frac{h_1 - h_0}{T_1 - T_0}$$

$$T = (T_0 + 273)e^{\frac{s - s_0}{c}} - 273$$

$$v = v_0 + \frac{v_1 - v_0}{T_1 - T_0}(T - T_0)$$

$$h = h_0 + c(T - T_0) + v_0(P - P_0)$$

Fig. 10.24 Compressed liquid interpolation using compressed liquid approximation expanded about compressed liquid table data

Fig. 10.25 Compressed liquid interpolation for state 4 of reheat cycle

COMPRESSED LIQUID DATA

REHEAT CYCLE STATE 4

P_0 (MPa)	30		
T (°C)	v (m³/kg)	h (kJ/kg)	s (kJ/kg/K)
20	0.0009886	111.77	0.2897
40	0.0009951	193.9	--

COMPRESSED LIQUID INTERPOLATION

INPUT

s (kJ/kg/K)	0.5051
P (MPa)	30.376

INTERPOLATED

c (kJ/kg/K)	4.107
T (oC)	35.779
v (m³/kg)	0.0010002
h (kJ/kg)	176.94

Chapter 11
Combined Cycles

Combined cycles are the standard-bearer for modern natural gas-fired power plants. Two conventional thermodynamic heat engines (a topping cycle and a bottoming cycle) are combined in a manner that addresses thermodynamic shortcomings of both. An internal combustion engine (usually a gas turbine for power plants) serves as the topping cycle and a steam turbine (Rankine-type) cycle as the bottoming cycle. High pressure steam is generated for the Rankine cycle from the hot exhaust of the topping cycle in a heat exchanger called a Heat Recovery Steam Generator (HRSG). There is an optimal operating pressure on the H_2O side of the HRSG for thermodynamic performance, unlike a stand-alone Rankine cycle (with a high temperature external combustion source) where the thermodynamic performance increases monotonically with boiler pressure. In this chapter, thermodynamic performance of a HRSG/steam turbine combination is investigated using a pinch point analysis that demonstrates how the saturation temperature, hence H_2O-side pressure, is related to the available inlet gas temperature from the topping cycle. The condenser is also analyzed as a heat exchanger, with cooling water that must ultimately exchange heat with the ambient through a cooling tower (studied more fully in the next chapter). Finally, a series of capstone workshops are conducted to simulate the performance of an industrial scale combined cycle power plant.

The topping cycle is an internal combustion engine and the bottoming cycle an external combustion engine for good reason. On the one hand, the working fluid of an internal combustion engine is higher than its containment vessel, and therefore, the equivalent heat addition takes place at a high temperature. However, the exhaust is at a considerably higher temperature than ambient. Being an open system by necessity, an internal combustion engine acting alone vents a high temperature (but nominal ambient pressure) gas to the environment. On the other hand, the working fluid of a Rankine cycle (a true closed cycle) is at a lower temperature in the boiler than the containment vessel. However, the temperature of the working fluid can be made to approach the ambient in the condenser by expanding in the turbine to a partial vacuum. Conceptually, then, a combined cycle power plant starts with a gas

G. Sidebotham, *An Inductive Approach to Engineering Thermodynamics*, Mechanical Engineering Series, https://doi.org/10.1007/978-3-030-20430-3_11

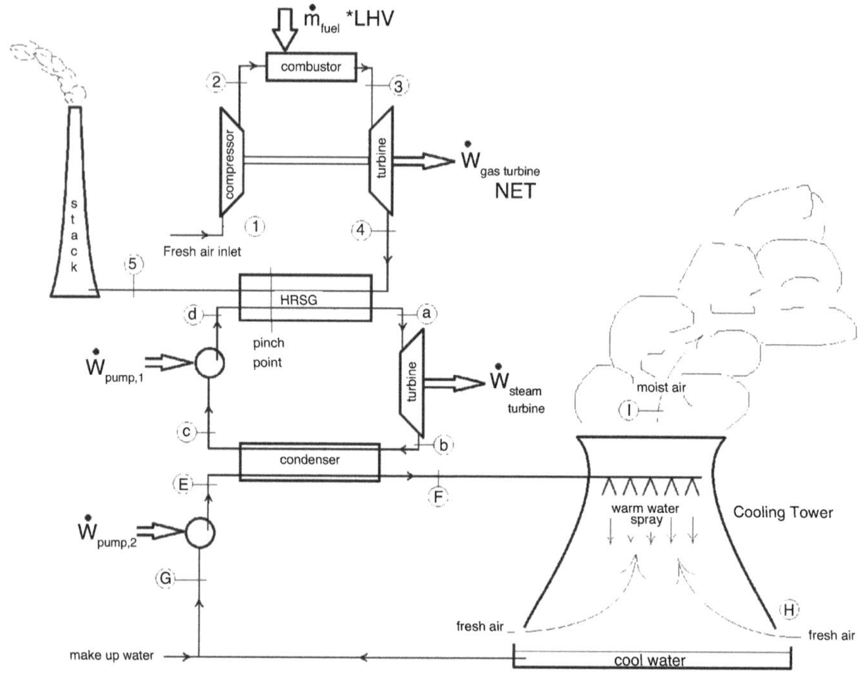

Fig. 11.1 Schematic of a combined cycle with gas turbine topping cycle and Rankine bottoming cycle

turbine as a stationary power plant, but rather than venting its exhaust, a cycle is designed to extract as much power as possible from its hot exhaust.

A schematic of a combined cycle is shown in Fig. 11.1 with a gas turbine as a topping cycle and a Rankine bottoming cycle. A single gas turbine is shown, but several gas turbines are commonly used in parallel to increase the total output. The exhaust of each gas turbine is combined into a single stream of hot gas (which has the properties of air, to a good engineering approximation) which serves as the heating source for the Rankine bottoming cycle. Heat is transferred internally in a heat recovery steam generator (HRSG) from the hot gas turbine exhaust to pressurized H_2O. The gas exit of the HRSG is warmer than ambient (and may be determined by the pinch point inside the HRSG, to be discussed) and is ultimately vented to the environment (with pollution control equipment at various points, not shown). The mass flow rate of H_2O produced for a given H_2O-side pressure is determined by the mass flow rate and entering temperature of the gas stream. The high pressure steam is fed to a steam turbine whose exit is maintained at a low pressure. The condenser is a heat exchanger between the condensing working fluid of the Rankine cycle and cooling water. Heat picked up by the cooling water is ultimately transferred to ambient air in a cooling tower where the warm water is sprayed directly in contact with ambient air. A portion of the cooling water is evaporated, which has an energy consequence, but cannot be analyzed fully with thermodynamics of pure substances (constant chemical composition). An approximate overall analysis will be conducted in this chapter, however, neglecting evaporation. Evaporative effects will be formally considered in Chap. 12.

Maximum Possible Power of Bottoming Cycle

The maximum possible power that can be produced from a hot exhaust stream can be determined by viewing the bottoming cycle as an open system on the hot side (hot exhaust in, warm exhaust out), and with mechanical power extracted and heat rejected to the local ambient (dead state). Figure 11.2 shows a schematic with energy terms (with ... representing kinetic and potential energy), and Fig. 11.3 shows entropy terms.

Steady-state energy and entropy balances, neglecting kinetic and potential energy are:

$$0 = -\dot{Q}_C - \dot{W}_{\text{bottom}} + \dot{m}_{\text{exh}}(h_{\text{exh}} - h_{\text{stack}})$$

$$0 = \frac{-\dot{Q}_C}{T_\infty} + \underset{\text{bottom}}{\overset{\dots}{\Sigma}} + \dot{m}_{\text{exh}}(s_{\text{exh}} - s_{\text{stack}})$$

Combining energy and entropy balances, by eliminating the heat rejection:

$$\dot{W}_{\text{bottom}} = \dot{m}_{\text{exh}}[(h_{\text{exh}} - T_\infty s_{\text{exh}}) - (h_{\text{stack}} - T_\infty s_{\text{stack}})] - e\underset{\text{D,bottom}}{\overset{\dots}{X}}$$

The thermodynamic pseudo-property $h - T_\infty s$ is the eXergy (open system form), so the maximum possible mechanical power extracted from the bottoming cycle

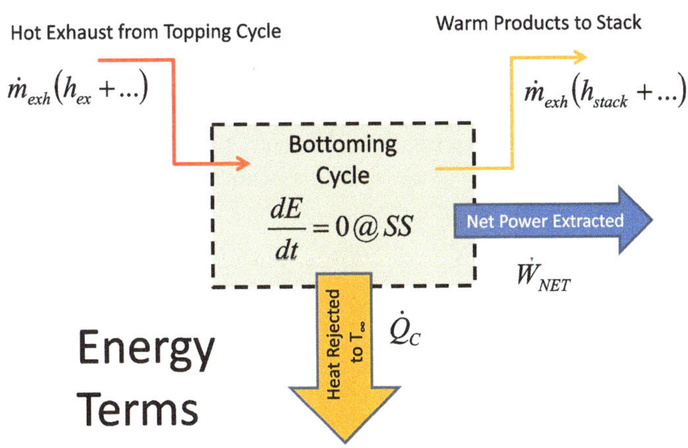

Fig. 11.2 Control volume of bottoming cycle, with energy terms

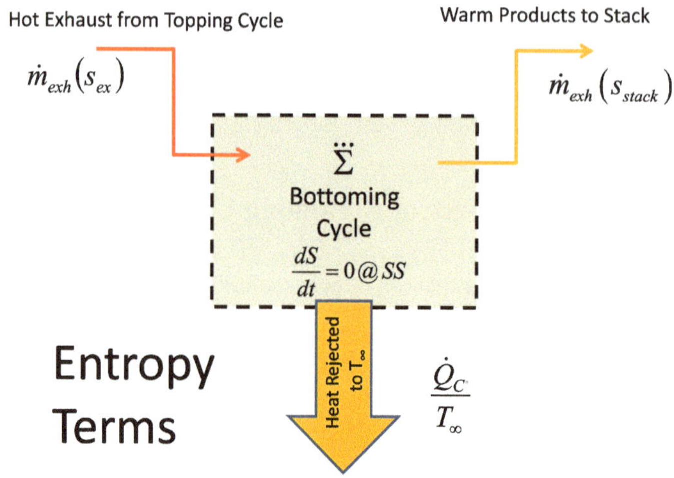

Fig. 11.3 Control volume of bottoming cycle, with entropy terms

equals the mass flow rate of exhaust times its change in eXergy. Any eXergy destruction inside the control volume reduces the maximum power extracted.

Since the exhaust will be a gas for the combined cycles investigated here, the enthalpy and entropy of the exhaust and stack gases can be related to their temperature and pressure, and the mechanical power extracted by the bottoming cycle per unit mass of the topping cycle flow rate:

$$\frac{\dot{W}_{\text{bottom}}}{\dot{m}_{\text{exh}}} = c_{\text{p}}\left[(T_{\text{exh}} - T_{\text{stack}}) - T_{\infty}\ln\left(\frac{T_{\text{exh}}}{T_{\text{stack}}}\right)\right.$$
$$\left.+\frac{RT_{\infty}}{c_{\text{p}}}\ln\left(\frac{P_{\text{exh}}}{P_{\text{stack}}}\right)\right] - e\ \dddot{X}_{\text{D,bottom}}$$

The maximum bottom cycle power divided by the exhaust mass flow rate occurs when there is no eXergy destruction and is plotted in Figs. 11.4 and 11.5 (a detail) as a function of stack temperature for several exhaust temperatures. The exhaust and stack pressures are equal in these cases. The maximum possible power that can be extracted increases strongly with exhaust temperature, not surprisingly. It also decreases with stack temperature, with elevated stack temperatures representing thermal energy vented to the environment. However, especially for higher exhaust temperatures typical of internal combustion engine exhaust, there is minimal penalty for stack temperatures that are somewhat warm.

Stack temperatures below the dead state temperature were deliberately plotted in Figs. 11.4 and 11.5 to make a point about eXergy calculations. It would be possible in theory for the exit of the exhaust to be below ambient, but only if an internal heat pump system were employed to do so, and there would be an eXergy cost (and

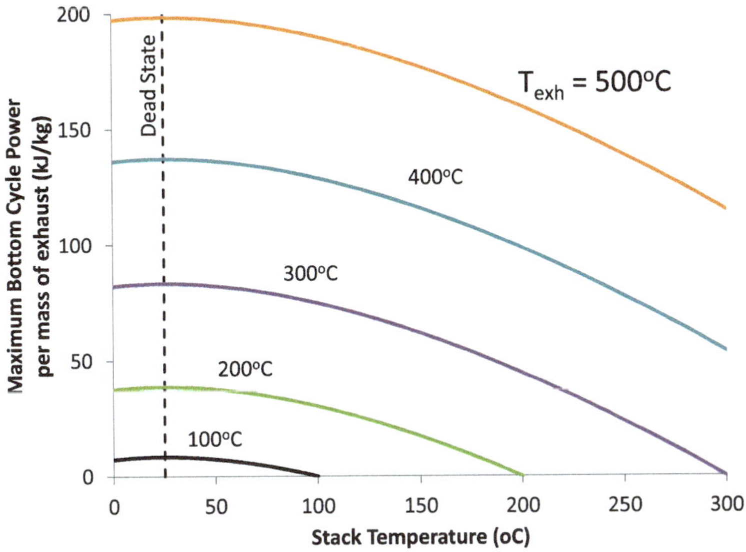

Fig. 11.4 Maximum bottom cycle power per mass flow rate of exhaust as a function of stack temperature for several topping cycle exhaust temperatures. The dead state temperature is 25 °C and the pressure does not change across the bottoming cycle. The enthalpy-specific heat is set to $c_p = 1.04$ kJ/kg/K and $R = 0.287$ kJ/kg/K

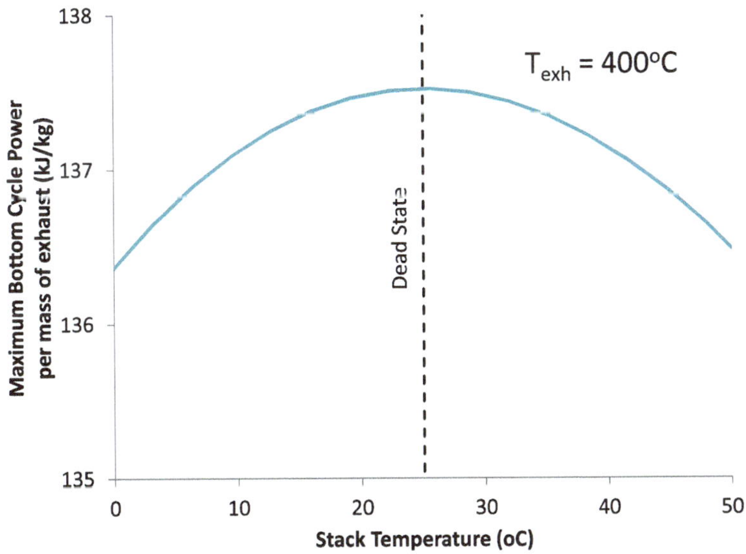

Fig. 11.5 Detail of previous plot demonstrating a peak when the stack temperature equals the dead state

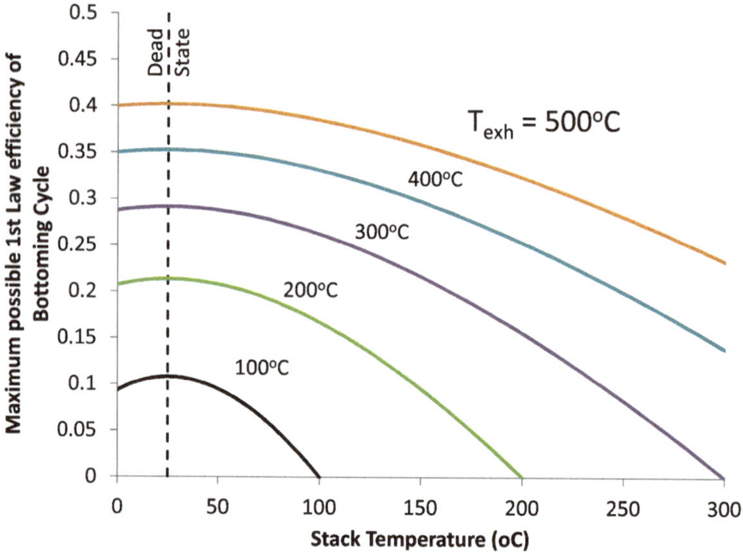

Fig. 11.6 Maximum possible 1st Law efficiency of bottoming cycle as a function of stack temperature. There is no pressure change across the exhaust stream

substantial capital cost) associated with that. So the best possible performance is obtained if the ultimate exhaust were delivered to the stack at the dead state temperature (and pressure).

A 1st Law efficiency in this case is defined as the net power output divided by the rate of enthalpy change of the exhaust gas stream:

$$\frac{\dot{W}_{\text{bottom}}}{\dot{m}_{\text{exh}} c_{\text{p}} (T_{\text{exh}} - T_{\infty})} = \left[\frac{(T_{\text{exh}} - T_{\text{stack}})}{(T_{\text{exh}} - T_{\infty})} - \frac{T_{\infty}}{(T_{\text{exh}} - T_{\infty})} \ln \left(\frac{T_{\text{exh}}}{T_{\text{stack}}} \right) + \frac{R T_{\infty}}{c_{\text{p}} (T_{\text{exh}} - T_{\infty})} \ln \left(\frac{P_{\text{exh}}}{P_{\text{stack}}} \right) \right] - \frac{e\dot{X}_{\text{D,bottom}}}{c_{\text{p}} (T_{\text{exh}} - T_{\infty})}$$

The maximum 1st Law efficiency (with eXergy destruction rate set to zero) is plotted in Figure 11.6 for several exhaust gas temperatures. The case of the stack temperature equal to the dead state corresponds to the Carnot efficiency.

Heat Recovery Steam Generator/Turbine: A Case Study

A specific example that demonstrates how the hot exhaust (at nominal ambient pressure) from an internal combustion engine can be used to produce output power is shown schematically in Fig. 11.7 (note that there is different labelling than Fig. 11.1). The hot exhaust (state "a") is fed to a heat exchanger (called a heat

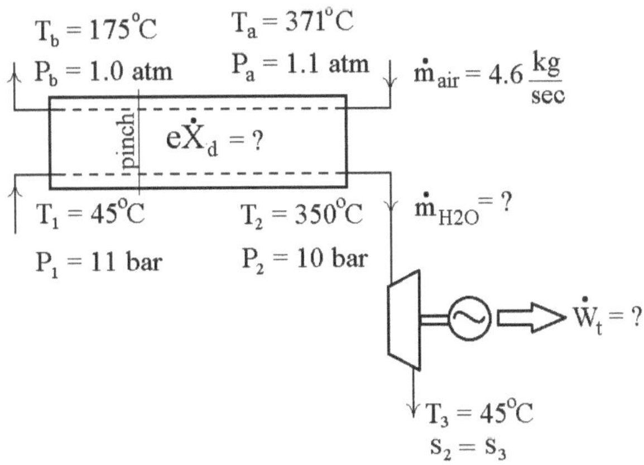

Fig. 11.7 Schematic of a Heat Recovery Steam Generator followed by a turbine

recovery steam generator, HRSG) designed to generate steam to drive a Rankine cycle ("bottoming" cycle). The gas exits at state "b." H_2O enters the HRSG at state 1 and exits at state 2, then enters an adiabatic, reversible turbine and exits (state 3) with a temperature above ambient (as required by the condenser that would follow the turbine). A pinch point (a deeper dive, Appendix 1) is indicated within the HRSG where the gas temperature must exceed the temperature of the saturated liquid.

For the conditions shown, calculate the mass flow rate of steam generated (in kg/sec), the power generated in the turbine (in kW), eXergy destruction rate in the HRSG (in kW, with $T_\infty = 25$ °C as the dead state temperature) and the equivalent high side reservoir temperature of the heat exchanger (a deeper dive, Appendix 2). Assume the gas stream is a calorically perfect gas with properties of air and an enthalpy-specific heat $c_p = 1.04$ kJ/kg/K (evaluated at the average gas temperature). Assume steady-state operation, neglect stray heat, kinetic and potential energy changes in both devices. For the gas side, define ambient temperature and pressure (the dead state) as the reference state. A blank worksheet template is provided and a detailed analysis follows (Fig. 11.8).

The formulas that go into the State and Results Table of the spreadsheet are listed in Table 11.1. The numerical values in these formulas and other cells are from saturated or superheat tables for H_2O. Note that the pressure on the air side is specified in atm, and the unit conversion 1 atm = 1.013 bar is used where needed. State b is at atmospheric pressure and, therefore, the pressure term of the entropy is not included.

The energy balance is applied to the HRSG as a control volume with 2 inlets and 2 outlets, with stray heat exchange between the external housing and the environment being:

$$\frac{dE}{dt} = -\dot{Q}_{\text{stray}} - \dot{W}_{\text{out}} + \dot{m}_{\text{air}}(h_{\text{a}} - h_{\text{b}})$$
$$+\dot{m}_{\text{H2O}}(h_1 - h_2) + \text{KE} + \text{PE} \text{ for the HRSG}$$

INPUT PARAMETERS

m'$_{air}$	4.6	kg/sec
$c_{p,air}$	1.04	kJ/kg/K, air specific heat, evaluated at average temp
R_{air}	0.287	kJ/kg/K, air gas constant
T_{inf}	25	oC, dead state temperature
P_{inf}	1	atm, ambient pressure

STATE PROPERTIES

state	P (bar)	T (oC)	h (kJ/kg)	s (kJ/kg/K)	quality or phase	
a	1.1143	371			GAS	air in
b	1.013	175			GAS	air out
1	11	45				H2O in HRSG
2	10	350				H2O out of HRSG
3		45				H2O out of turbine

RESULTS

m'$_{H2O}$		kg/sec, H2O flow rate
W'$_t$		kW, turbine power
eX'$_D$		kW, eXergy destruction rate in HRSG
$T_{H,equiv}$		oC, equivalent thermal reservoir temperature

Fig. 11.8 Blank spreadsheet template for HRSG/turbine workshop

The assumptions for simplification are: steady-state, negligible stray heat, kinetic and potential energy, and power output is identically zero. The energy balance then yields the mass flow rate of H_2O produced from a hot air stream, with inlet and exit conditions specified.

$$\dot{m}_{\mathrm{H2O}} = \dot{m}_{\mathrm{air}} \frac{(h_{\mathrm{a}} - h_{\mathrm{b}})}{(h_1 - h_2)}$$

For the problem at hand, this hot exhaust produces 0.316 kg/sec of steam at the superheated H_2O conditions specified much lower than the air flow rate feeding the HRSG (4.6 kg/s), because the H_2O changes phase (latent energy) while the air changes temperature only (sensible energy).

The power output of the turbine is obtained from an energy balance on the turbine:

$$\frac{dE}{dt} = \dot{Q}_{\mathrm{in}} - \dot{W}_{\mathrm{out}} + \dot{m}_{\mathrm{H2O}}(h_2 - h_3) + \mathrm{KE} + \mathrm{PE} \text{ for the turbine}$$

At steady-state, neglecting stray heat, kinetic and potential energy:

Table 11.1 Spreadsheet formulas for HRSG/turbine workshop

Cell	Quantity	Spreadsheet formula/comment
D14	h_a	=$B$7*(C14-$B$9)
E14	s_a	=$B$7*LN((C14 + 273)/($B$9 + 273))-$B$8*LN(B14/($B$10*1.013))
D15	h_b	=$B$7*(C15-$B$9)
E15	s_b	=$B$7*LN((C15+273)/($B$9+273))
D16	h_1	=188.44+0.00101*(B16-B18)*100, note unit conversion 1 bar = 100 kPa
D18	h_3	=188.44+F18*2394
F18	x_3	=(E18–0.6386)/(8.1633–0.6386)
Results table		
	$\dot{m}_{H2O}$	=B6*(D14-D15)/(D17-D16)
	$\dot{W}_{turbine}$	=B21*(D17-D18)
	$e\dot{X}_D$	=(B21*(E17-E16) + B6*(E15-E14))*(B9 + 273)
	$T_{H,\ equiv}$	=(D14-D15)/(E14-E15)-273
Pinch point analysis (Appendix 2, a deeper dive)		
A30	Q1f	=$B$21*(E30-$E$29)
D30	T1f	=179.88, saturation temperature
E30	H1f	=762.51, enthalpy of saturated liquid
H30	Tair@pp	=$H$29 + A30/($B$6*$B$7)
A31	Q1g	=$B$21*(E31-$E$29)
H31	Tair@ SH	=$H$29 + A31/($B$6*$B$7)
A32	Q2	=$B$21*(E32-$E$29)

$$\dot{W}_{out} = \dot{m}_{H2O}(h_2 - h_3)$$

The steam from the HRSG can produce 268 kW of power from this isentropic steam turbine whose exit pressure is 0.095 bar.

Many students will be confused by the notion that there is no heat transfer in a heat exchanger. For the control volume used, the heat transfer is internal. The HRSG as a whole is treated as adiabatic. Stray heat between the external housing and the environment could be estimated (tending to reduce the mass flow rate of H_2O produced), but that is a secondary effect. Students may inadvertently calculate the internal heat transfer rate by considering one stream or the other as the control volume. This approach is not wrong but unnecessary. The real problem with that comes in the 2nd Law analysis.

An entropy balance for the HRSG is:

$$\frac{dS}{dt} = \frac{-\dot{Q}_{stray}}{T_\infty} + \dot{m}_{air}(s_a - s_b) + \dot{m}_{H2O}(s_1 - s_2) + \dddot{\Sigma}_{HRSG}$$

At steady-state, neglecting stray heat (note the boundary temperature is the dead state) and multiplying by the dead state temperature, the eXergy destruction rate in the HRSG is:

$$e\dddot{X}_{\mathrm{D}} = \dot{m}_{\mathrm{air}} T_{\infty}(s_{\mathrm{b}} - s_{\mathrm{a}}) + \dot{m}_{\mathrm{H2O}} T_{\infty}(s_2 - s_1)$$

Inserting values:

$$e\dddot{X}_{\mathrm{D}} = \left(4.6\frac{\mathrm{kg}}{\mathrm{s}}\right)(298\,\mathrm{K})(0.4240 - 0.7741)\frac{\mathrm{kJ}}{\mathrm{kg}\cdot\mathrm{K}} + \left(0.316\frac{\mathrm{kg}}{\mathrm{s}}\right)(298\,\mathrm{K})$$
$$\times (7.3029 - 0.6386)\frac{\mathrm{kJ}}{\mathrm{kg}\cdot\mathrm{K}}$$

$$e\dddot{X}_{\mathrm{D}} = -479.9 + 627.6 = 147.7\,\mathrm{kW}$$

Notice that the entropy of the gas decreases, while the entropy of the H_2O increases more, resulting in a net positive entropy generation (hence eXergy destruction). Note also that the numbers are evaluated here with values truncated to 4 sig. figs, resulting in a net value of 147.7 kW. The result in the spreadsheet is 147.4 kW. The difference is that the spreadsheet carries extra significant figures. The truncation error is small in this example, but there can be cases where the net eXergy destruction rate is the difference between numbers which are close and a large truncation error can result in hand calculations. Therefore, follow the adage "carry extra significant figures in calculations, but report 3 or 4." Working in a spreadsheet helps facilitate that.

The source of the eXergy destruction is the internal heat transfer that takes place across a finite temperature difference. An overall energy and entropy balance allows it to be quantified without analyzing the internal effect in detail (an impossible task).

The exit gas temperature of the HRSG is still hot compared to ambient, and therefore, not all the available thermal energy is extracted in the HRSG. To see why that is necessary, a "pinch point" analysis is conducted in the deeper dive of Appendix 1 by considering the preheating section of the HRSG, with saturated liquid achieved at the pinch point that occurs between the preheater and boiler sections. A plot of the temperature versus the heat transferred is shown in Fig. 11.9 that reveals internal temperature differences that drive heat transfer, ultimately giving rise to the eXergy destruction within the boundary of the HRSG.

If the high side source for the HSRG were replaced by a thermal reservoir at a temperature T_{a}, steady-state energy and entropy balances (neglecting kinetic and potential energy, and multiply entropy by the dead state) are:

$$0 = \dot{Q}_{\mathrm{H}} + \dot{m}_{\mathrm{H2O}}(h_1 - h_2)$$

$$0 = T_{\infty}\left[\frac{\dot{Q}_{\mathrm{H}}}{T_{\mathrm{a}}} + \dot{m}_{\mathrm{H2O}}(s_1 - s_2) + \dddot{\Sigma}\right]$$

Combining and solving for the eXergy destruction rate:

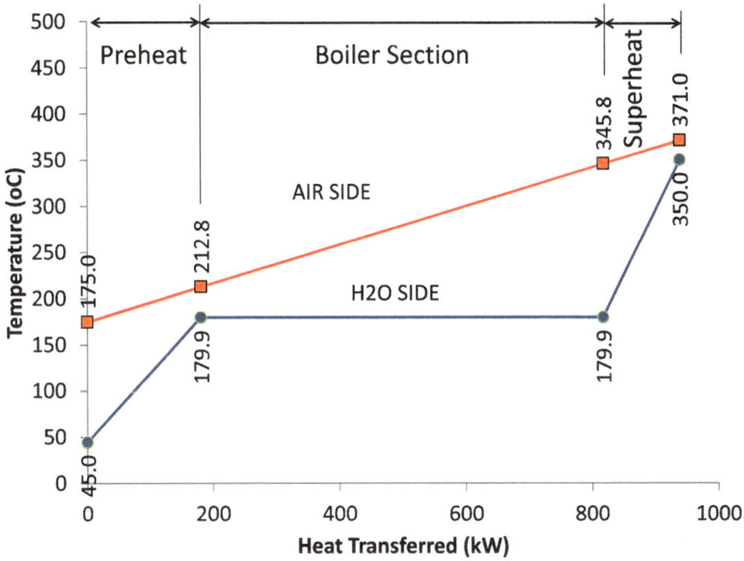

Fig. 11.9 HRSG temperatures versus heat transferred

$$e\dddot{X}_{\text{D}} = T_{\infty}\dot{\Sigma} = \dot{m}_{\text{H2O}}T_{\infty}\left(\frac{h_1 - h_2}{T_{\text{a}}} + (s_2 - s_1)\right)$$

$$e\dddot{X}_{\text{D}} = \left(0.316\frac{\text{kg}}{\text{s}}\right)(298\,\text{K})\left(\frac{189.54 - 3158.3}{371 + 273} + (7.3029 - 0.6386)\right)\frac{\text{kJ}}{\text{kg}\cdot\text{K}}$$

$e\dddot{X}_{\text{D}} = -434.1 + 627.6 = 193.5\,\text{kW}$ eXergy destruction rate from a thermal reservoir @ T_{a}. This eXergy destruction rate is 23.8% higher than it was from the HRSG. Physically, since the high temperature "source" fluid decreases as it flows through, the temperature difference driving the heat is lower, resulting in less eXergy destruction.

Appendix 2 derives a simple expression for an "equivalent thermal reservoir" associated with a heat exchanger. This hypothetic reservoir transfers heat at the same rate and has the same eXergy destruction rate as the heat exchanger. In this case, the equivalent thermal reservoir temperature is 309.2 °C (lower than the inlet air temperature of 371 °C).

Condenser

The condensation process in a Rankine cycle is accomplished in another heat exchanger, the condenser, shown schematically in the inset of Fig. 11.11 as an extension of the HRSG/turbine case study. In this case, heat is to be transferred out of

	A	B	C	D	E	F	G	H
5	**INPUT PARAMETERS**							
6	m'_{air}	**4.6**	kg/sec					
7	$c_{p,air}$	**1.04**	kJ/kg/K, air specific heat, evaluated at average temp					
8	R_{air}	**0.287**	kJ/kg/K, air gas constant					
9	T_{inf}	**25**	oC, dead state temperature					
10	P_{inf}	**1**	atm, ambient pressure					
11								
12	**STATE PROPERTIES**							
13	*state*	*P (bar)*	*T (oC)*	*h (kJ/kg)*	*s (kJ/kg/K)*	*quality or phase*		
14	a	**1.1143**	**371**	359.84	0.774075	**GAS**	air in	
15	b	**1.013**	**175**	156	0.424008	**GAS**	air out	
16	1	**11**	**45**	189.54	0.6386	Comp. Liq	H2O in HRSG	
17	2	**10**	**350**	3158.2	7.3029	SH Vapor	H2O out of HRSG	
18	3	0.09595	**45**	2308.70	7.3029	0.88566	H2O out of turbine	
19								
20	**RESULTS**							
21	m'_{H2O}	0.31585	kg/sec, H2O flow rate					
22	W'_t	268.318	kW, turbine power					
23	eX'_D	147.402	kW, eXergy destruction rate in HRSG					
24	$T_{H,equiv}$	309.287	oC, equivalent thermal reservoir temperature					
25								
26	**PINCH POINT ANALYSIS OF HRSG**							
27		WATER SIDE					AIR SIDE	
28	$Q_{transferred}$ (kW)	state	*P (bar)*	*T (oC)*	*h (kJ/kg)*	*quality*	*state*	*T*
29	0	1	11	45	189.541309	NA, CL	b	175
30	181.0	1f	10	179.88	762.51	0		212.8292
31	817.3	1g	10	179.88	2777.1	1		345.8386
32	937.7	2	10	350	3158.2	NA, SH	a	371
33								
34	**ΔT, PP**	**32.949**	oC, Pinch Point temperature difference (must be positive)					

Fig. 11.10 Solution to HRSG/turbine case study

the working fluid to a colder fluid (coolant) that acts as a thermal sink. In practice, it is most effective and economical to use liquid water as the cooling fluid, either directly taken from a natural source like a river or the return from a cooling tower. If river water is used, it will be returned to the river warm, a form of thermal pollution. On the other hand, a cooling tower (next section) brings ambient air in direct contact with the warmed coolant. In practice, ambient air cannot be used directly to condense the working fluid due to poor heat transfer effectiveness. As will be clear, because there is only a very modest change in temperature of a single phase substance, while the working fluid of the Rankine cycle undergoes phase change, the required flow rate of the coolant will be much larger than that of cycle working fluid.

Energy and entropy balances applied to the condenser, with the cooling water (CW) treated as a compressed liquid, are:

$$\frac{dE}{dt} = -\dot{Q}_{\text{stray}} + \dot{m}_{\text{H2O}}(h_{\text{in}} - h_{\text{out}})_{\text{Rankine}} + \dot{m}_{\text{CW}} c_{\text{liquid}}(T_{\text{CW,in}} - T_{\text{CW,out}}) + \text{KE} + \text{PE}$$

$$\frac{dS}{dt} = \frac{-\dot{Q}_{\text{stray}}}{T_\infty} + \dot{m}_{\text{H2O}}(s_{\text{in}} - s_{\text{out}})_{\text{Rankine}} + \dot{m}_{\text{CW}} c_{\text{liquid}} \ln\left(\frac{T_{\text{CW,in}}}{T_{\text{CW,out}}}\right) + \underset{\text{condenser}}{\dddot{\Sigma}}$$

At steady-state, neglecting stray heat and kinetic and potential energy, the energy balance can be used to express the required mass flow of the coolant in terms of its temperatures, with the change in cooling water temperature expressed as $\Delta T_{\text{CW}} = T_{\text{CW, out}} - T_{\text{CW, in}}$:

$$\frac{\dot{m}_{\text{CW}}}{\dot{m}_{\text{H2O}}} = \frac{(h_{\text{in}} - h_{\text{out}})_{\text{Rankine}}}{c_{\text{liquid}} \Delta T_{\text{CW}}}$$

The numerator is the change in specific enthalpy of the working fluid of the Rankine cycle, and the denominator is the change in specific enthalpy of the coolant which depends on the temperature change of the cooling water. The highest possible exit temperature of cooling water is the temperature of the condensing cycle water, 45 °C in this case study. For this problem then,

$$\left(\frac{\dot{m}_{\text{CW}}}{\dot{m}_{\text{H2O}}}\right)_{\text{minimum}} = \frac{2308.7 - 188.44 \text{ kJ/kg}}{4.2 \text{ kJ/kg}/\Delta°\text{C}(45 - 25)\Delta°\text{C}} = 31.2$$

This mass flow rate is plotted on semilogy scale as a function of cooling water temperature change in Fig. 11.11 for the case study with an inlet cooling water temperature equal to the ambient temperature of 25 °C. For a given cycle H_2O flow rate, the required cooling water increases sharply as the temperature change decreases. The required cooling flow rate is more than 100 times as large as the cycle H_2O for a temperature rise of 5 °C, 600 times as large for a cooling water temperature rise of 1 °C. As a practical matter, it is difficult to pin down a specific temperature rise of a coolant water that would be problematic as a thermal pollutant, and would involve regulatory activity.

The entropy balance becomes

$$\underset{\text{condenser}}{\dddot{\Sigma}} = -\dot{m}_{\text{H2O}}(s_{\text{in}} - s_{\text{out}})_{\text{Rankine}} + \dot{m}_{\text{CW}} c_{\text{liquid}} \ln\left(\frac{T_{\text{CW,out}}}{T_{\text{CW,in}}}\right)$$

Combining the energy balance to eliminate cooling water flow rate:

$$\underset{\text{condenser}}{\dddot{\Sigma}} = -\dot{m}_{\text{H2O}}(s_{\text{in}} - s_{\text{out}})_{\text{Rankine}} + \frac{\dot{m}_{\text{H2O}}(h_{\text{in}} - h_{\text{out}})_{\text{Rankine}}}{\Delta T_{\text{CW}}} \ln\left(1 + \frac{\Delta T_{\text{CW}}}{T_{\text{CW,in}}}\right)$$

A dimensionless eXergy destruction rate is obtained by dividing the eXergy destruction rate by the internal heat transfer rate from the cycle H_2O:

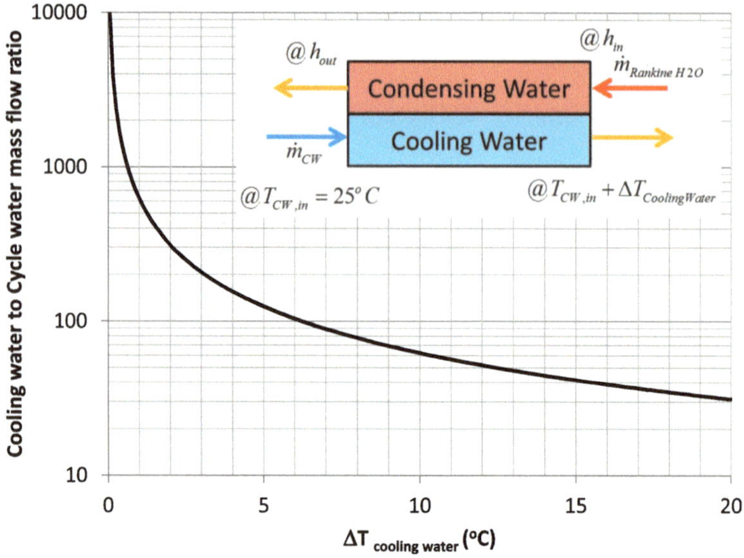

Fig. 11.11 Effect of cooling water temperature change on cooling water flow rate

$$\frac{\dddot{eX}_{\mathrm{D}}}{\dot{m}_{\mathrm{H2O}}(h_{\mathrm{in}}-h_{\mathrm{out}})_{\mathrm{Rankine}}}=\frac{-T_{\infty}(s_{\mathrm{in}}-s_{\mathrm{out}})}{(h_{\mathrm{in}}-h_{\mathrm{out}})}+\frac{T_{\infty}}{\Delta T_{\mathrm{CW}}}\ln\left(1+\frac{\Delta T_{\mathrm{CW}}}{T_{\mathrm{CW,in}}}\right)$$

The entropy and enthalpy for the cycle H_2O are expressed in terms of the quality and assuming the exit is a saturated liquid:

$$\frac{(s_{\mathrm{in}}-s_{\mathrm{out}})}{(h_{\mathrm{in}}-h_{\mathrm{out}})}=\frac{(s_{\mathrm{f}}+xs_{\mathrm{fg}}-s_{\mathrm{f}})}{(h_{\mathrm{f}}+xh_{\mathrm{fg}}-h_{\mathrm{f}})}=\frac{s_{\mathrm{fg}}}{h_{\mathrm{fg}}}$$
$$=\frac{1}{T_{\mathrm{SAT}}}\ \text{with}\ T_{\mathrm{SAT}}\ \text{in absolute units(Kelvin)}$$

The latter expression for the saturation temperature of the Rankine cycle H_2O is a fundamental relation. The dimensionless eXergy destruction rate can be expressed as:

$$\frac{\dddot{eX}_{\mathrm{D}}}{\dot{m}_{\mathrm{H2O}}(h_{\mathrm{in}}-h_{\mathrm{out}})_{\mathrm{Rankine}}}=\frac{T_{\infty}}{T_{\mathrm{SAT}}}\left(-1+\frac{T_{\mathrm{SAT}}}{\Delta T_{\mathrm{CW}}}\ln\left(1+\frac{\Delta T_{\mathrm{CW}}}{T_{\mathrm{CW,in}}}\right)\right)$$

This function is plotted versus cooling water temperature change in Fig. 11.12 for three different values of the saturation temperature of the Rankine cycle condenser H_2O. In the high coolant flow limit (where the heat to condense the cycle H_2O does not change the temperature of the coolant), the dimensionless eXergy destruction

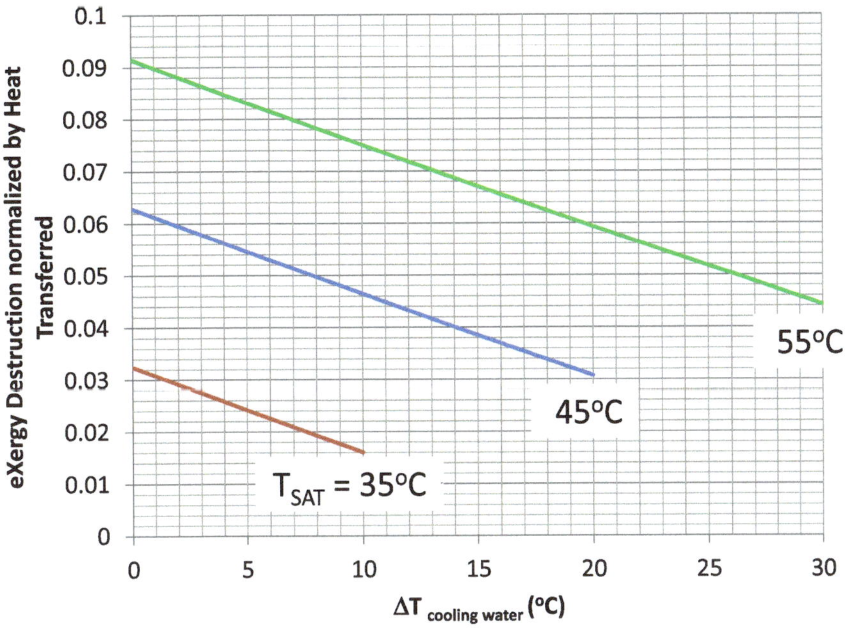

Fig. 11.12 Dimensionless eXergy destruction rate versus cooling water temperature change for three different Rankine cycle condenser H_2O temperatures. The inlet cooling water temperature equals the dead state temperature (25 °C). Results for three saturation temperatures of the Rankine cycle H_2O are shown

evaluates to 0.063 for the case of a saturation temperature of 45 °C (corresponding to the HRSG/turbine case study). In this limit, the coolant behaves like a thermal reservoir at the inlet temperature. The eXergy destruction rate decreases as the coolant flow rate decreases (resulting in a higher exit coolant temperature). This trend is due to the lower average temperature difference between the cycle H_2O and cooling water.

The eXergy destruction rate is relatively low in the condenser (relative to the desired heat that is transferred) but increases with increasing saturation temperature. This reinforces the incentive to lower the saturation temperature, which has the added benefit of increasing the power output of the turbine (by lowering the turbine exit pressure).

In designing heat exchangers, there is a trade-off between thermodynamic performance (i.e., minimizing eXergy destruction while transferring heat at the required rate) and capital cost. The heat transfer rate can be expressed as:

$$\dot{Q} = (\Omega A)_{\mathrm{HEX}}\left(\overline{T}_{\mathrm{H}} - \overline{T}_{\mathrm{C}}\right)$$

Where Ω is the effective heat transfer coefficient (a function of the both fluid types and the nature of the internal fluid mechanics and thermal resistance across the wall), A is the appropriate surface area of the physical wall that separates the two fluids, and the temperatures are average temperatures of the hot side and cold side fluids. These are not geometric averages, and can be defined more specifically based on inlet and outlet temperatures for certain heat exchanger designs (such as a Log-Mean Temperature Difference, or LMTD, for coflow or counterflow heat exchangers). The main point here is that the larger the change in the cooling water temperature, the lower the eXergy destruction rate (due to a lower average temperature difference driving heat) but the larger the surface area and capital cost. This type of trade-off between thermodynamic performance and physical size is typical. It translates to an economic trade-off between operating (fuel) cost and capital cost to install equipment.

The heat transfer coefficient depends on the fluid type, but can be increased moderately by design of the flow path, with baffles to increase turbulence, for example, or creating high local velocities with small flow channels. These increases come at the expense of higher pressure drops so there are trade-offs there as well.

Cooling Tower

There may not be a suitable natural source of cooling water, like a river, for a particular power plant application. In that case, the necessary heat rejection from the cycle H_2O in the condenser can be made to ambient air, but not directly. A secondary loop of cooling water is needed as it is much more effective at heat transfer than air. The heat picked up by the cooling water in the condenser is rejected to ambient air in a separate heat exchanger, called a cooling tower.

An open-type cooling tower is shown in the schematic of the complete combined cycle power plant (Fig. 11.1). The cooling water enters the cooling tower after picking up heat from the condensing water and is brought into direct contact with air (usually by spraying it as a mist). The air enters at ambient conditions and exits at a temperature below the cooling water inlet. The cooling water exits at a temperature above ambient and enters a pump that drives the secondary loop. Determination of the power required by this pump involves a detailed fluid mechanical analysis that will not be considered here. Similarly, power may be needed to drive a fan to force air through the cooling tower and is not considered. These auxiliary power requirements would reduce the net power output of the plant somewhat, but is considered of secondary importance.

The evaporation of air is an important effect on the thermodynamic performance of an open cooling tower. A small fraction of the water sprayed into the air stream evaporates, and the humidity of the air stream changes. There is a favorable energy consequence (called evaporative cooling). An additional advantage of the cooling tower compared to a closed heat exchanger is that the direct contact between the two fluids yields the lowest possible resistance to heat transfer. The water that evaporates must be added somewhere (make up water), shown before the pump in Fig. 11.1. These evaporative effects are neglected in this chapter, but will be revisited in the

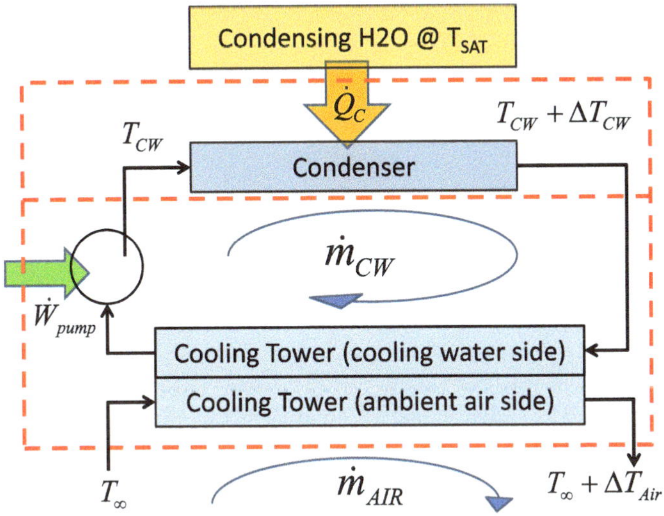

Fig. 11.13 Process diagram of a cooling tower with a cooling water loop removing heat from the condenser of a Rankine cycle. Control volume boundaries are shown as dotted red lines

next. The humid air that exits the cooling tower often creates the most visible part of a power plant, a white plume or cloud that forms as the exiting humid air is cooled by its ambient. This plume is sometimes mistaken for stack gases (combustion products), which can also form a visible plume.

The process diagram of Fig. 11.13 shows a cooling water loop in which the source of heat is a thermal reservoir at the saturation temperature of the H_2O in the condenser of the Rankine cycle. Two control volumes are defined to analyze the condenser and cooling tower separately, or equivalently, the condenser and the combined condenser and cooling tower. Cooling water (subscript cw) enters the condenser at a temperature T_{cw} and exits at a higher temperature $T_{\text{cw}} + \Delta T_{\text{cw}}$, which is the temperature it enters the cooling tower. The rate of heat transfer to the cooling water is set by the requirement of the condenser. The cooling tower is shown schematically as a closed-type heat exchanger (where the two fluids are separated by a wall) to facilitate analysis and emphasize the dominant thermal effect of mixing the warm cooling water with ambient air. Ambient air enters at the dead state temperature (T_∞)and picks up heat from the cooling water and exits at a higher temperature ($T_\infty + \Delta T_{\text{Air}}$). The cooling water loop requires a pump, whose power input which will be approximated. All quantities are shown as being inherently positive and consistent with the arrows for this application.

2nd Law constraints on the exit temperatures dictate their best theoretical performance. The temperature of the cooling water that exits the cooling tower must be greater than ambient, and its temperature leaving the condenser must be below the saturation temperature of the Rankine cycle condenser. That is:

$$T_\infty < T_{\text{cw}} \text{ and } T_{\text{cw}} + \Delta T_{\text{cw}} < T_{\text{SAT}} \text{ and or}$$
$$T_\infty < T_{\text{cw}} < (T_{\text{SAT}} - \Delta T_{\text{cw}})$$

The cooling water temperature is between ambient and saturation.

A constraint on the exit air temperature is that it must be below that of the cooling water that enters the tower:

$$T_\infty + \Delta T_{\text{air}} < T_{\text{cw}} + \Delta T_{\text{cw}}$$

Since all practical heat exchangers require a finite temperature difference to drive heat, and heat transfer across a finite difference destroys eXergy, it is clear that the addition of a cooling water loop comes with a thermodynamic penalty. In practice, Rankine cycle power plants that use a natural source of water can be designed with lower saturation temperatures than those that require a cooling tower for two reasons: the river water is usually colder than the ambient air and cooling water from a cooling tower is always higher than ambient air.

Control Volume around Condenser

The energy and entropy balances across the condenser only are (assuming steady-state and neglecting kinetic and potential energy):

$$0 = \dot{Q}_{\text{C}} + \dot{m}_{\text{cw}}(h_{\text{cw,in}} - h_{\text{cw,out}})$$
$$0 = \frac{\dot{Q}_{\text{C}}}{T_{\text{SAT}}} + \dot{m}_{\text{cw}}(s_{\text{cw,in}} - s_{\text{cw,out}}) + \dddot{\Sigma}_{\text{condenser}}$$

The cooling water is a compressed liquid, and the change in enthalpy is given by:

$$(h_{\text{cw,in}} - h_{\text{cw,out}}) = c_{\text{L}}(T_{\text{in}} - T_{\text{out}}) + v\Delta P$$

where c_{L} is the specific heat of liquid water (4.2 kJ/kg/K as a working value). The change in entropy of a compressed liquid is given by:

$$(s_{\text{cw,in}} - s_{\text{cw,out}}) = c_{\text{L}} \ln\left(\frac{T_{\text{cw,in}}}{T_{\text{cw,out}}}\right)$$

Neglecting the effect of the pressure drop across the cooling water side (equivalent to treating the cooling water as being isobaric across the condenser), the energy balance becomes:

$$0 = \dot{Q}_C + \dot{m}_{cw} c_L [T_{cw} - (T_{cw} + \Delta T_{cw})]$$

The cooling rate, considered fixed by the Rankine cycle, is:

$$\dot{Q}_C = \dot{m}_{cw} c_L \Delta T_{cw}$$

The minimum possible flow rate of cooling water occurs when the cooling water exits at the saturation temperature:

$$\dot{m}_{cw,\,min} = \frac{\dot{Q}_C}{c_L (T_{SAT} - T_{cw})}$$

A dimensionless cooling water mass flow rate is defined as a ratio of the actual to minimum flow rate:

$$\frac{\dot{m}_{cw}}{\dot{m}_{cw,\,min}} = \frac{T_{SAT} - T_{cw}}{\Delta T_{cw}}$$

The entropy balance becomes:

$$0 = \frac{\dot{Q}_C}{T_{SAT}} + \dot{m}_{cw} c_L \ln\left(\frac{T_{cw}}{T_{cw} + \Delta T_{cw}}\right) + \dddot{\Sigma}_{condenser}$$

Combining with the energy balance, multiplying by the dead state temperature and creating a nondimensional eXergy destruction rate (eXergy destruction rate divided by the condenser heat transfer rate):

$$\frac{e\dddot{X}_{D,condenser}}{\dot{Q}_C} = -\frac{T_\infty}{T_{SAT}} + \frac{T_\infty}{\Delta T_{cw}} \ln\left(1 + \frac{\Delta T_{cw}}{T_{cw}}\right)$$

This result is the same as obtained by analyzing the condenser as 2 inlet, 2 outlet stream, plotted in Fig. 11.12.

Control Volume around Cooling Tower

The energy and entropy balances on the control volume around the cooling tower are (assuming steady-state, neglecting stray heat, kinetic and potential energy changes and no evaporation of water between the air and cooling water stream):

$$0 = \dot{m}_{\mathrm{air}}(h_{\mathrm{air,in}} - h_{\mathrm{air,out}}) + \dot{m}_{\mathrm{cw}}c_{\mathrm{L}}(T_{\mathrm{cw}} + \Delta T_{\mathrm{cw}} - T_{\mathrm{cw}}) + \dot{W}_{\mathrm{pump}}$$

$$0 = \dot{m}_{\mathrm{air}}(s_{\mathrm{air,in}} - s_{\mathrm{air,out}}) + \dot{m}_{\mathrm{cw}}c_{\mathrm{L}}\ln\left(\frac{T_{\mathrm{cw}} + \Delta T_{\mathrm{cw}}}{T_{\mathrm{cw}}}\right) + \dddot{\Sigma}_{\mathrm{tower}}$$

If evaporative effects were included, the mass flow rates of all four streams would be different, and the enthalpy and entropy property relations would require treating the gas as a humid air stream with changing composition. These effects are studied in Chap. 12, and are considered to have a significant, but secondary effect on the cooling tower. The results by treating the cooling tower as a closed heat exchanger yield a good first approximation to required flow rates.

$$(h_{\mathrm{air,in}} - h_{\mathrm{air,out}}) = c_{\mathrm{p,air}}(T_{\infty} + \Delta T_{\mathrm{Air}})$$

$$(s_{\mathrm{air,in}} - s_{\mathrm{air,out}}) = c_{\mathrm{p,air}}\ln\left(\frac{T_{\infty}}{T_{\infty} + \Delta T_{\mathrm{Air}}}\right) - R_{\mathrm{air}}\ln\left(\frac{P_{\mathrm{Air,in}}}{P_{\mathrm{Air,out}}}\right)$$

Both air streams are open to the environment, so there is no entropy change due to pressure. The pump is required to overcome pressure flow losses associated with flow through the extensive piping and the condenser needed to move cooling water through the cooling water loop. Assuming a reversible pump and a liquid-specific volume v_{L}, the power required is:

$$\dot{W}_{\mathrm{pump}} = \dot{m}_{\mathrm{cw}}v_{\mathrm{L}}\Delta P_{\mathrm{pump}}$$

$$\frac{\dot{m}_{\mathrm{air}}}{\dot{m}_{\mathrm{cw}}} = \frac{c_{\mathrm{L}}\Delta T_{\mathrm{cw}} + v_{\mathrm{L}}\Delta P_{\mathrm{pump}}}{c_{\mathrm{p,air}}\Delta T_{\mathrm{Air}}}$$

The eXergy destruction rate in the cooling tower is:

$$e\dddot{X}_{\mathrm{D,tower}} = \dot{m}_{\mathrm{air}}c_{\mathrm{p,air}}T_{\infty}\ln\left(1 + \frac{\Delta T_{\mathrm{Air}}}{T_{\infty}}\right) - \dot{m}_{\mathrm{cw}}c_{\mathrm{L}}T_{\infty}\ln\left(1 + \frac{\Delta T_{\mathrm{cw}}}{T_{\mathrm{cw}}}\right)$$

Normalized by the condenser cooling rate, and manipulating:

$$\frac{e\dddot{X}_{\mathrm{D,tower}}}{\dot{Q}_C} = \left(1 + \frac{v_{\mathrm{L}}\Delta P}{c_{\mathrm{L}}\Delta T_{\mathrm{cw}}}\right)\frac{T_{\infty}}{\Delta T_{\mathrm{air}}}\ln\left(1 + \frac{\Delta T_{\mathrm{Air}}}{T_{\infty}}\right) - \frac{T_{\infty}}{\Delta T_{\mathrm{cw}}}\ln\left(1 + \frac{\Delta T_{\mathrm{cw}}}{T_{\mathrm{cw}}}\right)$$

These results are plotted in Figs. 11.14 and 11.15 for a case with a saturation temperature of 45 °C, an ambient of 25 °C and a fixed cooling water temperature change of 3 °C. A pressure drop of 0.5 atm is set. The dimensionless term where the pump work comes in evaluates to:

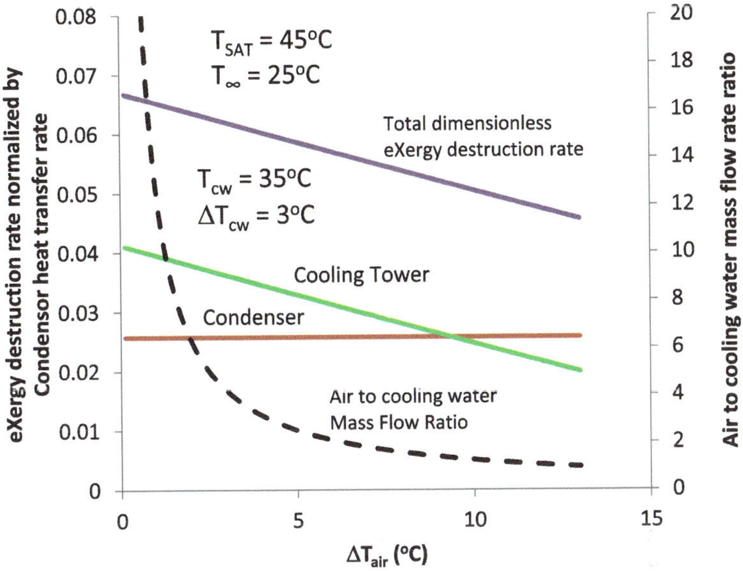

Fig. 11.14 Dimensionless eXergy destruction rates and air flow rate (normalized by Rankine cycle mass flow rate) as a function of air temperature change in cooling tower for a fixed cooling water temperature and change across condenser

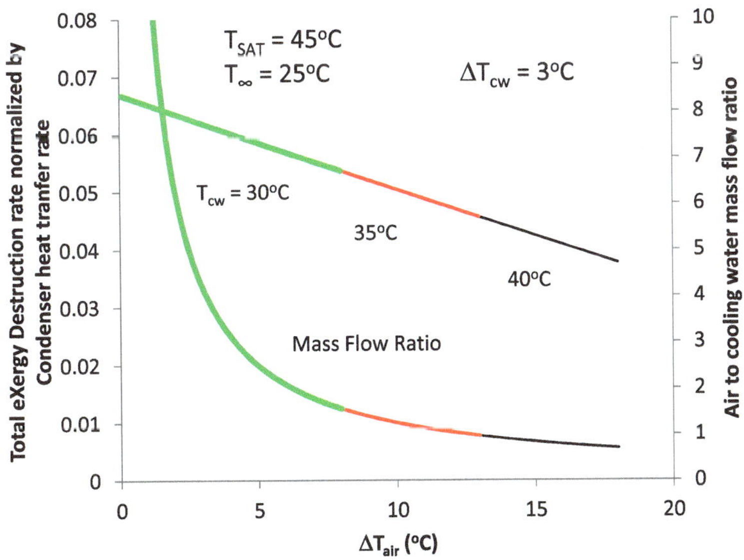

Fig. 11.15 Combined condenser/cooling tower dimensionless eXergy destruction rates and air to cooling water mass flow rates as a function of change in air temperature for three values of cooling water temperature (entering condenser)

$$\frac{v_L \Delta P}{c_L \Delta T_{cw}} = \frac{(0.001\,\mathrm{m^3/kg})(0.5\,\mathrm{atm}) \cdot |101.3\,\mathrm{kPa/atm}|}{(4.2\,\mathrm{kJ/kg/\Delta^\circ C})(3^\circ\mathrm{C})} = 0.00402$$

The pump work has a negligible effect on the thermodynamics for this cooling water flow rate (set by its temperature rise). A large pump is required however as part of the design. A lower change in cooling water temperature than set here would increase that requirement.

Figure 11.14 shows the dimensionless eXergy destruction rate and air to cooling water mass flow rate ratio as a function of the change in air temperature for a set cooling water temperature of 35 °C (midway between ambient and the saturation temperature). The eXergy destruction rate in the condenser is fixed, and the eXergy destruction rate decreases with air temperature change (due to the lower average temperature difference between the two fluids).

Figure 11.15 shows the results for three different values of the cooling water temperature (but all with the same 3 °C change in cooling water temperature). The eXergy destruction rate and mass flow rate fall on the same curve, but a lower overall eXergy destruction rate can achieved with the higher cooling water temperature and maximizing the air temperature change. The trade-off would be a more difficult condenser to design (due to lower temperature difference between cooling water and saturation temperature).

Workshop 11, Part 1: Gas Turbine Topping Cycle

In a stationary power plant application, a combined cycle that consists of a natural gas-fired gas turbine as a topping cycle and a Rankine bottoming cycle will be analyzed in four parts. This first part focuses on the topping cycle. The engine being modeled is an aero-derivative gas turbine (based on operating parameters similar to the GE-F110, the engine used in the F16 Fighting Falcon, a supersonic military aircraft), modified to burn natural gas, with the afterburner and nozzle replaced by a turbine to generate output shaft power. The exhaust gas will feed a Heat Recovery Steam Generator (HRSG) in Part 2, which produces steam fed to a steam turbine in a Rankine cycle (Part 3). A cooling tower is analyzed in Part 4 and the results are compiled. A single gas turbine is analyzed in this part, but the exhaust of three gas turbines will be combined in Parts 2–4 to increase the scale of the power plant.

A schematic of the basic gas turbine engine is shown in Fig. 11.16, which consists of a compressor, combustor and turbine. Air at ambient conditions (state 1) enters the compressor with a mass flow rate of $\dot{m}_{\mathrm{air}} = 122\ \mathrm{kg/s}$. The compressor has a specified pressure ratio of 30. Fuel with a lower heating value LHV = 50,000 kJ/kg (representing the chemical energy converted to thermal energy per unit mass of fuel burned) is injected in a combustor and burned with the compressed air (state 2). This process involves a conversion of one form of internal energy (chemical) into another (thermal), but will be modeled as an "equivalent heat addition" from a

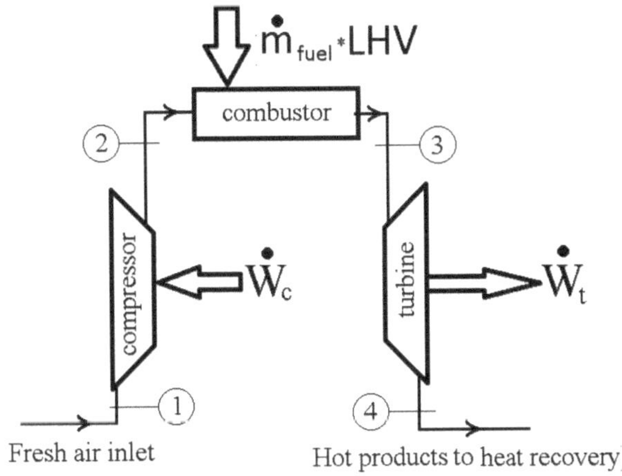

Fig. 11.16 Schematic of gas turbine topping cycle

thermal reservoir at the maximum temperature of the working fluid. A modest pressure drop occurs in the combustor. The products enter a turbine with a specified turbine inlet temperature of 1800 K (state 3) and are expanded to a pressure of 1.5 atm (state 4), higher than ambient to account for a pressure drop across the HRSG to follow.

Complete the State, Process and Results Tables, and sketch on the *T-s* diagram (which is to scale), assuming:

- Steady-state operation, with negligible stray heat, kinetic and potential energy changes in all devices.
- Isentropic efficiencies of 0.85 and 0.9 for the compressor and turbine, respectively. The hypothetical isentropic states are included in the state table to facilitate analysis.
- The working fluid is a calorically perfect gas with properties of air (i.e., enthalpy-specific heat $c_p = 1.14$ kJ/kg/K, and $R = 0.287$ kJ/kg/K). The specific heat is evaluated at 1000 K, a representative average temperature in the cycle.
- The internal combustion process ($\dot{m}_{fuel}$ LHV) is treated like an equivalent heat addition from a thermal reservoir at the turbine inlet temperature.

Once a well-designed spreadsheet is implemented, a parametric study can be conducted. A template to guide a sensitivity analysis is included, designed to incrementally observe the effect of nonideal behavior. That is, first change the compressor efficiency to unity, and copy the VALUE for 1st Law efficiency obtained. Then change the turbine efficiency to unity (keeping compressor efficiency at unity), then eliminate pressure drops in combustor and HRSG. The final case will be an ideal Brayton cycle. In the last row, as a check, enter the Brayton cycle formula $\eta_{Brayton} = 1 - (P_2/P_2)^{R/c_p}$ which should match the result from analyzing each process.

INPUT PARAMETERS

m'_{air}	122	kg/sec, mass flow rate of air
LHV	50000	kJ/kg, lower heating value of the fuel
cp	1.14	kJ/kg/K, specific of air at 1000K
R	0.287	kJ/kg/K, gas constant for air
η_t	0.9	isentropic efficiency of turbine
η_t	0.85	isentropic efficiency of compressor

STATE TABLE

state	T (K)	P (atm)	h (kJ/kg)	s (kJ/kg/K)
1	**298**	**1**	**0**	**0**
2s		**30**		
2		**30**		
3	**1800**	**28**		
4s		**1.5**		
4		**1.5**		

PROCESS TABLE (all quantities in MW)

PROCESS	DEVICE	$m'(h_{out}-h_{in})$	m'_{fuel}*LHV	W'_{out}	eX_d'
1-2	Compressor				
2-3	Combustor				
3-4	Turbine				

RESULTS

QUANTITY	VALUE	UNIT
Density of intake Air		kg/m^3
Inlet Volumetric Flow Rate		cfm
Back Work Ratio (Compressor to Turbine)		
Fuel mass flow rate		kg/sec
1st Law thermal efficiency of Topping Cycle		
Carnot Efficiency = 1-T1/T3		
2nd Law efficiency (1st Law over Carnot)		

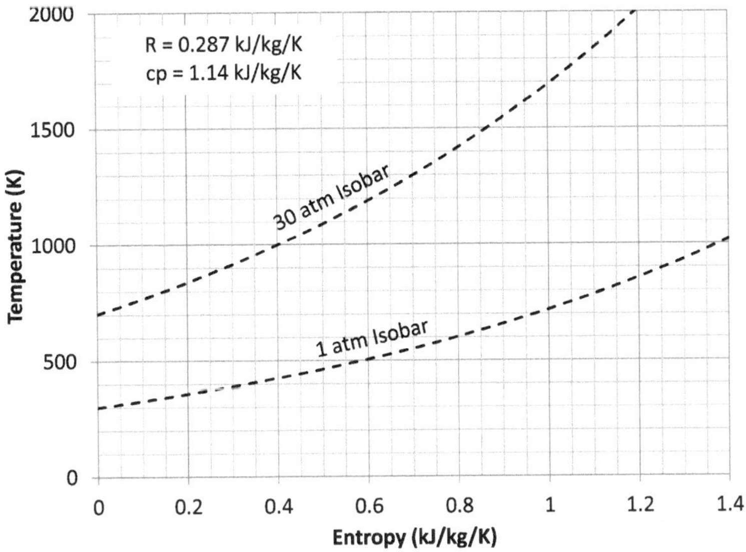

SENSITIVITY ANALYSIS

	change	CYCLE EFFICIENCY
BASE CASE	NA	
compressor efficiency	etac = 1	
turbine efficiency	etat = 1	
combustor pressure dropo	P3 = 30	
heat recovery pressure drop	P4 = 1	
BRAYTON CYCLE EFFICIENCY	NA	

Workshop 11, Part 2: Heat Recovery Steam Generator

The exhaust of the gas turbine analyzed in Part 1 is hot, and therefore contains thermal energy (relative to ambient). However, it is near atmospheric pressure and cannot be used directly to extract useful mechanical power. It can be used as a heating source for a bottoming Rankine cycle, the subject of this and the following workshop. A counterflow heat exchanger (called a Heat Recovery Steam Generator, or HRSG) is the link between the topping and bottoming cycle. The heat rejection from a topping cycle is used as the heat addition to a bottoming cycle. Heat is transferred internally from the hot gas to pressurized H_2O, generating a continuous stream of pressurized superheated H_2O (which will feed a steam turbine in Part 3).

In the schematic of Fig. 11.17, the exhausts of 3 of the gas turbines from Part 1 are combined to feed the HRSG on the air side. Mixing the 3 equivalent streams does not change their temperature or pressure, but the total mass flow rate sums together. H_2O

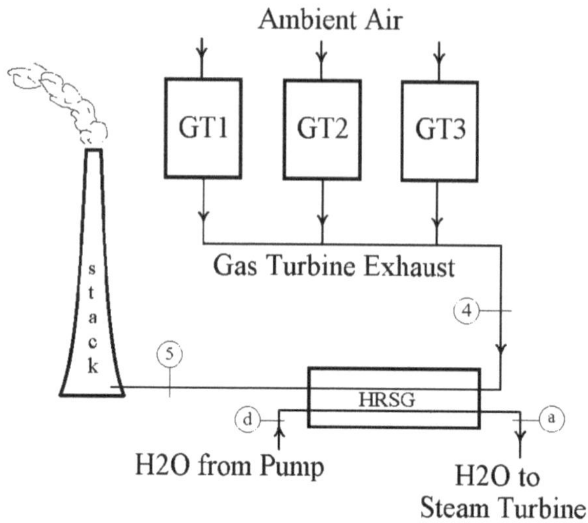

Fig. 11.17 Schematic of heat recovery steam generator (HRSG)

from a pump is fed at elevated pressure in a separate stream (state d). Note that the gas and H_2O streams can be at different pressures, but the pressure drop within them can be minimized by careful fluid mechanics design. In practice, the exit temperature of the H_2O must be below the inlet temperature of the gas stream ($T_a < T_4$), and the exit of the gas stream must be sufficiently high to allow proper function of the HRSG (from a "pinch point" analysis). A good exercise is to confirm that the value set of 85 °C for this example satisfies that requirement. The results in the form of a temperature versus heat transferred of that analysis for this case are shown in the Workshop solution Appendix. From this plot, it can be inferred from the temperature difference driving heat that significant eXergy destruction rates occur in the superheating and boiling sections. Improved performance could be attained with a more complicated system with two pressure levels, reheat and regeneration.

The control volume used for the HRSG is a two-inlet, two-outlet device. The heat transfer that takes place from the hot gas to the cold H_2O is an internal process, and does not appear in an energy or entropy balance. In Part 3, a different control volume on this same device will be chosen that focusses on just the H_2O side.

Complete the two State and the Results Tables (Fig. 11.18), and add the process 4–5 on the *T-s* diagram from Part 1, assuming:

- Steady-state operation, with negligible kinetic and potential energy changes in all devices.
- The working fluid is a calorically perfect gas with properties of air (i.e., constant specific heat, $c_p = 1.03$ kJ/kg/K, and $R = 0.287$ kJ/kg/K). Note that a lower value of c_p is used here, since the gas temperature is lower than in the gas turbine processes.
- The stray heat loss from the HRSG is negligible (the heat transfer is internal).
- The pressure drop across both streams is negligible.

INPUT PARAMETERS

n_{gt}	3	number of gas turbines
T1		oC, ambient temperature
m',air		kg/sec, gas-side mass flow rate
cp	1.03	kJ/kg/K, specific of air at 600K
R	0.287	kJ/kg/K, gas constant for air

STATE TABLE FOR GAS (AIR PROPERTIES)

STATE	P (atm)	T (oC)	h (kJ/kg)	s (kJ/kg/K)
4				
5	**1**	**85**		

STATE TABLE FOR H2O

STATE	P (bar)	T (oC)	h (kJ/kg)	s (kJ/kg/K)	x
d	**10**	**40**			
a	**10**	**350**			

RESULTS

m'_{H2O}		kg/s, mass flow rate of H2O
eX'_D		MW, Dissipation rate in HRSG

Fig. 11.18 Spreadsheet template for combined cycle workshop, part 2. The yellow highlighted cells are inputs for part 2 (HRSG). The blue cells involve inputs from part 1 (topping cycle)

Workshop 11, Part 3: Rankine Bottoming Cycle

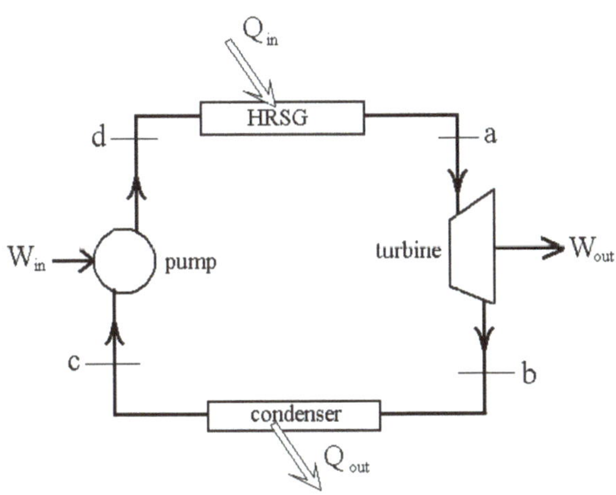

The main part of the bottoming cycle is analyzed in this part of the workshop. The high pressure H_2O vapor (steam) generated in the HRSG (state a) is used to generate additional output power in a steam turbine. To operate continuously in a closed loop (Rankine cycle) steam turbine exhaust (state b) is condensed (at constant pressure) to a saturated liquid (state c), then pressurized in a pump to the HRSG inlet (state d).

Note that the control volume used for the HRSG here is a single inlet, single outlet device on just the H_2O side, so that there is heat transfer from the hot gas side to the H_2O (which was an internal process in the control volume used for Part 2). The energy balance (1st Law analysis) is straightforward for this control volume, but the entropy balance (2nd Law analysis) is not, since the boundary temperature at which the heat transfer occurs is not well defined. The total entropy generation rate within the HRSG was calculated in Part 2 by treating it as a 2 inlet, 2 outlet device, and that value can be added to this worksheet directly.

Similarly, the heat transfer rate from the H_2O in the condenser generally takes place in a heat exchanger (2 inlet, 2 outlet) from the H_2O in the Rankine cycle to a cooling liquid (usually a separate water stream). If drawn from a river, the cooling liquid can be returned to the river a bit warmer (a form of thermal pollution). Alternatively, it can be sent to a cooling tower, where heat is transferred to ambient air. For this part of the Workshop, the heat transfer from the H_2O is considered to take place directly to ambient (equal to T_1 from Part 1) for the purposes of 2nd Law analysis. The next and final part of the Workshop will incorporate a cooling water loop and cooling tower.

Complete the State, Process and Results Tables (Fig. 11.19), and sketch on the *T-s* diagram for H_2O (Fig. 11.20, which is to scale), assuming:

- Steady-state operation, with negligible stray heat, kinetic and potential energy changes in all devices.
- The pump is adiabatic and reversible (hence isentropic).
- The steam turbine is adiabatic with an irreversibility defined by an isentropic efficiency (0.85).
- Pressure drops across the HRSG and condenser are negligible.

Workshop 11, Part 4: Cooling Tower and Compiled Results

This final part analyzes a simplified version of the elaborate system needed to accomplish the condensation process in the Rankine cycle. A complete schematic of the power plant that includes a cooling water loop and a cooling tower was shown in Fig. 11.1 and repeated here.

In part 3, the chosen control volume for the condenser included just the H_2O that circulates through the Rankine cycle. From this view, energy is removed in the form of heat and a 1st Law (energy) analysis is straightforward. However, a 2nd Law analysis is cumbersome, because the temperature of boundary where heat crosses is

INPUT PARAMETERS

T∞		oC, dead state temperature
m'H2O		kg/sec, mass flow rate of H2O
nt	0.85	steam turbine isentropic efficiency

STATE TABLE

State	P (bar)	T (oC)	h (kJ/kg)	s (kJ/kg/K)	x
a					
bs					
b					
c					0
d					

PROCESS TABLE (all quantities in MW)

PROCESS	DEVICE	$m'(h_{out}-h_{in})$	Q'_{in}	W'_{out}	eX'_D
a-b	Steam Turbine				
b-c	Condenser				
c-d	Pump				
d-a	HRSG				
	NET				

Fig. 11.19 Spreadsheet template for Combined Cycle workshop, part 3. The yellow highlighted cells are inputs for part 3 (steam turbine cycle). The blue cells involve inputs from parts 1 and 2 (topping cycle and HRSG)

not well defined. Therefore, the entropy transfer associated with the heat transfer is not readily obtained. However, in this part, the condenser is recognized as a heat exchanger, and the control volume is chosen to include both the Rankine H_2O and a separate stream of compressed liquid water (called cooling water). The heat transfer process between them is internal. Stray heat is generally neglected, justifiably, but if included, it occurs with the local ambient temperature and that entropy transfer can be calculated.

High quality steam from the steam turbine (state b) enters the condenser and exits as a saturated liquid (state c) at the same pressure. Cooling water enters in a separate stream at a specified temperature and exits at a temperature that must be below the temperature at which the steam condenses. For this case study, the cooling water enters at 30 °C and an absolute pressure of 1.5 atm (state E) and exits at 33 °C at atmospheric pressure (state F). The cooling water temperature is everywhere below the saturation temperature and is therefore a compressed liquid for all states. The elevated pressure at state E is needed to estimate the power required by the cooling water pump to push cooling water through the system.

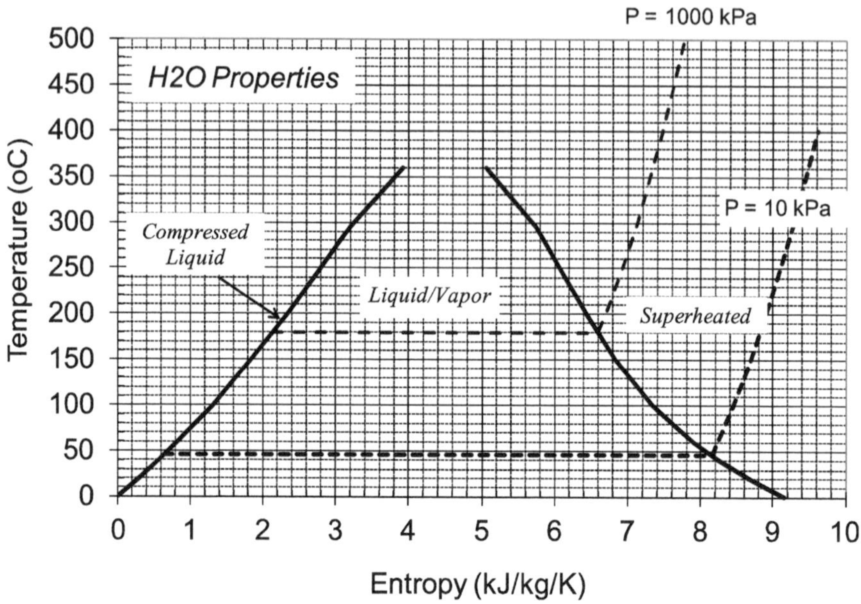

Fig. 11.20 Blank T_s diagram (to scale) for H_2O, with 2 representative isobars

After leaving the condenser at an elevated temperature, the cooling water can be brought into direct contact with ambient air to lower its temperature (in a cooling tower) and then pressurized by a pump. In the schematic of the power plant, an open-type heat exchanger called a cooling tower is shown. The cooling water enters the cooling tower (at $T_F = 33$ °C in this case study) and is brought into direct contact with air. The air enters at ambient conditions and exits at a temperature below the water inlet ($T_I = 28$ °C for this case study). The cooling water exits at a temperature above ambient ($T_G = 30$ °C for this case study) and enters a pump that drives the secondary loop. A pressure rise across the pump (equal to the pressure drop across the cooling water loop) is set to 0.5 atm, somewhat arbitrarily, but could be considered to be a design objective for the fluid mechanical aspect of the piping system that will not be considered further here. Similarly, power may be needed to drive a fan to force air through the cooling tower and is not considered. These auxiliary power requirements reduce the net power output of the plant somewhat.

An important (but not dominant) effect on the thermodynamic performance of an open cooling tower is the evaporation of air. A small fraction of the water sprayed into the air stream evaporates, and the humidity of the air stream changes. The water that evaporates is added prior to the pump (make up water). In this case study, these evaporative effects are neglected, but will be revisited in the next chapter.

The following assumptions apply to these components:

- Steady-state operation.
- Stray heat, kinetic energy and potential energy are negligible in all devices.

- The cooling water is a compressed liquid, with a specific heat of 4.2 kJ/kg/K.
- Neglect evaporative effects and pressure changes in the cooling tower and treat the air as a calorically perfect gas with $c_p = 1.00$ kJ/kg/K and $R = 0.287$ kJ/kg/K (properties of air at 298 K, 1 atm).

Complete the State Tables and Results Table (Fig. 11.21) for the cooling water loop and cooling tower. Then complete the Compiled Results Table (Fig. 11.22) that summarizes key metrics for all four parts of the Workshop. Be mindful of the scaling with a number of gas turbines. That is, Part 1 analyzed a single gas turbine, and the power plant has 3 of them. For volumetric flowrates, use the locations indicated in parenthesis, since specific volume can change within a given mass flow stream.

Think about the significance of the numbers as they are being generated. You are encouraged to work in teams and develop a spreadsheet with a separate worksheet for each part, and designed to allow you to conduct parametric studies by changing input parameters. That may require some manual entry of H_2O property data within the tables.

Appendix 1: Pinch Point Analysis

The heat recovery steam generator (HRSG) is viewed as a counterflow heat exchanger with three distinct sections, shown schematically in Fig. 11.23. That is, hot exhaust gases enter the HRSG at the gas turbine exit conditions and a known mass flow rate (from the right side of the schematic). Water enters in a separate stream (state d, top left) at the fixed operating pressure of the Rankine cycle. Heat is transferred (internally) and the gas turbine exhaust gases exit at ambient pressure and the stack temperature while the water exits at the turbine inlet conditions. The three main subsections of the HRSG are preheater (where compressed liquid is heated to its boiling point), boiler (where phase change occurs) and superheater (where a saturated vapor is heated to a superheated condition, suitable for the turbine). A section for pollution control is shown between the preheater and boiler (for removing nitrogen oxides, NOx), but not studied any further here, and assumed to have minimal impact on the thermodynamics of a HRSG.

The exit condition of the exhaust gas is determined by the "pinch point" that occurs between the preheater and boiler sections. The gas temperature at that point must exceed the boiling point of the H_2O at its operating pressure. For this case study in which all inlet and outlet temperatures (and mass flow rate of air) are specified, this pinch point temperature difference will be calculated to confirm it is acceptable. It is possible to specify inlet and outlet temperatures that are reasonable (exit H_2O below inlet air temperature and exit air above inlet H_2O temperature), but would result in a negative pinch point temperature (air temperature below H_2O saturation temperature at the inlet to the boiler). The spreadsheet calculations to determine the internal temperatures were detailed in Table 11.1 and plotted in Fig. 11.9 (repeated here) versus the heat transfer rate (to the H_2O) at each section. The temperature of the

INPUT PARAMETERS

T_∞		oC, dead state temperature
m'_{H2O}		kg/sec, mass flow rate of Rankine cycle H2O
$v_{H2OL)}$	0.001	m3/kg, specific volume of all liquid H2O streams
c_L	4.2	kJ/kg/K, specific heat of all liquid H2O streams
$c_{p,air}$	1.00	kJ/kg/K, enthalpy specific heat of cooling air
R_{air}	0.287	kJ/kg/K, gas constant for air

Rankine cycle H2O State Table

state	P (bar)	T (oC)	h (kJ/kg)	s (kJ/kg/K)	x
b					
c					

Cooling Water State Table (state G as reference)

state	P(atm)	T (oC)	h (kJ/kg)	s (kJ/kg/K)
E	1.5	30		
F	1	33		
G	1	30	**0**	**0**

Cooling Tower Air State Table (state H as reference)

state	P(atm)	T (oC)	h (kJ/kg)	s (kJ/kg/K)
H	1	0	**0**	**0**
I	1	28		

RESULTS

m'cw		kg/s, mass flow rate of cooling water
m'air		kg/s, mass flow rate of air into tower
W'cw_pump		MW, power into cooling water pump
$eX'_{D,condenser}$		MW, eXergy destruction rate in condenser
$eX'_{D,cooling\ tower}$		MW, eXergy destruction rate in condenser

Fig. 11.21 Spreadsheet template for Part 4 (Cooling Tower). Shaded yellow cells are inputs to Part 4, blue shaded cells are results or inputs from parts 1, 2, or 3

Compiled Results

FLOW RATES	Mass (kg/sec)	Volumetric (m3/sec)	
Gas Turbine Air (total)			(at inlet)
Fuel			(CH4 @ dead state)
H2O, Rankine Cycle			(at pump inlet)
Cooling Water			(anywhere in cw loop)
Cooling Tower Air			(inlet)

ENERGY RATES (in MW)

Chemical Energy Addition	
Gas Turbine Power Output (NET)	
Steam Turbine Output	
Rankine Pump Input	
Cooling Water Pump Input	
TOTAL POWER OUTPUT	

EXERGY DESTRUCTION RATES (in MW)

Topping Cycle Compressor	
Toppin Cycle Combustor	
Topping Cycle Turbine	
HRSG	
Steam Turbine	
Condenser	
Cooling Tower	
TOTAL EXERGY DESTRUCTION	

EFFICIENCIES

Overall First Law Efficiency	
2nd Law Efficiency*	

*Actual Power/(Actual Power + Total eXergy Destruction)
different than Carnot based (because cooling tower air is warm)

Fig. 11.22 Compiled results for all parts of the workshop

air decreases linearly (from right to left), a consequence of constant specific heats. The H_2O temperature increases linearly where no phase change occurs (preheater and superheater) and is constant in the boiler section. The pinch point (212.8–179.9 = 32.9 °C) occurs between the preheating and boiling sections. This value is considered reasonable. The problem could be specified with a somewhat lower exit gas temperature to generate a bit more steam, but not much.

It is clear that most of the heat transferred is needed to change the phase, and very little heat is associated with superheating. Yet superheating is vital for the Rankine cycle to maintain the steam quality of the turbine exit sufficiently high to avoid mechanical erosion.

In the next example, the exit gas temperature will not be known, but the pinch point temperature difference will be set (to 20 °C, a design specification). An energy balance between the exhaust gas inlet and this point with the pinch point constraints will yield the mass flow rate of water. Further cooling of the exhaust gases occurs in the preheater, and another constraint on the HRSG is that the stack temperature must

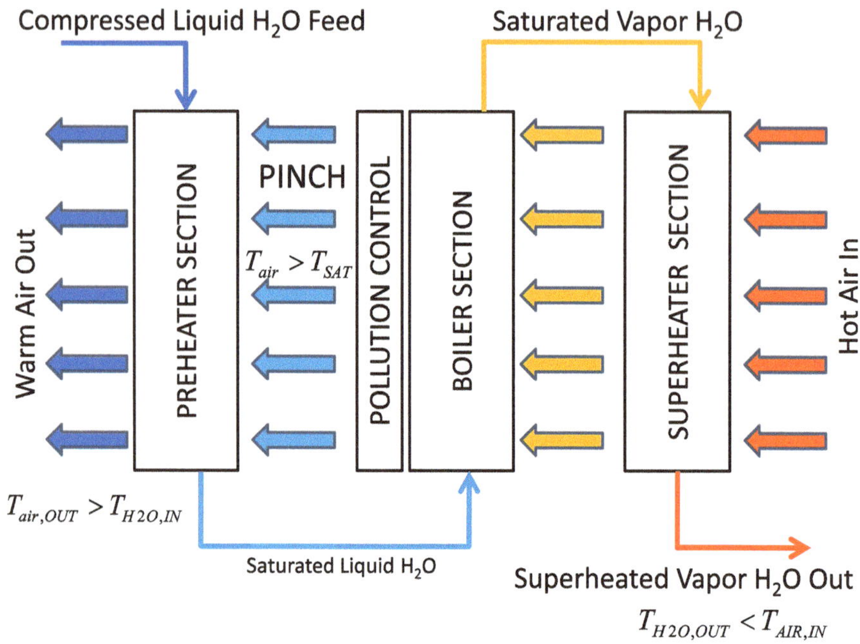

Fig. 11.23 Schematic of heat recovery steam generator

be warm enough for the chimney to achieve sufficient draft. In this case study, the stack temperature cannot be below 50 °C. One of these two constraints will uniquely define the design operating point of the HRSG, and thereby the mass flow rate of water that circulates in the Rankine cycle (the primary loop).

Workshop: Optimizing HRSG/Turbine Performance

This workshop is similar to the HRSG/turbine case study, with an important difference to demonstrate design optimization principles. The inlet hot gas stream from an internal combustion engine is the same (mass flow rate, temperature and pressure) but the exit air temperature is not specified. Rather, it is determined by a pinch-point constraint. Also, a different value of the Rankine cycle high side pressure is assigned to different student teams. The goal of this workshop is therefore two-fold: to optimize the performance of a basic HRSG with the constraint of the pinch point for a specified Rankine cycle pressure, and determine (by parametric study) the Rankine cycle pressure that maximizes power output of the bottoming cycle driven by the exhaust of an internal combustion engine.

The schematic for this workshop (Fig. 11.24) is modified from Fig. 11.7 to indicate these two effects. A parametric study of the effect of water side pressure

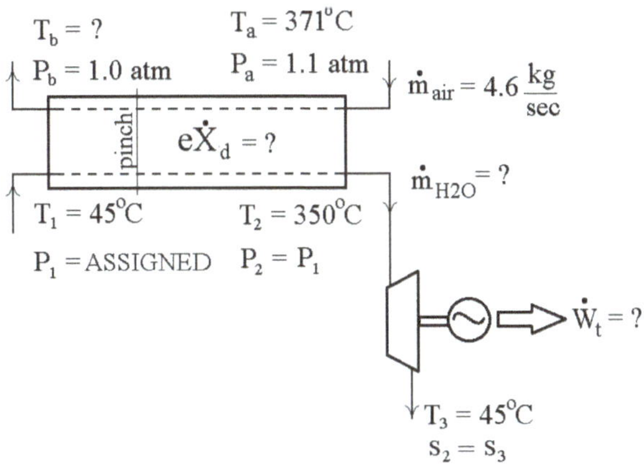

Fig. 11.24 Schematic of HRSG plus turbine

INPUT PARAMETERS

mdot,air	**4.6**	kg/sec
cp,air	**1.04**	kJ/kg/K, air specific heat, evaluated at average temp
R,air	**0.287**	kJ/kg/K, air gas constant
Tinf	**25**	oC, dead state temperature
Pinf	**1**	atm, ambient pressure

STATE PROPERTIES

state	*P (bar)*	*T (oC)*	*h (kJ/kg)*	*s (kJ/kg/K)*	*quality or phase*	
a	**1.5195**	**371**	359.84	0.68506096	**GAS**	air in
b	**1.013**	**158.947**	139.3045	0.38605687	**GAS**	air out
1	**10**	**45.81**	189.4403	0.6386	Comp. Liq	H2O in HRSG
2	**10**	**350**	3158.2	7.3029	SH Vapor	H2O out of HRSG
3	0.09595	**45**	2308.702	7.3029	0.885656571	H2O out of turbine

RESULTS

mdot,H2O	0.34171	kg/sec, H2O flow rate
Wdot,t	290.284	kW, turbine power
Xdot,D	268.754	kW, eXergy destruction rate in HRSG

PINCH POINT ANALYSIS OF HRSG

	WATER SIDE					AIR SIDE	
Qtransferred	**state**	***P (bar)***	***T (oC)***	***h (kJ/kg)***	***quality***	***state***	***T***
0	1	10	45.81	189.440309	NA, CL	b	158.9466
195.8252828	1f	10	179.88	762.51	0		199.88
884.2365921	1g	10	179.88	2777.1	1		343.7787
1014.463363	2	10	350	3158.2	NA, SH	a	371

ΔT, PP	**20**	oC, Pinch Point temperature difference (must be positive)

Fig. 11.25 Spreadsheet template for HRSG Optimization Workshop. Orange cells require input for the specific assigned Rankine high side pressure. Yellow cell represents an input gas temperature exit. The value shown is obtained using the "goal seek" feature to set pinch point temperature to a value of 20 °C

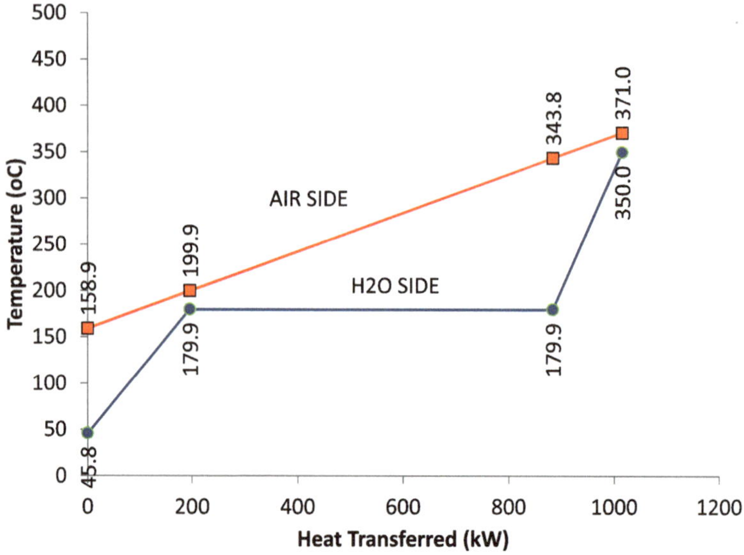

Fig. 11.26 Temperature versus heat transferred plot for the case of Rankine cycle high pressure of 10 bar and a pinch point temperature difference set to 20 °C

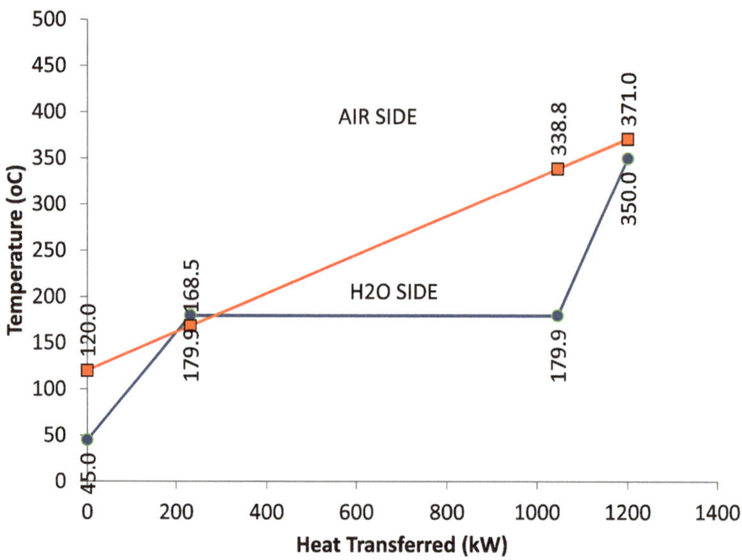

Fig. 11.27 Temperature versus heat transferred plot for the case of Rankine cycle high pressure of 10 bar and an exit air temperature set to 120 °C. The resulting pinch point temperature difference is −11.43 °C, the negative indicating that the 2nd Law is violated at the pinch point, and an exit air temperature this low would not be possible

on the overall performance will be conducted. To properly design the HRSG, a pinch point analysis will be conducted in order to determine the lowest allowable air temperature that can be obtained (extracting the most thermal energy from the air stream) given the 2nd Law constraint that the temperature of the air stream must everywhere be greater than that of the water stream.

A spreadsheet with the solution for an assigned pressure of 10 bar is shown in Fig. 11.25, with the corresponding temperature versus heat transferred plot in Fig. 11.26. The structure is similar, but not identical, to Fig. 11.10. The workshop consists of modifying the spreadsheet created for the HRSG/turbine case study for the assigned pressure (orange cells) and adding a plot for the temperatures. Everything should automatically update when the outlet air temperature is changed (T_b, the yellow cell). Once working, review the plot, then change T_b to a new value, and observe the changes in the output (temperature plot, mass flow rate of water, power developed in the turbine, eXergy destruction rate, and the pinch point temperature difference). Finally, use GOAL SEEK to set the pinch point to 20 °C by changing the inlet air temperature.

Results

Results from this workshop are reported and discussed in the remainder of this Appendix. First, the effect of pinch point temperature difference with the case of 10 bar is investigated. Figure 11.26 shows the air and H_2O temperatures inside the HRSG as a function of heat transferred for the design case (pinch point temperature difference set to 20 °C, with a Rankine cycle pressure of 10 bar. This case is considered to give near optimal performance, subject to a pinch point constraint. The air exit temperature of 158.9 °C shows that there is considerable thermal energy vented out the stack, a consequence of the pinch point.

Figure 11.27 shows a case where the exit temperature is set to a value (120 °C) that would appear to be reasonable for a stack temperature, but it would not be possible to design a HRSG with a single Rankine pressure because of the pinch point. Figure 11.28 shows a case where the exit air temperature is close to the inlet temperature. This case would produce minimal H_2O flow rate.

Figure 11.29 shows a compiled plot of the effect of pinch point temperature difference on the turbine power output and exit air temperature.

Figure 11.30 shows the effect of Rankine cycle high pressure on the mass flow rate of H_2O produced and the exit air temperature. Figure 11.31 shows the effect on the power produced in the steam turbine and the eXergy destruction rate in the HRSG. These results are compiled numbers submitted by students from an actual course, the few outlier points are indications of students that made some error in their calculations.

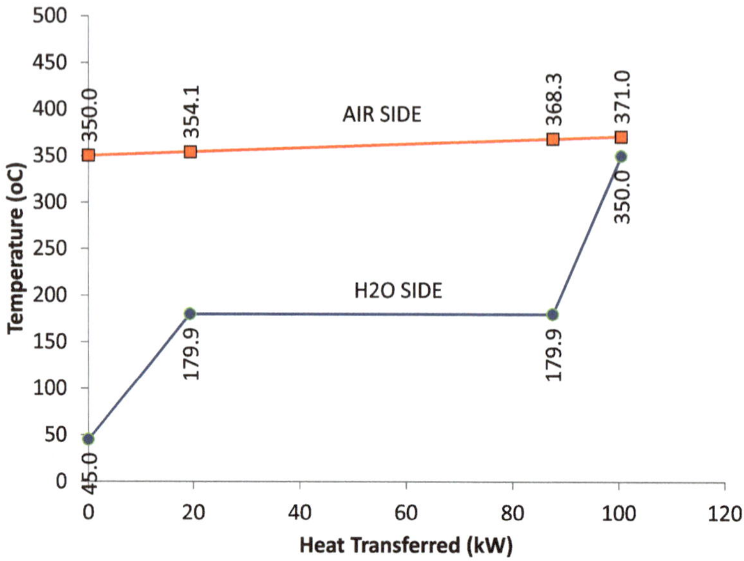

Fig. 11.28 Temperature versus heat transferred plot for the case of Rankine cycle high pressure of 10 bar and an exit air temperature set to 350 °C. The resulting pinch point temperature difference is 174.2 °C. The mass flow rate of H_2O produced is very low in this case (0.034 kg/sec)

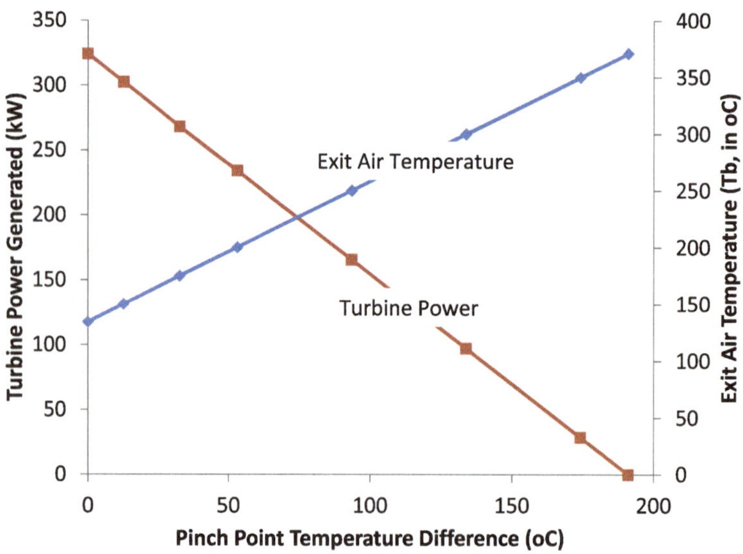

Fig. 11.29 Effect of pinch point temperature difference on steam turbine power and exit air temperature

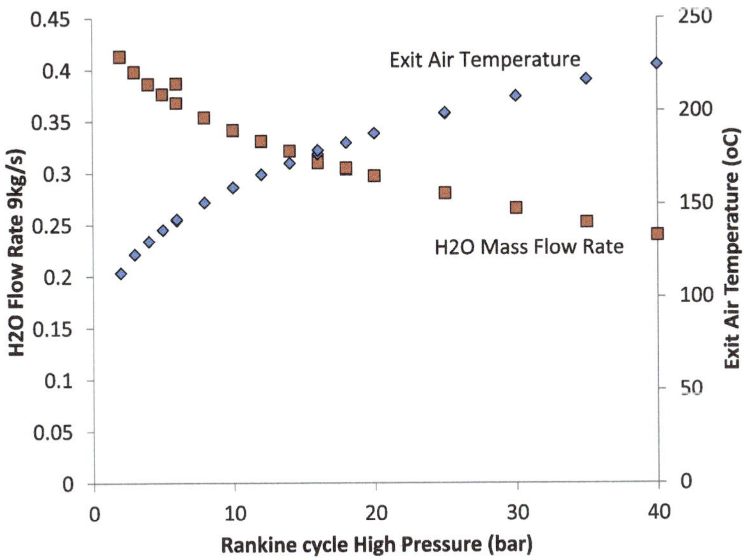

Fig. 11.30 H_2O mass flow rate produced in HRSG and exit air temperature as a function of Rankine cycle high pressure

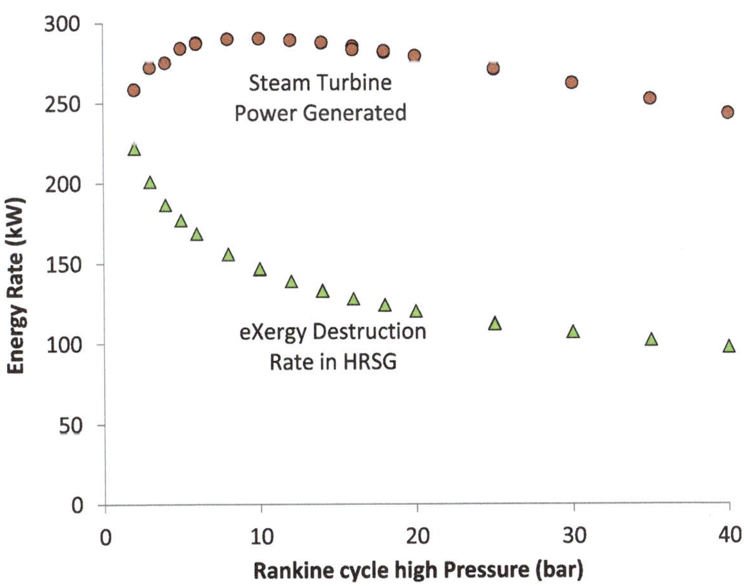

Fig. 11.31 Steam turbine power produced and eXergy destruction versus Rankine cycle high pressure

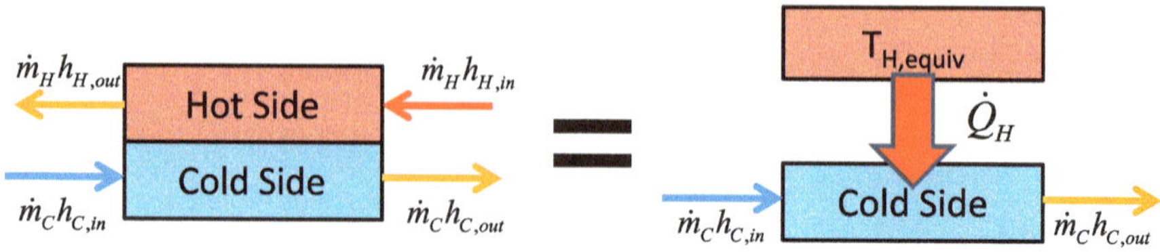

$$\dot{m}_H\left(h_{H,out}-h_{H,in}\right)=\dot{m}_C\left(h_{C,in}-h_{C,out}\right)$$

$$\dot{\Sigma}=\dot{m}_H\left(s_{H,out}-s_{H,in}\right)+\dot{m}_C\left(s_{C,out}-s_{C,in}\right)$$

$$\dot{Q}_H=\dot{m}_C\left(h_{C,in}-h_{C,out}\right)$$

$$\dot{\Sigma}=\frac{-\dot{Q}_H}{T_{H,equiv}}+\dot{m}_C\left(s_{C,in}-s_{C,out}\right)$$

Fig. 11.32 Representation of a heat exchanger with the hot fluid as an equivalent thermal reservoir source. Energy and entropy balances are shown (at steady-state, with negligible stray heat and kinetic and potential energy changes)

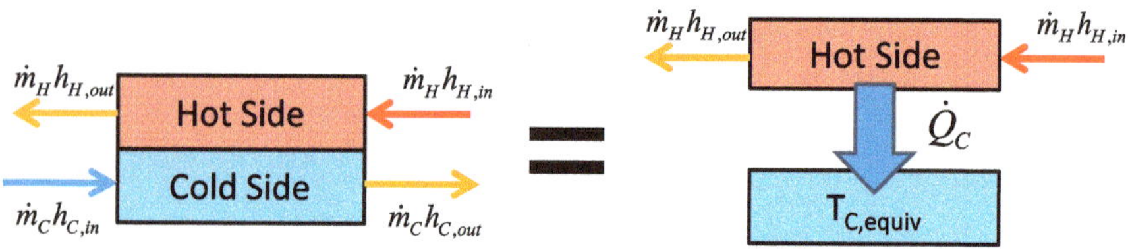

$$\dot{m}_H\left(h_{H,out}-h_{H,in}\right)=\dot{m}_C\left(h_{C,in}-h_{C,out}\right)$$

$$\dot{\Sigma}=\dot{m}_H\left(s_{H,out}-s_{H,in}\right)+\dot{m}_C\left(s_{C,out}-s_{C,in}\right)$$

$$\dot{Q}_C=\dot{m}_H\left(h_{H,in}-h_{H,out}\right)$$

$$\dot{\Sigma}=\frac{\dot{Q}_C}{T_{C,equiv}}+\dot{m}_H\left(s_{H,in}-s_{H,out}\right)$$

Fig. 11.33 Representation of a heat exchanger with the cold fluid as an equivalent thermal reservoir sink

Appendix 2: Equivalent Reservoir Temperature of a Heat Exchanger

The objective of a heat exchanger is to transfer heat from one fluid to another across a physical wall, either to change the temperature of a stream or to change its enthalpy. Generally, the temperature of both streams changes, and neither stream behaves like a thermal reservoir. However, it may be instructive for 2nd Law analysis to treat one of the streams as a thermal reservoir with a temperature that satisfies both 1st and 2nd Laws. Figure 11.32 shows the case for which the hot side fluid is treated as a thermal reservoir. For example, in the HRSG of a combined cycle, the hot side would be the exhaust gases, and the cold side is the H_2O in the Rankine cycle. The equivalent reservoir temperature would be an intermediate temperature between the hot gas inlet and the warm gas exit. Figure 11.33 shows the case for which the cold side is treated as a thermal reservoir. For example, the condenser of a Rankine cycle, where the hot side is the condensing Rankine H_2O, and the cold side is cooling water.

Combining the energy and entropy balances and equating the entropy generation rate for the case of the hot fluid as an equivalent thermal reservoir (Fig. 11.32), the equivalent thermal reservoir temperature is:

$$T_{\mathrm{H,equiv}} = \frac{h_{\mathrm{H,in}} - h_{\mathrm{H,out}}}{s_{\mathrm{H,in}} - s_{\mathrm{H,out}}}$$

Consider the case where the hot fluid is an ideal gas:

$$T_{\mathrm{H,equiv}} = \frac{c_p(T_{\mathrm{H,in}} - T_{\mathrm{H,out}})}{c_\mathrm{p} \ln\left(T_{\mathrm{H,in}}/T_{\mathrm{H,out}}\right) - R \ln\left(P_{\mathrm{H,in}}/P_{\mathrm{H,out}}\right)} \quad \text{for ideal gases on hot side}$$

If the process is isobaric, with $T_{\mathrm{H,\ out}} = T_{\mathrm{H,\ in}} - \Delta T_\mathrm{H}$:

$$T_{\mathrm{H,equiv}} = \frac{(T_{\mathrm{H,in}} - T_{\mathrm{H,out}})}{\ln\left(T_{\mathrm{H,in}}/T_{\mathrm{H,out}}\right)} = \frac{\Delta T_\mathrm{H}}{-\ln\left(1 - \Delta T_\mathrm{H}/T_{\mathrm{H,in}}\right)}$$

Consider a high flow limit for the hot fluid, meaning that the flow rate is sufficiently high that the change in temperature of is small. Expanding the ln term:

$$T_{\mathrm{H,equiv}} = \frac{\Delta T_\mathrm{H}}{-\left(1 - \Delta T_\mathrm{H}/T_{\mathrm{H,in}} + \left(\Delta T_\mathrm{H}/T_{\mathrm{H,in}}\right)^2/2 - \left(\Delta T_\mathrm{H}/T_{\mathrm{H,in}}\right)^3/3 + \ldots\right)}$$

Keeping one term:

$$T_{\mathrm{H,equiv}} \approx T_{\mathrm{H,in}} \quad \text{for isobaric ideal gases and } \Delta T_\mathrm{H} \ll T_{\mathrm{H,in}}$$

On the other hand, consider a low flow limit for the hot fluid, in which case it exits in thermal equilibrium with entering cold-side fluid ($T_{\mathrm{H,\ out}} = T_{\mathrm{C,\ in}}$)

$$T_{\mathrm{H,equiv}} = \frac{(T_{\mathrm{H,in}} - T_{\mathrm{C,in}})}{\ln\left(T_{\mathrm{H,in}}/T_{\mathrm{C,in}}\right)}$$

For example, if the hot fluid enters at 100 °C and the cold fluid enters at 0 °C, and the hot fluid is an ideal gas (with negligible pressure drop) and in a low flow limit, the equivalent hot reservoir temperature evaluates to 47.4 °C.

A similar result holds for liquids on the hot side (with no phase change):

$$T_{\mathrm{H,equiv}} = \frac{(T_{\mathrm{H,in}} - T_{\mathrm{H,out}})}{\ln\left(\frac{T_{\mathrm{H,in}}}{T_{\mathrm{H,out}}}\right)} \quad \text{for liquids on hot side}$$

Consider a phase changing case where the hot side fluid enters as a saturated vapor and exits as a saturated liquid at the same pressure. The equivalent thermal reservoir temperature is:

$$T_{\mathrm{H,equiv}} = \frac{h_{\mathrm{fg}}}{s_{\mathrm{fg}}} \quad \text{for isobaric condensation on hot side}$$

This analysis is therefore a formal derivation of the fundamental relationship between saturation temperature and the enthalpy and entropy of phase change, since the hot side fluid is fixed at the saturation temperature (corresponding to the constant pressure). That is:

$$\frac{h_{\mathrm{fg}}}{s_{\mathrm{fg}}} = T_{\mathrm{SAT@}P}$$

A similar analysis holds for the case where an equivalent cold side thermal reservoir temperature is sought (Fig. 11.33). In that case:

$$T_{\mathrm{C,equiv}} = \frac{h_{\mathrm{C,in}} - h_{\mathrm{C,out}}}{s_{\mathrm{C,in}} - s_{\mathrm{C,out}}}$$

Chapter 12
Psychrometrics

The field of psychrometrics pertains to engineering systems that involve humidity changes, or air/H_2O vapor mixtures, the subject of this final chapter. The theoretical methods developed in this chapter are launching points for more detailed studies in fields as diverse as combustion theory, chemical separation, meteorology, and climate studies. Mass-related concepts associated with gaseous mixtures, humidity (absolute and relative), and dew point are developed and then applied to determine the rate at which condensing water from a Rankine cycle evaporates in a cooling tower. Energy and entropy effects are developed for open systems with humidity changes (humidification and dehumidification applications). The difference between sensible energy (required to change temperature) and latent energy (required to change phase) is emphasized. Isothermal humidification involves only latent energy changes and demonstrates how energy (heat and/or work) must be added to change phase (and how eXergy is destroyed in the process). Adiabatic humidification demonstrates how a change in latent energy gives rise to an equal and opposite change in sensible energy, the result of which is air that is cooled by evaporation, up to a point. Evaporative cooling and air conditioning principles are thereby considered. All the major elements of a detailed psychrometric chart are developed, including adiabatic mixing of two humid streams. Workshops apply the methods to determine the biological mass and energy requirements associated with breathing, humidification needs of a household in winter, dehumidification needs in summer, and the energy cost associated with running a dehumidifier in a space that is simultaneously air conditioned.

Interpretation and use of a psychrometric chart is an important part of professional practice in the HVAC field. An example of a detailed psychrometric chart from the American Society of Heating, Refrigeration and Air Conditioning Engineers (ASHRAE) can be found at https://www.ashrae.org/File%20Library/Technical%20Resources/Bookstore/UP3/IP-1.pdf.

The basic elements of this imposing (and difficult to read without a magnifying glass) chart are developed methodically in this chapter.

G. Sidebotham, *An Inductive Approach to Engineering Thermodynamics*,
Mechanical Engineering Series, https://doi.org/10.1007/978-3-030-20430-3_12

Psychrometric Thought Experiment

Conduct this simple thought experiment, a warm-up to psychrometrics. Place a cup of water on a table and let it sit there. Is it in equilibrium? If you filled it with cold tap water (say at 15 °C), you'd say no. It quickly attains mechanical equilibrium, since the pressure of the liquid equals that of the room it is in and there is minimal motion (after a period of time during which any stirring settles down). However, it is not in thermal equilibrium. Given time, say an hour or so, heat will flow, driven by the temperature difference. The heat transferred will increase the internal energy, the H_2O will lose its eXergy and its temperature will become equal to the temperature in the room. That is, it will achieve thermal equilibrium as well as mechanical equilibrium. Is it now in equilibrium? Leave the cup there overnight and return the next day. Who took a sip during the night? Leave it another day, and another. A week later and there is no more water. It evaporated. There must be some other kind of equilibrium that is not established. Something related to phase change.

Repeat the thought experiment, only this time place a sealed container (like a 2 L soda bottle) approximately half full of water on a table. You find that this situation will reach mechanical and thermal equilibrium within an hour or so, but now, the water will not evaporate over time. What is the difference? In the open container experiment, the room is not a closed system but is ventilated. The open cup is not in phase equilibrium with the air in the room. Some of the molecules near the liquid/air interface have energy to overcome the intermolecular attraction and enter into the gas phase, never to return. In the closed container, molecules evaporate, but the number of molecules trapped in the air builds up over time. As they do, a portion of them strike the liquid/air interface and return to the liquid phase. Eventually, a dynamic equilibrium is established for which the rate at which molecules leave the liquid phase and enter the gas phase becomes equal to the rate at which molecules from the gas phase become absorbed in the liquid. As will be discussed, phase equilibrium is established when the partial pressure of H_2O in the gas phase equals the saturation pressure of H_2O at the corresponding temperature.

The liquid water is in direct contact with the humid air in the room at the free surface of the water. The chemical composition is different in the liquid phase and the gas phase as the air/water interface is crossed. The combined air/water system is not a pure substance (defined as two phases having the same chemical composition). Up to this point, whenever there were two phases present (liquid and vapor), there was a single chemical component, for example, R134a for refrigeration systems and H_2O for steam turbine cycles. However, there are important applications in engineering practice in which two phases are present but the chemical composition of the liquid differs from that of the gas. Psychrometrics applies when the liquid is nearly pure H_2O, the gas is humid air, and the conditions of pressure and temperature are in the general vicinity of normal weather and those desired for human comfort in interior spaces. The principles developed are applicable to other situations as well (e.g. fractional distillation of crude oil), but not developed in detail here.

In chemical engineering language, the psychrometric situation is classified as a two-phase system with a single noncondensable component. The air is always in the gas phase (noncondensable), while the H_2O can pass between gas and liquid phases. The energy changes associated with temperature changes are called "sensible energy," and those related to phase change are called "latent energy."

Humidity is an important measure associated with the comfort of air. Historically, a device called a sling psychrometer was the go-to method and language from that device persists, namely the term dry bulb and wet bulb temperature. Humidity is now generally measured with electronic sensors. A sling psychrometer consists of two liquid in glass thermometers. One is wrapped in a moistened cloth-like material and the other is not. The device is slung around the user like a sling-shot. Air passes through the moistened cloth and evaporates some liquid, which requires energy and lowers the temperature of the air. The wet bulb will give a temperature reading that is lower than the dry bulb. The difference between the wet bulb and dry bulb temperatures is related to the humidity in the air being measured, the theory of which is developed in this chapter.

A problem statement that introduces psychrometric principles related to a cooling tower is shown in Fig. 12.1. This problem will be solved once the appropriate background is developed. Ambient air enters from the bottom at a stated relative humidity (from a weather report, for example). Warm liquid H_2O (temperature not

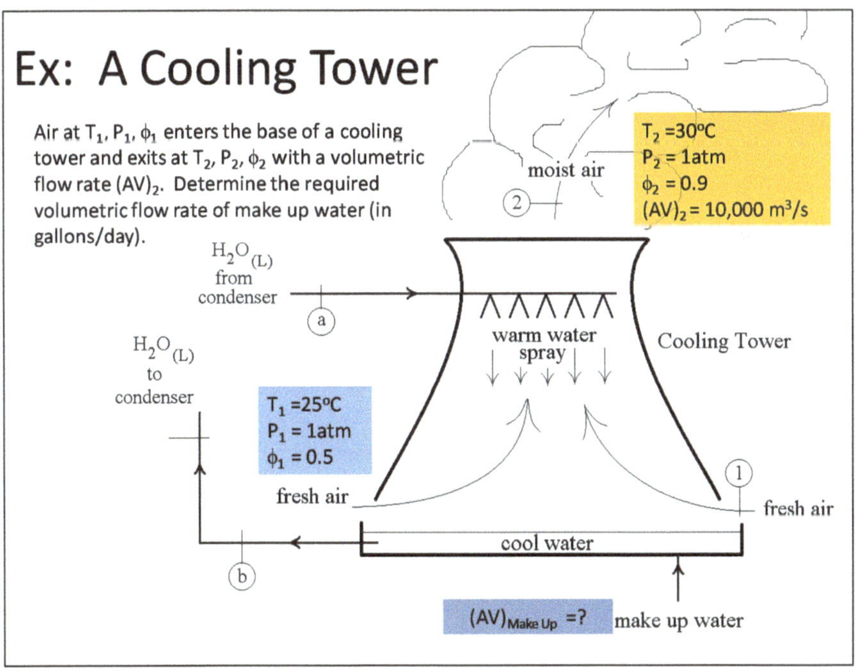

Fig. 12.1 Case study: Cooling tower problem statement that introduces psychrometric terminology related to mass principles

Fig. 12.2 Cooling tower plume from a coal power plant from a few miles distance

specified for this problem) from the condenser of a steam turbine cycle enters from the top and is sprayed into the air. Some of the H_2O evaporates into the air, and there is internal heat exchange. Therefore, the air exits warmer and with a higher relative humidity, with a specified volumetric flow rate. The goal of this mass-based calculation is to determine the required flow rate of make-up H_2O (liquid) needed to return the same flow rate to the condenser. The energy consequences of this problem will be revisited, which is needed to determine the flow rate of condenser cooling H_2O associated with this tower.

Some photographs of the plume from the cooler towers of a coal power plant (Drake power plant in Colorado Springs, Colorado) are shown in Figs. 12.2, 12.3, and 12.4. There are two plumes that may be visible due to psychrometric effects: the stack gas and the cooling tower plume. Figure 12.2 is a view of these plumes from a few miles away (in front of the coffee shop in which much of this book was written). The stack gas (from the tall chimney in Fig. 12.3) is from the products of the combustion of coal with air. Most of the O_2 from the air has been converted to CO_2 (an invisible greenhouse gas pollutant) and H_2O vapor, which is made visible as a cloud when it condenses as it cools in the ambient air. There are other minor chemical species (NO_x, SO_x, and particulates) at lower levels. On the other hand, the cooling tower plume is a humid air mixture, consisting of H_2O vapor in the main components of air (O_2 and N_2). The mole fraction of H_2O vapor in both plumes is

Fig. 12.3 Closer view of cooling towers (5 in a row). Note that stack gas (from the tall chimney behind the vapor cloud) is not visible under the conditions when this photo was taken

comparable and both plumes exit the power plant warm. The H_2O vapor condenses as the plume cools to ambient temperature if the ambient temperature is below the dew point temperature of the exiting plume (as will be explained). A similar process takes place in the visible vapor trail of jet engines.

Humid Air: Mass Considerations for Gas Mixtures

After some warm-up questions aimed at building a fundamental understanding of humidity, a methodology for analyzing gas mixtures is developed based on mass conservation, 1st Law and 2nd Law balances. The focus of this section is on mass-related effects, with qualitative discussions of energy effects. Quantitative methods that involve energy are considered in the following section. Then several workshops are introduced to investigate various applications of engineering importance.

Warm-Up Questions: In this Room...

What Is the Mass of Air in this Room?

Of course, that depends on what room you're in. Consider an average sized living room, with dimension 12 ft × 20 ft × 8 ft (240 ft^2 floor space, volume of

Fig. 12.4 Stack gas (from tall chimney) and cooling tower plume from a coal power plant. Approximate chemical composition is shown (in mole fractions, *x*, not to be confused with eXergy, here)

1920 ft^3 = 54.4 m^3). Assume that the pressure is nominal room conditions (1 atm, 25 °C). From what we've learned so far, the ideal gas law applies (with the gas constant of air being 0.287 kJ/kg/K), and the mass of air in the room is:

$$M_{\text{air}} = \frac{PV}{R_{\text{air}}T} = \frac{(1\ \text{atm})(12*20*8\ \text{ft}^3)}{(0.287\ \text{kJ/kg/K})(298\ \text{K})} \cdot \left|\frac{101.3\ \text{kN/m}^2}{\text{atm}}\right| \cdot \left|\frac{0.3048\ \text{m}}{\text{ft}}\right|^3$$

$$M_{\text{air}} = 64.4\ \text{kg}$$

Some people might be surprised by that simple calculation. It does not seem like the air in a typical room has a mass that would "weighs" about 150 pounds. But it does. It does not register that because its buoyant force equals its weight.

What Is the Mass of Water Vapor in this Room?

This question reveals that there is something incomplete in how "air" has been treated thus far. For the purposes of conducting engineering analysis with devices that use air as the working fluid (such as reciprocating engines and gas turbines), it has been acceptable to treat air as a calorically perfect gas with a gas constant $R = 0.287$ kJ/kg/K (corresponding to a molecular weight of 29.0 kg/kmol) and specific heat value selected at an appropriate average temperature. That approximation is not helpful here and a closer look at the precise composition of air is warranted. It will be shown shortly that if the relative humidity (ϕ) is 50%, then the absolute humidity (ω, the ratio of mass of vapor to mass of dry air) under these conditions (T and P) is approximately 0.01. That is

$$\omega = \frac{\text{Mass of } H_2O \text{ vapor}}{\text{Mass of bone dry air}} = \frac{M_v}{M_{BDA}} = 0.01$$

Air with H_2O vapor (denoted as H_2O_v) is called moist air, or humid air. The total mass of humid air is the sum of the H_2O_v and the rest of the air (called bone dry air, BDA). That is,

$$M_{air} = M_{BDA} + M_v$$

Manipulating in terms of the absolute humidity:

$$M_{air} = M_v(M_{BDA}/M_v + 1) = M_v(1/\omega + 1)$$

Rearranging for the mass of vapor

$$M_v = M_{air}\frac{\omega}{1+\omega} \approx \omega M$$

Evaluating for this case:

$$M_v = 0.644 \text{ kg}$$

A typical room has on the order of a kg of H_2O_v as part of the humid air.

What Volume Would the Water Vapor Occupy if It Were Condensed?
If all that H_2O were condensed to a pure liquid, its volume would be:

$$V = vM_v = \left(0.001\frac{\text{m}^3}{\text{kg}}\right)0.644 \text{ kg} \cdot \left|\frac{1000 \text{ L}}{\text{m}^3}\right|$$

$$V = 0.644 \text{ L}$$

Note, 1 L of liquid H_2O has a mass of approximately 1 kg, a convenient and often used conversion.

Why Is it a Vapor and Not a Liquid?
The temperature is 25 °C and the pressure is 1 atm. Shouldn't H_2O be a compressed liquid under those conditions? The saturation pressure at 25 °C is 0.0313 bar, and the

total pressure is much higher than that. However, the phase is gaseous, and the pressure is principally due to the presence of air (primarily N_2 and O_2), not H_2O. It will be shown shortly that a condensable chemical species in a gas mixture (H_2O_v in air in this case) exerts a *partial* pressure which depends on its relative amount. If the partial pressure of H_2O in moist air is less than the corresponding saturation temperature, it will be a superheated vapor, mixed with the air. A way to state that is to specify a relative humidity less than 100%.

Saturation Curve Fit for H_2O

Phenomena associated with humidity depend heavily on the saturation curve of H_2O in the temperature and pressure ranges associated with human living conditions. It is useful to obtain a working formula for the saturation curve (relationship between saturation pressure and temperature). A curve fit using the Clausius–Clapeyron relationship (derived in Chap. 4) over a temperature range of 0 to 60 °C is detailed in Fig. 12.5. Four points are obtained from a saturation table and the saturation pressure is plotted against inverse absolute temperature on semilog axes. The power law fit is shown to a high precision. The resulting curve is plotted directly to show the accuracy involved. This curve forms the basis of the psychrometric charts to be developed. The final working formula (using 4 significant figures) is

$$P_{\mathrm{SAT@}T} = 1.567(10)^6 e^{\left(\frac{-5283}{T+273}\right)} \ [=]\mathrm{bar} \ \text{with} \ T \ \text{in} \ {}^\circ\mathrm{C}$$

Rearranging for the saturation temperature in terms of the pressure:

$$T_{\mathrm{SAT@}P} = \frac{5283}{\ln\left(1.567(10)^6/P\right)} - 273 \ [=]{}^\circ\mathrm{C} \ \text{with} \ P \ \text{in bar}$$

Bone Dry Air (BDA), Partial Pressures, and Mole Fractions

As in the warm-up question, it is useful to separate humid air into two components, H_2O vapor (which is a condensable component) and bone dry air (BDA) which contains all the other components of air. For engineering systems that involve humidity changes (the field of Psychrometrics), the composition of BDA is adequately modeled to consist of two nonreactive chemical species, 79 mol% N_2 and 21 mol% O_2, and therefore, the properties of BDA are those of a single component with a molecular weight of 28.8 kg/kmol. The deeper dive of Appendix 1 briefly discusses more precise BDA, which consists of other chemical species at lower concentrations. These species do not play a significant role in the behavior of engineering systems, but can be important as pollutants (both global and local).

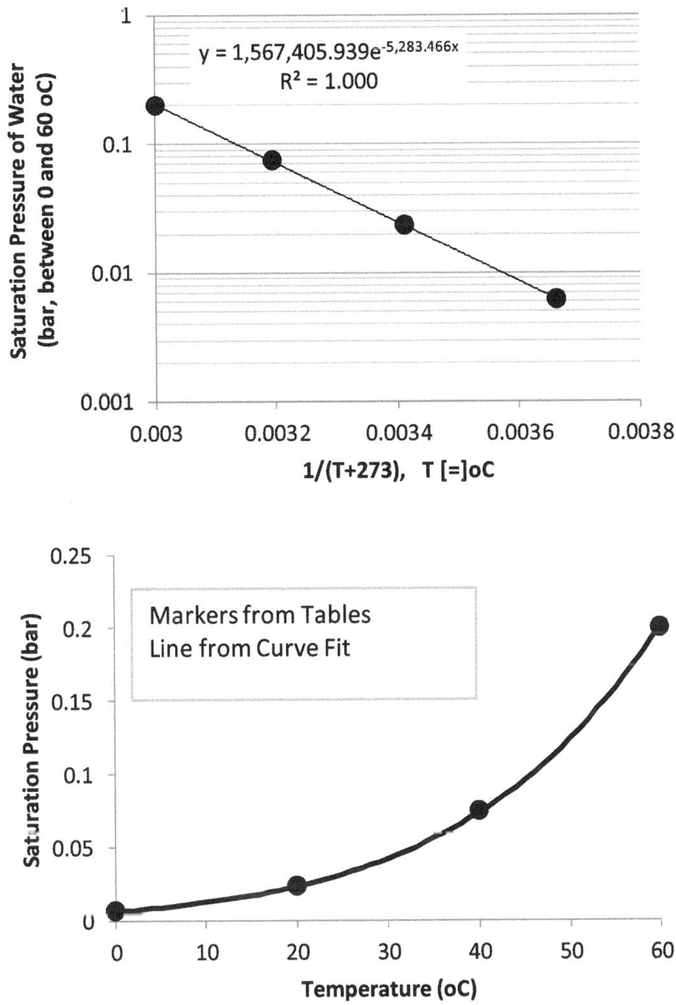

Fig. 12.5 Curve fit for saturation curve for H_2O over temperature range from 0 to 60 °C

Dalton's Law of Partial Pressures forms the foundation of gas mixture analysis, explained in Fig. 12.6. A sample of BDA is shown on the left (with two molecules of O_2 and eight of N_2). The molecules move in random directions with random speeds (the average of which increases with temperature). They do not influence each other except when they collide (the basis for perfect gas behavior). The momentum of a molecule is reversed when it strikes the wall. The collective impulse of all the wall collisions gives rise to the pressure exerted on the walls, and the ideal gas law in its molar form $\left(PV = n\widehat{R}T\right)$ is determined from kinetic theory (or experiment). There is no differentiation between N_2 and O_2 in this regard, and "n" is the total number of moles. However, there are approximately four times as many wall collisions of N_2 as

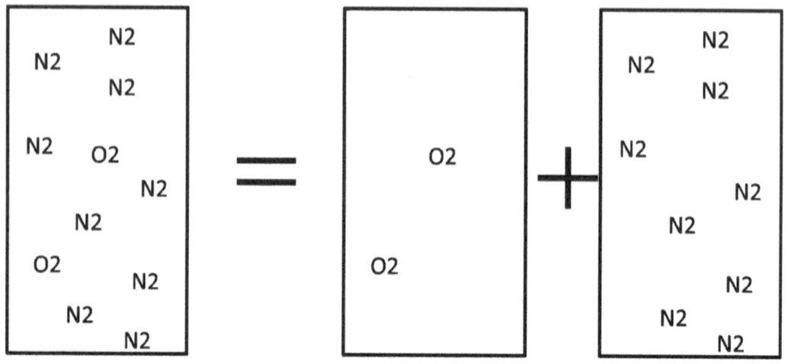

$$\frac{(n_{O2}+n_{N2})\hat{R}T}{V}=\frac{(n_{O2})\hat{R}T}{V}+\frac{(n_{N2})\hat{R}T}{V}$$

$$P=\frac{(n_{O2})}{(n_{O2}+n_{N2})}\frac{(n_{O2}+n_{N2})\hat{R}T}{V}+\frac{(n_{N2})}{(n_{O2}+n_{N2})}\frac{(n_{O2}+n_{N2})\hat{R}T}{V}$$

$$P=x_{O2}P+x_{N2}P$$

$$P=P_{O2}+P_{N2}$$

Fig. 12.6 Representation of Dalton's Law of Partial Pressures for bone dry air (BDA)

there are for O_2. The two right boxes show a hypothetical situation in which the O_2 and N_2 can be separated, as if they were independent. Both boxes have the same volume and the temperature of the two pure gases is the same. However, the pressure in each box is lower than the total, but they sum to the total pressure.

The equations underneath the boxes show the pressure expressed by the ideal gas law in various forms. First, the total number of moles (n) is split into the sum of O_2 and N_2. Then, each term is multiplied and divided by the total number of moles. This manipulation defines the mole fraction of each species (the moles of a species divided by the total number of moles):

$$\text{Mole fraction of } O_2=\left(\frac{n_{O_2}}{n}\right)=x_{O_2}$$

The mole fraction of a general species "*i*" is given the symbol "*x*," but many sources give it the symbol "*y*," so it important to know the nomenclature of a given source. Of course, "x" here is a totally different quantity than "x" used for eXergy.

Each component of a gas mixture exerts a partial pressure, defined by, and expressed in different equivalent ways:

$$\text{Partial pressure of } O_2=\left(\frac{n_{O_2}}{n}\right)P=x_{O_2}P=P_{O_2}$$

The molecular weight (MW) of a gas mixture is related to mole fractions and the molecular weights of individual components as follows:

$$\mathrm{MW}_{\mathrm{mix}} = \frac{\mathrm{Mass}_{\mathrm{mix}}}{Mol_{\mathrm{mix}}} = \frac{\mathrm{Mass}_{\mathrm{N_2}} + \mathrm{Mass}_{\mathrm{O_2}}}{\mathrm{Mol}_{\mathrm{N_2}} + \mathrm{Mol}_{\mathrm{O_2}}} = \frac{\mathrm{Mol}_{\mathrm{N_2}}\left(\frac{\mathrm{Mass}_{\mathrm{N_2}}}{\mathrm{Mol}_{\mathrm{N_2}}}\right) + \mathrm{Mol}_{\mathrm{N_2}}\left(\frac{\mathrm{Mass}_{\mathrm{N_2}}}{\mathrm{Mol}_{\mathrm{N_2}}}\right)}{\mathrm{Mol}_{\mathrm{N_2}} + \mathrm{Mol}_{\mathrm{O_2}}}$$

The last form is obtained by multiplying and dividing by the appropriate moles of the individual components, which brings in both mole fractions and molecular weights of individual components. The result is:

$$\mathrm{MW}_{\mathrm{mix}} = x_{\mathrm{N_2}}\mathrm{MW}_{\mathrm{N_2}} + x_{\mathrm{O_2}}\mathrm{MW}_{\mathrm{O_2}} = \sum_{\text{component } i} x_i \mathrm{MW}_i$$

The last form (summation) is a generalization to gas mixtures with components "*i*." Evaluating for BDA:

$$\mathrm{MW}_{\mathrm{BDA}} = 0.79\left(28\,\frac{\mathrm{kgN_2}}{\mathrm{kmolN_2}}\right) + 0.21\left(32\,\frac{\mathrm{kgO_2}}{\mathrm{kmolO_2}}\right) = 28.8\,\frac{\mathrm{kg}}{\mathrm{kmol}}$$

Dalton's Law of partial pressures is demonstrated on a molar (or number) basis because the universal gas constant is common to all species. However, in engineering systems, it is often more convenient to work in mass units. The relationship between mass fractions (mass of a component divided by the total mass, given the symbol "*y*") and mole fractions is occasionally required, derived as:

$$y_i = \frac{\mathrm{Mass}_i}{\mathrm{Mass}_{\mathrm{mix}}} = \frac{\mathrm{Mol}_i(\mathrm{Mass}_i/\mathrm{Mol}_i)}{\mathrm{Mol}_{\mathrm{mix}}(\mathrm{Mass}_{\mathrm{mix}}/\mathrm{Mol}_{\mathrm{mix}})} = x_i\left(\frac{\mathrm{MW}_i}{\mathrm{MW}_{\mathrm{mix}}}\right)$$

The mass fractions of BDA are:

$$y_{\mathrm{N_2}} = 0.79\left(\frac{28}{28.8}\right) = 0.768\,\frac{\mathrm{kg}_{\mathrm{N_2}}}{\mathrm{kg}_{\mathrm{mix}}} \quad \text{and} \quad y_{\mathrm{O_2}} = 0.21\left(\frac{32}{28.8}\right) = 0.233\,\frac{\mathrm{kg}_{\mathrm{O_2}}}{\mathrm{kg}_{\mathrm{mix}}}$$

The mass fraction of O2 is higher than its mole fraction because its molecular weight is higher than the mixture average.

Note that the molecular weight of a mixture can be determined from mass fractions, but not by the following mass fraction form:

$$\mathrm{MW}_{\mathrm{mix}} \neq \sum_{i \text{ component}} y_i \mathrm{MW}_i$$

Rather, as derived from a word statement, the mass fractions and molecular weights of individual components is related to the mixture by:

$$\frac{1}{\mathrm{MW}_{\mathrm{miz}}}=\frac{\mathrm{mol}_{\mathrm{mix}}}{\mathrm{mass}_{\mathrm{mix}}}=\frac{\left(\frac{\mathrm{mol}_{\mathrm{O}_2}}{\mathrm{mass}_{\mathrm{O}_2}}\right)\mathrm{mass}_{\mathrm{O}_2}+\left(\frac{\mathrm{mol}_{\mathrm{N}_2}}{\mathrm{mass}_{\mathrm{N}_2}}\right)\mathrm{mass}_{\mathrm{N}_2}}{\mathrm{mass}_{\mathrm{mix}}}=\frac{y_{\mathrm{O}_2}}{\mathrm{MW}_{\mathrm{O}_2}}+\frac{y_{\mathrm{N}_2}}{\mathrm{MW}_{\mathrm{N}_2}}$$

Or in general for a gas mixture with chemical components "*i*":

$$\frac{1}{\mathrm{MW}_{\mathrm{mix}}}=\sum_{i\ \mathrm{component}}\frac{y_i}{\mathrm{MW}_i}$$

Absolute and Relative Humidity

Consider the experimental arrangement and series of processes depicted in Fig. 12.7. A closed container is fitted with a syringe (from which liquid H_2O can be injected) and a piston/cylinder which allows for control of the total volume. This closed arrangement will be used to apply mass principles and then extended to open systems for energy and entropy principles. This section will focus on mass principles. The colored dots in the vessel signify H_2O vapor molecules (not drops of liquid). Heat and work can be exchanged in a manner that sets the temperature and pressure at any state to any desired values, depending on the type of process specified.

Initially (state 0), the container is filled with bone dry air (a difficult task to achieve in practice). Going forward, state 1 will be considered the starting point for various processes. A small amount of liquid is injected in process 0–1. At

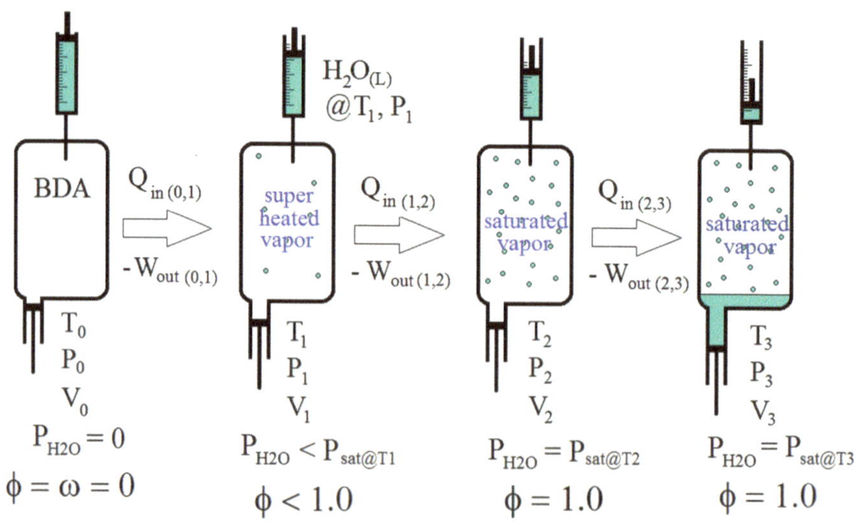

Fig. 12.7 Schematic of thought experiment to demonstrate humidity principles

equilibrium, all of the injected liquid evaporates, and the H_2O vapor exerts a partial pressure at state 1. However, the partial pressure of H_2O is less than the saturation pressure of H_2O at the corresponding temperature. Therefore, the H_2O is in the gas phase and it is superheated. From state 1 to 2, more liquid is injected, and it all evaporates. However, the partial pressure of H_2O vapor is equal to the saturation pressure at the corresponding temperature (T_2). The humid air at state 2 is saturated, and the air cannot hold any more H_2O vapor. If more liquid is injected, no further evaporation will occur and the injected liquid will simply fall to the bottom of the container. There is no real change of state from 2 to 3.

The relative humidity is a measure of the degree of superheat in humid air and depends strongly on air temperature. It is defined as

$$\phi = \frac{\text{Partial pressure of H2O vapor}}{\text{Saturation pressure at corresponding temperature}} = \frac{P_{H_2Ov}}{P_{SAT@T}}$$

The relative humidity can range from 0 to 1, and is often reported as a percent, not a fraction.

The partial pressure of H_2O vapor is plotted against temperature for fixed values of the relative humidity (as a fraction, not a percent) in Fig. 12.8. The temperature range (0–40 °C) spans that of typical outside air conditions, with desired interior temperatures in the middle. A relative humidity of 1 (100% relative humidity) corresponds to the saturation pressure (pressure at which H_2O would boil), and is the curve fit from Fig. 12.5. Lower relative humidity curves are simply fractions of

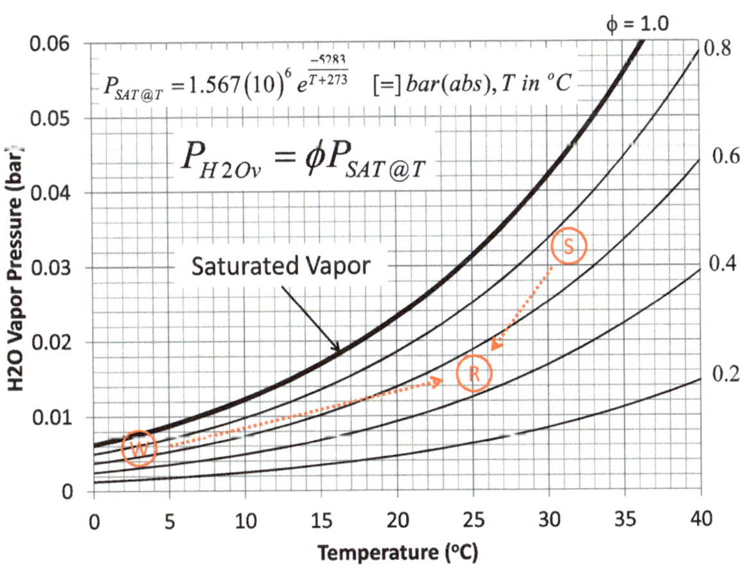

Fig. 12.8 H_2O vapor pressure as a function of temperature for fixed values of relative humidity. The states indicated are representative values for desired interior room conditions (*R*), outside winter conditions (*W*) and outside summer (*S*) in New York City

the saturation pressure. For humid air at 1 atmosphere pressure (1.013 bar), at the highest temperature and relative humidity, the partial pressure is only 0.06 bar (= 0.0592 atm), and the vapor pressure exerted by H_2O is much less than the partial pressure exerted by the BDA (0.94 bar). H_2O vapor is quite dilute in humid air.

Three representative states are shown. The circled "*R*" (for room) indicates a typical value of temperature and relative humidity that is desired for interior spaces. The "*W*" and "*S*" circles indicate typical outdoor conditions in winter and summer in New York City (or other mid-latitude regions). The dotted arrows show the overall change in air conditions that would be required of an HVAC system.

Whether controlled by HVAC equipment or not, living spaces "breathe" in that they require that outside air to be brought in and "conditioned" (like an inhale) and then interior air must be returned to the outside (an exhale). In winter, if the temperature of outdoor air was simply raised with no addition of H_2O vapor to increase its partial pressure (move horizontally from "*W*" to the right), the relative humidity inside would be quite low (about 20% in this case). This result explains why indoor air in winter tends to be very dry, and people are inclined to use humidifiers in winter. Air conditioning in winter requires both temperature change and addition of H_2O vapor to the air.

In contrast, if the temperature of outdoor air in summer were simply lowered with no removal of H_2O vapor (move horizontally from "*S*" to the left), the relative humidity would increase, and possibly become saturated, beyond which condensation of liquid H_2O would occur. This result explains why the development of mold is so problematic, particularly in warmer climates. The combination of high temperatures and moist conditions is a favorable breeding ground for bacteria. Air conditioning in summer requires both lowering the temperature (sensible energy change) and the removal of H_2O from the air (latent energy change).

The absolute humidity (sometimes called humidity ratio) is a measure of the quantity of H_2O vapor in air. It is defined as the ratio:

$$\omega = \frac{\text{Mass of H}_2\text{O vapor}}{\text{Mass of bone dry air}} = \frac{M_{\mathrm{H_2Ov}}}{M_{\mathrm{BDA}}} \quad \text{Absolute humidity}$$

It is not the mass fraction of H_2O vapor (which would be $y_{\mathrm{H_2Ov}} = M_{\mathrm{H_2Ov}}/(M_{\mathrm{H_2Ov}} + M_{\mathrm{BDA}})$ but it is close, since humid air is dilute in H_2O vapor. It will be apparent when conducting mass, energy, and entropy balances why the absolute humidity is better defined in this way.

At state 1 in Fig. 12.7, all the liquid H_2O injected has been evaporated into the gas phase, and the absolute humidity depends only on the mass of liquid injected and mass of BDA present, regardless of the temperature (provided it is sufficiently high to ensure complete evaporation). On the other hand, the relative humidity at state 1 depends strongly on its temperature. If process 1–2 were defined such that the temperature were increased without injecting more liquid the partial pressures of vapor and BDA and absolute humidity would remain constant. However, the saturation pressure would increase, and therefore, the relative humidity would decrease. This process would move the "winter air" horizontally to the right.

A number of relationships between mole fractions, partial pressures, total pressures, and between the two humidity measures can be obtained by suitable manipulation. The absolute humidity can be related to mole fractions of H_2O vapor as follows:

$$\omega = \frac{\text{mass}_{\text{H}_2\text{Ov}}}{\text{mass}_{\text{BDA}}} = \frac{\left(\frac{\text{mass}_{\text{H}_2\text{Ov}}}{\text{mol}_{\text{H}_2\text{Ov}}}\right)\left(\frac{\text{mol}_{\text{H}_2\text{Ov}}}{\text{mol}_{\text{mix}}}\right)}{\left(\frac{\text{mass}_{\text{BDA}}}{\text{mol}_{\text{BDA}}}\right)\left(\frac{\text{mol}_{\text{BDA}}}{\text{mol}_{\text{mix}}}\right)} = \left(\frac{M_{\text{H}_2\text{Ov}}}{M_{\text{BDA}}}\right)\left(\frac{x_{\text{H}_2\text{Ov}}}{x_{BDA}}\right) = \left(\frac{18}{28.8}\right)\left(\frac{x_{\text{H}_2\text{Ov}}}{1 - x_{\text{H}_2\text{Ov}}}\right)$$

$$\omega = 0.625\left(\frac{x_{\text{H}_2\text{Ov}}}{1 - x_{\text{H}_2\text{Ov}}}\right)$$

Expressing the H_2O mole fraction in terms of the absolute humidity:

$$x_{\text{H}_2\text{Ov}} = \frac{\omega}{0.625 + \omega}$$

Mole fractions are related to partial and total pressures by

$$P_{\text{H}_2\text{Ov}} = x_{\text{H}_2\text{O}} P$$

Therefore, the absolute humidity can be expressed as:

$$\omega = 0.625\left(\frac{P_{\text{H}_2\text{Ov}}}{P - P_{\text{H}_2\text{Ov}}}\right)$$

Or in terms of relative humidity and saturation pressure:

$$\omega = 0.625\left(\frac{\phi P_{\text{SAT@}T}}{P - \phi P_{\text{SAT@}T}}\right)$$

This expression can be rearranged for the relative humidity in terms of absolute humidity:

$$\phi = \frac{\omega}{(0.625 + \omega)} \frac{P}{P_{\text{SAT@}T}}$$

A psychrometric chart is a plot of the absolute humidity (or humidity ratio) against temperature. There are a lot of lines shown in published charts that will be explained one by one. Two basic psychrometric charts are shown in Fig. 12.9 for two values of the total pressure with lines of constant relative humidity (ϕ). Since the partial pressure of H_2O vapor is much less than the total pressure for this temperature range, the absolute humidity and partial pressure (Fig. 12.8) are nearly identical in shape. However, the absolute humidity has a mild dependence on total pressure.

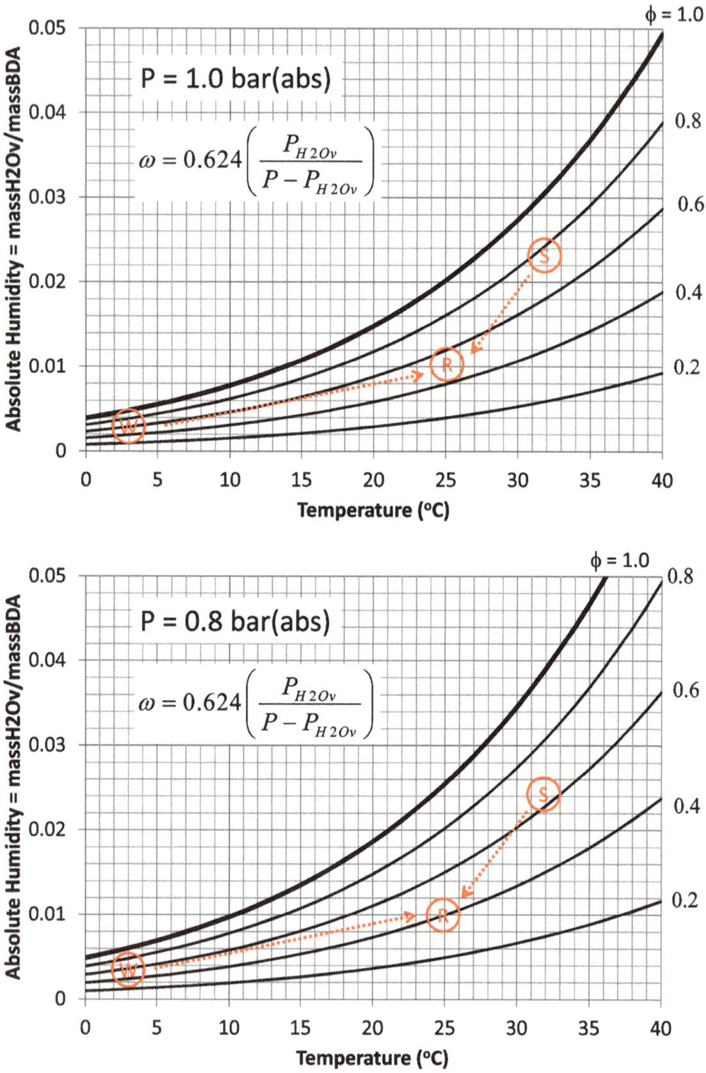

Fig. 12.9 Basic psychrometric chart at two different total pressures (1 bar, sea level, and 0.8 bar, a high elevation)

Use of the absolute humidity is more convenient than partial pressures or mole fractions to conduct mass and energy balances, and that is why psychrometric charts are prepared with absolute humidity. The mass of humid air is expressed as the sum of BDA and H_2O_v, and manipulated:

$$M_{\text{HumidAir}} = M_{\text{BDA}} + M_{\text{H}_2\text{Ov}} = M_{\text{BDA}}\left(1 + \frac{M_{\text{H}_2\text{Ov}}}{M_{\text{BDA}}}\right)$$

$$M_{\text{HumidAir}} = M_{\text{BDA}}(1 + \omega)$$

The mass of BDA is a constant for closed system processes (like Fig. 12.7) and a similar form will be used for open systems. When energy balances are conducted, the absolute humidity gives rise to a minor correction for changes in temperature (sensible energy), but energy changes associated with phase change (latent energy) are directly related to absolute humidity.

Isobaric Cooling/Heating and the Dew Point Temperature

Most people understand that air conditioning in the summer involves lowering the temperature of air to make it more comfortable. Fewer people are aware that air conditioning also involves removing H_2O vapor from the air, that is, dehumidifying the air. Air conditioning then involves a simultaneous lowering of the air temperature and its humidity. This case study investigates the basic process involved in most air conditioning applications.

Follow-Up "in this Room" Questions

How many liters of H_2O liquid would condense out if the air in the room were cooled to 5 °C? What would its relative humidity be if warmed back to a bit below room temperature?

This scenario involves the concept of the dew point and dehumidification and is depicted in Fig. 12.10. A closed system sequence of isobaric cooling processes is shown in which room air, initially at nominal room conditions and a relative humidity less than 100%, is cooled at constant pressure and then reheated to a

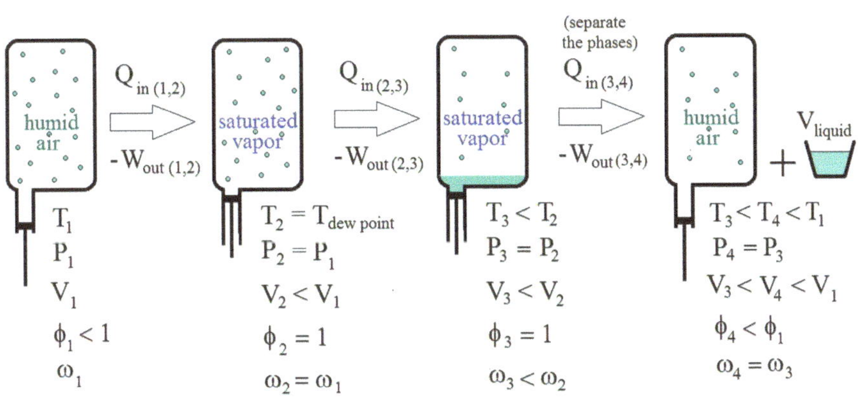

Fig. 12.10 Isobaric cooling of humid air to dew point (1–2) and below (2–3, with dehumidification) and reheating to original temperature (3–4)

INPUT PARAMETERS

V1	**54.4**	m3, volume of room
T1	**25**	oC, initial room temperature
ϕ1	**0.5**	initial relative humidity
T3	**5**	oC
T4	**20**	oC, final room temperature
P1=P2=P3	**1**	atm, total pressure for entire sequence

STATE TABLE **INPUTS**

state	P (bar)	T (oC)	ϕ	P_{H2Ov} (bar)	ω (kg_{H2O}/kg_{BDA})	MW_{mix}	R_{mix} (J/kg/K)	V_{gas} (m^3)	V_{liquid} (m^3)
1	**1.013**	**25**	**0.5**	0.01566	0.009798	28.63	290.4	54.4	0
2	**1.013**	13.787	**1**	0.01566	0.009798	28.63	290.4	52.35	0
3	**1.013**	**5**	**1**	0.00875	0.005436	28.71	289.6	50.40	0.000343
4	**1.013**	**20**	0.37799	0.00875	0.005436	28.71	289.6	53.06	NA

DERIVED PARAMETERS

Mair,1	63.687	kg, initial humid air mass
MBDA	63.069	kg, mass of bone dry air
MH2Ov,1	0.618	kg, initial mass of H2O vapor
Mair,3	63.412	kg, final humid air mass
MH2Ov,3	0.275	kg, final mass of H2O vapor
MH2OL,3	0.343	kg, mass of liquid condensed
ML/Min	55.5%	percent of H2O condensed

Fig. 12.11 Spreadsheet for isobaric cooling and reheating case study

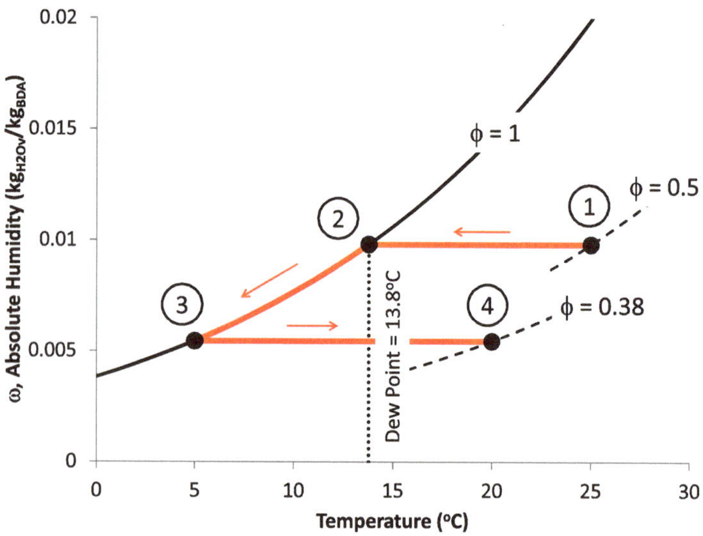

Fig. 12.12 Psychrometric plot for isobaric cooling below the dew point and reheating to initial temperature. The total pressure is 1 atm - = 1.013 bar

final temperature (a bit lower than the initial temperature). A spreadsheet solution is shown in Fig. 12.11, and the process path is traced on the psychrometric chart of Fig. 12.12. The yellow shaded cells in the State table represent input quantities from the input parameter tables (with a pressure unit conversion from atm to bar). The other cells require psychrometric calculations.

For state 1 (at 25 °C, 1 atm and 50% relative humidity), the partial pressure of H_2O vapor is calculated from

$$P_{H_2Ov,1} = \phi_1 P_{SAT@T1} = 0.0157 \text{ bar}$$

Note that the curve fit formula is used for the saturation pressure.

The initial absolute humidity is calculated from

$$\omega_1 = 0.624\left(\frac{P_{H_2Ov1}}{P - P_{H_2Ov1}}\right) = 0.0098 \frac{\text{kg}_{H_2Ov}}{\text{kg}_{air}}$$

The absolute humidity of approximately 1% reinforces the notion that humid air is truly a dilute mixture. Humidity effects would be negligible were it not for the role that it plays in human comfort and the energy associated with condensing or evaporating H_2O.

The next columns in Fig. 12.11 show the precise calculation of the molecular weight and gas constant for the humid air. The molecular weight of state 1 is given by:

$$\begin{aligned} \text{MW}_1 &= x_{H_2Ov1}\text{MW}_{H_2O} + x_{BDA1}\text{MW}_{BDA} = \frac{P_{H_2Ov1}}{P}\text{MW}_{H_2O} + \frac{P - P_{H_2Ov1}}{P}\text{MW}_{BDA} \\ &= 28.63 \frac{\text{kg}_{mix}}{\text{kmol}_{mix}} \end{aligned}$$

The molecular weight of the mixture is ever so slightly lower than that of BDA (28.8 kg/kmol) due to the presence of the lower molecular weight component H_2O. From here, the gas constant is calculated

$$R_1 = \frac{\widehat{R}}{MW_1} = \frac{8314 \text{ J/kmol/K}}{28.63 \text{ kg/kmol}} = 290.4 \frac{\text{J}}{\text{kg} \cdot \text{K}}$$

and then the total mass of humid air is

$$\begin{aligned} M_{air,1} &= \frac{P_1 V_1}{R_1 T_1} = \frac{(1 \text{ atm})(54.4 \text{ m}^3)}{(290.4 \text{ N} \cdot \text{m/kg/K})(25 + 273) \text{ K}} \cdot \left|\frac{1.013(10)^5 \text{ N/m}^2}{\text{atm}}\right| \\ &= 63.687 \text{ kg} \end{aligned}$$

This value differs slightly from that calculated in the warm-up questions (64.4 kg), where the molecular weight was taken to be that of BDA. The difference is in the third significant figure.

From here, the mass of BDA is calculated from

$$M_{\mathrm{BDA},1} = \frac{M_{\mathrm{air1}}}{1+\omega_1} = 63.069\ \mathrm{kg}$$

The mass of H_2O vapor in the air at state 1 is the difference, namely,

$$M_{\mathrm{H_2Ov},1} = M_{\mathrm{air},1} - M_{\mathrm{BDA}} = 0.618\ \mathrm{kg}$$

Process 1–2 in Fig. 12.10 involves cooling the air in the room at constant total pressure. To do so, the total volume must be adjusted (with the piston). There is work associated with that, and when open systems are considered, the flow work will play the role of this piston work. Initially, while the H_2O vapor is superheated, cooling lowers the temperature of the air but there is no change in chemical composition of the humid air. The partial pressure of H_2O vapor and the absolute humidity remain constant. However, as the temperature lowers, the H_2O saturation pressure lowers, and the ability of air to "hold" H_2O vapor is diminished. That is, the relative humidity increases as the absolute humidity stay constant. Eventually, the partial pressure exerted by the H_2O vapor equals the saturation temperature, and the relative humidity reaches 100%.

This condition is called the dew point, because further cooling results in the condensation of H_2O onto the cold surface that is causing the heat transfer. In nature, during the night, the temperature of the ground becomes lower than the temperature of the air because of radiative heat loss to outer space. If the ground temperature is below the dew point temperature of the air, then H_2O will condense onto the ground. That is, dew will form.

Weather reports frequently report the dew point temperature as a measure of human comfort. The closer the actual and dew point temperatures, the higher the relative humidity. The reason for the comfort effect has to do with the effectiveness of sweating. Sweating is only effective if the liquid H_2O that flows out of pores in the skin evaporates into the surrounding air. This effect is called evaporative cooling and will be addressed once the energy balances are developed.

It is easy to locate the dew point on a psychrometric chart, but its precise calculation requires some analysis. The condition for achieving the dew point is that the partial pressure exerted by H_2O vapor of the humid air equals the saturation pressure. That is:

$$P_{\mathrm{H_2Ov},2} = P_{\mathrm{H_2Ov},1} = \phi_1 P_{\mathrm{SAT}@T_1} = P_{\mathrm{SAT}@T_{\mathrm{DP}}}$$

Inserting the Clausius-Clapeyron curve fit (with coefficients A and B, for now):

$$\phi_1 A e^{\frac{-B}{T_1+273}} = A e^{\frac{-B}{T_{\mathrm{DP}}+273}}$$

The "*A*" cancels, and taking natural logarithms [with ln(*a***b*) = ln(*a*) + ln(*b*)]

$$\ln(\phi_1) + \frac{-B}{T_1 + 273} = \frac{-B}{T_{\mathrm{DP}} + 273}$$

Rearranging for the dew point temperature at state 1, and inserting $B = 5283$ K:

$$T_{\mathrm{DP@1}} = \frac{T_1 + 273}{1 - (T_1 + 273)\ln(\phi_1)/5283} - 273 \quad [=]^o\mathrm{C} \text{ with } T_1 \text{ in } ^\circ\mathrm{C}$$

When entered into the spreadsheet for this example, the dew point temperature evaluates to

$$T_{\mathrm{DP@1}} = 13.787 \ ^\circ\mathrm{C}$$

It is a good idea at this time to study process 1–2 on the psychrometric chart and observe what happens to the dew point temperature at higher or lower initial relative humidity levels and/or higher or lower initial air temperatures. What is the dew point in the Sahara desert at noon, for example? Or in London at midnight?

The volume of the air mass decreases during process 1–2, but not very much. Since there is no change in composition, the mass and gas constant remain constant, and the ideal gas law can be used in the form

$$\mathrm{MR} = \frac{P_1 V_1}{T_1} = \frac{P_2 V_2}{T_2}$$

Rearranging for the volume at state 2 and evaluating:

$$V_2 = \left(\frac{P_1}{P_2}\right)\left(\frac{T_2}{T_1}\right)V_1 = \left(\frac{1\ \mathrm{atm}}{1\ \mathrm{atm}}\right)\left(\frac{13.8 + 273}{25 + 273}\right)54.4\ \mathrm{m}^3 = 52.4\ \mathrm{m}^3$$

There is no change in the pressure and the change in absolute temperature is rather small. A common mistake in execution is to use temperature in Celsius, giving rise to large errors, both conceptual and in the result.

In Process 2–3, the now saturated humid air is cooled below the original the dew point. In practice, heat transfer must be driven by a temperature change, in this case, the walls of the container must be somewhat below that of the average air temperature. How much lower is a question of heat transfer theory, not developed here. It could be incrementally lower. Regardless of the magnitude of the temperature difference driving heat, if the walls are at a lower temperature than saturated air, H_2O liquid will condense on the walls. If equilibrium is maintained as the system temperature is decreased, the air will remain saturated, but since the saturation pressure is being lowered, the ability of the air to "hold" vapor will decrease, and liquid H_2O will continue to condense. The path taken on the psychrometric chart is along the saturation line. At the final state 3 (set to 5 °C in this case study), the H_2O_v partial pressure, and therefore, absolute humidity can be calculated (and/or read off the psychrometric chart). From there, the new gas constant, final humid air mass,

mass of vapor remaining and mass of liquid that condensed can all be calculated. The result is that 0.343 kg of H_2O liquid are condensed out of the air in the room. That is, 55% of the original H_2O vapor.

In the process sequence of this case study, once the air has been cooled below the dew point, and a liquid phase forms and settles on the bottom of the vessel, the phases can be physically separated. Wall unit air conditioners generally provide a means to drip this condensed water away. Then the air can be reheated to a comfortable temperature, but below its original temperature, and the liquid can be used to water your plants, flush your toilet, or drip on your head as you pass underneath a wall mounted air conditioner. That is, two different useful products have been created; air at a more comfortable temperature and lower humidity and liquid H_2O. This sequence is very much like what takes place in most air conditioning systems. It requires cooling the air to a value much lower than the delivered air temperature, which seems silly, but is necessary to remove H_2O in this way (as opposed to some sort of chemical desiccation, which is not widely used). The reheating can be accomplished in theory with a regeneration process, whereby the inlet humid air is precooled by the cold air prior to its return.

Humid Air Flow in Open Systems

Most applications in psychrometrics involve open systems. The analysis on closed systems has served to introduce important fundamental principles. Extending to open systems is relatively straightforward. Figure 12.13 is a representation of a flowing humid air stream as consisting of two components: bone dry air (BDA) and H_2O vapor. This form of mass flow rate is useful because the flow rate of BDA is a constant, but the flow rate of humid air changes with humidification or dehumidification.

Cooling Tower Results

With all that background into humidity, the pieces are in place to complete the cooling tower problem statement from the beginning of this chapter. A control volume is shown in Fig. 12.14 that indicates mass flows. The boundary is drawn outside the condenser water loop. Heat crosses this boundary but is not shown, as the current focus is on mass exchange. Humid air at ambient conditions (25 °C, 1 atm, 50% relative humidity) enters from the bottom and exits at 30 °C, 1 atm, and 90% relative humidity. Liquid H_2O (make-up water) enters from the bottom to ensure the flow rate of liquid H_2O that leaves the cooling tower as a vapor is the same that entered it as a liquid. This H_2O evaporates within the cooling tower. The H_2O that entered as a vapor in the inlet humid air stream also exits as a vapor (and does not change phase).

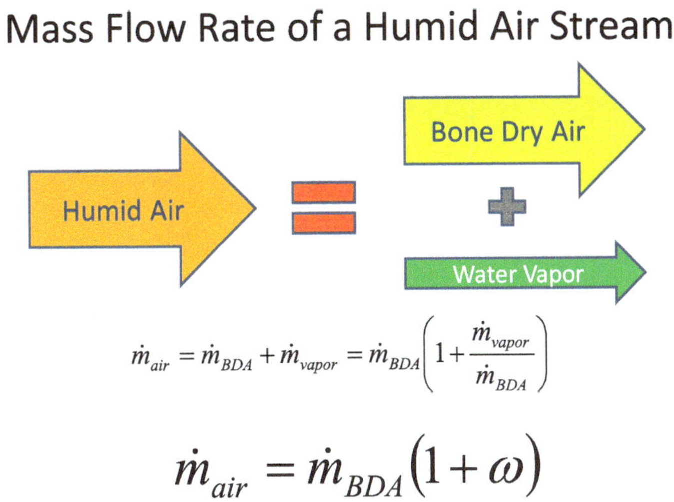

Fig. 12.13 Representation of a flowing humid air stream as two components: bone dry air (BDA) and H_2O vapor

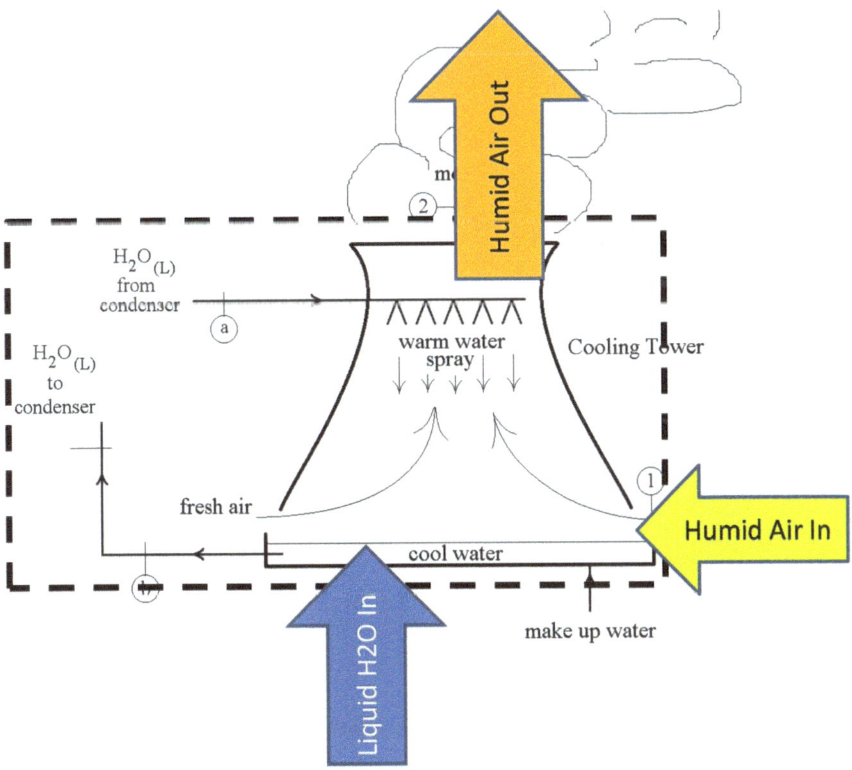

Fig. 12.14 Control volume for cooling tower showing mass crossing the boundary (not energy)

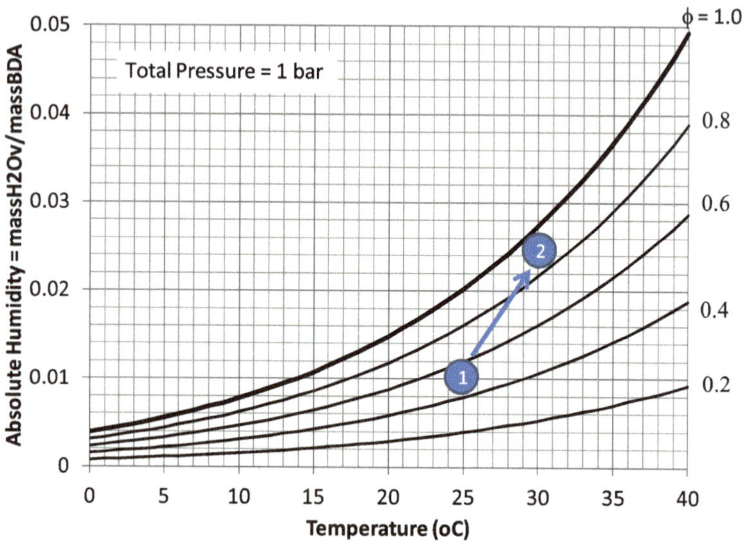

Fig. 12.15 Psychrometric chart for cooling tower case study

Figure 12.15 shows the process on a psychrometric chart, which can be drawn directly from the problem statement. A spreadsheet solution is shown in Fig. 12.16. A mass balance applied to the control volume is:

$$\frac{dM_{CV}}{dt} = \sum_{\text{in}} \dot{m} - \sum_{\text{out}} \dot{m}$$

At steady state, with three mass flow streams (two entering, one exiting)

$$0 = \dot{m}_{\text{make up}} + \dot{m}_1 - \dot{m}_2$$

Rearranging for the desired make-up water:

$$\dot{m}_{\text{make up}} = \dot{m}_2 - \dot{m}_1$$

Expressing the two humid air streams in terms of bone dry air and the absolute humidity:

$$\dot{m}_{\text{make up}} = \dot{m}_{\text{BDA}}(1 + \omega_2) - \dot{m}_{\text{BDA}}(1 + \omega_1)$$

An important concept for this single condensable component system is that while the mass flow rate of the humid air stream changes, the mass flow rate of the BDA does not. Therefore, the mass flow rate of the make-up water becomes:

INPUT PARAMETERS

T1	**25**	oC, temperature of inlet air
P1	**1**	atm, atmospheric pressure
ϕ1	**0.5**	relative humidity of ambient
T2	**30**	oC, temperature of exit air
P2	**1**	atm, atmospheric pressure
ϕ2	**0.9**	relative humidity of exit air
(AV)2	**10000**	m^3/sec, volumetric flow rate of exit
ρ_L	**1000**	kg/m^3, density of liquid water

STATE	T (oC)	ϕ	P (bar)	P_{SAT} (bar)	ω	y_{vapor}	MW	ρ (kg/m^3)
1 (Air In)	**25**	**0.5**	1.0130	0.0313	0.0098	0.0155	28.63	1.171
2 (Air out)	**30**	**0.9**	1.0130	0.0420	0.0242	0.0373	28.40	1.142

DERIVED PARAMETERS

m'2	11419	kg/sec, mass flow rate of exit stream
m'BDA	11150	kg/sec, mass flow rate of BDA
m'MAKE-UP	160.19	kg/sec, mass flow rate of Make Up water
(AV)MAKE-UP	0.16019	m^3/sec, volumetric flow rate of Make-Up
(AV)MAKE-UP	3,656,207	gallons/day, volumetric flow rate of Make-Up

Fig. 12.16 Spreadsheet solution for cooling tower case study. Numbers in bold are inputs. Shaded yellow in the state table refer to inputs. The spreadsheet is designed so that all calculations update with a change in an input parameter

$$\dot{m}_{\text{make up}} = \dot{m}_{\text{BDA}}(\omega_2 - \omega_1)$$

The volumetric flow rate of the make-up water is determined from

$$\rho_{\text{H}_2\text{O,liquid}}(AV)_{\text{make up}} = \dot{m}_{\text{make up}}$$

The molecular weight of a humid air stream is

$$\begin{aligned}\text{MW}_{\text{humid air}} &= (1 - x_{\text{H}_2\text{Ov}})\text{MW}_{\text{BDA}} + x_{\text{H}_2\text{Ov}}\text{MW}_{\text{H}_2\text{O}} \\ &= 28.8\left(1 - \phi\frac{P_{\text{SAT@}T}}{P}\right) + 18\phi\frac{P_{\text{SAT@}T}}{P}\end{aligned}$$

All the details are incorporated into the spreadsheet solution, and the result is a whopping 3.7 million gallons of make-up water per day, the rate at which condenser H_2O evaporates in the cooling tower. Determination of the total flow rate of condensing H_2O will require an energy balance, the topic of the next section. This section continues to focus on mass considerations.

Lines of Constant Specific Volume

Detailed psychrometric charts show lines of constant specific volume. In general, specific volume depends on temperature and pressure, but also on the humidity level, which affects the molecular weight of a humid air stream. These effects can be important for engineering design in sizing fans to move air in HVAC systems. Fan specifications published by manufacturers are for performance under a standard set of conditions. If the conditions where a system is to be used differ substantially, corrections can be made by appropriate use of the psychrometric chart. Recall that in open systems, the work per unit mass is "vee dePee" work, and the specific volume is a direct measure of the work required to move fluid.

To establish lines of constant specific volume, consider the specific volume of a humid air mixture (with T in Celsius and P in absolute units):

$$v_{\text{air}} = \frac{R_{\text{air}}(T+273)}{P} = \frac{\widehat{R}(T+273)}{\text{MW}_{\text{air}}P}$$

The molecular weight of air, in terms of absolute humidity is:

$$\begin{aligned}\text{MW}_{\text{air}} &= (1 - x_{\text{H}_2\text{Ov}})\text{MW}_{\text{BDA}} + x_{\text{H}_2\text{Ov}}\text{MW}_{\text{H}_2\text{O}} \\ &= 28.8\left(1 - \frac{\omega}{0.625+\omega}\right) + 18\frac{\omega}{0.625+\omega}\end{aligned}$$

Rearranging for the absolute humidity as a function of temperature (for fixed values of specific volume and pressure):

$$\omega = \frac{\left(\frac{18}{28.8}\right)\frac{\widehat{R}(T+273)}{Pv_{\text{air}}} - 18}{18 - \frac{\widehat{R}(T+273)}{Pv_{\text{ar}}}}$$

Figure 12.17 shows a psychrometric chart with lines of constant specific volume. The slant in the lines is due to the change in specific volume (or density) due to the change in molecular weight.

Energy and Entropy Considerations

The general forms of the open system mass, energy, and entropy balances for a control volume (CV) are:

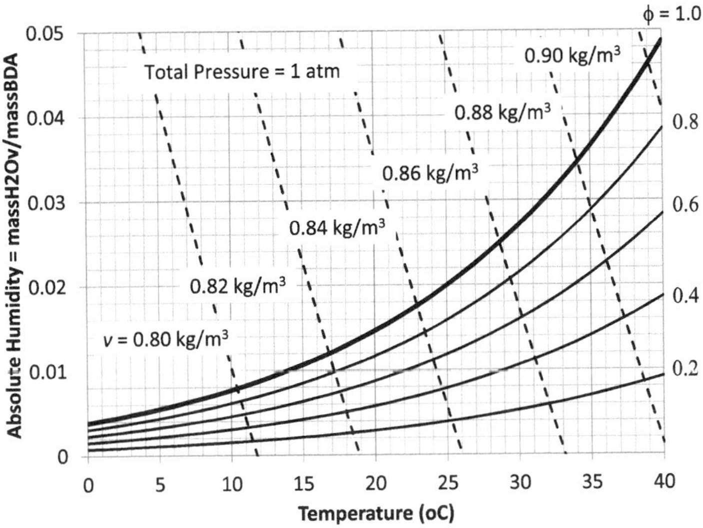

Fig. 12.17 Psychrometric chart with lines of constant specific volume

$$\frac{dM_{\mathrm{CV}}}{dt} = \sum_{\mathrm{in}} \dot{m} - \sum_{\mathrm{out}} \dot{m}$$

$$\frac{dE_{\mathrm{CV}}}{dt} = \dot{Q}_{in} - \dot{W}_{\mathrm{shaft,out}} + \sum_{\mathrm{in}} \dot{m}\left(h + \frac{V^2}{2} + gz\right) - \sum_{\mathrm{out}} \dot{m}\left(h + \frac{V^2}{2} + gz\right)$$

$$\frac{dS_{\mathrm{CV}}}{dt} = \frac{\dot{Q}_{\mathrm{in}}}{T_{\mathrm{D}}} + \ldots \Sigma_{\mathrm{CV}} + \sum_{\mathrm{in}} \dot{m}s - \sum_{\mathrm{out}} \dot{m}s$$

All cases considered in this chapter will be assumed to be at steady state, with negligible kinetic and potential energy changes. With that, and rearranging the mass balances so that the inlets equal the outlets, the energy balance so that the net energy added (heat and work) equals the difference in the outlet and inlet enthalpy flows, and the entropy balance for the eXergy destruction rate (by multiplying by the dead state temperature, T_∞:

$$\sum_{\mathrm{in}} \dot{m} = \sum_{\mathrm{out}} \dot{m}$$

$$\dot{Q}_{\mathrm{in}} - \dot{W}_{\mathrm{shaft,out}} = \sum_{\mathrm{out}} \dot{m}h - \sum_{\mathrm{in}} \dot{m}h$$

$$e\ldots X_{\mathrm{D}} = T_\infty \ldots \Sigma_{\mathrm{CV}} = -\frac{T_\infty}{T_{\mathrm{B}}} \dot{Q}_{\mathrm{in}} + \sum_{\mathrm{out}} \dot{m}T_\infty s - \sum_{\mathrm{in}} \dot{m}T_\infty s$$

Enthalpy Flow Rate of a Humid Air Stream

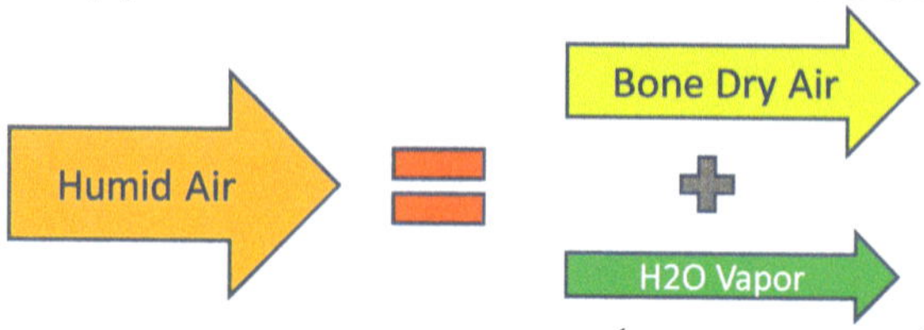

$$(\dot{m}h)_{air} = (\dot{m}h)_{BDA} + (\dot{m}h)_{H2Ov} = \dot{m}_{BDA}\left(h_{BDA} + \frac{\dot{m}_{vapor}}{\dot{m}_{BDA}} h_{H2Ov}\right)$$

$$h - h_{ref} = c_p\left(T - T_{ref}\right)$$

$$(\dot{m}h)_{air} = \dot{m}_{BDA}\left(c_{p,BDA} + \omega c_{p,H2Ov}\right)\left(T - T_{ref}\right)$$

Entropy Flow Rate of a Humid Air Stream

$$(\dot{m}s)_{air} = (\dot{m}s)_{BDA} + (\dot{m}s)_{H2Ov} = \dot{m}_{BDA}\left(s_{BDA} + \frac{\dot{m}_{vapor}}{\dot{m}_{BDA}} s_{H2Ov}\right)$$

$$\left(s - s_{ref}\right)_{BDA} = c_{p,BDA}\ln\left(\frac{T}{T_{ref}}\right) - R_{BDA}\ln\left(\frac{(1 - x_{H2Ov})P}{P_{ref}}\right)$$

$$\left(s - s_{ref}\right)_{H2Ov} = c_{p,H2Ov}\ln\left(\frac{T}{T_{ref}}\right) - R_{H2Ov}\ln\left(\frac{x_{H2Ov}P}{\mathrm{P}_{ref}}\right)$$

$$(\dot{m}s)_{air} = \dot{m}_{BDA}\left[\left(c_{p,BDA} + \omega c_{p,H2Ov}\right)\ln\left(\frac{T}{T_{ref}}\right) - R_{BDA}\ln\left(\frac{(1 - x_{H2Ov})P}{P_{ref}}\right) - \omega R_{H2Ov}\ln\left(\frac{x_{H2Ov}P}{P_{ref}}\right)\right]$$

Fig. 12.18 Representation of the enthalpy and entropy flows of a humid air stream separated into its components

A key in dealing with gas systems is to recognize that a single stream can be broken up into its components; in this case, BDA and H_2O vapor. Figure 12.18 is a representation of the enthalpy flow, with each component as an ideal gas (relative to a reference state). The entropy flow rate is more involved (due to the partial pressure effect). Formal 2nd Law analysis is not typically performed in traditional textbooks, and is mostly treated in deeper dives in this one. All calculations involve the differences between two states, and the reference state will cancel.

It is often convenient within a given stream to differentiate H_2O vapor that undergoes phase change from that which is carried with the air and does not change phase. Figure 12.19 shows a control volume and conceptual diagram (with enthalpy fluxes) for a general humidification process. Figure 12.20 shows the same thing for a

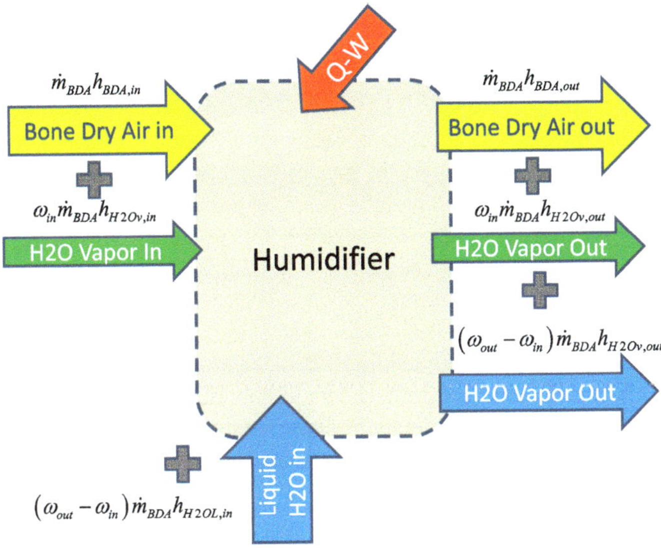

Fig. 12.19 Control volume and conceptual diagram of a humidification process. H_2O mass balances have been applied, such that the green inlet H_2O vapor is the same as the green exit H_2O vapor

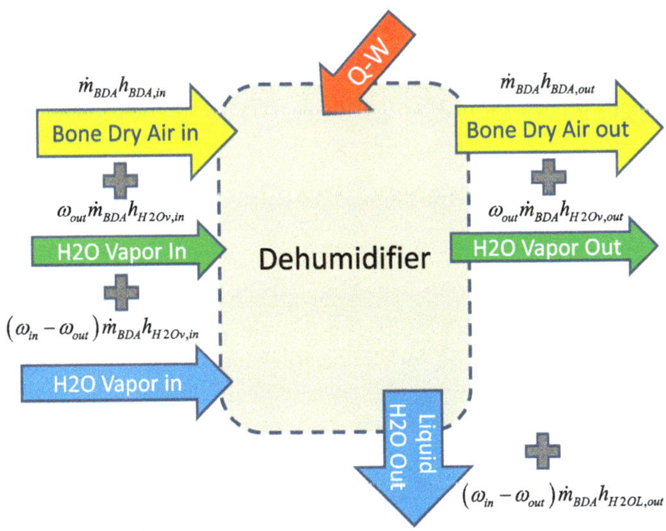

Fig. 12.20 Control volume and conceptual diagram of a dehumidification process

dehumidification process. Mass balances have been applied in a manner that defines three separate streams, each of constant mass flow rate, with some of them part of a gas mixture.

In the humidification case (Fig. 12.19), the H_2O vapor that enters with the BDA is mixed with the BDA (and is at the same temperature) and the liquid is in a separate stream (at a different temperature, in general). All three components of the exit stream (BDA, vapor that entered with BDA and vapor that entered as liquid) are part of a single stream, and at the same temperature. It is not a typo that the mass flow of the H_2O vapor out is based on the absolute humidity of the inlet stream. A mass balance has been applied. Separating the vapor this way helps clearly differentiate sensible energy changes (the H_2O vapor that enters with the BDA and changes temperature but not phase) from latent energy changes (the H_2O liquid that enters and changes phase and temperature). The mass flow rate of H_2O that changes phase is the difference between the outlet and inlet absolute humidity.

The general energy balance for the dehumidification cases can be expressed in terms of enthalpies as:

$$\dot{Q}_{\mathrm{in}} - \dot{W}_{\mathrm{shaft,out}} = \dot{m}_{\mathrm{BDA}}\left(h_{\mathrm{BDA,out}} - h_{\mathrm{BDA,in}}\right) + \omega_{\mathrm{in}}\dot{m}_{\mathrm{BDA}}\left(h_{\mathrm{H_2Ov,out}} - h_{\mathrm{H_2Ov,in}}\right)$$
$$+\left(\omega_{\mathrm{out}} - \omega_{\mathrm{in}}\right)\dot{m}_{\mathrm{BDA}}\left(h_{\mathrm{H_2Ov,out}} - h_{\mathrm{H_2OL,in}}\right)$$

The enthalpy changes of the BDA and H_2O vapor that enters with it are sensible enthalpy changes, and are expressed with the enthalpy-specific heats and combined (since they share the same inlet and exit temperatures):

$$\left(h_{\mathrm{BDA,out}} - h_{\mathrm{BDA,in}}\right) + \omega_{\mathrm{in}}\left(h_{\mathrm{H_2Ov,out}} - h_{\mathrm{H_2Ov,in}}\right) = \left(c_{\mathrm{P,BDA}} + \omega_{\mathrm{in}} c_{\mathrm{p,H_2Ov}}\right)\left(T_{\mathrm{out}} - T_{\mathrm{in}}\right)$$

The term for the H_2O vapor is a small effect, since the relative humidity is on the order of 0.01 for psychrometric applications. The change in sensible energy is dominated by the BDA.

The change in enthalpy of the H_2O that enters as liquid and exits as vapor undergoes both a temperature (sensible energy) change and phase (latent energy) change. A useful technique for evaluating thermodynamic properties like enthalpy and entropy that undergo phase (or chemical) change is to construct an equivalent thermodynamic path that changes phase where phase data is available. To do so in this case, the saturated vapor and liquid states of pure H_2O at the inlet temperature can be added and subtracted. That is:

$$\left(h_{\mathrm{H_2Ov,out}} - h_{\mathrm{H_2OL,in}}\right) = \left(h_{\mathrm{H_2Ov,out}} - h_{\mathrm{g@}T_{\mathrm{Lin}}}\right) + \left(h_{\mathrm{g@}T_{\mathrm{Lin}}} - h_{\mathrm{f@}T_{\mathrm{Lin}}}\right)$$
$$+ \left(h_{\mathrm{f@}T_{\mathrm{Lin}}} - h_{\mathrm{H_2OL,in}}\right)$$

The first term involves a change in temperature in the vapor phase between the exit temperature and inlet H_2O liquid temperature. The second term is the enthalpy of vaporization at the inlet liquid temperature. The third term represents the change in enthalpy of a compressed liquid due to pressure change from the corresponding saturation pressure at the inlet liquid H_2O temperature to the total pressure of the pure liquid, that is

$$\left(h_{f@T_{Lin}} - h_{H_2OL,in}\right) = v_{f@T_{L,in}}\left(P_{SAT@T_{L,in}} - P\right)$$

This term will be shown to be dwarfed by the enthalpy of vaporization, but retained for completeness, for now.

Defining the net energy exchange, per unit mass of BDA as $\left(\dot{Q}_{in} - \dot{W}_{shaft,out}\right)/\dot{m}_{BDA} = q - w$ and inserting all these enthalpy effects results in a general energy equation (at steady state, neglecting kinetic and potential energy) for humidification problems:

$$\begin{aligned} q - w = {} & \left(c_{p,BDA} + \omega_{in} c_{p,H_2Ov}\right)\left(T_{out} - T_{in}\right) + \left(\omega_{out} - \omega_{in}\right) \\ & \times \left(c_{p,H_2Ov}\left(T_{out} - T_{L,in}\right) + h_{fg@T_{L,in}} + v_{f@T_{L,in}}\left(P_{SAT@T_{L,in}} - P\right)\right) \end{aligned}$$

The choice to evaluate the phase change at the inlet liquid temperature is somewhat arbitrary, but convenient in this case. For substances other than water, where less data is available, thermodynamic data may only be available as the heat of vaporization at the normal boiling point (saturation temperature corresponding to 1 atm pressure). The specific heats of liquid and vapor would also be needed. Workshop 12.5 has an example of that, where the evaporative cooling effect in the intake of a spark ignition engine due to fuel injection is investigated.

In the dehumidification case (Fig. 12.20), the inlet stream has three components (BDA, H_2Ov that remains H_2Ov mixed with BDA and H_2Ov that condenses). There are two separate exit streams, a humid air stream and a liquid H_2O stream. A similar analysis can be done for dehumidification problems, resulting in:

$$\begin{aligned} q - w = {} & \left(c_{p,BDA} + \omega_{out} c_{p,H_2Ov}\right)\left(T_{out} - T_{in}\right) + \left(\omega_{in} - \omega_{out}\right) \\ & \times \left(c_{p,H_2Ov}\left(T_{L,out} - T_{in}\right) - h_{fg@T_{L,out}} + v_{f@T_{L,out}}\left(P - P_{SAT@T_{L,out}}\right)\right) \end{aligned}$$

A general equation for the eXergy destruction rate is developed in the deeper dive of Appendix 2.

An isothermal humidification case is considered in the next section. The dehumidification case is considered as a deeper dive in Appendix 6, as it involves a somewhat involved 2nd Law analysis aimed at demonstrating that there is a much larger energy consequence of a stand-alone dehumidifier appliance that the electricity that is required to run it. The necessary heat rejection of a dehumidifier results in a much larger electricity consumption of the air conditioner used to maintain the temperature of the interior space at its desired level.

Thermodynamic property data for H_2O in the range for psychrometric application is given in Table 12.1.

Table 12.1 Thermodynamic property data for H_2O

T (°C)	C_{Liq} (kJ/kg)	$C_{p,vapor}$ (kJ/kg)	h_{fg} (kJ/kg)	S_{fg} (kJ/kg/K)
10	4.20	1.89	2477.2	8.7488
20	4.18	1.91	2453.5	8.3696
30	4.18	1.92	2429.8	8.0152
40	4.18	1.93	2406.0	7.6832
100	4.22	2.08	2256.4	6.0470

Psychrometric charts generally show lines of constant enthalpy. The enthalpy of a humid air stream is given by:

$$h_{air} - h_{ref} = (h_{BDA} - h_{BDA,ref}) + \omega(h_{H_2Ov} + h_{H_2Ov,ref})$$

With the vapor phase as an ideal gas and the phase change at the reference temperature:

$$h_{air} - h_{ref} = (c_{p,BDA} + \omega c_{p,H_2Ov}) + \omega h_{fg@T_{ref}}$$

Rearranging for the humidity as a function of temperature:

$$\omega = \frac{(h - h_{ref}) - c_{p,BDA}(T - T_{ref})}{c_{p,H_2Ov}(T - T_{ref}) + h_{fg@T_{ref}}}$$

A psychrometric chart with lines of constant total enthalpy is shown in Fig. 12.21. Each line crosses the saturation line at the wet bulb temperature (and states to the left are not obtainable). These curves are approximate in that the enthalpy-specific heats are treated as constant, which is quite reasonable over this temperature range.

Isothermal Humidification of Bone Dry Air

The first example considered that involves energy and entropy calculations is an isothermal humidifier, shown schematically and on a psychrometric chart in Fig. 12.22. Only latent energy changes are involved with an isothermal process. Practical devices to operate in a similar manner are commonly used as stand-alone appliances in winter months. However, they do not start with BDA, the extreme case investigated here, intended to introduce the basic ideas of humidification. As will be shown, heat or work must be added to evaporate the liquid and maintain the temperature. In practice, it would be difficult to design a heat exchanger (a low temperature difference to drive heat), and much easier to use an electrical resistance heating coil. From a 1st Law perspective, there is not much difference between the two. However, the electrical option is shown to cause a much greater eXergy destruction in Appendix 3.

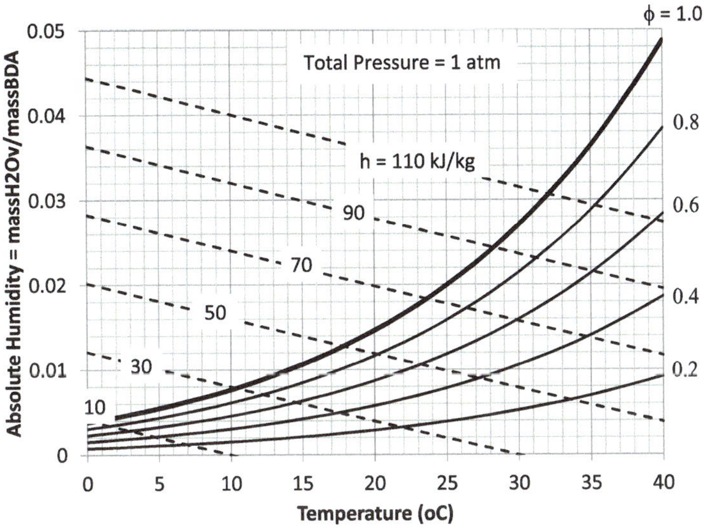

Fig. 12.21 Psychrometric chart showing lines of constant total enthalpy, using thermodynamic properties evaluated at 10 °C and a total pressure of 1 atm, and a reference temperature of 0 °C

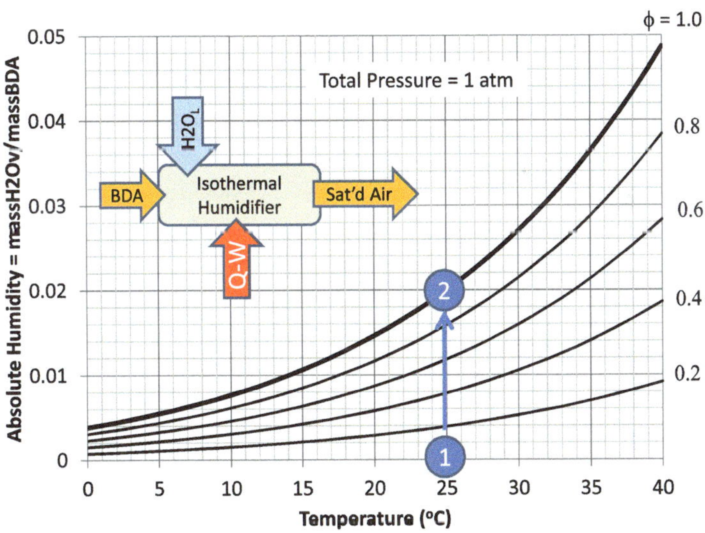

Fig. 12.22 Schematic and psychrometric chart for an isothermal, isobaric humidification process

In this case, liquid H_2O is sprayed into bone dry air (BDA), both at nominal room conditions. Just enough liquid is sprayed so that the exiting humid air is saturated. Any heat exchange is assumed to take place with the dead state temperature, equal to the temperature of all streams. In practice, a finite temperature difference would be required to drive the heat. The process considered here is a theoretical limit. Both cases (heat and work) will be considered. The goal of the calculation is to determine the energy required per unit mass of air and the associated eXergy destruction rate.

Consider first the steady-state mass balance:

$$\dot{m}_{\mathrm{H_2OL}} + \dot{m}_{\mathrm{BDA}}(1 + \omega_{\mathrm{in}}) = \dot{m}_{\mathrm{BDA}}(1 + \omega_{\mathrm{out}})$$

In this case, the absolute humidity entering is zero. Then,

$$\dot{m}_{\mathrm{H_2OL}} = \omega_{\mathrm{out}}\dot{m}_{\mathrm{BDA}}$$

The energy balance from the previous section, with all temperatures equal and zero inlet humidity, simplifies in this case to:

$$q - w = \omega_{\mathrm{out}}\left(h_{\mathrm{fg@}T_{\mathrm{L,in}}} + v_{f@T_{\mathrm{L,in}}}\left(P_{\mathrm{SAT@}T_{\mathrm{L,in}}} - P\right)\right)$$

The exiting absolute humidity is:

$$\omega_{\mathrm{out}} = 0.625\left(\frac{\phi P_{\mathrm{SAT@}T}}{P - \phi P_{\mathrm{SAT@}T}}\right) = 0.625\left(\frac{(1)0.0313}{1.013 - (1)0.0313}\right) = 0.020$$

The enthalpy change of the H_2O, keeping the two effects separate, is:

$$\begin{aligned} h_{\mathrm{fg@}T} + v_{\mathrm{f@}T}(P_{\mathrm{SAT@}T} - P) = &\quad 2441.7\frac{\mathrm{kJ}}{\mathrm{kg}} + 0.001003\frac{\mathrm{m}^3}{\mathrm{kg}}(3.1698 - 101.3)\frac{\mathrm{kN}}{\mathrm{m}^2} \\ = &\quad 2441.7 - 0.084\frac{\mathrm{kJ}}{\mathrm{kg}} = 2441.6\frac{\mathrm{kJ}}{\mathrm{kg}} \end{aligned}$$

Clearly, the pressure effect on the liquid enthalpy is negligible in comparison with the enthalpy associated with the change of phase. This pressure effect can be neglected safely in virtually all psychrometric cases going forward.

Therefore, the rate of energy addition required to simultaneously evaporate the liquid and maintain a constant temperature per unit flow of BDA is:

$$q - w = \omega_{\mathrm{out}}h_{\mathrm{fg@}T} = \left(0.02\frac{\mathrm{kg_{H_2Ov}}}{\mathrm{kg_{BDA}}}\right)\left(2441.7\frac{\mathrm{kJ}}{\mathrm{kg_{H_2O}}}\right) = 48.83\frac{\mathrm{kJ}}{\mathrm{kg_{BDA}}}$$

So clearly, energy must be added to evaporate the liquid, by heat addition and/or work input. From an energy perspective, it does not matter whether heat or work (negative work out) is added.

But, how much eXergy destruction is associated with this isothermal humidification process? Maybe the process is reversible, and a humid air stream can be separated into a dry air and liquid stream. Hmmm. A formal 2nd Law analysis is conducted in the deeper dive of Appendix 2 which demonstrates that there is relatively little eXergy destruction (2.68 kJ/kg$_{BDA}$) compared to the energy addition (48.8 kJ/kg$_{BDA}$) if the humidification is accomplished by heat transfer from ambient. However, that is difficult to achieve in practice, since heat addition requires a finite temperature difference. It is much easier to construct a device that uses electrical heating (working actually) to achieve the humidification. Doing so comes at a large thermodynamic penalty, however, as the eXergy destruction is 51.4 kJ/kg$_{BDA}$ if the humidification is accomplished by a work addition, rather than heat. It is greater than the input energy requirement because of the irreversibility associated with the decrease in partial pressure of the BDA.

The next example demonstrates the response of an evaporation process if neither heat nor work are added.

Adiabatic Humidification: Evaporative Cooling and the Wet Bulb Temperature

Humidifying air and maintaining its temperature has been shown to require an input of energy (heat or work). How does an air stream respond if that energy is not provided? In this case, if a fine mist of liquid H_2O is simply sprayed into air that is not saturated, the liquid will tend to evaporate, but where does the energy to do so come from? From the air, and then the internal energy of the air must decrease by the 1st Law. A schematic and psychrometric chart of this adiabatic humidification process is shown in Fig. 12.23.

Much of the analysis has been set up in the previous example. The general mass, energy, and entropy (converted to eXergy destruction) balance results are:

$$\dot{m}_{\mathrm{H_2OL}} = \omega_{\mathrm{out}} \dot{m}_{\mathrm{BDA}}$$

$$\left(\dot{Q}_{\mathrm{in}} - \dot{W}_{\mathrm{out}}\right) = \dot{m}_{\mathrm{BDA}}\left[\left(h_{\mathrm{BDA,out}} - h_{\mathrm{BDA,in}}\right) + \omega_{\mathrm{out}}\left(h_{\mathrm{H_2Ov,out}} - h_L\right)\right]$$

$$e\ldots X_{\mathrm{D}} = -\dot{Q}_{\mathrm{in}} + \dot{m}_{\mathrm{BDA}} T_{\infty}\left(s_{\mathrm{BDA,out}} - s_{\mathrm{BDA,in}}\right) + \dot{m}_{\mathrm{BDA}} \omega_{\mathrm{out}} T_{\infty}\left(s_{\mathrm{H_2Ov,out}} - s_{\mathrm{H_2Ov,in}}\right)$$

The heat transfer and work terms will be retained, but set to zero to evaluate the adiabatic humidification case. Since the BDA is a nonphase changing ideal gas, the change in enthalpy of the BDA (which was zero in the isothermal case) is:

$$\left(h_{\mathrm{BDA,out}} - h_{\mathrm{BDA,in}}\right) = c_{\mathrm{p,BDA}}\left(T_{\mathrm{out}} - T_{\mathrm{in}}\right)$$

The change in enthalpy of the H_2O is determined by adding and subtracting the intermediate states where phase change is fixed at the input temperature.

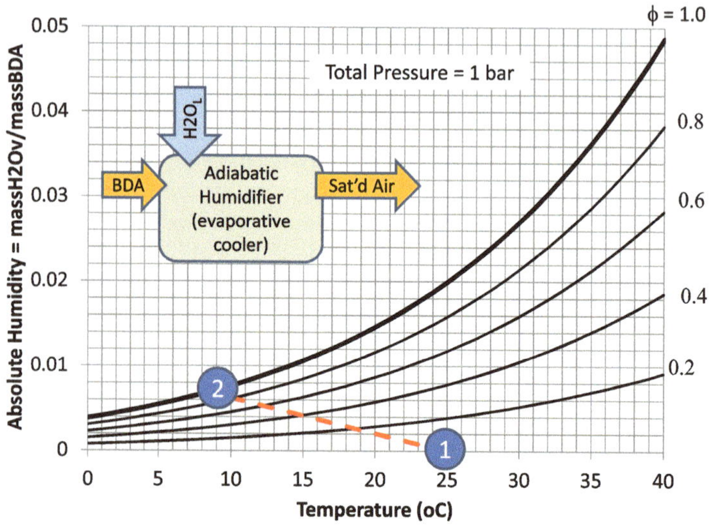

Fig. 12.23 Schematic and psychrometric chart for an adiabatic saturation process, or evaporative cooling, starting with bone dry air (BDA) at state 1 and terminating with saturated air at state 2 (the wet bulb temperature of state 1)

$$\left(h_{H_2Ov,out} - h_L\right) = \left(h_{H_2Ov,out} - h_{g@T_{in}}\right) + \left(h_{g@T_{in}} - h_{f@T_{in}}\right) + \left(h_{f@T_{in}} - h_L\right)$$

The first term is the change in enthalpy of the vapor, and is related to temperature change by the enthalpy-specific heat of H_2O_v. The 2nd term is the enthalpy of vaporization, and the last term is the negligible pressure effect on the liquid $v_{f@T_{in}}(P_{SAT@T_{in}} - P) \approx 0$. That is,

$$\left(h_{H_2Ov,out} - h_L\right) = c_{p,H_2Ov}(T_{out} - T_{in}) + h_{fg@T_{in}}$$

With the energy addition rate divided by BDA mass flow rate defined as $(q - w) = \left(\dot{Q} - \dot{W}\right)/\dot{m}_{BDA}$, the energy equation can be expressed as:

$$(q - w) = \left(c_{P,BDA} + \omega_{out}c_{p,H_2Ov}\right)(T_{out} - T_{in}) + \omega_{out}h_{fg@T_{in}}$$

The first term on the right represents the sensible energy change (i.e., temperature) while the second term is the latent energy change (phase). In the isothermal humidification case, the sensible energy change was identically zero, and the net energy exchange (heat and/or work) was entirely due to latent energy. In the adiabatic, no work case considered here, the change in sensible energy is equal and opposite to the latent energy change.

The evaporative cooling effect is analogous to the effect of an endothermic chemical reaction, one that absorbs rather than releases chemical energy. In that case, there is a conversion of thermal (sensible) energy into chemical energy. In adiabatic humidification, there is a conversion of sensible (thermal) energy into latent (phase) energy. An exothermic chemical reaction (like combustion), would raise the temperature as chemical energy is converted to thermal (sensible).

Rearranging for the outlet temperature:

$$T_{\text{out}} = T_{\text{in}} - \frac{\omega_{\text{out}} h_{\text{fg@}T_{\text{in}}} - (q - w)}{c_{\text{P,BDA}} + \omega_{\text{out}} c_{\text{p,H}_2\text{Ov}}}$$

Note that if the energy addition $(q - w)$ equals the outlet absolute humidity times enthalpy of vaporization, the outlet temperature equals the inlet temperature. That is the isothermal limiting case. On the other hand, if the process is adiabatic $(q - w) = 0$ and the outlet temperature decreases almost linearly with absolute humidity, the absolute humidity in the denominator has a small effect (since $\omega_{\text{out}} \ll 1$). A constant enthalpy line is followed on a psychrometric chart. Working values for the thermodynamic properties are obtained from Table 12.1:

Starting with BDA and adding H_2O results in an increase in both absolute and relative humidity, and a decrease in temperature, hence the term "evaporative cooling." The formula derived for the temperature is a simple closed form as a function of absolute humidity. The relative humidity is then determined from

$$\phi_{\text{out}} = \frac{P_{\text{H}_2\text{Ov,out}}}{P_{\text{SAT@}T_{\text{out}}}} = \frac{\omega_{\text{out}}}{(0.625 + \omega_{\text{out}})} \frac{P}{P_{\text{SAT@}T_{\text{out}}}}$$

Eventually, however, the air will become saturated, and the relative humidity will be 100%. The corresponding temperature at which saturation occurs is called the wet bulb temperature, so named from the reading obtained from the wet bulb of a sling psychrometer.

The exit temperature and relative humidity from an adiabatic humidifier as a function of the absolute humidity is shown in Fig. 12.24. The inlet is bone dry air (BDA) at 25 °C. For a fixed mass flow rate of BDA, the absolute humidity is proportional to the flow rate of liquid sprayed into it. The temperature decreases almost linearly (with an unnoticeable curvature due to the small effect of the absolute humidity term in the denominator). The relative humidity increases for two reasons; the increase in H_2O vapor (the numerator increases) and the saturation pressure lowers with the drop in temperature (the numerator decreases). Eventually, the exit air becomes saturated. This point can be determined iteratively (or by using a goal seek function in a worksheet, for example), but not in closed form.

Calculation of the wet bulb temperature is not straightforward, however, and requires the solution of a nonlinear algebraic equation, detailed in the deeper dive of Appendix 2. On the other hand, reading the wet bulb temperature off of a psychrometric chart is straightforward. In Fig. 12.23, the wet bulb temperature corresponding to BDA at 25 °C is approximately 8 °C, corresponding to an absolute humidity of approximately 0.007. Detailed psychrometric charts have lines of

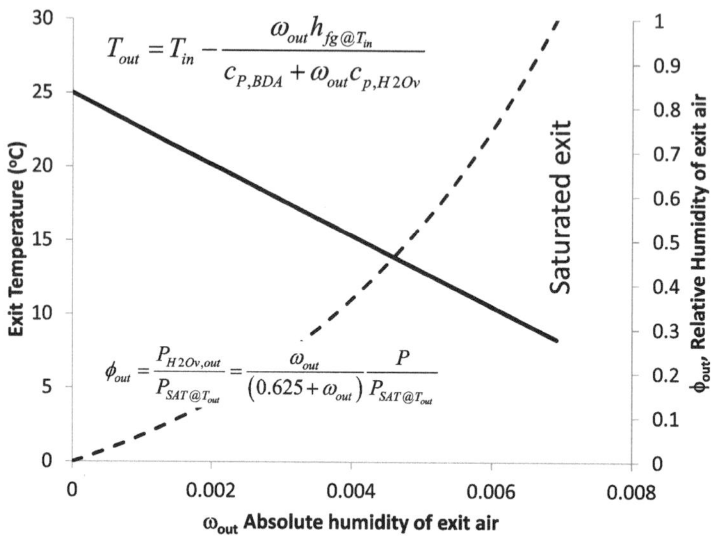

Fig. 12.24 Exit temperature and relative humidity as a function of absolute humidity. Initial state is bone dry air at 25 °C and a total pressure of 1 bar

constant enthalpy, like the dashed red line of Fig. 12.23, called lines of constant enthalpy, that terminate on the saturated line at the wet bulb temperature.

There is, of course, eXergy destruction associated with the adiabatic humidification process. The details are considered in the deeper dive of Appendix 3.

There are practical and commercially available devices called "Evaporative Coolers" or sometimes "Swamp Coolers" that put adiabatic humidification to useful purposes. They are able to achieve significant lowering of temperature of air by simply spraying liquid H_2O into it, without the complication and cost of a refrigeration system. Figure 12.25 shows adiabatic humidification lines for two representative locations in summer: New York City and Arizona. This plot shows partial pressure of H_2Ov (not absolute humidity) which is independent of total pressure. The temperature (dry bulb temperature) in NYC is shown a bit lower than in Arizona, but the relative humidity is higher. The wet bulb temperature represents the lowest temperature that can be achieved by adiabatic humidification, and is much lower for Arizona than NYC. Therefore, swamp coolers are attractive for hot, dry climates, but not for humid climates.

There is also an important biological role associated with adiabatic humidification, namely, sweating. There is an evaporative cooling effect when liquid H_2O is expelled through pores in the skin, *and the H_2O evaporates as it leaves*. The latter point is emphasized as it is easy to equate sweating with liquid dripping off (which does no good). The physical mechanism for evaporation is the transport process of diffusion of H_2O vapor away from the skin surface, which is driven by the difference in partial pressure at the surface (the saturation pressure at skin temperature) and the partial pressure in the surrounding air (dependent on its relative humidity and

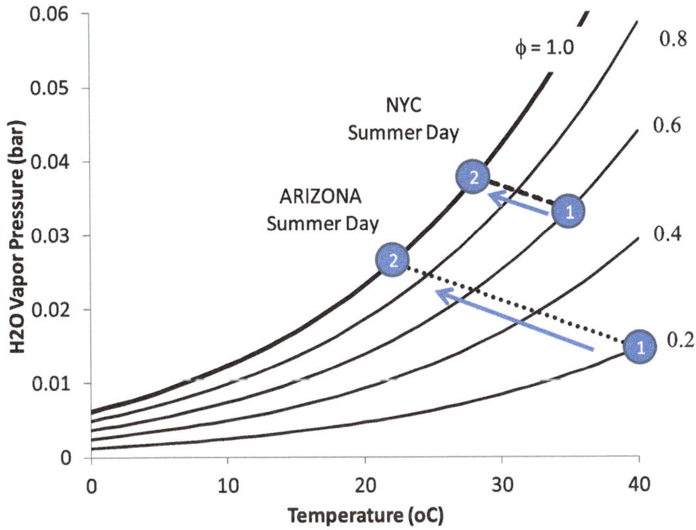

Fig. 12.25 H_2O vapor pressure plot showing constant enthalpy lines for a humid (NYC) and dry (Arizona) summer day

temperature). If the rate of H_2O liquid expelled exceeds the ability of the diffusion to evaporate it, excess H_2O will drip off as a liquid (achieving no significant cooling effect). It is for these reasons that the comfort level of air is related more to the dew point temperature than the actual temperature. The dew point temperature for the Arizona case is approximately 12 °C (by following a horizontal line to the saturation curve), while it is approximately 25 °C for NYC. The closer the dew point temperature is to the actual temperature (or dry bulb temperature), the more humid the air, and the less is its ability to evaporate sweat.

It is easier to become dehydrated in a dry climate than a humid one. In a dry climate, a large quantity of H_2O can be expelled from the pores, but it evaporates so the skin seems dry. You are not conscious of the amount of H_2O leaving. In a humid climate, you are dripping sweat and you know you are losing H_2O by sweating.

Mixing of Humid Air Streams

It is common in HVAC systems to mix two different streams of humid air that are at different temperatures and humidity levels. If the inlet streams are at a sufficiently low relative humidity, the streams will simply mix with no phase change. There can be condensation if both streams are very humid and at different temperatures. This case is not formally considered (but could be a good workshop). Figure 12.26 is a psychrometric chart with a schematic for a case study with a cold inlet with high

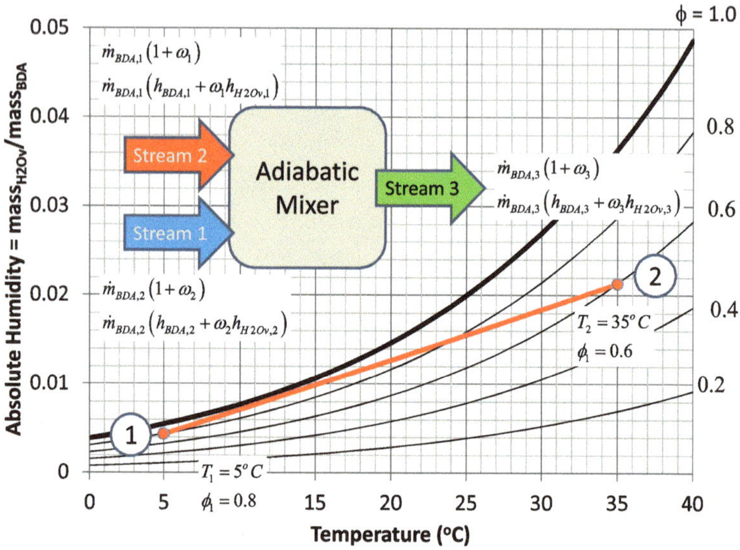

Fig. 12.26 Psychrometric chart showing the mixing line between two humid air streams. The exit state 3 lies on the red line, depending on the relative flow rates of the inlet streams. Total pressure = 1 atm

relative humidity mixing with a warm inlet with a lower relative humidity. The exit state will lie along the red line between the two states, but depends on the relative flow rates of the inlet streams, closer to the conditions of the stream with the higher flow rate. The equation of this line (which is not quite linear) is developed from manipulation of the mass and energy balances, as follows.

The overall mass balance for the mixing of two streams of humid air (states 1 and 2) to form a single exit stream (state 3) at steady state is:

$$\dot{m}_1 + \dot{m}_2 = \dot{m}_3$$

In terms of the BDA and absolute humidity, and recognizing that the exit BDA flow is the sum of the two inputs:

$$\dot{m}_{\text{BDA},1}(1+\omega_1) + \dot{m}_{\text{BDA},2}(1+\omega_2) = (\dot{m}_{\text{BDA},1} + \dot{m}_{\text{BDA},2})(1+\omega_3)$$

A H_2O mass balance is:

$$\dot{m}_{\text{BDA},1}\omega_1 + \dot{m}_{\text{BDA},2}\omega_2 = \dot{m}_{\text{BDA},3}\omega_3 = (\dot{m}_{\text{BDA},1} + \dot{m}_{\text{BDA},2})\omega_3$$

In terms of the ratio of the inlet stream 1 to the outlet stream BDA mass flow rate ratio, the exiting absolute humidity is a mass flow rate weighted average of the inlet and exit streams:

$$\omega_3 = \frac{(A_1/A_2)\omega_1 + \omega_2}{(A_1/A_2) + 1}$$

where

$$(A_1/A_2) = \frac{\dot{m}_{BDA,1}}{\dot{m}_{BDA,2}}$$

Is the ratio of the bone dry air mass flow rates.

Alternatively, a lever-rule type relationship between the change in absolute humidity and the ratio of the inlet BDA flow rates is:

$$\frac{\omega_3 - \omega_2}{\omega_1 - \omega_3} = (A_1/A_2)$$

Either form can be used, depending on how a given problem is formulated. It is emphasized that these results assume that there is no condensation, or that the exit stream has a relative humidity less than 100%. Condensation would occur if a line draw between the initial states on a psychrometric diagram crosses the saturation line (and depending on the ratio of air flows). The mass and energy balances would need to be revised to include an exiting liquid stream. Most cases of mixing would not fall in that category, however. The extreme case of Figs. 12.26 and 12.27 shows the line getting close to saturation, but not reaching it. A more humid stream 2 would result in some mass flow ratios with condensation.

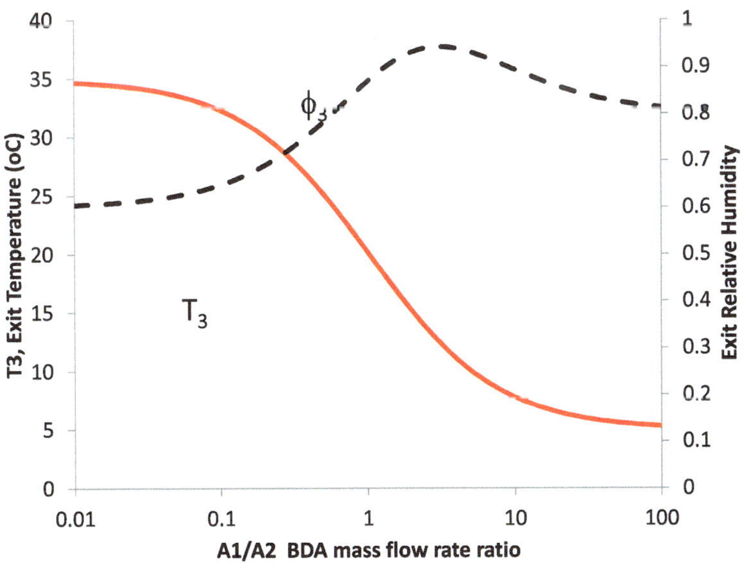

Fig. 12.27 Exit relative humidity and temperature as a function of the ratio of bone dry air mass flow rates of inlet streams corresponding to previous figure

The energy balance for adiabatic mixing is:

$$\dot{m}_1 h_1 + \dot{m}_2 h_2 = \dot{m}_3 h_3 = (\dot{m}_1 + \dot{m}_2) h_3$$

A lever rule expression based on total mass flow rates is:

$$\frac{h_1 - h_3}{h_3 - h_2} = \frac{\dot{m}_2}{\dot{m}_1}$$

This form is useful with a psychrometric chart with lines of constant enthalpy. Alternatively, an energy balance on the mixer can be treated more like a thermally-long heat exchanger. That is, the exit stream is expressed in two components, as if the two inlet streams are separated, but come to thermal equilibrium. If there is stray heat into the mixer from ambient (not the internal heat), the energy balance can be expressed as:

$$\dot{Q}_{\text{stray,in}} = \dot{m}_{\text{BDA,1}}\left(c_{\text{p,BDA}} + \omega_1 c_{\text{p,H}_2\text{Ov}}\right)(T_3 - T_1) + \dot{m}_{\text{BDA,2}}\left(c_{\text{p,BDA}} + \omega_2 c_{\text{p,H}_2\text{Ov}}\right) \times (T_3 - T_2)$$

The first term is the change in enthalpy flow of stream 1, and second term is the change in stream 2. For an adiabatic mixer (no stray heat), they are equal and opposite (like a heat exchanger). Rearranging for the exit temperature in terms of inlet conditions and the ratio of bone dry air mass flow rates:

$$T_3 = \frac{\left(c_{\text{p,BDA}} + \omega_1 c_{\text{p,H}_2\text{Ov}}\right)(A_1/A_2)T_1 + \left(c_{\text{p,BDA}} + \omega_2 c_{\text{p,H}_2\text{Ov}}\right)T_2}{\left(c_{\text{p,BDA}} + \omega_1 c_{\text{p,H}_2\text{Ov}}\right)(A_1/A_2) + \left(c_{\text{p,BDA}} + \omega_2 c_{\text{p,H}_2\text{Ov}}\right)}$$

The case study shown in Fig. 12.26 is constructed by calculating the exit absolute humidity and temperature as a function of bone dry air mass flow rates and plotting the exit absolute humidity against temperature. The relative humidity for each case is calculated and plotted (along with temperature) as a function of bone dry air mass flow ratio in Fig. 12.27. The relative humidity of the mixture increases for cases where the colder stream has the higher mass flow rate.

Cooling Tower Revisited: Condenser H_2O Flow Rate

In the cooling tower case study considered in the first part of this chapter, the control volume was chosen with the condensing H_2O internal (Fig. 12.14). The energy consequences were not considered, as the problem was formulated with specified air temperature and relative humidity. The required flow rate of make-up water could be calculated from mass balances alone. This case study problem statement is expanded

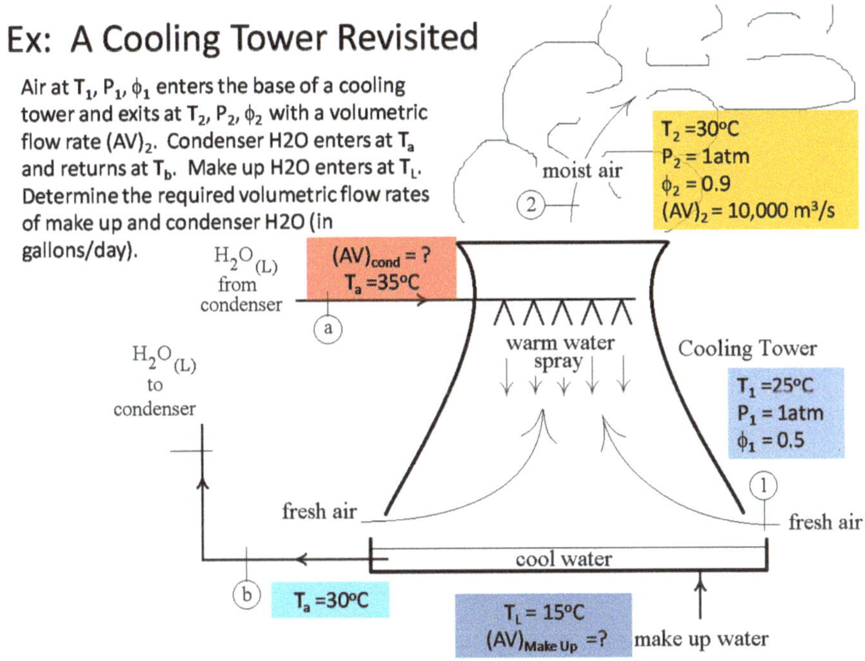

Fig. 12.28 Revised problem statement for cooling tower case study. The temperatures of the condenser H_2O and make-up water are specified

here in Fig. 12.28 to include inlet and exit condenser H_2O temperatures and make-up H_2O temperature (assumed to come from a ground source at 15 °C). The goal is to calculate the flow rate of condenser H_2O that can be cooled in the tower.

The control volume is defined in Fig. 12.29 for this calculation, across which both inlet and exit condenser H_2O streams flow, in addition to the same inlet and exit humid air and the make-up water as in the case study at the beginning of this chapter. The energy balance for this control volume, in a word statement, with the net energy input per unit mass of BDA (heat and work) on the left, and the enthalpy change in each stream on the right is:

$$q - w = \begin{pmatrix} \text{Sensible change of} \\ \text{BDA\&H}_2\text{Ov in air} \end{pmatrix} + \begin{pmatrix} \text{Sensible+} \\ \text{latent change} \\ \text{of make-up H}_2\text{O} \end{pmatrix} + \begin{pmatrix} \text{Sensible change} \\ \text{of condenser H2O} \end{pmatrix}$$

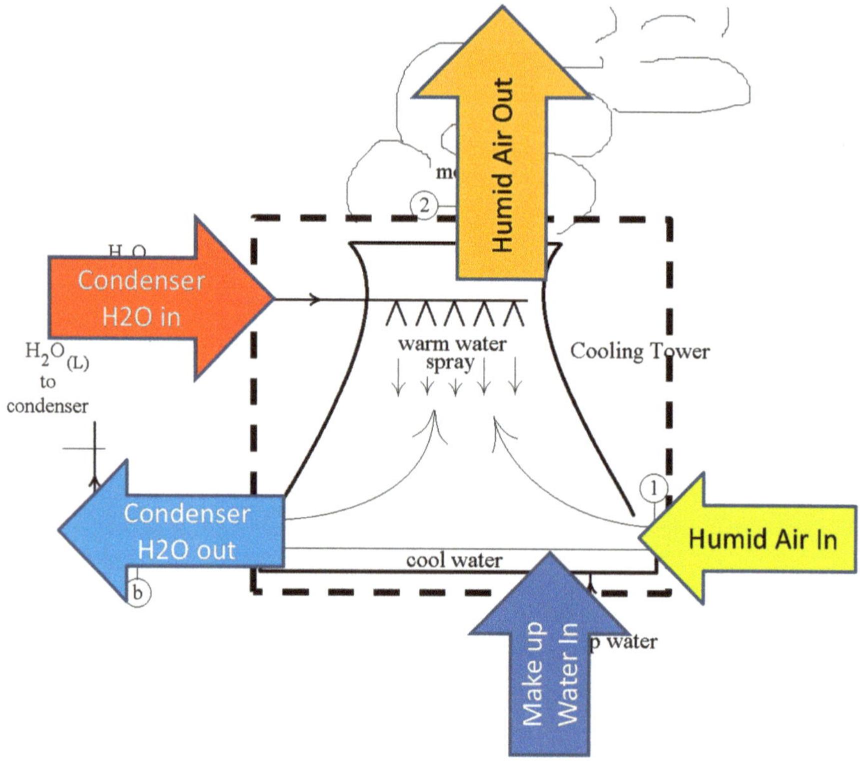

Fig. 12.29 Control volume for cooling tower case study showing mass (not energy) flows. Stray heat and work to drive fans and pumps could exist and are not shown

$$q - w = \left(c_{\mathrm{p,BDA}} + \omega_1 c_{\mathrm{p,H_2Ov}}\right)(T_2 - T_1) + (\omega_2 - \omega_1)\left(c_{\mathrm{p,H_2Ov}}(T_2 - T_{\mathrm{L}}) + h_{\mathrm{fg@}T_{\mathrm{L}}}\right) + \left(\frac{\dot{m}_{\mathrm{COND}}}{\dot{m}_{\mathrm{BDA}}}\right) c_{\mathrm{L}}(T_{\mathrm{b}} - T_{\mathrm{a}})$$

The system is analyzed neglecting stray heat (q) and input work associated with pumps and fans ($w = 0$), and the change in enthalpy of the humid air and makeup water is equal and opposite to the enthalpy change of the condenser water. The mass flow rate ratio, with $(q - w) = 0$ is:

$$\left(\frac{\dot{m}_{\mathrm{COND}}}{\dot{m}_{\mathrm{BDA}}}\right) = \frac{\left(c_{\mathrm{p,BDA}} + \omega_1 c_{\mathrm{p,H_2Ov}}\right)(T_2 - T_1) + (\omega_2 - \omega_1)\left(c_{\mathrm{p,H_2Ov}}(T_2 - T_L) + h_{\mathrm{fg@}T_{\mathrm{L}}}\right)}{c_{\mathrm{L}}(T_{\mathrm{b}} - T_{\mathrm{a}})}$$

The results of the case study are shown in the spreadsheet of Fig. 12.30. The energy terms are calculated separately in order to differentiate the relative importance of sensible and latent energy effects. The latent energy change of the make-up water accounts for 86.6% of the change, the rest being the sensible energy changes of

INPUT PARAMETERS

T1	**25**	oC, temperature of inlet air
P1	**1**	atm, atmospheric pressure
ϕ1	**0.5**	relative humidity of ambient
T2	**30**	oC, temperature of exit air
P2	**1**	atm, atmospheric pressure
ϕ2	**0.9**	relative humidity of exit air
Ta	**35**	oC, temperature of inlet condenser H2O
Tb	**30**	oC, temperature of outlet condenser H2O
TL	**15**	oC, temperature of make-up water
(AV)2	**10000**	m^3/sec, volumetric flow rate of exit
ρ_L	**1000**	kg/m^3, density of liquid water
cp,BDA	**1.005**	kJ/kg/K, specific heat of bone dry air
cp,H2Ov	**1.8**	kJ/kg/K, specific heat of H2O vapor
cp,L	**4.2**	kJ/kg/K, specific heat of liquid H2O
hfg@TL	**2465.5**	kJ/kg/K, enthalpy of evaporation at TL
$\eta_{Rankine}$	**0.3**	thermal efficiency of Rankine cycle

STATE	T (oC)	ϕ	P (bar)	P_{SAT} (bar)	ω	y_{vapor}	MW	ρ (kg/m^3)
1 (Air In)	25	0.5	1.0130	0.0313	0.0098	0.0155	28.63	1.171
2 (Air out)	30	0.9	1.0130	0.0420	0.0242	0.0373	28.40	1.142

DERIVED PARAMETERS, MASS CALCULATIONS

m'2	11419	kg/sec, mass flow rate of exit stream
m'BDA	11150	kg/sec, mass flow rate of BDA
m'MAKE-UP	160.19	kg/sec, mass flow rate of Make Up water
(AV)MAKE-UP	0.16019	m^3/sec, volumetric flow rate of Make-Up
(AV)MAKE-UP	3,656,207	gallons/day, volumetric flow rate of Make-Up

ENERGY TERMS (kJ/kgBDA)

	BDA (sensible)	H2O vapor (sensible)	Make-up H2O (sensible)	Make-up H2O (latent)	Condenser H2O
enthalpy change	5.03	0.088	0.388	35.42	-40.92
percent change	12.3%	0.22%	0.95%	86.6%	

DERIVED PARAMETERS, ENERGY CALCULATIONS

m'_{cond}/m'_{BDA}	1.9487	ratio of condenser H2O to bone dry air
m'_{cond}	21728	kg/sec, mass flow rate of condenser water
$(AV)_{cond}$	21.73	m^3/sec, volumetric flow rate of condenser H2O
$(AV)_{cond}$	495,920,141	gallons/day, volumetric flow rate of condenser H2O
$W'_{Rankine}$	195.5	MW, power output of associated Rankine cycle

Fig. 12.30 Spreadsheet solution for cooling tower case study. Yellow cells are inputs to the air state table

the BDA (12.3%), the H_2O vapor that entered with the humid air (0.22%), and the H_2O vapor of the make-up water after its phase change (0.95%). Clearly, the effectiveness of the cooling tower is due primarily to latent energy effects. Not only does the direct fluid/fluid interaction result in the minimal possible resistance to heat flow (as opposed to a heat exchanger), but the evaporation of the make-up water needed to humidify the air dominates the energy exchange between the air and condensing water streams.

The mass flow ratio (condenser water to BDA) evaluates to 1.95. The case study is defined with the same change in temperature for the air and condenser H_2O. If the cooling tower were replaced by a heat exchanger (no mixing of streams, no latent energy change), the energy balance would give a mass flow ratio related to the specific heats, as:

$$\frac{\dot{m}_{H_2O}}{\dot{m}_{BDA}} = \frac{c_{p,BDA}(T_{out} - T_{in})_{BDA}}{c_{p,H_2OL}(T_{in} - T_{out})_{H_2O}} = \frac{c_{p,BDA}}{c_{p,H_2OL}} = \frac{1.0}{4.2} = 0.24$$

The mass flow ratio for the cooling tower is more than 8 times higher than this, showing in another way how important latent energy changes are in cooling towers.

For the scale of this problem, which was set by the volumetric flow rate of exiting humid air, the volumetric flow rate of the condenser water is an astounding 496 million gallons of H_2O per day. The make-up water (and the amount evaporated) is 3.6 million gallons per day (0.7% of condenser water). Working backwards, what scale of a power plant would require that flow of condenser water? Let's see.

The thermal efficiency of a Rankine cycle of 30% is specified as an input, typical of a straight Rankine cycle from a coal power plant. The rate of heat exchange between the Rankine H_2O is equal and opposite to that of the condenser H_2O (neglecting stray heat) and that is calculated in this cooling tower problem (evaluating to 40.92 kJ/kg$_{BDA}$ and with a BDA mass flow rate of 11,150 kg/s, the cooling rate is 456 MW). The 1st Law efficiency of the Rankine cycle (that receives thermal input $\dot{Q}_{H,Rankine}$) is:

$$\eta_{Rankine} = \frac{\dot{W}_{Rankine}}{\dot{Q}_{H,Rankine}} = \frac{\dot{W}_{Rankine}}{\dot{W}_{Rankine} + \dot{Q}_{C,Rankine}}$$

The latter expression uses an overall energy balance for the Rankine cycle. Rearranging for the power output in terms of the condenser heat transfer rate and 1st Law efficiency:

$$\dot{W}_{Rankine} = \dot{Q}_{C,Rankine}\left(\frac{\eta_{Rankine}}{1 - \eta_{Rankine}}\right)$$

Evaluating, and converting units, the power output for the Rankine cycle that requires this scale of cooling tower is 195 MW, a moderately sized power plant, and approximately that whose plume is visible in Fig. 12.2.

Workshop 12.1: Humidification by Breathing

Every breath a human being takes consists of an inhale in which some H_2O vapor is taken into the lungs with the humid air and a greater amount of H_2O vapor that is expelled with the warmer and nearly saturated exhale. The goal of this workshop is to apply the principles developed in this chapter to quantify how much liquid H_2O a typical human being needs to drink daily to supply that biological requirement. Part 1 focuses on the mass loss. Part 2 focuses on the energy that must be supplied from the body to the gases in the lungs, and can be done as a separate exercise that builds on Part 1. In it, the relative importance of the sensible energy (required to raise air temperature) from the latent energy (to change phase) can be observed for this application.

Part 1: Mass Considerations

Estimate the total daily consumption rate of liquid H_2O (in L/day) of a typical human required to compensate for H_2O loss associated with breathing. The air being breathed is at $T_{\text{room}} = 25$ °C with a relative humidity of $\phi_{\text{room}} = 0.6$. Assume that the volume change of the lungs for each breath is $V_{\text{inhale}} = 0.2$ L, that the breathing rate is $N = 20$ breaths per minute, that each exhale is at a temperature $T_{\text{body}} = 37$ °C and is saturated (relative humidity of $\phi = 1$). Derive a closed form expression in terms of symbols, but evaluate for the default values given.

Develop a user-friendly worksheet (or program). Include at the top of each sheet or code a block for "Input Parameters," which can be changed by the user, with default values from the problem statement. Include a separate block for "Derived Parameters" which could include simple unit conversions. Display the final results prominently. Copy and paste a blank psychrometric chart (Fig. 12.31) into the spreadsheet and add graphics to indicate the process.

A curve fit (based on the Clausius–Clapeyron relation) that can be used for the saturation pressure of H_2O that is reasonably accurate over the range of conditions considered here is:

$$P_{\text{SAT}} = 1.567(10)^6 e^{\frac{-5283}{T+273}} \quad [=] \text{ bar, absolute,} \quad \text{where } T[=]^\circ\text{C}$$

Part 2: Energy Consequences

Estimate the total rate of energy loss (in Watts) associated with breathing, using the breath parameters from Part 1. Clearly compare the sensible (temperature change) and latent (phase change) components. The temperature of the liquid H_2O ingested can be taken to be $T_{\text{H}_2\text{OL,in}} - 5$ °C. Modify the spreadsheet (or program) from Part 1 to account for these energy effects. Relevant thermodynamic properties are:

$$h_{\text{fg@25 °C}} = 2441.4 \text{ kJ/kg}$$
$$c_{\text{p,BDA}} = 1.0 \text{ kJ/kg/}\Delta K$$
$$c_{\text{p,H}_2\text{Ov}} = 1.41 \text{ kJ/kg/}\Delta \text{K}$$

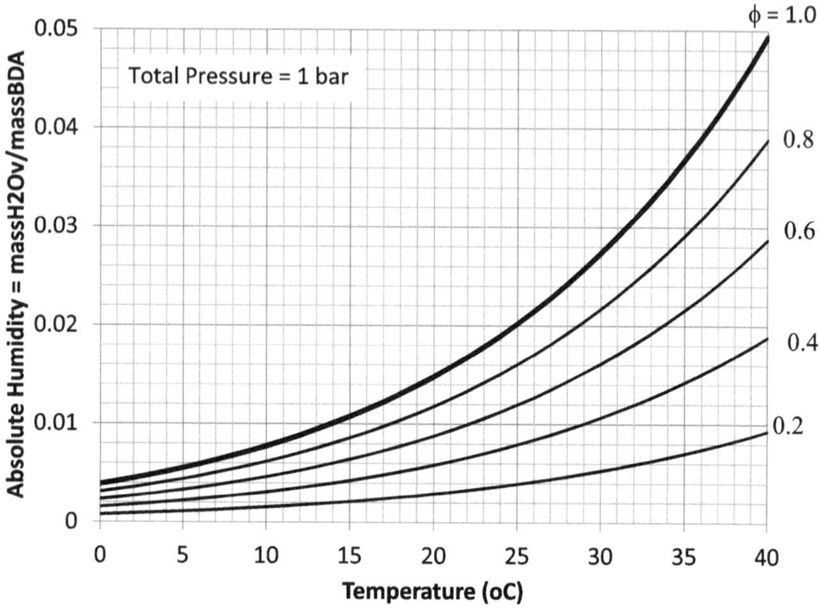

Fig. 12.31 Blank psychrometric chart

Workshop 12.2: Humidification Needs of Household in Winter

Think of your home during winter as a breathing entity. You need O_2, and therefore, fresh air from outside must enter for you to breathe (that is, there is a ventilation requirement needed for occupants of an interior space). Outside ambient air taken into the home is like an inhale, with temperature and relative humidity of outside air. By mass conservation, there is a nearly equal exiting mass flow, but at the average temperature and relative humidity of the indoor air (sustained by the HVAC system), so it is like an exhale. This workshop quantifies the required supply of liquid H_2O and the energy associated with changing both the temperature (sensible) and evaporation of the H_2O (latent) for a typical home.

Part 1: Mass Considerations

Estimate the total daily consumption rate of liquid H_2O (in L/day) of a typical household needed to maintain the humidity of the interior spaces at a comfortable level. Consider a day when the outside temperature is $T_\infty = -15\ °C$ with a relative humidity of $\phi_\infty = 0.9$, and a home with a floor area $A = 250\ m^2$ and average ceiling heights of $H = 2.5$ m. The interior space is maintained at $T = 23\ °C$ with a relative humidity of $\phi = 0.5$, with an air change rate of $n_{change} = 5$ per hour.

Develop a user-friendly worksheet (or program). Include at the top of each sheet or code a block for "Input Parameters," which can be changed by the user, with default values from the problem statement. Include a separate block for "Derived Parameters" which could include simple unit conversions. Display the final results prominently. Copy and paste a blank psychrometric chart (Fig. 12.31) into the spreadsheet and add graphics to indicate the process.

A curve fit (based on the Clausius–Clapeyron relation) that can be used for the saturation pressure of H_2O that is reasonably accurate over the range of conditions considered here is:

$$P_{SAT} = 1.567(10)^6 e^{\frac{-5283}{T+273}} \quad [=] \text{ bar, absolute,} \quad \text{where } T\,[=]\ ^\circ\text{C}$$

Part 2: Energy Consequences

Estimate the total input energy rate (in kW) of a typical household needed to maintain the temperature and humidity of the interior spaces at a comfortable level, using the parameters of Part 1. Clearly compare the sensible and latent components. The temperature of the liquid H_2O is $T_{H_2O,L} = 15\ ^\circ\text{C}$. Modify the spreadsheet (or program) from Part 1 to account for these energy effects. Relevant thermodynamic properties are:

$$h_{fg@25\ ^\circ\text{C}} = 2441.4\ \text{kJ/kg}$$
$$c_{p,BDA} = 1.0\ \text{kJ/kg/}\Delta\text{K}$$
$$c_{p,H_2Ov} = 1.41\ \text{kJ/kg/}\Delta K$$

Workshop 12.3: Dehumidification Needs of a Household in Summer

During summer, if the temperature of outside air were lowered by a cooling system without removing H_2O vapor, the relative humidity of the interior space would be too high for comfort and a breeding ground for mold. This workshop quantifies the amount of liquid H_2O that must be condensed and the associated energy removal (both sensible to change temperature and latent to condense H_2O) for a typical home.

Part 1: Mass Considerations

Estimate the total daily removal rate of liquid H_2O (in liters/day) of a typical household needed to maintain the humidity of the interior spaces at a comfortable level. Consider a day when the outside temperature is $T_\infty = 32.0\ ^\circ\text{C}$ with a relative humidity of $\phi_\infty = 0.7$, and a home with a floor area $A = 250\ \text{m}^2$ and average ceiling heights of $H = 2.5$ m. The interior space is maintained at $T = 25\ ^\circ\text{C}$ with a relative humidity of $\phi = 0.5$, with an air change rate of $n_{change} = 5$ per hour.

Develop a user-friendly worksheet (or program). Include at the top of each sheet or code a block for "Input Parameters," which can be changed by the user, with default values from the problem statement. Include a separate block for "Derived Parameters" which could include simple unit conversions. Display the final results prominently. Copy and paste a blank psychrometric chart (Fig. 12.31) into the spreadsheet and add graphics to indicate the process.

A curve fit (based on the Clausius–Clapeyron relation) that can be used for the saturation pressure of H_2O that is reasonably accurate over the range of conditions considered here is:

$$P_{SAT} = 1.567(10)^6 e^{\frac{-5283}{T+273}} \ [=] \text{ bar, absolute,} \quad \text{where } T \ [=] \ ^\circ\text{C}$$

Part 2: Energy Consequences

Estimate the total rate of energy removal rate (in kW) of a typical household needed to maintain the temperature and humidity of the interior spaces at a comfortable level, using the parameters of Part 1. Clearly compare the sensible and latent components. The temperature of the H_2O removed can be taken to equal to the interior space temperature. Modify the spreadsheet (or program) from Part 1 to account for these energy effects. Relevant thermodynamic properties are:

$$h_{fg@25\ ^\circ C} = 2441.4 \ \text{kJ/kg}$$
$$c_{p,BDA} = 1.0 \ \text{kJ/kg/}\Delta\text{K}$$
$$c_{p,H_2Ov} = 1.41 \ \text{kJ/kg/}\Delta\text{K}$$

Workshop 12.4: A Dehumidifier

A dehumidifier uses a conventional vapor/compression refrigeration cycle with R134a as the refrigerant. The cycle is assumed to operate on an ideal vapor/compression cycle at a steady state. That is, saturated vapor enters the compressor (state 1) and the compression process is isentropic (although the motor driving the compressor is assumed to be 90% efficient). The condenser process of the refrigeration cycle is isobaric and exchanges heat with the room air at $T_\infty = 25\ ^\circ$C, considered to be the dead state. The refrigerant enters the adiabatic valve as a saturated liquid (state 3). Refrigerant (state 4) and room air (state A1) enter the adiabatic evaporator in separate streams. The air is cooled below its dew point and returned to the room at $T_{A2} = 5\ ^\circ$C as a saturated humid air stream. Stray heat losses, kinetic and potential energy changes are negligible in all devices. The rate of liquid H_2O removal is $(AV)_L = 50$ L/day when the relative humidity of the room is $\phi_1 = 0.7$. The liquid drains at room temperature (state L). Complete the state and process tables and sketch the refrigeration process on the T-s and P-h diagrams, and the humid air process on the psychrometric chart provided.

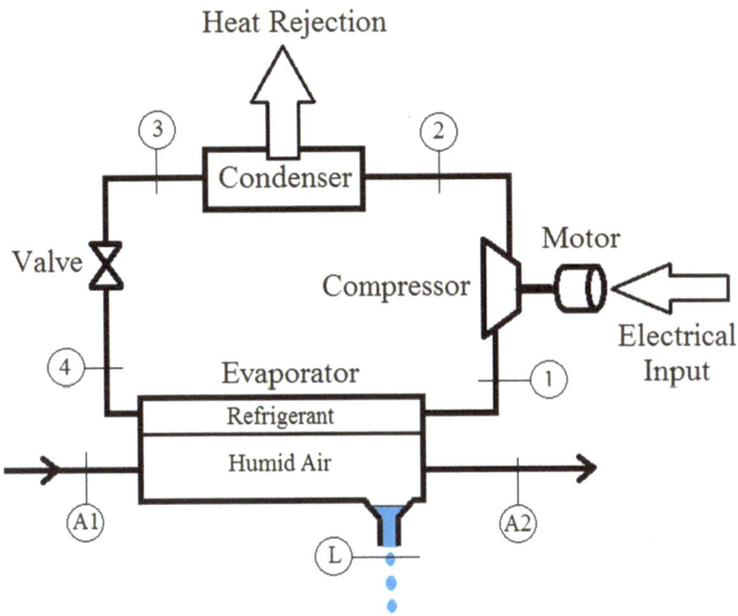

INPUT PARAMETERS

(AV)L	50	liters/day, H2O condensation rate
Pinf	1.013	bar, ambient pressure
ηm	0.9	motor efficiency
Tdead	25	oC, dead state temperature

REFRIGERANT STATE TABLE

State	P(kPa)	T(oC)	h(kJ/kg)	s(kJ/kg/K)	quality
1		0			1
2					
3		31.31			0
4					

HUMID AIR STATE TABLE

State	T(oC)	ϕ	ω	pv (bar)
1A	25	0.7		
2A	5	1		
L	25	NA	NA	NA

RESULTS TABLE

Quantity	Value	Unit
Dew Point of Room Air		oC
Bone Dry Air Mass Flow Rate		kg/sec
Inlet Humid Air Mass Flow Rate		kg/sec
Inlet Humid Air Volumetric Flow Rate		CFM
Refrigerant Mass Flow Rate		kg/sec
Compressor Shaft Power		Watts
Input Electrical Power		Watts
Heat Removal Rate from Air Stream		Watts
Heat Rejection Rate (Condenser to Room)		Watts
Net Heat Addition Rate to Room		Watts
System Coefficient of Performance		
eXergy Destruction Rate in condenser		Watts
eXergy Destruction Rate in valve		Watts
eXergy Destruction Rate in evaporator		Watts
eXergy Destruction Rate in motor		Watts

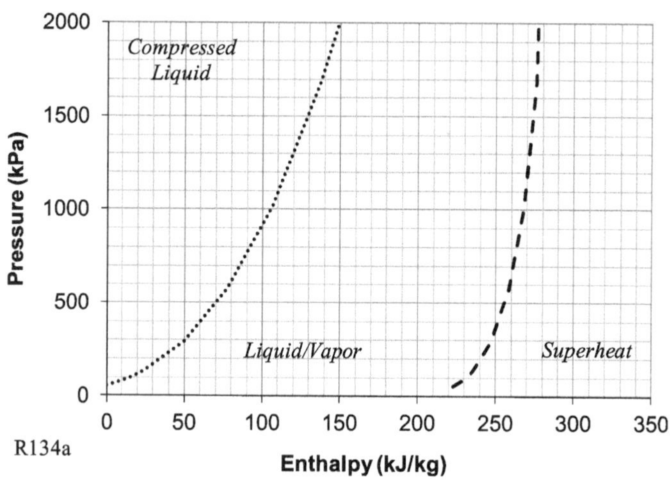

R134a

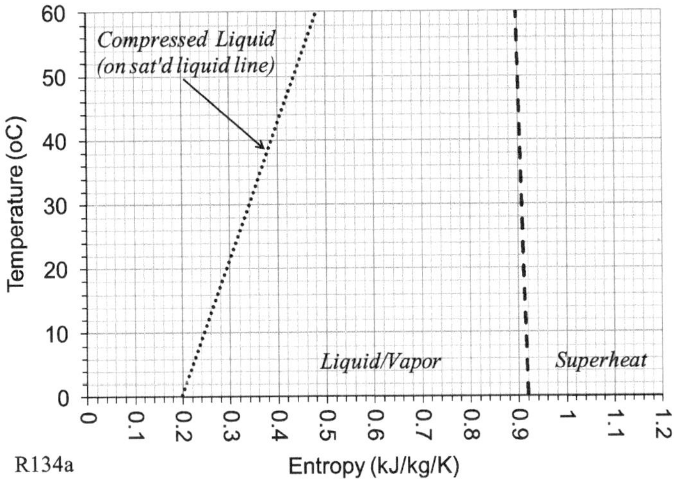

Workshop 12.5: Evaporative Cooling Due to Fuel Injection

This workshop applies the psychrometric principles developed in this chapter to an application that involves air as a noncondensable component but a liquid fuel as a condensable component (instead of H_2O). The readily available thermodynamic data is more limited than it is for H_2O.

Consider the fuel injection in a spark-ignition engine, in which liquid fuel is injected into an ambient air stream. The liquid will evaporate, and therefore, an evaporative cooling effect will take place. In practice, there will be significant heat transfer, but the goal of this workshop is more of a thought experiment, and to compare two liquid fuels, methanol and iso-octane (2–2-4 tri-methyl pentane, actually). The ratio of fuel to air injected is set by the combustion stoichiometry. What will the fuel/air adiabatic mixture temperature be? Will all of the injected liquid evaporate, or will the gas be saturated, with only a portion of the liquid evaporated?

Problem Statement

At a fixed total pressure, liquid fuel is injected into air in stoichiometric proportions and they mix adiabatically. Assume all of the liquid evaporates and determine the exit temperature, then check the relative humidity to determine whether the gas is saturated. Consider two pure hydrocarbon fuels: methanol and iso-octane (2,2,4 tri-methyl pentane). A spreadsheet template with input parameters and thermodynamic property data is shown. A deeper dive would be required if the exit is saturated.

Model Development

The thermodynamic data typically available is the normal boiling point (saturation temperature corresponding to 1 atm pressure) and the enthalpy of vaporization at the normal boiling point and the enthalpy-specific heats of liquid and vapor. Data from various on-line sources was used to obtain the input parameter values in the spreadsheet template. From Appendix 1 of Chap. 4, an approximate formula for vapor pressure is:

$$P = e^{\frac{h_{fg@1atm}}{R}\left(\frac{1}{T_{bp}}-\frac{1}{T}\right)}$$

where the pressure is in atm, and temperatures in Celsius. *R* is the gas constant for the liquid in question (universal gas constant divided by molecular weight).

Appendix 2 of Chap. 7 introduced combustion stoichiometry. For an oxygenated hydrocarbon with chemical formula $C_nH_mO_p$ mixed with air with an equivalence ratio ϕ, the chemical formula is:

$$C_nH_mO_p + \frac{1}{\phi}\left(n + \frac{m}{4} - \frac{p}{2}\right)(O_2 + 3.76N_2) \rightarrow \text{Products}$$

Therefore, the mole fraction of fuel in the exiting mixture is:

$$y_{\text{fuel}} = \frac{1}{1 + (4.76/\phi)(n + m/4 - p/2)}$$

The energy balance for the mixing process is:

$$\dot{Q}_{\text{in}} = \dot{m}_{\text{air}}\left(h_{\text{air,out}} - h_{\text{air,in}}\right) + \dot{m}_{\text{fuel}}\left(h_{\text{fuel,out}} - h_{\text{fuel,in}}\right)$$

The enthalpy change of the air is:

$$\left(h_{\text{air,out}} - h_{\text{air,in}}\right) = c_{\text{p,air}}\left(T_{\text{exit}} - T_{\text{A1}}\right)$$

A thermodynamic path through the normal boiling point is used to determine the enthalpy change of the fuel:

$$\left(h_{\text{fuel,out}} - h_{\text{fuel,in}}\right) = \left(h_{\text{fuel,out}} - h_{g@T_{\text{BP}}}\right) + \left(h_{g@T_{\text{BP}}} - h_{f@T_{\text{BP}}}\right) + \left(h_{f@T_{\text{BP}}} - h_{\text{fuel,in}}\right)$$

$$\left(h_{\text{fuel,out}} - h_{\text{fuel,in}}\right) = c_{\text{p,vapor}}\left(T_{\text{exit}} - T_{\text{BP}}\right) + h_{fg@T_{\text{BP}}} + c_{\text{L}}\left(T_{\text{BP}} - T_{\text{F1}}\right)$$

Adding to the energy equation, neglecting heat transfer and solving for the exit temperature (assuming complete vaporization):

$$T_{\text{exit}} = \frac{c_{\text{p,air}}T_{\text{A1}} + (\text{FAR})_{\text{mass}}\left(c_{\text{L}}\left(T_{\text{F1}} - T_{\text{BP}}\right) - h_{fg@T_{\text{BP}}} + c_{\text{p,vapor}}T_{\text{BP}}\right)}{c_{\text{p,air}} + (\text{FAR})_{\text{mass}}c_{\text{p,vapor}}}$$

Where $(\text{FAR})_{\text{mass}} = \dot{m}_{\text{fuel}}/\dot{m}_{\text{air}} = \text{MW}_{\text{fuel}}/(\text{MW}_{\text{air}}(4.76/\phi)(n + m/4 - p/2))$ is the fuel to air mass flow rate ratio (fuel to air). Methanol is chosen in part because it has both a higher than average FAR (due to oxygen in the fuel) and enthalpy of vaporization. Iso-octane has properties close to gasoline used in SI engines.

INPUT PARAMETERS

	METHANOL	ISO-OCTANE	
P	1	1	atm, total pressure
TA1	25	25	oC, inlet air temperature
TF1	15	15	oC, inlet fuel temperature
ϕ	1	1	equivalence ratio
n	1	8	carbon atoms in fuel
m	4	18	hydrogen atoms in fuel
p	1	0	oxygen atoms in fuel
TBP	64.5	99.3	oC, normal boiling point
hfg@Tbp	1100	308	kJ/kg, enthalpy of vaporization at 1 atm
cL	2.55	2.12	kJ/kg/K, specific heat of liquid fuel
cp,v	1.4	1.66	kJ/kg/K, specific heat of fuel vapor
cp,AIR	1.0	1.0	kJ/kg/K, specific heat of air

DERIVED PARAMETERS

MW			kg/kmol, molecular weight
R			kJ/kg/K, gas constant
FAR			fuel to air mass ratio

RESULTS

Texit			oC, exit temperature, assuming unsaturated exit
CHECK			return "IMPOSSIBLE" if exit is saturated

CHECK FOR SATURATION

yfuel			mol fraction of fuel in mixture
Pfuel			atm, partial pressure of fuel
Psat@Texit			atm, saturation pressure at Texit
hr			relative "humidity" - Pfuel/Psat

Appendix 1: Precise Bone Dry Air

A sample of air from the atmosphere would consist of the chemical species listed in Table 12.2 in decreasing order of mole fractions. The composition of an actual air sample will vary on a short time scale because of the change of H_2O_v associated with weather. This variation will not change the total amount of the other components, but it changes their relative amounts (but only at the 3rd significant figure, or so). Therefore, it is conventional to list the composition of the other components of air by removing the H_2O_v completely, resulting in bone dry air (BDA), the noncondensable component of humid air.

Reporting to five significant figures, the composition of precise BDA shows that N_2 and O_2 are dominant. All other species are at levels below a reasonable average

Table 12.2 Composition of Earth's air, with quantities listed in decreasing order in mole fractions. The value of CO2 is listed as a value to make the sum of mole fractions equal to unity (including trace gases at 0.000025)

Chemical species	Actual air sample	Precise bone dry air	Approximate bone dry air
N_2	Variable	0.78084	0.79
O_2	Variable	0.20946	0.21
H_2O_v	Variable with weather	0.00000	0
Argon	Variable	0.00934	0
CO_2	Variable (and rising)	$0.00033x$	0
Trace gases	Variable	<0.000025	0

for H_2O_v, with Argon being about 1% (making it actually a very large amount geologically). CO_2 is at the level of $0.00033x$. The "x" indicates that it varies with time. It has a natural seasonal variability as plants absorb CO_2 from the atmosphere during summer and release it steadily due to decay of dead plants. Historically, before 1950 or so, CO_2 peaked seasonally at levels below 0.000300 (or 300 ppm, parts per million). After that, there has been a steady rise superimposed on the natural variability due to the combustion of fossil fuels. The level at the time of writing exceeds 400 ppm. This man-made change in the atmosphere does not have a significant influence on the use of air in engineering systems, but it is generally accepted to be responsible for significant warming of the planet Earth, the greenhouse effect.

Trace gases include Neon, Helium, and other chemical species, many of which are direct pollutants (Nitrogen oxides, sulfur dioxides, particulates, unburned hydrocarbons, carbon monoxide) or greenhouse gases (methane, nitrous oxide). These all can vary with position and time. They are at sufficiently low levels that their presence has a negligible impact on the performance of most engineering systems, but require careful scrutiny for their role in air quality.

Appendix 2: Second Law Analysis in Psychrometrics

The eXergy destruction rate for a control volume at steady state is given by:

$$e\dddot{X}_{\mathrm{D}} = T_{\infty}\dddot{\Sigma}_{\mathrm{CV}} = -\frac{T_{\infty}}{T_{\mathrm{B}}}\dot{Q}_{\mathrm{in}} + \sum_{\mathrm{out}} \dot{m}T_{\infty}s - \sum_{\mathrm{in}} \dot{m}T_{\infty}s$$

For the general humidification case (Fig. 12.19), this can be expressed as:

$$e\dddot{X}_{\mathrm{D}} = -\frac{T_\infty}{T_{\mathrm{B}}}\dot{Q}_{\mathrm{in}} + \dot{m}_{\mathrm{BDA}}T_\infty\left[\left(s_{\mathrm{BDA,out}} - s_{\mathrm{BDA,in}}\right) + \omega_{\mathrm{in}}\left(s_{\mathrm{H_2Ov,out}} - s_{\mathrm{H_2Ov,in}}\right)\right.$$
$$\left. + \left(\omega_{\mathrm{out}} - \omega_{\mathrm{in}}\right)\left(s_{\mathrm{H_2Ov,out}} - s_{\mathrm{L,in}}\right)\right]$$

where the H_2O fluxes are broken into that which enters with the BDA and undergoes a sensible enthalpy change only, and the H_2O that undergoes both a sensible and latent (phase) change, The entropy of the BDA is determined from the calorically perfect gas relationship, namely:

$$\left(s_{\mathrm{BDA,out}} - s_{\mathrm{BDA,in}}\right) = c_{\mathrm{p,BDA}}\ln\left(\frac{T_{\mathrm{out}} + 273}{T_{\mathrm{in}} + 273}\right) - R_{\mathrm{BDA}}\ln\left(\frac{P - P_{\mathrm{H_2Ov,out}}}{P - P_{\mathrm{H_2Ov,in}}}\right)$$

There is both a temperature and a pressure effect (based on the partial pressure of BDA, which is the total pressure minus the partial pressure of H_2O vapor). The temperature is in °C in this expression. The entropy change of the vapor that is carried in with BDA is similar, as it does not undergo phase change:

$$\left(s_{\mathrm{H_2Ov,out}} - s_{\mathrm{H_2Ov,in}}\right) = c_{\mathrm{p,H_2Ov}}\ln\left(\frac{T_{\mathrm{out}} + 273}{T_{\mathrm{in}} + 273}\right) - R_{\mathrm{H_2O}}\ln\left(\frac{P_{\mathrm{H_2Ov,out}}}{P_{\mathrm{H_2Ov,in}}}\right)$$

The entropy change of the H_2O that enters as a liquid and exits as a vapor, the last term, is obtained by creating a thermodynamic path for which the phase change takes place at a specific temperature. If the inlet liquid temperature is used, the change in entropy can be expressed (by adding and subtracting the saturated vapor and liquid states) as:

$$\left(s_{\mathrm{H_2Ov,out}} - s_{\mathrm{L,in}}\right) = \left(s_{\mathrm{H_2Ov,out}} - s_{\mathrm{g@}T_{\mathrm{L,in}}}\right) + \left(s_{\mathrm{g@}T_{\mathrm{L,in}}} - s_{\mathrm{f@}T_{\mathrm{L,in}}}\right) + \left(s_{\mathrm{f@}T_{\mathrm{L,in}}} - s_{\mathrm{L,in}}\right)$$

The 1st term is the entropy change between the vapor in the outlet and the saturated vapor at the inlet liquid H_2O temperature and corresponding saturation pressure. That is:

$$\left(s_{\mathrm{H_2Ov,out}} - s_{\mathrm{g@}T_{\mathrm{L,in}}}\right) = c_{\mathrm{p,H_2Ov}}\ln\left(\frac{T_{\mathrm{out}} + 273}{T_{\mathrm{L,in}} + 273}\right) - R_{\mathrm{H_2O}}\ln\left(\frac{P_{\mathrm{H_2Ov,out}}}{P_{\mathrm{SAT@}T_{\mathrm{L,in}}}}\right)$$

The second term is recognized as the entropy change of evaporation at the inlet liquid temperature $\left(s_{\mathrm{fg@}T_{\mathrm{L,in}}}\right)$.

The third term represents the entropy change between a saturated liquid (at the saturation pressure) and the inlet liquid (at pressure P). Since the entropy of a compressed liquid is independent of pressure, this term is identically zero.

All the pieces are in place, and a final form of the eXergy destruction rate for a humidification process is:

$$\frac{e\dddot{X}_{\rm D}}{\dot{m}_{\rm BDA}} = -\frac{T_\infty}{T_{\rm B}}q +$$

$$\left(c_{\rm p,BDA} + \omega_{\rm in}c_{\rm p,H_2Ov}\right)T_\infty \ln\left(\frac{T_{\rm out}+273}{T_{\rm in}+273}\right)$$

$$-R_{\rm BDA}T_\infty \ln\left(\frac{P - P_{\rm H_2Ov,out}}{P - P_{\rm H_2Ov,in}}\right) - \omega_{\rm in}R_{\rm H_2O}T_\infty \ln\left(\frac{P_{\rm H_2Ov,out}}{P_{\rm H_2Ov,in}}\right)$$

$$+(\omega_{\rm out} - \omega_{\rm in})\left(c_{\rm p,H_2Ov}T_\infty \ln\left(\frac{T_{\rm out}+273}{T_{\rm L,in}+273}\right)\right.$$

$$\left.-R_{\rm H_2O}T_\infty \ln\left(\frac{P_{\rm H_2Ov,out}}{P_{\rm SAT@\mathit{T}_{L,in}}}\right) + T_\infty s_{\rm fg@\mathit{T}_{L,in}}\right)$$

A similar analysis for a dehumidification process (Fig. 12.20) results in:

$$\frac{e\dddot{X}_{\rm D}}{\dot{m}_{\rm BDA}} = -\frac{T_\infty}{T_{\rm B}}q +$$

$$\left(c_{\rm p,BDA} + \omega_{\rm out}c_{\rm p,H_2Ov}\right)T_\infty \ln\left(\frac{T_{\rm out}+273}{T_{\rm in}+273}\right)$$

$$-R_{\rm BDA}T_\infty \ln\left(\frac{P - P_{\rm H_2Ov,out}}{P - P_{\rm H_2Ov,in}}\right) - \omega_{\rm out}R_{\rm H_2O}T_\infty \ln\left(\frac{P_{\rm H_2Ov,out}}{P_{\rm H_2Ov,in}}\right)$$

$$+(\omega_{\rm in} - \omega_{\rm out})\left(c_{\rm p,H_2Ov}T_\infty \ln\left(\frac{T_{\rm out}+273}{T_{\rm in}+273}\right)\right.$$

$$\left.-R_{\rm H_2O}T_\infty \ln\left(\frac{P_{\rm SAT@\mathit{T}_{out}}}{P_{\rm H_2Ov,in}}\right) - T_\infty s_{\rm fg@\mathit{T}_{out}}\right)$$

Appendix 3: eXergy Destruction During Isothermal Humidification

Recall the eXergy destruction rate associated with a humidification process, per unit of bone dry air mass flow rate:

$$\frac{\dddot{eX}_D}{\dot{m}_{BDA}} = -\frac{T_\infty}{T_B}q +$$
$$\left(c_{p,BDA} + \omega_{in} c_{p,H_2Ov}\right) T_\infty \ln\left(\frac{T_{out}+273}{T_{in}+273}\right)$$
$$-R_{BDA} T_\infty \ln\left(\frac{P - P_{H_2Ov,out}}{P - P_{H_2Ov,in}}\right) - \omega_{in} R_{H_2O} T_\infty \ln\left(\frac{P_{H_2Ov,out}}{P_{H_2Ov,in}}\right)$$
$$+(\omega_{out} - \omega_{in})\left(c_{p,H_2Ov} T_\infty \ln\left(\frac{T_{out}+273}{T_{L,in}+273}\right) - R_{H_2O} T_\infty \ln\left(\frac{P_{H_2Ov,out}}{P_{SAT@T_{L,in}}}\right) + T_\infty s_{fg@T_{L,in}}\right)$$

Consider the example of an isothermal humidifier with BDA as the input and saturated air as the output, with the boundary temperature equal to the dead state temperature. The inlet temperature (liquid and humid air) equals the outlet temperature, the partial pressure of the exit H_2O vapor equals the saturation pressure at the corresponding temperature and the inlet absolute humidity is zero. The eXergy destruction rate simplifies to:

$$\frac{\dddot{eX}_D}{\dot{m}_{BDA}} = -q - R_{BDA} T_\infty \ln\left(1 - \frac{P_{SAT@T}}{P}\right) + \omega_{out} T_\infty s_{fg@T}$$

For the case at nominal room conditions (25 °C, 1 atm), using psychrometric formulas and saturation pressure curve fit,

$$P_{SAT@T}/P = 0.031$$
$$\omega = 0.02 \ \ \mathrm{kg_{H_2Ov}/kg_{BDA}}$$

Inserting values for the case study:

$$\frac{\dddot{eX}_D}{\dot{m}_{BDA}} = -q - \left(0.287 \frac{\mathrm{kJ}}{\mathrm{kg \cdot K}}\right)(298\ \ \mathrm{K}) \ln(1 - 0.031) + 0.02(298\ \ \mathrm{K}) \times \left(8.1895 \frac{\mathrm{kJ}}{\mathrm{kg}}\right)$$

$$\frac{\dddot{eX}_D}{\dot{m}_{BDA}} = -q + 2.701 + 48.663 = -q + 51.364 \frac{\mathrm{kJ}}{\mathrm{kg_{BDA}}}$$

The first numerical term is due to the increase in entropy of the BDA as its partial pressure is reduced. The much larger 2nd term is due to the entropy change associated with the phase change. The eXergy destruction is reduced by any heat transfer into the system.

To differentiate between the effects of heat and work, recall the energy balance associated with an isothermal humidification process:

$$q - w = \omega_{\text{out}} h_{\text{fg}@T_{\text{L,in}}}$$

Combining the energy and entropy balance equations by eliminating the heat transfer:

$$\frac{\dddot{eX}_{\text{D}}}{\dot{m}_{\text{BDA}}} = -w - R_{\text{BDA}} T_{\infty} \ln\left(1 - \frac{P_{\text{SAT}@T}}{P}\right) - \omega_{\text{out}}\left(h_{\text{fg}@T} - T_{\infty} s_{\text{fg}@T}\right)$$

$$\frac{\dddot{eX}_{\text{D}}}{\dot{m}_{\text{BDA}}} = -w - \left(0.287 \frac{\text{kJ}}{\text{kg} \cdot \text{K}}\right)(298\ \text{K}) \ln(1 - 0.031)$$
$$- 0.02\left[2441.4 \frac{\text{kJ}}{\text{kg}} - (298\ \text{K})\left(8.1895 \frac{\text{kJ}}{\text{kg}}\right)\right]$$

$$\frac{\dddot{eX}_{\text{D}}}{\dot{m}_{\text{BDA}}} = -w + 2.701 - 48.781 + 48.663 = -w + 2.683 \frac{\text{kJ}}{\text{kg}_{\text{BDA}}}$$

The first numerical term is associated with the entropy change of the BDA as before increase. The second term is negative and can be traced to the entropy transfer into the system associated with heat transfer. The last term is due to the phase change, as before. So, in the absence of work, there is eXergy destruction associated with the evaporation process, as there must be.

What is up with the output power term, w? In theory, a device could be constructed to extract useful output power (up to 2.683 kJ/kg$_{\text{BDA}}$ by setting the eXergy destruction to zero). I cannot envision a device that could, but maybe you can!

If the isothermal humidification were accomplished by heat transfer from ambient to the working fluid (at an incrementally lower temperature), and the work input identically zero, the eXergy destruction rate would be 2.683 kJ/kg. On the other hand, if the heat transfer rate were minimal, and the energy input was accomplished by an electrical resistance heating coil, then the power input would equal the change in enthalpy. The eXergy destruction rate would be:

$$\frac{\dddot{eX}_{\text{D}}}{\dot{m}_{\text{BDA}}} = 48.781 + 2.701 - 48.781 + 48.663$$
$$= 51.364 \frac{\text{kJ}}{\text{kg}_{\text{BDA}}} \quad \text{Adiabatic, with electrical work in}$$

Clearly, there is a large incentive thermodynamically to design a humidifier to operate by the transfer of heat instead of work, but that is difficult to accomplish in practice, at least for a stand-alone home appliance.

Consider now the reverse process, an isothermal dehumidification process, starting with saturated air at nominal room conditions, and dehumidifying it to bone dry air. The general expression for the eXergy destruction is:

$$\frac{\dddot{eX}_{\mathrm{D}}}{\dot{m}_{\mathrm{BDA}}} = -\frac{T_{\infty}}{T_{\mathrm{B}}} q +$$
$$\left(c_{\mathrm{p,BDA}} + \omega_{\mathrm{out}} c_{\mathrm{p,H_2Ov}}\right) T_{\infty} \ln\left(\frac{T_{\mathrm{out}} + 273}{T_{\mathrm{in}} + 273}\right)$$
$$-R_{\mathrm{BDA}} T_{\infty} \ln\left(\frac{P - P_{\mathrm{H_2Ov,out}}}{P - P_{\mathrm{H_2Ov,in}}}\right) - \omega_{\mathrm{out}} R_{\mathrm{H_2O}} T_{\infty} \ln\left(\frac{P_{\mathrm{H_2Ov,out}}}{P_{\mathrm{H_2Ov,in}}}\right)$$
$$+(\omega_{\mathrm{in}} - \omega_{\mathrm{out}})\left(c_{\mathrm{p,H_2Ov}} T_{\infty} \ln\left(\frac{T_{\mathrm{out}} + 273}{T_{\mathrm{in}} + 273}\right) - R_{\mathrm{H_2O}} T_{\infty} \ln\left(\frac{P_{\mathrm{SAT@}T_{\mathrm{out}}}}{P_{\mathrm{H_2Ov,in}}}\right) - T_{\infty} s_{\mathrm{fg@}T_{\mathrm{out}}}\right)$$

In this case, all the temperature terms and the exiting humidity evaluate to zero, reducing the eXergy destruction to:

$$\frac{\dddot{eX}_{\mathrm{D}}}{\dot{m}_{\mathrm{BDA}}} = -q - R_{\mathrm{BDA}} T_{\infty} \ln\left(\frac{1}{1 - P_{\mathrm{SAT@}T}/P}\right) - \omega_{\mathrm{in}} T_{\infty} s_{\mathrm{fg@}T}$$

Using the identity $\ln(1/x) = -\ln(x)$:

$$\frac{\dddot{eX}_{\mathrm{D}}}{\dot{m}_{\mathrm{BDA}}} = -q + R_{\mathrm{BDA}} T_{\infty} \ln\left(1 - \frac{P_{\mathrm{SAT@}T}}{P}\right) - \omega_{\mathrm{in}} T_{\infty} s_{\mathrm{fg@}T}$$

$$\frac{\dddot{eX}_{\mathrm{D}}}{\dot{m}_{\mathrm{BDA}}} = -q + R_{\mathrm{BDA}} T_{\infty} \ln\left(1 - \frac{P_{\mathrm{SAT@}T}}{P}\right) - \omega_{\mathrm{in}} T_{\infty} s_{\mathrm{fg@}T}$$

$$\frac{\dddot{eX}_{\mathrm{D}}}{\dot{m}_{\mathrm{BDA}}} = -q - 2.701 - 48.663 = -q - 51.364 \frac{\mathrm{kJ}}{\mathrm{kg_{BDA}}}$$

Unlike the humidification process, which could be adiabatic (with input work changing the phase, at great eXergy destruction cost), an isothermal dehumidification process must have a large heat transfer out of the working fluid.

To differentiate the effect of heat and work, consider the energy balance for this dehumidification process:

$$q - w = -\omega_{\mathrm{in}} h_{\mathrm{fg}}$$

Combining the energy and entropy balances

$$\frac{\dddot{eX}_{\mathrm{D}}}{\dot{m}_{\mathrm{BDA}}} = -w + R_{\mathrm{BDA}} T_{\infty} \ln\left(1 - \frac{P_{\mathrm{SAT}}}{P}\right) + \omega_{\mathrm{in}}\left(h_{\mathrm{fg}} - T_{\infty} s_{\mathrm{fg}}\right)$$

All the terms (except the work out) are of the opposite sign as the humidification case, as the process is just the reverse.

$$\frac{e\dddot{X}_{D}}{\dot{m}_{BDA}} = -w - 2.701 + 48.781 - 48.663 = -w - 2.683 \frac{\text{kJ}}{\text{kg}_{BDA}}$$

This result demonstrates that there must be a minimum of 2.683 kJ/kg$_{BDA}$ of work INTO the system in order to accomplish dehumidification. So work must be put in and a much larger amount of heat must be removed from a dehumidifier. Any internal irreversibility will increase both the required work input and heat rejection.

In Appendix 6, the energy consequence of using a stand-alone dehumidifier is considered. In that case, a large quantity of heat from the appliance is dumped into the interior space. An air-conditioning system is therefore required to remove that heat from the interior space. Dehumidification should involve heat rejection to the outside environment whenever possible. However, like a household refrigerator, it is much easier to simply dump heat into the interior space. This type of practice that satisfies convenience at the expense of energy costs may eventually become obsolete as a consequence of human evolution.

Appendix 4: Numerical Calculation of Wet Bulb Temperature

A closed-form solution for the wet bulb temperature is not possible due to the nonlinear nature of the saturation curve. There are several methods to solve for it iteratively, however. This Appendix develops the Regula–Falsi (or 2 point bracketing) method, a powerful method that works well for this type of problem and many others. A workshop format with a specific problem statement is developed. A general spreadsheet is developed that can be used for any other input conditions.

Problem Statement A humid air stream enters an evaporative (or swamp) cooler with a volumetric flow rate $(AV)_1 = 1000$ cfm (cubic feet per minute) at pressure $P = 1$ bar, temperature $T_{A1} = 30\ °C$ and a relative humidity $\phi_1 = 0.3$. Liquid H_2O is fed at a temperature $T_{W1} = 15\ °C$. The exiting humid air stream is saturated. Assume the device operates at steady state, with negligible kinetic and potential energy changes, and is adiabatic, with negligible input power required to operate. Determine the exit temperature and volumetric flow rate of the feed water (in gallons/day).

Model Development

The energy equation and definition of absolute humidity can be combined to yield a nonlinear algebraic equation for the exit temperature, which requires some form of iterative scheme to solve. The steady-state energy equation (neglecting kinetic and potential energy, but retaining heat and work exchanges, which can be set to zero for the adiabatic case, with no shaft work) is:

$$\dot{m}_{\text{BDA}} h_{\text{A1}} + \dot{m}_{\text{v1}} h_{\text{v1}} + \dot{m}_{\text{L}} h_{\text{H}_2\text{O,Lin}} + \dot{Q} - \dot{W} = \dot{m}_{\text{BDA}} h_{\text{A2}} + \dot{m}_{\text{v2}} h_{\text{v2}}$$

The H_2O balance is $\dot{m}_{\text{v1}} + \dot{m}_{\text{H}_2\text{OL}} = \dot{m}_{\text{v2}}$.

Dividing through by the dry air, and introducing the definition of absolute humidity (mass of H_2O vapor divided by mass of dry air):

$$\omega_1 + \dot{m}_{\text{H}_2\text{OL}} / \dot{m}_{\text{BDA}} = \omega_2$$

The energy equation becomes:

$$h_{\text{A1}} + \omega_1 h_{\text{v1}} + (\omega_2 - \omega_1) h_{\text{H}_2\text{O,Lin}} + \left(\dot{Q} - \dot{W}\right) / \dot{m}_{\text{BDA}} = h_{\text{A2}} + \omega_2 h_{\text{v2}}$$

Rearranging into a form that separates H_2O vapor that flows in with the inlet air stream from the H_2O liquid that flows in with the liquid stream (note, add and subtract $\omega_1 h_{\text{v2}}$ and add and subtract the enthalpy of vapor at the inlet liquid temperature):

$$\begin{aligned}(h_{\text{A2}} - h_{\text{A1}}) &+ \omega_1 (h_{\text{v2}} - h_{\text{v1}}) + (\omega_2 - \omega_1) \\ &\times \left(h_{\text{v2}} - h_{\text{H}_2\text{Ov@}T1\text{L}} + h_{\text{H}_2\text{Ov@}T1\text{L}} - h_{\text{H}_2\text{O,Lin}}\right) - \left(\dot{Q} - \dot{W}\right) / \dot{m}_{\text{BDA}} \\ &= 0\end{aligned}$$

Invoking the perfect gas approximation for air and H_2O vapor, and recognizing the heat of vaporization ($h_{\text{fg@T1L}}$) and defining the heat addition as $q - w = \left(\dot{Q} - \dot{W}\right) / \dot{m}_{\text{BDA}}$

$$\begin{aligned}\left(c_{\text{p,BDA}} + \omega_1 c_{\text{p,H}_2\text{Ov}}\right)(T_2 - T_1) &+ (\omega_2 - \omega_1)\left(c_{\text{p,H}_2\text{Ov}}(T_2 - T_{1\text{L}}) + h_{\text{fg@}T_{1\text{L}}}\right) \\ &- (q - w) \\ &= 0\end{aligned}$$

The first term is the change in sensible energy of the inlet humid air stream. The 2nd term is the change in energy of the liquid that evaporates (and is dominated by the latent energy compared to its change in sensible energy). The last term is the energy transfer (heat in and work out). This equation can be rearranged for the outlet temperature as a function of inlet conditions and outlet absolute humidity:

$$T_2 - \frac{\left(c_{\text{p,BDA}} + \omega_1 c_{\text{p,H}_2\text{Ov}}\right) T_1 + (\omega_2 - \omega_1) c_{\text{p,H}_2\text{Ov}} T_{1\text{L}} - h_{\text{fg@}T_{1\text{L}}} + q}{c_{\text{p,BDA}} + \omega_2 c_{\text{p,H}_2\text{Ov}}}$$

Given an absolute humidity of the exit stream, this energy equation yields an explicit expression for the exit temperature. However, there is an upper limit (defined by the saturation condition) that depends on the exit temperature, and finding that humidity makes the equation implicit (not explicit).

The definition of absolute humidity, expressed in terms of the relative humidity and saturation pressures is:

$$\omega_2 = \left(\frac{MW_{H_2O}}{MW_{BDA}}\right)\frac{\phi_2 P_{SAT@T2}}{P - \phi_2 P_{SAT@T2}}$$

The wet bulb temperature is obtained when the relative humidity at the exit temperature is 100%. However, the value of the wet bulb temperature and therefore the saturation pressure is unknown.

A recursion relation suitable for a powerful iterative method to solve certain types of nonlinear algebraic equations called the Regula–Falsi method (or 2 point bracketing) can be obtained by defining a function "f" that evaluates to zero when the energy balance is satisfied. That is:

$$\begin{aligned} f(T_2) &= \left(c_{p,BDA} + \omega_1 c_{p,H_2Ov}\right)(T_2 - T_1) + (\omega_2 - \omega_1) \\ &\quad \times \left(c_{p,H_2Ov}(T_2 - T_{1L}) + h_{fg@T_{1L}}\right) - (q - w) \\ &= 0 \end{aligned}$$

The Regula–Falsi method involves evaluating the function "*f*" for 2 values of T_2, one for which "*f*" evaluates to a negative number (called TLOW) and the other a positive number (THIGH). That is, the correct solution is bracketed. The next iterative guess is taken at an intermediate point (it could be midway, but does not need to be) and the function evaluated at that point. If "*f*" evaluates negative, then TLOW is replaced by the new value. Otherwise, THIGH is replaced.

A spreadsheet solution is shown in Fig. 12.32 (input and derived parameters), Fig. 12.33 (iteration table that implements Regula–Falsi method) and Fig. 12.34 showing results after the method converges to find the wet bulb temperature. The iteration table has three sections of four columns, one to evaluate the function "*f*" at THIGH, one for TLOW and one for TNEXT on each row. An in-then-else statement is entered into the next row to determine whether THIGH or TLOW is replaced by TNEXT. Ten iterations are shown, but 100 iterations are built in. In the RESULTS section, an if-then-else statement is implemented at the top to confirm that the method has converged (by testing whether the last value of "*f*" is below a convergence limit (set at 10^{-9}). It is clear from only 10 iterations that the method is converging to a wet bulb temperature of 17.9 °C for humid air at 30 °C and 30% relative humidity.

The spreadsheet allows for an exit relative humidity less than 100% to be specified.

INPUT PARAMETERS

AV1	1000	ft^3/min	Inlet air volumetric flow rate
P	1	bar	total pressure
TA1	30	oC	inlet air temperature
TL1	15	oC	inlet liquid H2O temperature
f1	0.3		inlet air relative humidity
f2	1		outlet air relative humidity

THERMO DATA

cp,air	1.005	kJ/kg/K	specific heat of air
cp,v	1.8	kJ/kg/K	specific heat of vapor
cL	4.2	kJ/kg/K	specific heat of liquid
hfg	2465.4	kJ/kg	heat of vaporization@15oC
rhoL	1000	kg/m^3	liquid density

DERIVED PARAMETERS

pv1	0.0126	bar	inlet vapor pressure
w1	0.0080	kgH2Ov/kga	inlet absolute humidity
M1	28.6640	kg/kmol	inlet molecular weight
rho1	1.1378	kg/m^3	inlet air density
AV1	0.4719	m^3/sec	inlet air volumetric flow rate
m'1	0.5370	kg/sec	inlet mass flow rate
m'BDA	0.5328	kgBDA/sec	mass flow rate of air

Fig. 12.32 Input and derived parameters for adiabatic saturation problem

ITERATION TABLE

k	THIGH	Psat@T2	w2	f(THIGH)	TLOW	Psat@T2	w2	f(TLOW)	TNEXT	Psat@T2	w2	f(TNEXT)
0	30	0.0419629	0.027376	48.37051	0	0.006177	0.003884	-40.53855815	15	0.016923087	0.010759	-8.41007
1	30	0.0410620	0.027376	48.37061	15	0.016923	0.010759	8.410066886	22.5	0.026957256	0.017315	15.52463
2	22.5	0.0269574	0.017315	15.52463	15	0.016923	0.010759	-8.410066886	18.75	0.021422883	0.013682	2.658423
3	18.75	0.0214229	0.013682	2.658423	15	0.016923	0.010759	-8.410066886	16.875	0.019055043	0.012141	-3.07818
4	18.75	0.0214229	0.013682	2.658423	16.875	0.019055	0.012141	-3.078177632	17.8125	0.02020812	0.012891	-0.26309
5	18.75	0.0214229	0.013682	2.658423	17.8125	0.020208	0.012891	-0.26308781	18.28125	0.020807615	0.013281	1.184023
6	18.28125	0.0208076	0.013281	1.184023	17.8125	0.020208	0.012891	-0.26308781	18.04688	0.020505918	0.013085	0.4571
7	18.046875	0.0205059	0.013085	0.4571	17.8125	0.020208	0.012891	-0.26308781	17.92969	0.020356535	0.012987	0.096169
8	17.9296875	0.0203565	0.012987	0.096169	17.8125	0.020208	0.012891	-0.26308781	17.87109	0.020282207	0.012939	-0.08367
9	17.9296875	0.0203565	0.012987	0.096169	17.87109	0.020282	0.012939	-0.083667663	17.90039	0.02031934	0.012963	0.006199
10	17.9003906	0.0203193	0.012963	0.006199	17.87109	0.020282	0.012939	-0.083667663	17.88574	0.020300766	0.012951	-0.03875

Fig. 12.33 Iteration table that implements Regula–Falsi method for adiabatic saturation problem

RESULTS

CONVERGENCE? YES

T2	17.89837	oC	exit air temperature
w2	0.012961	kgH2O/kgBDA	exit air absolute humidity
pv2	0.020317	bar	exit vapor pressure
M2	28.58058	kg/kmol	exit molecular weight
rho2	1.181734	kg/m^3	exit density
m'H2OL	0.00266	kg/sec	liquid water mass flow rate
(AV)H2OL	60.4812	gallons/day	liquid water volumetric flow rate

Fig. 12.34 Results table for adiabatic saturation problem

Appendix 5: eXergy Destruction Associated with Adiabatic Humidification

The general form (for steady-state open systems) of the eXergy destruction rate is:

$$\dddot{eX}_{\mathrm{D}} = T_{\infty}\dddot{\Sigma}_{\mathrm{CV}} = -\frac{T_{\infty}}{T_{\mathrm{B}}}\dot{Q}_{\mathrm{in}} + \sum_{\mathrm{out}} \dot{m}T_{\infty}s - \sum_{\mathrm{in}} \dot{m}T_{\infty}s$$

For the case of adiabatic humidification of BDA (Fig. 12.23), the eXergy destruction rate per unit of BDA mass flow rate is:

$$\frac{\dddot{eX}_{\mathrm{D}}}{\dot{m}_{\mathrm{BDA}}} = T_{\infty}\left(s_{\mathrm{BDA,out}} - s_{\mathrm{BDA,in}}\right) + \omega_{\mathrm{out}}T_{\infty}\left(s_{\mathrm{H_2O,out}} - s_{\mathrm{H_2Oin}}\right)$$

The entropy change of the BDA is given by, with the temperatures in absolute (Kelvin) units:

$$\left(s_{\mathrm{BDA,out}} - s_{\mathrm{BDA,in}}\right) = c_{\mathrm{p,BDA}} \ln\left(\frac{T_{\mathrm{out}}}{T_{\mathrm{in}}}\right) - R_{\mathrm{BDA}} \ln\left(\frac{\left(1 - x_{\mathrm{H_2Ov,out}}\right)P}{P}\right)$$

The entropy change of the H_2O is obtained by adding and subtracting intermediate saturated vapor and liquid states:

$$\left(s_{\mathrm{H_2O,out}} - s_{\mathrm{H_2Oin}}\right) = \left(s_{\mathrm{H_2O,out}} - s_{\mathrm{g@}T_{\mathrm{in}}}\right) + \left(s_{\mathrm{g@}T_{\mathrm{in}}} - s_{\mathrm{f@}T_{\mathrm{in}}}\right) + \left(s_{\mathrm{f@}T_{\mathrm{in}}} - s_{\mathrm{L}}\right)$$

The first term has a temperature and pressure effect.

$$\left(s_{\mathrm{H_2O,out}} - s_{\mathrm{H_2Oin}}\right) = c_{\mathrm{p,H_2Ov}} \ln\left(\frac{T_{\mathrm{out}}}{T_{\mathrm{in}}}\right) - R_{\mathrm{H_2O}} \ln\left(\frac{x_{\mathrm{H_2O,out}}P}{P_{\mathrm{SAT@}T_{\mathrm{in}}}}\right) + s_{\mathrm{fg@}T_{\mathrm{in}}} + c_{\mathrm{L}} \ln\left(\frac{T_{\mathrm{in}}}{T_{\mathrm{liquid,in}}}\right)$$

$$\frac{\dddot{eX}_{\mathrm{D}}}{\dot{m}_{\mathrm{BDA}}} = \left(c_{\mathrm{p,BDA}} + \omega_{\mathrm{out}}c_{\mathrm{p,H_2Ov}}\right)T_{\infty} \ln\left(\frac{T_{\mathrm{out}}}{T_{\mathrm{in}}}\right) - R_{\mathrm{BDA}}T_{\infty} \ln\left(1 - x_{\mathrm{H_2Ov,out}}\right) - \omega_{\mathrm{out}}R_{\mathrm{H_2O}}T_{\infty} \ln\left(\frac{P_{\mathrm{H_2Ov,out}}}{P_{\mathrm{SAT@}T_{\mathrm{in}}}}\right) + \omega_{\mathrm{out}}T_{\infty}s_{\mathrm{fg@}T_{\mathrm{in}}}$$

The results of the various terms and the net effect is plotted as a function of outlet absolute humidity in Fig. 12.35. The phase change effect and the temperature change of BDA are the dominant effects, of opposite sign.

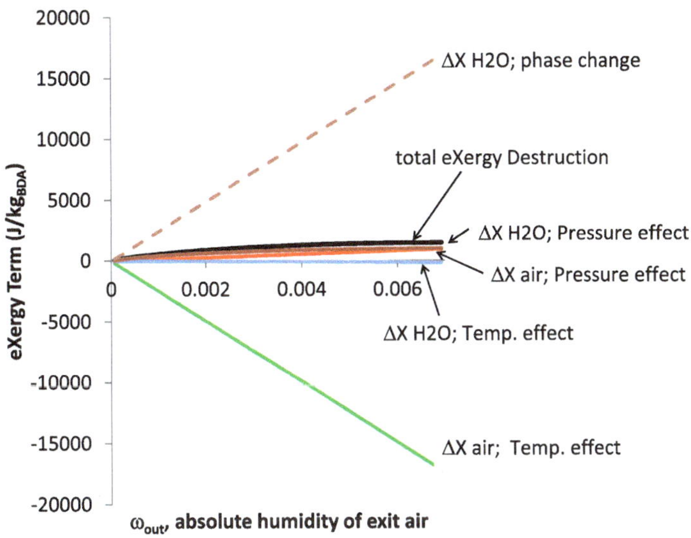

Fig. 12.35 eXergy destruction associated with adiabatic humidification. Individual terms and their sum are shown as a function of absolute humidity of exit air. Inlet is BDA at 25 °C

Appendix 6: Energy Requirements of a Stand-Alone Dehumidifier

Besides reinforcing mass, energy and eXergy principles in problems that involve humidity, the main goal of this extensive Appendix is to demonstrate that there is a large energy consequence associated with the use of a stand-alone dehumidifier appliance, a kind of unintended consequence. The unit draws electrical energy to operate, but dumps a large amount of heat into the interior space, which must be removed by an air conditioning system. The electrical energy associated with operating the dehumidifier will be shown to be much less than the electrical input energy required by the air conditioning system.

During the cooling season in humid climates, if ambient air is brought into a dwelling and the temperature lowered by an HVAC system without removing water vapor, the humid air will undergo a horizontal path to lower temperature on a psychrometric chart and therefore an increased relative humidity. This dampness in interior spaces can be a breeding ground for mold. Therefore, it is common practice in these climates to operate dehumidifiers (stand-alone electrically driven appliances) through the cooling season. The energy implications of that practice are considered in this section in a general way, without addressing the specific design of dehumidifiers. Workshop 12.4 is focused more on the way a dehumidifier could operate with a conventional vapor/compression refrigeration cycle.

The inherent energy cost of a dehumidifier should not be restricted to the electrical consumption of the appliance alone. As will be shown, there must be a

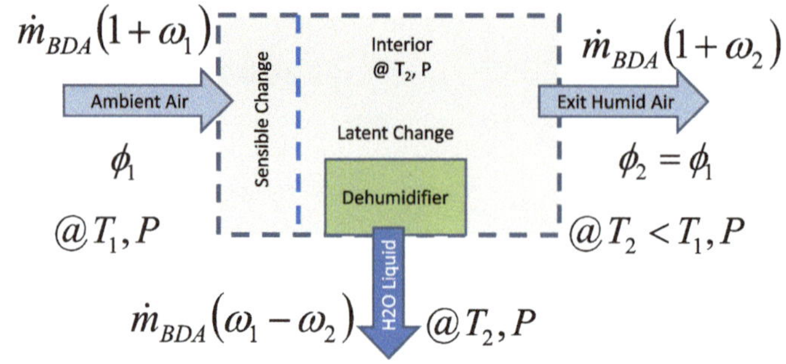

Fig. 12.36 Mass flows associated with an interior space equipped with a stand-alone dehumidifier to remove water vapor

heat rejection from the humidifier (the exiting enthalpy of the humid air is decreased by the latent energy change) to the interior space it is in. That is, the dehumidifier also acts as a heater (and the heat is much more than the work input). Therefore, since dehumidifiers are used in summer, the air conditioner that is used to maintain the temperature comfortable must work harder. A lot harder.

The scenario designed to focus on the effect of the dehumidifier is shown in Fig. 12.36, first from a purely mass-based point of view. Input and derived mass-related quantities are shown in the spreadsheet of Fig. 12.38. Energy and entropy flows will be considered shortly. The scale of this case study is an average single family house in 2010 in the USA (2400 square feet floor space, 8 foot high ceilings) with an air exchange rate set a 5 per hour, a typical value. Outside humid air is set at a value of $T_1 = 34$ °C (92.6 °F) and a relative humidity of $\phi_1 = 0.6$. The air inside the home is maintained at a set temperature ($T_2 = 25$ °C) and the study is designed so that the exit relative humidity is equal to the inlet. A thermodynamic pathway that differentiates sensible from latent enthalpy changes is shown in the psychrometric chart of Fig. 12.37. An intermediate state is shown where the sensible change is made first, followed by an isothermal dehumidification process.

This case study investigates only the energy requirements associated with the exchange of air, and not with the thermal load associated with heat transfer through the walls, windows and doors. The total requirements of the air conditioning system would have to include that effect as well.

The mass-related calculations for this case study are shown in the spreadsheet of Fig. 12.38. The H_2O liquid removal rate is almost 600 L/day.

The energy consequences of this scenario are shown in the energy flow diagram of Fig. 12.39. Entropy flows are shown in Fig. 12.40, for just the dehumidification section. Two heat pump cycles are indicated (which act as refrigerators, that is, the desired heat is removed from the cold space), which pump heat from the interior space to the outside ambient. The first heat pump cycle lowers the temperature (sensible energy) while the second is associated with the phase change (latent energy). A coefficient of performance (COP) of 3 is set for both. The combination of the heat pumps and the dehumidifier performs the function of an air conditioning

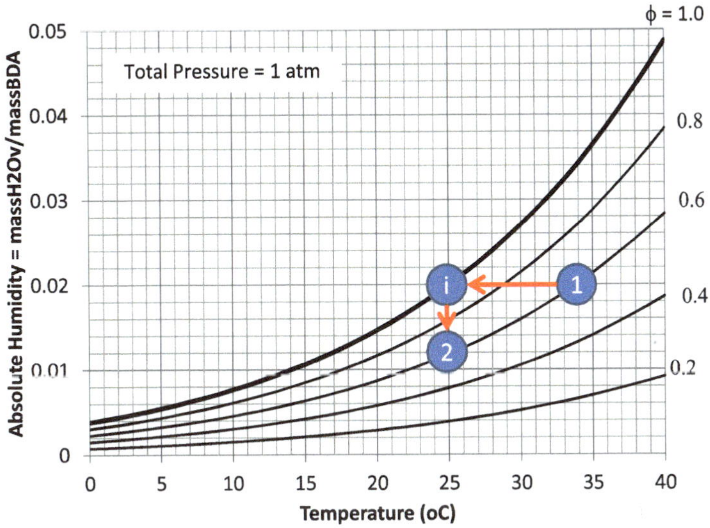

Fig. 12.37 Path of cooling and dehumidification process for case study. The path is through an intermediate state that is saturated. The inlet and exit relative humidities are set equal

unit, namely, to lower the temperature and absolute humidity. Their functions are separated in this case study in order to investigate the energy implications of the stand-alone dehumidifier.

Recall the general equation for a dehumidification process, on a per unit mass of BDA basis:

$$q_{\text{in}} - w_{\text{out}} = \left(c_{\text{p,BDA}} + \omega_{\text{out}} c_{\text{p,H}_2\text{Ov}}\right)\left(T_{\text{out}} - T_{\text{in}}\right) + \left(\omega_{\text{in}} - \omega_{\text{out}}\right)$$
$$\times \left(c_{\text{p,H}_2\text{Ov}}\left(T_{\text{L,out}} - T_{\text{in}}\right) - h_{\text{fg}@T_{\text{L,out}}} + v_{\text{f}@T_{\text{L,out}}}\left(P - P_{\text{SAT}@T_{\text{L,out}}}\right)\right)$$

The signs of the heat and work (q-w) from the generic form are reversed from the case shown in Fig. 12.39. The pressure effect on the liquid can be safely neglected, also. Applied to the first heat pump:

$$\dot{Q}_{\text{C1}} = \dot{m}_{\text{BDA}}\left(c_{\text{p,BDA}} + \omega_2 c_{\text{p,H}_2\text{Ov}}\right)\left(T_1 - T_2\right)$$

For the case study investigated, this heat transfer rate evaluates to 7623 W, which in common thermal units is 26,000 Btu/h (1 W = 3.41 Btu/h) or 2.17 tons of refrigeration (1 ton = 12,000 Btu/h). This represents the rate of cooling needed to lower the temperature of air that enters an average sized home from 34 °C to 25 °C.

The electrical input to the heat pump cycle is simply

INPUT PARAMETERS

A_{floor}	2400	ft^2, floor area
H	8	ft, ceiling height
exchange rate	5	per hour, exchange rate
T1	34	°C, outside temperature
T2	25	°C, room temperature
P	1.013	bar, room pressure
$\phi 1$	0.6	inlet relative humidity
$\phi 2$	0.6	exit relative humidity
COP_{AC}	3	Coef of performance of Heat Pump cycles

THERMO DATA

$c_{p,BDA}$	1005	J/kg/K, specific heat of air
$c_{p,H2Ov}$	1800	J/kg/K, specific heat of H2Ov
h_{fg}	2441700	J/kg, enthalpy of vaporization @ T2
s_{fg}	8189.5	J/kg/K, entropy of vaporatization @ T2
R_{BDA}	288.7	J/kg/K, gas constant for BDA
R_{H2O}	461.9	J/kg/K, gas constant for H2O
ρ_L	1000	kg/m^3, liquid density

DERIVED QUANTITIES

$p_{H2Ov,1}$	0.03160	bar, inlet vapor pressure
$p_{H2Ov,2}$	0.01879	bar, exit vapor pressure
$\omega 1$	0.02012	kgH2Ov/kgBDA, inlet absolute humidity
$\omega 2$	0.01181	kgH2Ov/kgBDA, exit absolute humidity
(AV)1	1600.00	CFM, inlet volumetric flow rate
(AV)1	0.75512	m^3/sec, inlet volumetric flow rate
MW1	28.09	kg/kmol, inlet molecular weight
ρ_1	1.115	kg/m^3, inlet density
m'_1	0.8419	kg/sec, inlet mass flow rate
m'_{BDA}	0.8253	kgBDA/sec, inlet BDA mass flow rate
m'_L	0.006859	kg/sec, condensate mass flow rate
$(AV)_L$	592.62	liters/day, condensate volumetric flow rate

Fig. 12.38 Spreadsheet with input parameters, thermo data, and derived quantities (mass related)

$$\dot{W}_{\text{e,HPcycle1}} = \frac{\dot{Q}_{\text{C1}}}{\text{COP}_{\text{R}}}$$

which evaluates to 2541 W.

The energy balance for the section that dehumidifies (isothermally) is:

$$\dot{Q}_{\text{C2}} - \dot{W}_{\text{e}} = \dot{m}_{\text{BDA}}(\omega_1 - \omega_2)h_{\text{fg@}T_2} = 16,750 \text{ W}$$

This energy represents the energy removal associated with the phase change (the latent energy). So the ratio of the latent to sensible energy change for this case study is 16,750/7623 = 2.2. This basic psychrometric energy calculation demonstrates that

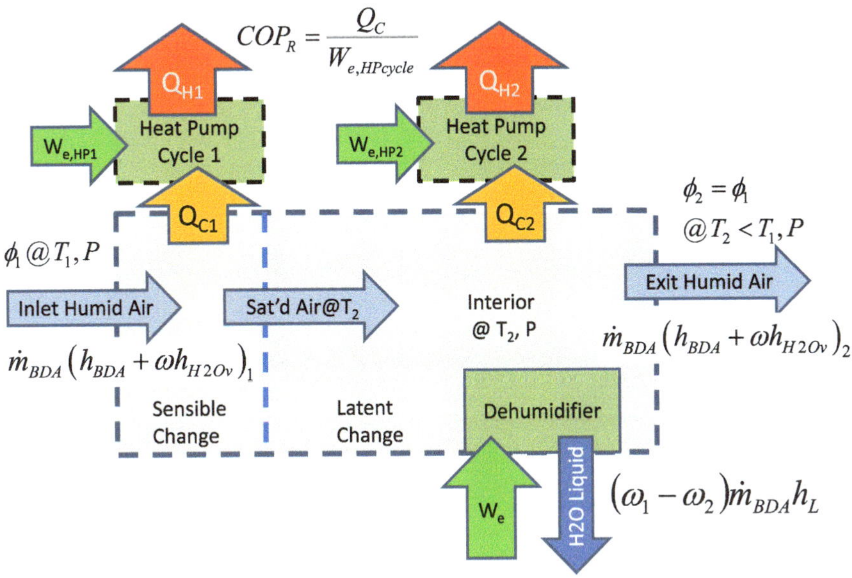

Fig. 12.39 Energy flows associated with a room equipped with a dehumidifier and two air conditioners; one to remove heat associated with changing the temperature (sensible change) and one to remove H_2O vapor (latent change)

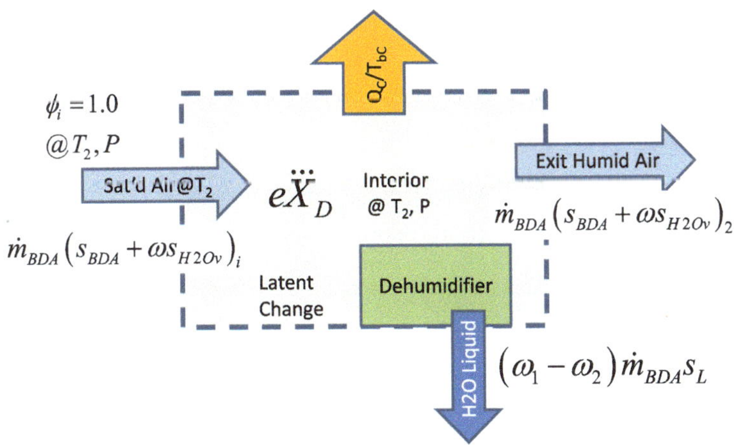

Fig. 12.40 Entropy flows in the dehumidifier section of the case study

latent energy changes are rather significant, despite the small change in air composition. A small relative amount of H_2O changes phase, but the energy associated with condensation is large.

At this point, it is not possible to differentiate heat from work, just that 16.7 kW of energy must be removed in order to dehumidify the air. One might even be tempted to say that you could make the process adiabatic, and then GENERATE 16.7 kW of electricity. That would be consistent with the 1st Law. But not the 2nd Law. A steady-state entropy balance on just the isothermal dehumidification process (Fig. 12.40) between state i (at T_2) is:

$$0 = -\frac{\dot{Q}_{C2}}{T_B} + \dot{m}_{BDA}[(s_i - s_2)_{BDA} + \omega_2(s_i - s_2)_{H_2Ov} + (\omega_1 - \omega_2)(s_{H_2Ov,i} - s_{2L})] + \dddot{\Sigma}$$

The boundary temperature for heat transfer is between the interior temperature and the temperature of the heat pump cycle. It will be evaluated at the interior temperature. The direction of the entropy transfer is out of the control volume. Note that the H_2O flows have been expressed with one stream that enters with the BDA as vapor and leaves as vapor (with exit humidity) and a stream that condenses.

Rearranging for the heat transfer term and normalizing by the bone dry air mass flow rate:

$$\frac{\dot{Q}_C}{\dot{m}_{BDA}} = T_B[(s_i - s_2)_{BDA} + \omega_2(s_i - s_2)_{H_2Ov} + (\omega_1 - \omega_2)(s_{H_2Ov,i} - s_{2L})] + \frac{T_B\,\dddot{\Sigma}}{\dot{m}_{BDA}}$$

Each stream has the same inlet and exit temperature, but there are entropy changes associated with partial pressure changes. Each of the entropy change terms is expressed, and then the result for the case study listed as follows. The entropy change (multiplied by the boundary temperature) of the BDA is:

$$T_B(s_i - s_2)_{BDA} = -R_{BDA}T_B \ln\left(\frac{P - P_{H_2Ov,1}}{P - P_{H_2Ov,2}}\right) = 1115\ \frac{J}{kg_{BDA}}$$

Note that the partial pressure of H_2O_v (and therefore BDA) at the intermediate state between the sensible and latent change is the same as for the inlet state 1. The term for the H_2O_v that does not condense is:

$$T_B\omega_2(s_i - s_2)_{H_2Ov} = -\omega_2 R_{H_2O}T_B \ln\left(\frac{P_{H_2Ov,i}}{P_{H_2Ov,2}}\right) = -845.1\frac{J}{kg_{BDA}}$$

This term is negative, and of comparable magnitude to the BDA change. The absolute humidity is small, but the ratio of inlet to outlet partial pressure of H_2O is much larger than the ratio for BDA (which is near unity in both cases).

The entropy change of the H_2O that condenses is obtained by creating a thermodynamic path through the phase change at the exit temperature:

$$\left(s_{\mathrm{H_2Ov},i} - s_{2\mathrm{L}}\right) = \left[\left(s_{\mathrm{H_2Ov},i} - s_{\mathrm{g}2}\right) + \left(s_{\mathrm{g}2} - s_{\mathrm{f}2}\right) + \left(s_{\mathrm{f}2} - s_{2\mathrm{L}}\right)\right]$$
$$= -R_{\mathrm{H_2O}} \ln\left(\frac{P_{\mathrm{H_2Ov},1}}{P_{\mathrm{SAT}@T_2}}\right) + s_{\mathrm{fg}@T_2}$$

The first term is the entropy change associated with changing the pressure from the partial pressure to the corresponding saturation pressure. The 2nd term is the phase change, and the last term is the change in pressure of the liquid to the pressure of the liquid (total pressure), which is zero. Evaluating:

$$(\omega_1 - \omega_2) T_{\mathrm{B}}\left(s_{\mathrm{H_2Ov,i}} - s_{2\mathrm{L}}\right) = (\omega_1 - \omega_2)\left[-R_{\mathrm{H_2O}} T_{\mathrm{B}} \ln\left(\frac{P_{\mathrm{H2Ov},1}}{P_{\mathrm{SAT}@T_2}}\right) + T_{\mathrm{B}} s_{\mathrm{fg}@T_2}\right]$$
$$= 584.3 + 20,280 \frac{\mathrm{J}}{\mathrm{kg_{BDA}}}$$

The pressure effect is small compared to the phase change, but the phase change is by far the largest effect of all, not surprisingly. Combining all the effects results in a heat transfer rate associated with the isothermal dehumidification of:

$$\frac{\dot{Q}_{\mathrm{C2}}}{\dot{m}_{\mathrm{BDA}}} = 21,136 \frac{\mathrm{J}}{\mathrm{kg}} + \frac{T_{\mathrm{B}} \dddot{\Sigma}}{\dot{m}_{\mathrm{BDA}}}$$

So heat and work are not equivalent. Heat must be removed in order to condense. Any irreversibilities within the dehumidifier will require greater heat transfer.

Combining the energy and entropy balances and rearranging for the input electrical power required:

$$\frac{\dot{W}_{\mathrm{e}}}{\dot{m}_{\mathrm{BDA}}} = 21,136 \frac{\mathrm{J}}{\mathrm{kg}} + \frac{T_{\mathrm{B}} \dddot{\Sigma}}{\dot{m}_{\mathrm{BDA}}} - (\omega_1 - \omega_2) h_{\mathrm{fg}@T_2}$$

Evaluating the enthalpy change:

$$\frac{\dot{W}_{\mathrm{e}}}{\dot{m}_{\mathrm{BDA}}} = 21,136 \frac{\mathrm{J}}{\mathrm{kg}} + \frac{T_B \dddot{\Sigma}}{\dot{m}_{\mathrm{BDA}}} - 20,292$$

The net result is

$$\frac{\dot{W}_{\mathrm{e}}}{\dot{m}_{\mathrm{BDA}}} = 844.5 \frac{\mathrm{J}}{\mathrm{kg}} + \frac{T_{\mathrm{B}} \dddot{\Sigma}}{\dot{m}_{\mathrm{BDA}}}$$

For the case study here ($\dot{m}_{\mathrm{BDA}} = 0.825\ \mathrm{kg/s}$), the energy rates, introducing the eXergy destruction in the dehumidifier section (noting that the boundary temperature and dead state are not equal):

$$\dot{Q}_{C2} = 17,445 + \left(\frac{T_B}{T_\infty}\right) e\dddot{X}_D \text{ W}$$

$$\dot{W}_e = 697 + \left(\frac{T_B}{T_\infty}\right) e\dddot{X}_D \text{ W}$$

The minimum heat transfer and electrical input rates are achieved if there is no eXergy destruction within the dehumidifier. That is:

$$\dot{Q}_{C2,\,min} = 17,445 \text{ W}$$

$$\dot{W}_{e,\,min} = 697 \text{ W}$$

The minimum electrical power represents the minimum power required to remove all the water vapor for this average house on a hot summer day. The heat transfer rate represents the heat rejection required. The way the case study is set up, with a stand-alone dehumidifier, this would be the heat rejection from the dehumidifier which is dumped into the interior space. This represents the unintended energy consequence of such an appliance. It is a very powerful heater. For every unit of input electrical power, approximately 25 units of heat are added to the interior space.

Contrast that with a household refrigerator, which also acts like a heater. The difference there is that the NET heat to the room exactly equals the net electrical input power. The refrigerator space is cold and heat from the room pours into it. In the dehumidifier, there is no such offsetting.

If all of the dehumidification is accomplished by the stand-alone dehumidifier (not as part of the heat pump cycles, which would occur to some degree), then the electrical energy to operate the heat pump cycles would be determined by the coefficient of performance, that is:

$$\dot{W}_{e,HP2} = \frac{\dot{Q}_{c2}}{COP_R}$$

which evaluates to 5815 W for the minimum value.

The energy results in Fig. 12.41 are before the inevitable irreversibilities within the dehumidifier are considered. A 2nd Law efficiency can be defined as the minimum input power divided by the minimum power plus whatever eXergy destruction occurs. That is:

$$\eta_{II} = \frac{\dot{W}_{e,\,min}}{\dot{W}_{e,\,min} + e\dddot{X}_D}$$

Rearranging for the eXergy destruction rate:

ENERGY RESULTS

Q'_{C1}	7623	W, heat removed from sensible energy section
$Q'_{C2} - W'_{e}$	16748	W, net rate of energy removal from dehumidification section
$Q'_{c2,min}$	17445	W, minimum required Qc2
$W'_{e,HP1}$	2541	W, electrical input to heat pump cycle 1
$W'_{e,min}$	697	W, minimum required input electricity to dehumidifier
$W'_{e,HP2}$	5815	W, minimum required input electricity to heat pump cycle 2

Fig. 12.41 Results of 1st and 2nd Law calculations

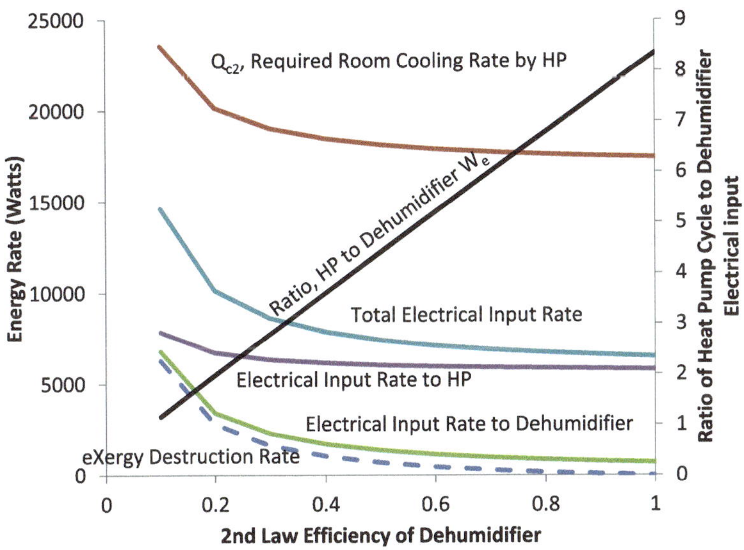

Fig. 12.42 Energy rates versus 2nd Law efficiency of dehumidifier. This extreme case assumes all H_2O condensation is accomplished by the dehumidifier. In practice, the heat pump cycles would remove a large portion of the H_2O vapor

$$e\dddot{X}_D = \dot{W}_{e,\,min}\left(\frac{1-\eta_{II}}{\eta_{II}}\right)$$

Figure 12.42 shows the effect of 2nd Law efficiency of the dehumidifier on energy terms. The rates shown for the heat pump electrical power are for the extreme case that all of the dehumidification takes place by the stand-alone dehumidifier. In reality, the air-conditioning units in a home would remove a large amount of the H_2O_v. These values scale, then, to the percentage of the H_2O_v removed by the heat pump cycles and that by the dehumidifier. The solid line in Fig. 12.42 shows the ratio of the electrical work that would be required of the heat pump to remove the heat rejection from the dehumidifier to the electrical work of the dehumidifier. For a 100% 2nd Law efficiency, the heat pump would require more than 8 units of electricity for every unit of input electricity to the dehumidifier. That ratio decreases with decreasing 2nd Law efficiency of the dehumidifier, but the total electrical consumption increases.

Appendix 1: Property Data

G. Sidebotham, *An Inductive Approach to Engineering Thermodynamics*,
Mechanical Engineering Series, https://doi.org/10.1007/978-3-030-20430-3

Specific Heats of Solids (Fig. 1)

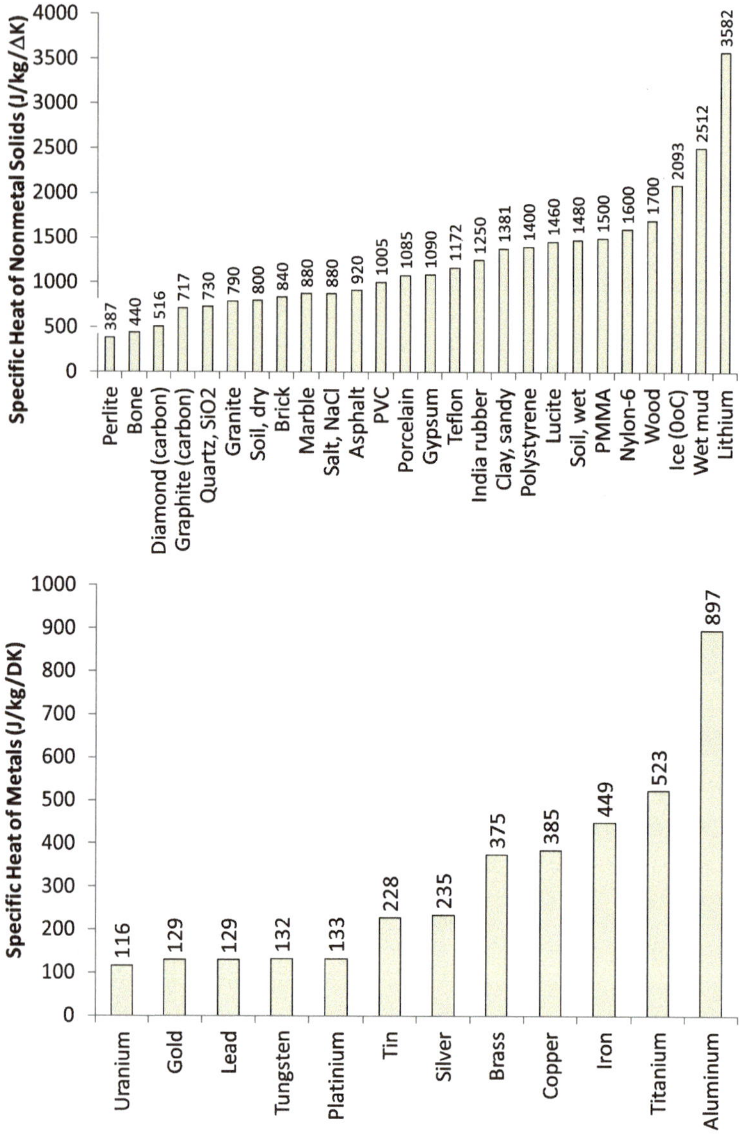

Fig. 1 Specific heat of nonmetals (top) and metals (bottom) at nominal room temperature. (Data adapted from http://www.engineeringtoolbox.com/specific-heat-capacity-d_391.html, accessed 6/26/2015)

Liquid Properties (Table 1)

Table 1 Liquid properties at nominal room temperature and their normal boiling points (sorted by increasing specific heat). Kerosene and engine oil are not well-defined mixtures of components and any property value is approximate

Liquid	Normal Boiling Point (°C)	Density (kg/m^3)	Specific Heat (J/kg/K)	Coef. of Thermal Expansion (1/K)
engine oil	>150	~900	~2000	~0.0007
kerosene	>150	~800	2010	0.00099
acetone	50.5	784.6	2150	0.00143
n-octane	125.6	699	2150	0.00114
dodecane	216	754.6	2210	
ethanol	79	789	2300	0.00109
ethylene glycol	197	1097	2360	0.00057
glycerin	290	1259	2430	0.0005
propylene glycol	187	965	2500	
water	100	997	4180	0.000257

Gas Properties (Tables 2 and 3)

Table 2 Properties of selected substances that are gases at nominal room conditions (25 °C, 1 atm). The compressibility factor (Z) are all 1.0 (by the low-pressure limit)

Substance	Chemical Formula	Molecular Weight	R (J/kg/K)	C_V (J/kg/ΔK) @ room conditions	Normal Boiling Point (°C)	T_{CRIT} (K)	P_{CRIT} (bar)	T/T_{CRIT} @ room conditions	P/P_{CRIT} @ room conditions	Classification
Helium	He	4	2079	3120	-269.0	5.3	2.3	56.2	0.440	gas
Air	mixture	29.0	287	718	-215.0	132.5	37.7	2.25	0.027	gas
Carbon Monoxide	CO	28	297	720	-191.5	133.0	35.0	2.24	0.029	gas
Carbon Dioxide	CO2	33	252	655	-78.5	304.2	73.9	0.980	0.014	vapor
Ethane	C2H6	30	277	1480	-88.0	305.5	44.8	0.975	0.023	vapor
Propane	C3H8	44	189	1480	-43.0	370.0	42.6	0.805	0.024	vapor
R-134a	CF3CH2F	102	81.5	2040	-26.3	374.2	40.6	0.796	0.025	vapor
Ammonia	NH3	17	489	1660	-35.5	405.5	112.8	0.735	0.009	vapor

Table 3 Ideal gas properties for a selection of common gases (specific heat values at nominal room temperature)

GAS	Molecular Weight (kg/kmol)	R (J/kg/K)	$C_{V,mass}$ @ 298K (J/kg/K)	$C_{V,molar}/R_u$ (@ 298K)	$k = C_p/C_V$ (@298K)
Hydrogen (H2)	2	4157	10,160	2.44	1.41
Helium	4	2079	3,120	1.50	1.67
Methane (CH4)	16	520	1,700	3.27	1.31
H_2O (vapor)	16	520	1,460	2.81	1.36
Nitrogen (N2)	28	297	743	2.50	1.40
2 component Dry Air	28.84	288	718	2.49	1.40
4 component Dry Air	28.95	287	718	2.50	1.40
Oxygen (O2)	32	260	659	2.54	1.39
Argon	40	208	312	1.50	1.67
CO_2	44	189	655	3.47	1.29

Two different models for dry air are shown, a 2 component (21 mol%O_2 in N_2) and a 4 component (78.09 mol%N_2, 20.95% O_2, 0.93%Argon, 0.03% CO_2). Water vapor is for pressures below the corresponding saturation pressure

H_2O Properties (Figs. 2 and 3)

Temperature Table

T (° C)	P(kPa)	v_f (m3/kg)	v_g (m^3/kg)	u_f (kJ/kg)	u_g (kJ/kg)	s_f (kJ/kg/K)	s_g (kJ/kg/K)
0.01	0.6117	0.001000	206.00	0	2374.9	0	9.1562
20	2.3392	0.001002	57.762	83.913	2402.3	0.2965	8.6661
40	7.3851	0.001008	19.515	167.53	2429.4	0.5724	8.2556
60	19.947	0.001017	7.667	251.16	2455.9	0.8313	7.9082
80	47.416	0.001029	3.4053	334.97	2481.6	1.0756	7.6111
100	101.42	0.001043	1.6720	419.06	2506.0	1.3072	7.3542
150	476.16	0.001091	0.392480	631.66	2559.1	1.8418	6.8371
200	1554.9	0.001157	0.127210	850.46	2594.2	2.3305	6.4303
250	3976.2	0.001252	0.050085	1080.7	2601.8	2.7933	6.0721
300	8587.9	0.001404	0.021659	1332.7	2563.6	3.2068	5.7432
350	16529	0.001741	0.008806	1642.4	2418.3	3.7788	5.2114
373.95	22064	0.003016	0.003016	2015.7	2015.7	4.407	4.407

Pressure Table

P(kPa)	T (° C)	v_f (m3/kg)	v_g (m^3/kg)	u_f (kJ/kg)	u_g (kJ/kg)	s_f (kJ/kg/K)	s_g (kJ/kg/K)
1	6.97	0.001000	129.19	29.302	2384.5	0.1059	8.9749
10	45.81	0.001010	14.67	191.79	2437.2	0.6492	8.1488
100	99.61	0.001043	1.6941	417.4	2505.6	1.3028	7.3589
1000	179.88	0.001127	0.19436	761.39	2582.8	2.1381	6.5850
10000	311.00	0.001452	0.018028	1393.3	2545.2	3.3603	5.6159

Fig. 2 Saturation property data for H_2O. (Adapted from the Appendices of *Thermodynamics, an Engineering Approach*, 7th ed., by Yunus A. Cengel and Michael A. Boles, McGraw-Hill. These tables are deliberately coarse. More detailed tables are available from a variety of textbooks and online sources. Notice that the reference state for these tables is the saturated liquid at the triple point of H_2O ($T_{ref} = 0.01$ °C, $P_{ref} = 0.6117$ kPa)

T (°C)	P (bar)	v_f (m^3/kg)	v_g (m^3/kg)	u_f (kJ/kg)	u_g (kJ/kg)	h_f (kJ/kg)	h_g (kJ/kg)	s_f (kJ/kg/K)	s_g (kJ/kg/K)
0.01	0.0061165	0.0010002	205.990	0.00	2374.9	0.000612	2500.9	0.00	9.1555
10	0.012282	0.0010003	106.300	42.02	2388.6	42.02	2519.2	0.15109	8.8998
20	0.023393	0.0010018	57.7570	83.91	2402.3	83.91	2537.4	0.29648	8.666
30	0.04247	0.0010044	32.8780	125.73	2415.9	125.73	2555.5	0.43675	8.452
40	0.073849	0.0010079	19.5150	167.53	2429.4	167.53	2573.5	0.5724	8.2555
50	0.12352	0.0010121	12.0270	209.33	2442.7	209.34	2591.3	0.70381	8.0748
60	0.19946	0.0010171	7.66720	251.16	2455.9	251.18	2608.8	0.83129	7.9081
70	0.31201	0.0010228	5.03950	293.03	2468.9	293.07	2626.1	0.95513	7.754
80	0.47414	0.0010291	3.40520	334.96	2481.6	335.01	2643	1.0756	7.6111
90	0.70182	0.0010360	2.35910	376.97	2494.0	377.04	2659.5	1.1929	7.4781
100	1.0142	0.0010435	1.67180	419.06	2506.0	419.17	2675.6	1.3072	7.3541
110	1.4338	0.0010516	1.20930	461.26	2517.7	461.42	2691.1	1.4188	7.2381
120	1.9867	0.0010603	0.891210	503.60	2528.9	503.81	2705.9	1.5279	7.1291
130	2.7028	0.0010697	0.668000	546.09	2539.5	546.38	2720.1	1.6346	7.0264
140	3.6154	0.0010798	0.508450	588.77	2549.6	589.16	2733.4	1.7392	6.9293
150	4.7616	0.0010905	0.392450	631.66	2559.1	632.18	2745.9	1.8418	6.8371
160	6.1823	0.0011020	0.306780	674.79	2567.8	675.47	2757.4	1.9426	6.7491
170	7.9219	0.0011143	0.242590	718.20	2575.7	719.08	2767.9	2.0417	6.665
180	10.028	0.0011274	0.193840	761.92	2582.8	763.05	2777.2	2.1392	6.584
190	12.552	0.0011415	0.156360	806.00	2589.0	807.43	2785.3	2.2355	6.5059
200	15.549	0.0011565	0.127210	850.47	2594.2	852.27	2792	2.3305	6.4302
210	19.077	0.0011727	0.104290	895.39	2598.3	897.63	2797.3	2.4245	6.3563
220	23.196	0.0011902	0.086092	940.82	2601.2	943.58	2800.9	2.5177	6.284
230	27.971	0.0012090	0.071503	986.81	2602.9	990.19	2802.9	2.6101	6.2128
240	33.469	0.0012295	0.059705	1033.4	2603.1	1037.60	2803	2.702	6.1423
250	39.762	0.0012517	0.050083	1080.8	2601.8	1085.80	2800.9	2.7935	6.0721
260	46.923	0.0012761	0.042173	1129.0	2598.7	1135.00	2796.6	2.8849	6.0016
270	55.030	0.0013030	0.035621	1178.1	2593.7	1185.30	2789.7	2.9765	5.9304
280	64.166	0.0013328	0.030153	1228.3	2586.4	1236.90	2779.9	3.0685	5.8579
290	74.418	0.0013663	0.025555	1279.9	2576.5	1290.00	2766.7	3.1612	5.7834
300	85.879	0.0014042	0.021660	1332.9	2563.6	1345.00	2749.6	3.2552	5.7059
310	98.651	0.0014479	0.018335	1387.9	2547.1	1402.20	2727.9	3.351	5.6244
320	112.84	0.0014990	0.015471	1445.3	2526.0	1462.20	2700.6	3.4494	5.5372
330	128.58	0.0015606	0.012979	1505.8	2499.2	1525.90	2666	3.5518	5.4422
340	146.01	0.0016376	0.010781	1570.6	2464.4	1594.50	2621.8	3.6601	5.3356
350	165.29	0.0017400	0.008802	1642.1	2418.1	1670.90	2563.6	3.7784	5.211
360	186.66	0.0018954	0.006949	1726.3	2351.8	1761.70	2481.5	3.9167	5.0536
370	210.44	0.0022152	0.004954	1844.1	2230.3	1890.70	2334.5	4.1112	4.8012
373.37	219.11	0.0025895	0.003868	1930.9	2120.7	1987.70	2205.5	4.2583	4.5952

Fig. 3 H_2O saturation data: temperature table. (Data downloaded from https://webbook.nist.gov/chemistry/fluid/ and formatted)

R134a Properties (Figs. 4, 5, and 6)

Pressure = 100 kPa (T_{sat} = 99.61°C)

T (° C)	v (m³/kg)	u (kJ/kg)	s (kJ/kg/K)
100	1.6959	2506.2	7.3611
200	2.1724	2658.2	7.8356
300	2.6389	2810.7	8.2172
400	3.1027	2968.3	8.5452
500	3.5655	3132.2	8.8362
600	4.0279	3302.8	9.0999
800	4.9519	3665.0	9.5682
1000	5.8755	4055.0	9.9800
1200	6.7988	4470.7	10.3504

Pressure = 10,000 kPa (T_{sat} = 311.00°C)

T (° C)	v (m³/kg)	u (kJ/kg)	s (kJ/kg/K)
400	0.026436	2833.1	6.0433
500	0.032811	3047.0	6.5995
600	0.038378	3242.0	6.9045
700	0.043597	3434.0	7.1693
800	0.048629	3628.2	7.4085
900	0.053547	3826.5	7.6290
1000	0.058391	4029.9	7.8349
1100	0.063183	4238.5	8.0289
1200	0.067938	4452.4	8.2126

Fig. 4 Superheated vapor properties of H_2O at 2 pressures. (Adapted from the Appendices of *Thermodynamics, an Engineering Approach*, 7th ed., by Yunus A. Cengel and Michael A. Boles, McGraw-Hill)

R134a Saturation Properties

T (oC)	P (bar)	v_f(m³/kg)	v_g(m³/kg)	h_f (kJ/kg)	h_g (kJ/kg)	s_f (kJ/kg/K)	s_g (kJ/kg/K)
-40	0.5164	0.0007054	0.360810	0.00	222.88	0.0000	0.9560
-28	0.9305	0.0007234	0.206660	14.37	230.38	0.0600	0.9411
-20	1.330	0.0007362	0.147290	24.26	235.31	0.0996	0.9332
0	2.928	0.0007723	0.069255	50.02	247.23	0.1970	0.9190
20	5.716	0.0008161	0.035969	77.26	258.36	0.2924	0.9102
40	10.16	0.0008720	0.019952	106.19	268.24	0.3866	0.9041
60	16.81	0.0009498	0.011434	137.42	275.99	0.4814	0.8973
80	26.32	0.0010772	0.006436	172.71	279.12	0.5814	0.8827
90	32.44	0.0011932	0.004599	193.69	276.32	0.6380	0.8656
100	39.74	0.0015269	0.002630	224.74	259.13	0.7196	0.8117

Fig. 5 Coarse listing of saturation properties

R134a Superheated Vapor Properties

P = 100 kPa (Tsat = -26.37°C)

T (°C)	v (m³/kg)	h (kJ/kg)	s (kJ/kg/K)	c_p(kJ/kg/K
-26.37	0.19254	234.44	0.9518	0.794
-20	0.19841	239.50	0.9721	0.799
-10	0.20743	247.49	1.0030	0.809
0	0.2163	255.58	1.0332	0.823
10	0.22506	263.81	1.0628	0.836
20	0.23373	272.17	1.0918	0.851
30	0.24233	280.68	1.1203	0.866
40	0.25088	289.34	1.1484	0.882
50	0.25937	298.16	1.1762	0.897
60	0.26783	307.13	1.2035	0.913
70	0.27626	316.26	1.2305	0.929
80	0.28465	325.55	1.2572	0.944
90	0.29303	334.99	1.2836	0.961
100	0.30138	344.60	1.3096	

note: c_p based on first order forward difference approximation

R134a Superheated Vapor Properties

P = 320 kPa (Tsat = 2.46°C)

T (°C)	v (m³/kg)	h (kJ/kg)	s (kJ/kg/K)	c_p(kJ/kg/K
2.46	0.06360	251.88	0.9301	0.903
10	0.06609	258.69	0.9544	0.897
20	0.06925	267.66	0.9856	0.899
30	0.07231	276.65	1.0157	0.905
40	0.07530	285.70	1.0451	0.915
50	0.07823	294.85	1.0739	0.926
60	0.08111	304.11	1.1021	0.937
70	0.08395	313.48	1.1298	0.950
80	0.08675	322.98	1.1571	0.964
90	0.08953	332.62	1.1840	0.977
100	0.09229	342.39	1.2105	0.991
110	0.09503	352.30	1.2367	1.005
120	0.09775	362.35	1.2626	

R134a Superheated Vapor Properties

P = 800 kPa (Tsat = 31.31°C)

T (°C)	v (m³/kg)	h (kJ/kg)	s (kJ/kg/K)	c_p(kJ/kg/K
31.31	0.025621	267.29	0.9183	1.054
40	0.027035	276.45	0.9480	1.024
50	0.028547	286.69	0.9802	1.012
60	0.029973	296.81	1.0110	1.007
70	0.031340	306.88	1.0408	1.009
80	0.032659	316.97	1.0698	1.013
90	0.033941	327.10	1.0981	1.020
100	0.035193	337.30	1.1258	1.029
110	0.036420	347.59	1.1530	1.038
120	0.037625	357.97	1.1798	1.048
130	0.038813	368.45	1.2061	1.060
140	0.039985	379.05	1.2321	1.071
150	0.041143	389.76	1.2577	1.083
160	0.042290	400.59	1.2830	

note: c_p based on first order forward difference approximation

R134a Superheated Vapor Properties

P = 1000 kPa (Tsat = 39.37°C)

T (°C)	v (m³/kg)	h (kJ/kg)	s (kJ/kg/K)	c_p(kJ/kg/K
39.37	0.020313	270.99	0.9156	1.143
40	0.020406	271.71	0.9179	1.103
50	0.021796	282.74	0.9525	1.064
60	0.023068	293.38	0.9850	1.047
70	0.024261	303.85	1.0160	1.040
80	0.025398	314.25	1.0458	1.039
90	0.026492	324.64	1.0748	1.042
100	0.027552	335.06	1.1031	1.047
110	0.028584	345.53	1.1308	1.053
120	0.029592	356.06	1.1580	1.063
130	0.030581	366.69	1.1846	1.071
140	0.031554	377.40	1.2109	1.082
150	0.032512	388.22	1.2368	1.093
160	0.033457	399.15	1.2623	

Fig. 6 Superheated vapor properties (at four pressures that correspond to typical condensing and evaporating temperatures)

Fuel Properties (Table 4)

Table 4 Chemical properties of typical fuels used in engineering systems

FUEL	primary chemical species	Chemical formula	Stoichiometric AFR_{mass}	LHV (MJ/kg_{FUEL})	Δu_{comb} (MJ/kg_{MIX})/ϕ	Density at room conditions (kg/m^3)	Volumetric Energy Density (MJ/liter)	kg_{CO2}/GJ
Hydrogen	Hydrogen	H_2	34.4	120.9	3.42	0.082	0.010	0
"Coal"	carbon	CH	13.2	35.2	2.86	800	28.2	85.5
:"Natural Gas"	methane	CH_4	17.2	50.1	2.76	0.654	0.033	54.8
"Gasoline"	octane	C_8H_{18}	15.1	42.4	2.73	750	31.8	74.7
"Diesel/Jet Fuel"	dodecane	$C_{12}H_{26}$	14.7	41.6	2.69	800	33.3	76.6
"Wood"	cellulose	$C_6H_{10}O_5$	4.9	16.1	2.64	380	6.1	101.3

Fuels are listed in parentheses because actual fuels are generally complex and not well-defined mixtures of many chemical components. The chemical formula given is an approximate equivalent pure species to represent the fuel

Appendix 2: Workshop Solutions

G. Sidebotham, *An Inductive Approach to Engineering Thermodynamics*,
Mechanical Engineering Series, https://doi.org/10.1007/978-3-030-20430-3

Workshop 1.1: Frequency of a Balloon Pendulum

RAW DATA

$M_{unfilled}$	3.7	grams, unfilled balloon mass
d	0.165	m, filled balloon diameter
M_{filled}	3.7	grams, filled balloon mass

trial	L (m)	revs	time (sec)
1	1.4	10	40
2	1.2	10	37
3	1	10	35
4	0.9	10	33
5	0.8	10	31.6
6	0.72	10	30.4
7	0.57	10	28.6
8	0.42	10	25
9	0.29	10	21
10	0.11	10	16

DATA ANALYSIS

$1/(L+d/2)^{0.5}$	frequency (Hz)
0.821	0.250
0.883	0.270
0.961	0.286
1.009	0.303
1.064	0.316
1.116	0.329
1.238	0.350
1.411	0.400
1.638	0.476
2.279	0.625

INPUT PARAMETERS

g	9.81	m/s^2	standard gravity value
mrubber	0.0037	kg	mass of empty balloon
rho,in	1.31	kg/m^3	density of gas inside balloon
rho,infinity	1.21	kg/m^3	density of ambient air
d	0.165	m	diameter of filled balloon

DERIVED PARAMETERS

Volume	0.00235	m^3	volume of balloon
mdisp	0.00285	kg	displaced mass
mtotal	0.00678	kg	total mass of filled balloon
me	0.00394	kg	effective mass of filled balloon

THEORETICAL PLOT

Lmin	0.11	m	smallest value of string length in data set
Lmax	1.4	m	largest value of string length in data set
n	100		number of steps to take for plottiing function
step size	0.0129	m	increment for plotting function

L	ftheory (Hz)
0.11	0.86550
0.1229	0.83789
0.1358	0.81275
0.1487	0.78975
0.1616	0.76860
0.1745	0.74906

(continue down for "n+1" entries

$$f = \frac{1}{2\pi}\sqrt{\frac{g(m_e/m)}{(L+d/2)}}$$

Workshop 2.1: Bouncing Ball

INPUT PARAMETERS

g	9.81	m/s^2, gravity value
me	0.044	kg, effective mass
mdisp	0.00005	kg, displaced mass of air
c	1500	J/kg/K, specific heat
H1	1.92	m, height of release
H4	1.56	m, height of capture
T_∞	25	oC, dead state and initial temperature

COLOR CODE

- INPUTS
- 1st Law
- 2nd Law
- Property Relation
- Process Relation

STATE TABLE

state	Height (m)	Speed (m/s)	Potential Energy (J)	Kinetic Energy (J)	Internal Energy (J)	Temperature (oC)	Entropy (J/K)	eXergy (J)
1	**1.92**	**0**	0.828749	0	**0**	**25**	**0**	0
2	**0**	6.134137	0	0.828749	0	25	0	0
3	**0**	5.529237	0	0.673358	0.15539	25.00235	0.0005214	6.13E-07
4	**1.56**	**0**	0.673358	0	0.15539	25.00235	0.0005214	6.13E-07
4'	**1.56**	**0**	0.673358	0	0	**0**	0	0

note: dead state is reference state for internal energy and entropy

PROCESS TABLE (all in J)

			1st LAW TERMS			2nd LAW TERMS		
Process	Description	Assumptions	ΔE	Q_{in}	W_{out}	$T_\infty \Delta S$	$Q_{in}(T_\infty/T_b)$	$eX_D = T_\infty \sigma$
1-2	Free Fall	Adiabatic, reversible, neglect fluid drag	0	0	0	0	0	0
2-3	Collision	Adiabatic, unchanged surroundings	0	0	0	0.15539	0	0.15539
3-4	Free Rise	Adiabatic, reversible, neglect fluid drag	0	0	0	0	0	0
4=4'	Wait	thermal equilibrium achieved with ambient	-0.15539	-0.15539	0	-0.15539	-0.15539	6.13E-07
4'-1	Lift	Adiabatic, neglect KE during process	0.15539	0	-0.15539	0	0	0
		NET	0	-0.15539	-0.15539	0	-0.15539	0.15539

Workshop 2.2: Measuring the Specific Heat of a Solid

INPUT PARAMETERS

D	0.1	m, diameter of sample
M	4	kg, mass of sample
R	100	Ohms, electrical resistance
V	12	Volts
T1	25	oC, initial and ambient temperature
T2	35	oC, final temperature
t	93	min, elapsed time

DERIVED PARAMETERS

Volume	0.000524	m^3, volume of material
Wdot,e	1.44	Watts, input electrical power
ρ	7639.4	kg/m^3, density of material
c	200.9	J/kg/ΔK, specific heat of material

STATE TABLE

State	T (oC)	u (J/kg)	s (J/kg/K)	eX (J/kg)
1	25	0	0	0
2	35	2008.8	6.6303	32.97

note: set reference state for u and s as the dead state

PROCESS TABLE (all in J)

Process	ΔU	Q_{in}	W_{out}	$T_{\infty}\Delta S$	$Q_{in}\,(T_{\infty}/T_B)$	$eX_D = T_{\infty}\Sigma$
1-2	8035	0	-8035	7903	0	7903
2-1	-8035	-8035	0	-7903	-8035	132
NET	0	-8035	-8035	0	-8035	8035

Workshop 4.2: Isothermal Expansion of H_2O

INPUT PARAMETERS

MASS	3.2	kg, mass of H2O
Tref	273.1434706	K, reference temperature for tables

GAS CONSTANT FOR H2O, USED FOR STATE 7		
RH2O	0.4619	kJ/kg/K

STATE TABLE

state	T(oC)	P(kPa)	v (m^3/kg)	u (kJ/kg)	s (kJ/kg/K)	x(quality)	PHASE (CL, L/V, or SH)
1	**25**	**101.3**	0.001003	104.83	0.3672	NA	CL
2	**25**	3.1698	0.001003	104.83	0.3672	**0**	L/V
3	**25**	3.1698	**1.00**	157.9451844	0.555974233	0.023050764	L/V
4	**25**	3.1698	19.22585291	1126.986671	**4**	0.443592405	L/V
5	**25**	3.1698	39.0061003	2178.673	7.73775	**0.9**	L/V
6	**25**	3.1698	43.34	2409.1	8.5567	**1**	L/V
7	**25**	1.37379132	**100**	2409.1	8.942882616	NA	SH

PROCESS TABLE (all in kJ)

process	ΔU	Q_{in}	W_{out}	$T_\infty \Delta S$	$Q_{in}(T_\infty/T_B)$	$eX_D = T_\infty \Sigma$
1-2	0	0	0	0	0	0.00E+00
2-3	169.969	180.102	10.133	180.102	180.102	0.00E+00
3-4	3100.933	3285.804	184.871	3285.804	3285.804	0.00E+00
4-5	3365.396	3566.034	200.638	3566.034	3566.034	0.00E+00
5-6	737.366	781.327	43.960	781.327	781.327	1.36E-12
6-7	0.000	367.558	367.558	368.441	367.558	8.83E-01
NET	7373.664	8180.825	807.161	8181.708	8180.825	0.883

$$W_{67} = \int PdV = MRT\int \frac{dV}{V} = P_6 V_6 \ln\left(\frac{V_7}{V_6}\right)$$

Workshop 4.3: Closed Rankine Cycle

INPUT PARAMETERS

M	**1**	kg, mass of H2O
D	**0.05**	m, cylinder diameter
T_∞	**25**	oC, dead state temperature (ambient)
T_H	**400**	oC, high temperature source temperature

ATE TABLE

state	*P(kPa)*	*T (° C)*	*v (m³/kg)*	*u (kJ/kg)*	*s (kJ/kg/K)*	*quality or phase*
1	**50**	81.32	0.00103	340.49	1.0912	**0**
2	8000	81.32	0.00103	340.49	1.0912	CL
3	**8000**	**400**	0.034344	2864.6	6.3658	SHV
4	50	81.32	2.62885	2078.7	6.3658	0.811239791

PROCESS TABLE (all in kJ)

Process	**ΔU**	**Q_{in} (kJ)**	**W_{out}**	**$T_\infty \Delta S$**	**$Q_{in}(T_\infty/T_b)$**	**$T_\infty \Sigma$ = eX_D**	***Process type***
1-2	0	0	0	0	0	0	adiabatic, reversible compression
2-3	2524	2791	267	3551	1236	2315	Isobaric heating (from "oven")
3-4	-786	0	786	0	0	0	Isentropic expansion
4-1	-1738	-1870	-131	-1441	-1870	429	Isobaric cooling (to ambient)
net	0	921	921	2110	-634	2743	

RESULTS TABLE

H_3	16.97	m	height of piston at state 3
H_4	1321.4	m	height of piston at state 4
W_{net}	921.0	kJ	Net work done by H2O
Q_{23}	2790.6	kJ	Heat added in "oven"
eX_D	2743.5	kJ	Total eXergy destruction
η_1	0.330		1st Law Efficiency (defined as net work over heat added)
η_2	0.251		a 2nd Law Efficiency (defined as net work over net work + eXergy destruction)
η_{CARNOT}	0.557		Carnot efficiency between "reservoirs"

Workshop 4.4: Refrigeration 101

INPUT PARAMETERS

M	**1**	kg, mass of refrigerant
Tspace	**0**	oC, temperature of cold space
Tinf	**25**	oC, dead state temperature

STATE	P (kPa)	T(oC)	V_{R134a} (m^3)	u (kJ/kg)	h (kJ/kg)	s(kJ//kg/K)	quality or phase
1	**1600**	28	0.0008313	**90**	91.42	0.33846	CL
2	132.82	**-20**	0.0496834	90	96.60	0.38552	0.333988
3	**280**	-1.25	0.0496834	172.626	186.53	0.69972	0.683321

all quantities in kJ

PROCESS	ΔU	Q_{in}	W_{out}	ΔS^*T_{inf}	$(Q/T_B)^*\ T_{inf}$	eX_d
1-2	0.00	0.00	0.00	14.03	0.00	14.03
2-3	82.63	82.63	0.00	93.68	90.19	3.49

Workshop 4.5: Closed Refrigeration Cycle

INPUT PARAMETERS

M	**2.5**	kg, mass of refrigerant
Tspace	**0**	oC, temperature of cold space
Pinf	**1**	atm
Tinf	**25**	oC, ambient (and dead state) temperature

R-134a STATE TABLE — results from Workshop 4.4

STATE	P (kPa)	T(oC)	v(m^3/kg)	u (kJ/kg)	s(kJ//kg/K)	quality, or phase
1	**1600**	28	0.000831	**90**	0.33846	NA
2	132.82	**-20**	0.04968	90	0.38552	0.33399
3	**280**	-1.25	0.04968	172.63	0.69972	0.68332
4	412.6	9.83	0.04968	235.58	0.92649	**1**
5	**1600**	62.74	0.01266	263.87	0.92649	superheated
6	**1600**	28	0.000831	**90**	0.33846	NA

all quantities in kJ

PROCESS	ΔU	Q_{in}	$W_{out,R134a}$	$W_{ambient}$	$\Delta S{*}T_{inf}$	$(Q/T_B){*}T_{inf}$	eX_d
1-2	0	0	0	0	35.08	0	35.075
2-3	206.57	206.57	0	0	234.20	225.47	8.726
3-4	157.4	157.38	0	0	169.03	157.38	11.65
4-5	70.7	**0**	-70.72	-9.376	0	0	0
5-6	-434.7	-481.99	-47.32	-2.996	-438.30	-481.99	43.69
6-1	0	0	0	12.372	0	0	0
net	0	-118.04	-118.04	0	0	-99.13	99.13

RESULTS

COP	1.750	Heat Removed from cold space/Net work In
Carnot Efficiency	10.926	= TC/(TH-TC), maximum theoretical COP of a refrigerator between cold space and ambient
2nd Law effy	0.160	= COP/CarnotCOP

Workshop 7.1: Otto and Diesel Cycle Comparison

OTTO CYCLE

INPUT PARAMS

Vd	1	liter
Rotto	10	
cv	0.72	kJ/kg/K
R	0.287	kJ/kg/K
Q/M	2730	
Ncyl	8	
Nengine	2000	RPM

STATES	V(liters)	P (atm)	T (K)
1	1.11111	**1**	**298**
2	0.11111	25.0387	746.152
3	0.11111	152.276	4537.82
4	1.11111	6.08163	1812.32

DIESEL CYCLE

INPUT PARAMS

Vd	1	liter
r	20	
cv	0.72	kJ/kg/K
R	0.287	kJ/kg/K
Q/M	583	kJ/kg
Ncyl	8	
Nengine	2000	RPM

STATES	V(liters)	P (atm)	T (K)
1	1.05263	**1**	**298**
2	0.05263	66.0138	983.606
3	0.08361	66.0138	1562.55
4	1.05263	1.91047	569.32

	OTTO	DIESEL
Mass (kg)	1.32E-03	1.25E-03
W12 (kJ)	-0.425	-0.615
W23 (kJ)	0.000	0.207
W34 (kJ)	2.583	0.892
W41 (kJ)	0.000	0.000
NET WORK (kJ)	2.158	0.483
Q23 (kJ)	3.593	0.727
1st Law Efficiency	0.601	0.665
Power Output (kW)	287.7	64.4
mep (atm)	21.3	4.8

STEADY-STATE ENGINE

Torque (kN-m)	1.3738	0.3077
mdot (kg/sec)	0.1579	0.1579
Heat Rate (kW)	479.0	96.9
Exit Enthalpy Rate (kW)	191.3	32.5
Exit Temperature (K)	1501.0	502.2

Workshop 7.2: Adding a Compressor and/or Turbine

ENGINE TYPE	Added Component	Mass Flow Rate (kg/sec)	Total Power (kW)	1st Law Efficiency	Exit Temperature (K)	COMMENTS?
Ideal 4 stoke SI Engine (OTTO)	NONE	0.158	259	0.601	1381	Base case. Note how hot the exhaust is (a candidate for cogeneration or a combined cycle for stationary applications
SI Engine w/SuperCharging	Compressor	0.259	422	0.596	1394	mass flow, power and torque increase compared to base case. Slight decrease in efficiency and increase in exit temperature
SI Engine w/SuperCharging and Intercooling	Compressor and Intercooler	0.316	511	0.592	1338	Intercooling increases the mass inducted, and therefore the mass flow rate, power and torque. There is a slight penalty in efficiency, and the exit temperature is lower (as there is cooling that didn't exist before).
SI Engine w/Turbocompounding	Turbine	0.158	299	0.695	1126	The main cycle is not affected (mass flow rate), but there is additional power generated, and therefore a higher efficiency. Power is extracted from the blow down which was dissipated in the base cycle. More power is derived in the turbine than is required by the additional pumping work.
SI Engine w/TURBO-CHARGING	Turbine and Compressor	0.259	520	0.735	1017	Combines advantages of both, namely, higher mass through put from compressor, and extraction of work from blowdown. Higher power and efficiency compared to base cycle.
SI Engine w/TURBO-CHARGING and INTERCOOLING	Turbine, Compressor and Intercooler	0.316	636	0.738	944	Intercooling gives an added boost in power, with a slight efficiency penalty

Workshop 8.1: A Gas Refrigerator/Water Heater, a Refreater

Argon State Table

STATE	T(oC)	P(atm)	h(kJ/kg)	s(kJ/kg/K)
1	**-10**	**10**	8.4729969	0.03325733
2	**50**	16.71623	39.690997	0.03325733
3	**30**	16.71623	29.284997	0
4	-26.2848	**10**	**0**	**0**

Results Table

Mass flow rate of Argon in Brayton cycle	0.0118022	kg/sec
Mass flow rate of hot water	0.0009747	kg/sec
Volumetric flow rate of hot water	84.2151	liters/day
Net Input Power to Cycle	22.813688	Watts
eXergy Destruction Rate in Water Heater	3.918021	Watts
eXergy Destruction Rate in Space Chiller	7.155117	Watts
1st Law Efficiency, COP_R	4.3833333	
2nd Law Efficiency, COP_R/COP_{CARNOT}	0.4014042	

Workshop 8.2: A Central Air Conditioning Application

INPUT PARAMETERS

$m'_{air}c_{p,air}(T_{send} - T_{return})$	**18**	kW, cooling rate of air
ΔT_{sub}	**5**	oC, degrees of subcooling
η_c	**0.85**	isentropic efficiency of compressor
$T_{H\infty}$	**35**	oC, outdoor (dead state) temperature
cp,air	**1**	kJ/kg/K, specific heat of air
R,air	**0.287**	kJ/kg/K, gas constant for air

STATE TABLE FOR REFRIGERATION CYCLE

state	P (kPa)	T (oC)	h (kJ/kg)	s (kJ/kg/K)	x or phase
1	2.928	**0**	247.23	0.919	**1**
2s (isentropic state)	**1000**	**40.3179**	**272.061**	0.919	SHV
2	**1000**	44.29061	276.4425	0.932745523	SHV
3	1000	34.37	98.04621	0.3600827	CL
4	2.928	0	98.04621	0.372827392	0.243528

STATE TABLE FOR CONDITIONED AIR

state	P (atm)	T (oC)	h (kJ/kg)	s (kJ/kg/K)
a	**1**	**25**	0	0
b	**1.003**	25.2563	0.256304	-9.87708E-17
c	**1.002**	**10**	-15	-0.052220016
d	**1.001**	4.743696	-20.2563	-0.070681612
e	**1**	**20**	-5	-0.016920877

MASS FLOW RATE RESULTS (in kg/sec)

m',air	3.6	kg/s, mass flow rate of air
m',R	0.126841	kg/s, mass flow rate of refrigerant

ENERGY RATE RESULTS (in kW)

W',c	3.705363	power input to compressor
Q',H	22.62806	heat transfer rate to outside ambient
W', f	0.922695	power input to fan

EXERGY DESTRUCTION RATES (in kW)

Compressor	0.536999
Condenser	0.792736
Valve	0.497899
Evaporator	0.867205
Regenerator	1.708349

EFFICIENCIES

COP	3.889321	Cooling Rate over total power in
COP,CARNOT	29.8	Carnot efficiency (interior to dead state)
$\eta 2$	0.130514	2nd Law eff'y (COP/COPcarnot)

Workshop 8.3: Water Chiller

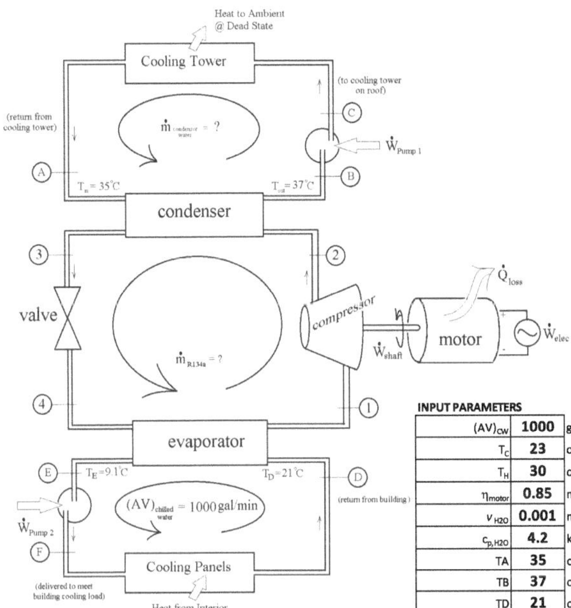

INPUT PARAMETERS

$(AV)_{cw}$	**1000**	gallons/minute, cooling water volumetric flow rate
T_C	**23**	oC, temperature of interior spaces
T_H	**30**	oC, temperature of outside ambient (dead state)
η_{motor}	**0.85**	motor efficiency (output shaft to input electric)
v_{H2O}	**0.001**	m3/kg, specific volume of all water streams
$c_{p,H2O}$	**4.2**	kJ/kg/K, specific heat of all water streams
TA	**35**	oC, water entering condenser
TB	**37**	oC, water exiting condenser
TD	**21**	oC, water entering evaporator
TE	**9.1**	oC, water exiting evaporator

R134a STATE TABLE

state	P (kPa)	T (oC)	h (kJ/kg)	s (kJ/kg/K)	quality
1	**320**	**10**	258.69	0.9544	NA
2	**1000**	50.58162	283.362	0.9544	NA
3	1000	**36**	102.4053	0.37609	NA
4	320	2.46	102.4053	0.3877823	0.240178

RESULTS TABLES

m',CW	63.1	kg/sec, mass flow rate of chilled water
Q',C	3153.2	kW, rate of heat transfer to the cooling panels
m',R	20.2	kg/sec, mass flow rate of R134a
m',cond	434.6	kg/sec, mass flow rate of condenser water
W',shaft	497.8	kW, shaft power to compressor
W',elec	585.6	kW, electric power to motor
Q'tower	3651.0	kW, heat rejected by cooling tower
COP	5.4	Qdot,C/Wdot,elec: overall coefficient of performance
COP,CARNOT	42.3	maximum possible COP to pump heat from inside to out
2nd Law eff'y	0.127	COP/COP,CARNOT

eXergy Destruction rates:		Tinf=	303
cooling tower	70.9	kW	
condenser	44.7	kW	
valve	71.5	kW	
evaporator	146.6	kW	
cooling panels	89.6	kW	
motor	89.9	kW	
TOTAL	513.1		

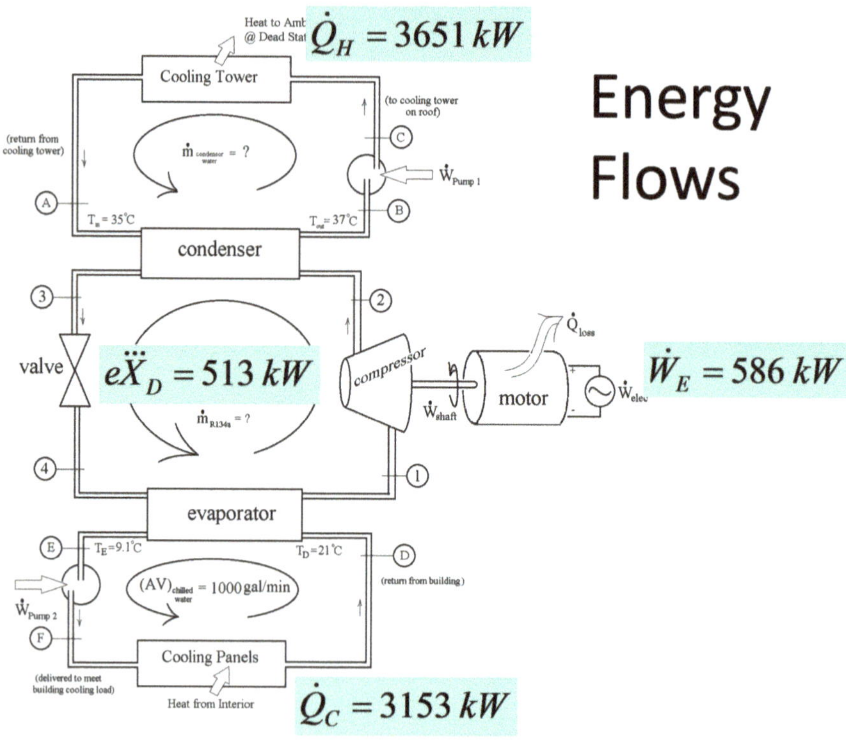

eXergy Destruction

$$e\dddot{X}_{D,TOTAL} = 513\ kW$$

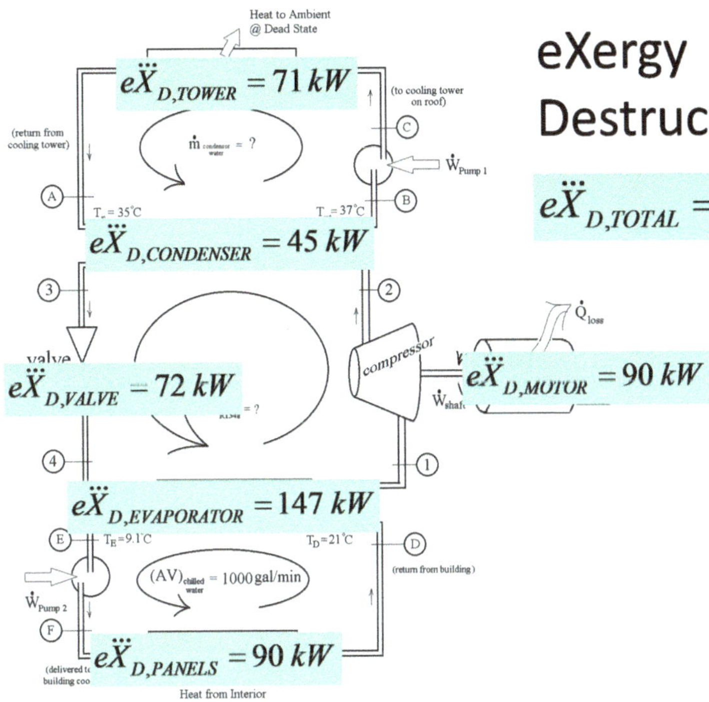

Workshop 9.1: Electric Hand Dryer

INPUT PARAMETERS

Cp=	**1005**	J/kg/K
R=	**287**	J/kg/K
D1=D2=	**0.05**	m
D3=	**0.015**	m

STATE TABLE (note: state 0 is set as a reference state)

state	T	P(gage)	h (J/kg)	V (m/s)	s(J/kg/K)	v(m^3/kg)
0	**25**	**0**	**0**	**0**	**0**	0.844
1	29.05	4909.2	4074.9	7.46	0	0.816
2	61.49	4909.2	36672.5	8.27	102.5	0.904
3	**57**	**0**	32160	**95**	102.5	0.935

RESULTS

air mass flow rate	0.0180	kg/s
Total input power (Ws + We)	658.5	Watts
eXergy destruction	548.5	Watts
Power to Heater	585.3	Watts
Power to Fan	73.2	Watts

Workshop 9.2: Brayton Cycle with Intercooling, Reheat and Regeneration

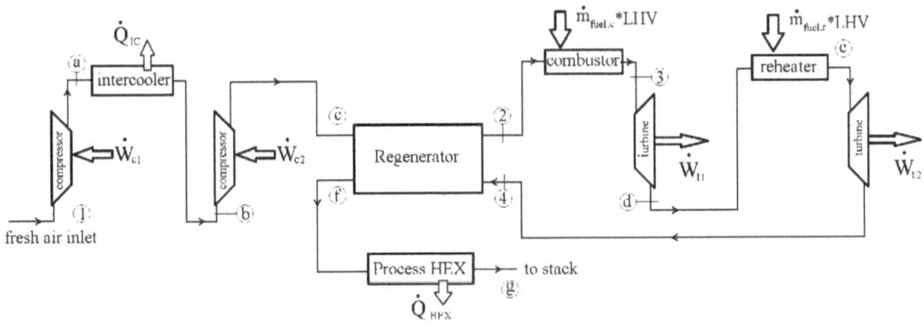

INPUT PARAMETERS

cp	**1.14**	kJ/kg/K, enthalpy specific heat
R	**0.287**	kJ/kg/K, gas constant
Tinf	**298**	K, dead state temperature
Pinf	**1**	atm, dead state pressure
T3 = Tc	**1800**	K, turbine inlet temperatures
Pa = Pb	**1**	atm, intercooler pressure
P2 = P3	**30**	atm, combustor pressure
Pd = Pe	**30**	atm, reheater pressure

STATE TABLE

STATE	P (atm)	T(K)	h(kJ/kg/K)	s(kJ/kg/K)
1	**1**	**298**	0.0	0.000
a	**1**	298.0	0.0	0.000
b	**1**	**298**	0.0	0.000
c	**30**	701.6	460.1	0.000
2	**30**	701.6	460.1	0.000
3	**30**	**1800**	1712.3	1.074
d	**30**	1800.0	1712.3	1.074
e	**30**	**1800**	1712.3	1.074
4	**1**	764.5	531.9	1.074
f	**1**	764.5	531.9	1.074
g	**1**	**298**	0.0	0.000

PROCESS TABLE (all in kJ/kg)

device	process	"q"in	wout	eXD
compressor 1	1-a	0.0	0.0	0.0
intercooler	a-b	0.0	0.0	0.0
compressor 2	b-c	0.0	-460.1	0.0
combustor	2-3	1252.2	0.0	112.8
turbine 1	3-d	0.0	0.0	0.0
reheater	'd-e	0.0	0.0	0.0
turbine 2	e-4	0.0	1180.4	0.0
process HEX	f-g	-531.9	0.0	211.8
REGENERATOR	c-2/4-f	0.0	0.0	0.0
	net	720.3	720.3	324.6

RESULTS

1st Law efficiency	0.575255
Carnot efficiency	0.834444
2nd Law efficiency	0.689387

INPUT PARAMETERS

cp	**1.14**	kJ/kg/K, enthalpy specific heat
R	**0.287**	kJ/kg/K, gas constant
Tinf	**298**	K, dead state temperature
Pinf	**1**	atm, dead state pressure
T3 = Tc	**1800**	K, turbine inlet temperatures
Pa = Pb	**5.5**	atm, intercooler pressure
P2 = P3	**30**	atm, combustor pressure
Pd = Pe	**30**	atm, reheater pressure

STATE TABLE

STATE	P (atm)	T(K)	h(kJ/kg/K)	s(kJ/kg/K)
1	**1**	**298**	0.0	0.000
a	**5.5**	457.7	182.1	0.000
b	**5.5**	**298**	0.0	-0.489
c	**30**	456.8	181.0	-0.489
2	**30**	456.8	181.0	-0.489
3	**30**	**1800**	1712.3	1.074
d	**30**	1800.0	1712.3	1.074
e	**30**	**1800**	1712.3	1.074
4	**1**	764.5	531.9	1.074
f	**1**	764.5	531.9	1.074
g	**1**	**298**	0.0	0.000

PROCESS TABLE (all in kJ/kg)

device	process	"q"in	wout	eXD
compressor 1	1-a	0.0	-182.1	0.0
intercooler	a-b	-182.1	0.0	36.3
compressor 2	b-c	0.0	-181.0	0.0
combustor	2-3	1531.3	0.0	212.4
turbine 1	3-d	0.0	0.0	0.0
reheater	'd-e	0.0	0.0	0.0
turbine 2	e-4	0.0	1180.4	0.0
process HEX	f-g	531.9	0.0	211.8
REGENERATOR	c-2/4-f	0.0	0.0	0.0
	net	817.3	817.3	460.4

RESULTS

1st Law efficiency	0.53376
Carnot efficiency	0.834444
2nd Law efficiency	0.639659

INPUT PARAMETERS

cp	**1.14**	kJ/kg/K, enthalpy specific heat
R	**0.287**	kJ/kg/K, gas constant
Tinf	**298**	K, dead state temperature
Pinf	**1**	atm, dead state pressure
T3 = Tc	**1800**	K, turbine inlet temperatures
Pa = Pb	**1**	atm, intercooler pressure
P2 = P3	**30**	atm, combustor pressure
Pd = Pe	**5.5**	atm, reheater pressure

STATE TABLE

STATE	P (atm)	T(K)	h(kJ/kg/K)	s(kJ/kg/K)
1	**1**	**298**	0.0	0.000
a	**1**	298.0	0.0	0.000
b	**1**	**298**	0.0	0.000
c	**30**	701.6	460.1	0.000
2	**30**	701.6	460.1	0.000
3	**30**	**1800**	1712.3	1.074
d	**5.5**	1174.3	999.0	1.074
e	**5.5**	**1800**	1712.3	1.561
4	**1**	1171.9	996.2	1.561
f	**1**	1171.9	996.2	1.561
g	**1**	**298**	0.0	0.000

PROCESS TABLE (all in kJ/kg)

device	process	"q"in	wout	eXD
compressor 1	1-a	0.0	0.0	0.0
intercooler	a-b	0.0	0.0	0.0
compressor 2	b-c	0.0	-460.1	0.0
combustor	2-3	1252.2	0.0	112.8
turbine 1	3-d	0.0	713.3	0.0
reheater	'd-e	713.3	0.0	27.0
turbine 2	e-4	0.0	716.1	0.0
process HEX	f-g	-996.2	0.0	531.1
REGENERATOR	c-2/4-f	0.0	0.0	0.0
	net	969.2	969.2	670.8

RESULTS

1st Law efficiency	0.493131
Carnot efficiency	0.834444
2nd Law efficiency	0.590969

INPUT PARAMETERS

cp	**1.14**	kJ/kg/K, enthalpy specific heat
R	**0.287**	kJ/kg/K, gas constant
Tinf	**298**	K, dead state temperature
Pinf	**1**	atm, dead state pressure
T3 = Tc	**1800**	K, turbine inlet temperatures
Pa = Pb	**5.5**	atm, intercooler pressure
P2 = P3	**30**	atm, combustor pressure
Pd = Pe	**5.5**	atm, reheater pressure

STATE TABLE

STATE	P (atm)	T(K)	h(kJ/kg/K)	s(kJ/kg/K)
1	**1**	**298**	0.0	0.000
a	**5.5**	457.7	182.1	0.000
b	**5.5**	**298**	0.0	-0.489
c	**30**	456.8	181.0	-0.489
2	**30**	456.8	181.0	-0.489
3	**30**	**1800**	1712.3	1.074
d	**5.5**	1174.3	999.0	1.074
e	**5.5**	**1800**	1712.3	1.561
4	**1**	1171.9	996.2	1.561
f	**1**	1171.9	996.2	1.561
g	**1**	**298**	0.0	0.000

PROCESS TABLE (all in kJ/kg)

device	process	"q"in	wout	eXD
compressor 1	1-a	0.0	-182.1	0.0
intercooler	a-b	-182.1	0.0	36.3
compressor 2	b-c	0.0	-181.0	0.0
combustor	2-3	1531.3	0.0	212.4
turbine 1	3-d	0.0	713.3	0.0
reheater	'd-e	713.3	0.0	27.0
turbine 2	e-4	0.0	716.1	0.0
process HEX	f-g	-996.2	0.0	531.1
REGENERATOR	c-2/4-f	0.0	0.0	0.0
	net	1066.2	1066.2	806.7

RESULTS

1st Law efficiency	0.475034
Carnot efficiency	0.834444
2nd Law efficiency	0.569281

INPUT PARAMETERS

cp	**1.14**	kJ/kg/K, enthalpy specific heat
R	**0.287**	kJ/kg/K, gas constant
Tinf	**298**	K, dead state temperature
Pinf	**1**	atm, dead state pressure
T3 = Tc	**1800**	K, turbine inlet temperatures
Pa = Pb	**5.5**	atm, intercooler pressure
P2 = P3	**30**	atm, combustor pressure
Pd = Pe	**5.5**	atm, reheater pressure

STATE TABLE

STATE	P (atm)	T(K)	h(kJ/kg/K)	s(kJ/kg/K)
1	**1**	**298**	0.0	0.000
a	**5.5**	457.7	182.1	0.000
b	**5.5**	**298**	0.0	-0.489
c	**30**	456.8	181.0	-0.489
2	**30**	1171.9	996.2	0.585
3	**30**	**1800**	1712.3	1.074
d	**5.5**	1174.3	999.0	1.074
e	**5.5**	**1800**	1712.3	1.561
4	**1**	1171.9	996.2	1.561
f	**1**	456.8	181.0	0.487
g	**1**	**298**	0.0	0.000

PROCESS TABLE (all in kJ/kg)

device	process	"q"in	wout	eXD
compressor 1	1-a	0.0	-182.1	0.0
intercooler	a-b	-182.1	0.0	36.3
compressor 2	b-c	0.0	-181.0	0.0
combustor	2-3	716.1	0.0	27.3
turbine 1	3-d	0.0	713.3	0.0
reheater	'd-e	713.3	0.0	27.0
turbine 2	e-4	0.0	716.1	0.0
process HEX	f-g	-181.0	0.0	35.9
REGENERATOR	c-2/4-f	0.0	0.0	0.0
	net	1066.2	1066.2	126.5

RESULTS

1st Law efficiency	0.745972
Carnot efficiency	0.834444
2nd Law efficiency	0.893975

Cycle	qin (kJ/kg)	wnet (kJ/kg)	qprocess (kJ/kg)	eXd (kJ/kg)	ηI	ηII
Brayton	1252	720	532	325	0.575	0.689
Intercooling	1531	817	532	460	0.534	0.640
Reheat	1965	969	996	671	0.493	0.591
Intercooling and Reheat	2245	1066	996	807	0.475	0.569
Intercooling, reheat and regeneration	1429	1066	181	126	0.746	0.894

Workshop 9.3: Jet Engine with Afterburner

INPUT PARAMETERS

m',air	100	kg/sec, mass flow rate of air
u0	300	m/s, cruise speed
P2/P1	25	pressure ratio across compressor
cp	1.14	kJ/kg/K, specific of air at 1000K
R	0.287	kJ/kg/K, gas constant for air

STATE PROPERTIES OF GAS GENERATOR

state	T (K)	P (atm)	h (kJ/kg)	$V^2/2$ (kJ/kg)	s (kJ/kg/K)
0	**230**	**0.3**	**0**	**45**	**0**
1	269.47	0.5628	45	0	0
2	605.97	14.07	428.61	0	0
3	**1800**	14.07	1789.8	0	1.24112
4	1463.5	6.1842	1406.19	0	1.24112

NOZZLE ONLY

state	T (K)	P (atm)	h (kJ/kg)	$V^2/2$ (kJ/kg)	s (kJ/kg/K)
5	683.2	**0.3**	516.6	889.5	1.24112

AFTERBURNER + NOZZLE

state	T (K)	P (atm)	h (kJ/kg)	$V^2/2$ (kJ/kg)	s (kJ/kg/K)
5	**2000**	6.1842	2017.8	0	1.59716
6	933.65	**0.3**	802.2	1215.6	1.59716

RESULTS

	BASIC CYCLE	w/Afterburner
Exit Temperature (K)	683	934
Exit Velocity (m/s)	1,334	1,559
Thrust (kN)	103,382	125,926
Compressor Power (kW)	38,361	38,361
m'*ΔKE (kW)	84,454	117,064
m',fuel*LHV (kW)	136,119	197,280
Thermal Efficiency	0.620	0.593

Workshop 9.4: Gas Turbine-Based Cogeneration with Supplemental Firing

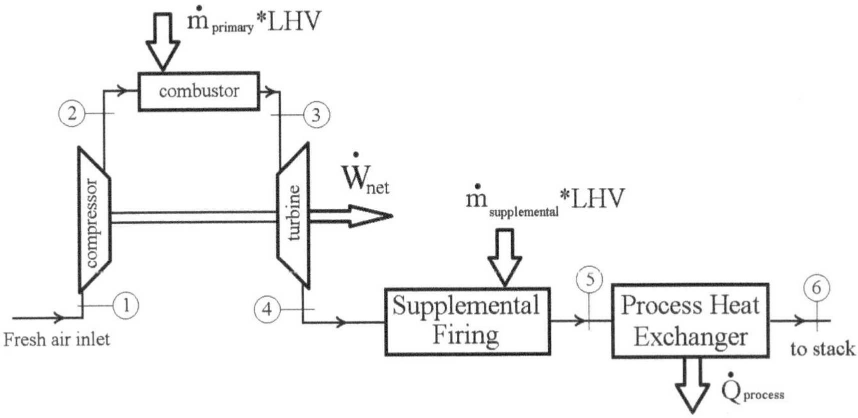

INPUT PARAMETERS

m'air	**100**	kg/s, flow rate of working fluid
cp	**1.14**	kJ/kg/K, enthalpy specific heat
R	**0.287**	kJ/kg/K, gas constant
Tinf	**298**	K, dead state temperature
Pinf	**1**	atm, dead state pressure
T3	**1800**	K, turbine inlet temperatures
P2 = P3	**15**	atm, combustor pressure
T6	**100**	oC, stack temperature
ηc	**0.85**	compressor isentropic efficiency
ηt	**0.9**	turbine isentropic efficiency
LHV	**50100**	kJ/kgfuel, lower heat value of fuel
Δhmix,comb	**2760**	kJ/kgmix, stoich misture heating value

STATE TABLE

state	P (atm)	T (K)	h (kJ/kg)	s (kJ/kg/K)
1	**1**	**298**	0	0
2	**15**	640.7	390.6	0.0953
3	**15**	**1800**	1712.3	1.2730
4	**1**	999.3	799.5	1.3793
5	**1**	999.3	799.5	1.3793
6	**1**	373.0	85.5	0.2559

PROCESS TABLE (kJ/kg of working fluid)

device	process	"qin"	wout	eXd
compressor	1-2	0.0	-390.6	28.4
combustor	2-3	1321.7	0.0	132.1
turbine	3-4	0.0	912.8	31.7
supplemental firing	4-5	0.0	0.0	0.0
Process Heater	5-6	-714.0	0.0	379.2
	NET	607.7	522.2	571.4

RESULTS

m'primary	2.64	kg/s, primary fuel flow rate
m'supplemental	0.00	kg/s, secondary fuel flow rate
W'net	52.2	MW, output power
m'fuel*LHV	132.2	MW, total chemical energy input rate
Q'process	71.4	MW, total process heat addition
eX'D	57.1	MW, total eXergy destruction rate
Cogen 1st Law	0.935	(Q'process+W'net)/m'fuel*LHV
2nd Law eff'y	0.684	(Q'p + W'n)/(Q'p + W'n +eX'D)

INPUT PARAMETERS

m'air	**100**	kg/s, flow rate of working fluid
cp	**1.14**	kJ/kg/K, enthalpy specific heat
R	**0.287**	kJ/kg/K, gas constant
Tinf	**298**	K, dead state temperature
Pinf	**1**	atm, dead state pressure
T3	**1800**	K, turbine inlet temperatures
P2 = P3	**15**	atm, combustor pressure
T6	**100**	oC, stack temperature
ηc	**0.85**	compressor isentropic efficiency
ηt	**0.9**	turbine isentropic efficiency
LHV	**50100**	kJ/kgfuel, lower heat value of fuel
Δhmix,comb	**2760**	kJ/kgmix, stoich misture heating value

STATE TABLE

state	P (atm)	T (K)	h (kJ/kg)	s (kJ/kg/K)
1	**1**	**298**	0	0
2	**15**	640.7	390.6	0.0953
3	**15**	**1800**	1712.3	1.2730
4	**1**	999.3	799.5	1.3793
5	**1**	2261.0	2237.8	2.3102
6	**1**	373.0	85.5	0.2559

PROCESS TABLE (kJ/kg of working fluid)

device	process	"qin"	wout	eXd
compressor	1-2	0.0	-390.6	28.4
combustor	2-3	1321.7	0.0	132.1
turbine	3-4	0.0	912.8	31.7
supplemental firing	4-5	1438.3	0.0	87.8
Process Heater	5-6	-2152.3	0.0	1540.1
	NET	607.7	522.2	1820.2

RESULTS

m'primary	2.64	kg/s, primary fuel flow rate
m'supplemental	2.87	kg/s, secondary fuel flow rate
W'net	52.2	MW, output power
m'fuel*LHV	276.0	MW, total chemical energy input rate
Q'process	215.2	MW, total process heat addition
eX'D	182.0	MW, total eXergy destruction rate
Cogen 1st Law	0.969	(Q'process+W'net)/m'fuel*LHV
2nd Law eff'y	0.595	(Q'p + W'n)/(Q'p + W'n +eX'D)

Workshop 10.1: Steam Turbine with Extraction

SUPERHEATED VAPOR DATA

STATE 1

	R (kJ/kg/K)	0.462	gas constant (8.314/MW)	
	P_0 (MPa)	20		
	T (°C)	v (m^3/kg)	h (kJ/kg)	s (kJ/kg/K)
T_0	550	0.016571	3396.2	6.339
T_1	600	--	3539.0	--

IDEAL GAS INTERPOLATION

INPUT

P (MPa)	18.9000
T (°C)	574

INTERPOLATED

cp (kJ/kg/K)	2.856
s(kJ/kg/K)	6.447
v (m^3/kg)	0.0180
h (kJ/kg)	3464.7

SUPERHEATED VAPOR DATA

STATE 2

	R (kJ/kg/K)	0.462	gas constant (8.314/MW)	
	P_0 (MPa)	3		
	T (°C)	v (m^3/kg)	h (kJ/kg)	s (kJ/kg/K)
T_0	350	0.09056	3116.1	6.745
T_1	400	--	3231.7	--

IDEAL GAS INTERPOLATION

INPUT

T (oC)	374.0000
s (kJ/kg/K)	6.8

INTERPOLATED

cp (kJ/kg/K)	2.312
P (Mpa)	3.218
v (m^3/kg)	0.0877
h (kJ/kg)	3171.6

INPUT PARAMETERS

Tinf	**25**	oC, dead state temperature
f	**0.21**	fraction of mass extracted at state 2
D1	**1**	m, inlet pipe diameter
D2	**1.5**	m, pipe diameter of extracted stream
D3	**12**	m, exit pipe diameter

State	P (bar)	T (oC)	v (m^3/kg)	h (kJ/kg)	s(kJ/kg/K)	quality	V (m/s)	z (m)	$V^2/2$ (kJ/kg)	gz (kJ/kg)
1	**189**	**574**	0.018047	3464.7	6.4472	NA	**20**	**100**	0.200	0.981
2	32.17949	**342**	0.087679	3171.59	**6.8000**	NA	43.2	**50**	0.933	0.4905
3	0.056291	**35**	22.6847	2322.8	7.5670	**0.9**	174.6	**0**	15.239	0

RESULTS

m'1	870.4	kg/s, mass flow rate into turbine
W'net	829.1	MW, net power extracted from turbine
W'net, neglecting PE	828.3	MW, net power neglecting PE
%difference	-0.092%	Error in neglecting PE
W'net, neglecting KE	838.8	MW, net power neglecting KE and PE
%difference	1.17%	Error in neglecting KE
eXd'	248.7	MW, Total eXergy destruction rate
2nd Law efficiency	0.769	W'/(W'+eX'd)

Workshop 10.2: Effect of Boiler Pressure on Ideal Rankine Cycle

Case	High Pressure (MPa)	Turbine Power (MW)	Pump Power (MW)	Heat Addition (MW)	Heat Rejection (MW)	1st Law Efficiency	Max. Theoretical Efficiency	2nd Law Efficiency	Wdot,net/(Wdot,net+eXd)	eXdot,D in boiler	eXdot,D in condenser
	0.1	22.2	0.0103	460.1	437.9	0.048	0.200	0.241	0.241	0.300	69.600
1	0.2	46.76	0.0309	484.6	437.9	0.096	0.291	0.331	0.331	24.820	69.570
2	0.2	46.77	0.00309	484.6	437.9	0.097	0.291	0.331	0.331	24.660	69.700
3	0.3	63.14	0.0515	501	437.9	0.1259	0.3567	0.353	0.353	46.06	69.57
4	0.3	63.13	0.0515	501	437.9	0.1259	0.3567	0.353	0.353	46.06	69.57
5	0.4	75.708	0.072	513.5	437.9	0.1473	0.399	0.3691	0.3691	59.73	69.57
6	0.5	85.75		523.6	437.9	0.1638	0.479	0.3419	0.3419	95.29	69.76
7	0.6	94.79	65.97	466.7	437.9	0.0618	0.3501	0.1764	0.1362	113	69.75
8	0.6	85.75	0.1133	523.5	437.9	0.1636	0.4289	0.3814	0.3814	69.3	69.57
9	0.8	109.4	0.1545	547.1	437.9	0.1996	0.4883	0.4088	0.4088	88.36	69.57
10	1	121.5	0.1957	559.2	437.9	0.2169	0.5139	0.422	0.422	96.52	69.57
11	1.5	144	0.298	581.6	437.9	0.2471	0.5558	0.4446	0.4446	109.9	69.57
12	1.2	132	0.262	569.6	437.9	0.2313	0.4462	0.518	0.433	102.8	69.65
13	1.4	140.9	0.3102	578.5	437.9	0.243	0.5497	0.442	0.442	107.9	69.6
14	1.6	148.9	0.3193	586.5	437.9	0.2534	0.563	0.4501	0.4501	111.94	69.64
15	1.8	156.8	0.3605	594.3	437.9	0.2632	0.575	0.4577	0.4577	115.8	69.5
16	2	163.7	0.402	601.2	437.9	0.2716	0.5852	0.4642	0.4642	118.9	69.57
17	2	163.7	0.4017	601.2	437.9	0.2717	0.5852	0.4642	0.4642	118.9	69.56
18	2.5	178	0.5047	615.4	437.9	0.2884	0.6045	0.4771	0.4771	124.9	69.5
19	3	190.7	0.6077	628	437.9	0.3027	0.5484	0.5521	0.4882	129.8	69.6
20	3.5	54.54	0.711	491.7	437.9	0.1095	0.3126	0	0.1488	713.7	
21	4	212.1	0.8367	649.2	437.9	0.6435	0.6435	0.5057	0.5057	136.88	69.58
22	4	212.1	0.8137	649.2	437.9	0.3267	0.6434	0.5077	0.5077	136.9	69.57
23	4.5	220.8	0.921	657.8	437.9	0.3343	0.6522	0.5128	0.5128	139.53	69.50
24	5	228.8	1.056	665.6	437.9	0.3421	0.6597	0.5186	0.5186	141.8	69.56
25	5	228.8	1.056	665.6	437.9	0.3421	0.5159	0.6632	0.5186	141.8	69.57
26	6	244	1.57	680.3	437.9	0.3563	0.673	0.5294	0.5294	145.9	69.57
27	7	256.8	1.432	693.2	437.9	0.3683	0.6834	0.5389	0.5389	148.9	69.57
28	8	267.8	10.213	695.5	437.9	0.3704	0.6919	0.5353	0.9854	3.568	0.233
29	9	278.1	1.844	714.1	437.9	0.3868	0.6993	0.5532	0.5532	153.6	69.57
30	10	286.6	2.085	722.3	437.9	0.3938	,7051	0.5585	0.5585		436.6
31	10	288.2	2.085	724	437.9	0.3952	0.7061	0.56	0.56	155.5	69.56
32	12.5	308.7	2.565	744	437.9	0.4114	0.7191	0.572	0.572	159.3	69.6
33	15	326.6	3.08	761.5	437.9	0.4249	0.7294	0.5826	0.5826	162.3	69.57
34	17.5	342.3	0.357	779.8	437.9	0.4385	0.7378	0.5943	0.5943	163.7	69.7
35	20										
	22.064	365.6	4.535	799	437.9	0.4519	0.7492	0.6033	0.6033	167.9	69.57

Note: These results were entered manually by students that analyzed a particular case. There are some duplicate cases, blanks and errors. Reported efficiencies are plotted

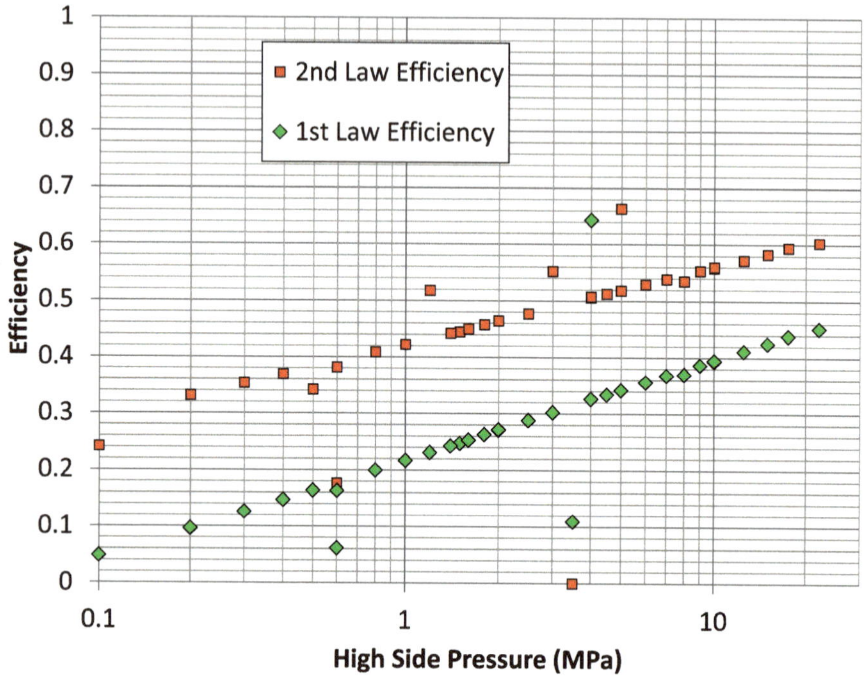
2nd Law Efficiency
1st Law Efficiency
Efficiency
1
0.9
0.8
0.7
0.6
0.5
0.4
0.3
0.2
0.1
0
0.1
1
10
High Side Pressure (MPa)

Workshop 11: Part 1: Gas Turbine Topping Cycle

STATE TABLE

state	T (K)	P (atm)	h (kJ/kg)	s (kJ/kg/K)
1	**298**	**1**	**0**	**0**
2s	701.6	**30**	460.1	0
2	772.82	**30**	541.3	0.1102
3	**1800**	**28**	1712.3	1.0939
4s	861.54	**1.5**	642.44	1.0939
4	955.39	**1.5**	749.42	1.2118

PROCESS TABLE (all quantities in MW)

PROCESS	DEVICE	$m'(h_{out}-h_{in})$	m'_{fuel}*LHV	W'_{out}	eX_d'
1-2	Compressor	66.0	0	-66.0	4.01
2-3	Combustor	142.9	142.9	0	12.1
3-4	Turbine	-117.5	0	117.5	4.29
		91.4	142.9	51.4	20.4

RESULTS

QUANTITY	*VALUE*	*UNIT*
Density of intake Air	1.184	kg/m^3
Inlet Volumetric Flow Rate	218250	cfm
Back Work Ratio (Compressor to Turbine)	0.562	
Fuel mass flow rate	2.86	kg/sec
1st Law thermal efficiency of Topping Cycle	0.360	
Carnot Efficiency	0.834	
2nd Law efficiency (1st Law over Carnot)	0.431	

Workshop 11: Part 2: Heat Recovery Steam Generator (Fig. 7)

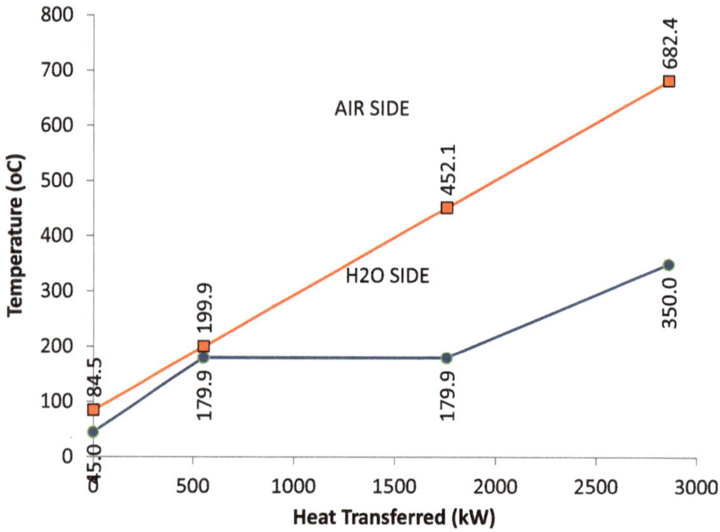

Fig. 7 Results of pinch point analysis for the Combined Cycle HRSG

Workshop 11: Part 3: Rankine Bottoming Cycle

INPUT PARAMETERS

T_∞	25	oC, dead state temperature
m'_{H2O}	75.328	kg/sec, mass flow rate of H2O
nt	0.85	steam turbine isentropic efficiency

STATE TABLE

State	P (bar)	T (oC)	h (kJ/kg)	s (kJ/kg/K)	x
a	10	350	3158.2	7.3029	NA
bs	0.073851	40	2275.191	7.3029	0.876002
b	0.073851	40	2407.643	7.7258634	0.931053
c	0.073851	40	167.53	0.5724	0
d	10	40	168.57	0.5724	NA

PROCESS TABLE (all quantities in MW)

PROCESS	DEVICE	$m'(h_{out}-h_{in})$	Q'_{in}	W'_{out}	eX'_D
a-b	Steam Turbine	-56.54	0.00	56.54	9.49
b-c	Condenser	-168.74	-168.74	0.00	8.16
c-d	Pump	0.0784	0.00	-0.0784	0.00
d-a	HRSG	225.20	225.20	0.00	53.51
	NET	0.0	56.5	56.46	71.2

Workshop 11: Part 4: Cooling Tower and Compiled Results

INPUT PARAMETERS

T_∞	25	oC, dead state temperature
m'_{H2O}	75.328	kg/sec, mass flow rate of Rankine cycle H2O
$v_{H2OL)}$	0.001	m3/kg, specific volume of all liquid H2O streams
c_L	4.2	kJ/kg/K, specific heat of all liquid H2O streams
$c_{p,air}$	1	kJ/kg/K, enthalpy specific heat of cooling air
R_{air}	0.287	kJ/kg/K, gas constant for air

Rankine cycle H2O State Table

state	P (bar)	T (oC)	h (kJ/kg)	s (kJ/kg/K)	x
b	0.07385	40	2407.6	7.7259	0.931
c	0.07385	40	167.53	0.5724	0

Cooling Water State Table (state G as reference)

state	P(atm)	T (oC)	h (kJ/kg)	s (kJ/kg/K)
E	1.5	30	0.05065	0
F	1	33	12.6	0.04138
G	1	30	**0**	**0**

Cooling Tower Air State Table (state H as reference)

state	P(atm)	T (oC)	h (kJ/kg)	s (kJ/kg/K)
H	1	25	**0**	**0**
I	1	28	3	0.010016778

RESULTS

m'cw	13446.4	kg/s, mass flow rate of cooling water
m'air	56247.7	kg/s, mass flow rate of air into tower
W'cw_pump	0.6811	MW, power into cooling water pump
$eX'_{D,condenser}$	5.230	MW, eXergy destruction rate in condenser
$eX'_{D,cooling\ tower}$	2.090	MW, eXergy destruction rate in condenser

Compiled Results

FLOW RATES	Mass (kg/sec)	Volumetric (m3/sec)	
Gas Turbine Air (total)	366	309.0	(at inlet)
Fuel	8.57	11.66	(CH4 @ dead state)
H2O, Rankine Cycle	75.33	0.0753	(at pump inlet)
Cooling Water	13,446	13.45	(anywhere in cw loop)
Cooling Tower Air	56,248	47,489	(inlet)

ENERGY RATES (in MW)

Chemical Energy Addition	429
Gas Turbine Power Output (NET)	154
Steam Turbine Output	56.5
Rankine Pump Input	0.0784
Cooling Water Pump Input	0.6811
TOTAL POWER OUTPUT	210

EXERGY DESTRUCTION RATES (in MW)

Topping Cycle Compressor	12.02
Toppin Cycle Combustor	36.33
Topping Cycle Turbine	12.86
HRSG	53.51
Steam Turbine	9.49
Condenser	5.23
Cooling Tower	2.09
TOTAL EXERGY DESTRUCTION	132

EFFICIENCIES

Overall First Law Efficiency	0.490
2nd Law Efficiency*	0.615

*Actual Power/(Actual Power + Total eXergy Destruction)
different than Carnot based (because cooling tower air is warm)

Workshop 12.1: Humidification by Breathing

FINAL RESULT

(AV)L	0.172	liters/day	Volumetric flow rate of liquid water
Sensible	0.95	Watts	
Latent	5.05	Watts	
Total	6.00	Watts	

INPUT PARAMETERS

Troom	25	oC	room temperature
Tlin	5	oC	temperature of ingested water
phi,room	0.6		relative humidity of room
Vinhale	0.2	liters	volume of lung expansion
N	20	per min	breathing rate
Texhale	37	oC	exhale temperature
phi,exit	1		exhale relative humidity
P	1.013	bar	total pressure
rhoL	1000	kg/m^3	liquid water density
cpair	1005	J/kg/K	air specific heat
cpv	1800	J/kg/K	H2Ov specific heat
hfg	2410000	J/kg	heat of vaporization@Tbody
cL	4200	J/kg/K	H2OL specific heat

DERIVED PARAMETERS

pv1	0.0188	bar	vapor pressure of inhale
pv2	0.0622	bar	vapor pressure of exhale
w1	0.0118	kgH2Ov/kgBDA	inhale absolute humidity
w2	0.0409	kgH2Ov/kgBDA	exhale absolute humidity
Mv1	2.73E-06	kg/breath	mass of H2Ov in inhale
Mv2	8.69E-06	kg/breath	mass of H2Ov in exhale
mdotL	1.99E-06	kg/sec	liquid mass flow rate
(AV)L	1.99E-09	m^3/sec	liquid volumetric flow rate
MBDA	2.31E-04	kg/breath	mass of BDA inhale (and exhale)
mdotBDA	7.70E-05	kg/sec	mass flow rate of BDA

Workshop 12.2 Humidification Needs of Household in Winter

FINAL RESULT

(AV)L	154.3	Liters/day
Sensible	9.02	kW
Latent	4.43	kW
Total	13.45	kW

INPUT PARAMETERS

Tinf	-15	oC	outside temperature
phi,inf	0.9		outside relative humidity
Tinside	23	oC	inside temperature
phi,inside	0.5		inside relative humidity
Tlin	15	oC	temperature of feed water
A	250	m^2	surface area of floor
H	2.5	m	average ceiling height
texchange	1	hr	time for a complete exchange
rhoL	1000	kg/m^3	liquid density
P	1.013	bar	atmospheric pressure
cpair	1.005	kJ/kg/K	air specific heat
cpv	1.8	kJ/kg/K	H2Ov specific heat
hfg	2441.7	kJ/kg	heat of vaporization@25oC
cL	4.2	kJ/kg/K	H2OL specific heat

DERIVED PARAMETERS

pv,inf	0.00180	bar	outside vapor pressure
w,inf	0.00112	kgH2Ov/kgBDA	outside absolute humidity
pv,inside	0.01389	bar	inside vapor pressure
w,inside	0.00869	kgH2Ov/kgBDA	inside absolute humidity
(AV)1	0.1736	m^3/sec	inlet volumetric flow rate
M1	28.78	kgMIX/kmolMIX	molecular weight of ambient
rho1	1.359	kg/m^3	density of ambient
mdot	0.236	kg/sec	mass flow rate of inlet air
mdotBDA	0.236	kg/sec	mass flow rate of dry air
mdotL	0.00179	kg/sec	mass flow rate of liquid water
(AV)L	1.785E-06	m^3/sec	volumetric flow rate of liquid water

Workshop 12.3: Dehumidification Needs of a Household in Summer

FINAL RESULT

(AV)L	186.8	Liters/day
Sensible	1.41	kW
Latent	5.31	kW
Total	6.72	kW

INPUT PARAMETERS

Tinf	32	oC	outside temperature
phi,inf	0.7		outside relative humidity
Tinside	25	oC	inside temperature
phi,inside	0.5		inside relative humidity
Tlout	25	oC	condensate temperature
A	250	m^2	surface area of floor
H	2.5	m	average ceiling height
texchange	1	hr	time for a complete exchange
rhoL	1000	kg/m^3	liquid density
P	1.013	bar	atmospheric pressure
cpair	1.005	kJ/kg/K	air specific heat
cpv	1.8	kJ/kg/K	H2Ov specific heat
hfg	2441.7	kJ/kg	heat of vaporization@25oC
cL	4.2	kJ/kg/K	H2OL specific heat

DERIVED PARAMETERS

pv,inf	0.0329	bar	outside vapor pressure
w,inf	0.0210	kgH2Ov/kgBDA	outside absolute humidity
pv,inside	0.0157	bar	inside vapor pressure
w,inside	0.00981	kgH2Ov/kgBDA	inside absolute humidity
(AV)1	0.17361	m^3/sec	inlet volumetric flow rate
M1	28.45	kgMIX/kmolMIX	molecular weight of ambient
rho1	1.136	kg/m^3	density of ambient
mdot	0.1973	kg/sec	mass flow rate of inlet air
mdotBDA	0.1932	kg/sec	mass flow rate of dry air
mdotL	0.002162	kg/sec	mass flow rate of liquid water
(AV)L	2.16E-06	m^3/sec	volumetric flow rate of liquid water

Workshop 12.4: A Dehumidifier

INPUT PARAMETERS

(AV)L	50	liters/day, H2O condensation rate
Pinf	1	atm, ambient pressure
Tdead	25	oC, dead state temperature

REFRIGERANT STATE TABLE

State	*P(kPa)*	*T(oC)*	*h(kJ/kg)*	*s(kJ/kg/K)*	*quality*
1	293.01	**0**	250.45	0.93139	**1**
2	800	35.14004	271.3272	0.93139	NA
3	800	**31.31**	95.47	0.35404	**0**
4	293.01	0	95.47	0.364032	0.2195871

HUMID AIR STATE TABLE

State	*T(oC)*	ϕ	ω	*pv (atm)*
1A	**25**	**0.7**	0.013826	0.021924
2A	**5**	**1**	0.005444	0.008748
L	**25**	**NA**	**NA**	**NA**

RESULTS TABLE

Quantity	***Value***	***Unit***
Dew Point of Room Air	19	oC
Bone Dry Air Mass Flow Rate	0.0690	kg/sec
Inlet Humid Air Mass Flow Rate	0.0696	kg/sec
Inlet Humid Air Volumetric Flow Rate	126.3	CFM
Refrigerant Mass Flow Rate	0.0182	kg/sec
Input Electrical Power	379.1	Watts
Heat Removal Rate from Air Stream	2814.4	Watts
Heat Rejection Rate (Condenser to Room)	3193.5	Watts
Net Heat Addition Rate to Room	379.1	Watts
Coefficient of Performance of Cycle	7.4	
eXergy Destruction Rate in condenser	69.1	Watts
eXergy Destruction Rate in valve	54.1	Watts
eXergy Destruction Rate in evaporator	57.9	Watts

Workshop 12.5 Evaporative Cooling due to Fuel Injection

INPUT PARAMETERS

	METHANOL	ISO-OCTANE	
P	1	1	atm, total pressure
TA1	25	25	oC, inlet air temperature
TF1	15	15	oC, inlet fuel temperature
ϕ	1	1	equivalence ratio
n	1	8	carbon atoms in fuel
m	4	18	hydrogen atoms in fuel
o	1	0	oxygen atoms in fuel
TBP	64.5	99.3	oC, normal boiling point
hfg@Tbp	1100	308	kJ/kg, enthalpy of vaporization at 1 atm
cL	2.55	2.12	kJ/kg/K, specific heat of liquid fuel
cp,v	1.4	1.66	kJ/kg/K, specific heat of fuel vapor
cp,AIR	1.0	1.0	kJ/kg/K, specific heat of air

DERIVED PARAMETERS

MW	32	114	kg/kmol, molecular weight
R	0.2598	0.0729	kJ/kg/K, gas constant
FAR	0.1556	0.0665	fuel to air mass ratio

RESULTS

Texit	-124.620	3.230	oC, exit temperature, assuming unsaturated exit
CHECK	IMPOSSIBLE	POSSIBLE	return "IMPOSSIBLE" if exit is saturated

CHECK FOR SATURATION

yfuel	0.12285	0.01653	mol fraction of fuel in mixture
Pfuel	0.12285	0.01653	atm, partial pressure of fuel
Psat@Texit	1.13784E-07	0.01935	atm, saturation pressure at Texit
hr	IMPOSSIBLE	0.854	relative humidity

Appendix 3: Nomenclature

Notes

- Units are shown as if written for computer code, without parentheses. For example, force is listed as kg/m/s^2 rather than kg/(m s^2) or kg/m s^2 which a computer would interpret as having the s^2 in the numerator.
- Temperature units are shown either as K or ΔK, depending on whether the absolute temperature or a temperature change is relevant to the quantity in question.
- Electrical units are used sparingly in this text, and not included in the strict SI unit definitions.
- It is inevitable that the same symbol is used for more than one quantity. "V" could refer to velocity and volume. Or h could be elevation or enthalpy. Or the practicing engineer will develop the ability to determine the appropriate symbol from the context.

The following initials are used as common combination units.

- °C = Celsius
- J = Joule
- N = Newton
- Pa = Pascal
- W = Watt

G. Sidebotham, *An Inductive Approach to Engineering Thermodynamics*,
Mechanical Engineering Series, https://doi.org/10.1007/978-3-030-20430-3

Roman Alphabet

Symbol	Quantity	Strict thermal/ fluid SI unit (kg, m, s, K)	Combination SI units
c_p	Enthalpy energy specific heat	$m^2/s^2/\Delta K$	J/kg/Δ°C
c_v	Internal energy specific heat	$m^2/s^2/\Delta K$	J/kg/Δ°C
c	Specific heat of solids or incompressible liquids	$m^2/s^2/\Delta K$	J/kg/Δ°C
F	Force	$kg/m/s^2$	N
h_f	Specific enthalpy of saturated liquid	m^2/s^2	J/kg
h_{fg}	Specific enthalpy of evaporation	m^2/s^2	J/kg
h_g	Specific enthalpy of saturated vapor	m^2/s^2	J/kg
h	Specific enthalpy (or elevation in m)	m^2/s^2	J/kg
HHV	Higher heating value	m^2/s^2	J/kg
I	Mass moment of inertia	$kg*m^2$	
k	Linear spring constant	kg/s^2	N/m
LHV	Lower heating value	m^2/s^2	J/kg
$\dot{m}$	Mass flow rate	kg/s	
m	Mass	kg	
P	Pressure	$kg/m/s^2$	N/m^2 or Pa
$\dot{Q}$	Heat transfer rate	$kg*m^2/s^3$	J/s = W
Q	Heat transfer	$kg*m^2/s^2$	N*m = J
q	Heat transfer rate per unit area	kg/s^3	W/m^2
s_f	Specific entropy of saturated liquid	$m^2/s^2/K$	J/kg/K
s_{fg}	Specific entropy of evaporation	$m^2/s^2/K$	J/kg/K
s_g	Specific entropy of saturated vapor	$m^2/s^2/K$	J/kg/K
S	Entropy (a property)	$kg*m^2/s^2/K$	J/K
s	Specific entropy	$m^2/s^2/K$	J/kg/K
T	Temperature	K	°C
t	Time	S	
u_f	Specific internal energy of saturated liquid	m^2/s^2	J/kg
u_{fg}	Specific internal energy of evaporation	m^2/s^2	J/kg
u_g	Specific internal energy of saturated vapor	m^2/s^2	J/kg
u	Specific internal energy	m^2/s^2	J/kg
$\forall$	Volume	m^3	= 1000 liter
V	Velocity	m/s	
v	Specific volume	m^3/kg	
$\dot{W}$	Work transfer rate	$kg*m^2/s^3$	J/s = W
W	Work	$kg*m^2/s^2$	J
eX_D	eXergy destruction	$kg*m^2/s^2$	J
eX	eXergy	$kg*m^2/s^2$	J
$e\ldots X_D$	eXergy destruction rate	$kg*m^2/s^3$	J/s = W

Greek Alphabet

Symbol	Quantity	Strict thermal/fluid SI Unit (kg, m, s, K)	Combination SI Units
∅	Relative humidity		
∅	Equivalence ratio		
$\mathbb{R}$	Thermal resistance	ΔK*s^3/kg/m^2	ΔK/W
ρ	Density	kg/m^3	
. . . Σ	Entropy generation rate	kg*m^2/s^3/K	W/K
Σ	Entropy production	kg*m^2/s^2/K	J/K
σ	Specific entropy production	m^2/s^2/K	J/kg/K
Σ	Entropy generation	kg*m^2/s^2/K	J/K
θ	Angular displacement		rad
τ	Torque	kg*m^2/s^2	N-m = J
Ψ	Battery voltage		Volt
ω	Absolute humidity		
ω	Angular velocity		rad/s
Ω	Overall-heat transfer coefficient	kg/s^3/ΔK	W/m^2/ΔK

Index

G. Sidebotham, *An Inductive Approach to Engineering Thermodynamics*, Mechanical Engineering Series, https://doi.org/10.1007/978-3-030-20430-3

F

H

I

J

K

L

FSC
www.fsc.org
MIX
Papier | Fördert gute Waldnutzung
FSC® C083411

Kolibri 360 GmbH
Ferdinand-Jühlke-Straße 7
99095 Erfurt, Deutschland
produktsicherheit@kolibri360.de